Lecture Notes in Computer Science 5092

Commenced Publication in 1973
Founding and Former Series Editors:
Gerhard Goos, Juris Hartmanis, and Jan van Leeuwen

Xiaodong Hu   Jie Wang (Eds.)

Computing and Combinatorics

14th Annual International Conference, COCOON 2008
Dalian, China, June 27-29, 2008
Proceedings

 Springer

Volume Editors

Xiaodong Hu
Chinese Academy of Sciences
Institute of Applied Mathematics
Zhong Guan Cun Dong Lu 55
Beijing 100190, China
E-mail: xdhu@amss.ac.cn

Jie Wang
University of Massachusetts–Lowell
Department of Computer Science
Center for Network and Information Security
Lowell, MA 01854, USA
E-mail: wang@cs.uml.edu

Library of Congress Control Number: 2008929349

CR Subject Classification (1998): F.2, G.2, I.3.5, C.2.3-4, E.1, E.5, E.4

LNCS Sublibrary: SL 1 – Theoretical Computer Science and General Issues

ISSN   0302-9743
ISBN-10   3-540-69732-2 Springer Berlin Heidelberg New York
ISBN-13   978-3-540-69732-9 Springer Berlin Heidelberg New York

Springer is a part of Springer Science+Business Media

springer.com

© Springer-Verlag Berlin Heidelberg 2008

Typesetting: Camera-ready by author, data conversion by Scientific Publishing Services, Chennai, India
Printed on acid-free paper   SPIN: 12279302   06/3180   5 4 3 2 1 0

Preface

The 14th Annual International Computing and Combinatorics Conference, CO-COON 2008, took place in Dalian, China, June 27–29, 2008. Past COCOON conferences were held in Xi'an (1995), Hong Kong (1996), Shanghai (1997), Taipei (1998), Tokyo (1999), Sydney (2000), Guilin (2001), Singapore (2002), Montana (2003), Jeju Island (2004), Kunming (2005), Taipei (2006), and Alberta (2007).

COCOON 2008 provided a forum for researchers working in the areas of algorithms, theory of computation, computational complexity, and combinatorics related to computing. The Program Committee received 172 submissions from 26 countries and regions: Canada, Australia, Austria, Brazil, China, Denmark, France, Germany, Hong Kong, Hungary, India, Iran, Israel, Japan, Korea, Mexico, Morocco, Netherlands, Pakistan, Singapore, Spain, Switzerland, Taiwan, Ukraine, the UK, and USA.

With the help of 116 external referees, each submission was reviewed by at least two Program Committee members or external referees. Of the 172 submissions, 66 papers were selected for presentation in the conference and are included in this volume. Some of these will be selected for publication in a special issue of *Algorithmica* and a special issue of the *Journal of Combinatorial Optimization* under the standard refereeing procedure. In addition to the selected papers, the conference also included two invited presentations by Der-Tsai Lee (Academia Sinica, Taiwan) and Takao Nishizeki (Tohoku University, Japan).

The best paper awards were given for "Visual Cryptography on Graphs" to Lu, Manchala and Ostrovsky, and for "A Linear Programming Duality Approach to Analyzing Strictly Nonblocking d-ary Multilog Networks under General Crosstalk Constraints" to Ngo, Wang and Le.

We thank the authors for submitting their papers to the conference. We are grateful to the members of the Program Committee and the external referees for their work within demanding time constraints. We thank the Organizing Committee for their contribution to making the conference a success. We also thank Jinrong Wu for helping us create conference Web pages, Andrei Voronkov for letting us use the Easychair conference system, and Dingzhu Du for offering guidance.

Finally, we thank the conference sponsors and supporting organizations for their support and assistance. COCOON 2008 was supported in part by the National Natural Science Foundation of China under Grant No. 70221001, 10531070, 10771209, 10721101, and was held at Mercure Teda Dalian Hotel, Dalian, China.

June 2008

Xiaodong Hu
Jie Wang

Organization

Executive Committee

Conference Chair: Dingzhu Du (University of Texas at Dallas, USA)

Program Committee

Eric Allender (Rutgers University, USA)
Dan Brown (University of Waterloo, Canada)
Danny Z. Chen (University of Notre Dame, USA)
John Crossley (Monash University, Australia)
Karen Daniels (University of Massachusetts, USA)
Lane Hemaspaandra (University of Rochester, USA)
Steven Homer (Boston University, USA)
Xiaodong Hu (Chinese Academy of Sciences, China) (Co-chair)
Toshihide Ibaraki (Kwansei Gakuin University, Japan)
Sanjay Jain (University of Singapore, Singapore)
Hiro Ito (Kyoto University, Japan)
Ming-Yang Kao (Northwestern University, USA)
Valentine Kabanets (Simon Fraser University, Canada)
Ker-I Ko (State University of New York at Stony Brook, USA)
Xiangyang Li (Illinois Institute of Technology, USA)
Chi-Jen Lu (Academia Sinica, Taiwan)
Shueh-I Lu (National Taiwan University, Taiwan)
Shinichi Nakano (Gunma University, Japan)
Rudiger Reischuk (University of Luebeck, Germany)
Georg Schnitgerm (University of Frankfurt, Germany)
Krishnaiyan Thulasiraman (University of Oklahoma, USA)
Biing-Feng Wang (National Tsing Hua University, Taiwan)
Jie Wang (University of Massachusetts, USA) (Co-chair)
Yinfeng Xu (Xi'an Jiaotong University, China)
Wenan Zang (Hong Kong University, Hong Kong)
Guochuan Zhang (Zhejiang University, China)
Xiao Zhou (Tohoku University, Japan)

Organization Committee

Xujin Chen (Chinese Academy of Sciences, China)
Jie Hu (Operations Research Society of China, China)
Xudong Hu (Chinese Academy of Sciences, China) (Chair)
Aimin Jiang (Chinese Academy of Sciences, China)
Degang Liu (Operations Research Society of China, China)
Yunbin Zhao (Chinese Academy of Sciences, China)

Referees

Akihiro Matsuura
Akihiro Uejima
Alexander K. Hudek
Alexander Russell
Amitabh Chaudhary
Andreas Jakoby
Benjamin Hescott
Bing Su
Binhai Zhu
Bodo Manthey
Boting Yang
Chang-Biau Yang
Chih-Jen Lin
Christian Hundt
Craig Kaplan
Cyril Allauzen
Dan Gusfield
Debajyoti Bera
Decheng Dai
Deshi Ye
Dingzhu Du
Edith Elkind
Eng Wee Chionh
Evanthia Papadopoulou
Feifeng Zheng
Frank Balbach
Frank Stephan
Frank Stephan
Fukuhito Ooshita
Gruia Calinescu
Haitao Wang
Heribert Vollmer
Hiroshi Hirai
Hirotaka Ono
Hsu-Chun Yen
Hu Zhang
Jeremy Avigad
Jesper Jansson
Jian Xia

Jin-Yi Cai
Johannes Textor
Johannes Uhlmann
Jun Tarui
Kazuyuki Miura
Ken-ichi Kawarabayashi
Kiyoshi Yoshimoto
Klaus Jansen
Koichi Yamazaki
Lance Fortnow
Liang Zhao
Lin Lin
Magnus Halldorsson
Maik Weinard
Mario Szegedy
Mark Braverman
Markus Hinkelmann
Masashi Kiyomi
Maw-Shang Chang
Max Alekseyev
Maxim Sviridenko
Min Xu
Ming Zhong
Mirela Damian
Mitsuo Motoki
Nirmalya Roy
Pangfeng Liu
Paul Goldberg
Petros Drineas
Ping Xu
Roger Grinde
Rolf Harren
Rupert Hartung
Ryuhei Uehara
Ryuhei Uehara
Sergey Bereg
Shang-Ju Liu
Sheung-Hung Poon
Shietung Peng

Shin-ichi Nakayama
Stasys Jukna
Stefan Langerman
Steven Skiena
Sun-Yuan Hsieh
Takaaki Mizuki
Takeaki Uno
Takehiro Ito
Takeyuki Tamura
Takuro Fukunaga
Takuya Yoshihiro
Thanh Minh Hoang
Tibor Jordan
Till Tantau
Timothy Chan
Toru Araki
Toru Hasunuma
Toshimasa Ishii
Toshinori Yamada
Toshiya Mashima
Troy Lee
Uri Zwick
Victor Milenkovic
Werner Kuich
Wing Kai Hon
Xiaohua Xu
Xuli Han
Yajun Wang
Yasuhito Asano
Yaw-Ling Lin
Yijie Han
Ying Xiao
Yoshio Okamoto
Yoshiyuki Kusakari
Yucheng Dong
Yufeng Wu
Yuichi Asahiro
Zhou Xu

Sponsoring Institutions

Academy of Mathematics and System Science, Chinese Academy of Sciences
Operations Research Society of China

Table of Contents

Algorithms and Data Structures

Algorithmic Game Theory and Online Algorithms

Automata, Languages, Logic, and Computability

Combinatorics Related to Algorithms and Complexity

Complexity Theory

Cryptography, Reliability and Security, and Database Theory

Computational Biology and Bioinformatics – Model

Computational Biology and Bioinformatics – Algorithms

Computational Algebra, Geometry, and Number Theory

Graph Drawing and Information Visualization

Graph Theory and Algorithms

Communication Networks and Optimization

Wireless Network

Network Optimization

Scheduling Problem

Efficient Compression of Web Graphs

Yasuhito Asano[1], Yuya Miyawaki[2], and Takao Nishizeki[2]

[1] Graduate School Informatics, Kyoto University, Yoshidahonmachi,
Sakyo-ku, Kyoto, 606-8051, Japan
`asano@i.kyoto-u.ac.jp`
[2] Graduate School of Information Sciences, Tohoku University, Aza-Aoba 6-6-05,
Aramaki, Aoba-ku, Sendai, 980-8579, Japan
`miyawaki@nishizeki.ecei.tohoku.ac.jp, nishi@ecei.tohoku.ac.jp`

Abstract. Several methods have been proposed for compressing the linkage data of a Web graph. Among them, the method proposed by Boldi and Vigna is known as the most efficient one. In the paper, we propose a new method to compress a Web graph. Our method is more efficient than theirs with respect to the size of the compressed data. For example, our method needs only 1.99 bits per link to compress a Web graph containing 3,216,152 links connecting 325,557 pages, while the method of Boldi and Vigna needs 2.84 bits per link to compress the same Web graph.

1 Introduction

A *Web graph* is a directed graph, whose vertex set consists of Web pages, and whose edge set consists of hyperlinks connecting these pages. A Web graph plays a central role in data mining on the Web. For example, search engines, including Google and Yahoo!, score Web pages by analyzing a Web graph containing billions of links [3],[6],[14]. Some other algorithms cluster Web pages by finding dense subgraphs in a Web graph [2],[12],[16],[23]. Such a Web graph has too many pages and links to be stored in the main memory of a computer. Thus, compressing a Web graph is indispensable for many applications, including search engines and clustering algorithms.

Several methods have been proposed for compressing a Web graph [1],[4],[5], [7],[9],[13],[19],[20],[22]. The previously known most efficient method, proposed by Boldi and Vigna [7], compresses the link data of a Web graph as little as about three bits per link. The so-called "localities of a Web graph" are utilized by most of the existing methods including that of Boldi and Vigna. The localities stem mainly from the fact that there are much more "intra-host links" than "inter-host links;" an *intra-host link* is a link between two pages in the same host computer, while an *inter-host link* is a link between two pages in distinct hosts. However, the fact has not been fully utilized by previous methods.

In the paper, we first propose a new method to compress a Web graph. Our method fully utilizes the localities of a Web graph together with the fact that there are much more intra-host links than inter-host links. The main idea of

X. Hu and J. Wang (Eds.): COCOON 2008, LNCS 5092, pp. 1–11, 2008.

our method is twofold: one is to deal with intra-host links separately for each host; the other is to use six types of "blocks" to cover all 1's in an adjacency matrix representing intra-host links for a host; each block consists of consecutive 1's in the matrix. (See Figure 1 in Section 4.2) Each type of blocks corresponds to some locality of intra-host links. The matrix can be represented by a sequence of blocks, each of which is represented by the "type," "beginning element" and "dimension" of a block. Thus the data of intra-host links can be efficiently compressed. We regard inter-host links as intra-host links of special type, and compress intra-host links and inter-host links all together. We then compare our method with that of Boldi and Vigna with respect to the size of compressed data and the retrieval speed. The size of data compressed by our method is smaller than 79% of that by their method. For example, our method compresses a Web graph containing 3,216,152 links and 325,557 pages as little as 1.99 bits per link, while their method compresses the same Web graph as little as 2.84 bits per link. Our method retrieves all the links of an original Web graph faster than their method, although our method could be slower than their method when retrieving only the links emanating from a specified page.

2   Localities of a Web Graph

Most of the existing methods to compress a Web graph pagenate all pages with integers in lexicographic order of their URIs, and hence two pages have close page numbers if their URIs share a long common prefix. We call the page number of a page the *index* of the page, and often call a page with index i simply *page i*. The following three facts, called the *localities of a Web graph*, hold true.

Locality (A): Two pages connected by a link often have close indices, that is, the difference of their indices are small.

Locality (B): Pages linked from the same page often have close indices.

Locality (C): Pages with close indices often have "similar" links. More precisely, if page i has a link to page j, then page h such that $|i - h|$ is small often has a link to the same page j.

These three kinds of localities stem from the following two facts on the Web: (1) all the pages in the same host have close indices; and (2) there are much more intra-host links than inter-host links in the Web.

The URIs of pages in the same host share a long common prefix, and hence these pages have close indices. An intra-host link and an inter-host link are formally defined as follows.

Definition 1. *A link between two pages is called an intra-host link if the pages belong to the same host; otherwise, the link is called an inter-host link.*

Intra-host links occupy more than 89% of all the links for the three data sets used in Section 5.

3   Previously Known Methods

In this section, we explain the ideas of two previous methods to compress a Web graph, one by Boldi and Vigna [7], and the other by Claude and Navarro [9].

We first present several definitions. If a link emanates from page p and enters page q, then the pages p and q are called the *source* and *destination* of the link, respectively.

Definition 2. *The destination list L_p of page p contains all indices of pages linked from page p, sorted in increasing order. The adjacency list of a Web graph is the set of all destination lists of pages in the Web graph.*

Methods of compressing a Web graph encode the adjacency list to a binary file, called the *compressed data*.

Boldi and Vigna's method [7] utilizes Localities (B) and (C), and has two positive integer parameters W and α. The parameter W is called a *refer range*. Their method represents the destination list L_p of page p by referring the destination list of one of the W pages preceding page p. They choose one of the W pages, say page q, such that L_q is most similar to L_p, that is, $|L_q \cap L_p|$ is maximum among all q, $p - W \le q \le p - 1$. They say that page p *refers* page q and page q is *referred* by page p. The *copy list* is a binary string of $|L_q|$ bits; its i-th bit, $1 \le i \le |L_q|$, is set to 1 if the i-th element of L_q is contained in L_p; otherwise, it is set to 0. Their method efficiently represents the subset $L_q \cap L_p$ of L_p by the copy list, while the remaining subset $L_p \setminus L_q$ of L_p is represented by a "differential list," called the *remaining list*; the *differential list* of k integers $i_1 < i_2 < \cdots < i_k$ is defined as a list of k integers $i_1, i_2 - i_1, i_3 - i_2, \cdots, i_k - i_{k-1}$. Their method uses several other techniques to efficiently compress the copy lists and the remaining lists. In particular, they allow that page q refers another page r, page r refers another page s, and so on. A set of pages $p_1, p_2, \cdots, p_k$ for some integer k is called a *copy chain* if page p_i refers page p_{i-1} for each i, $k \ge i \ge 2$, and p_1 does not refer any page. The integer k is called the *length of this copy chain*. The maximum length of a copy chain is bounded above by the parameter α. If α becomes larger, then the size of compressed data tends to become smaller but the retrieval speed becomes slower.

Claude and Navarro's method [9] uses a uniform technique called Re-Pair [17] to compress a Web graph comparably as small as that of Boldi and Vigna, while the retrieval time of links emanating from a specified page is several times faster than that of Boldi and Vigna.

4   Our Method

4.1   Classification of Links

We first partition the set of all links of a Web graph into several subsets, each consisting of all the links whose sources belong to the same host.

For example, consider a Web graph whose pages have URIs and indices written in Table 1, and whose adjacency list is represented in Table 2. The "URI"

Table 1. An example of URIs and indices of pages

URI	Original Index
abc.com/index.html	0
abc.com/link.html	1
abc.com/t0.html	2
abc.com/t1.html	3
ace.com/index.html	4
...	...
ace.com/pic300.html	314
add.com/a1.html	315
add.com/index.html	316
add.com/adv.html	317

Table 2. An example of an adjacency list

Source	Destinations
0	1, 2, 315
1	0, 2, 3, 315, 316
2	3
3	1
4	314
...	...
314	4
315	0, 316, 317
316	1, 317
317	316

column of Table 1 represents the URIs of pages, and the "Original Index" column represents the indices of pages. Let H_0 be a host whose name is abc.com, and let H_2 be a host whose name is add.com. Pages 0, 1, 2 and 3 belong to H_0, and pages 315, 316 and 317 belong to H_2. The "Destinations" column of each row in Table 2 represents the destination list of the page written in the "Source" column of the same row.

We then partition the set of all links whose sources belong to each host into two subsets: the set of all intra-host links and the set of all inter-host links. Table 3 depicts intra-host links and inter-host links for H_2. The destinations of intra-host links are written in the "Intra" column, and those of inter-host links are in the "Inter" column.

Table 3. Intra-host links and inter-host links for H_2

Source	Intra	Inter
315	316, 317	0
316	317	1
317	316	-

Table 4. Intra-destination lists for H_2

Source	Destinations
0	1, 2
1	2
2	1

Finally, for each host, we assign an integer, called a "local index," to each page in the host, and represent every intra-host link by a pair of local indices. The *original index* of a page is the index in lexicographic order of its URI, as described in Section 2. Tables 1–3 above use original indices. The *local index* of a page with original index p is defined to be the difference between p and the smallest original index of pages in the host containing page p. For example, the smallest original index in H_2 is 315, and hence the local index of the page with original index 317 is 2. Thus, a local index is much smaller than an original index. Table 4 represents the local indices of pages belonging to H_2. The "Source" column of each row represents the local index of a source of intra-host links, and the "Destinations" column represents the local indices of the destinations.

Comparing Tables 3 and 4, one can immediately realize the advantage of local indices in compressing intra-destination lists. From now on we call a page with local index i simply page i. The *intra-destination list* of page i is a list of all local indices of the destinations of intra-host links whose sources are page i.

4.2   Compression of Intra-host Links

Our method compresses the data of intra-host links in a Web graph, separately for each host, by extending a technique used for the compression of bi-level images [15].

Only for the sake of explanation, we employ an adjacency matrix A of intra-host links for a host. It should be noted that we use an adjacency list, in place of an adjacency matrix, for implementing our method. Let n be the number of pages in a host. Then the adjacency matrix A of the host is an $n \times n$ matrix such that, for $0 \leq i,\ j \leq n-1$, an (i,j)-element $A_{i,j} = 1$ if page i has a link to page j, and otherwise, $A_{i,j} = 0$. We say that an element $A_{i,j}$ is a *1-element* if $A_{i,j} = 1$; otherwise, it is called a *0-element*.

Our method finds the following six types of blocks in A, each consisting of 1-elements consecutive in A in some sense.

Definition 3. *(1) A* **singleton block** *consists of an "isolated" 1-element. (2) A* **horizontal block** *consists of two or more horizontally consecutive 1-elements. (3) A* **vertical block** *consists of two or more vertically consecutive 1-elements. (4) An* **L-shaped block** *is a union of a horizontal block and a vertical block sharing the upper leftmost 1-element. (5) A* **rectangular block** *is a submatrix of A such that all the elements in the submatrix are 1's and the submatrix has more than one consecutive rows and more than one consecutive columns. (6)*

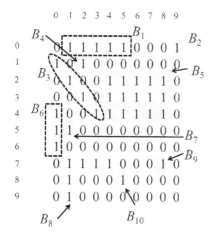

Fig. 1. Adjacency matrix A and blocks $B_1, B_2, \cdots, B_{10}$ in A

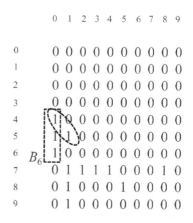

Fig. 2. Matrix A after B_5 is found

A **diagonal block** *consists of two or more 1-elements downward diagonally consecutive from upper left to lower right.*

In Figure 1, each block is either shaded or surrounded by dotted lines. For example, a horizontal block B_1 consists of five 1-elements $A_{0,1}, A_{0,2}, \cdots, A_{0,5}$. Block B_3 is diagonal, B_5 is rectangluar, B_6 is vertical, and B_8 is L-shaped. Blocks B_2, B_4, B_7, B_9, and B_{10} are singletons.

We represent a block by the "beginning element," the "type," and the "dimension" of the block. We denote by $type(B)$ the type of a block B. For example, $type(B_1) = Horizontal$. The upper leftmost element of a block B is called the *beginning element* of B, and is denoted by $b(B)$. Let $br(B)$ be the row number of $b(B)$, and let $bc(B)$ be the column number, then $b(B) = A_{br(B),bc(B)}$. Let $er(B)$ be the row number of the lowest element in B, and let $ec(B)$ be the column number of the rightmost element in B. We call $A_{er(B),ec(B)}$ the *ending element* of a block B unless B is L-shaped. An L-shaped block has two ending elements, the rightmost one $A_{br(B),ec(B)}$ and the lowermost one $A_{er(B),bc(B)}$. The *dimension* $d(B)$ of an L-shaped or rectangular block B is definded to be an ordered pair $(er(B) - br(B) + 1, ec(B) - bc(B) + 1)$. The *dimension* $d(B)$ of block B of the other types is defined to be the number of elements in B. For the example depicted in Figure 1, the beginning element of B_1 is $A_{0,1}$. The dimension of B_1 is 5, while the dimensions of B_5 and B_8 are $(3,5)$ and $(3,4)$, respectively. We can represent a block B in A by a quadraplet $(br(B), bc(B), type(B), d(B))$, called the *signature* $sig(B)$ of B. For example, $sig(B_5) = (2, 4, Rectangular, (3,5))$. A singleton block can be represented without the dimension, because the dimension of every singleton block is 1. For example, $sig(B_2) = (0, 9, Singleton)$. The *size* of a block B is the number of 1-elements in B, and is denoted by $size(B)$.

The five types of blocks, other than a singleton block, corresponds to localities of intra-host links; some are variants of the three localities mentioned in Section 2, and the remainder are newly found in the paper. We call pages with consecutive local indices simply *consecutive pages*. (i) A horizontal block corresponds to a variant of Locality (B); a page often has intra-host links to consecutive pages. (ii) A vertical block corresponds to a variant of Locality (C); in a host, consecutive pages often have an intra-host link to the same page. (Some of the previously known methods explicitly used (i) and (ii) [1],[4],[7],[13],[19],[20],[22].) (iii) A rectangular block corresponds to a newly found variant of Localities (B) and (C); in a host, several consecutive pages often have intra-host links to common consecutive pages. (iv) An L-shaped block also corresponds to a newly found variant of Localities (B) and (C); a page, say page p, often has intra-host links to several consecutive pages $q, q+1, \cdots, q+h$, and several consecutive pages $p + 1, p + 2, \cdots, p + k$ often have intra-host links to page q. For example, a site has such a locality if the site consists of ten pages **page0.html**, **page1.html**, **page2.html**, $\cdots$, **page9.html**, the top page **page0.html** has intra-host links to the remaining nine pages, and they have a link to **page0.html** for returning to the top page. There are a number of sites similar to this example. (v) A diagonal block corresponds to another newly found locality in intra-host links; if page p has an intra-host link to page q, then some consecutive pages $p + i$, $1 \le i \le k$, often have intra-host links to page $q+i$. For example, a site has such a locality if a site has ten pages page0.html, page1.html, page2.html, $\cdots$, page9.html and each page except the last page has an intra-host link to the next page. There are a number of sites similar to this example. We do not adopt an upward diagonal block from lower left to upper right, because there are very few such blocks.

We now explain how to find a set of blocks which cover all 1-elements in an adjacency matrix A. We first find the leftmost 1-element $A_{i,j}$ in the uppermost row containing 1-elements, then find a block B containing $A_{i,j}$ as its beginning element, then output the signature $sig(B)$, and finally replace all the 1-elements in B by 0's. We repeat the operation above from top to bottom and from left to right until there is no 1-element in A.

If there are two or more blocks whose beginning elements are $A_{i,j}$, then we choose B as follows. (One can easily find B using an adjacency list in place of an adjacency matrix.)

1. If there is a rectangular block whose beginning element is $A_{i,j}$, then we choose the *largest one*, that is, the block of the largest size, among these rectangular blocks.
2. Otherwise and if there is an L-shaped block whose beginning element is $A_{i,j}$, then we choose the largest one among them.
3. Otherwise, let B_h, B_v and B_d be the largest blocks with beginning element $A_{i,j}$ among the horizontal, vertical and diagonal blocks, respectively, and we choose the largest one among B_h, B_v and B_d. If there are no such blocks B_h, B_v and B_d, then we choose the singleton block B consisting only of the beginning element $A_{i,j}$.

For example, consider the adjacency matrix A in Figure 1, for which our algorithm finds $B_1, B_2, \cdots, B_{10}$ in this order. Figure 2 depicts matrix A just after the first five blocks $B_1, B_2, \cdots, B_5$ in Figure 1 are found and all the 1-elements in these blocks are replaced by 0's. $A_{4,0}$ is now the leftmost 1-element in the uppermost row containing 1-elements, and there are two blocks having $A_{4,0}$ as the beginning element: a vertical block B_6 and a diagonal block consisting of two 1-elements $A_{4,0}$ and $A_{5,1}$. Since the former is larger than the latter, we choose B_6 as the block containing $A_{4,0}$.

After all the blocks B covering all 1-elements in the adjacency matrix A are found, we encode $sig(B) = (br(B), bc(B), type(B), d(B))$ of all the blocks B in A. In order to encode an integer $br(B)$ to a binary string, we choose the best code among the following three kinds of variable-length codes: γ-code [11], δ-code [11] and ζ-code [8]. The best one depends on the distribution of the values of row numbers $br(B)$. Similarly, we choose the best code for encoding $bc(B)$ and $d(B)$.

We encode $type(B)$ to a binary string so that a type which often appears is encoded to a shorter string, but the details are omitted here.

4.3   Compression of Inter-host Links

For each host, we regard all the inter-host links for the host as intra-host links by giving "new local indices" to the destinations of inter-host links. We then compress the inter-host links and the intra-host links all together by the method described in Section 4.2.

For each host H, our method first gives "new local indices" to all the destinations of inter-host links whose sources belong to H. Let n be the number of pages in H. Let m be the number of destinations of inter-host links for H, and let $t(1) < t(2) < \cdots < t(m)$ be the original indices of all the destinations of inter-host links. The *new local index* $N(t(i), H)$ of a destination $t(i)$ is defined to be an integer $n + i - 1$.

Our method then constructs a "new intra-destination list" of each page p in host H. If $t(x_1) < t(x_2) < \cdots < t(x_k)$ are the original indices of destinations of inter-host links for page p, then the *new intra-destination list* of page p is the union of two lists: one is the intra-destination list of page p, and the other is the list of new local indices $N(t(x_1), H), N(t(x_2), H), \cdots, N(t(x_k), H)$.

We thus compress intra-host links and inter-host links all together for each host by the method in Section 4.2. The input to the method is the new intra-destination lists for the host. Let p be the local index of a page in host H, and let $t(x_i)$ be the i-th smallest orignal index of destinations of inter-host links of page p. Then, for most inter-host links, $|N(t(x_i), H) - p|$ is much smaller than $|t(x_i) - p|$. Thus, by using new local indices, we can cover all 1-elements in the adjacency matrix by a relatively small number of blocks, and can efficiently compress inter-host links together with intra-host links. Experimental analyses will be presented in Section 5.

For each host, it is necessary to store, in a table, a pair of the new local index and original index of the destination of each inter-host link. Using the table, one can retrieve orignal indices from new local indices. We efficiently represents the table by a differential list of the original indices.

5   Experiments

For the computational experiments, we use three data sets of Web graphs, named cnr-2000, in-2004, and eu-2005, collected by Boldi and Vigna [7]. These data sets can be downloaded from their site [21].

Table 5. The size of the compressed data

Data set	cnr-2000	in-2004	eu-2005
Pages	325, 557	1, 382, 908	862, 664
Links	3, 216, 152	16, 917, 053	19, 235, 140
Hosts	722	4, 409	417
BV, $\alpha = 3$(bit/link)	3.56	2.82	5.17
BV, $\alpha = \infty$(bit/link)	2.84	2.17	4.38
Ours (bit/link)	1.99	1.71	2.78
Ratio(%)	70.1	78.8	63.5

Table 5 depicts the size of each data set compressed by our method and that of Boldi and Vigna [7]. The numbers of pages and links in each data set are written in rows "Pages" and "Links," respectively. Each cell in the row "Ours" represents the size of the compressed data per link, obtained by our method. Similarly, the rows "BV, $\alpha = \infty$" and "BV, $\alpha = 3$" represent those by Boldi and Vigna's method with $\alpha = \infty$ and $\alpha = 3$, respectively, where α is the maximum length of a copy chain as described in Section 3. The "Ratio" row represents the ratio of the size of the data compressed by our method to that by the "Boldi and Vigna, $\alpha = \infty$." On average, the size of the data compressed by our method is smaller than 70.8% of theirs.

We use the method of Boldi and Vigna implemented by themselves, which is available on their site [21]. Our method is implemented with Java, and the experiments run on a PC with Core2 Duo E6600 (2.40GHz) and 2GB main memory.

Table 6. Retrieval time for the whole compressed data

Data set	cnr-2000	in-2004	eu-2005
BV, $\alpha = 3$	1.48s	5.87s	7.34s
BV, $\alpha = \infty$	$1.25\times$ 10^3s	$1.54\times$ 10^3s	$5.38\times$ 10^3s
Ours	0.68s	3.73s	1.94s

Table 7. Retrieval time for a specified page

Data set	cnr-2000	in-2004	eu-2005
BV, $\alpha = 3$	$3.51\times$ 10^{-2}ms	$6.07\times$ 10^{-2}ms	$4.07\times$ 10^{-2}ms
BV, $\alpha = \infty$	4.35ms	1.22ms	6.73ms
Ours	2.34ms	2.38ms	28.72ms

Table 6 depicts the time required to retrieve the whole compressed data. Table 7 depicts the time to retrieve the destination list of a specified page, which is the average time for randomly selected 10,000 pages for each data set. The retrieval time of our method is written in the column "Ours." Similarly, that of Boldi and Vigna's method with $\alpha = 3$ is written in the column "BV, $\alpha = 3$," which is much faster than their method with $\alpha = \infty$, written in the column "BV, $\alpha = \infty$."

Both our method and theirs take time $O(M)$ to retrieve the whole compressed data, where M is the number of links in a Web graph. However, our method is experimentally several times faster than theirs with $\alpha = 3$, and is about 1000 times faster than theirs with $\alpha = \infty$.

For a request to retrieve a specified page, our method must retrieve all the links whose sources belong to the same host as the specified page. On the other hand, their method must retrieve the destination lists of all the pages in a copy chain, whose length is at most α. Our method retrieves a specified page experimentally much slower than their method with $\alpha = 3$. For cnr-2000 data set, our method is faster than their method with $\alpha = \infty$, although our method is slower than theirs for in-2004 and eu-2005 data sets. Our method takes much time particularly for eu-2005 data set, because the data set have several hosts containing a huge number of links and pages.

6   Concluding Remarks

We have proposed a new efficient method of compressing a Web graph. We have introduced six types of blocks to cover all 1-elements in an adjacency matrix to fully utilize localities of intra-host links. We have also proposed a technique for compressing inter-host links and intra-host links all together by giving new local indices to the destinations of inter-host links. The size of data compressed by our method is about 70.8%, on average, of that by Boldi and Vigna's method which has been known as the most efficient method of compressing a Web graph [7]. The retrieval of our method for the whole compressed data is faster than their method, although that for a specified page could be slower than their method. Thus, one of the possible future works is to improve the retrieval speed for a specified page.

References

1. Asano, Y., Ito, T., Imai, H., Toyoda, M., Kitsuregawa, M.: Compact Encoding of the Web Graph Exploiting Various Power Laws: Statistical Reason Behind Link Database. In: Dong, G., Tang, C.-j., Wang, W. (eds.) WAIM 2003. LNCS, vol. 2762, pp. 37–46. Springer, Heidelberg (2003)
2. Asano, Y., Nishizeki, T., Toyoda, M., Kitsuregawa, M.: Mining Communities on the Web Using a Max-Flow and a Site-Oriented Framework. IEICE Trans. Inf. Syst. E89-D (10), 2606–2615 (2006)
3. Asano, Y., Tezuka, Y., Nishizeki, T.: Improvements of HITS Algorithms for Spam Links. In: Dong, G., Lin, X., Wang, W., Yang, Y., Yu, J.X. (eds.) APWeb/WAIM 2007. LNCS, vol. 4505, pp. 479–490. Springer, Heidelberg (2007)
4. Bharat, K., Broder, A., Henzinger, M., Kumar, P., Venkatasubramanian, S.: The Connectivity Server: Fast Access to Linkage Information on the Web. In: Proc. of the 7th WWW, pp. 469–477 (1998)
5. Blandford, D.K., Blelloch, G.E., Kash, I.A.: Compact Representation of Separable Graphs. In: Proc. of the 14th SODA, pp. 679–688 (2003)
6. Brin, S., Page, L.: The Anatomy of a Large-Scale Hypertextual Web Search Engine. In: Proc. of the 7th WWW, pp. 14–18 (1998)
7. Boldi, P., Vigna, S.: The Web Graph Framework I: Compression Techniques. In: Proc. of the 13th WWW, pp. 595–601 (2004)
8. Boldi, P., Vigna, S.: Codes for the World Wide Web. Internet Mathematics 2(4), 405–427 (2005)
9. Claude, F., Navarro, G.: A Fast and Compact Web Graph Representation. In: Ziviani, N., Baeza-Yates, R. (eds.) SPIRE 2007. LNCS, vol. 4726, pp. 118–129. Springer, Heidelberg (2007)
10. Cormen, T.H., Leiserson, C.E., Rivest, R., Stein, C.: Introduction to Algorithms. 2nd edn. MIT Press, Cambridge (2001)
11. Elias, P.: Universal Codeword Sets and Representaions of the Integers. IEEE Transactions on Information Theory 21, 194–203 (1975)
12. Flake, G.W., Lawrence, S., Giles, C.L.: Efficient Identification of Web Communities. In: Proc. of the 6th KDD, pp. 150–160 (2000)
13. Guillaume, J.L., Latapy, M., Viennot, L.: Efficient and Simple Encodings for the Web Graph. In: Meng, X., Su, J., Wang, Y. (eds.) WAIM 2002. LNCS, vol. 2419, pp. 328–337. Springer, Heidelberg (2002)

14. Kleinberg, J.: Authoritative Sources in a Hyperlinked Environment. In: Proc. of the 9th SODA, pp. 668–677 (1998)
15. Kou, W.: Digital Image Compression: Algorithms and Standards. Springer, Heidelberg (1995)
16. Kumar, R., Raghavan, P., Rajagopalan, S., Tomkins, A.: Trawling the Web for Emerging Cyber-Communities. Computer Networks 31(11-16), 1481–1493 (1999)
17. Larsson, N.J., Moffat, A.: Off-Line Dictionary-Based Compression. Proc. IEEE 88(11), 1722–1732 (2000)
18. Levenstein, V.E.: On the Redundancy and Delay of Separable Codes for the Natural numbers. Problems of Cybernetics 20, 173–179 (1968)
19. Randall, K., Stata, R., Wickremesinghe, R., Wiener, J.L.: The Link Database: Fast Access to Graphs of the Web. Research Report 175, Compaq Systems Research Center, Palo Alto, CA (2001)
20. Suel, T., Yuan, J.: Compressing the Graph Structure of the Web. In: Proc. of the Data Compression Conference, pp. 213–222 (2001)
21. WebGraph Homepage, http://webgraph.dsi.unimi.it/
22. Wickremesinghe, R., Stata, R., Wiener, J.: Link Compression in the Connectivity Server. Technical Report, Compaq Systems Research Center, Palo Alto, CA (2000)
23. Zhang, Y., Yu, J.X., Hou, J.: Web Communities: Analysis and Construction. Springer, Berlin (2006)

Damaged BZip Files Are Difficult to Repair

Christian Hundt and Ulf Ochsenfahrt

Fakultät für Informatik und Elektrotechnik, Universität Rostock, Germany
{christian.hundt,ulf.ochsenfahrt}@uni-rostock.de

Abstract. bzip is a program written by Julian Seward that is often used under Unix to compress single files. It splits the file into blocks which are compressed individually using a combination of the Burrows-Wheeler-Transformation, the Move-To-Front algorithm, Huffman and Runlength encoding. The author himself stated that compressed blocks that are damaged, i.e., part of which are lost, are essentially non-recoverable. This paper gives a formal proof that this is indeed true: focusing on the Burrows-Wheeler-Transformation, the problem of completing a transformed string, such that the decoded string obeys certain file format restrictions, is NP-hard.

1 Introduction

Consider an important Java source code file that is compressed with bzip [5] and sent to a backup server. During the transmission, the originating machine spontaneously bursts into flames and the connection is lost. How difficult is it to restore at least a part of the original source code from the incompletely transmitted file? In contrast to other compression algorithms, the bzip decoder does not process input data incrementally, but as a whole. Even if only a small fraction is lost, the decoder fails. Thus, the only option is to find a completion of the partial file with the hope that it can be subsequently decoded and results in valid Java source code.

This paper sets out to prove formally that this problem is in general NP-hard due to the use of the Burrows-Wheeler-Transformation (or BWT for short) in the bzip compression algorithm. The BWT deterministically calculates a permutation of the input data, which is consecutively easier to compress. Figure 1 shows the word 'spots' as an example. Part (a) shows the BWT matrix which consists of 'spots' rotated character-wise to every possible position, and then sorted lexicographically. The output of the algorithm is the last column of that matrix, namely 'pssto'. As shown in part (b), the inverse BWT first sorts the characters in the input string, resulting in two strings. It then infers a bipartite graph structure which forms a cycle. The output of the algorithm is every other character on that cycle, namely 'spots'.

Suppose that instead of the complete BWT output, 'pssto', only the first two characters 'ps' are known. Even if it is known that the decoded word must be a word in the English language consisting of the characters 'opsst', the solution is not unique. It could be either 'pssto', which decodes to 'spots', or it could

X. Hu and J. Wang (Eds.): COCOON 2008, LNCS 5092, pp. 12–21, 2008.

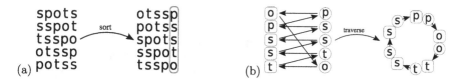

Fig. 1. (a) The Burrows-Wheeler-Transform matrix for 'spots' (b) the inverse transformation of 'pssto' to 'spots'

be 'pstos', which decodes to 'posts'. In general, given only the first part of the BWT output, it is difficult to complete this in any meaningful way, even if the structure of the original data is known, e.g., source code that conforms to the Java language grammar.

Proof Idea

Given a string of characters and a finite automaton, the BWT Reconstruction Problem consists of finding a valid completion of that string, such that the compound string can be BWT decoded and the result is accepted by the automaton. This paper proves its NP-completeness by reduction from the well-known directed Hamiltonian cycle problem. The reduction works as follows: Given a directed graph, we compute - in polynomial time - a string, a maximum length, and a finite automaton. There are many valid completions of that string for the given maximum length. But, when such a completion is transformed with the inverse BWT, then the automaton accepts the result if and only if it encodes a Hamiltonian cycle in the original graph.

Consider the graph in Figure 2(a) as an example. For the NP-completeness proof, this paper first constructs a finite automaton that accepts an encoding of any cycle in this graph. The automaton has one start state and a certain number of states for every vertex in the original graph. Each such group of states is arranged such that it accepts exactly the encoding of the corresponding vertex index. A vertex index encoding represents the index as a binary number using symbols 'b' and 'c' for 0 and 1, delimited by start and end markers 'd' and 'a'. For example, the vertex index 1 (binary: '01') is encoded as 'dbca'. Additionally, an encoding of a sequence of vertices starts with an 'e' and can have any number of 'f' characters inbetween index encodings. The sequence $(0, 3, 2)$ could then be encoded as 'e dbba ff dcca dcba f'.

Compared to the constructed automaton, which represents the structure of the original graph, the constructed string only depends on the number of vertices. It features an intricate pattern of letters 'b', 'c', and 'd', and ends with a single letter 'e'. The key to the proof is that the given pattern already strongly constrains the possible solutions, such that each transformed string contains substrings corresponding to all the index encodings, each and every one of them exactly once. By choosing different completions of the pattern with letters 'f' and 'a', these index encodings can be arranged in all possible permutations.

Combining the prefix string and the automaton representing the graph structure completes the proof. Every completion contains all vertex indices exactly

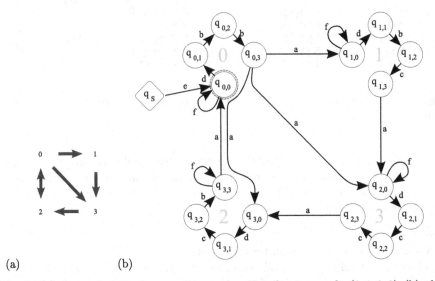

(a)   (b)

Fig. 2. (a) A graph with four vertices and a Hamiltonian cycle $(0, 1, 3, 2)$ (b) the corresponding automaton; the diamond represents the initial state, double circles are terminal states

once in some permutation. The automaton accepts a decoded string if and only if it forms a cycle in the original graph. If such a cycle contains all vertices exactly once, it is Hamiltonian. Since all permutations can be attained, solving this problem is equivalent to finding a Hamiltonian cycle.

2   Preliminaries

Throughout this paper we consider the Latin alphabet $\Sigma = \{\text{'a'}, \dots, \text{'z'}\}$. A string X over Σ is a finite ordered list of characters of length $|X|$. For all $i \in \{1, \dots, |X|\}$, $X[i]$ is the character at index i in X. If X and Y are strings then $X + Y$ is the string obtained by the concatenation of X and Y. Also, for any $c \in \Sigma$ and $i \in \mathbb{N}$, c^i is the string of length i which contains the character c i times.

We use the non-deterministic finite automaton model described by Hopcroft et al. [4], which is a 5-tuple $(Q, \Sigma, \delta, q_0, F)$, containing a set of states Q, an alphabet Σ, a transition relation $\delta \subset Q \times \Sigma \times Q$, an initial state q_0, and a set of accepting states F.

The graphs considered in this paper are finite, directed and contain neither self-loops nor multiple edges. Brandstädt et al. [1] performed a broad survey of graphs and their properties. Let $G = (V, E)$ be a graph. A sequence $P = (v_1, \dots, v_k)$ of pairwise distinct vertices is a path in G if $(v_1, v_2), \dots, (v_{k-1}, v_k) \in E$. If $(v_k, v_1) \in E$ holds, then P is called a cycle. A cycle is Hamiltonian if it contains every vertex of G exactly once. A graph is bipartite if V can be partitioned into two sets V_1, V_2 such that E contains only edges going from a vertex in V_1 to a vertex in V_2 or vice versa.

The DIRECTED HAMILTONIAN CYCLE PROBLEM is a known NP-complete problem [3] of determining whether a given graph G contains a directed Hamiltonian cycle. Garey and Johnson [3] performed a detailed survey on computational complexity.

The Burrows-Wheeler-Transformation is a fully invertible block sorting method used in data compression, originally introduced by Burrows and Wheeler [2]. For brevity, we only recapitulate the inverse BWT for a given string X (IBWT(X)).

The IBWT first takes the input string X of length n and sorts the characters in X to obtain a second string Z. It then builds a bipartite graph $B(X)$, as exemplified by Figure 1(b), with $2n$ nodes corresponding to all the characters in X and Z. Every such node has exactly one incoming and one outgoing edge. Every Character in X has an outgoing edge to the character in Z with the same index, but not necessarily the same character. Every character in Z has an outgoing edge to the same character in X, i.e., the 4th 'a' in Z has an outgoing edge to the 4th 'a' in X, irrespective of the actual indices of those characters. As Z and X are only permutations of the same list of characters, these two sets of edges denote one to one correspondences.

The resulting bipartite graph $B(X)$ must form a single cycle or the algorithm fails. It then simply outputs every other character on that cycle, a total of $\frac{2n}{2} = n$ characters, starting at a dedicated start character.

3   The BWT Reconstruction Problem Is NP-Complete

This section proves the NP-completeness of the following decision problem, which we call the BWT RECONSTRUCTION PROBLEM:

Instance: A string x over Σ, an automaton A and a natural number ℓ.
Question: Is there a string x' over Σ such that for $X = x + x'$ the inverse BWT on X yields a string Y of length ℓ which is accepted by A?

The proof is performed by reduction from the NP-complete directed Hamiltonian cycle problem [3]. Following the proof idea described in Section 1, this section first details construction of the finite automaton which encodes the graph structure, and the generic assembly of the prefix string depending on the vertex count in the original graph. As an intermediate step, this section argues that these constructions can all be performed on a deterministic Turing machine in polynomial time.

After detailing the construction scheme, this section proves that the described scheme is correct in four steps: (1) first it infers partial knowledge of the nodes and edges in the IBWT graph $B(X)$, then (2) it shows that every inversely transformed completion contains an encoding of every vertex index once, followed by (3) the demonstration that every possible permutation of vertex indices can be attained. Finally, (4) it summarizes the results in the theorem that the BWT Reconstruction Problem is NP-complete.

The Reduction Scheme

Given a graph $G = (V, E)$, this section provides instructions on how to construct an automaton $A(G)$, a string $x(G)$ and a length $\ell(G)$, such that G contains a Hamiltonian cycle if and only if a completion x' of length $\ell(G) - |x(G)|$ exists, where $Y = \text{IBWT}(x(G) + x')$ is accepted by $A(G)$.

We assume that $|V| = n = 2^w$ for $w \in \mathbb{N}$. If this is not the case, construct a graph G' which contains a Hamiltonian cycle if and only if G does, but consists of a number of vertices which is the next higher power of two. G' is identical to G except for one vertex v which is replaced by a directed path P of length $2^w - n + 1$ where $w = \lceil \log n \rceil$. In G' any in-edge of v leads to the first vertex of P and any out-edge comes from the last vertex of P. Then we perform the reduction on G' which has 2^w vertices.

Vertices in V are indexed by numbers from 0 to $n - 1$. For each vertex $v \in V$ the id-string s_v of length $w + 2$ encodes the index of v as a binary number using characters 'b' and 'c' for 0 and 1 additionally delimited by start and end markers 'd' and 'a'.

We construct the automaton $A(G)$ such that it only accepts sequences of identifications strings $s_{v_1}, s_{v_2}, \ldots, s_{v_k}$ which (1) give a cycle $(v_1, v_2, \ldots, v_k)$ in G, (2) have the first id-string s_{v_1} for the vertex v_1 with index 0, (3) are initiated by a single character 'e' and (4) contain arbitrarily long sequences of 'f' between id-strings. The formal definition $A(G) = (Q, \Sigma, \delta, q_S, F)$ of the automaton follows:

1. For every vertex $v \in V$, Q contains a set of states $\{q_{v,0}, \ldots, q_{v,w+1}\}$ which accept the id-string s_v. The transition relation between states of this set is:
   (a) $\forall v \in V : \delta(q_{v,0}, \text{'d'}, q_{v,1})$
   (b) $\forall v \in V, i \in \{1, \ldots, \log n\} : \delta(q_{v,i}, s_v[i + 1], q_{v,i+1})$
2. Q contains an initial state q_S and $\delta(q_S, \text{'e'}, q_{0,0})$
3. $\forall (u, v) \in E : \delta(q_{u,w+1}, \text{'a'}, q_{v,0})$
4. $F = \{q_{0,0}\}$
5. $\forall v \in V : \delta(q_{v,0}, \text{'f'}, q_{v,0})$

Checking that this construction of $A(G)$ is correct can be easily done and is omitted here. Note that $A(G)$ is non-deterministic. However, this is only for convenience and to make the technique as clear as possible. A deterministic version of $A(G)$ with $O(n^2)$ states can also be constructed in polynomial time.

In contrast to the automaton $A(G)$, the string $x(G)$ is independent of the structure of G and depends only on n. Define for all $i, j \in \mathbb{N}$ the strings $b_{i,j} = (\text{'b'})^{2^{i-1}}$, $c_{i,j} = (\text{'c'})^{2^{i-1}}$, $x_{0,j} = \text{'d'}$ as well as

$$x_{i,j} = b_{i,j} + c_{i,j} + x_{i-1,2j-1} + x_{i-1,2j}.$$

Then set $x(G) = x_{w,1} + \text{'e'}$. One can show by induction that $x(G)$ contains exactly $\frac{n}{2} \log n$ 'b'-, $\frac{n}{2} \log n$ 'c'-, n 'd'-, and a single 'e'-character. Finally, set $\ell(G) = n^2 + n \log n + 2n + 1$.

Lemma 1 (without proof). *For all graphs $G = (V, E)$ with $|V| = n = 2^w$ with $w \in \mathbb{N}$ the instance $(A(G), x(G), \ell(G))$ can be computed in polynomial time with respect to the size of G.*

The Burrows-Wheeler-Transform as described by the original authors [2] uses an explicit start index for the inverse transform. In this case, the start index is $1+2n+n \log n$, which is the position of the character 'e' in $x(G)$, and independent of the completion. However, since the character 'e' is unique in this case, the start index can be omitted. Therefore, the problem is not easier to solve if the start index is known.

The Nodes of the IBWT Graph $B(X)$
The prefix $x(G)$ already determines part of the string $Y = \text{IBWT}(x(G) + x')$ for any completion x', as every character of $x(G)$ must occur in Y. Additionally, $A(G)$ accepts only certain kinds of strings Y as described by the following four facts:

1. Y contains only characters 'a' to 'f',
2. Y contains only a single 'e' as the first character, followed by a 'd' or 'f' character,
3. the amount of 'a' in Y equals the amount of 'd', and
4. if α is the amount of 'a' and β the total amount of 'b' and 'c', then $\beta = \alpha \log n$.

These facts allow inferring a significant part of the IBWT graph $B(X)$, although only the prefix $x(G)$ of X is known. By the contents of $x(G)$ follows that Z has at least n 'a'-, n 'd'-, $\frac{n}{2} \log n$ 'b'- and $\frac{n}{2} \log n$ 'c'-nodes. Assume that the sum of 'a'-, 'b'- and 'c'-nodes is more than $n + n \log n$. In Z the characters are sorted, so this implies $Z[1 + n + n \log n]$ is either an 'a'-, 'b'- or 'c'-node. However, the construction of $x(G)$ gives that $X[1 + n + n \log n]$ is the 'e'-node, and the IBWT graph $B(X)$ contains an edge from $X[1 + n + n \log n]$ to $Z[1 + n + n \log n]$. It follows that such a decoded string Y would contain the character 'e' followed by a character 'a', 'b' or 'c', but not 'd' or 'f'. $A(G)$ only allows 'd' or 'f' after the initial 'e', which contradicts the assumption. Therefore, Z contains exactly n 'a'-, n 'd'-, $\frac{n}{2} \log n$ 'b'-, $\frac{n}{2} \log n$ 'c'-, and a single 'e'-node. The remaining $\ell(G) - 2n - n \log n - 1 = n^2$ characters then have to be 'f'. We summarize in the following lemma:

Lemma 2 (without proof). *For any completion x', $X = x(G) + x'$, $|X| = \ell(G)$ and $Y = IBWT(X)$ accepted by $A(G)$, x' contains n characters 'a' and n^2 characters 'f'. This gives for the IBWT graph $B(X)$ the nodes*

$$Z = (\text{'a'})^n + (\text{'b'})^{\frac{n}{2} \log n} + (\text{'c'})^{\frac{n}{2} \log n} + (\text{'d'})^n + (\text{'e'}) + (\text{'f'})^{n^2}$$

The above lemma states that $A(G)$, $x(G)$ and $\ell(G)$ give us complete knowledge on the nodes in Z and X, but not on the ordering of nodes in X except for the prefix $x(G)$.

The Edges of the IBWT Graph $B(X)$
Assuming that Y is accepted by $A(G)$, Lemma 2 gives complete knowledge of the nodes in $B(X)$. We now infer partial knowledge of the edges in $B(X)$.

Due to the IBWT mechanism, $B(X)$ contains an edge between the kth occurrence of a c-node in Z to the kth occurrence of a c-node in X for all $c \in \Sigma$ and each k. Remember that $x(G)$ is defined recursively in terms of sequences $b_{i,j}$ and $c_{i,j}$ of 'b' and 'c'. Each such sequence $b_{i,j}$ of 'b'-nodes in X corresponds to an equally long sequence $B_{i,j}$ of 'b'-nodes in Z such that there is a consecutive bundle of edges going from $B_{i,j}$ to $b_{i,j}$. Similarly, any sequence $c_{i,j}$ in $x(G)$ determines nodes $C_{i,j}$ in Z such that edges go from nodes in $C_{i,j}$ to nodes in $c_{i,j}$. Subsequently, we can find the recursive construction of $x(G)$ again in Z. We partition Z as

$$Z = Z\langle A\rangle + Z\langle B\rangle_{\log n,1} + Z\langle C\rangle_{\log n,1} + Z\langle D\rangle + \text{'e'} + (\text{'f'})^{n^2}$$

where $Z\langle A\rangle = (\text{'a'})^n$, $Z\langle D\rangle = (\text{'d'})^n$, and for all $j \in \mathbb{N}$ we define $Z\langle B\rangle_{0,j} = Z\langle C\rangle_{0,j}$ as the empty string, and for all $i \in \mathbb{N}$ we let

$$Z\langle B\rangle_{i,j} = B_{i,j} + Z\langle B\rangle_{i-1,2j-1} + Z\langle B\rangle_{i-1,2j} \text{ and}$$
$$Z\langle C\rangle_{i,j} = C_{i,j} + Z\langle C\rangle_{i-1,2j-1} + Z\langle C\rangle_{i-1,2j}.$$

Here each $B_{i,j} = (\text{'b'})^{2^{i-1}}$ is the sequence of 'b'-nodes which is connected by edges to the sequence $b_{i,j}$ in X. Similarly $C_{i,j} = (\text{'c'})^{2^{i-1}}$ is a sequence of 'c'-nodes connected to $c_{i,j}$ in X.

Apart from edges going from nodes in Z to nodes in X, some edges in the reverse direction are also predefined. In fact, the IBWT scheme tells us that for all $k \in \{1,\dots,n + n\log n + 1\}$ there is an edge going from $X[k]$ to $Z[k]$. By the following lemma we get this relation with respect to the recursively defined parts of Z and X.

Lemma 3 (without proof). *For all $i \in \{1,\dots,w\}$ and for all $j \in \{1,\dots,2^{w-i}\}$ there are edges in $B(X)$ going from the nodes in X defined by $x_{i-1,j}$ to the nodes in $Z\langle B\rangle_{i,j}$ and there are edges in $B(X)$ going from the nodes in X defined by $x_{i-1,j'}$ to the nodes in $Z\langle C\rangle_{i,j}$, where $j' = j + \frac{n}{2^i}$.*

The Result String Contains a Permutation of Index Encodings

Lemma 2 and Lemma 3 give us partial knowledge of the nodes and edges in the IBWT graph $B(X)$. As described in Section 2, the string $Y = \text{IBWT}(x(G)+x')$ is obtained by traversing the cycle of the bipartite graph $B(X)$. This implies that every directed path in $B(X)$ determines a substring of Y, as stated in the following lemma:

Lemma 4. *For each $v \in V$, $B(X)$ contains a path P_v determining the index encoding s_v as a substring of Y.*

Proof. We define the following kinds of path bundles: For all $i \in \{1,\dots,w\}$ and for all $j \in \{1,\dots,2^{w-i}\}$ the paths $B\langle i,j\rangle$, starting in nodes $B_{i,j}$ and the paths $C\langle i,j\rangle$ starting in nodes $C_{i,j}$.

We show by induction over $i \in \{w,\dots,1\}$ that

1. for all $j \in \{1,\dots,2^{w-i}\}$ any path of $B\langle i,j\rangle$ or respectively of $C\langle i,j\rangle$ describes the same string $s(B\langle i,j\rangle)$ or respectively $s(C\langle i,j\rangle)$ and

2. the subsumption of all paths $P\langle i\rangle = \bigcup_{j=1}^{2^{w-i}} B\langle i,j\rangle \cup C\langle i,j\rangle$ describe all possible strings of length $w - i + 2$ consisting of characters 'b' and 'c' and one terminal character 'a'.

Initially we have $i = w$. The paths $B\langle \log n, 1\rangle$ start in $B_{\log n,1}$ of Z and thus they all describe strings starting with character 'b'. Then, by definition, those paths go through $b_{w,1}$ in X to the nodes $Z[1], \ldots, Z[\frac{n}{2}] \in Z\langle A\rangle$ all associated to character 'a'. Hence, $s(B\langle w, 1\rangle) = $ 'ba'. Since $x(G)$ contains no character 'a' the paths end in $Z\langle A\rangle$. Accordingly the paths in $C\langle w, 1\rangle$ start in $C_{w,1}$, end in $Z[\frac{n}{2} + 1], \ldots, Z[n] \in Z\langle A\rangle$ and describe the string $s(C\langle w, 1\rangle) = $ 'ca'. Thus, $P\langle w\rangle = \{$'ba', 'ca'$\}$ describes all possible strings of length 2.

Now let $1 \le i < w$. Take first the paths $B\langle i+1, j\rangle$ for some $j \in \{1, \ldots, 2^{w-i-1}\}$ which by induction hypothesis start in $B_{i+1,j}$ and describe a unique string $s_1 = s(B\langle i + 1, j\rangle)$. As we have seen above in the proof of Lemma 3, the nodes of $B_{i+1,j}$ are seen from X by the sequence $(b_{i,j} + c_{i,j})$. The nodes $b_{i,j}$ are seen by $B_{i,j}$ and the nodes $c_{i,j}$ by $C_{i,j}$. Hence, $B\langle i, j\rangle$ and $C\langle i, j\rangle$ are sets of paths initiated in $B_{i,j}$ or $C_{i,j}$ and which contain as a tail the paths in $B\langle i+1, j\rangle$. Thus, the paths $B\langle i, j\rangle$ describe the unique string $s(B\langle i, j\rangle) = $ 'b' $+ s_1$ and similarly the paths $C\langle i, j\rangle$ describe the string $s(C\langle i, j\rangle) = $ 'c' $+ s_1$. The same can be shown for the paths $B\langle i, j'\rangle$ and $C\langle i, j'\rangle$ which go through $C_{i+1,j}$. They describe the strings $s(B\langle i, j'\rangle) = $ 'b' $+ s_2$ and $s(C\langle i, j'\rangle) = $ 'c' $+ s_2$. At the end of induction we obtain that $P\langle i\rangle$ contains all possible strings of length $w - i + 2$.

Now we have for all $j \in [\frac{n}{2}]$ paths in $B\langle 1, j\rangle$ and $C\langle 1, j\rangle$ starting at the nodes $B_{1,j}$ and $C_{1,j}$. The nodes $B_{1,j}$ and $C_{1,j}$ are seen from $x_{0,j} = $ 'd' or $x_{0,j+\frac{n}{2}} = $ 'd', respectively and thus, we get strings with initial character 'd'.   □

Note that the subsumption of all substrings $s_v, v \in V$ already consumes all available characters 'a' to 'd'. Therefore, apart from the id-strings, Y only contains one 'e' and n^2 characters 'f'. With other words, Y contains a permutation of all vertex index encodings, with one 'e' at the beginning and 'f' characters inbetween index encodings.

Every Permutation of Index Encodings can be Achieved
Building on Lemma 4, the following lemma states that every permutation of index encoding substrings in Y can be achieved by choosing for x' the appropriate order of characters 'a' and 'f'.

Lemma 5. *For any permutation $H = (v_1, \ldots, v_n)$ of the vertices V, where v_1 is the vertex with index 0, one can construct a completion x' from n characters 'a' and n^2 characters 'f' such that $Y = IBWT(x(G) + x')$ contains as substrings the index encodings s_v for all $v \in V$ in exactly the order given by H.*

Proof. As shown in the proof of Lemma 4, the IBWT graph $B(X)$ already contains paths corresponding to all n vertices. These paths begin at nodes $(Z[n + n \log n + 2], \ldots, Z[n + n \log n + n + 1])$ and end at nodes $(Z[1], \ldots, Z[n])$ in $B(X)$, with each path determining one index encoding s_v, except for the last which determines the string 'e' $+ s_0$.

Let s be the first start index $s = n + n \log n + 2$, let e be the first end index $e = 1$, and let W be the list of id-strings ordered with respect to the indices of their start nodes $(s, \ldots, s + n)$. We now iterate over the string x', adding one additional character in every step, either an 'a', or an 'f'. Adding an 'f' has the effect of moving the first string in the list W to the end of that list, and adding an 'a' has the effect of connecting two strings in W. As an iteration-invariant, we will maintain that every string in W corresponds to the path starting at node $Z[s + i]$, where i is the index of that string in W.

In particular, by adding an 'f', the path generated by the first string in W starting at index s is prolonged by two nodes 'f'. Its new start index is the index $s + |W|$. We update s and W by incrementing s, removing the first string p from W, and appending the string 'f'$+p$ to W. This maintains the iteration-invariant.

Adding an 'a' connects two paths, the path with end index e and the path with start index s. To maintain the iteration-invariant, we first determine the element q in W the path with end index e belongs to. Then we remove the first element p in W, and replace q in W with $q + p$. Then increment both s and e. Note that the list W is now one element shorter than before.

At each step, check whether the last node index encoded in q is adjacent to the first node index encoded in p in the permutation H. If so, add an 'a' to x'. Otherwise, add an 'f' to x' to move the string p to the end of the list W. The algorithm will always find a adjacent pair after at most n steps, because the end element q does not change when adding an 'f' to x'. After $n - 1$ 'a' characters have been placed, W only contains a single element. The algorithm then needs to add all remaining 'f' characters and the last 'a' character at the very end of x' to complete the cycle in $B(X)$. $\square$

NP-Completeness Theorem

By the use of the introduced construction and the related lemmas we will now show the connection between a Hamiltonian cycle in G and the possibility to find a completion for the instance $(A(G), x(G), \ell(G))$.

Theorem 1. *The BWT Reconstruction Problem is NP-complete.*

Proof. Membership: Given an instance (x, ℓ, A), a Turing machine can simply guess in a non-deterministic fashion all possibilities for $x' \in \Sigma^{\ell - |x|}$, compute in linear time whether $Y = \text{IBWT}(x + x')$ exists and, if it does, verify in linear time whether Y is accepted by the automaton A. Therefore, the BWT Reconstruction problem is in NP.

Completeness: For any graph $G = (V, E)$ with $|V| = n = 2^w$ and $w \in \mathbb{N}$ it is true: G contains a Hamiltonian cycle if and only if $(A(G), x(G), \ell(G))$ is a member of the BWT Reconstruction Problem.

$\Rightarrow$: If there is a Hamiltonian cycle H in G, then H is w.l.o.g. a permutation $(v_1, \ldots, v_n)$ of the vertices in V such that (1) v_1 is the vertex with index 0, (2) for all $i \in \{1, \ldots, n-1\}$ the edge (v_i, v_{i+1}) is in E, and (3) the edge (v_n, v_1) is in E. By Lemma 5 there is a completion x' such that in $Y = \text{IBWT}(x(G) + x')$ the substrings s_v occur exactly in the order given by H. The string Y would obviously be accepted by $A(G)$ since all consecutive substrings s_u and s_v in Y fulfill $(u, v) \in E$.

$\Leftarrow$: Reversely, if a completion x' exists such that $Y = \mathrm{IBWT}(x(G) + x')$, $|Y| = \ell(G)$ and $A(G)$ accepts Y then by Lemma 4 the string Y gives a permutation $(s_{v_1}, \ldots, s_{v_n})$ of the id-strings for the vertices V. Since Y is accepted by $A(G)$ it must be true that (1) v_1 is the vertex with index 0, (2) for consecutive strings s_u and s_v the edge (u, v) is in E, and (3) $(v_n, v_1) \in E$. Hence, $H = (v_1, \ldots, v_n)$ is a Hamiltonian cycle in G.

The reduction can be computed in polynomial time on a deterministic Turing machine, as shown in Lemma 1. □

4 Conclusions

This paper shows that recovering even partial content from damaged bzip-compressed files is generally not possible due to the use of the Burrows-Wheeler-Transform as part of the bzip compression algorithm. In practice, the current implementation of bzip splits its input into blocks and compresses each block separately. Therefore, undamaged blocks can still be recovered, and this partial recovery functionality is part of the implementation. Still, even minor damage to a bzip file can render large chunks of the original content essentially non-recoverable.

References

1. Brandstädt, A., Le, V.B., Spinrad, J.P.: Graph Classes; A Survey. SIAM Monographs on Discrete Mathematics and Applications (1999)
2. Burrows, M., Wheeler, D.J.: A Block-sorting Lossless Data Compression Algorithm. SRC Research Report (1994)
3. Garey, M.R., Johnson, D.S.: Computers and Intractability; A Guide to the Theory of NP-Completeness. W. H. Freeman & Co., New York (1979)
4. Hopcroft, J.E., Motwani, R., Ullman, J.D.: Introduction to Automata Theory, Languages, and Computation. Addison-Wesley, Reading (2000)
5. Seward, J.: The official BZip Homepage, http://www.bzip.org

Isoperimetric Problem and Meta-fibonacci Sequences*

B.V.S. Bharadwaj, L.S. Chandran, and Anita Das

Department of Computer Science and Automation,
Indian Institute of Science, Bangalore- 560012, India
{subramanya,sunil,anita}@csa.iisc.ernet.in

Abstract. Let $G = (V, E)$ be a simple, finite, undirected graph. For $S \subseteq V$, let $\delta(S, G) = \{(u, v) \in E : u \in S \text{ and } v \in V - S\}$ and $\phi(S, G) = \{v \in V - S : \exists u \in S, \text{ such that } (u, v) \in E\}$ be the edge and vertex boundary of S, respectively. Given an integer i, $1 \leq i \leq |V|$, the edge and vertex isoperimetric value at i is defined as $b_e(i, G) = \min_{S \subseteq V; \ |S|=i} |\delta(S, G)|$ and $b_v(i, G) = \min_{S \subseteq V; \ |S|=i} |\phi(S, G)|$, respectively. The edge (vertex) isoperimetric problem is to determine the value of $b_e(i, G)$ $(b_v(i, G))$ for each i, $1 \leq i \leq |V|$. If we have the further restriction that the set S should induce a connected subgraph of G, then the corresponding variation of the isoperimetric problem is known as the connected isoperimetric problem. The connected edge (vertex) isoperimetric values are defined in a corresponding way. It turns out that the connected edge isoperimetric and the connected vertex isoperimetric values are equal at each i, $1 \leq i \leq |V|$, if G is a tree. Therefore we use the notation $b_c(i, T)$ to denote the connected edge (vertex) isoperimetric value of T at i.

Hofstadter had introduced the interesting concept of meta-fibonacci sequences in his famous book "Gödel, Escher, Bach. An Eternal Golden Braid". The sequence he introduced is known as the Hofstadter sequences and most of the problems he raised regarding this sequence is still open. Since then mathematicians studied many other closely related meta-fibonacci sequences such as Tanny sequences, Conway sequences, Conolly sequences etc. Let T_2 be an infinite complete binary tree. In this paper we related the connected isoperimetric problem on T_2 with the Tanny sequences which is defined by the recurrence relation $a(i) = a(i - 1 - a(i - 1)) + a(i - 2 - a(i - 2))$, $a(0) = a(1) = a(2) = 1$. In particular, we show that $b_c(i, T_2) = i + 2 - 2a(i)$, for each $i \geq 1$.

We also propose efficient polynomial time algorithms to find vertex isoperimetric values at i of bounded pathwidth and bounded treewidth graphs.

1 Introduction

Let $G = (V, E)$ be a simple, finite, undirected graph. For $S \subseteq V$, $G[S]$ denotes the subgraph of G induced by the vertices in S. Let T be a rooted tree. The

---

* This research was funded by the DST grant SR/S3/EECE/62/2006.

X. Hu and J. Wang (Eds.): COCOON 2008, LNCS 5092, pp. 22–30, 2008.

depth of T is the number of nodes present in the longest path starting from the root and ending at a leaf. We assume the leaf to be at depth zero. Let $N(v) = \{w \in V(G)|vw \in E(G)\}$ be the set of neighbors of v and $N[v] = N(v) \cup \{v\}$.

Definition 1. *For $S \subseteq V$, the edge boundary $\delta(S, G)$ is the set of edges of G with exactly one end point in S. In other words,*

$$\delta(S, G) = \{(u, v) \in E : u \in S \text{ and } v \in V - S\}$$

Definition 2. *For $S \subseteq V$, the vertex boundary $\phi(S, G)$ is defined similarly.*

$$\phi(S, G) = \{v \in V - S : \exists u \in S, \text{ such that } (u, v) \in E\}$$

Definition 3. *Let i be an integer where $1 \le i \le |V|$. For each i define the edge isoperimetric value $b_e(i, G)$ and the vertex isoperimetric value $b_v(i, G)$ of G at i as follows*

$$b_e(i, G) = \min_{S \subseteq V; \ |S|=i} |\delta(S, G)|$$

$$b_v(i, G) = \min_{S \subseteq V; \ |S|=i} |\phi(S, G)|$$

It is easy to verify that $b_v(i, G) \le b_e(i, G)$, for each i. Note that, the set S which gives the minimum edge boundary need not be connected in general. Next we define a special kind of isoperimetric value by putting some restriction on the set S.

Definition 4. *Let i be an integer where $1 \le i \le |V|$. For each i define the connected edge isoperimetric value $b_{ce}(i, G)$ and the connected vertex isoperimetric value $b_{cv}(i, G)$ of G at i as follows:*

$$b_{ce}(i, G) = \min_{S \subseteq V; |S|=i; \ G[S] \text{ is connected}} |\delta(S, G)|$$

$$b_{cv}(i, G) = \min_{S \subseteq V; |S|=i; \ G[S] \text{ is connected}} |\phi(S, G)|$$

Lemma 1. *For a tree $T = (V, E)$, $b_{cv}(i, T) = b_{ce}(i, T)$, for each i, $1 \le i \le |V|$.*

Proof. It is easy to verify that $b_{cv}(i, T) \le b_{ce}(i, T)$. Now, if possible suppose $b_{cv}(i, T) < b_{ce}(i, T)$. Let $S_{cv} \subseteq V$ with $|S_{cv}| = i$ be such that $|\phi(S_{cv}, T)| = b_{cv}(i, T)$. As $|\phi(S_{cv}, T)| = b_{cv}(i, T) < b_{ce}(i, T) \le \delta(S_{cv}, T)$, there exists a vertex, say $x \in \phi(S_{cv}, T)$, such that at least two edges of $\delta(S_{cv}, T)$ are incident on x and clearly by the definition of $\delta(S_{cv}, T)$, the other end points of these edges belong to S_{cv}. Now, consider the subgraph induced by $A = S_{cv} \cup \{x\}$. Recalling that S_{cv} induces a connected subtree with i vertices and $i - 1$ edges, it is easy to see that A induces a connected subtree with $i + 1$ vertices but having $i + 1$ or more edges, which is clearly a contradiction. Hence the lemma. ∎

Let T_2 be an infinite complete binary tree. As connected vertex and edge isoperimetric problems are equivalent in T_2, we will call these problems to be *connected isoperimetric problem* of T_2 and the connected isoperimetric value of T_2 at i is denoted as $b_c(i, T_2)$.

1.1   Brief Literature Survey

Discrete isoperimetric inequalities form a very useful and important subject in graph theory and combinatorics. See [5], Chapter 16 for a brief introduction on isoperimetric problems. For a detailed treatment see the book by Harper [15]. See also the surveys by Leader [20] and by Bezrukov [2] for a comprehensive overview of work in the area. The edge (vertex) isoperimetric problem is NP-hard for an arbitrary graph.

The isoperimetric properties of graphs with respect to eigen values of their adjacency or Laplacian matrices is considered by many authors, for example see [1]. The isoperimetric properties of a graph is very closely related to its expansion properties. A graph G is called an expander graph if for every positive integer $i \leq \epsilon|V|$, $b_v(i, G) \geq \epsilon'i$, where ϵ and ϵ' are predefined constants.

The importance of isoperimetric inequalities lies in the fact that they can be used to give lower bounds for many useful graph parameters. For example it can be shown that $pathwidth(G) \geq b_v(G)$ [7], $bandwidth(G) \geq b_v(G)$ [12] and $cutwidth(G) \geq b_e(G)$ [3]. In [9], it is shown that given any j (where $1 \leq j \leq |V|$), $treewidth(G) \geq \min_{j/2 \leq i \leq j} b_v(i, G) - 1$ and in [8] it is shown that $carving\text{-}width(G) \geq \min_{j/2 \leq i \leq j} b_v(i, G)$, where $1 \leq j \leq |V|$ and in [12] it is shown that $wirelength(G) \geq \sum_{i=1}^{|V|} b_e(i, G)$.

A fibonacci sequence is a sequence of numbers defined by the recurrence relation $a(n) = a(n-1) + a(n-2)$ with $a(0) = 1$ and $a(1) = 1$. In [17], Hofstadter defined the sequence $Q(n)$ by

$$Q(n) = Q(n - Q(n-1)) + Q(n - Q(n-2)), \quad n > 2$$

with $Q(1) = Q(2) = 1$. He remarked on the apparent parallel between $Q(n)$ and the usual fibonacci recursion 'in that each new value is a sum of two previous values-but not of the immediately two values'. Motivated by this observation, Guy [11] reports that Malm calls $Q(n)$ a 'meta-fibonacci sequence'.

A meta-fibonacci sequence is given by the recurrence $a(n) = a(x_1(n) + a_1'(n - 1)) + a(x_2(n) + a_2'(n-2))$, where $x_i : \mathbb{Z}^+ \longrightarrow \mathbb{Z}^+$, $i = 1, 2$, is a linear function of n and $a_i'(j) = a(j)$ or $a_i'(j) = -a(j)$, for $i = 1, 2$. Hofstadter [17] had introduced the interesting concept of meta-fibonacci sequences in his famous book "Gödel, Escher, Bach. An Eternal Golden Braid". The sequence he introduced is known as the Hofstadter sequence and most of the problems he raised regarding this sequence is still open. Since then mathematicians studied many other closely related meta-fibonacci sequences such as Tanny sequences, Conway sequences, Conolly sequences etc. In [22], Tanny defined a sequence recursively as $a(i) = a(i - 1 - a(i-1)) + a(i - 2 - a(i-2))$, $a(0) = a(1) = a(2) = 1$. This sequence is known as Tanny sequence which is a close relative of the Hofstadter sequence.

1.2   Our Results

Let T_2 be an infinite complete binary tree. See Figure 1 for illustration. In this paper we related the connected isoperimetric problem on T_2 with the Tanny sequences. In particular we show the following:

Result 1. For a tree $T = (V, E)$, $b_{ce}(i, T) = b_{cv}(i, T)$, for each i, $1 \leq i \leq |V|$.

Result 2. For T_2, $b_c(i, T_2) = i + 2 - 2a(i)$, for each i, $i \geq 1$, where $a(i)$ is the Tanny sequence defined by the recurrence relation $a(i) = a(i - 1 - a(i - 1)) + a(i - 2 - a(i - 2))$, $a(0) = a(1) = a(2) = 1$.

Result 3. For a graph $G = (V, E)$ which has bounded pathwidth or bounded treewidth, we propose efficient polynomial time algorithms to find $b_v(i, G)$, for each i, $1 \leq i \leq |V|$.

Proof of theorems and lemmas are omitted due to page restriction and can be found in the journal version of the paper.

2   Connected Isoperimetric Problem on T_2

In this section we relate the connected isoperimetric problem on T_2 with some other problem on T_2 which enables us to find the connected isoperimetric value of T_2 at i, for each $i \geq 1$, easily.

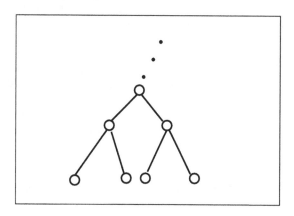

Fig. 1. Infinite complete binary tree

Definition 5. Let $S \subseteq V(T_2)$. Let $L(S) = \{u \in S \mid u \text{ is a leaf in } T_2\}$ and $l_{max}(i) = \max_{|S|=i;\ S \text{ is connected}} |L(S)|$. The leaf maximization problem is to find a subset S, $|S| = i$ such that $|L(S)| = l_{max}(i)$.

Lemma 2. Let T_2 be an infinite complete binary tree. If $S_c \subseteq V(T_2)$ with $|S_c| = i$ is such that $\delta(S_c, T_2) = \phi(S_c, T_2) = b_c(i, T_2)$, then $|L(S_c)| = l_{max}(i)$.

Definition 6. Let T be a tree rooted at r and has k children, say $c_1, c_2, \ldots, c_k$. Let $T_1, T_2, \ldots, T_k$ be the subtrees of T rooted at $c_1, c_2, \ldots, c_k$, respectively. The VLR-traversal of T (also known as pre-order traversal) is defined as follows: (1) Visit the root, (2) Visit T_1 in VLR-order, ..., $(k+1)$ Visit T_k in VLR-order.

The ordering of the vertices of T which is associated with the VLR-traversal is known as VLR-ordering.

Definition 7. *For a given integer i, $i > 0$, let d be the integer such that $2^{d-1} \leq i \leq 2^d - 1$ and T_2^d be the complete binary tree of depth d. Clearly T_2^d is a subtree of T_2. A VLR-tree of i nodes, denoted as $VLR(i)$, is the subtree of T_2 induced by the set of first i vertices in the VLR-ordering of T_2^d.*

Definition 8. *Let $S \subseteq V(T_2)$. A vertex $v \in S$ is called saturated in S if the set of children of v in T_2 is a subset of S, else v is called an unsaturated vertex.*

Given a tree $T = (V, E)$ with i nodes, it is easy to check whether it is isomorphic to a VLR-tree. Some properties of $VLR(i)$ are given below:

Property 1. At any depth $h \geq 0$ of T_2, $VLR(i)$ can contain at most one unsaturated vertex.

Property 2. If $h > 0$ is the minimum depth in T_2 such that $VLR(i)$ contains an unsaturated vertex, then all the vertices of $VLR(i)$ which are present at depth less than h are saturated in $VLR(i)$.

In this section we relate the connected isoperimetric problem on T_2 with the leaf maximization problem on T_2. Let $S \subseteq T_2$. For $h \geq 0$, let $N(S', h)$ denote the number of vertices of S present in depth h of T_2. The following lemma is easy to verify.

Lemma 3. *Let $VLR(i)$ be a VLR-tree with i vertices. Let S' induce a connected subtree of T_2. If $N(S', h) \geq N(VLR(i), h)$, then $N(S', h+1) \geq N(VLR(i), h+1) - 1$.*

Theorem 1. *Among all subtrees of T_2 with cardinality i, $VLR(i)$ contains maximum number of leaves.*

3   Meta-fibonacci Sequences and Connected Isoperimetric Problem on T_2

Let $VLR(i)$ be a VLR-tree. Number the leaves of T_2 which are present in $VLR(i)$ as follows: If two leaves of T_2 are present in $VLR(i)$ having same parent then number them as 1 and 2 else number as 1.

Definition 9. *A vertex v in $VLR(i)$ is called a i-leaf, $i = 1, 2$, if v is a leaf and is numbered as i.*

Definition 10. *A vertex in $VLR(i)$ which is the parent of some leaf is called a petiole. In particular, a vertex which is the parent of a j-leaf is called a j-petiole, $j = 1, 2$. Note that, a petiole can be 1-petiole and 2-petiole simultaneously.*

Let $T(i)$ denote the number of leaves present in $VLR(i)$ and set $T(0) = 1$. Let $T_j(i)$, denote the number of j-leaves present in $VLR(i)$, $j = 1, 2$. Clearly, for $0 \leq i < j$, $T_j(i) = 0$. The following lemma counts recursively the number of j-leaves, $j = 1, 2$, in $VLR(i)$.

Lemma 4. *In $VLR(i)$, $i \geq 3$, the number of j-leaves, $j = 1, 2$, is given by the following recurrence relation:*

$$T_j(i) = T(i - j - T(i - j)).$$

The following theorem gives a recurrence relation which finds the number of leaves present in a VLR-tree of i vertices.

Theorem 2. *The number of leaves in $VLR(n)$ is given by the following recurrence relation:*

$$T(n) = T(n - 1 - T(n - 1)) + T(n - 2 - T(n - 2)).$$

4   Primary Tree Decomposition of a Graph

For the definition of tree (path) decomposition and treewidth (pathwidth) we refer to [4].

Definition 11. *A **primary tree decomposition** of a graph $G = (V, E)$ is a triple $(\{X_i : i \in I\}, r \in I, T)$ where $(\{X_i : i \in I\}, T)$ is a tree decomposition of G that additionally satisfies the following properties.*

(1) T is a rooted binary tree and the subset X_r corresponds to the root node r.
(2) If a node $i \in I$ has two children j_1 and j_2, then $X_i = X_{j_1} = X_{j_2}$.
(3) If a node $i \in I$ has one child j, then $|X_i - X_j| = 1$.

It is well known that given a graph of treewidth at most k, for some constant k, a tree decomposition of optimum width can be constructed in linear time. Provided a tree decomposition of width k is given such a modified tree decomposition of the same width can be constructed in linear time whereby the new decomposition tree has at most $O(|V(G)|)$ nodes. We will assume in the reminder that such a modified tree decomposition T of G is given. A primary tree decomposition $(\{X_r : r \in I\}, r \in I, T)$ of G where T is a path is called a **primary path decomposition** of G. Note that as T is a path, the first and second condition in the definition of primary tree decomposition is trivially true.

5   Weighted Vertex Isoperimetric Problem in Bounded Pathwidth (Treewidth) Graphs

In this section, our intension is to develop efficient polynomial time algorithms for the vertex isoperimetric problem for graphs of bounded pathwidth and bounded treewidth. In fact we will solve a slightly more general problem, defined below:

Let $G = (V, E)$ be a weighted graph in which the vertices are weighted and $|V| = n$. Let $w : V(G) \longrightarrow R^+$ be the weight function defined on the vertices of G.

Definition 12. *For $S \subseteq V$, the weighted vertex boundary $\phi(S, G)$ is defined as*

$$\phi_w(S, G) = \sum_{v \in V-S \ : \ \exists u \in S \ such \ that \ (u,v) \in E} w(v).$$

Definition 13. *Let i be an integer where $1 \leq i \leq |V|$. For each i define the weighted vertex isoperimetric value $b_{wv}(i, G)$ at i as follows*

$$b_{wv}(i, G) = \min_{S \subseteq V; \ |S|=i} \phi_w(S, G)$$

The weighted vertex isoperimetric problem is to determine the value of $b_{wv}(i, G)$ for each i, $1 \leq i \leq |V|$. The reader may note that when the weight of each vertex is 1, then the weighted isoperimetric value problem is same as the usual vertex isoperimetric problem. Let $G = (V, E)$ and $|V| = n$ be a weighted graph in which the vertices are weighted and has pathwidth at most k, for some constant k.

Definition 14. *We define special vertex isoperimetric values as follows: Let $b_v^{S'}(i, G) = \min_{S \subseteq V, \ |S|=i, \ S' \subseteq S} \sum_{v \in V-S; \ \exists u \in S \ such \ that \ (u,v) \in E(G)} w(v)$. In other words, $b_v^{S'}$ is set to minimum of the sum of weight of the vertices which are in $V - S$ and an edge incident on these vertices whose another end point lies in S and S is made to contain a particular vertex set S'. If such a set S does not exist, that is when $|S| < |S'|$, then the value of $b_v^{S'}$ is set to ∞.*

Let T be a *primary path decomposition* of G which contains p bags, say X_1, $X_2, \ldots, X_p$ and rooted at a degree one node, say X_p. Note that, by definition of primary path decomposition, $p = O(n)$. Let $G_j = G[\bigcup_{i=1}^{j} X_i]$. Let i be a leaf node in T. As $|X_i| \leq k$, X_i can have at most 2^k possible subsets. The following lemma is easy to verify.

Lemma 5. *Let i be a leaf in T. Then for all $j \in \{0, 1, \ldots, |X_i|\}$, $b_{wv}(j, G[X_i])$ can be found in $O(k2^k)$ time.*

Definition 15. *The special vertex isoperimetric values of G_i at l, $l \leq |V(G_i)|$, with respect to the vertices of X_i is to find $b_v^S(l, G_i)$, for all $S \subseteq X_i$.*

Consider any bag X_i of T. Suppose we know all the special vertex isoperimetric values of G_{i-1} at j, $j \leq |V(G_{i-1})|$, with respect to the vertices of X_{i-1}. Now, our goal is to calculate the special vertex isoperimetric values of G_i at l, $l \leq |V(G_i)|$, with respect to the vertices of X_i. Note that there can be at most 2^k new special vertex isoperimetric values in G_i in comparison to the special vertex isoperimetric values of G_{i-1}.

Lemma 6. *Given all the special vertex isoperimetric values at i of G_r, $i \leq |V(G_r)|$, with respect to the vertices of X_r, all the special vertex isoperimetric values of G_{r+1} at j, $j \leq |V(G_{r+1})|$, with respect to the vertices of X_{r+1} (where $G_{r+1} = G[\bigcup_{i=1}^{r+1} X_i]$ and X_{r+1} is the parent of X_r in T) can be calculated in $O(n2^k)$ time.*

Now, suppose all the special vertex isoperimetric values of G_r at i, with respect to the vertices of X_r is known and the weighted vertex isoperimetric value of G_{r-1} is known for each i. Now, it is easy to see that the weighted vertex isoperimetric values of G_r at i can be found by taking the minimum of the weighted vertex isoperimetric value of G_{r-1} at i and all the special isoperimetric value of G_r at i with respect to the vertices of X_r. Recall that T can contain at most $O(n)$ nodes. So, to compute $b_{wv}(j, G_i)$, for each j, $j \leq |V(G_i)|$, it takes $O(n^2 2^k)$ time by Lemma 6. The following theorem follows from the above discussion.

Theorem 3. *There is a $O(2^k n^2)$ time algorithm to determine the weighted vertex isoperimetric value of a graph $G = (V, E)$ at i, $i \leq n$, having pathwidth at most k, for some constant k.*

Next suppose $G = (V, E)$ be a graph having treewidth at most k, for some constant k. Let $(\{X_i : i \in I\}, r \in I, T)$ be a primary tree decomposition of G. Let $G_i = G[\{\cup_j X_j, \text{ where } j \text{ is a descendent of } i \text{ in } T\}]$.

Let X_i has two children, say X_{i_1} and X_{i_2}, in T. Suppose all the special vertex isoperimetric values of G_{i_1} at j_1, $j_1 \leq |V(G_{i_1})|$, with respect to the vertices of X_{i_1} and G_{i_2} at j_2, $j_2 \leq |V(G_{i_2})|$, with respect to the vertices of X_{i_2} is known. As X_i contains at most k vertices, we need to compute only 2^k special vertex isoperimetric values.

Lemma 7. *Suppose X_i has two children, say X_{i_1} and X_{i_2}, in T. Given all the special vertex isoperimetric values of G_{i_1}, at j_1, $j_1 \leq |V(G_{i_1})|$, with respect to the vertices of X_{i_1} and all the special vertex isoperimetric values of G_{i_2} at j_2, $j_2 \leq |V(G_{i_2})|$, with respect to the vertices of X_{i_2}, all the special vertex isoperimetric values of G_i at j, $j \leq |V(G_i)|$, with respect to the vertices of X_i can be calculated in $O(n^2 2^k)$ time.*

Suppose all the special vertex isoperimetric values at j of G_i, $j \leq |V(G_i)|$, is given. Now, it is easy to see that the weighted vertex isoperimetric value of G_i at j, that is $b_{wv}(j, G_i)$, is nothing but $\min\limits_{S \subseteq V(G_i),\ |S| \leq j} b_v^S(j, G_i)$. As all the special vertex isoperimetric values of G_i at j is known, by using the above formula, $b_{wv}(j, G_i)$ can also be calculated. To compute $b_{wv}(j, G_i)$, for each j, $j \leq |V(G_i)|$, it takes $O(n^2 2^k)$ time by Lemma 7.

Theorem 4. *There is a $O(n^3 2^k)$ time algorithm to determine the weighted vertex isoperimetric values of a graph $G = (V, E)$ at i, $i \leq n$, with treewidth k, for some constant k.*

References

1. Alon, N., Millman, V.D.: λ_1, Isoperimetric Inequalities for Graphs and Super Concentrators. Journal of Combinatorial Theory Series B 38, 73–88 (1985)
2. Bezrukov, S.L.: Edge Isoperimetric Problems of Graphs. In: Graph Theory and Combinatorial Biology, vol. 7, pp. 157–197. Bolyai Soc. Math. Stud., Budapest (1999)

3. Bezrukov, S.L., Chavez, J.D., Harper, L.H., Röttger, M., Schroeder, U.-P.: The Congestion of n–cube Layout on a Rectangular Grid. Discrete Mathematics 213, 13–19 (2000)
4. Bodlaender, H.L.: A Tourist Guide Through Treewidth. Acta Cybernetica 11, 1–21 (1993)
5. Bollobás, B.: Combinatorics. Cambridge University Press, Cambridge (1986)
6. Bollobás, B., Leader, I.: Edge-Isoperimetric Inequalities in the Grid. Combinatorica 11, 299–314 (1991)
7. Chandran, L. S., Kavitha, T.: Treewidth and Pathwidth of Hypercubes. Special Issue of Discrete Mathematics on Minimal Separation and Chordal Completion (to appear, 2005)
8. Chandran, L.S., Kavitha, T.: The Carvingwidth of Hypercubes (2006)
9. Chandran, L.S., Subramanian, C.R.: Girth and Treewidth. Journal of Combinatorial Theory Series B 93, 23–32 (2005)
10. Diestel, R.: Graph Theory, 2nd edn. vol. 173. Springer, New York (2000)
11. Guy, R.K.: Some Suspiciously Simple Sequences. Amer. Math. Monthly 93, 186–190 (1986)
12. Harper, L.: Optimal Assignments of Numbers to Vertices. Jour. Soc. Indust. Appl. Math. 12, 131–135 (1964)
13. Harper, L.: Optimal Numberings and Isoperimetric Problems on Graphs. Journal of Combinatorial Theory 1, 385–393 (1966)
14. Harper, L.: On an Isoperimetric Problem for Hamming Graphs. Discrete Applied Mathematics 95, 285–309 (1999)
15. Harper, L.H.: Global Methods for Combinatorial Isoperimetric Problems. Cambridge University Press, Cambridge (2004)
16. Hart, S.: A Note on the Edges of the n–cube. Discrete Mathematics 14, 157–163 (1976)
17. Hofstadter, D., Gödel, Escher, Bach: An Eternal Golden Braid. Basic Books, New york (1979)
18. Houdré, C., Tetali, P.: Isoperimetric Invariants for Product Markov Chains and Graph Products. Combinatorica 24, 359–388 (2004)
19. Jackson, B., Ruskey, F.: Meta-Fibonacci Sequences, Binary Trees and Extremal Compact Codes. Electron. J. Combin. 13(1); Research Paper 26, p.13 (2006) (electronic)
20. Leader, I.: Discrete Isoperimetric Inequalities. In: Proc. Symp. Appl. Math, vol. 44, pp. 57–80 (1991)
21. Otachi, Y., Yamazaki, K.: A Lower Bound for the Vertex Boundary-Width of Complete k-ary Trees. Discrete Mathematics (in Press)
22. Tanny, S.M.: A Well-Behaved Cousin of the Hofstadter Sequence. Discrete Mathematics 105, 227–239 (1992)

On the Complexity of Equilibria Problems in Angel-Daemon Games*

Joaquim Gabarro, Alina García, and Maria Serna

ALBCOM Research Group
Universitat Politècnica de Catalunya
Edifici Ω, Campus Nord,
Jordi Girona, 1-3, Barcelona 08034, Spain
{gabarro,agarcia,mjserna}@lsi.upc.edu

Abstract. We analyze the complexity of equilibria problems for a class of strategic zero-sum games, called *Angel-Daemon* games. Those games were introduced to asses the goodness of a web or grid orchestration on a faulty environment with bounded amount of failures [6]. It turns out that *Angel-Daemon* games are, at the best of our knowledge, the first natural example of zero-sum succinct games in the sense of [1],[9]. We show that deciding the existence of a pure Nash equilibrium or a dominant strategy for a given player is Σ_2^p-complete. Furthermore, computing the value of an Angel-Daemon game is **EXP**-complete. Thus, matching the already known complexity results of the corresponding problems for the generic families of succinctly represented games with exponential number of actions.

1 Introduction

A Web or Grid application management system calls sites in order to perform sub-computations. There may be times when the user accepts the possibility of failure, but would like to have an estimate of the likelihood of success. Such an ex-ante analysis has been done using a subset of Orc [14],[15] to describe the orchestration of sites in the web application and game theory under some assumptions on the behavior of the failed sites and on the nature and amount of expected faults [6]. The study of systems under failure is not new [5],[16]; however, the analysis of orchestrations where control is *exercised by a single user* is different from the analysis of distributed systems under failure.

For doing such analysis in an environment with a bounded amount of failures the concept of *risk profile* was introduced in [7]. A risk profile specifies, for a set of sites participating in an orchestration, a partition into two sets $\mathcal{A}$ and $\mathcal{D}$ and two values $f_{\mathcal{A}}$ and $f_{\mathcal{D}}$. The set $\mathcal{A}$ represents those sites that fail due to

* Work partially supported by FET pro-active Integrated Project 15964 (AEOLUS) and by Spanish projects TIN2005-09198-C02-02 (ASCE), MEC-TIN2005-25859-E and TIN2007-66523 (FORMALISM). The first author was also partially supported by the FP6 Network of Excellence CoreGRID funded by the European Commission (Contract IST-2002-004265).

X. Hu and J. Wang (Eds.): COCOON 2008, LNCS 5092, pp. 31–40, 2008.

misbehavior or other reasons, but that essentially are not malicious, and thus called *angelic*. While the set $\mathcal{D}$ represent those sites with malicious behavior, and thus called *daemonic*. The numbers f_A and f_D constitute an estimation of the number of angelic and demonic failures. as a rough measure of the quality of the service. The final component is the *angel-daemon game* associated to a risk profile for a given Orc expression E. In an angel-daemon game there are two players the angel and the daemon. The angel selects f_A sites in $\mathcal{A}$ to fail and the daemon has to select f_D sites in $\mathcal{D}$ to fail. Once the set of failures is fixed we can compute the number of outputs of expression E when it is evaluated under such failures. The angel objective is to maximize this value while the daemon tries to minimize it. As such an angel-daemon game is a zero-sum game it follows from [22] that all the Nash equilibrium of the game have the same utility for the angel, which is known as the *game value*. The game value is used as the measure for assessing the expected behavior of expression E in the environment described by the risk profile [6],[7].

The main components of strategic games are players, their actions and payoff or utility functions. Each component of a game, that is a set or a function, can be described explicitly by means of a list or a table, or implicitly via computation models, for example a Turing machine. Depending on how we describe each component we can achieve different degree of succinctness [1],[10],[20]. Angel-daemon games constitute thus a natural class of strategic games with a succinct representation as the number of strategies is exponential in the size of the description. We are interested in analyzing the complexity of problems related to equilibria problems in such games. In particular the computational complexity of the following problems on angel-daemon games and on games with succinct representations.

Exists pure Nash equilibrium? (EPN). Given a game Γ decide whether Γ has a pure Nash equilibrium.

Exists dominant strategy? (EDS). Given a game Γ and a player i decide whether there is a dominant strategy for player i in Γ.

Game Value (GV). Given an zero-sum game Γ, compute its value.

We use standard terminology for classical complexity classes like **LOGSPACE**, **P**, **NP**, **coNP**, Σ_2^p and **EXP** [3],[19]. Basics on algorithmic game theory are in [17]. Our results include a characterization of the complexity of all the problems introduced above when the input is restricted to be an angel-daemon game. We show that deciding the PNE and the EDS problem are Σ_2^p-complete, and that GV is **EXP**-complete. This provides the first natural family of succinct games for which such complexity results can be established. Similar result for general families were already known for strategic games in implicit form [1] and succinct zero-sum games [9].

The paper provides first an introduction to strategic games, zero-sum games and their representations, followed by an introduction to the orchestration language Orc [14],[15].

2   Strategic Games and Succinct Representations

The following is the mathematical definition of strategic and perfect informa-tion extensive games borrowed form [18]. A *strategic game* Γ is a tuple $\Gamma = (N, (A_i)_{i \in N}, (u_i)_{i \in N})$ where $N = \{1, \dots, n\}$ is the set of players. For each player $i \in N$, A_i is a finite set of actions. For each player $i \in N$, u_i is a *utility* (or *payoff*) function, mapping $A_1 \times \dots \times A_n$ to the rationals.

Now we turn our attention to how the players play, given an strategic game $\Gamma = (N, (A_i)_{i \in N}, (u_i)_{i \in N})$ a strategy s_i for player i is a selection of an action $s_i \in A_i$. A *strategy profile* is a tuple $s = (s_1, \dots, s_n)$. Finally, after selecting independently the joint strategy profile s, player i gets utility $u_i(s)$. Given $s = (s_1, \dots, s_n)$ and a player i, we factorize s as (s_{-i}, s_i) where $s_{-i} = (s_j)_{j \neq i}$. A *zero-sum game* is a strategic two-player game in which for any strategy profile s holds $u_1(s) + u_2(s) = 0$.

A strategy profile $s = (s_1, s_2, \dots, s_n)$ is a *pure Nash equilibrium* if for any player i and any $s_i' \in A_i$ we have $u_i(s_{-i}, s_i) \geq u_i(s_{-i}, s_i')$. A *mixed strategy* σ_i for player i is a probability distribution on the set A_i. A *mixed strategy profile* is a tuple $\sigma = (\sigma_1, \dots, \sigma_n)$ and the utility of $u_i(\sigma)$ for player i is the expected utility. A mixed strategy profile $\sigma = (\sigma_1, \dots, \sigma_n)$ is a *Nash equilibrium* if for any player i and any other mixed strategy σ_i' for player i we have $u_i(\sigma_{-i}, \sigma_i) \geq u_i(\sigma_{-i}, \sigma_i')$. Any strategic game has a Nash equilibrium [18], in contraposition there are games without any pure Nash equilibrium. For the case of zero-sum games, a result by [22] states that in all the Nash equilibria the utility for player 1 is always the same, this value is called *the value* of the game [13]. A dominant strategy for player i is a strategy s_i^* such that, for any strategy profile s, $u_i(s_{-i}, s_i) \leq u_i(s_{-i}, s_i^*)$.

In order to study the computational complexity of problems on games it is fundamental to define how an input game Γ is represented. Notice that the complexity of a problem is always analyzed with respect to the size of the input. It is clear that the size of a game representation depends mainly on the number of players and on the size of the action set. And, in the same way, the utility function of each player, which is also part of the game description, depends on the number of strategy profiles. Because of these, we have to make clear how to describe the set of players, and for each player their set of actions and utility functions. Depending on the succinctness of the description of the action sets and depending on whether the pay-off functions are described implicitly by Turing machines (TM). A strategic game in *implicit form* [1] is represented by $\langle 1^n, 1^m, M, 1^t \rangle$. This game has n players. For each player i, their set of actions is $A_i = \Sigma^m$ and $\langle M, 1^t \rangle$ is the description of the utility functions. The utility of player i on strategy profile s is given by the output of the Turing machine M on input (s, i) after at most t steps.

Notice that in this case the length of the representation is proportional to n the number of players, to m the length of the actions (and then logarithmic to the number of different actions), and to t the computation time of the pay-off functions. Observe that this type of representation includes the *circuit games* considered in [10],[20]. In [1],[2] the complexity of several problems on games given in implicit form was analyzed. In particular they shown that the EPN problem is Σ_2^p-complete for games given in implicit form with four players. A *succinct* two-person zero-sum game [9] representation is a boolean circuit C such that the utilities are defined by $u_1(i,j) = C(i,j)$ and $u_2(i,j) = -C(i,j)$. It is well known that computing the value of a succinct zero-sum game is EXP-complete [8],[9]. To the best of our knowledge there are no results on the complexity of the EDS problem.

3   Orc and Angel-Daemon Games

An orchestration is a user-defined program that utilizes services on a web or grid. In Orc [14],[15] services are modelled by sites which have some predefined semantics. Typical examples of services are: an eigensolver, a search engine or a database [21]. A *site* accepts an argument and *publishes* a result value[1]. For example, a call to a search engine, $find(x)$, may publish the set of sites which *currently* offer service x. A site call is *silent* if it does not publish a result. Site calls may induce *side effects*. A site call can publish *at most one response*. An orchestration which composes a number of service calls into a complex computation can be represented by an *Orc expression*. An orchestrator may utilize any service that is available on the web. The simplest kind of Orc expression is a site (service) call. Orc has two special sites, site 1 and site 0. A call to 1 always returns exactly one signal. A call to 0 never returns and thus remains silent.

We define here the subset of Orc operations that we will use in our Orc expressions. A site call $s(v_1, \ldots, v_n)$ is called *non-blocking* if it must publish a result when $v_1, \ldots, v_n$ are well defined; otherwise s is *potentially blocking*. In Orc the predefined site *1* is non-blocking while $if(b)$ is potentially blocking. In this paper we will consider only some non-blocking expressions, to ensure that we use a subset of non-blocking operators. Two Orc expressions P and Q can be combined using the following operators:

- **Sequence** $P > x > Q(x)$: For each output, x, published by P an instance $Q(x)$ is executed. If P publishes the stream, $v_1, v_2, \ldots v_n$, then $P > x > Q(x)$ publishes some interleaved stream of the set $\{Q(v_1), Q(v_2), \ldots, Q(v_n)\}$. When the value of x is not needed we note $P \gg Q$.
- **Symmetric Parallelism** $P \mid Q$: $P \mid Q$ publishes *some* interleaving of the values published by P and Q.
- **Asymmetric parallelism** $P(x)$ where $x :\in Q$: P and Q are evaluated in parallel. Some subexpressions in P may become blocked by a dependency on x. The first result published by Q is bound to x, the remainder of Q's evaluation is terminated and evaluation of the blocked residue of P is resumed.

Observe that the two first constructions keep the structure of a regular expression but the second allows the introduction of variables to record partial results that can be used by other sites in the expression.

---

[1] The words "publishes","returns" and "outputs" are used interchangeably. The terms "site" and "service" are also used interchangeably.

We use $\alpha(E)$ be the set of sites that are referenced in orchestration E and $\alpha_+(E) = \alpha(E) \setminus \{0\}$. Let $\text{out}(E)$ be the number of outputs published by a non-blocking expression E. Given $E(z)$, depending on an input variable z, we denote by $E(\bot)$ the behavior of this expression when the value of z is undefined. This is equivalent to replace by 0 all subexpressions having a dependency on z. In [6] it was shown the following result. We keep $E(z)$ to represent the evaluation when z is defined.

Lemma 1 ([6]). *Let E_1, E_2 be non-blocking well formed Orc expression, and let s be a site, the number of publications, $\text{out}(E)$,*

$$\text{out}(0) = 0 \, , \ \text{out}(1) = 1$$

$$\text{out}(s(v_1, \ldots, v_k)) = \begin{cases} 1 & \text{if all the parameters are defined} \\ 0 & \text{otherwise} \end{cases}$$

$$\text{out}(E_1 | E_2) = \text{out}(E_1) + \text{out}(E_2)$$

$$\text{out}(E_1 > z > E_2(z)) = \text{out}(E_1) * \text{out}(E_2(z))$$

$$\text{out}(E_1(z) \text{ where } z :\in E_2) = \begin{cases} \text{out}(E_1(z)) & \text{if } \text{out}(E_2) > 0 \\ \text{out}(E_1(\bot)) & \text{if } \text{out}(E_2) = 0 \end{cases}$$

Given a non-blocking well formed Orc expression E, $\text{out}(E)$, can be computed in polynomial time with respect to the length of the expression E.

Although a site call may have a well-defined result it may be the case that a call to the site, in an untrusted environment, fails (silence). Given a non-blocking orchestration E it is unrealistic to assume that there will be no site failures during execution. Let $\mathcal{F} \subseteq \alpha_+(E)$ denote a set of sites that fail during an evaluation of E. The behavior of the evaluation of E in this environment corresponds to the orchestration in which all occurrences of sites $s \in \mathcal{F}$ are replaced by 0. Let $\varphi_{\mathcal{F}}(E)$ denote this expression. According to [6] and assuming that when a site fails it remains always silent; otherwise it publishes one result. The usefulness of the evaluation of E in failure environment $\mathcal{F}$ is measured by $\text{out}(\varphi_{\mathcal{F}}(E))$. Of course, according to Lemma 1, given E and the set $\mathcal{F}$, the value $\text{out}(\varphi_{\mathcal{F}}(E))$ can be computed in polynomial time in the length of E. Following we provide several examples of such non-blocking Orc expression.

Example 1. A call to site *CNN* publishes a digital version of the current newspaper. A call to *CNN* increases by one the site counter. Site *BBC* is similar. A call to *Email*(a, m) sends the email m to the address a. Also publishes the signal *sent* to denote completion. We consider restricted versions like *EmailAlice*(m) or *EmailBob*(m). Let us consider the *Two_Each* $= (CNN \mid BBC) > x >$ $(EmailAlice(x) \mid EmailBob(x))$ orchestration where sites *CNN* and *BBC* are called in parallel. Suppose that *CNN* responds first and the result is emailed to Alice and Bob. Later on *BBC* will answer and Alice and Bob get another newspaper. The orchestration publish 4 times *sent*, denoted as $\text{out}(\textit{Two\_Each}) = 4$. In *One_Each* $= (CNN > x > EmailAlice(x)) \mid (BBC > x > EmailBob(x))$ Alice receives the *CNN* and Bob the *BBC* and $\text{out}(\textit{One\_Each}) = 2$.

$$SEQ\_of\_PAR \triangleq (P \mid Q) \gg (R \mid S), \quad PAR\_of\_SEQ \triangleq (P \gg Q) \mid (R \gg S)$$

$$\mathcal{R}_1 = \langle SEQ\_of\_PAR, \{P, Q\}, \{R, S\}, 1, 1 \rangle, \mathcal{R}_2 = \langle SEQ\_of\_PAR, \{P, R\}, \{Q, S\}, 1, 1 \rangle$$

$$\mathcal{R}_3 = \langle PAR\_of\_SEQ, \{P, Q\}, \{R, S\}, 1, 1 \rangle, \mathcal{R}_4 = \langle PAR\_of\_SEQ, \{P, R\}, \{Q, S\}, 1, 1 \rangle$$

	$\mathcal{D}$			$\mathcal{D}$			$\mathcal{D}$			$\mathcal{D}$	
	R	S		Q	S		R	S		Q	S
$\mathcal{A}\ P$	1	1	$\mathcal{A}\ P$	0	1	$\mathcal{A}\ P$	0	0	$\mathcal{A}\ P$	1	0
Q	1	1	R	1	0	Q	0	0	R	0	1

$$\nu(\mathcal{R}_1) = 1 \qquad \nu(\mathcal{R}_2) = 1/2 \qquad \nu(\mathcal{R}_3) = 0 \qquad \nu(\mathcal{R}_4) = 1/2$$

Fig. 1. Expressions SEQ_of_PAR and PAR_of_SEQ are introduced in [4]. We consider angel daemon games for different risk profiles. The utility u_a is the number of outputs of the Orc expression. Games $\Gamma(\mathcal{R}_1)$ and $\Gamma(\mathcal{R}_3)$ have pure Nash equilibria. Games $\Gamma(\mathcal{R}_2)$ and $\Gamma(\mathcal{R}_4)$ have mixed Nash equilibria.

Example 2. When $\mathcal{F} = \{CNN\}$ we have $\varphi_{\{CNN\}}(CNN) = 0$, $\varphi_{\{CNN\}}(s) = s$ in all the other cases. Moreover as $0 \mid BBC = BBC$ we have

$$\varphi_{\{CNN\}}(Two\_Each) = BBC > x > (EmailAlice(x) \mid EmailBob(x))$$

and $\text{out}(\varphi_{CNN}(Two\_Each)) = 2$. Another example of faulty orchestration is

$$\text{out}(\varphi_{\{CNN, EmailAlice(x)\}}(Two\_Each)) = \text{out}(BBC > x > EmailBob(x)) = 1.$$

Our next definition introduced in [7] proposes a way to asses orchestrations in scenarios with a controlled environment in which some of the failures are benevolent while others are malicious.

Definition 1 ([7]). *Given E, the tuple $\mathcal{R} = \langle E, A, \mathcal{D}, f_A, f_D \rangle$ is a risk profile for E where $A \cup \mathcal{D} = \alpha_+(E)$, $A \cap \mathcal{D} = \emptyset$, $f_A \leq \#A$ and $f_D \leq \#\mathcal{D}$.*

Sites in A perform *as well as possible* and are called *angelic* while sites in $\mathcal{D}$ are expected to *maximize damage to the application* and are called *daemonic*. Given $\mathcal{R} = \langle E, A, \mathcal{D}, f_A, f_D \rangle$ a strategic situation occurs when E suffers the effects of two players $\mathfrak{a}$ and $\mathfrak{d}$, controling respectively A and $\mathcal{D}$ with opposing behaviors, called respectively the *angel* and the *daemon*. In particular consider the following zero-sum game:

Definition 2 ([6]). *The zero-sum angel-daemon game associated to risk profile $\mathcal{R}$ is the game $\Gamma(\mathcal{R}) = \langle N = \{\mathfrak{a}, \mathfrak{d}\}, A_\mathfrak{a}, A_\mathfrak{d}, u_\mathfrak{a}, u_\mathfrak{d} \rangle$ with two players, $\mathfrak{a}$ the angel and $\mathfrak{d}$ the daemon. The players have the following sets of actions, $A_\mathfrak{a} = \{a \subseteq A \mid \#a = f_A\}$ and $A_\mathfrak{d} = \{d \subseteq A \mid \#d = f_D\}$. As for an strategy profile $s = (a, d)$ the set of sites to fail associated with s is thus $a \cup d$, the utilities are $u_\mathfrak{a}(s) = \text{out}(\varphi_{a \cup d}(E))$ and $u_\mathfrak{d}(s) = -u_\mathfrak{a}(s)$.*

Figure 1 gives examples of angel-daemon games and contains cases in which the games have not pure Nash equilibrium. The value of the game $\nu(\mathcal{R})$ is a fundamental parameter as it is used to asses the viability of the execution of E under risk profile $\mathcal{R}$. Although it is easy to compute this value in some well behaved cases (as the ones in Figure 1) we will show that it is a computational hard problem.

Taking Lemma 1, given a risk profile $\mathcal{R}$ it is straightforward to obtain a description of the game $\Gamma(\mathcal{R})$ in implicit form or in zero-sum succinct form in polynomial time. Therefore, we have the following result.

Lemma 2. *Given a risk profile $\mathcal{R}$ for a non-blocking Orc expression E, a description of the angel-daemon game $\Gamma(R)$ in implicit form or zero-sum succinct form can be obtained in polynomial time.*

Thus angel-daemon games constitute a natural class of games with a succinct representation.

4 The Complexity of the EPN Problem

In this section we prove that deciding the existence of a pure Nash equilibrium for angel-daemon games is Σ_2^p-complete, thus this problem is as hard as for the general case of strategic games given in implicit form [1]. Due to the lack of space we state the reduction here and delay the proof to the full paper.

In order to prove the hardness for Σ_2^p we consider a restricted version of the Quantified Boolean Formula, the Q2SAT problem, which is Σ_2^p-complete [19]. We say that Φ is a Q2BF formula when Φ has the form

$$\Phi = \exists \alpha_1, \ldots, \alpha_n \forall \beta_1, \ldots \beta_m F(\alpha_1, \ldots, \alpha_n, \beta_1, \ldots \beta_m)$$

where F is a Boolean formula given in conjunctive normal form (CNF). Recall that Q2SAT is defined as follows: Given a Q2BF formula Φ over the boolean variables $\alpha_1, \ldots, \alpha_n, \beta_1, \ldots, \beta_m$, decide whether Φ is true.

We construct an Orc expression on $2(n + m) + 5$ sites from a given Q2BF formula $\Phi = \exists x_1, \ldots, x_n \forall y_1, \ldots, y_m F(x_1, \ldots, x_n, y_1, \ldots, y_m)$. For any i, $1 \leq i \leq n + 1$, we have two sites X_i and $\overline{X}_i$. For any j, $1 \leq j \leq m + 1$, we consider two sites Y_i and $\overline{Y}_i$. Let X be the following Orc expression $(X_1|\overline{X}_1) \gg \cdots \gg (X_{n+1}|\overline{X}_{n+1})$. Let Y be the expression $(Y_1 \gg \overline{Y}_1)|\cdots|(Y_{m+1} \gg \overline{Y}_{m+1})$.

Assume that $F = C_1 \wedge \cdots \wedge C_k$ where each C_λ, $1 \leq \lambda \leq k$ is a clause. We construct an expression O_λ for each clause $(1 \leq \lambda \leq k)$, this expression is formed by the parallel composition of the sites corresponding to the literals appearing in C_λ. The expression O associated to formula F is defined as $O_1 \gg \cdots \gg O_k$. The expression associated to the boolean formula is the following:

$$E_\Phi = Z(x) \text{ where } x :\in [(Y_{m+1} \gg X_{n+1})|(\overline{Y}_{m+1} \gg \overline{X}_{n+1})|(X \gg (O|Y))].$$

The associated risk profile is $\mathcal{R}(\Phi) =< E_\Phi, \mathcal{A}, \mathcal{B}, n + 1, m + 1 >$ with $\mathcal{A} = \{Z, X_1, \ldots, X_{n+1}, \overline{X}_1, \ldots, \overline{X}_{n+1}\}$ and $\mathcal{B} = \{Y_1, \ldots, Y_{m+1}, \overline{Y}_1, \ldots, \overline{Y}_{m+1}\}$. The proof of the following result will be given in the full paper.

Lemma 3. *The Q2BF formula Φ is true iff the angel-daemon game $\Gamma(\mathcal{R}(\Phi))$ has a pure Nash equilibrium.*

Putting together Lemma 2 and 3, and taking into account that the EPN problem for games given in implicit form belongs to Σ_2^p [1], we have.

Theorem 1. *Deciding whether an Angel-Daemon game has a pure Nash equilibrium is Σ_2^p-complete.*

As a consequence of the above result we can conclude also some results on the complexity of the EPN problem on succinct described games.

Corollary 1. *Deciding whether a zero-sum game given in implicit form has a pure Nash equilibrium is Σ_2^p-complete. Deciding whether a two-player strategic game given in implicit form has a pure Nash equilibrium is Σ_2^p-complete.*

The above results settles to a minimum the number of players, improving the already known results as in the reduction provided in [1] the number of players was four.

5   Computing the Value of an Angel-Daemon Game

In this section we prove that computing the number of outputs of a non-blocking Orc expression is P-complete and that computing the value of an angel-daemon game is **EXP**-complete. Thus, again the VG problem is as hard as for the general case of succinct zero-sum games [8],[9]. Due to the lack of space we sketch here the reduction and delay the complete construction to the full paper.

We provide a construction of an angel-daemon game associated to a succinct description of a zero-sum game. Assume that a boolean circuit C on $n + m$ inputs and $k + 1$ outputs describes a succinct zero-sum game Γ. Without loose of generality we assume that the circuit has only **and** and **or** gates and that all the negations are on the inputs.

We define first a non-blocking Orc expression S_C corresponding to a one output circuit C. For any i, $1 \le i \le n$, we have two sites X_i and $\overline{X}_i$. For any j, $1 \le j \le m$, we consider two sites Y_i and $\overline{Y}_i$. We refer to those sites as the *input sites*. The Orc expression uses a construction provided in [15] to express the orchestration of a work-flow on a dag. The idea is to associate a site with each gate and a variable to every wire in the circuit. The expression makes use of the **where** constructor to guarantee the appropriate work-flow. The input to the circuit is bound through variables to the input sites, so that when sites fail accordingly to a consistent assignment α, $\text{out}(\varphi_{\mathcal{F}_\alpha}(S_C)) = C(\alpha)$ where $\mathcal{F}_\alpha = \{X_i \mid x_i = 0\} \cup \{\overline{X}_i \mid x_i = 1\} \cup \{Y_i \mid y_i = 0\} \cup \{\overline{Y}_i \mid y_i = 1\}$. Thus the expression produces zero outputs iff and only if the circuit evaluates to zero. This construction is used to devise a reduction from the *Circuit value problem* which is P-complete [12],[11].

Theorem 2. *Given a non-blocking well formed Orc expression E, determining whether $\text{out}(E) \ne 0$ is P-complete. Given a set of failures $\mathcal{F} \subseteq \alpha_+(E)$, determining whether $\text{out}(\varphi_{\mathcal{F}}(E)) \ne 0$ is also P-complete.*

Turning again to the circuit C giving a succinct representation of a game. For each output gate we generate a different Orc expression, thus we have expressions

$S_C^0, \ldots, S_C^k$. We have to orchestrate those expressions in such a way that the final number of outputs correspond to the value represented by the circuit. To do so, for any $0 \leq j$ let us consider the Orc expression P_j that produces 2^j outputs, this can be obtained by repeatedly doubling the number of outputs, $P_0 = 1$ and $P_l = P_{l-1}|P_{l-1}$ for any $0 < l \leq j$. The combination $V_C = (S_C^0 \gg P_0)| \cdots |(S_C^k \gg P_k)$ produces $C(\alpha)$ outputs provided that the input variables fail according to α. The last step takes care of inconsistent sets of failures. Let X be the Orc expression $(X_1|\overline{X}_1) \gg \cdots \gg (X_n|\overline{X}_n)$ and let Y be the expression $(Y_1 \gg \overline{Y}_1)| \cdots |(Y_{m+1} \gg \overline{Y}_{m+1})$. Now we can define the expression associated to the circuit $E_C = X \gg (V_C|(Y \gg P_{k+1}))$, and the risk profile $\mathcal{R}_C = \langle E_C, \mathcal{A}, \mathcal{D}, n, m \rangle$ where $\mathcal{D} = \{Y_1, \ldots, Y_m, \overline{Y}_1, \ldots, \overline{Y}_m\}$ and $\mathcal{A} = \alpha^+(E_C) \setminus \mathcal{D}$. The proof of the following result will be given in the full paper.

Lemma 4. *Given a succinct description of a zero-sum game Γ by means of a circuit C, the angel-daemon game $\Gamma(\mathcal{R}_C)$ has the same value as Γ.*

According to Lemma 2 a succinct description of an angel-daemon game can be obtained in polynomial time. Furthermore, computing the value of a succinct zero-sum game is EXP-complete [9]. This together with Lemma 4 gives the following result.

Theorem 3. *Computing the value of an angel-daemon game is EXP-complete.*

6   Deciding the Existence of Dominant Strategies

In this section we conclude our contribution with some considerations on the last problem considered in the paper. The complexity of the EDN, to the best of our knowledge, has not been addressed before on succinctly represented games.

Observe that in the Orc expression constructed in the proof of Theorem 1, when the Q2BF formula Φ is true the associated angel-daemon game has a dominant strategy, but when formula Φ is not true the angel-daemon game has no pure Nash equilibrium and therefore has no dominant strategy. More formally, a player i has a dominant strategy if exists s_i^* such that for any other s_i and s_{-i} it holds $u_i(s_{-i}, s_i) \leq u_i(s_{-i}, s_i^*)$. From the definition is clear that EDS is in Σ_2^p. To prove completeness $\Phi = \exists \alpha_1, \ldots, \alpha_n \forall \beta_1, \ldots \beta_m F(\alpha_1, \ldots, \alpha_n, \beta_1, \ldots \beta_m)$ is mapped into two components. The first one is the game corresponding to $\mathcal{R}(\Phi) = <E_\Phi, \mathcal{A}, \mathcal{B}, n+1, m+1>$ given in in §4, the second one is $\mathfrak{a}$. We delay to the final paper the full proof of the following theorem.

Theorem 4. *Deciding whether a dominant strategy strategy exists for Angel-Daemon games, strategic games in implicit form and zero-sum games in implicit form is Σ_2^p-complete.*

References

1. Àlvarez, C., Gabarro, J., Serna, M.: Pure Nash Equilibria in Games with a Large Number of Actions. In: Jedrzejowicz, J., Szepietowski, A. (eds.) MFCS 2005. LNCS, vol. 3618, pp. 95–106. Springer, Heidelberg (2005)

2. Àlvarez, C., Gabarro, J., Serna, M.: Polynomial Space Suffices for Deciding Nash Equilibria Properties for Extensive Games with Large Trees. In: Deng, X., Du, D.-Z. (eds.) ISAAC 2005. LNCS, vol. 3827, pp. 634–643. Springer, Heidelberg (2005)

3. Balcazar, J.L., Díaz, J., Gabarro, J.: Structural Complexity I. 2nd edn. Springer, Heidelberg (1995)

4. Bougé, L.: Le mòdele de programmation à parallélisme de donés: une perspective sémantique. Techniques et science informatiques 12(5), 541–562 (1993)

5. Eliaz, K.: Fault Tolerant Implementation. Review of Economic Studies 69(3), 589–610 (2002)

6. Gabarro, J., García, A., Clint, M., Kilpatrick, P., Stewart, A.: Bounded Site Failures: an Approach to Unreliable Grid Environments. In: Danelutto, M., Frangopoulou, P., Getov, V. (eds.) Making Grids Work. Springer, Heidelberg (to appear, 2008)

7. Gabarro, J., García, A., Serna, M., Stewart, A., Kilpatrick, P.: Analysing Orchestrations with Risk Profiles and Angel-Daemon Games. In: CoreGRID Integration Workshop (April 2–4, 2008)

8. Feigenbaum, J., Koller, D., Shor, P.: A Game-Theoretic Classification of Interactive Complexity Classes. In: Proc. of the 10th Annual IEEE Conference on Structure in Complexity Theory, pp. 227–237 (1995)

9. Fortnow, L., Impagliazzo, R., Kabanets, V., Umans, C.: On the Complexity of Succinct Zero-Sum Games. In: Proc. of the 20th Annual IEEE Conference on Computational Complexity, pp. 323–332 (2005)

10. Gottlob, G., Greco, G., Scarcello, F.: Pure Nash Equilibria: Hard and Easy Games. J. Artif. Intell. Res. 24, 357–406 (2005)

11. Greenlaw, R., Hoover, J., Ruzzo, W.: Limits to Parallel Computation: P-Completeness Theory. Oxford University Press, Oxford (1995)

12. Ladner, R.: The Circuit Value Problem is Log Space Complete for P. ACM SIGACT News 7(1), 18–20 (1975)

13. von Neumann, J., Morgenstern, O.: Theory of Games and Economic Behavior. Princeton (1944)

14. Misra, J., Cook, W.: Computation Orchestration: A Basis for Wide-Area Computing. Software & Systems Modeling (2006) doi:10.1007/s10270-006-0012-1

15. Misra, J.: A Programming Model for the Orchestration of Web Services. In: Proc. of SEFM 2004, pp. 28–30 (2004)

16. Moscibroda, T., Schmid, S., Wattenhofer, R.: When Selfish Meets Evil: Byzantine Players in a Virus Inoculation Game. In: Proc. of PODC 2006, pp. 35–44 (2006)

17. Nisan, N., Roughgarden, T., Tardos, E., Vazirani, V.: Algorithmic Game Theory. Cambridge University Press, Cambridge (2007)

18. Osborne, M., Rubinstein, A.: A Course on Game Theory. MIT Press, Cambridge (1994)

19. Papadimitriou, C.: Computational Complexity. Addison-Wesley, Reading (1994)

20. Schoenebeck, G., Vadhan, S.: The Computational Complexity of Nash Equilibria in Concisely Represented Games. In: Proc. of the 7th ACM conference on Electronic commerce, pp. 270–279 (2006)

21. Stewart, A., Gabarro, J., Clint, M., Harmer, T., Kilpatrick, P., Perrott, R.: Managing Grid Computations: An ORC-Based Approach. In: Guo, M., Yang, L.T., Di Martino, B., Zima, H.P., Dongarra, J., Tang, F. (eds.) ISPA 2006. LNCS, vol. 4330, pp. 278–291. Springer, Heidelberg (2006)

22. Watson, J.: Strategy: An Introduction to Game Theory. W. W. Norton (2002)

Average-Case Competitive Analyses
for One-Way Trading

Hiroshi Fujiwara[1], Kazuo Iwama[2], and Yoshiyuki Sekiguchi[3]

[1] Department of Informatics, Kwansei Gakuin University
2-1 Gakuen, Sanda 669-1337, Japan
h-fujiwara@kwansei.ac.jp
[2] School of Informatics, Kyoto University
Yoshida-Honmachi, Kyoto 606-8501, Japan
iwama@kuis.kyoto-u.ac.jp
[3] Faculty of Marine Technology, Tokyo University of Marine Science and Technology
2-1-6 Etchujima, Koto-ku, Tokyo 135-8533, Japan
yoshi-s@kaiyodai.ac.jp

Abstract. Consider a trader who exchanges one dollar into yen and assume that the exchange rate fluctuates within the interval $[m, M]$. The game ends without advance notice, then the trader is forced to exchange all the remaining dollars at the minimum rate m. El-Yaniv et al presented the optimal *worst-case threat-based* strategy (WTB) for this game [4]. In this paper, under the assumption that the distribution of the maximum exchange rate is known, we provide average-case analyses using all the reasonable optimization measures and derive different optimal algorithms for each of them. Remarkable differences in behavior are as follows: Unlike other algorithms, *the average-case threat-based* strategy (ATB) that minimizes $E[\text{OPT}/\text{ALG}]$ exchanges *little by little*. The maximization of $E[\text{ALG}/\text{OPT}]$ and the minimization of $E[\text{OPT}]/E[\text{ALG}]$ lead to similar algorithms in that both exchange *all at once*. However, their timing is different.

1 Introduction

Since online competitive analysis was introduced, there has scarcely been any significant difference in attitude to worst-case performance evaluation of online algorithms. Most studies use, as their performance measure, $\max\left[\frac{\text{OPT}}{\text{ALG}}\right]$ or $\min\left[\frac{\text{ALG}}{\text{OPT}}\right]$ for maximization problems, where ALG and OPT denote the returns for maximization (the costs for minimization) of an online and an optimal offline algorithm, respectively. (The difference between the two is not essential, both are getting better with the quantity approaching to 1.0.) When it comes to average-case evaluation with an input distribution, what is an adequate measure? A natural extension based on expectations would be $E\left[\frac{\text{OPT}}{\text{ALG}}\right]$, $\frac{E[\text{OPT}]}{E[\text{ALG}]}$, $E\left[\frac{\text{ALG}}{\text{OPT}}\right]$, or $\frac{E[\text{ALG}]}{E[\text{OPT}]}$. It might seem that all of them are reasonable and which one to be used is just a matter of taste. In this paper we warn that such an

X. Hu and J. Wang (Eds.): COCOON 2008, LNCS 5092, pp. 41–51, 2008.
© Springer-Verlag Berlin Heidelberg 2008

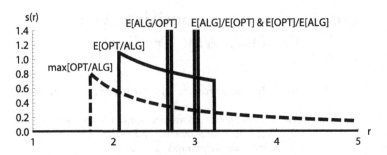

Fig. 1. Optimal exchange algorithms according to five different measures: The amount of dollars $s(r)dr$ to be exchanged when the exchange rate reaches r for the first time

easy choice is in fact quite dangerous by illustrating that the resulting optimal algorithm varies crucially depending on which measure is adopted.

Our problem in this paper is *one-way trading* [4]. In this problem the trader owns one dollar at the beginning and exchanges it into as much yen as possible, depending on the exchange rate. We assume the range of the exchange-rate fluctuation is guaranteed to be on $[m, M]$. Note that the rate does not necessarily reach either m or M. The trader is allowed to exchange an arbitrary amount of dollars to yen on each transaction but not allowed to get dollars back. We also assume that the cost of sampling the exchange rate and the transaction fees are negligible. There is a sudden end of the game at which the trader is obliged to exchange all the remaining dollars at the possible lowest rate m. The trader is not informed in advance when the game ends.

El-Yaniv et al [4] presented the *worst-case threat-based* algorithm WTB: (i) If the current exchange rate q is the highest thus far, then the trader should exchange $\int_{q_0}^{q} s_w(r)dr$ dollars, where $s_w(r) = \frac{1}{c_w(r-m)}$ for $mc_w \leq r \leq M$ and zero elsewhere, and q_0 is the highest rate on the previous transaction. (ii) Otherwise, the trader should not exchange. c_w is a constant determined by the equation $c = \ln \frac{M-m}{m(c-1)}$. Denoting the returns of an online and an optimal offline algorithms by ALG and OPT, respectively, WTB minimizes the *worst-case competitive ratio* $\max \left[\frac{\text{OPT}}{\text{ALG}}\right]$ (and maximizes $\min \left[\frac{\text{ALG}}{\text{OPT}}\right]$). They also proved that randomization does not help, i.e., the ratio cannot be improved even if the algorithm is randomized. Thus there does not seem any room for further improvement.

However, it is also true that worst-case analysis is often criticized as too pessimistic, and average-case analysis has always been an interesting research target. For online algorithms as well, several objective functions as presented above have recently appeared in the literature [6],[11],[1],[10],[7]. Our current problem involves a lot of human activities; it seems especially interesting to make an analysis under proper input distributions.

1.1   Our Contribution

Our goal in this paper is average-case competitive analyses for one-way trading under several input distributions and performance measures. Our first contribution is to show that the optimal algorithm becomes significantly different depending on the performance measure, such as $E\left[\frac{OPT}{ALG}\right]$, $\frac{E[OPT]}{E[ALG]}$, $E\left[\frac{ALG}{OPT}\right]$, or $\frac{E[ALG]}{E[OPT]}$. Figure 1 illustrates the optimal amount of dollars to be exchanged when the exchange rate reaches r for the first time, under the assumption that the maximum rate of a game is uniformly distributed from between one and five. For comparison, s_w of the algorithm WTB, independent of the distribution, is also drawn with a dashed line. (A) One can easily distinguish the difference between $E\left[\frac{OPT}{ALG}\right]$ and the others: Only $E\left[\frac{OPT}{ALG}\right]$ *exchanges little by little* on a certain rate range. In addition, this algorithm (ATB) that minimizes $E\left[\frac{OPT}{ALG}\right]$ differs from WTB: Whereas WTB waits for the possible maximum rate with keeping some dollars, ATB completes the transaction by the time the rate grows up to 3.24. (B) Note that the minimization of $E\left[\frac{OPT}{ALG}\right]$ and the maximization of $E\left[\frac{ALG}{OPT}\right]$ lead to different results. Recall that for worst-case analysis it does not make sense to change max and min with taking reciprocal. Unlike ATB, the algorithm that maximizes $E\left[\frac{ALG}{OPT}\right]$ is to *exchange all at once* at rate 2.67, which is drawn as a delta function in Figure 1. (C) Apparently, the minimization of $\frac{E[OPT]}{E[ALG]}$ and the maximization of $\frac{E[ALG]}{E[OPT]}$ are equivalent to the maximization of $E[ALG]$, since $E[OPT]$ is independent of the online algorithm. The result is to exchange all at once at rate 3.00. We will prove that for an arbitrary distribution both the maximization of $E\left[\frac{ALG}{OPT}\right]$ and that of $E[ALG]$ result in such one-time transaction. It is also shown that for IFR distributions, the exchange timing for $E\left[\frac{ALG}{OPT}\right]$ is earlier than that for $E[ALG]$.

We derive the performance of ATB and others all in closed form. This mathematically nontrivial analysis itself is another important contribution of this paper. Firstly, for one-way trading it is required to handle a function that describes an algorithm, i.e. *when and how much to exchange*, which obviously makes analysis involved. Secondly, unlike other average-case measures, this objective function is nonlinear: The return of an online algorithms appears in the denominator. El-Yaniv et al [4] gave some results on $E[ALG]$, which are derived implicitly based on the linearity. We constructively analyze $E\left[\frac{OPT}{ALG}\right]$ with the help of the calculus of variations. More specifically, a subproblem over a smaller solution space is firstly solved. Then we confirm the sufficient condition for the original problem.

1.2 Previous Work

Although the worst-case analysis introduced by Sleator and Tarjan [12] has provided us with beautiful results in online computation, its limitation has also been revealed: (a) The evaluation is often too pessimistic and therefore resulting optimal algorithms seem far from practical ones, for instance the optimal strategy for one-way trading is to wait for the possible maximum rate with some dollars left. (b) In some cases the worst-case competitive ratio cannot tell the performance difference properly, e.g. all of the algorithms LRU, FWF, and FIFO for paging have the same worst-case competitive ratio despite the clear difference in empirical performance [2]. For supplementing these weak points, average-case analysis using $E\left[\frac{ALG}{OPT}\right]$ and so on has been recently proposed. Note that the expectation $E[\cdot]$ is taken with respect to the input distribution.

To overcome the weak point (a), [6] first employed $E\left[\frac{\text{ALG}}{\text{OPT}}\right]$ as a criterion of optimality for designing algorithms for the ski-rental problem and presented a family of optimal online algorithms which is different from that for worst-case analysis. The significant difference in analysis between the ski-rent problem and one-way trading is as follows: Whereas an algorithm for the ski-rental problem is represented by a single real number which specifies when to buy skis, for one-way trading, as mentioned, we need to determine a function for trading.

As for the difficulty (b), paging and bin packing have been intensively studied. Becchetti showed the difference between FWF and LRU using $\frac{E[\text{ALG}]}{E[\text{OPT}]}$ [1]. Panagiotou and Souza investigated the performance of LRU for different cache size and mentioned that in this case $E\left[\frac{\text{ALG}}{\text{OPT}}\right]$ measures more adequately than $\frac{E[\text{ALG}]}{E[\text{OPT}]}$. Concerning bin packing, in contrast, $E\left[\frac{\text{ALG}}{\text{OPT}}\right]$ and $\frac{E[\text{ALG}]}{E[\text{OPT}]}$ make no difference as far as we consider the asymptotic ratio for a sufficiently long input sequence [10]. In the recent paper [7], Garg et al pointed out the difference between $E\left[\frac{\text{ALG}}{\text{OPT}}\right]$ and $\frac{E[\text{ALG}]}{E[\text{OPT}]}$ for the online Steiner tree problem: To obtain an upper bound of $E\left[\frac{\text{ALG}}{\text{OPT}}\right]$ requires a scenario-wise comparison independently of results for $\frac{E[\text{ALG}]}{E[\text{OPT}]}$.

El-Yaniv et al introduced one-way trading in the context of online computation [4]. They gave optimal algorithms using worst-case analysis, as well as some bounds of the worst-case competitive ratio. Chen et al [3] obtained the worst-case competitive ratio in explicit form under the assumption that the exchange rate follows a geometric Brownian motion. Lorenz et al [9] studied the problem under the setting that at each point of time the trader chooses whether to convert a fixed fraction of the initial wealth or to do nothing. Other models, including two-way trading and portfolio selection, are mentioned in [2].

2   Worst-Case Threat-Based Strategy

Recall that in our model the trader knows the possible range $[m, M]$ of the exchange-rate fluctuation and does not know when the game ends. (In [4] this model is referred to as Variant 2.) For convenience of calculus we assume that the exchange rate $q(t)$ is piecewise continuous with time $t \geq 0$. The trader can exchange $\int_t^{t+\Delta t} d\tilde{S}(t)$ dollars into $\int_t^{t+\Delta t} q(t)d\tilde{S}(t)$ yen in an arbitrarily small time interval Δt. Thus the algorithm of the trader is represented by a function $\tilde{S} : [0, \infty) \to [0, 1]$. The rule of one-way transaction forces $\tilde{S}$ to be a non-decreasing function. Suppose that the game is over at time τ. Then the trader has to exchange the remaining dollars at the minimum rate m. The total return is written as $\text{ALG}_{\tilde{S}}(\tau) = \int_0^\tau q(t)d\tilde{S}(t) + m\left(1 - \int_0^\tau d\tilde{S}(t)\right) = m + \int_0^\tau (q(t) - m)\, d\tilde{S}(t)$. On the other hand, the optimal offline algorithm will exchange the whole one dollar at the highest rate throughout the game and therefore its return is $\text{OPT}(\tau) = \max_{0 \leq x \leq \tau} q(x)$.

El-Yaniv et al stated that it is enough to consider algorithms that exchange only when the exchange rate is the highest so far [4]. Let $\bar{q}(t)$ denote $\max_{0 \leq x \leq t} q(x)$. Such an algorithm exchanges $\int_{\bar{q}(t)}^{q(t+\Delta t)} dS(r)$ dollars if $q(t+\Delta t) >$

$\bar{q}(t)$ and does not exchange otherwise, where $S : [m, M] \rightarrow [0, 1]$ is a non-decreasing function. We will describe the algorithm by $s(r) := \frac{dS(r)}{dr}$ if the derivative of S exists. Suppose that the highest exchange rate of the game is p. Without loss of generality, we assume $q(0) = m$. Let us denote simply by ALG the total return. We have $\text{ALG} = \int_m^p r dS(r) + m \left(1 - \int_m^p dS(r)\right) = m + \int_m^p (r - m) dS(r)$. The optimal offline algorithm gains $\text{OPT} = p$. Here is the (worst-case) threat-based algorithm proposed by El-Yaniv et al.

Algorithm WTB: Suppose that the exchange rate changes from $q(t)$ to $q(t + \Delta t)$. (i) If $q(t + \Delta t) > \bar{q}(t)$ then exchange $\int_{\bar{q}(t)}^{q(t+\Delta t)} s_w(r) dr$ dollars, where

$$s_w(r) = \begin{cases} \frac{1}{c_w(r-m)}, & mc_w \leq r \leq M; \\ 0, & m \leq r < mc_w. \end{cases}$$

(ii) If $q(t + \Delta t) \leq \bar{q}(t)$ then do not exchange.

Theorem 1 ([4]). *For any q, the algorithm WTB minimizes the worst-case competitive ratio $\max_\tau \left[\frac{OPT(\tau)}{ALG_S(\tau)}\right]$ to c_w, where c_w is a constant c determined by the equation $c = \ln \frac{M-m}{m(c-1)}$.*

3   Average-Case Threat-Based Strategy

We consider the model in which the highest exchange rate until a game is over follows a probability distribution and the trader devises a strategy with help of the distribution. The distribution is characterized by a cumulative distribution function $F : [m, M] \rightarrow [0, 1]$: $F(p)$ is the probability that the highest exchange rate of the game is equal to or less than p. Throughout this paper, $E[\cdot]$ denotes the expectation with respect to F unless otherwise specified as $E_H[\cdot]$ for a distribution H. We seek an optimal online algorithm that minimizes *the average-case competitive ratio*

$$E\left[\frac{\text{OPT}}{\text{ALG}}\right] = \int_m^M \frac{p}{m + \int_m^p (r - m) s(r) dr} dF(p)$$

among those that exchange only when the exchange rate is the highest so far.

The following lemma claims that it suffices to consider such algorithms that exchange only when the exchange rate is the highest thus far, as well as in the worst-case analysis. $\bar{q}(t)$ denotes $\max_{0 \leq x \leq t} q(x)$. The proof is omitted due to space limitation.

Lemma 1. *Given a distribution F and a rate function q, consider H such that $H(\tau) = F(\bar{q}(\tau))$ for all $\tau \geq 0$. (Note that $H(\tau)$ is the probability that the game ends by the time τ.) Suppose that there exists S^* such that $E_F[OPT/ALG_{S^*}] \leq E_F[OPT/ALG]$ holds for all S. Let us convert S^* into $\tilde{S}^*$ by $\tilde{S}^*(t) = S^*(\bar{q}(t))$ for all $t \geq 0$. Then, $E_H[OPT(\tau)/ALG_{\tilde{S}^*}(\tau)] \leq E_H[OPT(\tau)/ALG_{\tilde{S}}(\tau)]$ for all $\tilde{S}$.*

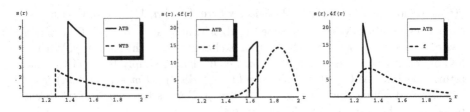

Fig. 2. ATB and WTB for a uniform distribution   **Fig. 3.** ATB for a Weibull-type distribution   **Fig. 4.** ATB for a Fréchet-type distribution

As a result of the analysis in Section 3.3 we obtain the optimal algorithm, which advises to exchange dollars only during a certain exchange-rate range of $[\alpha, \beta] \subset [m, M]$. In what follows we assume that the distribution F has a positive density function f.

Algorithm ATB: Suppose that the exchange rate changes from $q(t)$ to $q(t+\Delta t)$.
(i) If $q(t + \Delta t) > \bar{q}(t)$ then exchange $\int_{\bar{q}(t)}^{q(t+\Delta t)} s_a(r)dr$ dollars, where

$$
s_a(r) = \begin{cases} \frac{m}{(\alpha-m)\sqrt{\alpha f(\alpha)}} \left\{ \frac{3r-m}{2(r-m)} \sqrt{\frac{f(r)}{r}} + \frac{f'(r)}{2} \sqrt{\frac{r}{f(r)}} \right\}, & \alpha \le r \le \beta; \\ 0, & \text{otherwise.} \end{cases} \tag{1}
$$

(ii) If $q(t + \Delta t) \le \bar{q}(t)$ then do not exchange. α and β are constants given by

$$
\int_\alpha^\beta s_a(r)dr = 1 \text{ and } \beta(\beta - m)f(\beta) = \int_\beta^M pf(p)dp. \tag{2}
$$

Given f, one can have the value of β from the second equation in (2). The value of α is determined by the first equation in (2) after substituting s_a with α unknown. The optimality of ATB requires the condition (C). A simple sufficient condition for (C) is $(\tilde{C})$.

(C) $\alpha(\alpha - r)(\alpha - m)f(\alpha) - (r - m) \int_r^\alpha pf(p)dp \ge 0$ for all $r \in [m, \alpha]$, $(3r - m)f(r) + (r - m)rf'(r) > 0$ for all $r \in (\alpha, \beta)$, and $\beta(\beta - m)^2 f(\beta) - (r - m) \int_r^M pf(p)dp \ge 0$ for all $r \in [\beta, M]$.

$(\tilde{C})$ f is non-decreasing on $[m, \beta]$ and $(r - m)^2 rf(r) \ge (\beta - m)^2 \beta f(\beta)$ holds for all $r \in [\beta, M]$.

3.1   Uniform Distribution

Consider the probability density function $f(p) = \frac{1}{M-m}$, which means that the highest rate p is uniformly distributed on the range $[m, M]$. The optimal online algorithm ATB is then represented by

$$
s_a(r) = \begin{cases} \frac{m}{(\alpha-m)\sqrt{\alpha f(\alpha)}} \cdot \frac{3r-m}{(r-m)\sqrt{r}}, & \alpha \le r \le \beta; \\ 0, & \text{otherwise.} \end{cases}
$$

We derive $\beta = \frac{1}{3}\left(m + \sqrt{m^2 + 3M^2}\right) < m + \frac{2}{3}(M - m)$ from (2). One can easily confirm the condition $(\widetilde{C})$. Note that the uniform distribution means that we have a completely even chance of the highest exchange rate in the range. This allows us to concentrate on rather narrow range for trade, which is obviously impossible if we assume the worst-case adversary. Actually, the above inequality implies that ATB completes the conversion before the rate reaches 66% of the range $[m, M]$. Figure 2 illustrates the case of $m = 1$ and $M = 2$. When ATB finishes the entire transaction, WTB still retains 48% of the initial asset. Moreover, WTB waits with some dollars left until the exchange rate finally reaches M, no matter how scarcely the luck happens. The advantage of ATB appears in the performance measure: By exchanging dollars intensively on a narrow range of $[1.38, 1.53]$, $E\left[\frac{\text{OPT}}{\text{ALG}}\right]$ is improved down to 1.20 from 1.29 $(= c_w = E\left[\frac{\text{OPT}}{\text{WTB}}\right])$.

3.2 Extreme Value Distribution

If random variables are independent and identically distributed then the maxima under some affine transformation follow the generalized extreme value distribution [8,5] $f(p) = \frac{1}{\sigma}\left(1 + k\frac{p-\mu}{\sigma}\right)^{-1-\frac{1}{k}} \exp\left(\left(1 + k\frac{p-\mu}{\sigma}\right)^{-\frac{1}{k}}\right)$. Apparently, this distribution provides the trader with useful information on the exchange rate. Consequently, the degree of improvement is outstanding compared with that of the uniform distribution. Again, let $m = 1$ and $M = 2$. Recall that the value of c_w is 1.29 previously. (i) Weibull-type distribution, say, $k = -0.5, \sigma = 0.12, \mu = 1.77$ (see Figure 3): The density function has a peak at 1.84 and a long tail to the left. That is, the trader is informed that the rate is in an upward trend. ATB exchanges on the range $[1.59, 1.66]$ and $E\left[\frac{\text{OPT}}{\text{ALG}}\right]$ is reduced down to 1.15. This setting satisfies the condition $(\widetilde{C})$. (ii) Fréchet-type distribution, say, $k = 0.5, \sigma = 0.2, \mu = 1.39$ (see Figure 4): This setting yields a right-tailed distribution whose peak is at 1.32. The trader anticipates that the rate does not grow so much. ATB converts dollars on the range $[1.27, 1.34]$ and $E\left[\frac{\text{OPT}}{\text{ALG}}\right]$ becomes 1.19. One can confirm the condition (C).

3.3 Proof of Optimality

The proof of the optimality involves the calculus of variations. The minimization of $E\left[\frac{\text{OPT}}{\text{ALG}}\right]$ can be written as:

$$(\mathcal{P}) \text{ minimize } J(s) := \int_m^M \frac{pdF(p)}{m + \int_m^p (r - m)s(r)dr}$$

$$\text{subject to } \int_m^M s(r)dr = 1, \ s(r) \geq 0, \ s \in L^1[m, M].$$

Outline of the proof. We firstly provide a sufficient condition for global optimality for $(\mathcal{P})$. Next we consider a subproblem $(\mathcal{P}_0)$ over a smaller solution space. Starting from a necessary condition for $(\mathcal{P}_0)$, we derive a local optimal solution. Finally, it is confirmed that the obtained solution satisfies the sufficient condition for the original problem $(\mathcal{P})$.

The proof of Lemma 2 is omitted due to space limitation.

Lemma 2. *Suppose that there exist feasible $\bar{s} \in L^1[m, M]$, $\lambda \in R$ and $\mu \in L^\infty[m, M]$ such that $\mu(r) \leq 0$ almost everywhere, $\int_m^M \mu(r)\bar{s}(r)dr = 0$ and*

$$\int_m^M \frac{-p\int_m^p(r - m)v(r)dr}{\left(m + \int_m^p(r - m)\bar{s}(r)dr\right)^2}dF(p) + \int_m^M \lambda v(r)dr + \int_m^M \mu(r)v(r)dr = 0 \quad (3)$$

for all $v \in L^1[m, M]$. Then $\bar{s}$ is a unique solution to the problem $(\mathcal{P})$.

Theorem 2. *Suppose that a distribution F has a density function f which is positive, continuously differentiable, and satisfies the condition (C). Then the unique minimizer of the problem $(\mathcal{P})$ is s_a in (1). The constants are given by (2).*

Proof. First, we narrow the candidates of solutions to functions which are non-negative and continuous on a subinterval of $[m, M]$ and zero elsewhere. Then the problem finding a minimizer from such family of functions has the following form:

$(\mathcal{P}_0)$ minimize $J_0(s, \alpha, \beta)$

$$:= \int_m^\alpha \frac{pf(p)}{m}dp + \int_\alpha^\beta \frac{pf(p)dp}{m + \int_\alpha^p(r - m)s(r)dr} + \int_\beta^M \frac{pf(p)dp}{m + \int_\alpha^\beta(r - m)s(r)dr}$$

$$\text{subject to } G(s, \alpha, \beta) := \int_\alpha^\beta s(r)dr = 1, \ s(r) \geq 0, \ m < \alpha < \beta < M,$$

$$s \in C[m, M].$$

Observe the problem $(\mathcal{P}_0)$ is no longer a convex programming. Thus we find a function $\bar{s}$ which satisfies a necessary condition for local optimality in $(\mathcal{P}_0)$. Then we will confirm the condition in Lemma 2 by explicitly giving a constant λ and a function μ there.

Suppose $(\bar{s}, \alpha, \beta)$ is a local optimal of $(\mathcal{P}_0)$ and $\bar{s}$ is positive. Then there exists $\lambda \in R$ such that $DJ_0(\bar{s}, \alpha, \beta)(v, \xi, \eta) + \lambda DG(\bar{s}, \alpha, \beta)(v, \xi, \eta) = 0$ for all $\xi, \eta \in R$, $v \in C[m, M]$, where $DJ_0(\bar{s}, \alpha, \beta)$ and $DG(\bar{s}, \alpha, \beta)$ are the Gâteaux derivatives for J_0 and G, respectively. By substituting $\xi = \eta = 0$, we have $\int_\alpha^\beta \frac{-pf(p)\int_\alpha^p(r-m)v(r)dr}{\left(m+\int_\alpha^p(t-m)\bar{s}(t)dt\right)^2}dp + A \cdot \int_\alpha^\beta(r - m)v(r)dr + \lambda \int_\alpha^\beta v(p)dp = 0$, where $A = \frac{-\int_\beta^M pf(p)dp}{\left(m+\int_\alpha^\beta(t-m)\bar{s}(t)dr\right)^2}$.

We change the order of the integration of the first term and replace p with r in the third term. Then we obtain

$$\int_\alpha^\beta v(r)\left\{(r - m)\left(\int_r^\beta \frac{-pf(p)dp}{\left(m + \int_\alpha^p(t - m)\bar{s}(t)dt\right)^2} + A\right) + \lambda\right\}dr = 0.$$

DuBois-Reymond's lemma states that if the equation holds for every $v \in C[m, M]$, then

$$(r - m) \left(\int_r^\beta \frac{-pf(p)dp}{\left(m + \int_\alpha^p (t - m)\bar{s}(t)dt\right)^2} + A \right) + \lambda = 0 \tag{4}$$

for all $r \in [\alpha, \beta]$. By substituting $r = \beta$ and $r = \alpha$ in the equation (4), we have respectively $(\alpha - m)B + (\alpha - m)A + \lambda = 0$ and $(\beta - m)A + \lambda = 0$, where $B = \int_\alpha^\beta \frac{-pf(p)dp}{\left(m + \int_\alpha^p (t - m)\bar{s}(t)dt\right)^2}$. We also have

$$\int_r^\beta \frac{-pf(p)dp}{\left(m + \int_\alpha^p (t - m)\bar{s}(t)dt\right)^2} + (r - m)\frac{rf(r)}{\left(m + \int_\alpha^r (t - m)\bar{s}(t)dt\right)^2} + A = 0 \tag{5}$$

by differentiating the equation (4) and hence $B + \frac{(\alpha-m)\alpha f(\alpha)}{m^2} + A = 0$ by letting $r = \alpha$. Thus we obtain $\lambda = \frac{(\alpha-m)^2 \alpha f(\alpha)}{m^2}$. Then we eliminate the first term in the equations (4) and (5) and hence $(r - m)^2 \frac{rf(r)}{\left(m + \int_\alpha^r (t-m)\bar{s}(t)dt\right)^2} - \lambda = 0$. Thus we obtain $m + \int_\alpha^r (t - m)\bar{s}(t)dt = \frac{1}{\sqrt{\lambda}}(r - m)\sqrt{rf(r)}$. By differentiating we have $\bar{s}(r) = \frac{1}{\sqrt{\lambda}}\left\{ \frac{3r-m}{2(r-m)}\sqrt{\frac{f(r)}{r}} + \frac{f'(r)}{2}\sqrt{\frac{r}{f(r)}} \right\}$. The second inequality in the condition (C) guarantees its feasibility. In addition, by substituting β for r in the equation (5), we have the condition for β. Therefore we have shown that $\bar{s}$ is equal to s_a.

To complete the proof, we use Lemma 2 to show the function to be a minimizer of the original problem $(\mathcal{P})$. We substitute $\bar{s}$ for s in (3) and change the order of the integration. The condition (C) implies that we can choose $\mu(r) \leq 0$ almost everywhere to make each term zero and (3) is satisfied by obtained $\bar{s}$, λ and μ. □

3.4   Randomized Algorithms

The following result shows that randomization of one-way trading algorithms cannot improve E_G [RALG] nor therefore the average-case performance measure such as $E_F \left[\frac{\text{OPT}}{E_G[\text{RALG}]} \right]$, $E_F \left[\frac{E_G[\text{RALG}]}{\text{OPT}} \right]$, or $E_F [E_G [\text{RALG}]]$, where E_G [RALG] denotes the expected return of a randomized algorithm RALG.

Proposition 1 ([4]). *Let RALG be any randomized algorithm including mixed and behavioral strategies. Then, there exists a deterministic algorithm represented by S such that E_G [RALG] $= ALG_S$ for any scenario of the exchange-rate fluctuation. The reverse statement also holds true: For any deterministic algorithm that is represented by S, there exists a randomized algorithm RALG such that satisfies the above equality.*

4   Optimal Algorithms for Other Measures

In this section we consider $E\left[\frac{ALG}{OPT}\right]$ and $E\left[ALG\right]$. Unlike the minimization of $E\left[\frac{OPT}{ALG}\right]$, both the maximization of $E\left[\frac{ALG}{OPT}\right]$ and that of $E\left[ALG\right]$ lead to an algorithm that *exchanges the whole one dollar at a certain timing*. Such an algorithm is referred to as RPP in [4], in which the maximization of $E\left[ALG\right]$ is discussed. We remark that for a distribution $F \in$ IFR, the optimal timing of the one-time transaction for $E\left[\frac{ALG}{OPT}\right]$ is earlier than that of $E\left[ALG\right]$. As [8], we say that $F \in$ IFR if the function $f(p)/(1 - F(p))$ increases for all $p \in [m, M]$. IFR includes a comprehensive family of distributions such as the uniform, the normal, or a special case of Weibull distributions.

From the same argument as in Lemma 1, we can concentrate on algorithms that exchange only when the exchange rate is the highest so far. It is also seen that randomization does not help for either measures as in Proposition 1. The proofs of the following results are omitted due to lack of space.

Theorem 3. *An algorithm represented by* $S_1(r) = 1$ *for* $r \in [r_1, M]$ *and zero elsewhere, i.e. one that exchanges the whole one dollar when the rate reaches* r_1 *for the first time, maximizes* $E\left[\frac{ALG}{OPT}\right]$, *where* $r_1 < M$ *is a maximizer of* $(r - m) \int_r^M \frac{dF(p)}{p}$.

Theorem 4 ([4]). *An algorithm represented by* $S_0(r) = 1$ *for* $r \in [r_0, M]$ *and zero elsewhere, i.e. one that exchanges the whole one dollar when the rate reaches* r_0 *for the first time, maximizes* $E\left[ALG\right]$, *where* r_0 *is a maximizer of* $(r - m)(1 - F(r))$.

Proposition 2. *For a distribution* $F \in$ *IFR,* r_1 *is smaller than* r_0.

5   Concluding Remarks

If one analyzes the ski-rental problem using the average-case competitive ratio instead of the conventional worst-case competitive ratio, then the optimal timing of buying skis shifts earlier [6]. In one-way trading the same property appears: Whereas algorithm WTB waits until the exchange rate reaches the possible maximum, algorithm ATB quits the transaction at a certain timing. It might be interesting to investigate whether the same property holds for other online problems in general. We should also realize a limit of average-case analysis: To carry out an average-case analysis, it is strongly required that the input sequence has a simple structure and is explicitly parameterized, such as the total time of ski trips and the maximum exchange rate.

Acknowledgments. This work was partially supported by Grant-in-Aid for Scientific Research (No. 16092216 and 19700015) from Ministry of Education, Culture, Sports, Science and Technology of Japan.

References

1. Becchetti, L.: Modeling locality: A Probabilistic Analysis of LRU and FWF. In: Albers, S., Radzik, T. (eds.) ESA 2004. LNCS, vol. 3221, pp. 98–109. Springer, Heidelberg (2004)
2. Borodin, A., El-Yaniv, R.: Online Computation and Competitive Analysis. Cambridge University Press, Cambridge (1998)
3. Chen, G., Kao, M., Lyuu, Y., Wong, H.: Optimal Buy-and-Hold Strategies for Financial Markets with Bounded Daily Returns. SIAM J. Comput. 31(2), 447–459 (2001)
4. El-Yaniv, R., Fiat, A., Karp, R.M., Turpin, G.: Optimal Search and One-Way Trading Online Algorithms. Algorithmica 30(1), 101–139 (2001)
5. Embrechts, P., Kluppelberg, C., Mikosch, T.: Modelling Extremal Events: for Insurance and Finance. Springer, London (1997)
6. Fujiwara, H., Iwama, K.: Average-Case Competitive Analyses for Ski-Rental Problems. Algorithmica 42(1), 95–107 (2005)
7. Garg, N., Gupta, A., Leonardi, S., Sankowski, P.: Stochastic Analyses for Online Combinatorial Optimization Problems. In: Proc. SODA 2008, pp. 942–951 (2008)
8. Gertzbakh, I.B.: Statistical Reliability Theory. Marcel Dekker, New York (1989)
9. Lorenz, J., Panagiotou, K., Steger, A.: Optimal Algorithms for k-Search with Application in Option Pricing. In: Arge, L., Hoffmann, M., Welzl, E. (eds.) ESA 2007. LNCS, vol. 4698, pp. 275–286. Springer, Heidelberg (2007)
10. Naaman, N., Rom, R.: Average Case Analysis of Bounded Space Bin Packing Algorithms. Algorithmica 50(1), 72–97 (2008)
11. Panagiotou, K., Souza, A.: On Adequate Performance Measures for Paging. In: Proc. STOC 2006, pp. 487–496 (2006)
12. Sleator, D.D., Tarjan, R.E.: Amortized Efficiency of List Update and Paging Rules. Commun. ACM 28(2), 202–208 (1985)

On the Monotonicity of Weak Searching*

Boting Yang and Yi Cao

Department of Computer Science, University of Regina
{boting,caoyi200}@cs.uregina.ca

Abstract. In this paper, we propose and study two digraph searching models: weak searching and mixed weak searching. In these two searching models, each searcher must follow the edge directions when they move along edges, but the intruder can move from tail to head or from head to tail along edges. We prove the monotonicity of the mixed weak searching model by using Bienstock and Seymour's method, and prove the monotonicity of the weak searching model by using LaPaugh's method. We show that both searching problems are NP-complete.

1 Introduction

Given a graph/digraph that contains an intruder hiding on vertices or along edges, a graph/digraph searching problem, which serves as a mathematical model for many real world problems, is to find the minimum number of searchers required to capture this intruder. Throughout this paper, we assume that the intruder has complete knowledge of the location of every searcher and always takes the best strategy for him to avoid being captured, and we also assume that the speed of the intruder is significantly larger than the speed of searchers.

Megiddo et al. [8] proposed the edge searching model, in which there are three types of actions for searchers, i.e., placing, removing and sliding. Kirousis and Papadimitriou [6] proposed the node searching model, in which there are only two types of actions for searchers, i.e., placing and removing. Bienstock and Seymour [3] introduced the mixed searching model that combines the edge searching and node searching models.

An undirected graph is not always sufficient in representing all the information of a real-world problem. Yang and Cao [10] introduced the directed searching model in which all searchers and the intruder must follow the edge directions when they move along edges. Yang and Cao [11] also introduced the strong searching model in which the intruder must follow the edge directions but searchers need not when they move along edges. In [12] Yang and Cao introduced the directed vertex separation and investigated the relations between different digraph searching models, directed vertex separation, and directed pathwidth. In the above digraph searching models, there are three types of actions for searchers: placing, removing and sliding. Alspach et al. [1] proposed three digraph searching models in which searchers have only two actions: placing and sliding. In all of

---

* Research was supported in part by NSERC and MITACS.

X. Hu and J. Wang (Eds.): COCOON 2008, LNCS 5092, pp. 52–61, 2008.

the above graph/digraph searching models, the intruder is invisible, he may hide on vertices or along edges, and he can move at a great speed at any time along a path that contains no searchers. There are several other digraph searching models studied in [2],[5],[9], in which the intruder hides only on vertices.

In this paper, we introduce and study two searching models on digraphs: the weak searching model and the mixed weak searching model, in which searchers must follow the edge directions and the intruder need not. These two models are motivated by some applications in which the robber ignores the orientation of edges, while the law-abiding cops may only move along an edge in the direction of that edge. Notice that in the directed searching model of [10] and the strong searching model of [11], the intruder must follow the edge directions when he moves along edges. This difference makes the weak search number possibly far greater than the directed search number or the strong search number. For example, one searcher suffices to clear a transitive tournament in the directed or strong searching model, but we need n searchers to clear a transitive tournament in the weak searching model, where n is the number of vertices in the tournament. After Bienstock and Seymour [3] proved the monotonicity of the mixed searching model, their method has been successfully used to prove many monotonicity results (for example, see [4],[10],[11]). However, it seems difficult to use this method to prove the monotonicity of the weak searching model. We can prove the monotonicity of the mixed weak searching model using Bienstock and Seymour's method. But we cannot find a simple relation between the mixed weak searching model and the weak searching model so that we can use this relation to show the monotonicity of the weak searching model. Fortunately, we can use LaPaugh's method [7] to prove the monotonicity of the weak searching model. To the best of our knowledge, this is the first time to use LaPaugh's method to show monotonicity after her pioneer work in showing the monotonicity of the edge searching model. Due to the monotonicity of the weak and mixed weak searching models, we can show the NP-completeness results for these two models.

2 Definitions and Notation

All graphs and digraphs in this paper contain at least one edge. Throughout this paper, we use D to denote a digraph, and (u, v) to denote a directed edge with tail u and head v. An *undirected path* between two vertices $u, v \in V(D)$ is defined as a path between u and v in the underlying graph of D, and is denoted as $u \sim v$. We assume that D contains no searchers and all edges of D are *contaminated* before the first action is carried out in any searching model.

In the *weak/mixed weak searching model* on a digraph D, searchers must move in the edge directions but the intruder need not. The intruder can move at a great speed at any time from vertex $u \in V(D)$ to vertex $v \in V(D)$ along an undirected path $u \sim v$ that contains no searchers. There are three types of actions for searchers: (1) placing a searcher on a vertex, (2) removing a searcher from a vertex and (3) sliding a searcher along an edge from its tail to its head.

A *weak/mixed weak search strategy* is a sequence of actions such that the final action leaves all edges of D *cleared*. In a weak/mixed weak searching model, a cleared edge will be *recontaminated* if there is an undirected path containing this cleared edge and a contaminated edge such that there is no searchers stationing on any internal vertex of this undirected path.

We say that a vertex in D is *occupied* at some moment if at least one searcher is located on this vertex at this moment. Any searcher that is not on D at some moment is said *free* at this moment. Note that all searchers are initially free.

We now define the edge-clearing process of the two searching models in a different way so that they can be easily used in the two different proof methods described in Sections 3 and 4 respectively. We first define the three types of actions in the weak searching model formally.

Definition 1. Let $(s_1, s_2, \ldots, s_\ell)$ be a weak search strategy for a digraph D. For each action s_i, let A_i be the set of cleared edges and Z_i be the *multiset* of occupied vertices immediately after s_i, and let $A_0 = Z_0 = \emptyset$. Let E_u be the set of all edges incident with u (i.e., u is the tail or head of each edge in E_u). Each action s_i, $1 \leq i \leq \ell$, is one of the following three types:

(a) (*placing a searcher on v*) $Z_i = Z_{i-1} \cup \{v\}$ for some vertex $v \in V(D)$ and $A_i = A_{i-1}$;

(b) (*removing a searcher from v*) $Z_i = Z_{i-1} - \{v\}$ for some vertex $v \in Z_{i-1}$, and A_i is the set of edges $e \in A_{i-1}$ such that every undirected path containing e and an edge of $E(D) - A_{i-1}$ has an internal vertex in Z_i;

(c) (*sliding along e*) $Z_i = (Z_{i-1} - \{u\}) \cup \{v\}$ and $A_i = A_{i-1} \cup \{e\}$ for some edge $e = (u, v) \in E(D) - A_{i-1}$ such that the multiset Z_{i-1} contains at least two u, or Z_{i-1} contains one u and $E_u - \{e\} \subseteq A_{i-1}$.

From case (c) of Definition 1, we know that a contaminated edge (u, v) can be cleared in one of two ways by a sliding action: (1) sliding a searcher from u to v along (u, v) while at least one searcher is located on u, or (2) sliding a searcher from u to v along (u, v) while all edges incident with u except (u, v) are already cleared. In the mixed weak searching model, we replace the first way of clearing a contaminated edge by the *node-search-clearing* rule: a contaminated edge can be cleared if both of its end vertices are occupied. This rule is the same as that used in the node searching model [6]. We keep the second way in the mixed weak searching model and call it *edge-search-clearing*. By this modification, we can define the mixed weak searching so that each vertex contains at most one searcher at any step. More precisely, the four types of actions in the mixed weak searching model are defined as follows.

Definition 2. Let $(s_1, s_2, \ldots, s_\ell)$ be a mixed weak search strategy for a digraph D. For each action s_i, let A_i be the set of cleared edges and Z_i be the set of occupied vertices immediately after s_i, and let $A_0 = Z_0 = \emptyset$. Let E_u be the set of all edges incident with u. Each action s_i, $1 \leq i \leq \ell$, is one of the following four types:

(a) (*placing a searcher on v*) $Z_i = Z_{i-1} \cup \{v\}$ for some vertex $v \in V(D) - Z_{i-1}$ and $A_i = A_{i-1}$;

(b) (*removing a searcher from v*) $Z_i = Z_{i-1} - \{v\}$ for some vertex $v \in Z_{i-1}$, and A_i is the set of edges $e \in A_{i-1}$ such that every undirected path containing e and an edge of $E(D) - A_{i-1}$ has an internal vertex in Z_i;

(c) (*node-search-clearing e*) $Z_i = Z_{i-1}$ and $A_i = A_{i-1} \cup \{e\}$ for some edge $e = (u, v) \in E(D) - A_{i-1}$ with both ends u and v in Z_{i-1};

(d) (*edge-search-clearing e*) $Z_i = (Z_{i-1} - \{u\}) \cup \{v\}$ and $A_i = A_{i-1} \cup \{e\}$ for some edge $e = (u, v) \in E(D) - A_{i-1}$ with $u \in Z_{i-1}$, $v \in V(D) - Z_{i-1}$ and $E_u - \{e\} \subseteq A_{i-1}$.

From Definitions 1 and 2, we know that only removing actions may cause recontamination in the weak/mixed weak searching model. The digraph D is *cleared* if all of its edges are cleared. The minimum number of searchers needed to clear D in the weak searching model is the *weak search number* of D, denoted by $\mathrm{ws}(D)$. Similarly, the minimum number of searchers needed to clear D in the mixed weak searching model is the *mixed weak search number* of D, denoted by $\mathrm{mws}(D)$. In a searching model, let S be a search strategy and let A_i be the set of cleared edges immediately after the ith action. The strategy S is *monotonic* if $A_i \subseteq A_{i+1}$ for each i. We say that a searching model has the property of *monotonicity* (or is *monotonic*) if for any digraph D, there exists a monotonic search strategy that can clear D using k searchers, where k is the search number of D in this searching model.

3   Monotonicity of Mixed Weak Searching

In this section, we prove the monotonicity of the mixed weak searching model by using Bienstock and Seymour's method [3].

Definition 3. Let D be a digraph. For an edge set $X \subseteq E(D)$, the *boundary* of X, denoted by $\partial(X)$, is the set of vertices incident with both edges inside and outside X.

Definition 4. Let D be a digraph. For a weak or mixed weak search strategy that clears D, a vertex of $V(D)$ is *exposed* at some moment if it is incident with both a cleared edge and a contaminated edge at this moment.

In the weak/mixed weak searching model, it follows from Definitions 3 and 4 that at any moment, each vertex in the boundary of the cleared edge set is exposed and must be occupied by a searcher. The following lemma due to Bienstock and Seymour [3].

Lemma 1. *For any $X, Y \subseteq E(D)$, $|\partial(X \cap Y)| + |\partial(X \cup Y)| \le |\partial(X)| + |\partial(Y)|$.*

Definition 5. For a digraph D, let $\mathrm{mws}(D) = k$ and $S = (s_1, s_2, \ldots, s_\ell)$ be a mixed weak search strategy for D using k searchers. If s_i is a removing action and there are k exposed vertices immediately before s_i, then we say that s_i is an *avoidable removing action*. If S contains no avoidable removing actions, we say that S is a *normalized* mixed weak search strategy.

For any digraph D, we can always construct a normalized mixed weak search strategy that clears D using $\mathrm{mws}(D)$ searchers.

Lemma 2. *For any digraph D with $\mathrm{mws}(D) = k$, there exists a normalized mixed weak search strategy that clears D using k searchers.*

We now define the k-weak campaign that corresponds to the sequence of the cleared edge sets in a mixed weak search strategy.

Definition 6. Given a digraph D and a nonnegative integer k, a sequence $(X_0, X_1, \ldots, X_\ell)$ of subsets of $E(D)$ is called a *k-weak campaign* if it satisfies the following three conditions:

  (i) $X_0 = \emptyset$, $X_\ell = E(D)$, $|\partial(X_i)| \leq k$, and $|X_j - X_{j-1}| \leq 1$, for $1 \leq j \leq \ell$;
 (ii) if $|\partial(X_i)| = k$ for some i satisfying $0 \leq i \leq \ell - 1$, then $X_{i+1} = X_i \cup \{(u,v)\}$ for some edge $(u,v) \in E(D) - X_i$ such that $u \in \partial(X_i)$ and either $v \in \partial(X_i)$ or $u \notin \partial(X_{i+1})$; and
(iii) if $|\partial(X_i)| = k$ and $|\partial(X_{i+1})| < k$ for some i satisfying $0 \leq i \leq \ell - 2$, then $X_{i+1} \subset X_{i+2}$.

A k-weak campaign is *progressive* if $X_0 \subseteq X_1 \subseteq \cdots \subseteq X_\ell$ and $|X_j - X_{j-1}| = 1$, for $1 \leq j \leq \ell$.

Given a digraph D, a *weak extension* of D, denoted by D_m^+, is a digraph which contains two disjoint subgraphs D and F_m, such that the underlying graph of F_m is m isolated edges, $m \geq 0$. Thus, F_m has m vertices of indegree 0 and outdegree 1 and m vertices of indegree 1 and outdegree 0. Since D contains at least one edge and F_m can be cleared by one searcher, we have $\mathrm{mws}(D_m^+) = \mathrm{mws}(D)$.

Lemma 3. *For any digraph D with $\mathrm{mws}(D) = k$, there exists a weak extension D_m^+, $m \geq 0$, such that there is a k-weak campaign in D_m^+.*

We can now prove that there always exists a progressive k-weak campaign if there is a k-weak campaign in D.

Lemma 4. *Let D be a digraph. If there is a k-weak campaign in D, then there is a progressive k-weak campaign in D.*

Lemma 5. *Let D be such a digraph that except the end vertices of isolated edges, every vertex is incident with at least two edges. Let $(X_0, X_1, \ldots, X_\ell)$ be a progressive k-weak campaign in D, and let $X_i - X_{i-1} = e_i$ for $1 \leq i \leq \ell$. Then there is a monotonic mixed weak search strategy that clears D using k searchers such that the edges of D are cleared in the order $e_1, e_2, \ldots, e_\ell$.*

Lemma 6. *Let D be such a digraph that except the end vertices of isolated edges, every vertex is incident with at least two edges. If $\mathrm{mws}(D) = k$, then there is a monotonic mixed weak search strategy that clears D using k searchers.*

Finally we can prove the monotonicity of the mixed weak searching model.

Theorem 1. *Let D be a digraph. If $\mathrm{mws}(D) = k$, then there is a monotonic mixed weak search strategy that clears D using k searchers.*

4   Monotonicity of Weak Searching

In Section 3, we proved the monotonicity of the mixed weak searching model. However, we cannot find a transformation from a digraph D to another digraph D' so that $\mathrm{mws}(D) = \mathrm{ws}(D')$. It seems not easy to prove the monotonicity of the weak searching model from Theorem 1 using Bienstock and Seymour's method [3]. In this section we will prove this monotonicity result using LaPaugh's method [7].

LaPaugh proved the monotonicity of the edge searching model for undirected graphs. The difference between the weak searching of a digraph D and the edge searching of the underlying graph of D is that searchers must move in the edge directions in weak searching but need not in edge searching. In both models, the intruder can move along an undirected path at a great speed if there is no searcher on this path.

We first define a new model, *encapsulated weak searching*, in which any digraph has the same search number as in the weak searching model. In the encapsulated weak searching model, there are two types of moves: *recontamination moves* and *clearing moves*. Each move consists of a sequence of actions. During a clearing move, exactly one edge is cleared. During a recontamination move, at least one edge is recontaminated. We say that a vertex is *cleared* if all edges incident with it are cleared and *contaminated* if all edges incident with it are contaminated. From Definition 4, a vertex is *exposed* if there are both contaminated and cleared edges incident on it. Recall that a vertex is *occupied* if it is located by at least one searcher. At the end of each move, the following two conditions are satisfied:

(I) No cleared or contaminated vertex is occupied; and
(II) no vertex contains more than one searcher.

There are three phases in a *clearing move:* (1) Placing some number of searchers on vertices; (2) clearing a contaminated edge, say (u, v), by a sliding action, that is, sliding a searcher from u to v along (u, v) while at least one searcher is located on u, or sliding a searcher from u to v along (u, v) while all edges incident with u except (u, v) are already cleared; and (3) Removing the searchers that violate conditions (I) and (II).

There are four phases in a *recontamination move:* (1) Removing some number of searchers from exposed vertices; (2) placing some number of searchers on cleared vertices; (3) Recontaminating any cleared edge that lies on an undirected path satisfying that it contains a contaminated edge and there is no searcher on any internal vertex of this path; and (4) Removing the searchers that violate conditions (I) and (II).

Similar to Lemmas 1 and 2 in [7], we have the following two lemmas.

Lemma 7. *Let D be a digraph. If there is a weak search strategy that clears D using k searchers, then there is an encapsulated weak search strategy that clears D using k searchers.*

Lemma 8. *Let D be a digraph. If there is an encapsulated weak search strategy that clears D using k searchers and contains no recontamination moves, then there is a monotonic weak search strategy that clears D using k searchers.*

From Lemmas 7 and 8, we know that the monotonicity of the weak searching model follows from the monotonicity of the encapsulated weak searching model.

The remainder of this section is concerned with proving the monotonicity of the encapsulated weak searching model. We begin by presenting some notation. For a recontamination move r, let $R(r)$ denote the set of vertices from which searchers are removed in phase (1), and let $P(r)$ denote the set of vertices on which searchers are placed in phase (2). For a clearing or recontamination move s, let $V_{\text{before}}(s)$ denote the set of occupied vertices immediately before s and $V_{\text{after}}(s)$ denote the set of occupied vertices immediately after s; let $E_{\text{before}}(s)$ denote the set of contaminated edges immediately before s and $E_{\text{after}}(s)$ denote the set of contaminated edges immediately after s. Because both clearing moves and recontamination moves satisfy conditions (I) and (II), the occupied vertex set is the same as the exposed vertex set immediately before or after a clearing move or recontamination move. Given a digraph D, the *state of D* in the search process is described by the set of contaminated edges and the set of occupied vertices in D. Because of conditions (I) and (II), $E_{\text{after}}(s)$ is sufficient to specify the state of D just after the move s. The *state of a vertex* in D at some moment is whether this vertex is occupied or not at this moment. Recall that any searcher that is not on the digraph at some moment is called a *free* searcher at this moment.

Definition 7. Let r be a recontamination move in an encapsulated weak search strategy of a digraph D. We say that r is *irredundant* if no searchers are removed in phase (4), that is, phase (4) is unnecessary.

For a recontamination move r, $R(r)$ and $P(r)$ are disjoint, and any vertex in $R(r)$ is exposed just before r and contaminated just after r. Furthermore, from Definition 7, an irredundant recontamination move r has the following properties: $V_{\text{after}}(r) = (V_{\text{before}}(r) - R(r)) \cup P(r)$, and any vertex in $P(r)$ is cleared just before r and exposed just after r.

Note that the definition of the recontamination move in the encapsulated weak searching model is exactly the same as that in [7]. Hence, all the lemmas in [7] that only involve recontamination moves can be directly applied here.

From Lemma 4 in [7], any encapsulated weak search strategy can be modified to an encapsulated weak search strategy such that for any recontamination move, both its preceding move and its following move are clearing moves. In addition to combining recontamination moves, we can also decompose a recontamination move.

Definition 8. For a given irredundant recontamination move r in an encapsulated weak search strategy of digraph D, a set of vertices, denoted by $C(r)$, is a *separating set* of $R(r)$ and $P(r)$ if in digraph $D - (V_{\text{before}}(r) - R(r))$ every path from a vertex in $R(r)$ to a vertex in $P(r)$ contains a vertex in $C(r)$.

Note that $C(r)$ does not need to be disjoint from $R(r)$ or $P(r)$ but is disjoint from $V_{\text{before}}(r) - R(r)$.

Lemma 9. *Let s_r be the last recontamination move in an encapsulated weak search strategy S of a digraph D. Suppose that s_r is irredundant and the move immediately preceding s_r is a clearing move. If the number of free searchers just after s_r is less than or equal to the number just before s_r, and the sets $R(s_r)$ and $P(s_r)$ have no separating set $C(s_r)$ satisfying $|C(s_r)| < |R(s_r)|$, then there is another encapsulated weak search strategy S' for D using at most the same number of searchers as S such that the sequences of moves in S' and S are identical before s_r and S' contains one fewer recontamination move than S, that is, s_r is eliminated in S'.*

Lemma 10. *Let s_r be an irredundant recontamination move that is preceded by a clearing move s_c in an encapsulated weak search strategy for a digraph D. If the number of free searchers immediately after s_r is more than the number immediately before s_r, then moves s_c and s_r can be replaced by a new recontamination move t_r and a new clearing move t_c such that t_r precedes t_c, and the state of D immediately after the sequence of moves $t_r t_c$ is the same as the state of D immediately after $s_c s_r$. One or both of t_r and t_c may be empty moves; if t_c is not empty, it clears the same edge as s_c does.*

Theorem 2. *Let D be a digraph. If there is an encapsulated weak search strategy for D using k searchers, then there is an encapsulated weak search strategy that contains no recontamination moves and clears D using k searchers.*

Proof. Suppose that there is an encapsulated weak search strategy that contains at least one recontamination move and clears D using k searchers. From Lemmas 3 and 4 in [7], there is an encapsulated weak search strategy S for D using k searchers such that for each recontamination move in S, it is irredundant and both its preceding move and its following move are clearing moves. Let $(CM(S), ER(S))$ be an ordered pair such that $CM(S)$ is the number of clearing moves immediately before the last recontamination move in S and $ER(S)$ is the number of edges recontaminated in the last recontamination move in S. For two strategies S_1 and S_2, $(CM(S_1), ER(S_1)) < (CM(S_2), ER(S_2))$ if and only if $CM(S_1) < CM(S_2)$ or $CM(S_1) = CM(S_2)$ and $ER(S_1) < ER(S_2)$. Because S has at least one recontamination move, we know that $(CM(S), ER(S)) > (0,0)$. We will construct another encapsulated weak search strategy S' for D using k searchers such that $(CM(S'), ER(S')) < (CM(S), ER(S))$. Let r be the last recontamination move in S. We now have three cases.

Case 1. If $|R(r)| > |P(r)|$, then we can modify S by replacing r and its preceding clearing move by a new recontamination move followed by a new clearing move according to Lemma 10. Let S' be this modified strategy. It is easy to see that $CM(S') < CM(S)$.

Case 2. If $|R(r)| \leq |P(r)|$ and the sets $R(r)$ and $P(r)$ have no separating set $C(r)$ with $|C(r)| < |R(r)|$, then r can be eliminated according to Lemma 9. Let S' be this modified strategy. Thus, $CM(S') < CM(S)$.

Case 3. If $|R(r)| \leq |P(r)|$ and there is a minimal separating set $C(r)$ of $R(r)$ and $P(r)$ with $|C(r)| < |R(r)|$, then we can decompose r into two recontamination moves according to Lemma 5 in [7]. Recall that the first move removes searchers from $R(r) - C(r)$ and places searchers on $C(r) - R(r)$; the second move removes searchers from $C(r) - P(r)$ and places searchers on $P(r) - C(r)$. Since $|C(r)| < |R(r)|$, we know that $R(r) - C(r)$ is not empty, and some edge is recontaminated during the first move. Since $|R(r) - C(r)| > |C(r) - R(r)|$, there are more free searchers just after the first move than just before it. By Lemma 10, we can replace the first recontamination move and its preceding clearing move by a new recontamination move followed by a new clearing move. If the new clearing move is empty, then we combine the new recontamination move with the second recontamination move. Let S' be the updated strategy. We have $CM(S') < CM(S)$. If the new clearing move is not empty, the second recontamination move is the new last recontamination move. If we still use S' to denote the updated strategy, then we have $CM(S') = CM(S)$ and $ER(S') < ER(S)$.

In all the cases, we have $(CM(S'), ER(S')) < (CM(S), ER(S))$. By repeating the above procedure a finite number of times, we can obtain an encapsulated weak search strategy S^* for D using k searchers such that $(CM(S^*), ER(S^*)) = (0, 0)$, which means that S^* does not contain any recontamination move.   □

Finally we can prove the monotonicity of the weak searching model.

Theorem 3. *For a digraph D, if $\mathrm{ws}(D) = k$, then there is a monotonic weak search strategy that clears D using k searchers.*

5 NP-Completeness Results

Megiddo et al. [8] proved that the edge searching problem on undirected graphs is NP-hard. Kirousis and Papadimitriou [6] proved that the node searching problem on undirected graphs is NP-complete. In this section, we will prove that both mixed weak searching and weak searching problems are NP-complete. Given a digraph D and an integer k, the weak/mixed weak problem is to determine whether k searchers can clear D under the weak/mixed weak searching model. Similar to [10], we can prove the following relationship between the mixed weak searching and the node searching.

Lemma 11. *Let G be an undirected graph. If D_G is a digraph obtained from G by replacing each edge $uv \in E(G)$ with two directed edges (u, v) and (v, u), then $\mathrm{mws}(D_G) = \mathrm{ns}(G)$, where $\mathrm{ns}(G)$ is the node search number of G.*

From Theorem 1 and Lemma 11, we can prove the following result.

Theorem 4. *The Mixed Weak Searching problem is NP-complete.*

We can establish the following relationship between the weak searching and the node searching.

Lemma 12. *Let G be an undirected graph. If D'_G is a digraph obtained from G by replacing each edge $uv \in E(G)$ with three directed paths of length two, then $\mathrm{ws}(D'_G) = \mathrm{ns}(G) + 1$.*

We can also establish a relation between the weak search number and the edge search number as follows:

Lemma 13. *Let G be an undirected graph. If D''_G is a digraph obtained from G by replacing each edge $uv \in E(G)$ with four directed edges (u, u_1), (u_2, u_1), (u_2, u_3) and (v, u_3), then $\mathrm{ws}(D''_G) = \mathrm{es}(G) + 2$, where $\mathrm{es}(G)$ is the edge search number of G.*

From Theorem 3 we can prove that the Weak Searching problem belongs to NP class, and from Lemma 12 or 13, we can show it is NP-hard. Therefore, we have the following result.

Theorem 5. *The Weak Searching problem is NP-complete.*

References

1. Alspach, B., Dyer, D., Hanson, D., Yang, B.: Arc Searching Digraphs without Jumping. In: Dress, A.W.M., Xu, Y., Zhu, B. (eds.) COCOA. LNCS, vol. 4616, pp. 354–365. Springer, Heidelberg (2007)
2. Barat, J.: Directed Path-Width and Monotonicity in Digraph searching. Graphs and Combinatorics 22, 161–172 (2006)
3. Bienstock, D., Seymour, P.: Monotonicity in Graph Searching. Journal of Algorithms 12, 239–245 (1991)
4. Fomin, F., Thilikos, D.: On the Monotonicity of Games Generated by Symmetric Submodular Functions. Discrete Applied Mathematics 131, 323–335 (2003)
5. Hamidoune, Y.O.: On a pursuit game on Cayley graphs. European Journal of Combinatorics 8, 289–295 (1987)
6. Kirousis, L., Papadimitriou, C.: Searching and Pebbling. Theoretical Computer Science 47, 205–218 (1996)
7. LaPaugh, A.: Recontamination does not Help to Search a Graph. Journal of ACM 40, 224–245 (1993)
8. Megiddo, N., Hakimi, S., Garey, M., Johnson, D., Papadimitriou, C.: The Complexity of Searching a Graph. Journal of ACM 35, 18–44 (1998)
9. Nowakowski, R.J.: Search and Sweep Numbers of Finite Directed Acyclic Graphs. Discrete Applied Mathematics 41, 1–11 (1993)
10. Yang, B., Cao, Y.: Directed Searching Digraphs: Monotonicity and Complexity. In: Cai, J.-Y., Cooper, S.B., Zhu, H. (eds.) TAMC 2007. LNCS, vol. 4484, pp. 136–147. Springer, Heidelberg (2007)
11. Yang, B., Cao, Y.: Monotonicity of Strong Searching on Digraphs. Journal of Combinatorial Optimization 14, 411–425 (2007)
12. Yang, B., Cao, Y.: Digraph Searching, Directed Vertex Separation and Directed Pathwidth. Discrete Applied Mathematics (Accepted, 2007)

VC Dimension Bounds for Analytic Algebraic Computations

José Luis Montaña*, Luis Miguel Pardo**, and Mar Callau

Departamento de Matemáticas, Estadística y Computación
Universidad de Cantabria
{montanjl,pardol,callaum}@unican.es

Abstract. We study the Vapnik-Chervonenkis dimension of concept classes that are defined by computer programs using analytic algebraic functionals (Nash operators) as primitives. Such bounds are of interest in learning theory because of the fundamental role the Vapnik-Chervonenkis dimension plays in characterizing the sample complexity required to learn concept classes. We strengthen previous results by Goldberg and Jerrum giving upper bounds on the VC dimension of concept classes in which the membership test for whether an input belongs to a concept in the class can be performed either by an algebraic computation tree or by an algebraic circuit containing analytic algebraic gates. These new bounds are polynomial both in the height of the tree and in the depth of the circuit. This means in particular that VC dimension of computer programs using Nash operators is polynomial not only in the sequential complexity but also in the parallel complexity what ensures polynomial VC dimension for classes of concepts whose membership test can be defined by well-parallelizable sequential exponential time algorithms using analytic algebraic operators.

1 Introduction

In this paper we consider the Vapnik-Chervonenkis (VC) dimension as it applies to the distribution-free Probably Approximately Correct (PAC) learning model of Valiant (see [11]). The VC dimension was studied by Blumer et al. in [3] as a means of analyzing non-discrete concept classes (having a geometrical motivation) where concepts and instances are represented using real numbers. The assumption is that one unit is charged for representing and operating on a real number, which is used in the neural networks literature. In this situation one cannot use counting arguments to show that a consistent hypothesis of some limited complexity achieves PAC learning. However, Blumer et al. have shown that the VC dimension of a concept class determines how many examples are necessary for a learner to conjecture a PAC hypothesis. Moreover, Blumer et al. have shown that uniform learnability is characterized by finite VC dimension.

* Partially supported by the Spanish MCyT under project TIN2007-67466-C02-02.
** Partially supported by the Spanish MCyT under project MTM2007-62799.

1.1   Background

In PAC learning, the set of all objects that may be presented to the learner is called the instance domain, usually denoted by X. Members of X (instances) are classified according to membership or non-membership of an unknown subset C of X, called the target concept and the goal of the learner is to construct a hypothesis H that is a good approximation of C. The target concept is restricted to be a member of a known collection C of subsets of X, called the concept class. Examples are assumed to be generated according to a fixed but unknown probability distribution P on X. We say that a hypothesis H ϵ-approximates C if the probability that H and C disagree on a random instance drawn according to P is at most ϵ. The criterium for successful learning is that the hypothesis should reliably classify further instances drawn according to P. This criterion is captured by the notion of learning function.

A learning function takes positive real parameters ϵ and δ (representing error and confidence) and a sequence of classified instances (a sample) according to membership or non membership to C and drawn according to P and produces a hypothesis H that ϵ-approximates C with probability at least $1 - \delta$.

A learning algorithm, i.e. a program that implements a learning function, should ideally run in time polynomial in the accuracy ϵ^{-1} and in the confidence precision δ^{-1} and in the other parameters of the learning problem. Besides ϵ^{-1} and δ^{-1}, there are the domain dimension and the (syntactical) concept complexity that will be denoted by n and k respectively. Suppose that the instance domain X can be stratified as a disjoint union of subsets X_n (n representing the dimension of the instance). Assume also that the class of concepts is further stratified according to the complexity k of the concept, so that the concept class C is a disjoint union of concept classes $C_{k,n}$ where $C_{k,n}$ are families of subsets of X_n, for every k. Even though the VC dimension of C may be unbounded, the results by Blumer et al. ([3]) assures that the PAC learning of a concept in $C_{k,n}$ is possible from a learning set of size polynomial in k and n provided that the VC dimension of $C_{k,n}$ is bounded by a polynomial in k and n.

1.2   Main Results

We deal with general concept classes whose concepts and instances are represented by tuples of real numbers. For such a concept class C, let $C_{k,n}$ be C restricted to concepts represented by k real values and instances represented by n real values. Following Goldberg and Jerrum ([6]), the *membership test* of a concept class C over domain X takes as input a concept $C \in C$ and an instance $x \in X$, and returns the boolean value "$x \in C$". The membership test of a concept class can be thought of in two common ways: either as a formula or as an algorithm taking as input representations of a concept and an instance, and evaluating to the boolean value indicating membership.

Throughout this paper, the membership test for a concept class $C_{k,n}$ is assumed to be expressed as a computer problem $\mathcal{P}_{k,n}$ taking $k + n$ real inputs, representing a concept $C \in \mathbf{R}^k$ and an instance $x \in X = \mathbf{R}^n$. The program

$\mathcal{P}_{k,n}$ uses exact real arithmetic, analytic algebraic operators as primitives (this includes the standard arithmetic operators and other more sophisticated operators like series having fractional exponents), conditional statements, and returns the truth value $x \in \mathcal{C}$.

For classes defined by programs as described above we announce the following results.

- For a hierarchy of concept classes $\mathcal{C}_{k,n}$, defined by computer programs $\mathcal{P}_{k,n}$ which run in sequential time $t = t(k, n)$ the VC dimension of $\mathcal{C}_{k,n}$ is polynomial in k, n and t.
- For a hierarchy of concept classes $\mathcal{C}_{k,n}$, defined by algorithms $\mathcal{P}_{k,n}$ which run in parallel time $d = d(k, n)$ the VC dimension of $\mathcal{C}_{k,n}$ is also polynomial in d, k, n and in the number of analytic algebraic operators that the program contains.

Precise statements of these results are given in Theorems 3 and 4 below.

1.3 Related Work

I n [6] Goldberg and Jerrum exhibited sufficient conditions on the size of formulas in the first order theory of the real numbers $\Phi_{k,n}$, and on the sequential running time of programs $\mathcal{P}_{k,n}$, using the standard arithmetic operations as primitives, to define a class of polynomial VC dimension in k and n (see also Alonso and Montaña [1]). Special mention deserves the seminal paper by Karpinski and Macintyre ([7]), where polynomial bounds on the VC dimension of sigmoidal networks, and networks with general Pfaffian activation functions, are derived.

2 Relevant Results About VC Dimension of Formulas

The Vapnik-Chervonenkis dimension was developed as a statistical tool by Vapnik and Chervonenkis ([12]). It is defined as follows.

Definition 1. *Let $\mathcal{F}$ be a class of subsets of a set X. We say that $\mathcal{F}$ shatters a set $A \subset X$ if for every subset $E \subset A$ there exists $S \in \mathcal{F}$ such that $E = S \cap A$. The VC dimension of $\mathcal{F}$ is the cardinality of the largest set that is shattered by $\mathcal{F}$.*

Along this section we deal with concept classes $\mathcal{C}_{k,n}$ such that concepts are represented by k real numbers, $y = (y_1, \ldots, y_k)$, instances are represented by n real numbers, $x = (x_1, \ldots, x_n)$, and the membership test to the family $\mathcal{C}_{k,n}$ is expressed by a formula $\Phi_{k,n}(y, x)$ taking as inputs the pair concept/instance (y, x) and returning the value 1 if "x belongs to the concept represented by y" and 0 otherwise.

We can think of $\Phi_{k,n}$ as a function from $\mathbb{R}^{k+n}$ to $\{0, 1\}$. So for each concept $y \in \mathbb{R}^k$, define:

$$C_y := \{x \in \mathbb{R}^n : \Phi_{k,n}(y, x) = 1\}, \tag{1}$$

The objective is to obtain an upper bound on the VC dimension of the collection of sets

$$\mathcal{C}_{k,n} = \{C_y \, : \, y \in \mathbb{R}^k\}. \tag{2}$$

For boolean combinations of polynomial equalities and inequalities the following seminal result by Goldberg and Jerrum is known.

Theorem 1 ([6], Theorem 2.2). *Suppose $\mathcal{C}_{k,n}$ is a class of concepts whose membership test can be expressed by a boolean formula $\Phi_{k,n}$ involving a total of s polynomial equalities and inequalities, where each polynomial has degree no larger than d. Then the VC dimension V of $\mathcal{C}_{k,n}$ satisfies*

$$V \leq 2k \, \log_2(4eds) \tag{3}$$

Now assume that formula $\Phi_{k,n}$ is a boolean combination of s atomic formulas, each of them being of one of the following forms:

$$\tau_i(w, x) > 0 \tag{4}$$

or

$$\tau_i(w, x) = 0, \tag{5}$$

where $\{\tau_i(w,x)\}_{1 \leq i \leq s}$ are infinitely differentiable functions from $\mathbb{R}^{k+n}$ to $\mathbb{R}$. Next, make the following assumptions about the functions τ_i. Let $\alpha_1, \ldots, \alpha_v \in \mathbb{R}^n$. Form the $s.v$ functions $\tau_i(w, \alpha_j)$ from $\mathbb{R}^k$ to $\mathbb{R}$. Choose $\Theta_1, \ldots, \Theta_r$ among these, and let

$$\Theta : \mathbb{R}^k \to \mathbb{R}^r \tag{6}$$

be defined by

$$\Theta(w) := (\Theta_1(w), \ldots, \Theta_r(w)) \tag{7}$$

Assume there is a bound B independent of the α_i, r and $\epsilon_1, \ldots, \epsilon_r$ such that if $\Theta^{-1}(\epsilon_1, \ldots, \epsilon_r)$ is a $(k - r)$-dimensional C^∞-submanifold of $\mathbb{R}^k$ then $\Theta^{-1}(\epsilon_1, \ldots, \epsilon_r)$ has at most B connected components.

With the above set-up, the following result is proved by Karpinski and Macintyre ([7]).

Theorem 2. *The VC dimension V of a family of concepts $\mathcal{C}_{k,n}$ whose membership test can be expressed by a formula $\Phi_{k,n}$ satisfying the above conditions satisfies:*

$$V \leq 2 \log_2 B + 2k \log_2(2es) \tag{8}$$

3 Technical Statements for the VC Dimension of Formulas Involving Nash Functions

We study the VC dimension of formulas involving analytic algebraic functions. Such functions are called Nash functions in the mathematical literature (see Bochnak et al. [4]). A Nash function $f : \mathbb{R}^n \to \mathbb{R}$ is an analytic function satisfying a nontrivial polynomial equation $P(x, f(x)) = 0$ for all $x \in \mathbb{R}^n$. The degree of a Nash function is the minimal degree of nontrivial polynomials vanishing on its graph.

A sign assignment to a Nash function f is one of the (in)equalities: $f > 0$ or $f = 0$ or $f < 0$. A sign assignment to a set of s Nash functions is consistent if all s (in)equalities can be satisfied simultaneously by some assignment of real numbers to the variables. The following Lemma is an easy consequence of Bézout Theorem for Nash functions which is proved by Ramanakoraisina in [10].

Lemma 1. *Let $f_1, \ldots, f_s$ be n-variate Nash functions each f_i of degree bounded by d. Then, the subset of $\mathbb{R}^n$ defined by the equations:*

$$f_1 = 0, \ldots, f_s = 0 \qquad (9)$$

has at most $(2d)^{(s+1)(2n-1)}$ connected components.

We show for Nash functions a statement that bounds the number of consistent sign assignments of a finite family of such functions. The technical details of the proof are omitted.

Lemma 2. *Let $\mathcal{F}$ be a finite family of s n-variate Nash functions with degree bounded by $d \geq 1$. If $s \geq (n+1)(2n-1)$ the number of consistent sign assignments to functions of the family $\mathcal{F}$ is at most*

$$\left(\frac{8eds}{(n+1)(2n-1)}\right)^{(n+1)(2n-1)}. \qquad (10)$$

Next we state the following result concerning VC dimension of families of concepts defined by Nash functions. Its proof is a consequence of Theorem 2 and Lemma 1.

Proposition 1. *Let $x = (x_1, \ldots, x_n)$ and $y = (y_1, \ldots, y_k)$ denote vectors of real variables. Suppose $\mathcal{C}_{k,n}$ is a class of concepts whose membership test can be expressed by a boolean formula $\Phi_{k,n}$ involving a total of s (in)equalities of polynomials belonging to the polynomial ring $\mathbb{R}[x, y, f_1(x, y), \ldots, f_q(x, y)]$, where each polynomial has degree no larger than d, and each function f_i is Nash of degree bounded by d'. Then the VC dimension V of $\mathcal{C}_{k,n}$ satisfies*

$$V \leq 2(1 + \log_2 \max\{d, d'\})(k(q+1) + 1)(2k(q+1) - 1) + 2k \log_2(2es) \quad (11)$$

4   VC Dimension Bound for Sequential Nash Computations

We next show a development of the previous result by extending it to concept classes whose membership tests are programs described by bounded height algebraic computation trees. For various computational problems like sorting, searching and others coming from computational geometry, Steel and Yao, Ben-Or ([2]), Montaña et al. ([9]) and other authors have used results about the homology of semialgebraic sets (see Warren [13], and also Milnor [8]) to give lower bounds on the height of any algebraic computation tree that solves them.

Theorem 3. *Let $C_{k,n}$ be a family of concept classes for which the test for membership of an instance x in a concept C is program $\mathcal{P}_{k,n}$ taking $k + n$ real inputs representing C and x, whose running time is $t = t(k, n)$ and which returns the truth value $a \in C$. The algorithm is allowed to perform conditional branching (conditioned on equality and inequality) of real values, execute the standard arithmetic operations on real numbers $\{+, -, *, /\}$ and Nash operations f where each f is of degree bounded by $D \geq 2$. Then the VC dimension V of $C_{k,n}$ satisfies:*

$$V = O((\log_2 D)k^2(t+1)^2). \tag{12}$$

Proof. Let $t = t(k, n)$ be the running time of program $\mathcal{P} = \mathcal{P}_{k,n}$. Following standard approaches (Ben-Or [2]), the set of instructions performed by a program $\mathcal{P}$ on inputs accepted (or rejected) within running time t can be described by an algebraic computation tree $\mathcal{T}$ of height at most t whose nodes are of the following kinds:

(a) Arithmetic nodes v with a unique next node that have associated a computational instruction $z_v := z_{v_1} \, op_v \, z_{v_2}$ where z_{v_1}, z_{v_2} are either inputs in the set $\{x_1, \ldots, x_n\} \cup \{y_1, \ldots, y_k\}$ or v_1, v_2 are ancestors of v in the tree, and op_v is an arithmetic operation in $\{+, -, *, /\}$.

(b) Nash nodes v with a unique next node that have associated a computational instruction $z_v := f_v(z_{v_1}, \ldots, z_{v_r})$, where the v_i are ancestors of v in the tree (or z_{v_i} is an input) and f_v is the Nash operator associated to this node.

(c) Branching nodes v with two next nodes according to whether a condition $(z_{v_1} \sigma_i 0)$, $\sigma_i \in \{>, =, <\}$, is or is not satisfied, where v_1 is an ancestor of v in the tree (or z_{v_1} is an input).

(d) Leave nodes v labelled "True" or "False".

Each input causes the execution of program $\mathcal{P}$ to take some path trough this tree. In order to take a particular path, the input must satisfy at most t tests, consisting of (in)equalities of values z_v computed in the arithmetic or Nash nodes v of the tree $\mathcal{T}$. So the condition for taking a particular path is a conjunction of such (in)equalities. Then the output of program $\mathcal{P}$ on input (x, y) is given by the disjunction over all paths ending in a "true" of the conjunction of atomic predicates $\Theta(x, y) > 0$, $\Theta(x, y) = 0$ where $\Theta(x, y)$ is a term of the language L with symbols $+, -, *, /, f_1, \ldots, f_q, 0, 1$, where $f_1, \ldots, f_q$ denotes the Nash operators of the program $\mathcal{P}$.

We now apply a projection technique to get a bound on the VC dimension of class $C_{k,n}$. According to Theorem 2 we need to know a bound on the number of connected components of a manifold of dimension $k - r$ defined by conditions of the form

$$\Theta_i(\alpha_i, y) - \epsilon_i = 0, \quad 1 \leq i \leq r \ (\leq k) \tag{13}$$

where $\alpha_i \in \mathbb{R}^n$ and $\Theta_i(x, y)$ is an L-term computed at some computation node of $\mathcal{T}$. We appeal directly to the estimates given in Lemma 1 applied in a high-dimensional space.

For each computation node v of $\mathcal{T}$ let $\Theta_v(x, y)$ be the L-term computed at node v. For each i, $1 \leq i \leq r$, let v_i be the computation node satisfying

$\Theta_i(\alpha_i, y) = \Theta_{v_i}(\alpha_i, y)$. Let N_i be the set of computation nodes formed joining to $\{v_i\}$ all the computation nodes (arithmetic and Nash nodes) being ancestors of v_i in the tree.

Next, for each $v \in N_i$ we introduce variables $Z_{v,i}$ and equations

$$Z_{v,i} - Z_{v_1,i} op_v Z_{v_2,i} = 0, \tag{14}$$

if v is an arithmetic node and op_v is in $\{+, -, *\}$ and

$$Z_{v,i} Z_{v_2,i} - Z_{v_1,i} = 0, \tag{15}$$

if v is a division. In all cases, op_v is the operation performed at node v in the tree, $Z_{v,i}, Z_{v_1,i}, Z_{v_2,i}$ having the obvious meaning and consequently, the v_j, $j \in \{1,2\}$, are either ancestors of v or $Z_{v_j,i}$ are variables in $\{y_1, \ldots, y_k\}$ or coordinates of point $\alpha_i \in \mathbb{R}^n$.

We add variables $Z_{v,i}$ and equations

$$Z_{v,i} - f_v(Z_{v_1,i}, \ldots, Z_{v_l,i}) = 0 \tag{16}$$

if v is an Nash node. Here f_v is the Nash operation performed at node v and the v_j, $1 \leq j \leq l$ are ancestors of v or $Z_{v_j,i}$ is a variable in $\{y_1, \ldots, y_k\}$ or a coordinate of the point α_i.

Finally, add the equation

$$Z_{v_i,i} - \epsilon_i = 0 \tag{17}$$

Let $W(\alpha_1 \ldots, \alpha_r, y, \epsilon_1, \ldots, \epsilon_r, Z_{v,i})$ be the system obtained by Equations 14, 15, 16 and 17 for all i, $1 \leq i \leq r$. We make explicit the dependence from the α_i's and the y_i's because in our equations some of the $Z_{v,i}$ are variables y_i and coordinates α_i. In all other cases v is a node of the tree and $Z_{v,i}$ a new variable.

The key point is the following: the set defined in the variables $y_1, \ldots, y_k$ by the equations $\Theta_i(\alpha_i, y) = \epsilon_i$, $1 \leq i \leq r$, is the projection onto the variables $y_1, \ldots, y_k$ of the set defined in the variables $y_1, \ldots, y_k, Z_{v,i}$ by the system $W(\alpha_1 \ldots, \alpha_r, y, \epsilon_1, \ldots, \epsilon_r, Z_{v,i})$.

Note that there are at most $k + rt \leq k(t+1)$ variables among $y_1, \ldots, y_k, Z_{v,i}$ and system W contains at most $r(t+1) \leq k(t+1)$ equations. Accordingly, from Lemma 1 the value of B in Theorem 2 satisfies

$$\log_2 B \leq (k(t+1)+1)(2k(t+1)-1)(1+\log_2 D) \tag{18}$$

Note that since $\mathcal{T}$ is a binary tree the number of paths is bounded by 2^t and the number of atomic predicates is bounded by $t2^t$. So the result follows from Theorem 2 taking $s = t2^t$ and the bound in Equation 18.

5   VC Dimension Bounds for Parallel Nash Computations

Our model of parallel computation is that of Nash networks, that is an arithmetic network (von zur Gathen [5], Montaña et al. [9]) augmented with Nash operators.

A Nash network $\mathcal{N}$ is a circuit using Nash operators augmented with a special kind of nodes, called *sign nodes*. A sign node outputs 1 if their input is greater or equal than 0 and 0 otherwise. Next we provide a precise definition of a Nash network.

Definition 2. *A Nash (q, β)-network $\mathcal{N}$ over $\mathbb{R}$ is a directed acyclic graph having nodes with indegree 0 labeled as inputs or with elements of $\mathbb{R}$; nodes with indegree 2 labelled with a binary operation of $\mathbb{R}$, that is $+, -, *, /$; nodes of indegree 1 are sign gates; and q nodes with indegree at most β labelled with some Nash operator of arity at most β.*

To each node v we inductively associate a function as follows.

- If v is an input or constant node then f_v is the label of v.
- If v has indegree 2 and v_1 and v_2 are the ancestors of v then $f_v = f_{v_1} op_v f_{v_2}$ where $op_v \in \{+, -, *, /\}$ is the label of v.
- If v is labelled by a Nash operator f and $v_1, \ldots, v_k$ are the ancestors of v then $f_v = f(f_{v_1}, \ldots, f_{v_k})$ with $k \leq \beta$.
- If v is a sign node then $f_v = sign(f'_v)$ where v' is the ancestor of v in the graph.

In particular the function associated to the output node is the function computed by the network. We say that a subset W is accepted by a Nash network $\mathcal{N}$ if the function computed by $\mathcal{N}$ is the characteristic function of W. In this case it is assumed that the output node is a sign node.

Given a Nash network $\mathcal{N}$, the size $s(\mathcal{N})$ is the number of nodes in $\mathcal{N}$. The depth $d(\mathcal{N})$ is the length of the longest path from some input node to some output node. As usual, we shall refer to $d(\mathcal{N})$ as parallel time.

Remark 1. Observe that the combination of computation nodes with sign nodes may increase the number of terms involved in the computation (the size of the formula) up to a number which is doubly exponential in the depth (parallel time). If d is parallel time, sequential time could be, in the worst case, $t = 2^d$. On the other hand, if t is sequential time the formula size, could be at worst 2^t. Accordingly, we see that, using this straightforward argument, the best we can expect from Theorem 3 is an $O(k^2(2^d + 1)^2)$ upper bound for the VC dimension of concept classes $\mathcal{C}_{k,n}$ whose membership test is represented by an algorithm $\mathcal{N}_{n,k}$ working within parallel time $d = d(n, k)$. A formal explanation of this situation can be found in [1].

Lemma 3. *Let $\mathcal{C}_{k,n}$ be a family of concept classes whose membership test can be expressed by a family of Nash networks $\mathcal{N}_{k,n}$ having $k+n$ real variables representing the concept and the instance and depth $d = d(k, n)$. Assume that the network $\mathcal{N}_{k,n}$ has at most q Nash nodes of degree bounded by $D \geq 2$ and indegree bounded by β. Then, the membership test to $\mathcal{C}_{k,n}$ can be expressed by a family of formulas $\Phi_{k,n}$ in $k + n$ free variables having size at most $(\max\{\beta, 2\}D)^{O(((n+k+\beta q)d)^2)}$.*

Proof. (Sketch) We transform $\mathcal{N}_{k,n}$ into a formula $\Phi_{k,n}$ having the required formula size. Let $d = d(k,n)$ be the parallel time of $\mathcal{N}_{k,n}$. For each node v being an input of a Nash node let us introduce a variable z_v that contains the function value computed at this node. Call z the set of new variables z_v. We introduce at most $q\beta$ new variables. Let $v(i,1),\ldots,v(i,l_i)$ be the collection of sign nodes of the network $\mathcal{N}_{k,n}$ whose depth is $i \leq d = d(k,n)$. Now, for each pair (i,j), $1 \leq j \leq l_i$, let $f_{i,j}$ be the function that the sign node $v(i,j)$ receives as input.

Since the indegree of the arithmetic nodes is bounded by 2, it easily follows by induction that $f_{i,j}$ is a piecewise rational function of $(x,y,z,(f_l(x,y,z))_{1\leq l\leq q}$ of formal degree bounded by 2^i (the variables z can be eliminated by substitution to get $f_{i,j}$ as function of the input variables x, y). Note that at level i the number of non spurious (those connected with the output node) l_i is bounded above by $\max\{\beta,2\}^{d-i}$.

Now, for each sign assignment $\epsilon = (\epsilon_{i,j}) \in \{>,=,<,\}^{\sum_{1\leq i\leq d} l_i}$ let Φ_ϵ be the formula:

$$\Phi_\epsilon = \bigwedge_{1\leq i\leq d, 1\leq j\leq l_i} (f_{i,j}\epsilon_{i,j}0), \tag{19}$$

Using finite induction on the number of conjunctions in Equation 19 one arrives to the following.

Fact A. For every $\epsilon \in \{>,=,<,\}^{\sum_{1\leq i\leq d} l_i}$ there are rational functions $r_{i,j}$ of $(x,y,z,f_1,\ldots,f_q)$ of formal degree bounded by 2^i such that formula Φ_ϵ is equivalent to the formula

$$\bigwedge_{1\leq i\leq d, 1\leq j\leq l_i} (r_{i,j}\epsilon_{i,j}0) \tag{20}$$

In what follows formula in Equation 20 will be also denoted by Φ_ϵ. Notice that the set of inputs accepted by the arithmetic network $N_{k,n}$ can be described by a disjunction of some of the formulas Φ_ϵ. Hence, the proof of Lemma 3 finishes if we show the following.

Lemma 4. *The number of tuples ϵ such that formula Φ_ϵ represents a consistent sign assignment is bounded by* $(\max\{\beta,2\}D)^{O((k+n+\beta q)^2.d^2)}$.

The proof of Lemma 4 is a technical consequence of Lemma 2.

Theorem 4. *Let $\mathcal{C}_{k,n}$ be a family of concept classes whose membership test can be expressed by a family of Nash networks $\mathcal{N}_{k,n}$ having $k + n$ real variables representing the concept and the instance and depth $d = d(k,n)$. Assume that the network $\mathcal{N}_{k,n}$ has at most q Nash nodes of degree bounded by $D \geq 2$ and indegree bounded by β. Then, the VC dimension of $\mathcal{C}_{k,n}$ is in the class*

$$O((\log_2 D + \log_2 \max\{\beta,2\}) k(n + k + \beta q)^2 d^2) \tag{21}$$

Proof. (Sketch) In order to prove of Theorem 4 we appeal to Proposition 1 in a high dimensional space. From Fact A and Lemma 3 we conclude that $\mathcal{C}_{k,n}$ is a class of concepts whose membership test can be expressed by a boolean formula $\Phi_{k,n}$ involving a total of $s = (\max\{\beta,2\}D)^{O((k+n+\beta q)^2.d^2)}$ (in)equalities of polynomials belonging to the polynomial ring $\mathbb{R}[x, y, z, f_1(x, y, z), \ldots, f_q(x, y, z)]$ (where variables z can be eliminated by substitution). Each polynomial has degree no larger than 2^d, and each function f_i is Nash of degree bounded by D. Then, from Proposition 1 we conclude that the VC dimension V of $\mathcal{C}_{k,n}$ satisfies

$$V = O(\log_2 D + \log_2 \max\{\beta, 2\})k(n + k + \beta q)^2 d^2) \tag{22}$$

as wanted.

References

1. Alonso, C.L., Montaña, J.L.: VapnikChervonenkis Dimension of Parallel Arithmetic Computations. In: Proc. of Algorithmic Learning Theory, 18th International Conference, pp. 107–119 (2007)
2. Ben-Or, M.: Lower Bounds for Algebraic Computation Trees. In: Proc. of STOC 1983, pp. 80–86 (1983)
3. Blumer, A., Ehrenfeucht, A., Haussler, A., Warmuth, M.K.: Learnability and the Vapnik-Chervonenkis Dimension. Journal of the Association for Computing Machinery 36(4), 929–965 (1989)
4. Bochnak, J., Coste, M., Roy, M.-F.: Géométrie algébrique réelle (French) [Real algebraic geometry]. In: Ergebnisse der Mathematik und ihrer Grenzgebiete (3) [Results in Mathematics and Related Areas (3)], Berlin, vol. 12 (1987)
5. von zur Gathen, J.: Parallel Arithmetic Computations: A Survey. In: Wiedermann, J., Gruska, J., Rovan, B. (eds.) MFCS 1986. LNCS, vol. 233, pp. 93–112. Springer, Heidelberg (1986)
6. Goldberg, P., Jerrum, M.: Bounding the Vapnik-Chervonenkis Dimension of Concept Classes Parametrized by Real Numbers. Machine Learning 18, 131–148 (1995)
7. Karpinski, M., Macintyre, A.: Polynomial Bounds for VC Dimension of Sigmoidal and General Pffafian Neural Networks. Journal of Comput. System Sci. 54, 169–176 (1997)
8. Milnor, J.: On the Betti Numbers of Real Varieties. Proceedings of the American Mathematical Society 15, 275–280 (1964)
9. Montaña, J.L., Pardo, L.M.: Lower Bounds for Arithmetic Networks. Applicable Algebra in Engineering, Communication and Computing 4(1), 1–24 (1993)
10. Ramanakoraisina, R.: Bezout Theorem for Nash Functions. Comm. Algebra 17(6), 1395–1406 (1989)
11. Valiant, L.G.: A Theory of the Learnable. Communications of the ACM 27, 1134–1142 (1984)
12. Vapnik, V., Chervonenkis, A.: On the Uniform Convergence of Relative Frequencies of Events to their Probabilities. Theory of Probability and its applications 16, 264–280 (1971)
13. Warren, H.E.: Lower Bounds for Approximation by non Linear Manifolds. Trans. A.M.S. 133, 167–178 (1968)

Resource Bounded Frequency Computations
with Three Errors

Ulrich Hertrampf[1] and Christoph Minnameier[2]

[1] Abt. Theor. Informatik, University of Stuttgart, D-70569 Stuttgart, Germany
Hertrampf@informatik.uni-stuttgart.de
[2] Lst. Prakt. Informatik II, University of Mannheim, D-68131 Mannheim, Germany
cmm@informatik.uni-mannheim.de

Abstract. We deal with frequency computations in polynomial time, or more generally with resource bounded frequency computations. We investigate the first non-trivial case of the Hinrichs-Wechsung conjecture, which states that as soon as we have at least $2^d + d$ inputs to be queried, it does not become harder to get an answer with at most d errors, if we increase the number of inputs to be queried. This conjecture can easily be seen to hold for cases $d < 3$, and it seems very hard to prove in general. We solve the problem affirmatively in the case $d = 3$ by a combination of theoretical reasoning with a highly optimized computer search.

1   Introduction

The concept of frequency computation goes back to G. F. Rose [14]. The idea was as follows: If a function f is not computable in the usual sense it may still be possible to compute it in the following relaxed way: On n (pairwise different) inputs $x_1, \ldots, x_n$, a sequence of outputs $y_1, \ldots, y_n$ shall be produced, which approximates the function in such a way that at least m of the output values are correct, i.e. $\|\{i \in \{1, \ldots, n\} \mid y_i = f(x_i)\}\| \geq m$. The class of functions computable in this way is denoted by (m, n). The same idea can also be applied to sets instead of functions.

Essentially there are three versions of the frequency computation model: For the recursion theoretic setting, see e.g. [4],[9],[12],[16],[3]. The resource bounded case was investigated in [9],[5], and for the finite state model setting, see [8],[2],[1]. A lot of related work can be found in the literature (e.g. [6],[7],[11],[10],[13],[15]).

As in many other areas it turned out that the recursion theoretic world much resembles the finite state machine world, whereas the resource bounded (for convenience we will in this paper always speak of polynomially time bounded) world looks completely different.

In all models it is clear that $(m + 1, n + 1) \subseteq (m, n)$, because given an $(m + 1, n + 1)$-algorithm one can obtain an (m, n)-algorithm as follows: On input $x_1, \ldots, x_n$ choose any element not equal to any of the x_i and call it x_{n+1}. Now query $x_1, \ldots, x_{n+1}$ to the given $(m+1, n+1)$-algorithm and ignore its $(n+1)$-th output. Of the remaining n outputs no more than $(n + 1) - (m + 1) = n - m$ can be erroneous.

X. Hu and J. Wang (Eds.): COCOON 2008, LNCS 5092, pp. 72–81, 2008.
© Springer-Verlag Berlin Heidelberg 2008

In [5] it was shown that in the resource bounded case for all $m < 2^d$ (where $d = n - m$), the class $(m+1, n+1)$ is a proper subset of (m, n), thus especially for polynomial time $(m+1, n+1)P \subsetneq (m, n)P$. Furthermore Hinrichs and Wechsung conjectured that to the contrary, one always obtains equality for $m \geq 2^d$. In this paper we will call that conjecture *the Hinrichs-Wechsung conjecture.*

As a hint on the general computation power of frequency computations we should remark here, that even in the finite state machine setting the classes (m, n) with $2m \leq n$ contain non-countably many languages, i.e. especially they contain non-r.e. sets.

For $d < 3$ the Hinrichs-Wechsung conjecture can easily be proven (see the full paper for a proof), or it can be deduced from [9]:

Proposition 1. *Let $m > 2$. Then $(m, m + 1)P = (2, 3)P$. Let $m > 4$. Then $(m, m + 2)P = (4, 6)P$.*

However, for $d = 3$, the claim would be: Let $m > 8$. Then $(m, m+3)P = (8, 11)P$. No similar proof for this case is known. To provide a proof is the subject of this paper. We reach this goal using computer-aided search, where the search space is drastically reduced using theoretical arguments.

2   The Problem and Some Preliminaries

Definition 1. *The class $(m, n)P$ consists of all languages L, such that a polynomial time algorithm exists, which on every set of n different inputs produces an n-bit vector, one output bit for each of the n inputs, such that these outputs coincide with the characteristic function of L on at least m of the inputs.*

The problem we want to address is the following: Let $m \geq 9$. Let L be a language from $(m, m + 3)P$, witnessed by the algorithm A. Assume we have a set of $m+4$ different inputs $x_1, \ldots, x_{m+4}$, and assume we have obtained $m+4$ results by querying every combination of $m+3$ of these inputs to A. We want to show that, no matter what these results look like, we can find an $(m + 4)$-bit vector, which also makes at most 3 errors with respect to the characteristic function of L. Since our result will be obtained from the $m+4$ result vectors of the queries without further access to the inputs themselves, we can view the whole procedure as a polynomial time algorithm itself, thus proving that $(m, m+3)P \subseteq (m+1, m+4)P$. In other words, this will prove the Hinrichs-Wechsung conjecture for the case $d = 3$.

Remark 1: In fact this only proves $(9, 12)P = (10, 13)P = \ldots$ (due to the condition $m \geq 9$ above), whereas the conjecture starts with $(8, 11)P = (9, 12)P$. However, starting with the class $(9, 12)P$ is necessary to obtain the extendability of the investigation to higher values (c.f. the discussion in the proof). But, the case $(8, 11)P = (9, 12)P$ can be obtained by a simple adaptation of our proof, so we will state this case as a corollary.

Remark 2: Of course, it is well known that for any fixed value of m, the question whether $(m, m+3)P = (m+1, m+4)P$ can be considered to be a finite problem (by the characterization of Kummer and Stephan [9]). But, first of all the instance size already for $m = 9$ is rather big, and it is not at all clear how to investigate this case systematically. Moreover, as soon as we speak of variable m, we would have to consider infinitely many cases, and this considerably increases the difficulty in finding a solution.

The objects we deal with will be 13×13-matrices with entries in $\{0, 1, *\}$, asterisks being exactly in those positions, where row index plus column index equals 14, i.e. in the diagonal from bottom left to top right. All other entries will be 0 or 1. The rows of these matrices shall represent the results of the queries on 12 of the 13 inputs, in the i-th column the resulting bit for input x_i. The asterisk indicates that we have no answer for that input. This means, the first row represents the result on $x_1, \ldots, x_{12}$, the second row the results on $x_1, \ldots, x_{11}, x_{13}$, and so on.

Our proof is structured as follows: We present an algorithm, which checks that for all matrices of the described type, one of the following two properties will hold:

1. The matrix is not consistent, meaning: There is no possibility to obtain the answers from the matrix for any true characteristic sequence without making more than three errors in one row.

   Take e.g. a matrix with first row $(0,0,0,0,0,0,0,0,0,0,0,0,0,*)$ and second row $(1,1,1,1,1,1,1,0,0,0,0,*,0)$. No matter what the real characteristic sequence should be, one of the two rows has to make at least 4 errors on the first 7 inputs, because the rows differ on all seven.

   Of course, these matrices will never appear in a computation of the kind we are interested in, as long as the underlying algorithm is a correct $(9, 12)$P-algorithm for the language L to be decided.

2. We will be able to produce a 13-bit output, which coincides with the real characteristic vector on the 13 inputs in at least 10 components. Such an output will in the sequel always be called a *solution*.

In fact we will not try to enumerate all such matrices and investigate them – this would last by far too long. Instead, we consider partial matrices consisting of a few rows (always less than 13) and check, whether we can already show that one of the cases given will be true without knowing the other rows of the matrix. Because of this, the second possibility (proving the ability to produce a solution) will be split in two cases: Either we can explicitly give the solution already, or we can prove for a given i, that x_i is definitely in (or definitely not in, resp.) the language L. Then, using at most one more row of the given matrix (namely row $14 - i$), we can obtain a solution.

Moreover our case inspection will take advantage of occuring symmetries: Rearranging the input vector by application of a permutation results in (simultaneous) permutations of rows and columns, without changing the properties investigated: An inconsistent matrix will still be inconsistent after such a permutation, and a solution will be transformed to a solution of the new matrix, if we apply the same permutation to its bits. The second symmetry operation will be what we call bit-flipping: If we change all bits of a given column of our matrix (thus mapping 0 to 1 and 1 to 0, but leaving an asterisk unchanged), we also can see that the properties of inconsistency or having a solution remain the same (only in a solution, the according bit has to be flipped too). We will give details in Section 4.

Finally, we will show (inductively) that, whenever we know that $(9 + k, 12 + k)P = (10 + k, 13 + k)P$, then the same proof can be used to show that also $(10 + k, 13 + k)P = (11 + k, 14 + k)P$, which will complete the proof of our general result.

3   The Maxdist Technique

Dealing with the Hinrichs-Wechsung conjecture for quite some time, we found that a combination of theoretical arguments and machine power has to be used to make considerable progress in the direction of an affirmative solution. This means, we have to find a method of dividing the problem in subproblems, which may be easy enough to solve. Our approach does that by partitioning the set of all matrices to be considered according to a parameter called *maxdist*, the maximum distance between any two rows of the matrix. More formally:

Definition 2. *The* distance *of two (equal-sized) vectors over $\{0, 1, *\}$ is the number of places, where one vector carries value 0 and the other carries value 1.*

Remark 3: The distance between two rows of our matrices will not be changed, if a permutation as described above is applied. The same holds for bit-flipping, because we always flip all bits of a given column.

Definition 3. *The value* maxdist *for a 13×13-matrix of the kind considered is the maximum distance that appears between any two rows of the matrix.*

Lemma 1. *If a 13×13-matrix of the considered kind has a maxdist value of 0 or a maxdist value greater than 4, it always satisfies one of our desired properties (being not consistent or allowing a solution).*

Proof: We first look at maxdist at least 7. Then there are two rows in our matrix, which on 7 indices i give different answers to the question "$x_i \in L$?". As in a consistent matrix both should err on at most 3 inputs, this is a contradiction. Thus the matrix has to be inconsistent.

Now, look at maxdist either 5 or 6. We pick two rows with distance maxdist, say the i-th and the j-th row. Perform a permutation that maps i to 1 and j to 2 and the columns in such a way that the differences between these two rows appear in the first 5 (or 6) columns. Now, the first row is $(a_1, \ldots, a_{12}, *)$, the second is $(b_1, \ldots, b_{11}, *, b_{13})$, and $a_i = b_i$ for $i \in \{7, \ldots, 11\}$, while $a_i \neq b_i$ for $i \in \{1, \ldots, 5\}$. Then, on the first 5 inputs, both rows together make exactly 5 errors with respect to the real characteristic sequence on $x_1, \ldots, x_5$. It follows that all of $a_7, \ldots, a_{11}$ coincide with the characteristic sequence of $x_7, \ldots, x_{11}$, because otherwise the two given rows would make at least 7 errors in the sum, contradicting the assumption that each makes at most 3 errors. So we can choose for example row 3, where all but x_{11} are queried, to obtain answers for the other 12 inputs, of which at most 3 are erroneous. With the answer $a_{11} = b_{11}$ for x_{11} we have the desired solution.

Finally, let the matrix have maxdist value 0. Then all rows giving answer for x_i agree on that input (otherwise their distance would be greater than 0). We call the according answer a_i. If more than three of the a_is would be wrong, we could choose a row that answers at least four of them. Then, this row would make 4 errors, in contradiction to our assumptions. Thus the vector $a_1, \ldots, a_{13}$ is a solution.   □

The case of maxdist 1 is solved in the same way as the cases 2, 3, and 4. However, the number of cases to be considered is rather small. Thus, it is a good opportunity to get some insight, how our computer program has to work. That is why we now show, how to perform this case:

Lemma 2. *If a 13 × 13-matrix of the considered kind has a maxdist value of 1, then it always satisfies one of our desired properties (being not consistent or allowing a solution).*

Proof: First observe that our matrix has the following property: all pairs of rows have either distance 0 or distance 1.

By a suitable rearrangement (applying a permutation), we make sure that the first two rows have distance 1, and that the difference occurs in the first component. Now, by bit-flipping we change the matrix in such a way that the first two rows get the following form:

$$(0,0,0,0,0,0,0,0,0,0,0,0,*)$$
$$(1,0,0,0,0,0,0,0,0,0,0,*,0)$$

We call this partial matrix 1-2-1, which means: maxdist is 1, depth (number of rows) is 2, and it is the first (in this case the only) matrix with these values of maxdist and depth, which we have to consider. For the third row we get the following possible values:

a) $(0,0,0,0,0,0,0,0,0,0,*,0,0)$
b) $(1,0,0,0,0,0,0,0,0,0,*,0,0)$
c) $(0,0,0,0,0,0,0,0,0,0,*,1,0)$
d) $(1,0,0,0,0,0,0,0,0,0,*,0,1)$

All other possible rows would contradict our assumption of maxdist 1. Now, case b) can be transformed to case a) by permuting the first two rows (and thus also the last two columns) and bit-flipping in the first column. The same holds for case d) to case c). So we only need to consider two cases of partial matrices with 3 rows. Thus we now have two matrices to consider on depth 3, namely

$$(0,0,0,0,0,0,0,0,0,0,0,0,*)$$
$$(1,0,0,0,0,0,0,0,0,0,0,*,0)$$
$$(0,0,0,0,0,0,0,0,0,0,*,0,0)$$

(denoted as matrix 1-3-1), and

$$(0,0,0,0,0,0,0,0,0,0,0,0,*)$$
$$(1,0,0,0,0,0,0,0,0,0,0,*,0)$$
$$(0,0,0,0,0,0,0,0,0,0,*,1,0)$$

(denoted as matrix 1-3-2).

We proceed with matrix 1-3-1 and find the following three cases of possible next rows:

a) (0,0,0,0,0,0,0,0,0,*,0,0,0)
b) (1,0,0,0,0,0,0,0,0,*,0,0,0)
c) (0,0,0,0,0,0,0,0,0,*,0,1,0)

All three cases are different and will lead to matrices 1-4-1, 1-4-2, and 1-4-3. But, from matrix 1-3-2 we only have two possible next rows:

a) (0,0,0,0,0,0,0,0,0,*,0,0,0)
b) (0,0,0,0,0,0,0,0,0,*,0,1,0)

Here, case a) can be transformed to matrix 1-4-3 (which was case c) above), by swapping rows 3 and 4 (and of course also columns 11 and 10). And also case b) can be transformed to that matrix by a somewhat more complex operation: A circular swap of rows 1, 3, and 4, followed by a bit-flipping in column 12.

Now, we have to consider depth 4. However, as the principle should be clear now, we invite interested readers to perform the rest of the proof themselves.

4   Our Algorithm

Now we want to complete the proof of $(9, 12)P = (10, 13)P$. Therefor we have to consider the cases of maxdist value 2, 3, or 4. This now is definitely a case for the computer. But, in order to reduce computation time, we want to reduce the problem as far as possible.

In Section 2 we already mentioned that we will use transformations, and we did apply this technique in Section 3 for the easy cases. We justify this by the following discussion.

Let $x_1, \ldots, x_{13}$ be any sequence of (pairwise different) inputs, which are to be checked for membership in L, the given language from $(9, 12)P$. By 13 queries to the given $(9, 12)P$-algorithm we will obtain 13 times 12 output bits for the possible combinations of 12 out of the 13 inputs. We will know that at least 9 output bits in each answer sequence are correct. (But of course, we do not a priori know which ones!)

Thus the output could look like:

(0,1,*,0,1,1,0,0,1,1,0,0,1)
(0,0,1,0,1,0,0,0,1,0,0,1,*)
(0,0,0,0,0,1,*,0,1,0,0,0,1)
(0,0,1,0,1,1,1,*,1,1,1,0,1)
(0,0,0,0,0,1,1,1,1,0,0,*,1)
(0,0,1,0,1,1,1,1,1,0,*,1,1)
(0,0,1,0,1,0,1,0,1,*,0,1,1)
(0,0,1,0,0,0,1,0,*,0,0,1,1)

and so on. Clearly, we can choose to write any row first, then any other row second, and so forth. But it is obvious that there is only one way to arrange the rows, where the asterisks appear in the diagonal from top right to bottom left:

(0,0,1,0,1,0,0,0,1,0,0,1,*)
(0,0,0,0,0,1,1,1,1,0,0,*,1)
(0,0,1,0,1,1,1,1,1,0,*,1,1)

and so on. We always choose the sequence of rows in this way and call that a normalization. Now, whenever we swap two rows, or more generally whenever we perform a permutation on the rows, we have to swap columns too, in order to keep our normalized form. Note, that swapping of rows i and j has to be followed by swapping of columns $14 - i$ and $14 - j$ (for $1 \leq i < j \leq 13$), and similarly for general permutations. However, while the ordering of the rows technically means nothing, the ordering of the columns refers to the ordering of the queried inputs $x_1, \ldots, x_{13}$. Thus, one should keep in mind that such an operation always means a rearrangement of the inputs.

Lemma 3. *If the normalized matrix A can be transformed into the normalized matrix B by a sequence of row and column permutations, then B is consistent if and only if A is, and B allows a solution, if and only if A does.*

Proof: (omitted).

Corollary 1. *If the normalized partial matrix A (i.e. an upper part of a normalized complete matrix) can be transformed into the normalized partial matrix B by a sequence of row and column permutations, then B can be extended to a consistent matrix if and only if A can, and B already allows for a solution, if and only if A does.*

We will normalize the matrices further: We want to consider only such matrices, where the first row is $(0,0,0,0,0,0,0,0,0,0,0,0,*)$, and the second row has a 0 in the 13-th place. This can be arranged by bit-flipping of all columns, where the first row had a 1, resp. of column 13, if the second row had a 1 there. This operation may be done because of the following lemma:

Lemma 4. *If the normalized matrix A can be transformed to B by bit-flipping on one or more columns, then B is consistent if and only if A is, and B allows a solution, if and only if A does.*

Proof: It suffices to prove the claim for bit-flipping on one column. Let column i be flipped. If $a_1, \ldots, a_{13}$ is a possible truth for the original matrix A, then the sequence obtained by flipping a_i is a possible truth for the new matrix B, and vice versa. Thus, A is consistent if and only if B is. If a solution for A is given, then it can be changed to a solution for B by flipping the i-th bit, and also vice versa. Thus, a solution for A exists, if and only if a solution for B exists.   □

Notation: The relation between (partial) matrices, given in such a way that A is related to B if and only if A can be transformed to B by the above introduced transformation rules, is an equivalence relation. We will often call matrices related in this way to be *symmetric* to each other.

By the above normalization rules we will always be able to transform a given matrix to the following form:

- The asterisk of row i is in column $14 - i$.
- The first row consists only of 0 entries (and the asterisk in column 13).
- The second row has a 0 in column 13.
- The distance between rows 1 and 2 is exactly the value of maxdist.

- In each row, the following holds: If columns j and $j + 1$ are exactly equal in all rows above the current one, then in the current row we may not have a 0 in column j and a 1 in column $j + 1$. (We call that the monotonicity rule.)
- The monotonicity rule, together with the rule about the distance between rows 1 and 2 implies for row 2 that there are 1-entries exactly in columns 1 to maxdist, and all other entries in row 2 are 0 (except for the asterisk in column 12).

The discussion proved that in order to show that all possible (that is consistent) matrices have solutions, it is sufficient to only consider matrices in the described form.

Now we are ready to introduce our algorithm for the cases of maxdist values 2, 3, and 4. A pseudo-code formulation of the algorithm, as well as an exe-file with our implementation (and the source code in ADA) can be found on the web page

http://134.155.88.3/main/chair_de/03/cmm_download/index_de.html

Essentially, the algorithm works exactly as the explicit procedure for the case of maxdist value 1 in Lemma 2. For a given value of maxdist (2, 3, or 4) it starts with the first two rows, which are uniquely determined by the above described form and the value of maxdist. We call that stage "depth 2" and we initialize a list of matrices to be considered with that one matrix of two rows. Moreover we compute the set of possible truths, i.e. the set of all 13-bit vectors which have distance at most 3 to both rows of that matrix, and we attach that set to the matrix.

Now for a given depth (starting from depth 2), we take all matrices from the list, compute all candidates for next rows, which have a distance less than or equal to maxdist to all rows already in the matrix. For each of these candidate rows, we check, whether the row would obey the monotonicity rule. If it does, it is output as a new row to be considered.

The algorithm has to cancel out all so far possible truths, which contradict the new row (by having distance greater than 3 to it), and then examine the remaining set of possible truths, in order to find out, whether one of the following cases happens:

(a) The (partial) matrix becomes inconsistent (set of possible truths is empty), or
(b) All possible truths have the same value in one position (this will lead to a solution), or
(c) We can find a 13-bit vector, which has distance less than or equal to 3 for all possible truths, or
(d) None of these cases, but in our list for the next depth, we can already find a matrix which can be obtained from the current one by row permutation and bit-flipping operations, or
(e) None of the other cases, so we have to append the new matrix to the list for the next depth.

The most complex test to be performed here is for case d). We could try to apply all possible transformations in order to get any of the matrices already in the list. However, we do it the other way round: We feed all pairs, built of the current matrix and one matrix of the list, into a symmetry detecting procedure called *the matcher*, which uses additional structural properties of the matrices to speed up detection of nonsymmetry in many cases: The mainly used property is the vector of numbers of pairs of a given distance. See the full paper for details.

Now, if one of cases a), b), c) or d) happens, the algorithm produces an output telling exactly which case happened, and moreover in case b) the special position and the value that all truths have in that position, in case c) the solution, and in case d) the row permutation that leads to the other matrix, and the number of that matrix in our list (note, that the according column permutation and the bit-flipping operations are implicitly given by our normalization conditions). When all partial matrices of a given depth are done, the algorithm starts to examine the list for the next depth. If that is empty, the algorithm terminates.

The algorithm can be performed on maxdist values 2, 3 and 4, and it runs very fast. The output can also be found on the above named web page. By inspection of the output one can see that in all three cases the maximum depth to be considered is depth 8. At the maximum depth the algorithm terminates in each case, meaning there are no more matrices to consider. Thus we obtain:

Theorem 1. $(9, 12)P = (10, 13)P$.

One can observe that all (partial) matrices occuring in the execution of the algorithm have at least one column of all zeroes. If we omit one such column in every considered matrix, the whole procedure does exactly the same, only it now works on only 12 inputs. Thus, as a corollary we get:

Corollary 2. $(8, 11)P = (9, 12)P$.

5   Extendability

As we observed at the end of Section 4, every matrix used in the execution of our algorithm has at least one column of all zeroes. To be more exact: In cases maxdist value 3 or 4, we always have at least one such column, and in case maxdist value 2 (and maxdist value 1 as well, as investigated in Lemma 2), we always have at least two such columns. These zero columns will be the key in the proof, that the result of Theorem 1 can be extended to all cases with more inputs and up to three errors. Together with Corollary 2 this will complete the solution of the $d = 3$ instance of the Hinrichs-Wechsung conjecture. Unfortunately, the rigorous space restrictions in these lecture notes only allow us to state the final result. We refer the reader to the full paper for details.

Theorem 2. *For all* $k > 0$, *we have* $(8 + k, 11 + k)P = (8, 11)P$.

6   Further Work

One obvious next step would be an adaptation of the proof to case $d = 4$, which might be done by a simple extension of our algorithm. Or, certainly more interesting, a (theoretical) proof should be given for the following conjecture:

Conjecture: For any $m < n$ we have:

$$(m, n)\text{P} = (m + 1, n + 1)\text{P} \implies (m + 1, n + 1)\text{P} = (m + 2, n + 2)\text{P}$$

References

1. Austinat, H., Diekert, V., Hertrampf, U.: A Structural Property of Regular Frequency Computations. Theor. Comp. Sc. 292(1), 33–43 (2003)
2. Austinat, H., Diekert, V., Hertrampf, U., Petersen, H.: Regular Frequency Computations. In: RIMS Symposium on Algebraic Systems, Formal Languages and Computation, Kyoto, Japan, pp. 35–42 (2000)
3. Beigel, R., Gasarch, W.I., Kinber, E.B.: Frequency Computation and Bounded Queries. Theor. Comp. Sc. 163(1–2), 177–192 (1996)
4. Degtev, A.N.: On (m, n)-Computable Sets. Algebraic Systems, 88–99 (1981) (in Russian)
5. Hinrichs, M., Wechsung, G.: Time Bounded Frequency Computations. Information and Computation 139(2), 234–257 (1997)
6. Kinber, E.B.: Frequency Calculations of General Recursive Predicates and Frequency Enumeration of Sets. Soviet Mathematics Doklady 13, 873–876 (1972)
7. Kinber, E.B.: On Frequency-Enumerable Sets. Algebra i Logika 13, 398–419 (1974) (in Russian); English translation in Algebra and Logic 13, 226–237 (1974)
8. Kinber, E.B.: Frequency Computations in Finite Automata. Kibernetika 2, 7–15 (1976) (in Russian); English translation in Cybernetics 12, 179–187 (1976)
9. Kummer, M., Stephan, F.: The Power of Frequency Computation. In: Reichel, H. (ed.) FCT 1995. LNCS, vol. 969, pp. 323–332. Springer, Heidelberg (1995)
10. Kummer, M., Stephan, F.: Recursion Theoretic Properties of Frequency Computation and Bounded Queries. Information and Computation 120, 59–77 (1995)
11. Kummer, M.: A Proof of Beigel's Cardinality Conjecture. J. of Symbolic Logic 57(2), 677–681 (1992)
12. McNaughton, R.: The Theory of Automata, a Survey. Advances in Computers 2, 379–421 (1961)
13. McNicholl, T.: The Inclusion Problem for Generalized Frequency Classes. PhD thesis, George Washington University, Washington (1995)
14. Rose, G.F.: An Extended Notion of Computability, Abstracts Int. Congress for Logic, Methodology, and Philosophy of Science. Stanford, California, p. 14 (1960)
15. Tantau, T.: Towards a Cardinality Theorem for Finite Automata. In: Diks, K., Rytter, W. (eds.) MFCS 2002. LNCS, vol. 2420, pp. 625–636. Springer, Heidelberg (2002)
16. Trakhtenbrot, B.A.: On the Frequency Computation of Functions. Algebra i Logika 2, 25–32 (1963) (in Russian)

A Sublinear Time Randomized Algorithm for Coset Enumeration in the Black Box Model

Bin Fu and Zhixiang Chen

Dept. of Computer Science,
University of Texas - Pan American, TX 78539, USA
{binfu,chen}@cs.panam.edu

Abstract. Coset enumeration is for enumerating the cosets of a subgroup H of a finite index in a group G. We study coset enumeration algorithms by using two random sources to generate random elements in a finite group G and its subgroup H. For a finite set S and a real number $c > 0$, a random generator R_S is a c-random source for S if $c \cdot \min\{\Pr[a = R_S())|a \in S]\} \geq \max\{\Pr[a = R_S())|a \in S]\}$. Let c be an arbitrary constant. We present an $O(\frac{|G|}{\sqrt{|H|}}(\log |G|)^3)$-time randomized algorithm that, given two respective c-random sources R_G for a finite group G and R_H for a subgroup $H \subseteq G$, computes the index $t = \frac{|G|}{|H|}$ and a list of elements $a_1, a_2, \cdots, a_t \in G$ such that $a_i H \cap a_j H = \emptyset$ for all $i \neq j$, and $\cup_{i=1}^{t} a_i H = G$. This algorithm is sublinear time when $|H| = \Omega((\log |G|)^{6+\epsilon})$ for some constant $\epsilon > 0$.

1 Introduction

Coset enumeration is for enumerating the cosets of a subgroup H of a finite index in a group G, given a set of defining relations for G and words generating its subgroup H. It was first studied in Todd and Coxeter's pioneer paper [11]. This is a fundamental problem in computational group theory and has a long line of research [11],[7],[9],[3],[8],[6]. Some computer programs for coset enumeration have been developed and studied such as the implementation by Cannon, Dimino, Havas and Watson [3]. The work in [3] includes comprehensive references to earlier implementations. Later advancements in this area may be found in Neubüser [8], Leech [7] and Sims [9]. Neubüser and Leech gave useful introductions to coset enumeration. Sims gave a formal account of coset enumeration in terms of automata and proved interesting new results on coset enumeration behavior. More details on the experimental work which led to the selection of those methods was given by Havas and Lian [6].

In this paper, our work is on a different model for coset enumeration. We use a black box to access a finite group G and its subgroup H. The input is a random source R_G to generate random elements of group G and another random source R_H to generate random elements for a subgroup H of G. We allow the random generators for the groups to have some constant factor bias, which is defined below. For a finite set S and a real number $c > 0$, a random generator R_S is a

X. Hu and J. Wang (Eds.): COCOON 2008, LNCS 5092, pp. 82–91, 2008.

c-random source for S if $c \cdot \min\{\Pr[a = R_S()] | a \in S)\} \geq \max\{\Pr[a = R_S()] | a \in S)\}$. We assume that each random source has at most c-factor bias. We use a c-random source to generate a random element in a finite group.

There are important reasons that we do not use the uniform random generator to get a random element in two groups. In order to save storage, a finite group is often represented by a small number of generators (e.g. [9]). It is not trivial to generate a random elements from a finitely generated finite group [2],[4]. Theoretically, Babai showed that a random element can be generated in $O(\log |G|)^5$ steps. A more practical and widely used algorithm to generate random elements based on the replacement method was introduced by Celler, Leedham, Murray, and Niemeyer [4]. Celler et al's algorithm does not have theoretical warrant of uniform random distribution.

Let c be an arbitrary constant. We present an $O(\frac{|G|}{\sqrt{|H|}}(\log |G|)^3)$-time randomized algorithm that, given two respective c-random sources R_G for a finite group G and R_H for a subgroup $H \subseteq G$, computes the index $t = \frac{|G|}{|H|}$ and a list of elements $a_1, a_2, \cdots, a_t \in G$ such that $a_i H \cap a_j H = \emptyset$ for all $i \neq j$, and $\cup_{i=1}^t a_i H = G$.

The principle of Birthday Paradox [10] plays an important role in our algorithm. For a set S of n elements, if $\sqrt{n}$ elements are randomly selected independent from S, then with high probability two elements among those selected are identical.

2   Notations and Model of Computation

For two positive integers x and y, (x, y) represents the greatest common divisor (GCD) between them. For a set A, $|A|$ denotes the number of elements in A. For a real number x, $\lfloor x \rfloor$ is the largest integer $\leq x$ and $\lceil x \rceil$ is the smallest integer $\geq x$. For two integers x and y, $x|y$ means that $y = xc$ for some integer c.

A *group* is a nonempty set G with a binary operation "$\cdot$" that is closed in set G and satisfies the following properties (for simplicity, "ab" represents "$a \cdot b$"): 1)for every three elements a, b and c in G, $a(bc) = (ab)c$; 2)there exists an identity element $e \in G$ such that $ae = ea = a$ for every $a \in G$; 3)for every element $a \in G$, there exists $a^{-1} \in G$ with $aa^{-1} = a^{-1}a = e$. A group G is *finite* if G contains a finite number of elements. Let e be the identity element of G, i.e. $ae = a$ for each $a \in G$. For $a \in G$, ord(a), the order of a, is the least integer k such that $a^k = e$. For a subset S of the group G and an element $a \in G$, define $aS = \{as | s \in S\}$ and $Sa = \{sa | s \in S\}$. For a subgroup H of the group G and an element $a \in G$, aH (Ha) is called a *left coset* (*right coset*, respectively).

The computation model of our algorithm is to call two c-random sources R_G and R_H to generate elements in G and its subgroup H, respectively. We also need to access a black box for group multiplication and assume that the cost is one step for one multiplication. We assume that it takes one step to check if two elements in group G are identical. We further assume that each element of a group can be saved in one unit memory and is encoded as an integer. The inverse operation is not used in our proposed algorithm.

3   Algorithm Find-Cosets

In this section we describe the algorithm Find-Cosets. It has three subroutines Merge, Approximate and Find-Cosets.

Definition 1. *Given a series of sets $L = S_1, \cdots, S_k$, the intersection graph $G = (V, E)$ is defined as follows: each set S_i is a node in V, and there is an edge connecting two nodes S_i and S_j if $S_i \cap S_j \neq \emptyset$. Such a graph is called intersection graph of the sets $S_1, \cdots, S_k$. The intersection graph is decomposed into a series of connected components $(V_1, E_1), \cdots, (V_t, E_t)$. A representation of L is a list of elements $a_1, \cdots, a_t$ such that each a_i is taken from one component $a_i \in \cup_{S_j \in V_i} S_j$.*

The following algorithm returns a representation of a list of finite sets $S_1, \cdots, S_k$. It will be used to select one element from each coset of G/H in our algorithm.

> Merge($S_1, \cdots, S_k$)
> Input: $S_1, \cdots, S_k$ is a list of finite sets.
> Output: a representation of $S_1, \cdots, S_k$.
> Begin
> > For each S_i, select one element $a_i \in S_i$ to represent it.
> > Let $T = \emptyset$.
> > Let L be an empty list.
> > For $i = 1$ to k
> > Begin
> > > For every element a in S_i
> > > > If (a is not in T)
> > > > > Insert a to T.
> > > > > Let the linked list of a be empty.
> > > > Else (T contains a)
> > > > > Append a_i (which represents S_i) to the linked list for a.
> > End of For.
> > Build graph U with nodes $S_1, \cdots, S_k$.
> > For each $a \in T$
> > Begin
> > > Let S_i be the first node in the linked list of a.
> > > For every other node S_j in the linked list of a
> > > > Add an edge between S_i and S_j in the graph U.
> > > Find all connected components for the graph U.
> > > For each connected component, select one node
> > > a_i from it and put a_i into L.
> > End of For.
> > Output L.
> End of Merge

The following algorithm Approximate(R_H) is used to approximate the square root $\sqrt{|H|}$ of H. It is a randomized algorithm based on the Birthday Paradox.

Approximate(R_H)
Input: R_H is a c-random source of set H.
Output: an integer m to approximate $\sqrt{|H|}$.
Begin
    Let $m_1 = 1$.
    Let $found = false$.
    While $(found = false)$
    Begin
        Generate a list $L_i = a_1 a_2 \cdots a_{m_i}$ of m_i elements in H by calling R_H m_i times independently.
        If L_i contains two identical elements then $found = true$.
        Else Let $m_{i+1} = 2m_i$ and $i = i + 1$.
    End
    Output m_i.
End of Approximate.

Define $u_1(c, \epsilon) = \sqrt{\frac{\epsilon}{4c}}$ and $u_2(c, \epsilon) = 2(\sqrt{\frac{2\ln\frac{\epsilon}{2}}{c}} + 1)$. The algorithm Find-Cosets(R_G, R_H, ϵ) computes the index $z = \frac{|G|}{|H|}$ and a list of elements $v_1, \cdots, v_z$ of G such that $v_1 H, \cdots, v_z H$ form a partition of G. Let $\beta = \frac{1}{4}$. The algorithm Find-Cosets is described by the routine below, which uses the other routines defined before. The number α in Find-Cosets follows that in Lemma 5.

Find-Cosets(R_G, R_H, ϵ)
Input: R_G is a c-random source for group G, R_H is a c-random source for the subgroup H of G, and ϵ is a positive real number.
Output: elements $a_1, \cdots, a_t$ such that $a_1 H, \cdots, a_t H$ is a partition of G.
Begin
    Let $m = $ Approximate(R_H) and $n = $ Approximate(R_G).
    Let $k = (c \ln \frac{1}{\beta \epsilon}) \cdot (\frac{u_2(\epsilon, \beta)^2}{u_1(\epsilon, \beta)^2}) \cdot (\frac{n^2}{m^2}) \cdot (\log((\frac{u_2(\epsilon, \beta)^2}{u_1(\epsilon, \beta)^2}) \cdot (\frac{n^2}{m^2})))$.
    Let $m' = \max(1, \frac{2c}{\alpha} \cdot (\ln(\frac{1}{\beta \epsilon}))) \cdot \max(1, \frac{1}{u_1(c, \epsilon \beta)}, \frac{1}{\sqrt{u_1(c, \epsilon \beta)}}) \cdot m \cdot \max(1, 2\ln k)$.
    Sample k elements $a_1, a_2, \cdots, a_k$ with R_G.
    Sample m' elements $S = \{b_1, b_2, \cdots, b_{m'}\}$ with R_H.
    Compute $a_1 S, a_2 S, \cdots, a_k S$.
    $L = $Merge($a_1 S, a_2 S, \cdots, a_v S$).
    Output the number of elements in the list L and L.
End of Find-Cosets.

4 Analysis for Algorithm Find-Cosets

The following lemma is basic and well known in group theory. We include its proof for completeness.

Lemma 1. *Assume that G is a finite group and H is a subgroup of G. Then 1) For any a, b in G, either $aH \cap bH = \emptyset$ or $aH = bH$; and 2) For any a in G, $|aH| = |H|$.*

Proof. 1) Assume $aH \cap bH \neq \emptyset$. There are elements $h \in H$ and $h' \in H$ with $ah = bh'$. We have $(ah)h^{-1} = (bh')h^{-1} = b(h'h^{-1})$. Since H is a subgroup of G, $h'h^{-1} \in H$. Therefore, $a \in bH$. We have $aH \subseteq bH$. Similarly, we also have $bH \subseteq aH$. Thus, $aH = bH$.

2)For every two elements, $b_1, b_2 \in H$, if $ab_1 = ab_2$, then $b_1 = a^{-1}(ab_1) = a^{-1}(ab_2) = (a^{-1}a)b_2 = b_2$. Therefore, we have $|aH| = |H|$.

For a c-random source R_S for a set S with n elements, we have Lemma 2 that shows that every element has probability at least $\frac{1}{cn}$ and at most $\frac{c}{n}$ to be generated by R_S.

Lemma 2. *Let S be a set of n elements and R_S be a c-random source. Then for every element $a \in S$, $\Pr[R_S() = a] \in [\frac{1}{cn}, \frac{c}{n}]$.*

Proof. We have $1 = \sum_{a \in S} \Pr[R_S() = a]$. Since S have n elements, we have that $1 \geq n \min\{\Pr[R_S() = a]|a \in S\}$ and $1 \geq n \max\{\Pr[R_S() = a]|a \in S\}$. Therefore, $\frac{1}{n} \leq \min\{\Pr[R_S() = a]|a \in S\}$ and $\frac{1}{n} \leq \max\{\Pr[R_S() = a]|a \in S\}$.

Since $c \cdot \min\{\Pr[a = R_S()]|a \in S)\} \geq \max\{\Pr[a = R_S()]|a \in S)\}$, we have $\min\{\Pr[R_S() = a]|a \in S\} \geq \frac{1}{c} \max\{\Pr[a = R_S()]|a \in S)\} \geq \frac{1}{cn}$. On the other hand, $\max\{\Pr[R_S() = a]|a \in S\} \leq c \min\{\Pr[a = R_S()]|a \in S)\} \leq \frac{c}{n}$.

In the rest of this paper, if R_S is a c-random source for a set S of n elements, then we automatically use the condition that for each $a \in S$, $\Pr[R_S() = a] \in [\frac{1}{cn}, \frac{c}{n}]$ by Lemma 2.

Lemma 3. *Let c be a constant of at least 1. Assume m and n are two non-negative integer with $m \leq n$. Then there exists a constant $\alpha \in (0,1)$ such that for every integer m_1 with $0 \leq m_1 \leq \alpha m$, $\binom{n}{m_1}(\frac{cm_1}{n})^m \leq e^{-m}$.*

Proof. We first show that $m_1! \geq \frac{m_1^{m_1}}{e^{m_1}}$ for the integer $m_1 \geq 1$. This is equivalent to prove that $\frac{m_1!}{m_1^{m_1}} \geq e^{-m_1}$, which is implied by the following inequalities: $\ln \frac{m_1!}{m_1^{m_1}} = m_1 \cdot (\frac{1}{m_1} \cdot \sum_{i=1}^{m_1} \ln \frac{i}{m_1}) \geq m_1 \cdot \lim_{a \to +0} \int_a^1 (\ln x)dx = -m_1$. Therefore, we have that $\binom{n}{m_1} < \frac{n^{m_1}}{m_1!} \leq \frac{e^{m_1}n^{m_1}}{m_1^{m_1}}$. Let $\alpha \in (0,1)$ be a constant that satisfies $0 < \alpha < 1$ and $\alpha \leq \frac{1}{c} \cdot \frac{1}{e^{1/(1-\alpha)}} \cdot \frac{1}{c^{1/(1-\alpha)}} \cdot \frac{1}{e^{1/(1-\alpha)}}$. It is easy to see the existence of α when it is small enough since $2, c$ and e are all constants.

We have

$$\binom{n}{m_1}(\frac{cm_1}{n})^m \leq \frac{e^{m_1}n^{m_1}}{m_1^{m_1}}(\frac{cm_1}{n})^m = e^{m_1}c^m(\frac{m_1}{n})^{m-m_1} \tag{1}$$

$$\leq e^{m_1}c^m(\alpha)^{m-m_1} \leq e^{m_1}c^m\alpha^{(1-\alpha)m} \tag{2}$$

$$\leq e^{m_1}c^m(\frac{1}{e^{1/(1-\alpha)}} \cdot \frac{1}{c^{1/(1-\alpha)}} \cdot \frac{1}{e^{1/(1-\alpha)}})^{(1-\alpha)m} \tag{3}$$

$$= e^{m_1}c^m e^{-m}c^{-m}e^{-m} \leq e^{-m} \tag{4}$$

Basing on the well known Birthday Paradox, we develop Lemmas 4 through 7 so that they can be used for c-random source. We need to check if two elements

a and b are from the same coset ($aH = bH$). Those lemmas will support us to convert the checking of $aH = bH$ into the checking for $aS \cap aS \neq \emptyset$, where S is a set of $\sqrt{|H|}$ random samples from H.

Lemma 4. *Let S be a set of n elements and R_S be a c-random source. Assume that $x_1, \cdots, x_m$ be m elements in S generated by R_S independently. Then with probability at most $\binom{n}{m_1}(\frac{cm_1}{n})^m$, $x_1, \cdots, x_m$ contains at most m_1 different elements.*

Proof. For a subset $S' \subseteq S$ with $|S'| = m_1$, the probability is at most $(\frac{cm_1}{n})^m$ that all elements $x_1, \cdots, x_m$ are in S'. For every subset $X \subseteq S$ with $|X| \leq m_1$, there exists another subset $S' \subseteq S$ such that $|S'| = m_1$. We have that $\Pr[\text{the list } x_1, \cdots, x_m \text{ has no more than } m_1 \text{ distinct elements}] \leq \Pr[\text{all elements in the list } x_1, \cdots, x_m \text{ are in some } S' \subseteq S \text{ with } |S'| = m_1]$. There are $\binom{n}{m_1}$ subsets of S with size m_1. We have the probability at most $\binom{n}{m_1}(\frac{cm_1}{n})^m$ that $x_1, \cdots, x_m$ contains at most m_1 different elements.

Lemma 5. *Let c be a constant, S be a set of n elements, and R_S be a c-random source. Then there exists some constant $\alpha > 0$ such that the probability is at most $2e^{-\frac{\alpha m \min(m,n)}{cn}}$ to have $A \cap B = \emptyset$, where $A = \{x_1, \cdots, x_m\}$ and $B = \{y_1, \cdots, y_m\}$ are each generated by calling $R_S()$ m times.*

Proof. We prove the lemma by two cases.

Case 1: $m < n$. Let α_1 be the constant to be determined later. We assume that A has been generated by R_S and has at least $m_1 = \lceil \alpha_1 m \rceil$ different elements. We are going to give an upper bound about the probability that B does not contain any element in A. For each element $y_i \in B$, with probability at most $1 - \frac{m_1}{cn}$ that y_i is not in A. Therefore, the probability is at most $(1 - \frac{m_1}{cn})^m$ that B does not contain any element in A.

By Lemma 4, the probability is at most $\binom{n}{m_1}(\frac{cm_1}{n})^m$ that A contains at most m_1 elements. We have that

$$\Pr[A \cap B = \emptyset] = \Pr[B \cap A = \emptyset || A| \geq m_1] \cdot \Pr[|A| \geq m_1] + \tag{5}$$
$$\Pr[B \cap A = \emptyset || A| < m_1] \cdot \Pr[|A| < m_1] \tag{6}$$
$$\leq \Pr[B \cap A = \emptyset || A| \geq m_1] + \Pr[|A| < m_1] \tag{7}$$
$$\leq (1 - \frac{m_1}{cn})^m + \binom{n}{m_1}(\frac{cm_1}{n})^m \tag{8}$$
$$\leq e^{-\frac{m_1^2}{cn}} + \binom{n}{m_1}(\frac{cm_1}{n})^m. \tag{9}$$

The inequality $(1 - \frac{m_1}{cn})^m \leq e^{-\frac{\alpha_1 m^2}{cn}}$ follows from the fact that $1 - x \leq e^{-x}$. By Lemma 3, $\binom{n}{m_1}(\frac{cm_1}{n})^m \leq e^{-m} \leq e^{-\frac{\alpha_1 m^2}{cn}}$ for some constant $0 < \alpha_1 < 1$. Therefore, $\Pr[A \cap B = \emptyset] \leq 2e^{-\frac{\alpha_1 m^2}{cn}}$.

Case 2: $m \geq n$. Let α_2 be the constant to be determined later. We assume that A has been generated by R_S and has at least $n_1 = \lceil \alpha_2 n \rceil$ different elements.

We are going to give an upper bound about the probability that B does not contain any element in A. For each element $y_i \in B$, with probability at most $1 - \frac{n_1}{cn} \leq 1 - \frac{\alpha_2}{c}$ that y_i is not in A. Therefore, the probability is at most $(1 - \frac{\alpha_2}{c})^m$ that B does not contain any element in A.

By Lemma 4, the probability is at most $\binom{n}{n_1}(\frac{cn_1}{n})^m$ that A contains less than n_1 elements. We have that

$$\Pr[A \cap B = \emptyset] = \Pr[B \cap A = \emptyset \mid |A| \geq n_1] \cdot \Pr[|A| \geq n_1] + \tag{10}$$
$$\Pr[B \cap A = \emptyset \mid |A| < n_1] \cdot \Pr[|A| < n_1] \tag{11}$$
$$\leq \Pr[B \cap A = \emptyset \mid |A| \geq n_1] + \Pr[|A| < n_1] \tag{12}$$
$$\leq (1 - \frac{\alpha_2}{c})^m + \binom{n}{n_1}(\frac{cn_1}{n})^m \tag{13}$$
$$\leq e^{-\frac{\alpha_2 m}{c}} + \binom{n}{n_1}(\frac{cn_1}{n})^m \tag{14}$$

The inequality $(1 - \frac{\alpha_2}{c})^m \leq e^{-\frac{\alpha_2 m}{c}}$ follows from the fact that $1 - x \leq e^{-x}$. By Lemma 3, $\binom{n}{n_1}(\frac{cn_1}{n})^m \leq e^{-m} \leq e^{-\frac{\alpha_2 m^2}{cn}}$ for some constant α_2 with $0 < \alpha_2 < 1$. Therefore, $\Pr[A \cap B = \emptyset] \leq 2e^{-\frac{\alpha_2 m}{c}}$.

By combining Cases 1 and 2, we have that the probability is at most $2e^{-\frac{\alpha \cdot m \cdot \min(m,n)}{cn}}$ for $A \cap B = \emptyset$, where α is a constant greater than 0.

Lemma 6. *Assume that S is a set of n elements and is partitioned into m subsets of equal size $S_1, \cdots, S_m$. Let R_S be a c-random source of S and $U = \{a_1, \cdots, a_k\}$ is k elements, which is independently generated by calling $R_S()$ k times. Then with probability at most $me^{-\frac{k}{cm}}$, $U \cap S_i = \emptyset$ for some $i \in \{1, \cdots, m\}$.*

Proof. Since R_S is a c-random source of S, we have that $\min\{\Pr[R_S() = a] \mid a \in S\} \geq \frac{1}{cn}$ by Lemma 2. For each $a_i \in U$ and S_j, $\Pr[a_i \notin S_j] \leq 1 - \frac{|S_j|}{cn} = 1 - \frac{1}{cm}$. We have that $\Pr[U \cap S_j = \emptyset]$ is at most $(1 - \frac{1}{cm})^k \leq e^{-\frac{k}{cm}}$ since $1 - x \leq e^{-x}$. We have that with probability at most $me^{-\frac{k}{cm}}$, $U \cap S_i = \emptyset$ for some $i \in \{1, \cdots, m\}$.

The following Lemma 7 is essentially from the well known Birthday Paradox. We slightly transform it to fit the c-random source instead of uniform random source.

Lemma 7. *Let R_S be a c-random source of S and $a_1, \cdots, a_k$ be k elements independently generated by $R_S()$. Then 1) with probability at most $(1 - \frac{1}{cn})^{\frac{k(k-1)}{2}}$, all elements in the list $a_1, \cdots, a_k$ are different. 2) With probability at least $(1 - \frac{c}{n})^{\frac{k(k-1)}{2}}$, all elements in the list $a_1, \cdots, a_k$ are different.*

Proof. Since R_S is a c-random source of S, we have that $\min\{\Pr[R_S() = a] \mid a \in S\} \geq \frac{1}{cn}$ by Lemma 2. Let P_i be the probability that i elements generated by R_S are different. For i different elements, a_{i+1} is generated by R_S. The probability that $a_{i+1} \neq a_j$ for $j = 1, \cdots, i$ is at most $(1 - \frac{1}{cn})^i$. We have that $P_{i+1} \leq (1 - \frac{1}{cn})^i P_i$. Therefore, $P_k \leq (1 - \frac{1}{cn})^{\frac{k(k-1)}{2}}$.

Given i different elements $a_1, \cdots, a_i$, assume that a_{i+1} is generated by R_S. The probability that $a_{i+1} \neq a_j$ for $j = 1, \cdots, i$ is at least $(1 - \frac{c}{n})^i$ by Lemma 2. We also have that $P_{i+1} \geq (1 - \frac{c}{n})^i P_i$. Thus, we have that $P_k \geq (1 - \frac{c}{n})^{\frac{k(k-1)}{2}}$.

We have Lemma 8 about the complexity and correctness of subroutine Merge$(S_1, \cdots, S_k)$.

Lemma 8. *1. There exists an $O(n \log n)$ time algorithm to test if a list of n elements has two elements identical.*

2. There exists an $O(n \log n)$ time algorithm Merge such that given a series of sets $S_1, \cdots, S_k$, it outputs a series of elements $a_1, \cdots, a_t$ that is a representation of the intersection graph of $S_1, \cdots, S_k$, where $n = |S_1| + |S_2| + \cdots + |S_k|$ and t is the number of its connected components.

Proof. We need a data structure that can support query and insertion in $O(\log n)$ time. The Red-black tree or 2-3 tree is sufficient (see [5],[1]).

Statement 1. This part is trivial. All elements are added into a 2-3-tree one by one. The 2-3-tree T has $O(\log n)$ time complexity for querying and searching. Before inserting a new element into the tree, check if the tree T has already contained it.

Statement 2. In the algorithm Merge, it needs $O(n)$ queries and insertions to the 2-3-tree. Each insertion adds one element to a linked list in a node. It takes $O(n \log n)$ time to build up the tree. Finding all connected components takes $O(|V| + |E|)$ time for a graph (V, E) by using the depth-first search method in a undirected graph (see [5],[1]). The graph U built in the algorithm has $O(n)$ edges and $O(k)$ nodes.

Lemma 9 gives the computational time of subroutine Approximate(R_H) and guarantees that the value returned by Approximate(R_H) has only constant factor difference with $\sqrt{|H|}$ with high probability. Recall that we have defined $u_1(c, \epsilon) = \sqrt{\frac{\epsilon}{4c}}$ and $u_2(c, \epsilon) = 2(\sqrt{2c \ln \frac{\epsilon}{2}} + 1)$.

Lemma 9. *Let ϵ be a positive small real number. Algorithm Approximate(R_H) is a randomized algorithm such that given a set of H and a c-random generator R_H for H, the algorithm runs in time $O(\sqrt{|H|})$ and returns a number m with $u_1(c, \epsilon)\sqrt{|H|} \leq m \leq u_2(c, \epsilon)\sqrt{|H|}$ with probability at most ϵ to fail.*

Proof. Let $n = |H|$. Let i be the largest number with $m_i < u_1(c, \epsilon)\sqrt{n}$. If $m_j < u_1(c, \epsilon)\sqrt{n}$, then by Lemma 7, with probability at least $(1 - \frac{c}{n})^{\frac{m_j(m_j-1)}{2}}$, list L_j has no two identical elements. Since $m_j = 2m_{j+1}$, we have $\sum_{j=1}^i m_j^2 \leq m_i^2(1 + (\frac{1}{4}) + (\frac{1}{4})^2 + \cdots) \leq m_i^2 \cdot \frac{4}{3}$. Therefore, with probability at least $\prod_{j=1}^i (1 -$
$\frac{c}{n})^{\frac{m_j(m_j-1)}{2}} \geq \prod_{j=1}^i (1 - \frac{c}{n})^{\frac{m_j^2}{2}} \geq (1 - \frac{c}{n})^{\frac{1}{2}\sum_{j=1}^i m_j^2} \geq (1 - \frac{c}{n})^{m_i^2} \geq 1 - \frac{cm_i^2}{n}$, no m_j with $j \leq i$ will be outputted by Approximate(R_H). With probability at most $1 - (1 - \frac{cm_i^2}{n}) = \frac{cm_i^2}{n}$, m_j with $j \leq i$ is returned by Approximate(R_H). Therefore, With probability at most $P_1 = \sum_{j=1}^i \frac{cm_i^2}{n} \leq \frac{2cm_i^2}{n} \leq \frac{\epsilon}{2}$, one of m_j with $j \leq i$ is returned by Approximate(R_H).

Let i' be the largest number with $m_{i'} < u_2(c, \epsilon)\sqrt{n}$. It implies that $m_{i'+1} \geq u_2(c, \epsilon)\sqrt{n}$. Since $m_{i'+1} = 2m_{i'}$, we have $m_{i'} \geq \frac{u_2(c, \epsilon)\sqrt{n}}{2}$, which implies the following inequality:

$$m_{i'} \geq (\sqrt{2c\ln\frac{\epsilon}{2}} + 1)\sqrt{n} \geq \sqrt{2c\ln\frac{\epsilon}{2}}\sqrt{n} + 1. \tag{15}$$

By Lemma 7, with probability at most $(1 - \frac{1}{cn})^{\frac{m_{i'}(m_{i'}-1)}{2}}$, all elements in list $L_{i'}$ are distinct. Therefore, with probability at most $P_2 = (1 - \frac{1}{cn})^{\frac{m_{i'}(m_{i'}-1)}{2}} \leq e^{-\frac{1}{cn} \cdot \frac{m_{i'}(m_{i'}-1)}{2}} \leq e^{-\frac{1}{cn} \cdot \frac{(m_{i'}-1)(m_{i'}-1)}{2}} \leq \frac{\epsilon}{2}$ (by (15)), all elements in list $L_{i'}$ are distinct. Therefore, the probability is at most P_2 that the algorithm returns $m_i > m_{i'}$.

The probability is at most $P_1 + P_2 \leq \epsilon$ that the algorithm Approximate fails to return a number m with $u_1(c, \epsilon)\sqrt{|H|} \leq m \leq u_2(c, \epsilon)\sqrt{|H|}$.

Lemma 9 shows that with probability at least $1 - \epsilon$, Approximate(R_H) returns m with $u_1(c, \epsilon)\sqrt{|H|} \leq m \leq u_2(c, \epsilon)\sqrt{|H|}$. The following Lemma 10 shows the computational time and correctness of Find-Cosets.

Lemma 10. *Let ϵ be a positive small real number. The algorithm Find-Cosets (R_G) runs in $O(\frac{|G|}{\sqrt{|H|}}(\log |G|)^3)$ time and has probability at most ϵ to fail to return the correct $t = \frac{|G|}{|H|}$ and a list of elements $g_1, \cdots, g_t$ that $g_1 H, \cdots, g_t H$ form a partition of G.*

Proof. By Lemma 9, with probability at most $P_0 = \epsilon\beta$, the number m returned by Approximate(R_H) does not satisfy $u_1(c, \epsilon\beta)\sqrt{|H|} \leq m \leq u_2(c, \epsilon\beta)\sqrt{|H|}$. Similarly, we have probability at most $P_1 = \epsilon\beta$ that the number n returned from Approximate($R_H, \epsilon\beta$) does not satisfy $u_1(c, \epsilon\beta)\sqrt{|G|} \leq n \leq u_2(c, \epsilon\beta)\sqrt{|G|}$. We assume that $u_1(c, \epsilon\beta)\sqrt{|H|} \leq m \leq u_2(c, \epsilon\beta)\sqrt{|H|}$ and $u_1(c, \epsilon\beta)\sqrt{|G|} \leq n \leq u_2(c, \epsilon\beta)\sqrt{|G|}$ are true in the following analysis. Let $q = |G|/|H|$. Let $U = \{a_1, a_2, \cdots, a_k\}$.

For the m' elements in S, let X be the first $\frac{m'}{2}$ elements of S and Y be the last $\frac{m'}{2}$ elements of S. By Lemma 5, for each pair of elements a_j and $a_{j'}$ from the same coset ($a_j H = a_{j'} H$), the probability is at most $e^{-\alpha \frac{m'}{2} \cdot \frac{\min(\frac{m'}{2}, |H|)}{c|H|}} \leq \frac{\beta\epsilon}{k^2}$ (by the setting of m' in Find-Cosets) that $a_j X \cap a_{j'} Y = \emptyset$. With probability at most $P_2 = k^2 \frac{\beta\epsilon}{k^2} = \beta\epsilon$, there exists a pair of elements a_j and $a_{j'}$ in U from the same coset ($a_j H = a_{j'} H$) such that $a_j S \cap a_{j'} S = \emptyset$.

Since k is large enough, the set U shall have intersection with all the cosets in G. Let $t = \frac{|G|}{|H|}$. By Lemma 6, the probability is at most $P_3 = te^{-\frac{k}{ct}} \leq \beta\epsilon$ that $U \cap aH = \emptyset$ for some $a \in G$. Assume that the set U has non-empty intersection with all cosets. Merge computes the index $t = \frac{|G|}{|H|}$ and a list $L = g_1 g_2 \cdots g_t$ such that $g_1 H, \cdots, g_t H$ form a partition of G.

According to the analysis above, the total probability for Find-Cosets to fail is at most $P_0 + P_1 + P_2 + P_3 < \epsilon$. The sum of sizes in $a_1 S, \cdots, a_k S$ is $|U||S| \leq$

km'. The Merge function takes $O(km' \log(km'))$ time. Since $|G| \geq |H|$, $k = O(t \log t)$, and $m' = \sqrt{|H|} \log k$, the total computational time for Find-Cosets is $O(km' \log(km')) = O(\frac{|G|}{\sqrt{|H|}}(\log |G|)^3)$.

Theorem 1. *Let δ be a small fixed positive real number. There exists an $O(\frac{|G|}{\sqrt{|H|}}$ $(\log |G|)^3)$ time randomized algorithm such that given a c-random source R_G for a group G and a c-random source R_H for a subgroup H of G, it returns the index $\frac{|G|}{|H|}$ and the list of all cosets in G with probability at least $1 - \delta$.*

Proof. It follows from Lemma 10.

References

1. Aho, A.V., Hopcroft, J.E., Ullman, J.D.: The Design and Analysis of Computer Algorithms. Addison-Wesley, Reading (1974)
2. Babai, L.: Local Expansion of Vertex-Transitive Graphs and Random Generation in Finite Groups. In: Proceedings of the 23rd Annual ACM Symposium on Theory of Computing, pp. 164–174 (1991)
3. Cannon, J.J., Dimino, L.A., Havas, G., Watson, J.M.: Implementation and Analysis of the Todd-Coxeter Algorithm. Math. Comput. 27, 449–463 (1973)
4. Celler, F., Leedham-Green, C.R., Murray, S.H., Niemeyer, A.C., O'Brien, E.A.: Generating Random Elements of a Finite Group. Comm. Algebra 23, 4931–4948 (1995)
5. Cormen, T.H., Leiserson, C.E., Rivest, R.L., Stein, C.: Introduction to Algorithms. 2nd edn. The MIT Press, Cambridge (2001)
6. Havas, G., Lian, J.X.: New Implementations of Coset Enumeration. Technical Report, Department of Computer Science, University of Queensland (1991)
7. Leech, J.: Coset Enumeration on Digital Computers. Proc. Cambridge Philos. Soc. 59, 257–267 (1963)
8. Neubuser, J.: An Elementary Introduction to Coset-Table Methods in Computational Group Theory. London Math. Soc. Lecture Note Ser. 71, 1–45 (1982)
9. Sims, C.: Computation with Finitely Presented Groups. Cambridge University Press, Cambridge (1994)
10. Stallings, W.: Cryptography and Network Security. Prentice Hall, Englewood Cliffs (2005)
11. Todd, J.A., Coxeter, H.S.M.: A Practical Method for Enumerating Cosets of a Finite Abstract Group. Proc. Edinburg Math. Ann. 5, 584–594 (1936)

Smallest Formulas for Parity of 2^k Variables Are Essentially Unique

Jun Tarui

University of Electro-Comm, Chofu, Tokyo 182-8585, Japan
tarui@ice.uec.ac.jp

Abstract. For $n = 2^k$, we know that the size of a smallest AND/OR/NOT formula computing the Boolean function $\text{Parity}(x_1, \ldots, x_n) = \text{Odd}(x_1, \ldots, x_n)$ is exactly n^2: For any n, it is at least n^2 by classical Khrapchenko's bound, and for $n = 2^k$ we easily obtain a formula of size n^2 by writing and recursively expanding

$$\text{Odd}(x_1, \ldots, x_n) = [\, \text{Odd}(x_1, \ldots, x_{n/2}) \wedge \text{Even}(x_{n/2+1}, \ldots, x_n) \,]$$
$$\vee \,[\, \text{Even}(x_1, \ldots, x_{n/2}) \wedge \text{Odd}(x_{n/2+1}, \ldots, x_n) \,].$$

We show that for $n = 2^k$ the formula obtained above is an essentially unique one that computes $\text{Parity}(x_1, \ldots, x_n)$ with size n^2. In the equivalent framework of the Karchmer-Wigderson communication game, our result means that an optimal protocol for Parity of 2^k variables is essentially unique.

1 Introduction and Summary

A *formula* is an AND/OR/NOT circuit in which each gate has fan-out 1, i.e., the underlying graph is a tree. Such AND/OR/NOT formulas are sometimes called *DeMorgan formulas*. The *size* of a formula F is the number of leaves in F, which equals the number of ANDs and ORs in F plus one. We assume, without loss of generality for our purposes, that all the NOT gates appear as negated input literals $\neg x_i$'s. For a Boolean function f, let $L(f)$ denote the minimum possible size of a formula computing f. The largest known lower bound for $L(f)$ for an explicit function f is due to Hastad [1], and it is $n^{3-o(1)}$.

For $x \in \{0,1\}^n$, $\text{Parity}(x) = \text{Parity}(x_1, \ldots, x_n)$ is 1 if $\sum x_i$ is odd, and 0 otherwise. We also use $\text{Odd}(x)$ and $\text{Even}(x)$ to respectively denote $\text{Parity}(x)$ and its negation. Khrapchenko's theorem [2] (also explained in [5],[6],[7]) asserting that $L(\text{Parity}) \geq n^2$ is a nice classic in combinatorial complexity.

A formula F computing a Boolean function f bijectively corresponds to a two-party communication protocol P solving the Karchmer-Wigderson communication game for f [3],[4] (also explained in, e.g., [5],[6],[8]). In this game Alice and Bob are respectively given $x, y \in \{0,1\}^n$ such that $f(x) = 1$ and $f(y) = 0$. The task is to agree on some $i \in \{1, \ldots, n\}$ such that $x_i \neq y_i$. A leaf of a formula F corresponds to a leaf of a protocol P, or equivalently, to a *monochromatic rectangle* in the partition induced by P; thus the size of F equals the number of

X. Hu and J. Wang (Eds.): COCOON 2008, LNCS 5092, pp. 92–99, 2008.

leaves of P, which also equals the number of monochromatic rectangles in the partition induced by P.

We can obtain a formula for $\text{Parity}(x_1, \ldots, x_n) = \text{Odd}(x_1, \ldots, x_n)$ by writing and recursively expanding

$$\text{Odd}(x_1, \ldots, x_n) = [\, \text{Odd}(x_1, \ldots, x_{n/2}) \wedge \text{Even}(x_{n/2+1}, \ldots, x_n) \,]$$
$$\vee [\, \text{Even}(x_1, \ldots, x_{n/2}) \wedge \text{Odd}(x_{n/2+1}, \ldots, x_n) \,].$$

For $n = 2^k$, this yields a formula of size n^2 and depth $2\log_2 n$. We can also use the following dual expression.

$$\text{Odd}(x_1, \ldots, x_n) = [\, \text{Odd}(x_1, \ldots, x_{n/2}) \vee \text{Odd}(x_{n/2+1}, \ldots, x_n) \,]$$
$$\wedge [\, \text{Even}(x_1, \ldots, x_{n/2}) \vee \text{Even}(x_{n/2+1}, \ldots, x_n) \,].$$

For $n = 2^k$, using the two forms above alternately yields a formula of size n^2 and depth $2\log_2 n$ in which the number of AND/OR alternations along each path from the root to a leaf is $\log_2 n$. The number of AND/OR *alternations* corresponds to the number of *rounds* in a Karchmer-Wigderson protocol.

Thus for $n = 2^k$ we know the formula size complexity exactly; it is n^2. In this paper we show that a formula obtained above is an essentially unique smallest one: We can partition the set $\{1, \ldots, n\}$ into two sets of size $n/2$ in an arbitrary way; we can use either one of the dual forms above and we can "syntactically shuffle and rotate" gates of the same type (AND/OR) appearing at *consecutive* levels. But up to these variations, a smallest formula is unique.

While most important questions in computational complexity are *only* meaningful in asymptotic forms of $O(\cdot)$, $\Omega(\cdot)$, and $\Theta(\cdot)$, when an underlying model is a simple natural one such as formula, comparator network, or arithmetic formula/circuit, to determine the *exact* complexity is an interesting problem. In somewhat rare cases where we can determine the exact complexity, we can further raise the question about the *uniqueness of an optimal solution*. A famous example is Borodin's result [12] in arithmetic complexity that Horner's rule for evaluating a polynomial is uniquely optimal (also explained, e.g., in [13, p. 498]). But there are very few results establishing such uniqueness. Our result suggests that uniqueness can sometimes be established if one focuses on *special nice cases*, e.g., for n a power of 2, or if one appropriately defines *essential uniqueness*. We hope that our result encourages people to keep in mind questions of this type.

We mention some more previous works related to this one. Zwick [9] gave some extension of Khrapchenko's bound by considering the formula size in which variables have different costs. Koutsoupias [10] gave an extension and an alternative proof of Khraphcheko's bound by considering the eigenvalues of a certain matrix associated with a formula. Through a line of research on *quantum query complexity* of Boolean functions, Laplante, Lee, and Szegedy [11] have given a new method for establishing size lower bounds for (classical) formulas; the new method yields an interesting alternative proof for Khrapchenko's bound.

1.1   Results

Throughout the paper we do not want to distinguish, e.g., $F \wedge G$ and $G \wedge F$; all our statements are up to the interchange of left and right. Consider the formula F obtained by connecting k subformulas $F_1, \ldots, F_k$ by $(k-1)$ AND gates for $k \geq 3$; for example, consider

$$F_0 = [\,[\,[\,F_1 \wedge F_2\,] \wedge F_3\,] \wedge F_4\,].$$

We say that, e.g., F_0 is of the form $[\,F_1 \wedge F_2\,] \wedge [\,F_3 \wedge F_4\,]$ *up to AND rearrangements*. Similarly we speak of *OR rearrangements*. Note that AND rearrangements and OR rearrangements do *not* change the number of AND/OR alternations along any path from the root to a leaf. For a set $S \subseteq \{1, \ldots, n\}$ and $x \in \{0,1\}^n$, the function $\mathrm{Odd}_S(x_1, \ldots, x_n)$ is 1 if $\sum_{i \in S} x_i$ is odd, and 0 otherwise; we also use $\mathrm{Even}_S(x_1, \ldots, x_n)$ to denote $\neg\mathrm{Odd}_S(x_1, \ldots, x_n)$. We state our main result in the following form.

Theorem 1. Assume that $n = 2^k$.

(A) Let F be a formula of size n^2 that computes $\mathrm{Odd}(x_1, \ldots, x_n)$. Then, the following hold: If $n = 1$, F is a single literal x_1. For $n \geq 2$:

(A-1) If the top gate is OR, its two children at depth 1 must be AND gates, and up to AND rearrangement at depth 1, F is of the following form for some $S \subseteq \{1, \ldots, n\}$ with $|S| = n/2$: $F = [\,F_1 \wedge F_2\,] \vee [\,F_3 \wedge F_4\,]$, where

$$F_1(x) = \mathrm{Odd}_S(x), \ F_2(x) = \mathrm{Even}_{\overline{S}}(x), \ F_3(x) = \mathrm{Even}_S(x), \ F_4(x) = \mathrm{Odd}_{\overline{S}}(x).$$

(A-2) If the top gate is AND, its two children at depth 1 must be OR gates, and up to OR rearrangement at depth 1, F is of the following form for some $S \subseteq \{1, \ldots, n\}$ with $|S| = n/2$: $F = [\,F_1 \vee F_2\,] \wedge [\,F_3 \vee F_4\,]$, where

$$F_1(x) = \mathrm{Odd}_S(x), \ F_2(x) = \mathrm{Odd}_{\overline{S}}(x), \ F_3(x) = \mathrm{Even}_S(x), \ F_4(x) = \mathrm{Even}_{\overline{S}}(x).$$

(B) Similar statements hold for a size-n^2 formula computing $\mathrm{Even}(x_1, \ldots, x_n)$.

Our main result above immediately yields some new *quantitative* results about simultaneous optimality such as the following.

Corollary 2. For $n = 2^k$, the minimum number of AND/OR alternations in a smallest formula for $\mathrm{Parity}(x_1, \ldots, x_n)$ is $\log_2 n$. Equivalently, for $n = 2^k$, the minimum number of rounds in an optimal Karchmer-Wigderson protocol for $\mathrm{Parity}(x_1, \ldots, x_n)$ is $\log_2 n$.

1.2   Proof Outline

The outline of our proof of Theorem 1 is as follows. Our proof first closely follows the simplified proof of Khrapchenko's bound due to Mike Paterson (see [5]). In each inductive step, we analyze properties of a subformula whose size equals the lower bound; the notion of *certificate* will play a key role. Then we further analyze a whole formula. A proof of Khrapchenko's theorem appears as part of our analysis, making this paper self-contained.

2   Proof of Theorem 1

Following Wegener [7], for $a, b \in \{0, 1\}^n$ say that a and b are *neighbors* if the Hamming distance between a and b is 1. For $A, B \subseteq \{0, 1\}^n$, define the set $A \otimes B$ as follows.

$$A \otimes B = \{(a, b) : a \in A, b \in B, \ a \text{ and } b \text{ are neighbors.}\}$$

We will consider $A \otimes B$ in the case where A and B are disjoint nonempty sets; for such A and B, define $K(A, B)$ as follows.

$$K(A, B) = \frac{|A \otimes B|^2}{|A| \cdot |B|}.$$

In other words, consider the average, respectively over $a \in A$ and $b \in B$, of the number of neighbors in B and A. The product of these two averages is $K(A, B)$. For disjoint sets $A, B \subseteq \{0, 1\}^n$ and a formula F, say that F *separates* A and B if F is constant on A and B taking different 0/1 values, i.e., $f(A) = \{0\}$, $f(B) = \{1\}$ or $f(A) = \{1\}$, $f(B) = \{0\}$.

For $a \in \{0, 1\}^n$ and $i \in \{1, \ldots, n\}$, let $a^{(i)}$ denote the n-bit vector obtained by flipping the i-th bit of a. For $x \in \{0, 1\}^n$, a formula F, and a set $S \subseteq \{1, \ldots, n\}$, say that set S is a *certificate of x for F* if for any $x' \in \{0, 1\}^n$ such that $x'_i = x_i$ for all $i \in S$, $F(x') = F(x)$. Say that a formula F separating A and B is *K-optimal for A and B* if $\text{size}(F) = K(A, B)$.

Lemma 3. Let F be a formula separating disjoint nonempty $A, B \subseteq \{0, 1\}^n$.

(I) [KHRAPCHENKO-PATERSON] $\text{size}(F) \geq K(A, B)$.

(II) Let $s = |A \otimes B|/|A|$ and $t = |A \otimes B|/|B|$. Assume that F is K-optimal for A and B, i.e., $\text{size}(F) = K(A, B) = st$. Then, each $a \in A$ has s neighbors in B and each $b \in B$ has t neighbors in A. Furthermore, for each $a \in A$, if $a^{(i_1)}, a^{(i_2)}, \ldots, a^{(i_s)}$ are the s neighbors of a in B, then the set $\{i_1, i_2, \ldots, i_s\}$ is a certificate of a for F. Similarly, for each $b \in B$, if $b^{(j_1)}, b^{(j_2)}, \ldots, b^{(j_t)}$ are the t neighbors of b in A, then the set $\{j_1, j_2, \ldots, j_t\}$ is a certificate of b for F.

Proof. The proof is by induction on the size of a formula F. Assume that F separates nonempty disjoint sets A and B.

Base: $\text{size}(F) = 1$ and F is a single literal x_i or $\overline{x_i}$: Each $a \in A$ has at most one neighbor in B; the only candidate is $a^{(i)}$. Similarly each $b \in B$ has at most one neighbor in A; the only candidate is $b^{(i)}$. Thus (I) holds since we have

$$K(A, B) = \frac{|A \otimes B|}{|A|} \times \frac{|A \otimes B|}{|B|} \leq 1 \times 1 = 1 = \text{size}(F).$$

Formula F is K-optimal, i.e., $K(A, B) = 1$ if and only if each $a \in A$ and each $b \in B$ respectively has one neighbor $a^{(i)}$ in B and one neighbor $b^{(i)}$ in A. In this case the set $\{i\}$ is a certificate for a and for b and (II) holds.

Induction: Assume that $F = F_1 \vee F_2$ and $F(A) = \{0\}, F(B) = \{1\}$. The other dual case $F = F_1 \wedge F_2$ is similar. Put $B_1 = \{b \in B : F_1(b) = 1\}$ and $B_2 = B - B_1$. Then, (I) holds since we have

$$\text{size}(F) = \text{size}(F_1) + \text{size}(F_2) \geq K(A, B_1) + K(A, B_2) \geq K(A, B). \qquad (1)$$

The first equality is by definition of formula size and the first inequality is by induction. The second inequality is by the following calculation: Let

$$a = |A|, \quad b = |B_1|, \quad c = |B_2|, \quad d = |A \otimes B_1|, \quad e = |A \otimes B_2|$$

so that

$$K(A, B_1) = \frac{d^2}{ab}, \quad K(A, B_2) = \frac{e^2}{ac}, \quad K(A, B) = \frac{(d+e)^2}{a(b+c)}.$$

We can obtain the inequality by writing a series of equivalent inequalities:

$$\frac{d^2}{ab} + \frac{e^2}{ac} \geq \frac{(d+e)^2}{a(b+c)}.$$
$$(b+c)(cd^2 + be^2) \geq bc(d+e)^2.$$
$$c^2 d^2 + b^2 e^2 \geq 2bcde.$$
$$(cd - be)^2 \geq 0.$$

Note that equalities hold in place of the inequalites above if and only if $d/b = e/c = |A \otimes B|/|B|$.

To see that (II) holds, assume that F is K-optimal for A and B. Then, equalities hold in place of the inequalities in (1). Thus (i) F_1 is K-optimal for A and B_1, (ii) F_2 is K-optimal for A and B_2, and (iii) $|A \otimes B_1|/|B_1| = |A \otimes B_2|/|B_2| = |A \otimes B|/|B|$. Induction yields (II). $\qquad \square$

Remark. Assume that $A, B \subseteq \{0,1\}^n$ are disjoint and nonempty. Consider a *partition* of $A \times B$ into rectangles that are monochromatic in the sense of Karchmer-Wigderson. The number of monochromatic rectangles is at least $K(A, B)$; see [6, Sec 5.1]. This assertion is more general than Lemma 3-(I) since partitions properly contain protocol-induced partitions, which correspond to formulas. Call a partition *K-optimal* if the number of monochromatic rectangles equals $K(A, B)$. By inspecting the arguments in [6, Sec 5.1] we can also see that in a K-optimal partition of $A \times B$, all the monochormatic rectangles must be squares of the same size; it follows that each $a \in A$ and each $b \in B$ has the same number of neighbors in B and in A respectively. But one cannot define *certificates* for general partitions. In fact, we do not know the answer to the following question: For $n = 2^k$, it there a size-n^2 partition of Odd×Even that is *not* induced by a protocol?

The following is a strengthened version of an exercise in Wegener's book [7, Chapter 8, Exercise 1, p. 263]. (The exercise is to show that $L(H) = L(f) + L(g)$ for H, f, g below.) We include a proof in the appendix.

Lemma 4. Suppose that $X = \{x_1, \ldots, x_s\}$ and $Y = \{y_1, \ldots, y_t\}$ are disjoint sets of variables and $f(x_1, \ldots, x_s)$ and $g(y_1, \ldots, y_t)$ are nonconstant Boolean functions. Let H be a smallest formula that computes $f(x_1, \ldots, x_s) \wedge g(y_1, \ldots, y_t)$. Then, up to AND-rearrangements, H is of the form $H = F \wedge G$, where F is a smallest formula that computes f and G is a smallest formula that computes g. The dual statement with OR-rearrangements holds for a smallest formula computing $f \vee g$.

Proof of Theorem 1. Let $n = 2^k \geq 2$ and let F be a size-n^2 formula computing $\text{Odd}(x_1, \ldots, x_n)$. Let $A = \text{Odd}^{-1}(0)$ and $B = \text{Odd}^{-1}(1)$, i.e., A is the set of even vectors in $\{0, 1\}^n$ and B is the set of odd vectors in $\{0, 1\}^n$. Clearly, F is a K-optimal separation of A and B.

Assume that F is of the form $F = F_1 \vee F_2$. The other dual case is similar; we omit a proof. Put $B_1 = \{b \in B : F_1(b) = 1\}$ and $B_2 = B - B_1$. As explained in the proof of Lemma 3, F_1 is K-optimal for A and B_1 and F_2 is K-optimal for A and B_2. By Lemma 3, there exist s and t such that $n = s + t$ and that each $a \in A$ has s neighbors $a^{(i_1)}, \ldots, a^{(i_s)}$ in B_1 and t neighbors $a^{(j_1)}, \ldots, a^{(j_t)}$ in B_2 with the set $\{i_1, \ldots, i_s\}$ being a 0-certificate of a for F_1 and the set $\{j_1, \ldots, j_t\}$ being a 0-certificate of a for F_2.

Consider $a_0 = (0, 0, \ldots, 0) \in A$. The point a_0 has s neighbors $a_0^{(i_1)}, \ldots, a_0^{(i_s)}$ in B_1. Let $S = \{i_1, \ldots, i_s\}$ and $\overline{S} = \{1, \ldots, n\} - S$.

Claim. Formula F_1 computes function $\text{Odd}_S \wedge \text{Even}_{\overline{S}}$ and Formula F_2 computes function $\text{Even}_{\overline{S}} \wedge \text{Odd}_S$.

Proof of Claim. Renaming indexes, if necessary, assume that $S = \{1, \ldots, s\}$. Set $S = \{1, 2, \ldots, s\}$ is a 0-certificate of a_0 for F_1 and set $\overline{S} = \{s + 1, \ldots, n\}$ is a 0-certificate of a_0 for F_2. Thus F_1 is 0 for any n-bit vector of the form

$$(\underbrace{0, 0, \ldots, 0}_{s}, \underbrace{*, *, \ldots, *}_{n-s}),$$

and F_2 is 0 for any n-bit vector of the form

$$(\underbrace{*, *, \ldots, *}_{s}, \underbrace{0, 0, \ldots, 0}_{n-s}).$$

Consider n-bit vectors of the form

$$a = (\underbrace{0, 0, \ldots, 0}_{s}, \underbrace{a_{s+1}, \ldots, a_n}_{n-s}) \in A.$$

Note that since a is in A, $a_{s+1} + \cdots + a_n$ is even. For each such a, S must be a 0-certificate of a for F_1 since a agrees with a_0 on those coordinates in S and S is a 0-certificate for a_0. Each such a has a 0-certificate for F_2 of size $n - s$; but that certificate must be $\overline{S}$ because if a size-$(n - s)$ set T other than $\overline{S}$ is a 0-certificate for F_2, the set $S \cup T$ of size less than n is a 0-certificate of a for $F = F_1 \vee F_2$ contradicting the assumption that F computes $\text{Odd}(x_1, \ldots, x_n)$.

Thus, for every $(n-s)$-tuple $a_{s+1}, \ldots, a_n$ such that $a_{s+1} + \cdots + a_n$ is even, $(*, \ldots, *, a_{s+1}, \ldots, a_n)$ is a 0-certificate for F_2. Therefore, for each $b \in B$ such that $b_{s+1} + \cdots + b_n$ is even, $F_2(b) = 0$; but $F(b) = F_1(b) \vee F_2(b) = 1$; so $F_1(b) = 1$. By similarly arguing with $a = (a_1, a_2, \ldots, a_s, 0, \ldots, 0) \in A$, we conclude that for each $b \in B$ such that $b_1 + \cdots + b_s$ is even, $F_2(b) = 1$. Thic completes the proof of Claim.

We continue the proof of Theorem 1. Recall that $|S| = s$ and $|\overline{S}| = n - s$. By Lemma 4 and Khrapchenko's bound(Lemma 3-(I)), formula F_1 computing $\mathrm{Odd}_S \wedge \mathrm{Even}_{\overline{S}}$ has size at least $s^2 + (n-s)^2$ and the same applies to formula F_2. Hence $\mathrm{size}(F_1) + \mathrm{size}(F_2) \geq 2(s^2 + (n-s)^2)$. But the sum of two sizes equals n^2 by assumption, and thus we must have $s = n/2$. Now we know what F_1 and F_2 *compute*. By Lemma 4 we conclude that they are of the form as claimed in Theorem 1.   □

References

1. Håstad, J.: The Shrinkage Exponent of De Morgan Formulae is 2. SIAM Journal on Computing 27(1), 48–64 (1998)
2. Khrapchenko, V.M.: A Method of Determining Lower Bounds for the Complexity of Π-Schemes. Mat.Zametski 10(1), 83–92 (1971) (in Russian); English translation in: Math. Notes 10(1), 474–479 (1971)
3. Karchmer, M., Wigderson, A.: Monotone Circuits for Connectivity Require Super-Logarithmic Depth. SIAM J. Discrete Mathematics 3(2), 255–265 (1990)
4. Karchmer, M.: Communication Complexity: A New Approach to Circuit Depth. MIT Press, Cambridge (1989)
5. Boppana, R., Sipser, M.: The Complexity of Finite Functions. In: van Leeuwen, J. (ed.) Handbook of Theoretical Computer Science volume A, Algorithms and Complexity. MIT Press, Cambridge (1990)
6. Kushilevitz, E., Nisan, N.: Communication Complexity. Cambridge Univ. Press, Cambridge (1997)
7. Wegener, I.: The Complexity of Boolean Functions. Wiley, Chichester (1987) (on-line copy available at the web site of ECCC under Monographs)
8. Arora, S., Barak, B.: Complexity Theory: A Modern Approach (to be published, 2008)
9. Zwick, U.: An Extension of Khrapchenko's Theorem. Information Processing Letters 37, 215–217 (1991)
10. Koutsoupias, E.: Improvements on Khraphchenko's Theorem. Information Processing Letters 116, 399–403 (1993)
11. Laplante, S., Lee, T., Szegedy, M.: The Quantum Adversary Method and Classical Formula Size Lower Bounds. Computational Complexity 15(2), 163–196 (2006)
12. Borodin, A.: Horner's Rule is Uniquely Optimal. In: Kohavi, Z., Paz, A. (eds.) Theory of Machines and Computations, pp. 45–58. Academic Press, London (1971)
13. Knuth, D.: The Art of Computer Programming, 3rd edn. Seminumerical Algorithms, vol. 2. Addison-Wesley, Reading (1997)

Appendix: Proof of Lemma 4. Let $X = \{x_1, \ldots, x_s\}$ and $Y = \{y_1, \ldots, y_t\}$ be disjoint sets of variables and let $f(x_1, \ldots, x_s)$ and $g(y_1, \ldots, y_t)$ be nonconstant functions. Suppose that smallest DeMorgan formulas computing f and g have size l and m respectively. Let H be a smallest formula computing $f \wedge g$.

Claim 1. The formula H contains l X-variables and m Y-variables.

Proof. Since f is a nonconstant function, there is a 0/1-assignment a for X-variables that makes f to be 1. Fixing all the X-variables in H according to the assignment a yields a formula that computes g; thus the number of Y-variables in H is at least m. Similarly, the number of X-variables in H is at least l.

Claim 2. Let $P = Q \square R$ be a subformula of H, where $\square$ is a connective, the formula Q contains only X-variables, and the formula R contains at least one Y-variable. Then the connective $\square$ must be $\wedge$, and for every assignment a of X-variables, $f(a) = 1 \Longrightarrow Q(a) = 1$.

Proof. To get a contradiction, assume that the connective $\square$ is $\vee$. Further assume, for the sake of contradiction, that for every assignment a to X-variables, $f(a) = 1 \Longrightarrow Q(a) = 0$. Then, it is easy to see that the formula we obtain by fixing Q to be constant 0 in H still computes $f \wedge g$; this contradicts the minimum size assumption about H. Thus there is an assignment a to X-variables such that $f(a) = 1$ and $Q(a) = 1$. Fixing all X-variables in H according to a yields a formula for g; but since the connective is $\vee$ and $Q(a) = 1$, this formula contains less than m Y-variables; a contradiction. Hence the connective $\square$ must be $\wedge$. Suppose that there is an assignment a such that $f(a) = 1$ and $Q(a) = 0$. Then, fixing all X-variables as in a yields a formula computing g with size less than m; a contradiction. Thus the last assertion of Claim 2 holds.

Say that a subformula Q of H is *pure* if it is a maximal subformula containing only X-variables or only Y-variables. From Claim 2 and the symmetry, it follows that whenever $f \wedge g = 1$, all pure subformulas evaluate to 1. To get a contradiction to the assertion of Lemma 4, assume that there is a subformula P such that P contains both X and Y variables and has the form $P = Q \vee R$. Then the smaller formula obtained by setting Q (or R) to be 0 must still compute $f \wedge g$ since when $f \wedge g = 1$, all the pure subformulas evaluate to 1, and thus $R = 1$; this contradicts the minimum size assumption. Thus H connects all the pure subformulas by the connective $\wedge$. $\qquad\square$

Counting Polycubes without the Dimensionality Curse*

Gadi Aleksandrowicz and Gill Barequet

Center for Graphics and Geometric Computing
Dept. of Computer Science
Technion—Israel Institute of Technology
Haifa 32000, Israel
[gadial|barequet]@cs.technion.ac.il

Abstract. A d-D polycube of size n is a connected set of n cells (hypercubes) of an orthogonal d-dimensional lattice, where connectivity is through $(d-1)$-dimensional faces of the cells. Computing $A_d(n)$, the number of distinct d-dimensional polycubes of size n, is a long-standing elusive problem in discrete geometry. In a previous work we described the generalization from two to higher dimensions of a polyomino-counting algorithm of Redelmeier. The main deficiency of the algorithm is that it keeps the entire set of cells that appear in any possible polycube in memory at all times. Thus, the amount of required memory grows exponentially with the dimension. In this paper we present a method whose order of memory consumption is a (very low) *polynomial* in both n and d. Furthermore, we parallelized the algorithm and ran it through the Internet on dozens of computers simultaneously. This enables us to find $A_d(n)$ for values of d and n far beyond any previous attempt.

1 Introduction

A d-dimensional polycube of size n is a face-connected set of n d-dimensional cubes on the regular orthogonal lattice $\mathbb{Z}^d$. *Fixed* polycubes are considered distinct if they differ in their shapes *or* orientations. The number of fixed d-dimensional polycubes of size n is denoted by $A_d(n)$. For example, Figures 1(a,b) show the $A_3(2) = 3$ and $A_3(3) = 15$ dominoes and trominoes, respectively, in three dimensions. There are two main open problems related to polycubes: (i) The number of d-dimensional polycubes of size n, as a function of d and n; and (ii) For a fixed dimension d, the growth-rate limit of $A_d(n)$ (that is, $\lim_{n \to \infty} A_d(n+1)/A_d(n)$). Since no analytic formula for $A_d(n)$ is known, even for $d = 2$, a great portion of the research has so far focused on efficient algorithms for computing $A_2(n)$ for as high as possible values of n.

---

* Work on this paper has been supported in part by Jewish Communities of Germany Research Fund.

X. Hu and J. Wang (Eds.): COCOON 2008, LNCS 5092, pp. 100–109, 2008.
© Springer-Verlag Berlin Heidelberg 2008

Redelmeier [15] introduced the first algorithm for counting polyominoes (2-D polycubes), which generates all the polyominoes sequentially *without* repetitions. Thus, it only has to count generated polyominoes but does not have to compare every polyomino to all the previously-generated polyominoes. Since the algorithm generates each polyomino in $O(1)$ time, its total running time is $O(A_2(n))$.[1] Redelmeier implemented his algorithm in Algol W and in the PDP assembly language. The program required

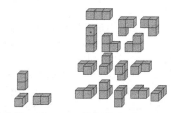

(a) Dominoes   (b) Trominoes

Fig. 1. Fixed three-dimensional dominoes and trominoes

about 10 months of CPU time on a PDP-11/70 for counting fixed polyominoes of up to size 24. A faster algorithm is known only in 2D. Jensen [6],[7] was able to to compute $A_2(n)$ up to $n = 56$ [7].

Polyominoes and polycubes have triggered the imagination of not only mathematicians. Extensive studies can also be found in statistical-physics literature, where 2- and 3-dimensional fixed polycubes are usually referred to as *lattice animals*. Animals play an important role in computing the mean cluster density in percolation processes, in particular those of fluid flow in random media [3].

There have only been a few attempts to count 3-dimensional fixed polycubes:

- Lunnon [10] analyzed in 1972 3-D polycubes by considering symmetry groups, and computed (manually!) $A_3(n)$ up to $n = 6$. Values of $A_3(n)$ up to $n = 12$ can be derived from a subsequent work of Lunnon [11] in 1975.
- Sykes et al. [16] used in 1976 a method of [13] in order to analyze series expansions on a 3-D lattice, but did not compute new values of $A_3(n)$.
- Gong [5] computed, in a series of attempts (in 1992, 1997, and 2004) $A_3(n)$ up to $n = 9$, 15, and 16, respectively.
- However, Flammenkamp [4] computed $A_3(17)$ already in 1999.
- The authors of this paper computed $A_3(18)$ in 2006 [1].

In higher dimensions, the only previous work on counting polycubes that we are aware of is that of Lunnon [11], in which he counted polycubes that could fit into restricted boxes. (In fact, Lunnon counted *proper* polycubes, that is, polycubes that cannot be embedded in lower-dimensional lattices, but the numbers of all polycubes can easily be deduced from the numbers of proper polycubes.) Lunnon computed values up to $A_4(11)$, $A_5(9)$, and $A_6(8)$ (with slight errors in $A_6(7)$ and $A_6(8)$). In [1] we generalized Redelmeier's algorithm in a rather naive way; this allowed us to compute values up to $A_4(15)$, $A_5(13)$, $A_6(9)$, $A_7(6)$, $A_8(5)$, and $A_9(4)$. Some formulae for the numbers of polycubes are given in [2].

---

[1] This is true under the assumption that operations on integers can be done in $O(1)$ time. This work is targeted at situations where the handled numbers are large enough to make this issue significant, so that each operation requires time that is linear in the number of bits required to represent the respective number. In this perspective, just counting the polyominoes takes $O(A_2(n) \log A_2(n)) = O(A_2(n)n)$ time.

Redelmeier's original algorithm [15] counts (2-D) polyominoes. Although it is not presented in this way, it is based on counting connected subgraphs in the underlying graph of the 2-D orthogonal lattice, that contain one particular node. However, his algorithm does not depend on any property of the graph. In the generalization of this algorithm to higher dimensions [1], we needed to compute the respective lattice graphs, clipped to the size of the sought-after polycubes, and to apply the same subgraph-counting algorithm. The main drawback of this approach in higher dimensions is that it keeps in memory at all times the entire set of "reachable" cubes—all cubes that are part of some polycube. Since the number of such cells is roughly $(2n)^d$, the algorithm becomes useless for relatively low values of d. In the current paper we significantly improve the algorithm by not keeping this set of reachable cells in memory *at all*. Instead, we maintain only the "current" polycube and a set of its immediate neighboring cells, entirely omitting the lattice graph. The latter is computed locally "on demand" in the course of the algorithm. Hence, we need to store in memory only $O(nd)$ cells. Furthermore, we parallelized the new version of the algorithm.

In the next sections we describe Redelmeier's original algorithm, its generalization to higher dimensions, our new method of eliminating the need to hold the entire underlying graph in memory, and how to parallelize the algorithm. This enabled us to compute $A_d(n)$ for values of d far beyond any previous attempt.

2   The Original Algorithm

Redelmeier's original algorithm [15] is a procedure for connected-subgraph counting, where the underlying graph is induced by the square lattice. Since translated copies of a fixed polyomino are considered identical, one must decide upon a canonical form. Redelmeier's fixed the leftmost square of the bottom row of a polyomino at the origin, that is, at the square (0,0). (Coordinates are associated with squares and not with their corners.) Thus, he needed to count the number of edge-connected sets of squares (that contain the origin) in

$$\{(x,y) \mid (y > 0) \text{ or } (y = 0 \text{ and } x \geq 0)\}.$$

The squares in this set are located above the thick line in Figure 2(a). The shaded area in this figure consists of all the *reachable* cells, (possible locations of cells) of pentominoes (polyominoes of size 5). Counting pentominoes amounts to counting all the connected subgraphs of the graph shown in Figure 2(b), that contain the vertex a_1. The algorithm [15] is shown in Figure 3. Step 4(a) deserves some attention. By "new neighbors" we mean only neighbors of the new cell c that was chosen in Step 2, which *were not neighbors* of any cells of the polyomino prior to adding c to the polyomino. This ensures that we will not count the same polyomino more than once. We elaborate more on this in Section 3.2.

This sequential subgraph-counting algorithm can be applied to any graph, and it has the property that it never produces the same subgraph twice.

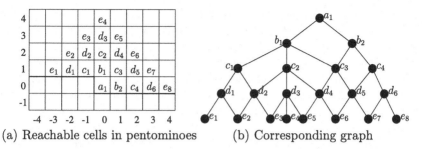

(a) Reachable cells in pentominoes   (b) Corresponding graph

Fig. 2. Pentominoes as subgraphs of some underlying graph

Initialize the parent to be the empty polyomino, and the untried set
to contain only the origin. The following steps are repeated until the
untried set is exhausted.

1. Remove an arbitrary element from the untried set.
2. Place a cell at this point.
3. Count this new polyomino.
4. If the size is less than n:
   (a) Add new neighbors to the untried set.
   (b) Call this algorithm recursively with the new parent being the
       current polyomino, and the new untried set being a copy of the
       current one.
   (c) Remove the new neighbors from the untried set.
5. Remove newest cell.

Fig. 3. Redelmeier's algorithm (taken verbatim from [15, p. 196])

3   Polycubes in Higher Dimensions

3.1   Naive Generalization

In [1] we generalized Redelmeier's algorithm to higher dimensions by reorganizing it into two stages: (1) Computing the cell-adjacency graph and marking the canonical cell; and (2) Applying the original procedure, i.e., counting the connected subgraphs containing the canonical cell. As noted above, this method does not depend on the structure of the graph. Thus, we only needed to rewrite the first stage to compute the neighborhood graph in the appropriate dimension (up to the sought size). Then, we invoked the subgraph-counting procedure.

Denote the coordinates of a cell as a vector $x = (x_1, x_2, \ldots, x_d)$, and regard it as a number in some base $t \in \mathbb{N}$ with d digits. We mapped the lattice cell x to the integer $\Gamma(x) = \sum_{k=1}^{d} x_k t^{k-1}$. The number t was chosen large enough so that no two cells were mapped to the same number. The minimum possible choice of t was the size of the range of coordinates attainable by reachable cells, i.e., $t = 2n - 2$, where n is the polycube size. A clear benefit of this representation was the ability to compute neighboring cells efficiently: The images under $\Gamma(\cdot)$ of the immediate neighbors of a cell x along the kth direction were $\Gamma(x) \pm t^{k-1}$.

Originally, the canonical cell was fixed as $(0, \ldots, 0)$. Except in the dth direction, the original range of reachable coordinates was $[-(n-2), \ldots, n-1]$, so it was shifted by $n-2$ in order to have only nonzero coordinates. In the dth direction the range of reachable coordinates is $[0, \ldots, n-1]$, and so no shift was needed. Thus, the canonical cell was shifted to $\bar{0} = (n-2, \ldots, n-2, 0)$. It is easy to verify that a cell $x = (x_1, \ldots, x_d)$ is reachable if and only if $x \geq \bar{0}$ lexicographically, and reachable cells are in the d-dimensional box defined by $(0, \ldots, 0)$ and $x_M = (2n-3, \ldots, 2n-3, n-1)$. The function $\Gamma(\cdot)$ maps reachable cells to numbers in the range 0 to $M = \Gamma(x_M) = \sum_{k=1}^{d-1}(2n-3)(2n-2)^{k-1} + (n-1)(2n-2)^{d-1} = (2n-2)^{d-1}n - 1$.[2] Figure 4 shows the algorithm for building the d-dimensional graph.

ALGORITHM GraphOrthodD(int n, int d)
 begin
  1. $o := \Gamma((n-2, \ldots, n-2, 0))$; $M := (2n-2)^{d-1}n - 1$;
  2. for $i = o, \ldots, M$ do
   2.1 $b := 1$; $counter := 0$;
   2.2 for $j = 1, \ldots, d$ do
    2.2.1 if $i + b \geq o$ then do
     2.2.1.1 neighbors$[i][counter] := i + b$;
     2.2.1.2 $counter := counter+1$;
    end if
    2.2.2 if $i - b \geq o$ then do
     2.2.2.1 neighbors$[i][counter] := i - b$;
     2.2.2.2 $counter := counter+1$;
    end if
    2.2.3 $b := b(2n-2)$;
   end for
   2.3 neighbors_num$[i] := counter$;
  end for
 end GraphOrthodD

Fig. 4. Computing the graph for a d-dimensional orthogonal lattice

The cell-adjacency graph contains, then, at most $(2n-2)^{d-1}n$ nodes. Actually, only roughly one half of these nodes, that is, about $2^{d-2}(n-1)^{d-1}n$, are really reachable (see Figure 2(a) for an illustration). In two dimensions, this number (n^2) is negligible for polyominoes whose counting is time-wise possible. However, in higher dimensions the "dimensionality curse" appears. The size of the graph, and not the number of distinct polycubes, becomes the bottleneck. For example, in order to count polycubes of size 7 in 7 dimensions, we need to maintain a graph with about 10^7 nodes. Recall that each node has $2d$ neighbors—14, in the last example. Computing $A_d(n)$ by this algorithm becomes here infeasible.

---

[2] In fact, x_M itself is not reachable. The reachable cell with the largest image under Γ is $y = (n-2, \ldots, n-2, n-1)$ (the top cell of a "stick" aligned with the dth direction). We have $\Gamma(y) = (t^{d+1} - 2t^{d-1} - t + 2)/(2(t-1))$, where $t = 2n - 2$.

3.2 Eliminating the Computation of the Graph

Our main goal is to eliminate the need to hold the entire graph in memory during the computation. Clearly, the distribution of cells of different statuses is extremely uneven: At one time, at most n cells are in the current polycube, at most $2nd$ cells are in the untried set (since each of the n used cells has at most $2d$ free neighbors), and all the other cells are free (unoccupied). Since the lattice graph has a well-defined structure, it is not necessary to create it in advance. Instead, one could create the graph locally "on demand." Given a cell, one could compute its neighbors on-line and always in the same order. Instead of maintaining a huge vertex set (lattice cells) of the graph, and keeping the status of each cell (occupied, in the untried set, or free), we could keep only the first two small sets of cells, and deduce that a cell is free from the fact that it does not belong to any of these two sets. The major obstacle is that we might need to distinguish between free cells that, in the course of the algorithm, were already part of some counted polycubes, and are thus "done with," and cells that were not yet part of any polycube. It turns out that this differentiation is redundant.

To show this, we formally proved the correctness of the algorithm.[3] That is, we proved that although the graph contains plenty of cycles, no subgraph is obtained twice with different orders of nodes. It then became clear that no information about the nodes should be maintained. From the definition of the algorithm, it was clear that every polycube would be counted at least once. The more delicate issue was to prove that every polycube is counted only once.

Theorem 1. *Redelmeier's algorithm counts every connected subgraph (polycube) that contains the origin exactly once.*

We provide the proof in the full version of the paper. In particular, it implies the following. Denote by P' the polycube to which a cell a_k is added by the algorithm. After a_k is removed from the current polycube, it will never again be added to the untried set (and hence neither to the polycube) as long as the recursion continues with P' as its root. When the recursion is unfolded, and the current polycube is shrunk to be a subset of P' plus some other cells, it is certainly possible that a_k will be added again to the untried set and later to the polycube. Nevertheless, no future polycube will have P' as a subset.

The proof also implies that we do not need to store any information about the free cells. This is because in order to avoid counting the same polyomino twice, all that is required is the current polyomino and its neighboring cells. To maintain this information we only need to be able to explore the graph locally "on demand," and for each neighbor c' of a cell c which is moved from the untried set to the polycube, determine whether or not it is now a *new* neighbor. The first task is easy to accomplish, since the structure of the orthogonal lattice is well defined, and so is the definition of the mapping $\Gamma(\cdot)$. The second task is also feasible: Such a cell c' is a new neighbor if and only if its only neighboring cell in the current polycube is c. This is easy to determine using the polycube's set of cells.

---

[3] Redelmeier [15, p. 195] only gave a brief description of the course of the algorithm.

3.3   Counting Proper Polycubes

As reported in Section 5, our counts of $A_6(7)$ and $A_6(8)$ did not match those of Lunnon [11]. Lunnon actually computed $DX(n, d)$, the number of *proper* polycubes of size n in d dimensions. (Proper d-dimensional polycubes span all the d dimensions, and, thus, cannot be embedded in $d - 1$ dimensions.) According to Lunnon's formula, $A_d(n) = \sum_{i \leq d} \binom{d}{i} DX(n, i)$. In order to trace the source of the discrepancies, we modified our program to also count proper polycubes.

To this aim, we should have maintained a "dimension status" that should have counted, for each dimension $1 \leq i \leq d$, how many cells of the current polycube have the ith coordinate different than $\bar{0}$, the cell at the origin. Each addition or deletion of a cell to/from the polycube should have been accompanied by updating the appropriate counter in the dimension-status record. To achieve this, we needed an efficient implementation of $\Gamma^{-1}(\cdot)$, a function that maps numbers (cell ids) back to their original source cells with d coordinates.

However, for efficiency, we wanted to avoid such a function $\Gamma^{-1}(\cdot)$. When we add a cell c to the polycube, instead of increasing by 1 the counters of all its dimensions relative to the origin (which are many, and not simple to identify), we increase only the counter of its dimension relative to the cell c' due to which it was previously put in the untried set (which is unique, and is known while c is added). This specific counter is decreased by 1 when cell c is removed from the polycube. During the entire period in which c belongs to the polycube, this dimension of the polycube is used. Thus, although the counter of dimension i does not really count the number of cells that occupy the ith dimension, it is easy to see that it is still positive whenever there is any cell that does so.

3.4   Parallelization

Since now the bottleneck of the algorithm is the running time, we also parallelized the new version of Redelmeier's algorithm. Mertens and Lautenbacher [14] did this for the two-dimensional version of the problem, while we do this for any dimension. Since, as is already observed above, the execution of any subtree of the recursion does not change the contents of the data structures at the top of the subtree, the computation of the subtree can be done independently and in parallel, while the main algorithm skips the subtree.

In practice, the main procedure recurses only until some level $k < n$ (for some fixed dimension d). Regardless of the value of n, this induces $A_d(k)$ subtrees, which can be assigned independently to the processors we have at hand. The results for all levels $k < m \leq n$, and, in particular, $m = n$, are collected from all the processors, and their summation yields the values of $A_d(m)$. This can be done for counting both proper and all (proper and improper) polycubes.

A description of our parallel implementation is given in the full version of the paper. The programs for the Internet server and client were written in Ruby, while versions of the polycube-counting program were written in C anc C++. The client can be run in either a Windows or Linux environment.

4   Complexity Analysis

We assume that $n > d$, otherwise the polycubes cannot span all the d dimensions. In d dimensions, there are $\Theta(2^{d-2}n^d)$ distinct reachable cells, so the amount of memory (and time) needed to store (and process) the identity of a single cell is $\Theta(d \log n)$. (In other studies this factor was considered as $\Theta(1)$ or $\Theta(d)$.)

We maintain two data structures: The set of currently-occupied cells and the set of untried cells (neighbors of the polycube). The size of the current polyomino is at most n, while the size of the untried set is at most $2dn - 2(n-1) = 2((d-1)n + 1)$. That is, the amount of cells stored by the algorithm is $\Theta(dn)$. For this we need $\Theta(d^2 n \log n)$ space. For a fixed value of d, this is $\Theta(n \log n)$.

Since we do not compute the lattice graph, its size does not directly affect the algorithm's running time. The important factor is the number of operations (additions and deletions of cells) on the untried set. This number is proportional to $A_d(n)$, the number of counted polycubes. (More precisely, the number of these operations is proportional to the total number of d-dimensional polycubes of *all* sizes up to n, but the polycubes of size n outnumber all the smaller polycubes.)

We maintain the current polycube as a balanced binary tree (sorted by $\Gamma(\cdot)$), and since it always contains up to n cells, each operation requires $O(\log n)$ steps (where each step requires $O(d \log n)$ time). The untried set should support two types of operations: (1) A queue-like structure, where new cells are added at the rear and cells are removed from the front; (2) Searching for a node, to avoid multiple additions of cells. An appropriate tree-like structure (or two trees with cross references) can do the job. Since the size of the untried set is $O(dn)$, each operation on the set requires $O(\log d + \log n) = O(\log n)$ steps.

When a cell c is moved from the untried set to the polycube, its $2d$ neighbors are computed, and each neighbor c' is tested to see if it is a *new* neighbor. Recall that c' is a new neighbor if it is free and among all its own $2d$ neighbors, only c belongs to the current polycube. Hence, each such test can be done in $O(d \log n)$ steps by using the polycube structure. This can be done more efficiently if we keep a secondary set of neighboring cells, which includes not only those in the untried set, but also cells that were removed from the polycube, i.e., neighboring cells that cannot become "new neighbors." (The size of this set is $\Theta(dn)$.) For each cell in this set we maintain the count of its neighbors that belong to the polycube. When a cell is removed from the polycube, not only it is put in this list, but also the neighbors counts of all its neighbors (in this set) are decreased by 1. This takes $O(d)$ steps. When the neighbors count of a cell reaches 0, the cell is removed from this set. Thus, we can test in only $O(\log n)$ steps whether the cell c' is new. Otherwise, it is not added to the untried set.

Overall, each operation results in $O(d \log n)$ steps. Since the algorithm performs $A_d(n)$ operations, and each step requires $O(d \log n)$ time, the total running time is $O(A_d(n)d^2 \log^2 n)$. The major factor, $A_d(n)$, is exponential in n; its base is an unknown constant that depends solely on d (e.g., ~ 4.06 for $d = 2$ [7]).

For computing only proper polycubes we invest $O(\log d)$ time per polycube to maintain the dimension-status record, and $O(d \log d)$ time for checking whether a polycube is proper. This does not change asymptotically the running time.

5   Results

We implemented the algorithm in C and C++, and ran the serial program locally in an MS Windows environment on an IBM X500 with four 2.4GHz XEON processors and 3.5GB of RAM. The parallel version ran over the Internet on dozens of computers simultaneously, offering a wide range of processor frequency and available amount of memory.

Table 1. Fixed polycubes (new values in boldface)

n	$A_6(n)$	$A_7(n)$	$A_8(n)$
1	1	1	1
2	6	7	8
3	66	91	120
4	901	1,484	2,276
5	13,881	27,468	49,204
6	231,008	551,313	**1,156,688**
7	4,057,660	**11,710,328**	**28,831,384**
8	74,174,927	**259,379,101**	**750,455,268**
9	1,398,295,989	**5,933,702,467**	
10	**27,012,396,022**	**139,272,913,892**	

Table 1 shows the values of $A_d(n)$ obtained in 6, 7, and 8 dimensions. New values, which are tabulated for the first time, are shown in bold font. The non-parallel version of the program computed the reported values of $A_6(\cdot)$ and $A_8(\cdot)$ in about 26 days and a little more than a week, respectively. The values of $A_7(\cdot)$ were computed by the parallel program in about a week, using a dozen computers for gathering a total of 91 days of CPU. All the old values in the table agree with previous publications, including ours [1], except for the values of $A_6(7)$ and $A_6(8)$ computed by Lunnon [11]. From his calculations of proper polyominoes, one can infer the values 4,038,205 and 71,976,512, respectively. However, the now-known explicit formulae for $DX(d, d + 1)$ and $DX(d, d + 2)$ (given in the reference [2]) confirm our counts.

6   Conclusion

In this paper we presented an efficient implementation of Redelmeier's serial algorithm for counting fixed high-dimensional polycubes. We used our program to compute new terms of the series $A_d(n)$, for a few values of d, which, to the best of our knowledge, have not been known before. The main contributions of this work is eliminating the need to store the entire lattice graph (in which polycubes are sought), and parallelizing the algorithm. The parallel version was actually implemented and run over the Internet. We plan to continue running it and compute even more yet unknown values of $A_d(n)$.

We end with a note about the growth rates of polycubes. It is known [8] that the limit $\lambda_2 := \lim_{n\to\infty} \sqrt[n]{A_2(n)}$ exists. A similar method shows that the limit $\lambda_d := \lim_{n\to\infty} \sqrt[n]{A_d(n)}$ exists for any fixed value of $d > 1$. Only less than a decade ago it was proven by Madras [12] that the limit $\lim_{n\to\infty}(A_2(n+1)/A_2(n))$ also exists and that it is equal to λ_2, and, similarly, that for any fixed value of $d > 1$, the limit $\lim_{n\to\infty}(A_d(n+1)/A_d(n))$ exists and is equal to λ_d. The constant λ_2 is estimated to be around 4.06 [7]. From inspecting the experimental growth rates of polycubes in higher dimensions (but for relatively small values of d), one may conclude that λ_d is roughly equal to $4(d-1)$. In fact, it is proven in [2] that λ_d is roughly $2ed$.

References

1. Aleksandrowicz, G., Barequet, G.: Counting d-dimensional Polycubes and Non-rectangular Planar Polyominoes. In: Chen, D.Z., Lee, D.T. (eds.) COCOON 2006. LNCS, vol. 4112, pp. 418–427. Springer, Heidelberg (2006)
2. Barequet, R., Barequet, G.: Formulae and Growth Rates of High-Dimensional Polycubes (manuscript)
3. Broadbent, S.R., Hammersley, J.M.: Percolation Processes: I. Crystals and mazes. Proc.Cambridge Philosophical Society 53, 629–641 (1957)
4. The on-line Encyclopedia of Integer Sequences, http://www.research.att.com/~njas/sequences
5. Gong, K.: http://kevingong.com/Polyominoes/Enumeration.html
6. Jensen, I.: Enumerations of Lattice Animals and Trees. J. of Statistical Physics 102, 865–881 (2001)
7. Jensen, I.: Counting Polyominoes: A Parallel Implementation for Cluster Computing. In: Sloot, P.M.A., Abramson, D., Bogdanov, A.V., Gorbachev, Y.E., Dongarra, J., Zomaya, A.Y. (eds.) ICCS 2003. LNCS, vol. 2659, pp. 203–212. Springer, Heidelberg (2003)
8. Klarner, D.A.: Cell Growth Problems. Canadian J. of Mathematics 19, 851–863 (1967)
9. Lunnon, W.F.: Counting Polyominoes. In: Atkin, Birch (eds.) Computers in Number Theory, pp. 347–372. Academic Press, London (1971)
10. Lunnon, W.F.: Symmetry of Cubical and General Polyominoes. In: Read (ed.) Graph Theory and Computing, pp. 101–108. Academic Press, NY (1972)
11. Lunnon, W.F.: Counting Multidimensional Polyominoes. The Computer Journal 18, 366–367 (1975)
12. Madras, N.: A Pattern Theorem for Lattice Clusters. Annals of Combinatorics 3, 357–384 (1999)
13. Martin, J.L.: Computer Techniques for Evaluating Lattice Constants. In: Domb, Green (eds.) Phase Transitions and Critical Phenomena, vol. 3, pp. 97–112. Academic Press, London (1974)
14. Mertens, S., Lautenbacher, M.E.: Counting Lattice Animals: A Parallel Attack. J. of Statistical Physics 66, 669–678 (1992)
15. Redelmeier, D.H.: Counting Polyominoes: Yet Another Attack. Discrete Mathematics 36, 191–203 (1981)
16. Sykes, M.F., Gaunt, D.S., Glen, M.: Percolation Processes in Three Dimensions. J. of Physics, A: Mathematical and General 10, 1705–1712 (1976)

Polychromatic Colorings of n-Dimensional Guillotine-Partitions

Balázs Keszegh[*]

Central European University, Budapest

Abstract. A *strong hyperbox-respecting coloring* of an n-dimensional hyperbox partition is a coloring of the corners of its hyperboxes with 2^n colors such that any hyperbox has all the colors appearing on its corners. A *guillotine-partition* is obtained by starting with a single axis-parallel hyperbox and recursively cutting a hyperbox of the partition into two hyperboxes by a hyperplane orthogonal to one of the n axes. We prove that there is a strong hyperbox-respecting coloring of any n-dimensional guillotine-partition. This theorem generalizes the result of Horev et al. [8] who proved the 2-dimensional case. This problem is a special case of the n-dimensional variant of *polychromatic colorings*. The proof gives an efficient coloring algorithm as well.

1 Introduction

A k-coloring of the vertices of a plane graph is *polychromatic* (or *face-respecting*) if on all its faces all k colors appear at least once (with the possible exception of the outer face). The polychromatic number of a plane graph G is the maximum number k such that G admits a polychromatic k-coloring, we denote this number by $\chi_f(G)$. For an introduction about polychromatic colorings see for example the introduction of [2] or [4]. We restrict ourselves to a brief introduction to this topic and list some results. Alon et al. [2] showed that if g is the length of a shortest face of a plane graph G, then $\chi_f(G) \geq \lfloor (3g - 5)/4 \rfloor$ (clearly $\chi_f(G) \leq g$), and showed that this bound is sufficiently tight. Mohar and Škrekovski [10] proved using the four-color theorem that every simple plane graph admits a polychromatic 2-coloring, later Bose et al. [3] proved that without using the four-color theorem. Horev and Krakovski [9] proved that every plane graph of degree at most 3, other than K_4 admits a polychromatic 3-coloring. Horev et al. [7] proved that every 2-connected cubic bipartite plane graph admits a polychromatic 4-coloring. This result is tight, since any such graph must contain a face of size four.

We define a *rectangular partition* as a partition of an axis-parallel rectangle into an arbitrary number of non-overlapping axis-parallel rectangles, such that no four rectangles meet at a common point. One may view a rectangular partition as a plane graph whose vertices are the corners of the rectangles and edges are the line segments connecting these corners. Dinitz et al. [6] proved that every rectangular partition admits a polychromatic 3-coloring. A *guillotine-partition* is obtained by recursively cutting a rectangle into two subrectangles by either

---

[*] This research was partially supported by Hungarian Science Fundation OTKA project 48826.

a vertical or a horizontal line. For this subclass of rectangular partitions Horev et al. [8] proved that they admit a polychromatic 4-coloring. Actually, they prove a stronger statement. We define a *strong rectangle-respecting coloring* of a rectangular partition R as a vertex coloring of R with four colors such that every rectangle of R has all four colors among the four corners defining it. This is clearly a polychromatic 4-coloring as well. For examples see Figure 1. They proved that such coloring exists for any guillotine-partition. Recently, Dimitrov et al. [4] proved that any rectangular partition admits a strong rectangle respecting coloring, using a theorem about plane graphs.

Our main result is a generalization of the result for guillotine-partitions for n dimensions. An *n-dimensional hyperbox* is an n-dimensional axis-parallel hyperbox. For us a *partition* of an n-dimensional hypercube or hyperbox is a partition to hyperboxes such that each corner vertex is a corner of 2 hyperboxes, except the corners of the original hypercube. Note that this definition differs a bit from the natural definition, where we would allow a vertex to be the corner of more than 2 hyperboxes. This is needed, as using the more natural definition even in the plane there are simple counterexamples for our main theorem. The hyperboxes of the partition are called the *basic hyperboxes*. A *guillotine-partition* is obtained by starting with a partition containing only one basic hyperbox and recursively cutting a basic hyperbox into two hyperboxes by a hyperplane orthogonal to one of the n axes. The structure of such partitions is widely investigated, used in the area of integrated circuit layouts and other areas. Guillotine-partitions are also the underlying structure of orthogonal *binary space partitions* (BSPs) which are widely used in computer graphics. In [1] Ackerman et al. determine the asymptotic number of structurally different guillotine-partitions, we refer to the introduction of the same paper for more on this topic.

A *strong hyperbox-respecting coloring* of a partition is a coloring of the corners of its basic hyperboxes with 2^n colors such that any basic hyperbox has all the colors appearing on its corners. Note that a corner belongs to two basic hyperboxes except the 2^n corners of the partitioned big hyperbox, which belong to only one basic hyperbox. The natural extension to n dimensions of a polychromatic coloring would be a coloring of the corners of its basic hyperboxes with 2^n colors such that any basic hyperbox has all the colors appearing on its boundary. Clearly, every strong hyperbox-respecting coloring has this property.

Theorem 1. *There is a strong hyperbox-respecting coloring of any n-dimensional guillotine-partition.*

2   Proof of the Main Theorem

First we start with some definitions to be able to phrase the theorem we will actually prove, implying Theorem 1. We can assume w.l.o.g. that every hyperbox is a hypercube. Let us formulate this more precisely. We begin by introducing some notations. From now on $x = (x_1, x_2, \ldots, x_n)$, y, a, b, etc. always refer to some n-long 0-1 vector. We define the sum of two such vectors (denoted simply by $+$) as summing independently all coordinates mod 2. The $(0, 0, \ldots, 0)$ vector

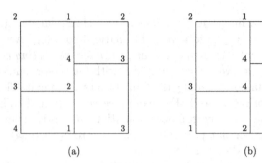

(a)   (b)

Fig. 1. a (a) polychromatic 4-coloring and (b) strong rectangle-respecting coloring of a guillotine partition

is denoted by $\mathbf{0}$ and the vector $(1, 0, 0, \ldots, 0)$ by e_1. A *face* is always an $(n-1)$-dimensional face of a hyperbox. Any axis-parallel n-dimensional hyperbox can be uniquely scaled and translated to be the hypercube with the two opposite corners being $(0, 0, \ldots, 0)$ and $(1, 1, \ldots, 1)$. We refer to the corner of the hyperbox which maps into the point with coordinates $x = (x_1, x_2, \ldots, x_n)$ by $C(x)$. For some fixed $x \neq \mathbf{0}$ we define the *reflection* R_x being the function on the set of corners for which $R_x(C(y)) = C(x + y)$ for all y. Observe that $R_x(R_x(C(y))) = C(y)$ for any y.

From now on when we speak about a coloring of some hyperbox then it is always a hyperbox-respecting coloring. We say that a coloring of the corners of a hyperbox is an R_x-*coloring* ($x \neq \mathbf{0}$) if two corners $C(y)$ and $C(z)$ have the same colors if and only if $R_x(C(y)) = C(z)$ (or equivalently $R_x(C(z)) = C(y)$). Observe that such a coloring will have 2^{n-1} different colors appearing on the corners of the hyperbox, each occuring twice. Further, we say that the coloring of the corners is an R_0-*coloring* if all corners are colored differently. Note that permuting the colors of an R_x-coloring gives another R_x-coloring for any x. From now on we will always restrict ourselves to these kinds of colorings. If any pair of such colorings could be put together along any axes to form another such coloring then it would already imply a recursive proof for the main theorem. As this is not the case we have to be more precise about our freedom of how to color a partition, making necessary to define sets of such colorings.

For any $x \neq \mathbf{0}$ S_x is defined as the union of all R_y for which $x \cdot y = 1$ (the scalar product of x and y mod 2). S_0 is the one element set of R_0. If for some $x \neq \mathbf{0}$ for all $y \in S_x$ the hyperbox partition has a strong hyperbox-respecting coloring which is an R_y-coloring on its corners, we say that the hyperbox partition can be colored by the *color-range* S_x. If it has a strong hyperbox-respecting coloring which is an R_0-coloring on its corners, we say that the hyperbox partiton can be colored by the (one element) *color-range* S_0.

We will prove the following theorem, which implies Theorem 1.

Theorem 2. *Any n-dimensional guillotine-partition can be colored by some color-range.*

Proof. We proceed by induction on the number of guillotine-cuts of the partiton. The corners of a hyperbox containing only one basic hyperbox (i.e. the partition has 0 cuts) can be colored trivially with all different colors, thus colorable by color-range S_0. In the general step we take a cut of the hyperbox B splitting it into two hyperboxes B_1 and B_2 with smaller number of cuts in them. Thus, by induction they can be colored by some color-ranges S_x and S_y for some x and y. We need to prove that there exists a z for which our hyperbox partition can be colored by S_z. First we prove this for the case when the cut is orthogonal to the first axis. Finally, we will prove that as the definition of R's and S's is symmetrical on every pair of axes, the claim follows for any kind of cut.

We regard the first axis (the one which corresponds to the first coordinate of points) as the usual x-axis, and so we can say that an object (corner, face, hyperbox etc.) is *left* from another if its first coordinates are smaller or equal than the other's (B_1 is left from B_2 for example). Similarly we can say *right* when its coordinates are bigger or equal than the other's.

We always do the following. Take an $R_a \in S_x$ and $R_b \in S_y$ and take a coloring of B_1 which is an R_a-coloring on its corners and a coloring of B_2 which is an R_b-coloring on its corners by induction such that the colors of the corners which should fit together (the right face of B_1 and the left face of B_2) have the same colors at the corners which will be identified. This is not always possible but when it is, it gives a coloring of B (the corners on the left face of B_1 and on the right face of B_2 are the corners of B). Note that we can permute the colors on the two hyperboxes in order to achieve such a fit of the colors. Clearly, the resulting coloring of B is a hyperbox-respecting coloring by induction. If the resulting coloring can be an R_c-coloring on the corners for some c then we write $R_a \cdot R_b \rightarrow R_c$. See Figures 2 and 3 for examples for 3 dimensions. The definition of $\rightarrow$ is good as the existence of such a fit depends only on the color of the corners. Observe that this operation is not commutative by definition and can hold for more than one c and has the hidden parameter that we put them together along the first axis (i.e. the two partitions are put together by the face which is orthogonal to the first axis). As we remarked earlier, if for any a and b there would be a c with $R_a \cdot R_b \rightarrow R_c$ then it would be enough to prove the main theorem by induction without defining color-ranges. As this is not the case we need to deal with color-ranges and define the function $\rightarrow$ on them as well.

We write $S_x \cdot S_y \rightarrow S_z$ if $\forall R_c \in S_z \; \exists R_a \in S_x$ and $R_b \in S_y$ such that $R_a \cdot R_b \rightarrow R_c$. Clearly, we need to prove that there exists such a z for any choice of x and y that $S_x \cdot S_y \rightarrow S_z$. Lemma 2 states this. For the proof of this lemma we will first need to prove Lemma 1 about the behaviour of $\rightarrow$ for R's.

Finally, we need to prove that we can put together color-ranges along any axis. One can argue that we can obviously do that as the definition of color-ranges is symmetrical on any pair of coordinates and because of that analogs of Lemma 1 and Lemma 2 are true for an arbitrary axis. For a more rigorous argument see Lemma 3 and its proof in the Appendix.

Lemma 1 (fitting together colorings). *For $a, b, c \neq 0$ we have*

(a) $R_0 \cdot R_0 \rightarrow R_c$, if the first coordinate of c is 1,
(b) $R_a \cdot R_0 \rightarrow R_0$ and $R_0 \cdot R_a \rightarrow R_0$, if the first coordinate of a is 1,

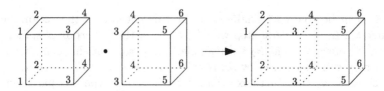

Fig. 2. Example to Lemma 1(c): $R_{010} \cdot R_{010} \rightarrow R_{010}$

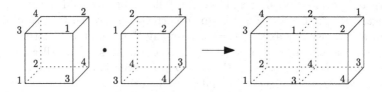

Fig. 3. Example to Lemma 1(d): $R_{110} \cdot R_{101} \rightarrow R_{111}$

(c) $R_a \cdot R_a \rightarrow R_a$, *if the first coordinate of a is 0,*
(d) $R_a \cdot R_b \rightarrow R_c$, *if the first coordinate of a and b is 1 and* $c = a + b + e_1$.

Proof. (a) Take an arbitrary c with its first coordinate being 1. For an R_c-coloring each color appearing on the corners of the hyperbox appears once on its left and once on its right face. We want to fit together two R_0-colorings to have an R_c-coloring. Take an arbitrary R_0-coloring of B_1. Take an R_0-coloring of B_2 and permute its colors such that the corners on its left face fit together with the corners on the right face of B_1. Now the set of colors on the right face of B_2 is the same set of colors as on the left face of B_1. After a possible permutation of these colors on B_2 we can get an R_c-coloring on B.

(b) Take an R_a-coloring of B_1 and an R_0-coloring of B_2, permute the colors on B_2 such that the needed faces fit together. This can be done as R_a does not have a color appearing twice on its right face. As on B_1's left face the same set of colors appear as on its right face, B has all the colors of B_2's coloring appearing on its corners, thus it is an R_0-coloring of B. The proof for the other claim is similar.

(c) Take an R_a-coloring of B_1. Take an R_a-coloring of B_2 and permute the colors on it such that the corners on its left face fit together with the corners on the right face of B_1 and on its right face all colors are different from the ones we used to color the corners of B_1. This can be done as both are R_a-colorings where the first coordinate of a is 0, so on the common face the same pair of corners need to have the same color. Similarly, we see that these fit together to form an R_a-coloring of B. For an illustration for 3 dimensions see Figure 2.

(d) Take an R_a-coloring of B_1. Take an R_b-coloring of B_2 and permute again the colors such that the corners on its left face fit together with the corners on the right face of B_1. This can be done as the corners on the right face of B_1 all have different colors and this is what we need on the left face of B_2 to make an R_b coloring (as the first coordinate of a and b is 1). Now it is enough to see that the resulting coloring of B is an R_c coloring with $c = a + b + e_1$ (recall that e_1 is the vector with all-0 coordinates except the first coordinate being 1). Take an arbitrary corner on its left face, $C(d)$ (thus d has first coordinate 0). In the coloring of B_1 its pair (the corner with the same color) is $C(d+a)$. This is on the right face of B_1, and so it is fitted together with the corner $C(d+a+e_1)$ of B_2 on B_2's left face. By the R_b-coloring of B_2 the corner $C(d+a+e_1+b)$ has the same color. This is also the $C(d + a + e_1 + b)$ corner of B. This holds for any corner of B on its left side and symmetrically on its right side as well, and so this is indeed an R_c-coloring of B. For an illustration for 3 dimensions see Figure 3. $\square$

Lemma 2 (fitting together color-ranges). *For $x, x', y \neq 0$ we have*

(a) $S_0 \cdot S_0 \rightarrow S_{e_1}$,
(b) $S_x \cdot S_0 \rightarrow S_0$ *and* $S_0 \cdot S_x \rightarrow S_0$,
(c) $S_x \cdot S_y \rightarrow S_{e_1}$, *if x and y differ somewhere which is not the first coordinate,*
(d) $S_x \cdot S_x \rightarrow S_{x'}$, *if x' is the same as x with the possible exception at the first coordinate, which is 1 in x'.*
(e) $S_x \cdot S_{x'} \rightarrow S_x$ *and* $S_{x'} \cdot S_x \rightarrow S_x$, *if x' is the same as x except at the first coordinate, which is 0 in x and 1 in x'.*

Proof. Let us recall first that S_0 is the one element set of R_0 and for any $x \neq 0$ S_x is defined as the union of all R_y for which $x \cdot y = 1$.

(a) by Lemma 1(a) $R_0 \cdot R_0 \rightarrow R_c$ for any $c \cdot e_1 = 1$.

(b) In S_x ($x \neq 0$) there is always an R_a where the first coordinate of a is 1. By Lemma 1(b) $R_a \cdot R_0 \rightarrow R_0$. The proof for the other claim is similar.

(c) We need to prove that for any $R_c \in S_{e_1}$ ($c \cdot e_1 = 1$) there is an $R_a \in S_x$ and $R_b \in S_y$ such that $R_a \cdot R_b \rightarrow R_c$.

Suppose x and y differ in the kth coordinate ($k \neq 1$). Define X as the set of coordinates l where $x_l = 1$, and Y the set of coordinates l where $y_l = 1$. We want to apply Lemma 1(d) which is symmetrical on a and b and so we can suppose that $k \notin X$ and $k \in Y$. The first coordinate of c is 1, so we choose a and b having the first coordinate 1 as well. We need that $a + b + e_1 = c$ to be able to apply Lemma 1(d). First define the coordinates of a being in X all zero except one (this is the first if $1 \in X$, some other otherwise), thus by any choice of the other coordinates we will have $R_a \in S_x$. Now define the coordinates of b being in $X \setminus \{1\}$ such that $a_l + b_l = c_l$ for all $l \in X$. Define the rest of the coordinates of b such that $b \cdot y = 1$, this can be done as we can choose the kth coordinate as we want. Thus, $R_b \in S_y$ as well. Finally, choose the coordinates of a not in X such that $a_l + b_l = c_l$ for all $l \notin X$. This way $a + b + e_1 = c$ as needed.

(d) We need to prove that for any $R_c \in S_{x'}$ there is an $R_a \in S_x$ and $R_b \in S_x$ such that $R_a \cdot R_b \rightarrow R_c$.

First we prove the case when the first coordinate of x is 1 and so $x' = x$. For a c with first coordinate 0 by Lemma 1(c) we have $R_c \cdot R_c \rightarrow R_c$, all in S_x as needed. For a c with first coordinate 1 take an arbitrary a with first coordinate 1 and $R_a \in S_x$. Choose b such that $a + b + e_1 = c$ and so by Lemma 1(d) $R_a \cdot R_b \rightarrow R_c$ holds. We need that R_b is in S_x, which is true as $b \cdot x = (a+c+e_1) \cdot x = 1+1+1 = 1$.

Now we prove the case when the first coordinate of x is 0 and so $x' = x+e_1$. For a c with first coordinate 0 by Lemma 1(c) we have $R_c \cdot R_c \rightarrow R_c$, all in S_x and in S'_x too (as for such a c we have $c \cdot x = c \cdot x' = 1$). For a c with first coordinate 1 take an arbitrary a with first coordinate 1 and $R_a \in S_x$. Choose b such that $a+b+e_1 = c$ and so by Lemma 1(d) $R_a \cdot R_b \to R_c$ holds. We need that R_b is in S_x, which is true as $c \cdot x' = 1$, $c \cdot e_1 = 1$ and so $b \cdot x = (a+c+e_1) \cdot x = 1 + c \cdot (x'+e_1) + 0 = 1$.

(e) For $S_x \cdot S_{x'} \to S_x$ we need to prove that for any $R_c \in S_x$ there is an $R_a \in S_x$ and $R_b \in S_{x'}$ such that $R_a \cdot R_b \to R_c$.

For a c with first coordinate 0 by Lemma 1(c) we have $R_c \cdot R_c \to R_c$, all in S_x and in S'_x too (as for such a c we have $c \cdot x = c \cdot x' = 1$). For a c with first coordinate 1 take an arbitrary a with first coordinate 1 and $R_a \in S_x$. Again, choose b such that $a + b + e_1 = c$ and so by Lemma 1(d) $R_a \cdot R_b \to R_c$ holds. We need that R_b is in S'_x, which is true as $b \cdot x' = (a+c+e_1) \cdot x' = (a+c+e_1) \cdot (x+e_1) = a \cdot x + c \cdot x + e_1 \cdot x + a \cdot e_1 + c \cdot e_1 + e_1 \cdot e_1 = 1 + 1 + 0 + 1 + 1 + 1 = 1$.

As Lemma 1(d) is symmetrical on a and b, $S_{x'} \cdot S_x \to S_x$ follows the same way.   □

The Lemmas above conclude the proof of Theorem 2.   □

3   Algorithm and Remarks

Assuming we know the cut-structure of the partition, the proof yields a simple linear time algorithm (in the number of cuts, regarding the dimension n as a fixed constant) to give a strong hyperbox-respecting coloring. First we determine the color-ranges and then the colorings of the hyperboxes using the lemmas. We will sketch how to do that.

First we construct the rooted binary tree with its root on the top representing our guillotine-cuts (each node corresponds to a hyperbox, the leaves are the basic boxes, the root is the original hyperbox). From bottom to top we can determine for each node v the unique $s(v)$ for which the corresponding hyperbox will have color-range $S_{s(v)}$ (leaves have color-range S_0, then it is easy to determine the rest going upwards using Lemma 2). Now from top to bottom we can give appropriate R_y-colorings to the hyperboxes. For the root w give arbitrary $R_{r(w)}$-coloring with $r(w) \in S_{s(w)}$. Then by induction if we gave an $R_{r(w)}$-coloring ($r_w \in S_{s(w)}$) to some hyperbox corresponding to the node w with children u and v then by Lemma 2 there exists $r(u) \in S_{s(u)}$ and $r(v) \in S_{s(v)}$ such that an $R_{r(u)}$ and an $R_{r(v)}$ can be put together (at the appropriate face) to form an $R_{r(w)}$-coloring. Such colorings can be found in the same way as in the proof of Lemma 2. Thus, we can give such colorings to the hyperboxes corresponding to u and v. Finishing the coloring this way the basic boxes will have R_0-colorings, i.e. the coloring will be a strong hyperbox-respecting coloring.

It is easy to see that using this algorithm any S_x color-range can appear with appropriate cuts.

It was observed by D. Dimitrov and R. Škrekovski [5] using a double-counting argument that when a (not necessary guillotine) partition contains an odd number of basic hyperboxes then a coloring of it must have all the corners colored differently. From Lemma 1 one can easily deduce that when the partition contains an odd number of basic hyperboxes then our algorithm will give an R_0-coloring thus having all corners colored differently indeed. Further it was also observed that when a partition contains an even number of basic hyperboxes then all the colors appear pair times on the corners of the hyperbox. In the even case our algorithm will give an R_a-coloring with $a \neq \mathbf{0}$ thus having all colors appearing zero times or twice on the corners.

As mentioned in the Introduction, the general case is solved in 2-dimensions, but it is still unknown for which other dimensions can it hold.

Problem 1. For which $n > 2$ do exist a strong hyperbox-respecting coloring of any n-dimensional partition.

Acknowledgments. This paper was done while the author visited in Berlin the Theoretical Computer Science Workgroup of Freie Universitaet and the Discrete Mathematics Workgroup of Technische Universitaet. The author is grateful to Darko Dimitrov for introducing this topic, further wish to thank Darko Dimitrov, Eyal Ackerman and Christian Knauer for the entertaining discussions about the proof. Finally, Eyal Ackerman and Dömötör Pálvölgyi and an anonymous reviewer for their several improvements on the presentation of the paper.

References

1. Ackerman, E., Barequet, G., Pinter, R.Y., Romik, D.: The Number of Guillotine Partitions in d Dimensions. Inf. Process. Lett. 98(4), 162–167 (2006)
2. Alon, N., Berke, R., Buchin, K., Buchin, M., Csorba, P., Shannigrahi, S., Speckmann, B., Zumstein, P.: Polychromatic Colorings of Plane Graphs. In: Proc. of 24th Annual ACM Symposium on Computational Geometry (2008)
3. Bose, P., Kirkpatrick, D., Li, Z.: Worst-Case-Optimal Algorithms for Guarding Planar Graphs and Polyhedral Surfaces. Comput. Geom. Theory Appl. 26(3), 209–219 (2003)
4. Dimitrov, D., Horev, E., Krakovski, R.: Polychromatic 4-Coloring of Rectangular Partitions. In: Proc. of 24th European Workshop on Computational Geometry (2008)
5. Dimitrov, D.: Personal Communication (2007)
6. Dinitz, Y., Katz, M.J., Krakovski, R.: Guarding Rectangular Partitions. In: Proc. of 23rd European Workshop on Computational Geometry (2007)
7. Horev, E., Katz, M.J., Krakovski, R.: Polychromatic Colorings of Cubic Bipartite Plane Graphs (submitted, 2007)
8. Horev, E., Katz, M.J., Krakovski, R., Löffler, M.: Polychromatic 4-Colorings of Guillotine Subdivision (submitted, 2007)
9. Horev, E., Krakovski, R.: Polychromatic Colorings of Bounded Degree Plane Graphs (submitted, 2007)
10. Mohar, B., Škrekovski, R.: The Grötzsch Theorem for the Hypergraph of Maximal Cliques. Electr. J. Comb. R26, 1–13 (1999)

Appendix

Define $\circ^i$ as the function on the 0-1 vectors which exchanges the first and the ith coordinates, i.e. for a vector x the vector x^i has the same coordinates except that

$x_1^i = x_i$ and $x_i^i = x_1$ (thus x^{ii} is the identity and x^i is a bijection). For vectors corresponding to corners of a hyperbox this is a reflection on a hyper-plane going through the corners having the same first and ith coordinate. Clearly, applying $\circ^i$ on an R_x-coloring of the corners we get an R_{x^i}-coloring of the corners. Lemma 3 states that the color-ranges S_x and S_y can be put together along the ith axis to give the color-range S_z if the color-ranges S_{x^i} and S_{y^i} can be put together along the first axis to give the color range S_{z^i}. We have seen this can be done for any x^i and y^i with some z^i, thus fitting along any other axis is also possible.

Lemma 3 (fitting together along a general axis). *If the color-ranges S_{x^i} and S_{y^i} can be put together along the first axis to give the color range S_{z^i} then the color-ranges S_x and S_y can be put together along the ith axis to give the color-range S_x.*

Proof. First we prove that if $R_{a^i} \cdot R_{b^i} \rightarrow R_{c^i}$ for some c then an appropriate R_a-coloring and R_b-coloring can be put together by the ith axis to form an R_c-coloring. For that take an R_{a^i}-coloring and an R_{b^i}-coloring which fit together along the first axis to form an R_{c^i}-coloring. Apply $\circ^i$ on these colorings. The original ones had the same colors on the pair of corners $C(v)$ on the first one and $C(v + e_1)$ on the second one for arbitrary v having first coordinate 1. Thus after applying $\circ^i$ their images, the pair of corners $C(w)$ and $C(w + e_i)$ (e_i is the vector with all-0 coordinates except the ith coordinate being 1), will have the same colors for arbitrary w with ith coordinate 1 and so we can put together the two colorings along the ith axis.

By assumption when putting together along the first axis, the result was an R_{c^i}-coloring. If $c = c^i = \mathbf{0}$ then it had all different colors on its corners, thus the same is true after applying $\circ^i$ and putting together along the ith axis, so the result is indeed an R_c-coloring.

Otherwise if c^i has first coordinate 0 then on the R_{a^i}-coloring the corners $C(v)$ and $C(v + c^i)$ had the same colors for any v with first coordinate 0 and on the R_{b^i}-coloring the corners $C(w)$ and $C(w + c^i)$ had the same colors for any w with first coordinate 1. Thus after applying $\circ^i$, the corners $C(v)$ and $C(v + c)$ of the R_a-coloring have the same colors for any v with ith coordinate 0 and the corners $C(w)$ and $C(w + c)$ of the R_b-coloring have the same colors for any w with ith coordinate 1. As in this case the ith coordinate of c is 0, the resulting coloring after fitting these two together along the ith axis is indeed an R_c-coloring.

If c^i has first coordinate 1 then the corner $C(v)$ of the $R(a^i)$-coloring and the corner $C(v + c^i)$ of the $R(b^i)$-coloring had the same color for any v with first coordinate 0. Thus after applying $\circ^i$, the corners $C(v)$ of the $R(a)$-coloring and the corner $C(v+c)$ of the $R(b)$-coloring have the same colors for any v with ith coordinate 0. Putting these together along the ith axis gives indeed an R_c-coloring.

Finally, back to the hyperboxes colorable with color-ranges S_x and S_y which need to be put together along the ith axis, applying x^i on all the colorings of S_x we get S_{x^i} and similarly from S_y we get the color-range S_{y^i} and we can put these together by the first axis to get the color-range S_{z^i} for some z and so S_x and S_y can be put together by the ith axis to get the color-range S_z. $\square$

The Computational Complexity of Link Building

Martin Olsen

MADALGO*
Department of Computer Science
University of Aarhus
Aabogade 34, DK 8200 Aarhus N, Denmark
mo@madalgo.au.dk

Abstract. We study the problem of adding k new links to a directed graph $G(V, E)$ in order to maximize the minimum PageRank value for a given subset of the nodes. We show that this problem is NP-hard if k is part of the input. We present a simple and *efficient* randomized algorithm for the simple case where the objective is to compute *one* new link pointing to a given node t producing the maximum increase in the PageRank value for t. The algorithm computes an approximation of the PageRank value for t in $G(V, E \cup \{(v, t)\})$ for all nodes v with a running time corresponding to a *small* and *constant* number of PageRank computations.

1 Introduction

Google uses the PageRank algorithm [3],[10] to calculate a universal measure of the popularity of the web pages. The PageRank algorithm assigns a measure of popularity to each page based on the link structure of the web graph. The PageRank algorithm – or variants of the algorithm – can be used to assign a measure of popularity to the nodes in any directed graph. As an example it can also be used to rank scientific journals and publications [2],[4],[9] based on citation graphs.

An organization controlling a set T of web pages might try to identify potential new links that would produce the maximum increase in the PageRank values for the pages in T. Subsequently the organization could try to make sure that these links were added to the web graph. If for example a link from a page p not controlled by the organization is considered beneficial then the organization could simply contact the people controlling p and offer them money for the new link[1]. The problem of obtaining optimal new – typically incoming – links

---

* Center for Massive Data Algorithmics, a Center of the Danish National Research Foundation.

[1] The author of this paper is fully aware that Google attempts to take counter measures to paid links as can be seen on the blog of Matt Cutts (www.mattcutts.com/blog/text-links-and-pagerank/). Matt Cutts is the head of Google's Web spam team. The subject causes much debate which justifies looking at it from a theoretical standpoint.

X. Hu and J. Wang (Eds.): COCOON 2008, LNCS 5092, pp. 119–129, 2008.

is known as *link building* and this problem attracts much attention from the *Search Engine Optimization* (SEO) industry. In this paper we will look at the computational complexity of link building and try to answer the following question in a formal context: How hard is it to identify optimal new links using only information of the link structure?

Langville and Meyer [8] deal with the problem of updating PageRank efficiently without starting from scratch. Avrachenkov and Litvak [1] study the effect on PageRank if a given page establishes one or more links *to* other pages. Avrachenkov and Litvak show that an optimal linking strategy for a page is to establish links only to pages in the *community* of the page. When Avrachenkov and Litvak speak about a web community they mean "... a set of Web pages that a surfer can reach from one to another in a relatively small number of steps". It should be stressed that Avrachenkov and Litvak look for optimal links in $\{p\} \times V$ for a given page p where V denotes the nodes in the directed graph under consideration and that they conclude that p "... cannot significantly manipulate its PageRank value by changing its outgoing links". In this paper we will look for optimal links in $V \times V$ and $V \times \{p\}$ respectively which could cause a significant increase in the PageRank value of p.

1.1   Contribution and Outline of the Paper

We briefly introduce the mathematical background and notation for the paper in Sect. 2 and present introductory examples in Sect. 3. A general formulation of the link building problem is considered in Sect. 4 where we show that this general variant of the problem is NP-hard. In Sect. 5 we look at the simplest case of the problem where we want to find *one* new link *pointing* to a given node t producing the maximum increase for the PageRank value of t. In contrast to the intractability of the general case we present a simple randomized algorithm solving the simplest case with a time complexity corresponding to a small and constant number of PageRank computations. Results of experiments with the algorithm on artificial computer generated graphs and a crawl of the Danish part of the web graph are also reported in Sect. 5.

2   Mathematical Background

This section gives the mathematical background for the PageRank algorithm. We refer to [6] for more details on Finite Markov Chains in general and to [7] for more details on the PageRank algorithm. All vectors throughout this paper are column vectors.

Let $G(V, E)$ denote a directed graph and let $|V| = n$ and $|E| = m$. We allow multiple occurrences of $(u, v) \in E$ implying a *weighted* version of the PageRank algorithm as described in [2]. A *random surfer* visits the nodes in V according to the following rules: When visiting u the surfer picks a link $(u, v) \in E$ uniformly at random and visits v. If u is a sink[2] then the next node to visit is chosen

---

[2] A sink is a node not linking to any node.

uniformly at random. The sequence of pages visited by the random surfer is a Finite Markov Chain with state space V and transition probability matrix $P = \{p_{uv}\}$ given by $p_{uv} = \frac{\text{m}(u,v)}{\text{outdeg}(u)}$ where $\text{m}(u,v)$ is the multiplicity of link (u,v) in E and $\text{outdeg}(u)$ is the out degree of u. If $\text{outdeg}(u) = 0$ then $p_{uv} = \frac{1}{n}$.

Now we modify the behavior of the random surfer so that he behaves as described above with probability α when visiting u but performs a *hyper jump* with probability $1 - \alpha$ to a node v chosen uniformly at random from V. If E is the matrix with all 1's then the transition probability matrix Q for the modified Markov Chain is given by $Q = \frac{1-\alpha}{n}E + \alpha P$. The powers $w^T Q^i$ converge to the same probability distribution π^T for any initial probability distribution w on V as i tends to infinity – implying $\pi^T Q = \pi^T$. The vector $\pi = \{\pi_v\}_{v \in V}$ is known as the PageRank vector. Computing $w^T Q^i$ can be done in time $O((n + m)i)$ and according to [7] 50 - 100 iterations provide a useful approximation for π for $\alpha = 0.85$. Two interpretations of π are the following:

- π_v is the probability that a random surfer visits v after i steps for large i.
- All nodes perform a vote to decide which node is the most popular and π is the result of the vote. The identity $\pi^T Q = \pi^T$ shows that a node is popular if it is pointed to by popular nodes.

The matrix $I - \alpha P$ is invertible and entry z_{uv} in $Z = (I - \alpha P)^{-1}$ is the *expected* number of visits – preceding the first hyper jump – on page v for a random surfer starting at page u. If $u = v$ then the initial visit is also included in the count.

In this paper we will typically look at the PageRank vector for the graph we obtain if we add a set of links E' to $G(V, E)$. We will let $\tilde{\pi}_v(E')$ denote the PageRank value of v in $G(V, E \cup E')$. The argument E' may be omitted if E' is clear from the context.

3   Introductory Examples

We now present two examples of link building problems involving a small graph where the nodes are organized as a hexagon connected with one link to a clique consisting of two nodes. For simplicity we only allow links with multiplicity 1 in this section. Our objective is to identify new links pointing to node 1 maximizing $\tilde{\pi}_1$ – the PageRank value for node 1 *after* insertion of the links. In this paper we will typically try to maximize the PageRank *value* for a node as opposed to try to achieve the maximum improvement in the *ranking* of the node in which case we also have to take the values of the competitors of the node into consideration.

Figure 1(a) shows an optimal new link if we only look for one new link and Fig. 1(b) shows an optimal set of two new links. The two most popular nodes in the set $\{3, \ldots, 7\}$ prior to the modification are the nodes 6 and 7. The examples show that adding links from the most popular nodes are not necessarily the optimal solution – even in the case where the most popular nodes have a low out degree. The examples show that the topology of the network has to be taken into consideration.

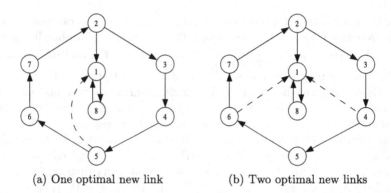

(a) One optimal new link   (b) Two optimal new links

Fig. 1. The dotted links produce the maximum value of $\tilde{\pi}_1$. The PageRank values prior to the update are $\pi_3 = 0.0595$, $\pi_4 = 0.0693$, $\pi_5 = 0.0777$, $\pi_6 = 0.0848$ and $\pi_7 = 0.0908$.

4   A Result on Intractability

A natural question to ask for a set of pages T and numbers x and k is the following: "Is it possible for all the pages in T to achieve a PageRank value greater than x by adding k new links anywhere in the web graph?". This is an informal way to phrase the decision version of the following optimization problem:

Definition 1. *MAX-MIN PAGERANK problem:*

- *Instance: A directed graph $G(V,E)$, a subset of nodes $T \subseteq V$ and a number $k \in \mathbb{Z}^+$.*
- *Solution: A set $S \subseteq \{(u,v) \in V \times V : u \neq v\}$ with $|S| = k$ maximizing $\min_{t \in T} \tilde{\pi}_t(S)$.*

We allow multiple occurrences of (u,v) in E and S.

The MAX-MIN PAGERANK problem is solvable in polynomial time if k is a fixed constant in which case we can simply calculate $\tilde{\pi}(S)$ for all possible S. If k is part of the input then the problem is NP-hard which is formally stated by the following theorem:

Theorem 1. *MAX-MIN PAGERANK is NP-hard.*

Theorem 1 is proved by reduction from the NP-complete balanced version of the PARTITION problem [5, page 223]. The rest of this section gives the proof in detail.

   In order to prove that MAX-MIN PAGERANK is NP-hard when k is part of the input we need three lemmas concerning the graph in Fig. 2 where the weight of a link is the number of occurrences in E. The intuition behind the lemmas and the proof is the following: The nodes A and B are identical twins devoted to each other – the number of links x between them is big – and they share the

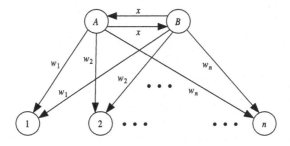

Fig. 2. A directed graph with weights indicating the number of occurrences of the links

same view on the world by assigning the same weight w_i to any other node i in the network. Suppose that you would like to maximize $\min(\tilde{\pi}_A, \tilde{\pi}_B)$ with n new links. The best you can do is to add one new link from every node in $\{1, \ldots, n\}$ to either A or B such that $\tilde{\pi}_A = \tilde{\pi}_B$. It turns out that we have to split the friends of A and B in two groups of equal cardinality and weight to achieve $\tilde{\pi}_A = \tilde{\pi}_B$ and let one group link to A and the other group link to B. Splitting the friends is a well known NP-complete problem [5, page 223].

In the following we let $N = \{1, \ldots, n\}$ and $W = \sum_{i=1}^{n} w_i$. We will write $\tilde{\pi}_{AB}(E')$ as a shorthand for $\tilde{\pi}_A(E') + \tilde{\pi}_B(E')$. We will now formally introduce the term *sum-optimal* and justify this definition in the two subsequent lemmas.

Definition 2. *A set of links E' is called* sum-optimal *if*

$$\forall i \in N : (i, A) \in E' \lor (i, B) \in E' \ .$$

In Lemma 1 we show that we achieve the same value for $\tilde{\pi}_A + \tilde{\pi}_B$ for all sum-optimal sets of n links. In Lemma 2 we show that we will achieve a lower value of $\tilde{\pi}_A + \tilde{\pi}_B$ for any other set of links.

In Lemma 3 we show that we can achieve $\tilde{\pi}_A = \tilde{\pi}_B$ for a sum-optimal set of n links *if and only if* we can split the friends of A and B in two groups of equal cardinality and weight. The three lemmas show that we can identify such a potential split by maximizing $\min(\tilde{\pi}_A, \tilde{\pi}_B)$.

Lemma 1. *Consider the graph in Fig. 2. If E_1' and E_2' denote two arbitrary sum-optimal sets of n links then we have the following:*

$$\tilde{\pi}_{AB}(E_1') = \tilde{\pi}_{AB}(E_2') \ . \tag{1}$$

Proof. Let E' be an arbitrary sum-optimal set of n links. The only nodes that link to the nodes in N are A and B and A and B both use a fraction of $\frac{W}{W+x}$ of their links on N. Since no node in N is a sink and the sum of PageRank values of the nodes in N is $1 - \tilde{\pi}_{AB}(E')$ we have the following:

$$1 - \tilde{\pi}_{AB}(E') = (1 - \alpha)\frac{n}{n+2} + \alpha\tilde{\pi}_{AB}(E')\frac{W}{W+x} \ . \tag{2}$$

From (2) we obtain an expression for $\tilde{\pi}_{AB}(E')$ that proves (1):

$$\tilde{\pi}_{AB}(E') = \frac{1 - (1-\alpha)\frac{n}{n+2}}{1 + \alpha\frac{W}{W+x}} \ . \qquad \square$$

Lemma 2. *Let x satisfy the following inequality:*

$$x > \frac{W(n+2)^2}{n(1-\alpha)} - W \ . \tag{3}$$

If E' is an arbitrary sum-optimal set of n links and L is an arbitrary set of links which is not sum-optimal then we have that

$$\tilde{\pi}_{AB}(E') > \tilde{\pi}_{AB}(L) \ . \tag{4}$$

Proof. There has to be at least one node $u \in N$ that does not link to A and does not link to B since L is not sum-optimal. A fraction of $1 - \alpha$ of the PageRank value of u is spread uniformly on all nodes. No matter whether u is a sink or not then it will spread at least a fraction $\frac{n}{n+2}$ of the remaining part of its PageRank value to the other nodes in N. The PageRank value of u is greater than $\frac{1-\alpha}{n+2}$ which enables us to establish the following inequality:

$$1 - \tilde{\pi}_{AB}(L) > (1-\alpha)\frac{n}{n+2} + \alpha\frac{1-\alpha}{n+2} \cdot \frac{n}{n+2} \ . \tag{5}$$

From (3) we get $\frac{(1-\alpha)n}{(n+2)^2} > \frac{W}{W+x}$. Now we use (2), (5) and $\tilde{\pi}_{AB}(E') < 1$ to conclude that $1 - \tilde{\pi}_{AB}(L) > 1 - \tilde{\pi}_{AB}(E')$ that proves (4). $\qquad \square$

Lemma 3. *Let E' denote an arbitrary sum-optimal set of n links and let x satisfy*

$$x > \frac{\alpha W(n+2)}{1-\alpha} - W \ . \tag{6}$$

Let $A_{\leftarrow} = \{i \in N : (i, A) \in E'\}$. The set $A_{\leftarrow}$ consists of the nodes in N that link to A. We define $W_{A_{\leftarrow}} = \sum_{i \in A_{\leftarrow}} w_i$. We also define $B_{\leftarrow}$ and $W_{B_{\leftarrow}}$ accordingly.

The following two statements are equivalent where E' is omitted as an argument for $\tilde{\pi}_A$ and $\tilde{\pi}_B$:

1. $W_{A_{\leftarrow}} = W_{B_{\leftarrow}} \wedge |A_{\leftarrow}| = |B_{\leftarrow}|$.
2. $\tilde{\pi}_A = \tilde{\pi}_B$.

Proof. Let $\tilde{\pi}_{A_{\leftarrow}}$ and $\tilde{\pi}_{B_{\leftarrow}}$ denote the sum of PageRank values for the two sets $A_{\leftarrow}$ and $B_{\leftarrow}$ respectively. Following the same line of reasoning as used in the proof of Lemma 1 we have the following:

$$\tilde{\pi}_A = \frac{1-\alpha}{n+2} + \alpha\tilde{\pi}_{A_{\leftarrow}} + \alpha\frac{x}{x+W}\tilde{\pi}_B \tag{7}$$

$$\tilde{\pi}_B = \frac{1-\alpha}{n+2} + \alpha\tilde{\pi}_{B_{\leftarrow}} + \alpha\frac{x}{x+W}\tilde{\pi}_A \tag{8}$$

$$\tilde{\pi}_{A_{\leftarrow}} = |A_{\leftarrow}|\frac{1-\alpha}{n+2} + \alpha\frac{W_{A_{\leftarrow}}}{W+x}(\tilde{\pi}_A + \tilde{\pi}_B) \tag{9}$$

$$\tilde{\pi}_{B_{\leftarrow}} = |B_{\leftarrow}|\frac{1-\alpha}{n+2} + \alpha\frac{W_{B_{\leftarrow}}}{W+x}(\tilde{\pi}_A + \tilde{\pi}_B) \ . \tag{10}$$

$1 \Rightarrow 2$: Assume that $W_{A_{\leftarrow}} = W_{B_{\leftarrow}}$ and $|A_{\leftarrow}| = |B_{\leftarrow}|$ for a sum-optimal set E' consisting of n links. By using (9) and (10) we conclude that $\tilde{\pi}_{A_{\leftarrow}} = \tilde{\pi}_{B_{\leftarrow}}$. By solving (7) and (8) we get that $\tilde{\pi}_A = \tilde{\pi}_B$.

$2 \Rightarrow 1$: Assume that $\tilde{\pi}_A = \tilde{\pi}_B$ for a sum-optimal set E' of n links. In this case we can conclude that $\tilde{\pi}_{A_{\leftarrow}} = \tilde{\pi}_{B_{\leftarrow}}$ by using (7) and (8). If $x > \frac{\alpha W(n+2)}{1-\alpha} - W$ then $\frac{1-\alpha}{n+2} > \alpha\frac{W}{W+x}$. This means that the last term in (9) and (10) are smaller than $\frac{1-\alpha}{n+2}$. We conclude that $|A_{\leftarrow}| = |B_{\leftarrow}|$ with $W_{A_{\leftarrow}} = W_{B_{\leftarrow}}$ as a consequence. $\qquad\square$

We are now in a position to prove Theorem 1.

Proof. We show how to solve an instance of the balanced version of the PARTITION problem [5, page 223] – which is known to be NP-complete – in polynomial time if we are allowed to consult an *oracle*[3] for solutions to the MAX-MIN PAGERANK problem.

For an instance of the balanced version of PARTITION we have a $w_i \in \mathbb{Z}^+$ for each $i \in N$. The question is whether a subset $N' \subset N$ exists such that $\sum_{i \in N'} w_i = \sum_{i \in N-N'} w_i$ and $|N'| = |N - N'|$.

In polynomial time we transform this instance into an instance of MAX-MIN PAGERANK given by the graph G in Fig. 2 with $x = \frac{W(n+2)^2}{n(1-\alpha)}$, $T = \{A, B\}$ and $k = n$. We claim that the following two statements are equivalent:

1. $N' \subset N$ exists such that $\sum_{i \in N'} w_i = \sum_{i \in N-N'} w_i$ and $|N'| = |N - N'|$.
2. The solution S to the MAX-MIN PAGERANK instance is a sum-optimal set of links with $W_{A_{\leftarrow}} = W_{B_{\leftarrow}}$ and $|A_{\leftarrow}| = |B_{\leftarrow}|$.

$1 \Rightarrow 2$: Let $E' = [N' \times \{A\}] \cup [(N - N') \times \{B\}]$. According to Lemma 1 and Lemma 2 then $\tilde{\pi}_{AB}(E')$ is at its maximum compared to any other set of n new links. According to Lemma 3 we also have that $\tilde{\pi}_A(E') = \tilde{\pi}_B(E')$. This means that $\min(\tilde{\pi}_A(E'), \tilde{\pi}_B(E'))$ is at its maximum. The solution S to the MAX-MIN PAGERANK instance must match this value so S must be sum-optimal (Lemma 2) with $\tilde{\pi}_A(S) = \tilde{\pi}_B(S)$. According to Lemma 3 then $W_{A_{\leftarrow}} = W_{B_{\leftarrow}}$ and $|A_{\leftarrow}| = |B_{\leftarrow}|$ for S.

$2 \Rightarrow 1$: Take $N' = A_{\leftarrow}$.

We can now solve the PARTITION instance by checking whether 2) is satisfied in the solution of the MAX-MIN PAGERANK instance. The checking procedure can be done in polynomial time. $\qquad\square$

5   An Efficient Algorithm for the Simplest Case

We now turn to the simplest variant of the link building problem where the objective is to pick *one* link pointing to *a given page* in order to achieve the maximum increase in the PageRank value for the page. This problem can be solved naively in polynomial time using n PageRank computations. We present

---

[3] An oracle is a hypothetical computing device that can compute a solution in a *single step* of computation.

an efficient randomized algorithm that solves this problem with a running time corresponding to a *small* and *constant* number of PageRank computations. The main message is that if we have the machinery capable of calculating the Page-Rank vector for the network we can also solve the simple link building problem.

If page $j \neq 1$ establishes a link to page 1 then we have the following according to [1, Theorem 3.1]:

$$\tilde{\pi}_1 = \pi_1 + \pi_j \frac{\alpha z_{11} - z_{j1}}{k_j + z_{jj} - \alpha z_{1j}} . \tag{11}$$

The central idea for the link building algorithm is to avoid an expensive matrix inversion and only calculate the entries of Z playing a role in (11) for all $j \neq 1$. We approximate z_{11}, z_{1j} and z_{j1} for *all* $j \neq 1$ by performing two calculations where each calculation has a running time comparable to one PageRank computation. The diagonal elements z_{jj} are approximated by a randomized scheme tracking a random surfer. When we have obtained approximations of all *relevant* entries of Z then we can calculate (11) in constant time for any given page j.

5.1   Approximating Rows and Columns of Z

We will use the following expression for Z [6]:

$$Z = (I - \alpha P)^{-1} = \sum_{i=0}^{+\infty} (\alpha P)^i . \tag{12}$$

In order to get row 1 from Z we multiply (12) with e_1^T from the left where e_1 is a vector with a 1 at coordinate 1 and 0's elsewhere:

$$e_1^T Z = \sum_{i=0}^{+\infty} e_1^T (\alpha P)^i = e_1^T + e_1^T \alpha P + (e_1^T \alpha P)\alpha P + \cdots . \tag{13}$$

Equation (13) shows how to *approximate* row 1 in Z with a simple iterative scheme using the fact that each term in (13) is a row vector obtained by multiplying αP with the previous term from the left. We simply track a group of random surfers starting at page 1 and count the number of hits they produce on other pages preceding the first hyper jump.

The elements appearing in a term are non negative and the sum of the elements in the $i'th$ term is α^{i-1} which can be shown by using the fact that $Pe = e$ where e is the vector with all 1's so the iterative scheme converges quickly for $\alpha = 0.85$. The iterative scheme has roughly the same running time as the power method for calculating PageRank and 50-100 iterations gives adequate precision for approximating the fraction in (11) since $z_{jj} \geq 1$ for all j.

By multiplying (12) with e_1 from the right we will obtain an iterative scheme for calculating the first column in Z with similar arguments for the convergence.

5.2   Approximating the Diagonal of Z

Now we only have to find a way to approximate z_{jj} for $j \neq 1$. In order to do this we will keep track of a *single* random surfer. Each time the surfer decides *not* to

follow a link the surfer changes identity and continues surfing from a new page
– we chose the new page to start from by adding 1 (cyclically) to the previous
start page. For each page p we record the identity of the surfer who made the
most recent visit, the total number of visits to p and the number of different
surfers who have visited p. The total number of visits divided by the number of
different surfers will most likely be close to z_{pp} if the number of visits is large.

If Z_{pp} denotes the stochastic variable denoting the number of visits on page
p for a random surfer starting at page p prior to the first hyper jump then we
have the following [6]:

$$Var(Z_{pp}) = z_{pp}^2 - z_{pp} = z_{pp}(z_{pp} - 1) \ . \tag{14}$$

where $Var(\cdot)$ denotes the variance. Since we will obtain the highest value of z_{pp}
if all the nodes pointed to by p had only one link back to p then we have that

$$z_{pp} \leq 1 + \alpha^2 + \alpha^4 + \cdots = \frac{1}{1 - \alpha^2} \ . \tag{15}$$

Combining (14) and (15) we have that $Var(Z_{pp}) = O(1)$ so according to *The
Central Limit Theorem* we roughly need a *constant* number of visits per node of
the random surfer to achieve a certain level of certainty of our approximation
of z_{pp}.

Our main interest is to calculate z_{pp} for pages with high values of π_p – luckily
$k\pi_p$ is the expected number of visits to page p if the random surfer visits k pages
for large k [6] so our approximation of z_{pp} tends to be more precise for pages
with high values of π_p. We also note that it is easy to parallelize the algorithm
described above simply by tracking several random surfers in parallel.

5.3   Experiments

Experiments with the algorithm were carried out on artificial computer generated graphs and on a crawl of the Danish part of the web graph. Running the
algorithm on a subgraph of the web graph might seem to be a bad idea but if
the subgraph is a community it actually makes sense. In this case we are trying
to find optimal link modifications only involving our direct competitors. Locating the community in question by cutting away irrelevant nodes seems to be a
reasonable prepossessing step for the algorithm.

Experiments on Artificial Graphs. The algorithm was tested on 10 computer generated graphs each with 500 nodes numbered from 1 to 500 and 5000
links with multiplicity 1 inserted totally at random. For each graph $G(V, E)$
and for each $v \in V$ such that $(v, 1) \notin E$ we computed $\tilde{\pi}_1(\{(v, 1)\})$. The new
PageRank value $\tilde{\pi}_1$ of node 1 was computed in two ways: 1) by the algorithm
described in this section and 2) by the power method. We used 50 terms when
calculating the rows and columns of the Z-matrix and 50 moves per edge for
the random surfer when calculating the diagonal of Z. For the PageRank power
method computation we used 50 iterations. For all graphs and all v the *relative
difference* of the two values of $\tilde{\pi}_1$ was less than 0.1%.

Experiments on the Web Graph. Experiments were also carried out on a crawl from spring 2005 of the Danish part of the web graph with approximately 9.2 million pages and 160 millions links. For each page v in the crawl we used the algorithm to compute the new PageRank value for www.daimi.au.dk – the home-page of the Department of Computer Science at University of Aarhus, Denmark – obtained after adding a link from v to www.daimi.au.dk. The list of potential new PageRank values was sorted in decreasing order.

The PageRank vector and the row and column of Z corresponding to www.daimi.au.dk was calculated using 50 iterations/terms and the diagonal of Z was computed using 300 moves of the random surfer per edge. The computation took a few hours on standard PC's using no effort on optimization. The links were stored on a file that was read for each iteration/term in the computation of the PageRank vector and the rows and columns of Z.

As can be seen from Equation (11) then the diagonal element of Z plays an important role for a potential source with a low out degree. As an example we will look at the pages www.kmdkv.dk/kdk.htm and news.sunsite.dk which we will denote as page a and b respectively in the following. The pages a and b are ranked 22 and 23 respectively in the crawl with π_a only approximately 3.5% bigger than π_b. Page a has out degree 2 and page b has out degree 1 so based on the information on π_a, π_b and the out degrees it would seem reasonable for www.daimi.au.dk to go for a link from page b because of the difference on the out degrees. The results from the experiment show that it is a better idea to go for a link from page a: If we obtain a link to www.daimi.au.dk from page a we will achieve a PageRank value approximately 32% bigger than if we obtain a link from page b. The reason is that z_{bb} is relatively big producing a relatively big denominator in the fraction in (11).

Acknowledgments. The author of this paper would like to thank Torsten Suel and his colleagues at Polytechnic University in New York for a crawl of the Danish part of the web graph and Gerth S. Brodal from University of Aarhus for valuable comments and constructive criticism.

References

1. Avrachenkov, K., Litvak, N.: The effect of new links on Google Pagerank. Stochastic Models 22(2), 319–331 (2006)
2. Bollen, J., Rodriguez, M.A., Van de Sompel, H.: Journal Status. Scientometrics 69(3), 669–687 (2006)
3. Brin, S., Page, L.: The anatomy of a large-scale hypertextual Web search engine. Computer Networks and ISDN Systems 30(1–7), 107–117 (1998)
4. Chen, P., Xie, H., Maslov, S., Redner, S.: Finding scientific gems with Googles PageRank algorithm. Informetrics 1(1), 8–15 (2007)
5. Garey, M.R., Johnson, D.S.: Computers and Intractability: A Guide to the Theory of NP-Completeness. W.H. Freeman, New York (1979)
6. Kemeny, J.G., Snell, J.L.: Finite Markov Chains. Van Nostrand (1960)
7. Langville, A.N., Meyer, C.D.: Google's PageRank and Beyond: The Science of Search Engine Rankings. Princeton University Press, Princeton (2006)

8. Langville, A.N., Meyer, C.D.: Updating Markov Chains with an Eye on Google's PageRank. SIAM J. Matrix Anal. Appl. 27(4), 968–987 (2006)
9. Meho, L.I., Yang, K.: Multi-faceted Approach to Citation-based Quality Assessment for Knowledge Management. In: World Library and Information Congress: 72nd IFLA General Conference and Council (2006)
10. Page, L., Brin, S., Motwani, R., Winograd, T.: The PageRank Citation Ranking: Bringing Order to the Web. Technical report, Stanford Digital Library Technologies Project (1998)

Improved Parameterized Algorithms for Weighted 3-Set Packing*

Jianxin Wang and Qilong Feng

School of Information Science and Engineering, Central South University,
Changsha 410083, P.R. China
jxwang@mail.csu.edu.cn

Abstract. Packing problems form an important class of NP-hard problems. For the weighted 3-Set Packing problem, we provide further theoretical study on the problem and present a deterministic algorithm of time $O^*(10.6^{3k})$. Based on the randomized divide-and-conquer method, the above result can be further reduced to $O^*(7.56^{3k})$, which significantly improves the previous best result $O^*(12.8^{3k})$.

1 Introduction

In the complexity theory, packing problems form an important class of NP-hard problems, which are widely used in scheduling and code optimization area. We first give some related definitions [12],[5],[9].

Assume all the elements used in this paper are from U.

Definition 1. Set Packing: Given a collection S of n sets, find the largest subset S' such that no two sets in S' have the common elements.

Definition 2. (Parameterized) Weighted 3-Set Packing: Given a pair (S, k), where S is a collection of n weighted sets, each of which contains 3 elements, and k is an integer, either construct a k-packing with the maximum weight or report that no such packing exists.

Definition 3. Weighted 3-Set Packing Augmentation: Given a pair (S, P_k), where S is a collection of n weighted sets and P_k is a k-packing with the maximum weight in S, either construct a $(k + 1)$-packing with the maximum weight or report that no such packing exists.

A lot of attention has been paid to the weighted m-Set Packing problem (each set contains m elements). From the approximation point of view, Hassin [2] and Bafna et al. [3] both analyzed the weighted m-Set Packing problem using local

---

* This work is supported by the National Basic Research 973 Program of China (2008CB317107), the National Natural Science Foundation of China under Grant (60433020, 60773111), the Program for New Century Excellent Talents in University of China under Grant (NCET-05-0683), the Program for Changjiang Scholars and Innovative Research Team in University of China under Grant (IRT0661).

X. Hu and J. Wang (Eds.): COCOON 2008, LNCS 5092, pp. 130–139, 2008.

search, and got the same approximation ratio $m - 1 + \varepsilon$. Combining greedy solution and local improvement, Chandra [6] presented an approximation algorithm with ratio $2(m + 1)/3$. Based on the local improvement, Berman [4] gave an approximation algorithm with ratio $(m + 1)/2$.

Recently, Y. Liu et al. [11] proposed a parameterized algorithm with time complexity $O^*(12.8^{mk})$, which is the current best result.

In this paper, we will discuss how to construct a maximum weighted $(k + 1)$-packing from a maximum weighted k-packing, so as to solve the weighted 3-Set Packing problem. After further analyzing the structure of the problem, we can get an improved parameterized algorithm of time $O^*(10.6^{3k})$ for the weighted 3-Set Packing problem using color coding. Based on the randomized divide and conquer, the above result can be further reduced to $O^*(7.56^{3k})$, which greatly improves the current best result $O^*(12.8^{3k})$.

2   Related Terminology and Lemmas

We first give some terminologies and lemmas.

Fellows et al. [10] gave that the 3-Set Packing problem has a kernel of size $O(27k^3)$. For the weighted 3-Set Packing problem, based on the same kernelization process, it is easy to get a kernel of size $O(27k^3)$ (When using pattern to reduce the sets in S, delete the sets with smaller weight).

Based on the Lemma 2.1 in [12], we can get the following lemma.

Lemma 1. *For any constant $c > 1$, the weighted 3-Set Packing Augmentation problem can be solved in time $O^*(c^k)$ if and only if the weighted 3-Set Packing problem can be solved in time $O^*(c^k)$.*

By Lemma 1, our objective is to reduce the time complexity of the weighted 3-Set Packing Augmentation problem. In this paper, color coding technology is efficiently used, which was first introduced by Alon et al. [1] and has the best runtime bounds of the form $O^*(6.1^k)$ obtained by Chen et al. [8].

3   An $O^*(10.6^{3k})$ Algorithm Based on Color Coding

For an instance (S, P_k) of weighed 3-Set Packing Augmentation problem, we divide S into S_1, S_2 two parts to handle: for each set ρ in S, if ρ and P_k has no common elements, put it into S_1, or else put it into S_2.

The general idea of our algorithm is that:

Divide S into S_1 and S_2. Find the set with the largest weight in S_1, plus the k sets in P_k, then a $(k + 1)$-packing can be constructed, denoted as $PW1$. Use color coding to find a maximum weighted $(k + 1)$-packing contained in S_2, denoted as $PW2$. Choose the one with larger weight from $PW1$ and $PW2$.

Lemma 2. *For an instance of the weighted 3-Set Packing Augmentation problem (S, P_k), where P_k is a maximum weighted k-packing in S. Either $PW1$ or $PW2$ is a maximum weighted $(k + 1)$-packing.*

Proof. $PW1$ is a maximum weighted $(k+1)$-packing over all the $(k+1)$-packing constructed by the sets in S_1 and the sets in P_k. $PW2$ is a maximum weighted $(k+1)$-packing in S_2. Assume that $PW2 \leq PW1$, we need to prove that $PW1$ is a maximum weighted $k+1$-packing in S.

Assume there is $(k+1)$-packing P constructed by the sets of S_1 and S_2. Assume set ρ has the largest weight over all the sets of S_1 contained in P. Thus, $P - \rho$ is a k-packing. Because P_k is a maximum weighted k-packing in S, the weight of $(k+1)$-packing constructed by P_k and ρ is no less than P. It is known that $PW1$ is the maximum weighted $(k+1)$-packing over all the $(k+1)$-packing constructed by the sets in S_1 and the sets in P_k. Thus, the weight of $P_k + \rho$ is no larger than $PW1$. Therefore, $PW1$ is a maximum weighted $(k+1)$-packing in S. For the case $PW2 > PW1$, the proof is similar. □

The remaining problem is to find a $(k+1)$-packing with the maximum weight in S_2. For the $(k+1)$-packing contained in S_2, we can get the following lemma.

Lemma 3. *For an instance of the weighted 3-Set Packing Augmentation problem (S_2, P_k). Assume there exists P_{k+1}, each set in P_{k+1} should contain at least one element of P_k, that is, the P_{k+1} contains at least $k+1$ elements of P_k.*

Proof. In the process of dividing S, if a set ρ in S has common elements with P_k, put it into S_2. Therefore, each set in S_2 contains at least one element of P_k. □

By Lemma 3, the $(k+1)$-packing contains at least $k+1$ elements of P_k. Since P_{k+1} contains $3k+3$ elements, the number of elements contained in P_{k+1} but not in P_k is at most $2k+2$. Use color coding technique to find those at most $2k+2$ elements in $S_2 - P_k$. We introduce another $3k$ colors to color the $3k$ elements of P_k. Based on the color coding, we will use dynamic programming to find a maximum weighted $(k+1)$-packing in S_2.

For a packing P, we call it is properly colored if P contains no two elements having the same color. Let $cl(P)$ denote the set of colors used by packing P. The algorithm of finding the maximum weighted $(k+1)$-packing in S_2 is given in Fig.1.

Theorem 1. *If there is a maximum weighted $(k+1)$-packing in S_2, algorithm $3SPDP(S_2, f, k)$ will definitely return a maximum weighted $(k+1)$-packing in $O^*(2^{5k})$.*

Proof. We get an induction for the i in the step 4 so as to prove the theorem. For any arbitrary i $(1 \leq i \leq n)$, assume that $X_i(X_i \subseteq S_2)$ denotes the i sets in S_2. Therefore, we only need to prove the following claim.

Claim 1. *If there exists a maximum weighted j-packing P_j in X_i, then after i-th execution of the for-loop in step 4, Q_1 must contains a j-packing P_j' which uses colors $cl(P_j)$ and has the same weight with P_j.*

In step 3, $Q_1 = \emptyset$. If there is a 0-packing in X_i, the claim is true.

When $i \geq 1$, assume there exists a maximum weighted j-packing $P_j = \{\varphi_{l_1}, \varphi_{l_2}, \cdots, \varphi_{l_j}\}$ in X_i, where $1 \leq l_1 < l_2 < \cdots < l_j \leq i$, then there

---

Algorithm 3SPDP(S_2, f, k)

Input: A collection S_2 of 3-sets, which is colored by coloring f, and an
      integer k

Output: a $(k+1)$-packing with the maximum weight in S_2 if it exists

1.   remove any sets in S_2 in which two elements have the same color;
2.   let the remaining 3-sets in S_2 be $\rho_1, \rho_2, \cdots, \rho_n$;
3. $Q_1 = \{\emptyset\}$;
4.   **for** $i = 1$ **to** n **do**
4.1.   **for** each packing C in Q_1 **do**
4.2   **if** $cl(C)$ has no same color with $cl(\rho_i)$ **then** $C' = C \cup \{\rho_i\}$;
4.3   **if** C' contains no more than $k+1$ sets and Q_1 contains no packing
           that uses exactly the same colors as that used by C'
           then add C' to Q_1;
4.4   **if** there exists a packing C_1 in Q_1 using exactly the same colors
           as $cl(C')$ and the weight of C_1 is less than C'
           then replace C_1 with C';
5.   return a $(k+1)$-packing with maximum weight in Q_1 if such packing exists;

Fig. 1. Find a maximum weight $(k+1)$-packing in S_2

must exist a $(j-1)$-packing $P_{j-1} = \{\varphi'_{l_1}, \varphi'_{l_2}, \cdots, \varphi'_{l_j-1}\}$ in X_{l_j-1} such that $cl(P_{j-1}) = cl(P_j - \varphi_{l_j})$ and P_{j-1} has the same weight with $P_j - \varphi_{l_j}$. By the induction assumption, after the $(l_j - 1)$-th execution of for-loop in step 4, Q_1 contains a maximum weighted $(j-1)$-packing P'_{j-1}, which uses the same colors with P_{j-1} and has the same weight. By the assumption, X_i contains a maximum weighted j-packing P_j. Therefore, when the set φ_{l_j} is considered in step 4.2, the colors in P'_{j-1} are totally different from the colors in φ_{l_j}. As a result, if there is no packing in Q_1 containing the same $cl(P'_{j-1} \cup \{\varphi_{l_j}\})$, the j-packing $P'_{j-1} \cup \{\varphi_{l_j}\}$ will be added into Q_1. If there exists a packing C_1 in Q_1 using exactly the same color as $cl(P'_{j-1} \cup \{\varphi_{l_j}\})$ and the weight of C_1 is not larger than $P'_{j-1} \cup \{\varphi_{l_j}\}$, then replace C_1 with $P'_{j-1} \cup \{\varphi_{l_j}\}$. Because all the packing in Q_1 are not removed from Q_1 and $l_j \leq i$, after the i-th execution of the for-loop in step 4, Q_1 must contains a j-packing using the same colors with P_j and having the same weight. When $i = n$, if there exists a maximum weighted $(k+1)$-packing in S_2, Q_1 must contain a $(k+1)$-packing with the maximum weight.

At last, we give the analysis of the time complexity. For each j $(0 \leq j \leq 5k+2)$ and any subset of containing $3j$ colors, Q_1 records at most one packing using those $3j$ colors, thus, Q_1 contains at most $\sum_{j=0}^{k+1} \binom{5k+2}{3j}$ packing. Therefore, the time complexity of algorithm 3SPDP is $O^*(\sum_{j=0}^{k+1} \binom{5k+2}{3j}) = O^*(2^{5k})$. $\qquad \square$

Based on Lemma 2 and Theorem 1, the algorithm solving the weighted 3-Set Packing Augmentation problem is given in Fig.2.

Algorithm 3SPG(S, P_k)
Input: A collection S of 3-sets, a maximum weighted k-packing P_k
Output: a $(k + 1)$-packing with the maximum weight in S if it exists
1. $PW1 = \{\emptyset\}$, and find those sets having no common elements with P_k,
   denoted as S_1, and let S_2 denote the sets in $S - S_1$;
2. **if** S_1 is not empty **then**
   use the set of the largest weight in S_1 to construct the P_{k+1} with the k
   sets in P_k, denoted as $PW1$;
3. use $2k + 2$ colors to find the at most $2k + 2$ elements in P_{k+1} but not in
   P_k, and use F to denote the color coding scheme;
4. introduce another $3k$ colors to color the $3k$ elements in P_k, such that each
   coloring f in F uses $5k + 2$ colors;
5. **for** each coloring f in F **do**
       $PW2 =$ 3SPDP(S_2, f, k);
       if the weight of $PW2$ is larger than $PW1$ **then** $PW1 = PW2$;
6. return $PW1$;

Fig. 2. Find a maximum weighted $(k + 1)$-packing in S

Theorem 2. *For an instance of the weighted 3-Set Packing Augmentation prob-lem (S, P_k), if there exists P_{k+1}, algorithm 3SPG(S, P_k) will definitely return a P_{k+1} with the maximum weight in time $O^*(10.6^{3k})$.*

Proof. For an instance of the weighted 3-Set Packing Augmentation problem (S, P_k), we divide S into two parts to handle: S_1, S_2. PW1 can be easily con-structed by the sets in S_1 and the k sets in P_k.

For the $(k + 1)$-packing in S_2, we use color coding to find that at most $2k + 2$ elements contained in P_{k+1} but not in P_k , and can get a color coding scheme F of size $O^*(6.1^{2k})$. After introducing another $3k$ colors to color the $3k$ elements in P_k, each coloring in F uses $5k + 2$ colors to color the elements in S_2. If there exists a maximum weighted $(k + 1)$-packing in S_2, the $(k + 1)$-packing must be properly colored by a coloring f in F. By Theorem 1, we know that if there is a maximum weighted $(k + 1)$-packing in S_2, algorithm 3SPDP(S_2, f, k) will definitely return a $(k + 1)$-packing with time complexity $O^*(2^{5k})$. At last, make a comparison between $PW1$ and the packing returned by algorithm 3SPDP, and choose the largest one.

The $(k+1)$-packing PW1 can be handled in polynomial time. For each coloring f in F, we need to call algorithm 3SPDP to find the maximum weighted $(k+1)$-packing. Therefore, the total time complexity of algorithm 3SPG is $O^*(6.1^{2k}) \times O^*(2^{5k}) = O^*(10.6^{3k})$.   □

Based on Lemma 1 and Theorem 2, we can get the following corollary.

Corollary 1. *The weighted 3-Set Packing problem can be solved in time $O^*(10.6^{3k})$.*

4 An $O^*(7.56^{3k})$ Algorithm Based on Randomized Divide and Conquer

In this section, we will use randomized divide and conquer to further reduce the time complexity of the weighted 3-Set Packing problem. Because the general idea used in this part is similar to the one in [14], we just omit the similar part and only give the necessary part of solving the weighted 3-Set Packing problem. All the lemmas and theorems in the following are related to the $(k + 1)$-packing in S_2. Based on Lemma 3, it is easy to get the following lemma.

Lemma 4. *Given an instance of weighted 3-Set Packing Augmentation problem* (S_2, P_k). *Assume there exists* P_{k+1}, *each* $(k + 1)$-*packing* P *has at most* r *sets such that each contains only one element of* P_k, $0 \le r \le k + 1$, *has at most* s *sets such that each contains only two elements of* P_k, $0 \le s \le k + 1$, *and has at most* t *sets such that each contains three elements of* P_k, $0 \le t \le k - 1$, *where* $r + s + t = k + 1$.

Let C_i $(1 \le i \le 3)$ denote all the sets having i common elements with P_k. Assume U_{P_k} denotes the $3k$ elements in P_k and $U_{S_2 - P_k}$ denotes the elements in $S_2 - P_k$.

By Lemma 4, if P_{k+1} exists, there are r sets in P_{k+1} such that each contains only one element of P_k, which are obviously included in C_1. To find the r elements from U_{P_k}, there are $\binom{3k}{r}$ enumerations. For one fixed enumeration, let H be the collection of sets in C_1 containing one of those r elements.

Assume $U_{P_{k+1} - P_k}$ contains all the elements in P_{k+1} but not in P_k, and the size of the $U_{P_{k+1} - P_k}$ is denoted by $y = |U_{P_{k+1} - P_k}|$. By Lemma 3, P_{k+1} contains at least $k + 1$ elements of U_{P_k}, thus, $U_{P_{k+1} - P_k}$ contains at most $2k + 2$ elements of $U_{S_2 - P_k}$, that is, $y \le 2k + 2$. It can be seen that the elements in $U_{P_{k+1} - P_k}$ are either in H or in $C_2 \cup C_3$. Assume that $U_{P_{k+1} - P_k}^H$ contains the elements of $U_{P_{k+1} - P_k}$ belonging to H. When the elements in $U_{P_{k+1} - P_k}$ are partitioned, the probability that the elements in $U_{P_{k+1} - P_k}^H$ are exactly partitioned into H is $\frac{1}{2^y}$.

The general ideal of our randomized algorithm is as follows:

Divide P_{k+1} into two parts to handle, one of which is in H and the other in $C_2 \cup C_3$. For the part contained in $C_2 \cup C_3$, we use dynamic programming to find a maximum weighted $(k + 1 - r)$-packing; For the part in H, we use randomized divide and conquer to handle.

4.1 Find a Maximum Weighted $(k + 1 - r)$-Packing in $C_2 \cup C_3$

The general idea of finding a maximum weighted $(k+1-r)$-packing in $C_2 \cup C_3$ is similar to the algorithm SP in [14]. We just point out the different part. Assume that we use set Q_3 to save the packings generated by the algorithm. When there exists a packing C_1 in Q_3 using exactly the same elements with a newly constructed packing C', we need to compare the weight of C_1 and C', and choose the larger one. We can get the following theorem from Theorem 1 in [14].

Theorem 3. *If there exists a maximum weighted* $(k+1-r)$-*packing in* $C_2 \cup C_3$, *a maximum weighted* $(k + 1 - r)$-*packing can be found in* $O^*(2^{3k})$.

4.2   Find a Maximum Weighted r-Packing in H

The general idea of finding a maximum weighted r-packing in H is similar to the one in [14]. However, we give a more efficient algorithm to find a maximum weighted r-packing in H, just as shown in figure 3. The proof of correctness and time complexity is also same with the Theorem 2 in [14]. Therefore, we can get the following theorem.

Algorithm RSP(H', D', r)
Input: H' is a subset of H, D' is a collection of r'-packing and there is no
      common elements between H' and D'
Output: return a collection D of packings and each packing in D is the
      combination of the packing in H' and D'
1. $D = \phi$;
2. **if** $r = 1$ **then**
2.1 **if** $D' = \phi$ **then** return all the 1-packing in H'
    else for each r'-packing in D' and each set ρ in H' **do**
2.2   **if** P has no common elements with ρ **then** $P' = P \cup \{\rho\}$;
2.3   **if** there is no $(r' + 1)$-packing containing ρ in D
             then put P' into D;
2.4   **if** there is a $(r' + 1)$-packing $\bar{P}$ containing ρ in D and the
             weight of $\bar{P}$ is no larger than P' **then** replace $\bar{P}$ with P';
2.5   return D;
3. randomly pick $\lceil \frac{r}{2} \rceil$ elements from the r elements, and let H_1 denote all
    the sets containing one of those $\lceil \frac{r}{2} \rceil$ elements, $H_2 = H - H_1$;
4. **for** $3 \cdot 2^{2r}$ times **do**
4.1 $H'_1 = H_1, H'_2 = H_2$;
4.2   mark all the elements from $U_{S_2 - P_k}$ in H'_1 and H'_2 with red and blue;
4.3   for each set ρ in H'_1, if the colors of the elements belonging to $U_{S_2 - P_k}$
      are not both red, delete the set ρ;
4.4   for each set ρ' in H'_2, if the colors of the elements belonging to $U_{S_2 - P_k}$
      are not both blue, delete the set ρ';
4.5 $D_1 =$RSP$(H'_1, D', \lceil \frac{r}{2} \rceil)$;
4.6   **if** $D_1 \neq \phi$ **then**
4.7 $D_2 =$RSP$(H'_2, D_1, r - \lceil \frac{r}{2} \rceil)$;
4.8   **for** each $(r' + r)$-packing α in D_2 **do**
        if there does not exist $(r' + r)$-packing in D **then** put α into D;
        if there exists $(r' + r)$-packing $\bar{\alpha}$ in D and the weight of $\bar{\alpha}$ is no
        larger than α **then** replace $\bar{\alpha}$ with α;
5. return D;

Fig. 3. Find a maximum weighted r-packing in H

Theorem 4. *If H has a maximum weighted r-packing, a maximum weighted r-packing will be returned with probability larger than 0.75, and the time complexity is bounded by $O^*(4^{2r})$.*

4.3 The General Algorithm for Weighted 3-Set Packing Augmentation

The algorithm GSP in [14] can be used to solve the weighted 3-Set Packing Augmentation problem. However, some modification should be made. For the randomized algorithm, in order to find a maximum weighted $(k + 1)$-packing in S, the whole process is iterated at least 2^{2k+2} times. We use a set Q to save the $(k + 1)$-packing constructed by each iteration, which is different from the unweighted case. Based on the algorithm GSP in [14], we can get the following theorem.

Theorem 5. *If S_2 has a maximum weighted $(k + 1)$-packing, a maximum weighted $(k + 1)$-packing can be found with probability larger than 0.75 in time $O^*(7.56^{3k})$.*

Proof. The proof of correctness is similar to the proof of Theorem 3 in [14]. We just give time complexity analysis. For each r, there are $\binom{3k}{r}$ ways to enumerate r elements from U_{P_k}. By theorem 3, the time complexity of finding a maximum weighted $(k + 1 - r)$-packing in $C_2 \cup C_3$ is bounded by $O^*(2^{3k})$. By theorem 4, the time complexity of finding a maximum weighted r-packing is $O^*(4^{2r})$, $0 \leq r \leq k + 1$. Therefore, the total time complexity of algorithm GSP is $\sum_{r=0}^{k+1} \binom{3k}{r}(2^{2k+2}(2^{3k-r} + 4^{2r})) = O^*(7.56^{3k})$. $\square$

4.4 Derandomization

When there exists a maximum weighted $(k+1)$-packing, in order to make failure impossible, we need to derandomize the above algorithm. We first give some related definitions [7]. A *splitting function* over set Z_n is a $\{0,1\}$ (i.e., Boolean) function over Z_n. Moreover, a splitting function can be interpreted as a partition (X_1, X_2) of the set Z_n (i.e., putting all x in Z_n such that $f(x) = 0$ in X_1 and putting all x in Z_n such that $f(x) = 1$ in X_2).

A set F of splitting functions over Z_n is an (n, k)-*universal set* if for every k-subset W of Z_n and any partition (W_1, W_2) of W, there is a splitting function f in F that implements partition (W_1, W_2). The size of an (n, k)-universal set F is the total number of splitting functions in F.

Chen and Lu [7] gave that the size of the (n, k)-universal set is $O(n2^{k+12\log^2 k + 12\log k + 6})$, which can be constructed in time $O(n2^{k+12\log^2 k + 12\log k + 6})$. In the constructing process, Chen and Lu [7] constructed a function $h(x)=((ix \bmod p)\bmod k^2)$ from $\{0, 1, \cdots, n-1\}$ to $\{0, 1, \cdots, k^2-1\}$, and used the fact that there are at most $2n$ $h(x)$. However, the bound $2n$ is not tight. Naor et al. [13] gave that the size of functions from $\{0, 1, \cdots, n-1\}$ to $\{0, 1, \cdots, k^2 - 1\}$ is bounded by $O(k^6 \log k \log n)$. Therefore, the size of the (n, k)-universal set is $O(\log n2^{k+12\log^2 k+18\log k})$.

If there exists a maximum weighted P_{k+1} in S_2, the P_{k+1} contains at most $2k + 2$ elements of $U_{S_2-P_k}$. After picking r elements from U_{P_k}, in order to get the P_{k+1}, the $2r$ elements in $U_{S_2-P_k}$ that belongs to H must be divided into H, and the others are in $C_2 \cup C_3$. We can construct the $(|U_{S_2-P_k}|, 2k+2)$-universal set, whose size is bounded by $O(\log n 2^{2k+2+12\log^2(2k+2)+18\log(2k+2)})$.

For each partition in the $(|U_{S_2-P_k}|, 2k+2)$-universal set, assume U_{H-P_k} contains the elements partitioned into H. If H has a maximum weighted r-packing P_r, there are $2r$ elements of $U_{S_2-P_k}$ in P_r, denoted by $U_{P_r-P_k}$. Assume $U'_{P_r-P_k}$ contains the elements of $U_{P_r-P_k}$ in H_1. In order to find P_r, $U'_{P_r-P_k}$ must be partitioned into H_1, and $U_{P_r-P_k} - U'_{P_r-P_k}$ should be in H_2. We can construct $(|U_{H-P_k}|, 2r)$-universal set, whose size is bounded by $O(\log n 2^{2k+2+12\log^2(2k+2)+18\log(2k+2)})$.

Based on the above result, we can get the following theorem.

Theorem 6. *Weighted 3-Set Packing Augmentation problem can be solved deterministically in time* $O^*(7.56^{3k})$.

Proof. According to the process of constructing the $(|U_{S_2-P_k}|, 2k+2)$-universal set and $(|U_{H-P_k}|, 2r)$-universal set, it is easy to prove the correctness of the theorem. We just analyze the time complexity of the algorithm.

In the derandomization of algorithm RSP, the time complexity is:

$$T_r \leq \log n 2^{2k+2+12\log^2(2k+2)+18\log(2k+2)}(T_{\lceil \frac{r}{2} \rceil} + T_{\lfloor \frac{r}{2} \rfloor})$$
$$\leq \log n 2^{2k+2+12\log^2(2k+2)+18\log(2k+2)+1}T_{\lceil \frac{r}{2} \rceil}$$
$$= O((2k+2)^{\log\log n} 2^{2(2k+2)+4\log^3(2k+2)+15\log^2(2k+2)+13\log(2k+2)}).$$

Because the weighted 3-Set Packing problem has a kernel of size $O(27k^3)$, the elements in S_2 is bounded by $O(81k^3)$. Therefore, the size of $(|U_{S_2-P_k}|, 2k+2)$-universal set is bounded by $O((4\log 3 + 3\log k) 2^{2k+2+12\log^2(2k+2)+18\log(2k+2)})$. T_r is bounded by: $O((2k+2)^{\log(4\log 3 + 3\log k)} 2^{2(2k+2)+4\log^3(2k+2)+15\log^2(2k+2)+11\log(2k+2)})$.

If there exists maximum weighted P_{k+1}, on the basis of $(|U_{S_2-P_k}|, 2k+2)$-universal set and T_r, we can get the P_{k+1} deterministically with time complexity

$$\sum_{r=0}^{k+1} \binom{3k}{r}((4\log 3 + 3\log k)2^{2k+2+12\log^2(2k+2)+18\log(2k+2)}(2^{3k-r} + T_r))$$
$$= O^*(7.56^{3k}). \qquad \square$$

By Lemma 1 and Theorem 6, we can get the following corollary.

Corollary 2. *Weighted 3-Set Packing problem can be solved in* $O^*(7.56^{3k})$.

5   Conclusions

In this paper, we discuss how to construct a maximum weighted $(k+1)$-packing from a maximum weighted k-packing, so as to solve the weighted 3-Set Packing problem. After a further analysis the structure of the problem, we present a deterministic algorithm of time $O^*(10.6^{3k})$ for the weighted 3-Set Packing problem. Based on the randomized divide-and-conquer method, the above result can be further reduced to $O^*(7.56^{3k})$, which significantly improves the previous best result $O^*(12.8^{3k})$.

References

1. Alon, N., Yuster, R., Zwick, U.: Color-Coding. Journal of the ACM 42, 844–856 (1995)
2. Arkin, E., Hassin, R.: Approximating Weighted Set Packing by Local Search. Math. Oper.Res. 24, 640–648 (1998)
3. Bafna, V., Narayan, B., Ravi, R.: Nonoverlapping Local Alignments (Weighted Independent Sets of Axis-Parallel Rectangles). Discrete Appl. Math., 41–53 (1996)
4. Berman, P.: A $d/2$ Approximation for Maximum Weight Independent Set in d-claw Free Graphs. In: Halldórsson, M.M. (ed.) SWAT 2000. LNCS, vol. 1851, pp. 214–219. Springer, Heidelberg (2000)
5. Chandra, B., Halldorsson, M.M.: Greedy Local Improvement and Weighted Set Packing Approximation. Journal of Algorithms 39, 223–240 (2001)
6. Chandra, B., Halldorsson, M.M.: Approximating Weighted Set Packing by Local Search. Journal of Algorithms, 223–240 (2001)
7. Chen, J., Lu, S.: Improved Parameterized Set Splitting Algorithms: a Probabilistic Approach. Algorithmica (to appear, 2008)
8. Chen, J., Lu, S., Zhang, F.: Improved Algorithms for Path, Matching, and Packing Problems. In: Proc. of SODA 2007, pp. 298–307 (2007)
9. Downey, R., Fellows, M.: Parameterized Complexity. Springer, New York (1999)
10. Fellows, M.R., Knauer, C., Nishimura, N., Ragde, P., Rosamond, F., Stege, U., Thilikos, D., Whitesides, S.: Faster Fixed-Parameter Tractable Algorithms for Matching and Packing Problems. In: Albers, S., Radzik, T. (eds.) ESA 2004. LNCS, vol. 3221, pp. 311–322. Springer, Heidelberg (2004)
11. Liu, Y., Chen, J., Wang, J.: On Efficient FPT Algorithms for Weighted Matching and Packing Problems. In: Cai, J.-Y., Cooper, S.B., Zhu, H. (eds.) TAMC 2007. LNCS, vol. 4484, pp. 692–702. Springer, Heidelberg (2007)
12. Liu, Y., Lu, S., Chen, J., Sze, S.H.: Greedy Localization and Color-Coding: Improved Matching and Packing Algorithms. In: Bodlaender, H.L., Langston, M.A. (eds.) IWPEC 2006. LNCS, vol. 4169, pp. 84–95. Springer, Heidelberg (2006)
13. Naor, M., Schulman, L., Srinivasan, A.: Splitters and Near-Optimal Derandomization. In: FOCS 1995, pp. 182–190 (1995)
14. Wang, J., Feng, Q.: An $O^*(3.52^{3k})$ Parameterized Algorithm for 3-Set Packing. In: Proc. of TAMC 2008 (to appear, 2008)

Structural Identifiability in Low-Rank Matrix Factorization

Epameinondas Fritzilas[1], Yasmin A. Rios-Solis[1,*], and Sven Rahmann[2]

[1] Faculty of Technology, Bielefeld University, Germany
efritzil@cebitec.uni-bielefeld.de, yasmin@yalma.fime.uanl.mx
[2] Computer Science 11, Technische Universität Dortmund, Germany
Sven.Rahmann@tu-dortmund.de

Abstract. In many signal processing and data mining applications, we need to approximate a given matrix Y of "sensor measurements" over several experiments by a low-rank product $Y \approx A \cdot X$, where X contains source signals for each experiment, A contains source-sensor mixing coefficients, and *both* A and X are unknown. We assume that the only a-priori information available is that A must have zeros at certain positions; this constrains the source-sensor network connectivity pattern.

In general, different AX factorizations approximate a given Y equally well, so a fundamental question is how the connectivity restricts the solution space. We present a combinatorial characterization of uniqueness up to diagonal scaling, called *structural identifiability* of the model, using the concept of structural rank from combinatorial matrix theory.

Next, we define an optimization problem that arises in the need for efficient experimental design: to minimize the number of sensors while maintaining structural identifiability. We prove its NP-hardness and present a mixed integer linear programming framework with two cutting-plane approaches. Finally, we experimentally compare these approaches on simulated instances of various sizes.

1 Introduction

The problem of explaining observed signals as linear mixtures of hidden signals is long standing and appears in many different areas of science and engineering. Applications can be found, for example, in text mining [11], image processing [5] and bioinformatics [2]. We consider the following generic setting. There is a bipartite network of m (hidden) signal sources and n (observable) sensors that are monitored over k discrete time points. The sensors are, in general, non-specific, i.e., each one of them measures a mixture of signals from more than one source. We assume that the sensor measurements depend linearly on the source signals, but the mixing coefficients are unknown. However, from the connectivity of the source-sensor network we know which of these coefficients are *fixed* to zero. Given the sensor measurements, our task is to infer the source signals and the non-zero mixing coefficients. Clearly, this problem is inherently ill-posed: since both the source signals and the non-zero mixing coefficients are

---

* Current affiliation: Graduate Program of Systems Engineering - UANL, México.

X. Hu and J. Wang (Eds.): COCOON 2008, LNCS 5092, pp. 140–148, 2008.
© Springer-Verlag Berlin Heidelberg 2008

unknown, many equally good linear explanations may exist for a given set of sensor measurements.

In formal terms, we are given an $n \times k$ matrix $Y = (y_{it})$ that contains the measurements at each sensor i at each time point t. We want to express it as a low-rank product of an $n \times m$ matrix $A = (a_{ij})$ that contains the mixing coefficients between each sensor i and each signal source j, and an $m \times k$ matrix $X = (x_{jt})$ that contains the source signal intensities. We assume that $m \leq \min\{n, k\}$ (low-rank factorization), and that the sparsity structure of A is known a priori. This knowledge is modeled with a $0/1$ matrix $\mathcal{Z}$ (called *zero pattern*). We say that a real-valued matrix A *satisfies* a zero pattern $\mathcal{Z}$ (denoted by $A \lhd \mathcal{Z}$), if A and $\mathcal{Z}$ have the same dimensions and $\mathcal{Z}_{ij} = 0$ implies $A_{ij} = 0$.

Since an exact low-rank factorization is impossible in general, we seek a matrix pair (A, X) such that $A \lhd \mathcal{Z}$ and an appropriately chosen error measure $\|Y - AX\|$ (e.g. the Frobenius norm) is minimized. We refer to this generic optimization problem as *Approximate Matrix Factorization* (AMF). What distinguishes AMF from other low-rank matrix approximation methods, e.g. based on SVD, is the a priori knowledge of $\mathcal{Z}$.

Related Work and Our Contributions. In this paper we do not focus on the computation of a global minimum for AMF, whose specifics depend on the properties of the chosen norm and on any constraints, other than the zero pattern, that apply to the factors A and X [7]. Instead, we focus on the inherent non-identifiability of the model. Assume that we have obtained a solution (A, X) for AMF. Many other pairs $(\widehat{A}, \widehat{X})$ may exist that satisfy $\widehat{A}\widehat{X} = AX$ and $\widehat{A} \lhd \mathcal{Z}$, and are therefore indistinguishable from (A, X) in terms of how well they approximate Y.

The zero pattern $\mathcal{Z}$ restricts the $(\widehat{A}, \widehat{X})$-space. In Sect. 2, we characterize the zero patterns $\mathcal{Z}$ that, under mild non-degeneracy assumptions on A and X, make the AMF model identifiable up to diagonal scaling; that is, we introduce the notion of *structural identifiability* of a zero pattern. It is based on the concept of structural rank from combinatorial matrix theory and relies on maximum bipartite matchings.

The basic fact that $\mathcal{Z}$ restricts the $(\widehat{A}, \widehat{X})$-space, as it is illustrated in the example of Fig. 1, was recognized in the context of network component analysis, a method for studying bipartite gene regulatory networks [6,1]. However, these papers missed the connection to structural rank that makes it possible to cast the discussion in a purely combinatorial framework. This connection was recently recognized in [10], a work that was done independently from ours.

In Sect. 3, we define the MINSENSOR combinatorial optimization problem and prove its NP-hardness: Given is a set of signal sources and a set of designed but not yet manufactured sensors. Which is the minimum subset of sensors that we have to manufacture, in order to monitor all sources with an identifiable design? For the exact solution of MINSENSOR, we develop a mixed integer linear programming (MILP) framework with two different cutting-plane strategies in Sect. 4. Finally, in Sect. 5, we experimentally test our approach on simulated problem instances.

$$
\overbrace{\begin{pmatrix} 0 & a_{12} & a_{13} & 0 \\ 0 & 0 & 0 & a_{24} \\ 0 & 0 & 0 & a_{34} \\ a_{41} & 0 & a_{43} & 0 \\ a_{51} & a_{52} & 0 & 0 \\ 0 & a_{62} & 0 & a_{64} \end{pmatrix}}^{A}
\overbrace{\begin{pmatrix} r_{11} & r_{12} & r_{13} & r_{14} \\ r_{21} & r_{22} & r_{23} & r_{24} \\ r_{31} & r_{32} & r_{33} & r_{34} \\ r_{41} & r_{42} & r_{43} & r_{44} \end{pmatrix}}^{R}
\vartriangleleft
\overbrace{\begin{pmatrix} 0 & 1 & 1 & 0 \\ 0 & 0 & 0 & 1 \\ 0 & 0 & 0 & 1 \\ 1 & 0 & 1 & 0 \\ 1 & 1 & 0 & 0 \\ 0 & 1 & 0 & 1 \end{pmatrix}}^{\mathcal{Z}}
\quad \text{leads to}
$$

$$
\overbrace{\begin{pmatrix} a_{12} & a_{13} & 0 \\ 0 & 0 & a_{24} \\ 0 & 0 & a_{34} \\ a_{62} & 0 & a_{64} \end{pmatrix}}^{\widetilde{A}_1}
\begin{pmatrix} r_{21} \\ r_{31} \\ r_{41} \end{pmatrix} = \begin{pmatrix} 0 \\ 0 \\ 0 \\ 0 \end{pmatrix} \;,
\quad
\overbrace{\begin{pmatrix} 0 & 0 & a_{24} \\ 0 & 0 & a_{34} \\ a_{41} & a_{43} & 0 \end{pmatrix}}^{\widetilde{A}_2}
\begin{pmatrix} r_{12} \\ r_{32} \\ r_{42} \end{pmatrix} = \begin{pmatrix} 0 \\ 0 \\ 0 \end{pmatrix} \;,
$$

$$
\overbrace{\begin{pmatrix} 0 & 0 & a_{24} \\ 0 & 0 & a_{34} \\ a_{51} & a_{52} & 0 \\ 0 & a_{62} & a_{64} \end{pmatrix}}^{\widetilde{A}_3}
\begin{pmatrix} r_{13} \\ r_{23} \\ r_{43} \end{pmatrix} = \begin{pmatrix} 0 \\ 0 \\ 0 \\ 0 \end{pmatrix} \;,
\quad
\overbrace{\begin{pmatrix} 0 & a_{12} & a_{13} \\ a_{41} & 0 & a_{43} \\ a_{51} & a_{52} & 0 \end{pmatrix}}^{\widetilde{A}_4}
\begin{pmatrix} r_{14} \\ r_{24} \\ r_{34} \end{pmatrix} = \begin{pmatrix} 0 \\ 0 \\ 0 \end{pmatrix} .
$$

Fig. 1. An example of a zero pattern $\mathcal{Z}$, a general matrix $A \vartriangleleft \mathcal{Z}$, and the four linear systems derived from A, $\mathcal{Z}$, and the condition $AR \vartriangleleft \mathcal{Z}$

2   The Combinatorial Flavor of Zeros

Given a matrix pair (A, X) and a zero pattern $\mathcal{Z}$ with $A \vartriangleleft \mathcal{Z}$, we characterize the class of matrix pairs $(\widehat{A}, \widehat{X})$, such that $\widehat{A}\widehat{X} = AX$ and $\widehat{A} \vartriangleleft \mathcal{Z}$. The proof of the following lemma is omitted.

Lemma 1. *Let A and $\widehat{A}$ be $n \times m$ matrices and, X and $\widehat{X}$ be $m \times k$ matrices with $AX = \widehat{A}\widehat{X}$ and $rank(A) = rank(X) = m$. Then there exists an invertible $m \times m$ matrix R such that $\widehat{A} = AR$ and $\widehat{X} = R^{-1}X$.*

Under the full-rank assumption of the lemma, the entries of R are restricted by the condition $\widehat{A} = AR \vartriangleleft \mathcal{Z}$. Only zeros in $\mathcal{Z}$ induce constraints: $\mathcal{Z}_{ij} = 0$ means $\sum_\ell a_{i\ell}r_{\ell j} = 0$. Since $A \vartriangleleft \mathcal{Z}$, the diagonal elements of R are not constrained. This leads to m linear systems, one for each column of R. An example is shown in Fig. 1.

We denote the set of integers $\{1, \ldots, n\}$ with $[n]$. Given an $n \times m$ zero pattern $\mathcal{Z}$, we define $\mathcal{Z}_j^0 = \{i \in [n] : \mathcal{Z}_{ij} = 0\}$. Given an $n \times m$ matrix (zero pattern) M, a set of row indices $I \subseteq [n]$ and a set of column indices $J \subseteq [m]$, we denote the correspondingly indexed submatrix (subpattern) with $M[I, J]$; when $J = [m]$, we write $M[I, *]$ as a shortcut. Finally, for a zero pattern $\mathcal{Z}$, we write $\widetilde{\mathcal{Z}}_j$ as a shortcut for $\mathcal{Z}[\mathcal{Z}_j^0, [m] \setminus \{j\}]$. With this notation, the linear system that restricts the j-th column of R can be written as $\widetilde{A}_j \cdot R[[m] \setminus \{j\}, j] = \mathbf{0}$, where $\widetilde{A}_j = A[\mathcal{Z}_j^0, [m] \setminus \{j\}]$.

The column ranks of the matrices $\widetilde{A}_j$ determine the degrees of freedom of R. If $rank(\widetilde{A}_j) = m - 1$ for all j, then all systems only allow the zero solution and all non-diagonal elements of R are constrained to be zero. In that case, R is a diagonal matrix, and $\widehat{A}$ differs from A only by column scaling (and $\widehat{X}$ from X by row scaling). In many applications, normalization of A with respect to a certain row (or of X with respect to a certain column) does not affect the interpretation of the results and, in this sense, the solution of AMF is essentially unique.

The ranks of the submatrices $\widetilde{A}_j$ depend on both the structure of $\mathcal{Z}$ and on the values of the free a_{ij} entries. In the example of Fig. 1, let us note that the rank of $\widetilde{A}_2$ can be at most 2, for any values of a_{ij}. On the other hand, the ranks of $\widetilde{A}_1$, $\widetilde{A}_3$ and $\widetilde{A}_4$ can be 3 or smaller, depending on the numerical values of a_{ij}. While the rank degeneracy of $\widetilde{A}_2$ is due to the "bad structure" of $\mathcal{Z}$, only adversarily chosen values of A would make the other $\widetilde{A}_j$ rank degenerate. Below we formalize the notion of a "badly structured" zero pattern.

We represent a zero pattern $\mathcal{Z}$ as a bipartite graph $G(\mathcal{Z}) = (R \cup C, E)$, where the vertex partitions R and C correspond to the rows and columns of $\mathcal{Z}$, respectively, and the edges E correspond to the ones in $\mathcal{Z}$.

Definition 1 (e.g. [9]). *Let $\mathcal{Z}$ be an $n \times m$ zero pattern. The structural rank of $\mathcal{Z}$, denoted by srank($\mathcal{Z}$), is the size of a maximum matching in $G(\mathcal{Z})$. If srank($\mathcal{Z}$) $= m$, then $\mathcal{Z}$ is called* column-perfect.

Lemma 2 (e.g. [9]). *If $\mathcal{Z}$ is a zero pattern and A is a matrix that satisfies $\mathcal{Z}$, then rank(A) $\leq$ srank($\mathcal{Z}$). Furthermore, if the non-zero elements of A are algebraically independent, then rank(A) $=$ srank($\mathcal{Z}$).*

In our example, the structure of $\mathcal{Z}$ immediately gives a certificate for the non-identifiability of AMF. More specifically, srank($\widetilde{\mathcal{Z}}_2$) $= 2$ and this is a structural reason that makes rank($\widetilde{A}_2$) < 3 and the AMF model non-identifiable. In applications, we would like to avoid source-sensor networks that give rise to such inherently problematic zero patterns. This observation motivates the following definition.

Definition 2. *Let $\mathcal{Z}$ be a $n \times m$ zero pattern. We call its j-th column structurally identifiable (for simplicity just identifiable), if $\widetilde{\mathcal{Z}}_j$ is column-perfect. We call $\mathcal{Z}$ identifiable, if all its columns are identifiable.*

We can efficiently check if $\mathcal{Z}$ is identifiable, by computing a maximum matching for $G(\widetilde{\mathcal{Z}}_j)$, for all $j \in [m]$ [4]. Let us emphasize that a structurally identifiable zero pattern *does not guarantee* the uniqueness (up to scaling) of AMF's solution, but rather avoids its non-uniqueness. On the other hand, the possible values for the non-zeros of A that violate the algebraic independence requirement of Lemma 2 form a set of measure zero. Intuitively, this means that rank(A) $=$ srank($\mathcal{Z}$), if the numerical entries at the non-zero positions of A are "generic enough".

3   Most Parsimonious Selection of Sensors

Assume that we are given the design for a large bipartite source-sensor network modeled by a zero pattern $\mathcal{H}$. Although it is easy to check if the whole network is identifiable, it becomes more difficult if, due to economical or spatial constraints, we want to monitor the sources with the *minimum number of used sensors*.

Problem 1. MINSENSOR
Given a $N \times M$ zero pattern $\mathcal{H}$, find a *minimum* subset of row indices $I \subseteq [N]$, such that the subpattern $\mathcal{Z} = \mathcal{H}[I, *]$ is identifiable.

We prove that MINSENSOR is NP-hard with a reduction from HITTING SET.

Problem 2. HITTING SET [3]
Given a family $\mathcal{F}$ of subsets of a finite set S and a positive integer $\widehat{k} \leq |S|$, is there a subset $S' \subseteq S$, with $|S'| \leq \widehat{k}$, that intersects all members of $\mathcal{F}$?

Theorem 1. *The decision version of MINSENSOR is NP-complete.*

Proof. Clearly, the decision version of MINSENSOR is in NP. Let $(\mathcal{F}, S, \widehat{k})$ be an arbitrary instance of HITTING SET with $\mathcal{F} = \{C_1, \ldots, C_m\}$; without loss of generality we assume that $S = [n]$. At first, we construct a $n \times m$ zero pattern $\widehat{\mathcal{H}}$, such that $\widehat{\mathcal{H}}_{ij} = 0$ iff $i \in C_j$. Then, HITTING SET is equivalent to asking if there is a subset of row indices $\widehat{I} \subseteq [n]$, with $|\widehat{I}| \leq \widehat{k}$, such that $\widehat{\mathcal{H}}[\widehat{I}, *]$ contains at least one zero per column.

Now, we construct a zero pattern $\mathcal{H}$ as follows: $\mathcal{H}$ has in total $2m$ columns $(c_1, \ldots, c_m, a_1, \ldots, a_m)$ and $2m^2 - m + n$ rows divided in $m+2$ blocks $(\mathcal{A}, \mathcal{B}_1, \ldots, \mathcal{B}_m, \mathcal{C})$. Block $\mathcal{A}$ consists of two $n \times m$ sub-blocks: $\mathcal{A}_1 = \widehat{\mathcal{H}}$ and $\mathcal{A}_2 = \mathbf{1}$. For $j \in [m]$, block $\mathcal{B}_j$ consists of two identical $(2m - 2) \times m$ sub-blocks, filled with ones except for the j-th column which is all zero. Finally, block $\mathcal{C}$ consists of two $m \times m$ sub-blocks: $\mathcal{C}_1 = \mathbf{1}$ and $\mathcal{C}_2 = \mathbf{1} - I$. For notational convenience we denote with L the subset of rows from index $n + 1$ to the end of $\mathcal{H}$. By construction, $\widehat{\mathcal{H}}$ has a row subset $\widehat{I}$, with $|\widehat{I}| \leq \widehat{k}$, such that $\widehat{\mathcal{H}}[\widehat{I}, *]$ contains at least one zero per column, iff $\mathcal{H}$ has a row subset I, with $|I| \leq \widehat{k} + |L|$, such that $\mathcal{H}[I, *]$ is identifiable.

Forward: For all columns c_j there exists a row index $\widehat{i}_j \in \widehat{I}$ with $c_j[\widehat{i}_j] = 0$. Let us consider $I = \widehat{I} \cup L$; clearly, $|I| \leq \widehat{k} + |L|$. We also show that $\mathcal{H}[I, *]$ is identifiable. For column c_j we consider the subpattern $\widetilde{\mathcal{H}}_{c_j}$: in $\widetilde{\mathcal{H}}_{c_j}$ all columns, except for a_j, can be matched to the rows of block $\mathcal{B}_j$ and column a_j is matched to row $\widehat{i}_j$. For column a_j, we consider the subpattern $\widetilde{\mathcal{H}}_{a_j}$: in $\widetilde{\mathcal{H}}_{a_j}$ all columns, except for c_j, can be matched to the rows of block $\mathcal{B}_j$ and column a_j is matched to the j-th row of block $\mathcal{C}$.

Reverse: A necessary condition for identifiability is that each column of $\mathcal{H}[I, *]$ contains at least $2m - 1$ zeros. For all columns a_j to contain $2m - 1$ zeros, it must be $I \supseteq L$. Since the columns c_j have exactly $2m - 2$ zeros in $\mathcal{H}[L, *]$, there must exist an $\widehat{I} \subseteq [n]$, such that $I = L \cup \widehat{I}$ and $\widehat{I}$ covers each column c_j with at least one zero. Clearly, $|\widehat{I}| \leq \widehat{k}$. $\qquad\square$

MILP 1. A complete formulation for MINSENSOR

$$\text{minimize:} \quad \sum_{i=1}^{N} y_i$$

subject to:

$$y_i \in \{0, 1\} \qquad \forall i \in [N] \tag{1}$$

$$z_{ij}^s \geq 0 \qquad \forall s \in [M], \forall (i, j) \in E^s \tag{2}$$

$$z_{ij}^s \leq y_i \qquad \forall s \in [M], \forall (i, j) \in E^s \tag{3}$$

$$\sum_{j:(i,j)\in E^s} z_{ij}^s \leq 1 \qquad \forall s \in [M], \forall i \in [N] \tag{4}$$

$$\sum_{i:(i,j)\in E^s} z_{ij}^s = 1 \qquad \forall s \in [M], \forall j \in [M] \setminus \{j\} \tag{5}$$

4   MILP Framework for MINSENSOR

We develop a mixed integer linear program (MILP) for the exact solution of MINSENSOR. For all $s \in [M]$, we define $E^s := \{(i, j) \in [N] \times [M] : (\mathcal{H}_{is} = 0) \wedge (j \neq s) \wedge (\mathcal{H}_{ij} = 1)\}$. Let us note that E^s is the edge set of $G(\widetilde{\mathcal{Z}}_s)$. MILP 1 contains N binary variables y_i, such that $y_i = 1$, if and only if $i \in I$. It also contains $\mathcal{O}(dNM^2)$ *real*-valued variables z_{ij}^s, in order to guarantee the identifiability of the subpattern $\mathcal{Z} = \mathcal{H}[I, *]$, where d is the fraction of ones in $\mathcal{H}$. For a column $s \in [M]$, the variables z_{ij}^s and the corresponding group of constraints (2)-(5) can be seen as an independent "unit" that guarantees the identifiability of the s-th column of $\mathcal{Z}$.

Correctness. A vector y defines a row subset I such that $\mathcal{Z} = \mathcal{H}[I, *]$ is identifiable, if and only if there exists a vector z such that (y, z) is a feasible solution of MILP 1.

Proof. At first, we prove that there exists a z, such that (y, z) is feasible iff there exists a *binary* $\widehat{z}$, such that $(y, \widehat{z})$ is feasible. The reverse direction is obvious; for the forward direction we need the concept of total unimodularity and some of its properties that can be found, e.g., in [12].

If (y, z) is feasible, then y is binary and z lies in the polyhedron $P(y)$ defined by (2)-(5), $P(y) := \left\{ z : 0 \leq z \leq y \text{ and } \begin{pmatrix} A \\ B \\ -B \end{pmatrix} z \leq \begin{pmatrix} 1 \\ 1 \\ -1 \end{pmatrix} \right\}$, where the matrices A and B contain the coefficients of the constraints (4) and (5), respectively. We observe that each variable z_{ij}^s appears with coefficient 1 exactly once in (4) and exactly once in (5); therefore, the matrix $\begin{pmatrix} A \\ B \end{pmatrix}$ is totally unimodular (TU). Then, it follows that the matrix defining $P(y)$ is also TU. Since $P(y)$ is non-empty and bounded, it has at least one vertex $\widehat{z}$. Because the matrix of $P(y)$ is TU and y is integer, $\widehat{z}$ is also integer. Furthermore, (2) and (4) imply that $\widehat{z}_{ij}^s \in \{0, 1\}$ for all (s, i, j). Therefore, $(y, \widehat{z})$ is a binary feasible solution of MILP 1.

Now, we show that y defines a row subset I, such that $\mathcal{Z} = \mathcal{H}[I, *]$ is identifiable, iff there exists a binary $\widehat{z}$, such that $(y, \widehat{z})$ is feasible. Let us fix a column $s \in [M]$ and consider the subpatterns $\widetilde{\mathcal{H}}_s$ and $\widetilde{\mathcal{Z}}_s$. The bipartite graph $G(\widetilde{\mathcal{H}}_s)$ has edge set E^s and for each $(i, j) \in E^s$, MILP 1 contains a variable z_{ij}^s. Furthermore, $G(\widetilde{\mathcal{Z}}_s)$ is a subgraph of $G(\widetilde{\mathcal{H}}_s)$: the row-vertex i and its incident edges (i, j) appear in $G(\widetilde{\mathcal{Z}}_s)$, iff $i \in I$. Due to (3), z_{ij}^s can be non-zero, iff the edge (i, j) appears in $G(\widetilde{\mathcal{Z}}_s)$. $\mathcal{Z}$ is identifiable, iff, for all $s \in [M]$, there exists a subset of edges in $G(\widetilde{\mathcal{Z}}_s)$ that covers all column-vertices exactly once and all row-vertices at most once. This happens, iff there is a 0/1 assignment of the variables z_{ij}^s that satisfies (3)-(5).   □

In practice, instead of solving the full-fledged MILP 1 at once, we can adopt a cutting-plane approach. We first solve a relaxation of MINSENSOR (called MINSENSOR-REL), where the requirement of $\mathcal{Z}$ being identifiable is replaced by two necessary (but not sufficient) conditions. Namely, $\mathcal{Z}$ must contain at least $M - 1$ zeros per column (C1) and, for all $(j_1, j_2) \subseteq [M]^2$ with $j_1 \neq j_2$, it must be $\mathcal{Z}_{j_1}^0 \not\subseteq \mathcal{Z}_{j_2}^0$ (C2). An argument similar to that of Sect. 3 easily implies that MINSENSOR-REL is also NP-hard. The corresponding ILP (not shown here) contains M^2 constraints, involving the y_i variables. Given a solution $\mathcal{Z}$ of MINSENSOR-REL, we can efficiently check a posteriori if all columns of $\mathcal{Z}$ are identifiable. If this is the case, then MINSENSOR has been optimally solved. Otherwise, for the non-identifiable columns, we add the corresponding variables z_{ij}^s and constraints (2)-(5) and reiterate. In principle, it is possible to iterate up to M times, each time solving a larger MILP.

As an alternative practical heuristic, we can add cutting-planes based on Hall's Theorem. Let $\mathcal{Z}$ be a $n \times m$ zero pattern, $X \subset [m]$ be a subset of columns and $s \notin X$ be a single column. With $N_s(X)$ we denote the set of neighbors of X in $G(\widetilde{\mathcal{Z}}_s)$. Then, from Hall's Theorem, the s-th column of $\mathcal{Z}$ is identifiable, iff $|N_s(X)| \geq |X|$ for all $X \subseteq [m] \setminus \{s\}$. If the s-th column is not identifiable, we can efficiently find a subset $X \subseteq [M] \setminus \{s\}$ that minimizes the surplus $|N_s(X)| - |X|$ [8]. For this X, which "maximally" violates Hall's condition, we add the constraints (6)-(11) (correctness proof omitted) to the MILP and reiterate. Theoretically, this approach can generate exponentially many constraints and iterations.

$$0 \leq \gamma_i^{(s,X)} \leq 1 \qquad \forall i \in [N] \tag{6}$$

$$\gamma_i^{(s,X)} \geq y_i \qquad \forall i \in [N] : (\mathcal{H}_{is} = 0) \wedge (\exists j \in X : \mathcal{H}_{ij} = 1) \tag{7}$$

$$\gamma_i^{(s,X)} \leq y_i \qquad \forall i \in [N] \tag{8}$$

$$\gamma_i^{(s,X)} \leq 1 - \mathcal{H}_{is} \qquad \forall i \in [N] \tag{9}$$

$$\gamma_i^{(s,X)} \leq \sum_{j \in X} \mathcal{H}_{ij} \qquad \forall i \in [N] \tag{10}$$

$$\sum_{i=1}^{N} \gamma_i^{(s,X)} \geq |X| \tag{11}$$

Correctness. $(y, \gamma^{(s,X)})$ satisfies (6)-(11), iff $|N_s(X)| \geq |X|$ in $\mathcal{Z}$.

5   Computational Experiments

We generated zero patterns with M columns (sources), N rows (sensors), and density d (fraction of ones), with 24 different parameter combinations: (M, N, d) $\in \{30, 60, 90\} \times \{2M, 3M\} \times \{10\%, 20\%, 30\%, 40\%\}$. For each combination, 100 random patterns are generated by setting each element in an $N \times M$ binary matrix to 1 with probability d independently. The integer programs are solved on a 64-bit Sparc machine (450 MHz, 4 GB of RAM), using the AMPL/CPLEX software platform (version 10.2) with default CPLEX parameters, and an upper limit of 20 CPU minutes for the solution of a single MILP. The ILPs are simplified by the presolver of AMPL, before being sent to the optimization engine.

We use the number of cutting-plane (outer) iterations and simplex (inner) iterations, and the final MILP size (number of variables and number of constraints), averaged over the solved instances for a given parameter combination, to compare the two cutting-plane approaches (Table 1).

In both cases, we first try to solve the relaxed MILP and add the matching constraints on demand, only if they are violated. In most cases, the relaxed

Table 1. Each cell corresponds to a certain parameter combination (M, N, d). In each cell, the first row shows the average number of cutting-plane iterations (ci) and the average total number of simplex iterations (si). The second row shows the average final number of variables (v) and constraints (c). The number of instances solved within the time limit is shown in parentheses (out of 100, all solved if number not given).

Straightforward Cutting Plane Approach					
	M=30		M=60		M=90	
d (%)	N=60	N=90	N=120	N=180	N=180	N=270
10	0.98ci 5890si	1.09ci 21694si	0.05ci 906si	0.08ci 4674si	0ci 23640si	0ci 35781si
	2113v 3461c	5571v 8483c	362v 3880c	1544v 5283c	180v 8100c	270v 8100c
					(95)	(71)
20	0.01ci 101si	0.08ci 2054si	0ci 1490si	0ci 28887si	0ci 32332si	0ci 44335si
	79v 86c	339v 1208c	120v 3600c	180v 3600c	180v 8100c	270v 8100c
				(95)	(77)	(55)
30	0ci 45si	0ci 129si	0ci 858si	0ci 57413si	0ci 22644si	0ci 49805si
	60v 892c	90v 900c	120v 3600c	180v 3600c	180v 8100c	270v 8100c
				(80)	(86)	(17)
40	0ci 23si	0ci 197si	0ci 61si	0ci 65078si	0ci 4144si	0ci 42516si
	45v 517c	90v 900c	103v 2623c	180v 3600c	175v 7625c	270v 8100c
				(76)		(11)

Cutting Plane Approach Based on Hall's Theorem			
	M=30		M=60	
d (%)	N=60	N=90	N=120	N=180
10	2.42ci 282si	5.98ci 1076si	0.05ci 225si	0.08ci 1467si
	918v 2364c	4725v 10208c	135v 3608c	344v 3930c
20	0.01ci 70si	0.08ci 122si		
	66v 883c	125v 971c		

We list only the cases, where the addition of any cutting-planes was necessary.

MILP returns an exact solution immediately. For the instances that are not immediately solved, a few cutting-plane iterations are enough in practice. This observation is in accordance with our intuition: In a *dense enough, randomly generated* zero pattern, if $\mathcal{Z}$ contains enough ($\geq M-1$) zeros per column, then most probably large matchings exist. Comparing the two cutting-plane approaches, we observe that the one based on Hall's Theorem takes more cutting-plane iterations, but finally leads to smaller ILPs. Of course, MINSENSOR-REL is itself a hard problem. As we face bigger and more dense zero patterns, moving towards the lower-right corner of Table 1, its solution is more and more computationally intensive and in several cases cannot be achieved within the time limit.

Acknowledgments. We thank Martin Milanič for his comments. E. Fritzilas is supported by the NRW International Graduate School in Bionformatics and Genome Research.

References

1. Boscolo, R., Sabatti, C., Liao, J., Roychowdhury, V.: A Generalized Framework for Network Component Analysis. IEEE Trans. Comp. Biol. Bioinf. 2, 289–301 (2005)
2. Brunet, J.P., Tamayo, P., Golub, T.R., Mesirov, J.P.: Metagenes and Molecular Pattern Discovery Using Matrix Factorization. PNAS 101, 4164–4169 (2004)
3. Garey, M., Johnson, D.: Computers and Intractability: A Guide to the Theory of NP-Completeness. W.H. Freeman and Company, New York (1979)
4. Hopcroft, J., Karp, R.: An $n^{5/2}$ Algorithm for Maximum Matchings in Bipartite Graphs. SIAM J. Comput. 2, 225–231 (1973)
5. Lee, D., Seung, H.: Learning the Parts of Objects by Non-negative Matrix Factorization. Nature 401, 788–791 (1999)
6. Liao, J., Boscolo, R., Yang, Y.L., Tran, L., Sabatti, C., Roychowdhury, V.: Network Component Analysis: Reconstruction of Regulatory Signals in Biological Systems. PNAS 100, 15522–15527 (2003)
7. Lin, C.J.: Projected Gradient Methods for Non-Negative Matrix Factorization. Neural Computation 19, 2756–2779 (2007)
8. Lovasz, L., Plummer, M.: Matching Theory. North-Holland, Amsterdam (1986)
9. Murota, K.: Matrices and Matroids for Systems Analysis. Springer, Heidelberg (2000)
10. Narasimhan, S., Rengaswamy, R., Vadigepalli, R.: Structural Properties of Gene Regulatory Networks: Definitions and Connections. IEEE Trans. Comp. Biol. Bioinf. (accepted, 2007)
11. Shahnaz, F., Berry, M., Pauca, V., Plemmons, R.: Document Clustering Using Nonnegative Matrix Factorization. Inf. Proc. & Manag. 42, 373–386 (2006)
12. Wolsey, L.: Integer Programming. Wiley Interscience, Chichester (1998)

Complexity of Counting the Optimal Solutions

Miki Hermann[1] and Reinhard Pichler[2]

[1] LIX (CNRS, UMR 7161), École Polytechnique, 91128 Palaiseau, France
hermann@lix.polytechnique.fr
[2] Institut für Informationssysteme, Technische Universität Wien,
A-1040 Wien, Austria
pichler@dbai.tuwien.ac.at

Abstract. Following the approach of Hemaspaandra and Vollmer, we can define counting complexity classes $\#\cdot\mathcal{C}$ for any complexity class $\mathcal{C}$ of decision problems. In particular, the classes $\#\cdot\Pi_k P$ with $k \geq 1$ corresponding to all levels of the polynomial hierarchy have thus been studied. However, for a large variety of counting problems arising from optimization problems, a precise complexity classification turns out to be impossible with these classes. In order to remedy this unsatisfactory situation, we introduce a hierarchy of new counting complexity classes $\#\cdot\mathrm{Opt}_k P$ and $\#\cdot\mathrm{Opt}_k P[\log n]$ with $k \geq 1$. We prove several important properties of these new classes, like closure properties and the relationship with the $\#\cdot\Pi_k P$-classes. Moreover, we establish the completeness of several natural counting complexity problems for these new classes.

1 Introduction

Many natural decision problems are known to be complete for the class $\Theta_k P = \Delta_k P[\log n]$, defined by Wagner in [19], or for $\Delta_k P$. In particular, they often occur in variants of $\Sigma_{k-1}P$-complete problems when cardinality-minimality or weight-minimality (or, likewise, cardinality-maximality or weight-maximality) is imposed as an additional constraint. Two prototypical representatives of such problems are as follows (The completeness of these problems in $\Theta_2 P$ and $\Delta_2 P$ is implicit in [6]).

Problem: MIN-CARD-SAT (MIN-WEIGHT-SAT)
Input: A propositional formula φ in conjunctive normal form over variables X (together with a weight function $w: X \to \mathbb{N}$) and a subset of variables $X' \subseteq X$.
Question: Are X' set to true in some cardinality-minimal (weight-minimal) model of φ?

A straightforward $\Theta_2 P$-algorithm for MIN-CARD-SAT computes the minimum cardinality of the models of φ by means of logarithmically many calls to an NP-oracle, asking questions of the type "Does φ have a model of size $\leq k$?". As soon as the minimum cardinality k_0 is known, we can proceed by a simple NP-algorithm, checking if the subset X' is true in some model of size k_0. Analogously, a $\Delta_2 P$-algorithm for MIN-WEIGHT-SAT first computes the minimum weight of all models of φ. In any reasonable representation, the weights are exponential with respect to their representation (e.g., they are represented in

X. Hu and J. Wang (Eds.): COCOON 2008, LNCS 5092, pp. 149–159, 2008.

binary notation). Hence, the straightforward algorithm for computing the minimum weight needs logarithmically many calls to an NP-oracle with respect to the total weight of all variables. This comes down to polynomially many calls with respect to the representation of the weights.

Note that the membership in $\Theta_2 P$ and $\Delta_2 P$ recalled above is in great contrast to *subset-minimality*, i.e., minimality with respect to set inclusion (or, likewise, *subset-maximality*), which often raises the complexity one level higher in the polynomial hierarchy. E.g., the following problem is well-known to be $\Sigma_2 P$-complete (cf. [13]).

Problem: MIN-SAT
Input: A propositional formula φ in conjunctive normal form over variables X and a subset $X' \subseteq X$.
Question: Are X' set to true in some subset-minimal model of φ?

As far as the complexity of the corresponding *counting problems* is concerned, only the counting problem corresponding to MIN-SAT has been satisfactorily classified so far. The following problem was shown to be #·coNP-complete in [3]: Given a propositional formula φ in conjunctive normal form, how many subset-minimal models does φ have? On the other hand, the counting complexity of the remaining aforementioned problems has remained obscure. The main goal of this paper is to introduce new counting complexity classes #·OptP and #·OptP[$\log n$], needed to pinpoint the precise complexity of these and many similar optimality counting problems. We will also show the relationship of these new classes with respect to the known classes in the counting hierarchy. Moreover, we will show that these new classes are not identical to already known ones, unless the polynomial hierarchy collapses. Finally, we will present several natural optimization counting problems, which turn out to be complete for one or the other introduced counting class. The definition of new natural counting complexity classes is by no means limited to the first level of the polynomial hierarchy. Indeed, we will show how the counting complexity classes #·OptP and #·OptP[$\log n$] can be generalized to #·Opt$_k$P and #·Opt$_k$P[$\log n$] for arbitrary $k \geq 1$ with #·OptP = #·Opt$_1$P and #·OptP[$\log n$] = #·Opt$_1$P[$\log n$].

Due to lack of space, proofs had to be omitted in this paper. A full version with detailed proofs of all results presented here is provided as a technical report.

2   Preliminaries

We recall the necessary concepts and definitions, but we assume that the reader is familiar with the basic notions in computational counting complexity. For more information, the interested reader is referred to Chapter 18 in the book [13] or the survey [4].

The study of *counting problems* was initiated by Valiant in [17], [18]. While decision problems ask if at least one solution of a given problem instance exists, counting problems ask for the number of different solutions. The most intensively studied counting complexity class is #P, which denotes the functions that count the number of accepting paths of a non-deterministic polynomial-time Turing machine. In other words, #P captures the counting problems corresponding to decision problems contained in NP. By allowing the non-deterministic polynomial-time Turing machine access to an oracle in NP, $\Sigma_2 P$, ..., we can define an infinite hierarchy of counting complexity classes.

Alternatively, a *counting problem* is presented using a suitable *witness* function which for every input x, returns a set of *witnesses* for x. Formally, a *witness* function is a function $A: \Sigma^* \to \mathcal{P}^{<\omega}(\Gamma^*)$, where Σ and Γ are two alphabets, and $\mathcal{P}^{<\omega}(\Gamma^*)$ is the collection of all finite subsets of Γ^*. Every such witness function gives rise to the following *counting problem*: given a string $x \in \Sigma^*$, find the cardinality $|A(x)|$ of the *witness* set $A(x)$. According to [7], if $\mathcal{C}$ is a complexity class of decision problems, we define #·$\mathcal{C}$ to be the class of all counting problems #·A whose witness function A satisfies the following conditions.

1. There is a polynomial $p(n)$ such that for every x and every $y \in A(x)$, we have that $|y| \le p(|x|)$, where $|x|$ is the length of x and $|y|$ is the length of y;
2. The decision problem "given x and y, is $y \in A(x)$?" is in $\mathcal{C}$.

It is easy to verify that #P = #·P. The counting hierarchy is ordered by linear inclusion [7]. In particular, we have that #P $\subseteq$ #·coNP $\subseteq$ #·Π_2P $\subseteq$ #·Π_3P, etc. Analogously, one can define the classes #·NP, #·Σ_2P, #·Σ_3P, etc. Toda and Ogiwara [15] determined the precise relationship between these classes as follows: #·Σ_kP $\subseteq$ #·$P^{\Sigma_k P}$ = #·Π_kP. Since the identity #·$P^{\Sigma_k P}$ = #·Δ_{k+1}P trivially holds, Toda and Ogiwara showed that there are no new Δ-classes in the counting hierarchy.

The prototypical #·Π_kP-complete problem for $k \in \mathbb{N}$ is #·Π_kSAT [3], defined as follows. Given a formula $\psi(X) = \forall Y_1 \exists Y_2 \cdots Q Y_k \; \varphi(X, Y_1, \ldots, Y_k)$, where φ is a Boolean formula and $X, Y_1, \ldots, Y_k$ are sets of propositional variables, count the number of truth assignments to the variables in X that satisfy ψ.

Completeness of counting problems is usually proved by means of polynomial-time Turing reductions, also called Cook reductions. However, these reductions do not preserve the counting classes #·Π_kP [16]. Hence, *parsimonious reductions* are usually considered instead. Consider two counting problems #·$A: \Sigma^* \to \mathbb{N}$ and #·$B: \Sigma^* \to \mathbb{N}$. We say that #·$A$ reduces to #·B via a parsimonious reduction if there exists a polynomial-time function $f \in$ FP, such that for each $x \in \Sigma^*$ we have #·$A(x)$ = #·$B(f(x))$. Parsimonious reductions are a special case of Karp reductions with a one-to-one relation between solutions for the corresponding instances of the problems #·A and #·B. However, parsimonious reductions are not always strong enough to prove completeness of well-known problems in counting complexity classes. E.g., the problem #POSITIVE 2SAT [2], [18] of counting satisfying assignments to a propositional formula with positive literals only and with two literals per clause cannot be #P-complete under parsimonious reductions, unless P = NP. Therefore Durand, Hermann, and Kolaitis [3] generalized parsimonious reductions to *subtractive reductions* and showed that all the classes #·Π_kP are closed under them. Subtractive reductions are defined as follows. The counting problem #·A reduces to #·B

via a *strong subtractive reduction* if there exist two polynomial-time computable functions f and g such that for each $x \in \Sigma^*$ we have $B(f(x)) \subseteq B(g(x))$ and $|A(x)| = |B(g(x))| - |B(f(x))|$.

A *subtractive reduction* is a composition (transitive closure) of a finite sequence of strong subtractive reductions. Thus, a *parsimonious* reduction corresponds to the special case of a strong subtractive reduction with $B(f(x)) = \emptyset$. In [3], subtractive reductions have been shown to be strong enough to prove completeness of many interesting problems in #P and other counting classes, but their power remains tame enough to preserve several interesting counting classes between #P and #PSPACE.

3   Optimization Counting Complexity Classes

Recall that, according to [7], a counting complexity class $\#{\cdot}\mathcal{C}$ can in principle be defined for any decision complexity class $\mathcal{C}$. However, as far as the polynomial hierarchy is concerned, this definition does not yield the desired diversity of counting complexity classes. In fact, if we simply consider $\#{\cdot}\mathcal{C}$ for either $\mathcal{C} = \Delta_k \mathrm{P}$ or $\mathcal{C} = \Theta_k \mathrm{P}$, then we do not get any new complexity classes, since the relationship $\#{\cdot}\Theta_k \mathrm{P} = \#{\cdot}\Delta_k \mathrm{P} = \#{\cdot}\Pi_{k-1}\mathrm{P}$ is an immediate consequence of the aforementioned result by Toda and Ogiwara [15].

Hence a different approach is necessary if we want to obtain a more fine grained stratification of the counting hierarchy. For this reason we introduce in the sequel the counting classes $\#{\cdot}\mathrm{Opt}_k\mathrm{P}[\log n]$ and $\#{\cdot}\mathrm{Opt}_k\mathrm{P}$ for each $k \in \mathbb{N}$, which will be appropriate for optimization counting problems. Of special interest will be the classes $\#{\cdot}\mathrm{OptP}[\log n] = \#{\cdot}\mathrm{Opt}_1\mathrm{P}[\log n]$ and $\#{\cdot}\mathrm{OptP} = \#{\cdot}\mathrm{Opt}_1\mathrm{P}$. We will define the new counting complexity classes via the nondeterministic transducer model (see [14]), as well as by an equivalent predicate based definition following the approach from [7]. The following definition generalizes the definition of nondeterministic transducers [14] to oracle machines.

Definition 1. *A* nondeterministic transducer *M is a nondeterministic polynomial-time bounded Turing machine, such that every accepting path writes a binary number. If M is equipped with an oracle from the complexity class $\mathcal{C}$, then it is called a* nondeterministic transducer with $\mathcal{C}$-oracle. *A $\Sigma_k\mathrm{P}$-transducer M is a nondeterministic transducer with a $\Sigma_{k-1}\mathrm{P}$ oracle. We identify nondeterministic transducers without oracle and $\Sigma_1\mathrm{P}$-transducers.*

For $x \in \Sigma^*$, we write $\mathrm{opt}_M(x)$ to denote the optimal value*, which can be either the* maximum *or the* minimum*, on any accepting path of the computation of M on x. If no accepting path exists then $\mathrm{opt}_M(x)$ is undefined.*

The above definition of a nondeterministic transducer is similar to a metric Turing machine defined in [10] and its generalization in [11]. However, our definition deviates from the machine models in [10], [11] in the following aspects:

1. We take the optimum value only over the *accepting* paths, while in [10] every path is accepting. Our ultimate goal is to count the number of optimal

solutions. Hence, above all, the objects that we want to count have to be *solutions*, i.e., correspond to an accepting computation, and only in the second place we are interested in the optimum.

2. In [10], only the maximum value is considered and it is mentioned that the minimum value is treated analogously. We prefer to make the applicability both to max and min explicit. The definition of the counting complexity classes below is not affected by this distinction.

3. In [11], NP-metric Turing machines were generalized to higher levels of the polynomial hierarchy by allowing alternations of minimum and maximum computations. However, for our purposes, in particular for the predicate-based characterization of the counting complexity classes below, the generalization via oracles is more convenient. Proving the equivalence of the two kinds of generalizations is straightforward.

It will be clear in the sequel that the generalization of nondeterministic transducers [14] to oracle machines is exactly the model we need. A similar idea but with a deterministic Turing transducer was used by Jenner and Torán in [8] to characterize the functional complexity classes $FP_{\parallel}^{NP}$, $FP_{\log}^{NP}$, and $FL_{\log}^{NP}$.

Definition 2. *We say that a counting problem $\#{\cdot}A\colon \Sigma^* \to \mathbb{N}$ is in the class $\#{\cdot}Opt_k P$ for some $k \in \mathbb{N}$, if there is a $\Sigma_k P$-transducer M, such that $\#{\cdot}A(x)$ is the number of accepting paths of the computation of M on x yielding the optimum value $opt_M(x)$. If no accepting path exists then $\#{\cdot}A(x) = 0$. If the length of the binary number written by M is bounded by $O(z(|x|))$ for some function $z(n)$, then $\#{\cdot}A$ is in the class $\#{\cdot}Opt_k P[z(n)]$.*

In this paper, we are only interested in $\#{\cdot}Opt_k P[z(n)]$ for two types of functions $z(n)$, namely the polynomial function $z(n) = n^{O(1)}$ and the logarithmic function $z(n) = \log n$. Clearly, $\#{\cdot}Opt_k P$ is the same as $\#{\cdot}Opt_k P[n^{O(1)}]$.

Distinguishing between max and min gives no additional computational power, as it is formalized by the following result.

Proposition 1. *Suppose that some counting problem $\#{\cdot}A\colon \Sigma^* \to \mathbb{N}$ is defined in terms of a $\Sigma_k P$-transducer M with the optimum being the maximum (minimum). Then there exists a parsimonious reduction to a counting problem $\#{\cdot}A'$ defined via a $\Sigma_k P$-transducer M' with the optimum value corresponding to the minimum (maximum).*

Krentel defined in [10] the class $OptP[z(n)]$ of optimization problems for a given function $z(n)$. He showed that $OptP[z(n)]$ essentially corresponds to $FP^{NP[z(n)]}$ (see also [9]). More precisely, for every "smooth" function[1] $z(n)$ (see [10]) we have $OptP[z(n)] \subseteq FP^{NP[z(n)]}$ and every function $f \in FP^{NP[z(n)]}$ can be represented as an $OptP[z(n)]$-problem followed by a polynomial-time function h. This correspondence between $OptP[z(n)]$ and $FP^{NP[z(n)]}$ can be generalized as

---

[1] A function $f\colon \mathbb{N} \to \mathbb{N}$ is *smooth* if it is nondecreasing and its unary representation is computable in polynomial time.

follows: replacing the $\Sigma_{k-1}P$ oracle in a $\Sigma_k P$-transducer by a $\Delta_k P$ oracle does not increase the expressive power.

We show next that the definition of $\#\cdot\mathrm{Opt}_k P[z(n)]$ via Turing machines (see Definition 1) has an equivalent definition via predicates. The basic idea is to decompose the computation of a $\Sigma_k P$-transducer M into a predicate B, which associates inputs x with computations y, and a function f which computes the number written by the transducer M following the computation path y.

Theorem 1. *For any function $z(n)$, a counting problem $\#\cdot A\colon \Sigma^* \to \mathbb{N}$ is in the class $\#\cdot\mathrm{Opt}_k P[z(n)]$ if and only if there exist an alphabet Γ, a predicate B on $\Sigma^* \times \Gamma^*$, and a polynomial-time computable function $f\colon \Gamma^* \to \mathbb{N}$ satisfying the following conditions.*

  (i) There is a polynomial $p(n)$ such that every pair of strings $(x, y) \in B$ satisfies the relation $|y| \leq p(|x|)$;
  (ii) The predicate B is decidable by a $\Delta_k P$ algorithm;
  (iii) The length of $f(y)$ is bounded by $O(z(|x|))$ for every (x, y);
  (iv) $\mathrm{opt}_f^B(x) = \mathrm{opt}(\{f(y) \mid (x, y) \in B\})$ with $\mathrm{opt} \in \{\max, \min\}$;
  (v) $A(x) = \{y \mid (x, y) \in B \wedge f(y) = \mathrm{opt}_f^B(x)\}$.

As far as complete problems for these new complexity classes are concerned, we propose the following natural generalizations of minimum cardinality and minimum weight counting satisfiability problems to quantified Boolean formulas.

Problem: $\#\mathrm{Min\text{-}Card\text{-}\Pi_k SAT}$ ($\#\mathrm{Min\text{-}Weight\text{-}\Pi_k SAT}$)
Input: A $\Pi_k \mathrm{SAT}$ formula $\psi(X) = \forall Y_1 \exists Y_2 \cdots Q Y_k\, \varphi(X, Y_1, \ldots, Y_k)$ with $k \in \mathbb{N}$, where φ is a quantifier-free formula and $X, Y_1, \ldots, Y_k$ are sets of propositional variables, (and a weight function $w\colon X \to \mathbb{N}$ assigning positive values to each variable $x \in X$) such that Q is either $\exists$ for k even or $\forall$ for k odd.
Output: Number of cardinality-minimal (weight-minimal) models of $\psi(X)$ or 0 if $\psi(X)$ is unsatisfiable.

We define the classes $\#\mathrm{Min\text{-}Card\text{-}SAT}$ and $\#\mathrm{Min\text{-}Weight\text{-}SAT}$ to be the $\#\mathrm{Min\text{-}Card\text{-}\Pi_0 SAT}$ and $\#\mathrm{Min\text{-}Weight\text{-}\Pi_0 SAT}$, respectively. Moreover, we can assume, following the ideas of Wrathal [20], that the formula φ is in CNF for k even and in DNF for k odd. Notice that for k *even* (*odd*), the formula φ has an *odd* (*even*) number of variable vectors, since the first variable block X remains always unquantified.

Theorem 2. *For every $k \in \mathbb{N}$, the following problems are complete via parsimonious reductions. $\#\mathrm{Min\text{-}Weight\text{-}\Pi_k SAT}$ is $\#\cdot\mathrm{Opt}_{k+1}P$-complete and $\#\mathrm{Min\text{-}Card\text{-}\Pi_k SAT}$ is $\#\cdot\mathrm{Opt}_{k+1}P[\log n]$-complete.*

As usual, also the versions of $\#\mathrm{Min\text{-}Weight\text{-}\Pi_k SAT}$ and $\#\mathrm{Min\text{-}Card\text{-}\Pi_k SAT}$ restricted to 3 literals per clause are $\#\cdot\mathrm{Opt}_{k+1}P$-complete and $\#\cdot\mathrm{Opt}_{k+1}P[\log n]$-complete, respectively, since there exists a parsimonious reduction to them.

Apart from containing natural complete problems, a complexity class should also be closed with respect to an appropriate type of reductions. We consider the closure of the considered counting classes under subtractive reductions. Note that we cannot expect the class $\#\text{·Opt}_k\text{P}[z(n)]$ to be closed under subtractive reductions for any function $z(n)$ since we can always get an arbitrary polynomial speed-up simply by padding the input. We show in the sequel that the two most interesting cases, namely $\#\text{·Opt}_k\text{P}$ and $\#\text{·Opt}_k\text{P}[\log n]$ for each $k \in \mathbb{N}$, are indeed closed under subtractive reductions.

Theorem 3. *The complexity classes $\#\text{·Opt}_k\text{P}$ and $\#\text{·Opt}_k\text{P}[\log n]$ are closed under subtractive reductions for all $k \in \mathbb{N}$.*

Our new considered classes $\#\text{·Opt}_k\text{P}$ and $\#\text{·Opt}_k\text{P}[\log n]$ need to be confronted with the already known counting hierarchy. We will present certain inclusions of the new classes with respect to already known counting complexity classes and show that the inclusions are proper, unless the polynomial hierarchy collapses.

Theorem 4. *We have $\#\text{·}\Pi_k\text{P} \subseteq \#\text{·Opt}_{k+1}\text{P}[\log n] \subseteq \#\text{·Opt}_{k+1}\text{P} \subseteq \#\text{·}\Pi_{k+1}\text{P}$ for each $k \in \mathbb{N}$.*

Finally, the following result shows that the new classes are robust.

Theorem 5. *If $\#\text{·Opt}_{k+1}\text{P}[\log n]$ or $\#\text{·Opt}_{k+1}\text{P}$ coincides with either $\#\text{·}\Pi_k\text{P}$ or $\#\text{·}\Pi_{k+1}\text{P}$ for some $k \in \mathbb{N}$, then the polynomial hierarchy collapses to the k-th or $(k+1)$-st level, respectively.*

4   Further Optimization Counting Problems

The most interesting optimization counting problems are of course those belonging to the classes on the first level of the optimization counting hierarchy, namely $\#\text{·OptP}$ and $\#\text{·OptP}[\log n]$. In this section we will focus on such problems of particular interest.

Gasarch *et al.* presented in [6] a plethora of optimization problems complete for OptP and OptP[$\log n$]. Either their lower bound is already proved by a parsimonious reduction or the presented reduction can be transformed into a parsimonious one similarly to Galil's construction in [5]. The counting version of virtually all these problems can therefore be proved to be complete for $\#\text{·OptP}$ or $\#\text{·OptP}[\log n]$. Likewise, Krentel presented in [11] several problems belonging to higher levels of the optimization hierarchy. They give rise to counting problems complete for $\#\text{·Opt}_k\text{P}$ or $\#\text{·Opt}_k\text{P}[\log n]$ with $k > 1$.

Problem: #MIN-CARD-SAT
Input: A propositional formula φ in conjunctive normal form over the variables X.
Output: Number of models of φ with minimal Hamming weight.

The dual problem #MAX-CARD-SAT asks for the number of models with *maximal* Hamming weight. The problems #MIN-WEIGHT-SAT and #MAX-WEIGHT-SAT are the corresponding weighted versions of the aforementioned problems.

Following Theorem 2, both counting problems #MIN-CARD-SAT and #MAX-CARD-SAT are #·OptP[log n]-complete, whereas #MIN-WEIGHT-SAT and #MAX-WEIGHT-SAT are #·OptP-complete. We consider only the cardinality-minimal problems in the sequel.

It is also interesting to investigate special cases of the optimization counting problems involving the following restrictions on the formula φ. As usual, a literal is a propositional variable (positive literal) or its negation (negative literal), whereas a clause is a disjunction of literals, and a formula in conjunctive normal form is a conjunction of clauses. We say that a clause c is **Horn** if it contains at most one positive literal, **dual Horn** if it contains at most one negative literal, **Krom** if it contains at most two literals. A formula $\varphi = c_1 \wedge \cdots \wedge c_n$ in conjunctive normal form is Horn, dual Horn, or Krom if all clauses c_i for $i = 1, \ldots, n$ satisfy the respective condition. Formulas restricted to conjunctions of Horn, dual Horn, or Krom clauses are often investigated in computational problems related to artificial intelligence, in particular to closed world reasoning [1]. We denote by the specification in brackets the restriction of the counting problem #MIN-CARD-SAT to the respective class of formulas.

The models of Horn formulas are closed under conjunction, i.e., for two models m and m' of a Horn formula φ, also the Boolean vector $m \wedge m' = (m[1] \wedge m'[1], \ldots, m[k] \wedge m'[k])$ is a model of φ. Hence there exists a unique model with minimal Hamming weight if and only if φ is satisfiable. Therefore a Horn formula φ has either one cardinality-minimal model or none, depending on the satisfiability of φ. A similar situation arises for #MIN-CARD-DNF, the problem of counting the number of assignments with minimal Hamming weight to a propositional formula in disjunctive normal form. These considerations imply the following results.

Proposition 2. #MIN-CARD-SAT[HORN] *and* #MIN-CARD-DNF *are in* FP.

Vertex covers, cliques, and independent sets have a particular relationship. The set X is a smallest vertex cover in $G = (V, E)$ if and only if $V \smallsetminus X$ is a largest independent set in G if and only if $V \smallsetminus X$ is a largest clique in the complement graph $\bar{G} = (V, V \times V \smallsetminus E)$. The size of the largest clique has been investigated by Krentel [10] and proved to be OptP[log n]-complete (the same proof is also given in [13]). Using this knowledge, we can determine the complexity of the following problems.

Problem: #MAX-CARD-INDEPENDENT SET
Input: Graph $G = (V, E)$.
Output: Number of independent sets in G with maximum cardinality, i.e., number of subsets $V' \subseteq V$ where $|V'|$ is maximal and for all $u, v \in V'$ we have $(u, v) \notin E$.

Problem: #MAX-CARD-CLIQUE
Input: Graph $G = (V, E)$.
Output: Number of cliques in G with maximum cardinality, i.e., number of subsets $V' \subseteq V$ where $|V'|$ is maximal and $(u, v) \in E$ holds for all $u, v \in V'$ such that $u \neq v$.

Problem: #MIN-CARD-VERTEX COVER
Input: Graph $G = (V, E)$.
Output: Number of vertex covers of G with minimal cardinality, i.e., number of subsets $V' \subseteq V$ where $|V'|$ is minimal and $(u, v) \in E$ implies $u \in V'$ or $v \in V'$.

Theorem 6. *The problems* #MAX-CARD-INDEPENDENT SET, #MAX-CARD-CLIQUE, *and* #MIN-CARD-VERTEX COVER *are* #·OptP[$\log n$]*-complete. Their weighted versions are* #·OptP*-complete.*

We can easily transform the counting problem #MIN-CARD-VERTEX COVER to both #MIN-CARD-SAT[DUAL HORN] and #MIN-CARD-SAT[KROM]. Indeed, we can represent an edge $(u, v) \in E$ of a graph $G = (V, E)$ by a clause $(u \lor v)$ which is both Krom and dual Horn. Hence a cardinality-minimal vertex cover of a graph $G = (V, E)$ corresponds to a cardinality-minimal model of the formula $\varphi_G = \bigwedge_{(u,v) \in E} (u \lor v)$.

Corollary 1. *The counting problems* #MIN-CARD-SAT[DUAL HORN] *and* #MIN-CARD-SAT[KROM] *are* #·OptP[$\log n$]*-complete via parsimonious reductions.*

The following problem is a classic in optimization theory. It is usually formulated as the maximal number of clauses that can be satisfied. We can also ask for the number of truth assignments that satisfy the maximal number of clauses.

Problem: #MAX2SAT
Input: A propositional formula φ in conjunctive normal form over the variables X with at most two variables per clause.
Output: Number of assignments to φ that satisfy the maximal number of clauses.

The optimization variant of the following counting problem is presented in [6] under the name CHEATING SAT. We can interpret it as a satisfiability problem in a 3-valued logic, where the middle value τ is a "don't-know". In this setting it is interesting to investigate the minimal size of uncertainty we need to satisfy a formula for the optimization variant, as well as the number of satisfying assignments with the minimal size of uncertainty.

Problem: #MIN-SIZE UNCERTAINTY SAT
Input: A propositional formula φ in conjunctive normal form over the variables X.
Output: Number of satisfying assignments $m \colon X \to \{0, \tau, 1\}$ of the formula φ, where $m(x) = \tau$ satisfies both literals x and $\neg x$, with minimal cardinality of the set $\{x \in X \mid m(x) = \tau\}$.

Theorem 7. #MAX2SAT   *and*   #MIN-SIZE   UNCERTAINTY   SAT   *are* #·OptP[$\log n$]*-complete.*

Even though the complete problems for the classes #·OptP and #·OptP[$\log n$] are the most interesting ones, there also exist some interesting complete problems in the classes $\#\cdot\mathrm{Opt}_k\mathrm{P}$ and $\#\cdot\mathrm{Opt}_k\mathrm{P}[\log n]$ for $k > 1$. The following problem is an example of such a case.

Problem: #MAXIMUM k-QUANTIFIED CIRCUIT
Input: A Boolean circuit $C(\boldsymbol{x}, \boldsymbol{y}_1, \ldots, \boldsymbol{y}_k)$ over variable vectors $\boldsymbol{x}, \boldsymbol{y}_1, \ldots, \boldsymbol{y}_k$.
Output: Number of maximum values $\boldsymbol{x} \in \{0,1\}^n$ in binary notation satisfying the quantified expression $\forall \boldsymbol{y}_1 \exists \boldsymbol{y}_2 \cdots Q\boldsymbol{y}_k \ (C(\boldsymbol{x}, \boldsymbol{y}_1, \ldots, \boldsymbol{y}_k) = 1)$, where Q is either $\forall$ or $\exists$ depending on the parity of k.

Theorem 8. #MAXIMUM k-QUANTIFIED CIRCUIT *is* $\#\cdot\mathrm{Opt}_k\mathrm{P}$-*complete.*

5   Concluding Remarks

In the scope of the result from [16] showing that all classes between #P and #PH, the counting equivalent of the polynomial hierarchy, collapse to #P under 1-Turing reductions, it is necessary (1) to find suitable reductions strong enough to prove completeness of well-known counting problems, but tame enough to preserve at least some counting classes, (2) to identify counting classes with interesting complete problems preserved under the aforementioned reduction. The first problem was mainly addressed in [3], whereas in this paper we focused on the second point. We introduced a new hierarchy of optimization counting complexity classes $\#\cdot\mathrm{Opt}_k\mathrm{P}$ and $\#\cdot\mathrm{Opt}_k\mathrm{P}[\log n]$. These classes allowed us to pinpoint the complexity of many natural optimization counting problems which had previously resisted a precise classification. Moreover, we have shown that these new complexity classes have several desirable properties and they interact well with the counting hierarchy defined by Hemaspaandra and Vollmer in [7]. Nevertheless, the Hemaspaandra-Vollmer counting hierarchy does not seem to be sufficiently detailed to capture all interesting counting problems. Therefore an even more fine-grained stratification of the counting complexity classes is necessary, which started with the contribution of Pagourtzis and Zachos [12] and has been pursued in this paper.

Finally, further decision problems in $\Delta_k\mathrm{P}$ (respectively $\Theta_k\mathrm{P}$) with $k \in \mathbb{N}$ and corresponding counting problems should be inspected. It should be investigated if the complexity of the latter can be precisely identified now that we have the new counting complexity classes $\#\cdot\mathrm{Opt}_k\mathrm{P}$ (respectively $\#\cdot\mathrm{Opt}_k\mathrm{P}[\log n]$) at hand. Moreover, we would also like to find out more about the nature of the problems that are complete for these new counting complexity classes. In particular, it would be very interesting to find out if there also exist "easy to decide, hard to count" problems, i.e., problems whose counting variant is complete for $\#\cdot\mathrm{Opt}_k\mathrm{P}$ (respectively $\#\cdot\mathrm{Opt}_k\mathrm{P}[\log n]$) while the corresponding decision problem is below $\Delta_k\mathrm{P}$ (respectively $\Theta_k\mathrm{P}$). Clearly, such a phenomenon can only exist if we consider completeness with respect to reductions stronger than the parsimonious ones. Hence, the closure of our new counting classes under *subtractive reductions* (rather than just under parsimonious reductions) in Theorem 3 is an indispensable prerequisite for further research in this direction.

References

1. Cadoli, M., Lenzerini, M.: The Complexity of Propositional Closed World Reasoning and Circumscription. Journal of Computer and System Sciences 48(2), 255–310 (1994)
2. Creignou, N., Hermann, M.: Complexity of Generalized Satisfiability Counting Problems. Information and Computation 125(1), 1–12 (1996)
3. Durand, A., Hermann, M., Kolaitis, P.G.: Subtractive Reductions and Complete Problems for Counting Complexity Classes. Theoretical Computer Science 340(3), 496–513 (2005)
4. Fortnow, L.: Counting complexity. In: Hemaspaandra, L.A., Selman, A.L. (eds.) Complexity Theory Retrospective II, pp. 81–107. Springer, Heidelberg (1997)
5. Galil, Z.: On Some Direct Encodings of Nondeterministic Turing Machines Operating in Polynomial Time into P-complete Problems. SIGACT News 6(1), 19–24 (1974)
6. Gasarch, W.I., Krentel, M.W., Rappoport, K.J.: OptP as the Normal Behavior of NP-complete Problems. Mathematical Systems Theory 28(6), 487–514 (1995)
7. Hemaspaandra, L.A., Vollmer, H.: The Satanic Notations: Counting Classes beyond #P and other Definitional Adventures. SIGACT News, Complexity Theory Column 8 26(1), 2–13 (1995)
8. Jenner, B., Torán, J.: Computing Functions with Parallel Queries to NP. Theoretical Computer Science 141(1-2), 175–193 (1995)
9. Köbler, J., Schöning, U., Torán, J.: On Counting and Approximation. Acta Informatica 26(4), 363–379 (1989)
10. Krentel, M.W.: The Complexity of Optimization Problems. Journal of Computer and System Sciences 36(3), 490–509 (1988)
11. Krentel, M.W.: Generalizations of OptP to the Polynomial Hierarchy. Theoretical Computer Science 97(2), 183–198 (1992)
12. Pagourtzis, A., Zachos, S.: The Complexity of Counting Functions with Easy Decision Version. In: Královič, R., Urzyczyn, P. (eds.) MFCS 2006. LNCS, vol. 4162, pp. 741–752. Springer, Heidelberg (2006)
13. Papadimitriou, C.H.: Computational Complexity. Addison-Wesley, Reading (1994)
14. Selman, A., Mei-Rui, X., Book, R.: Positive Relativizations of Complexity Classes. SIAM Journal on Computing 12(3), 565–579 (1983)
15. Toda, S., Ogiwara, M.: Counting Classes are at least as Hard as the Polynomial-Time Hierarchy. SIAM Journal on Computing 21(2), 316–328 (1992)
16. Toda, S., Watanabe, O.: Polynomial-Time 1-Turing Reductions from #PH to #P. Theoretical Computer Science 100(1), 205–221 (1992)
17. Valiant, L.G.: The Complexity of Computing the Permanent. Theoretical Computer Science 8(2), 189–201 (1979)
18. Valiant, L.G.: The Complexity of Enumeration and Reliability Problems. SIAM Journal on Computing 8(3), 410–421 (1979)
19. Wagner, K.: Bounded Query Classes. SIAM Journal on Computing 19(5), 833–846 (1990)
20. Wrathall, C.: Complete Sets and the Polynomial-Time Hierarchy. Theoretical Computer Science 3(1), 23–33 (1976)

The Orbit Problem Is in the GapL Hierarchy

V. Arvind[1] and T.C. Vijayaraghavan[2]

[1] The Institute of Mathematical Sciences, Chennai 600 113, India
[2] Chennai Mathematical Institute, SIPCOT IT Park Padur PO,
Siruseri 603 103, India

Abstract. The *Orbit problem* is defined as follows: Given a matrix $A \in \mathbb{Q}^{n \times n}$ and vectors $\mathbf{x}, \mathbf{y} \in \mathbb{Q}^n$, does there exist a non-negative integer i such that $A^i \mathbf{x} = \mathbf{y}$. This problem was shown to be in deterministic polynomial time by Kannan and Lipton in [7]. In this paper we put the problem in the logspace counting hierarchy GapLH. We also show that the problem is hard for $\mathrm{C}_{=}\mathrm{L}$.

1 Introduction

The *Orbit problem* is defined as follows.

> Given $A \in \mathbb{Q}^{n \times n}$ and $\mathbf{x}, \mathbf{y} \in \mathbb{Q}^n$, does there exist a non-negative integer i such that $A^i \mathbf{x} = \mathbf{y}$.

The goal of this paper is to give a new upper bound for the complexity of the orbit problem using logspace counting classes. We show that the orbit problem is in $\mathrm{AC}^0(\mathrm{GapL})$, and hence is in NC^2 (indeed, even in TC^1) as $\mathrm{AC}^0(\mathrm{GapL}) \subseteq \mathrm{TC}^1 \subseteq \mathrm{NC}^2$.

In a celebrated paper, Kannan and Lipton in [7] gave a polynomial time algorithm for the orbit problem. Their approach is to reduce it to the *Matrix power problem*. In the matrix power problem, we are given two matrices $B, D \in \mathbb{Q}^{n \times n}$ as input and we need to check if there exists a non-negative integer i such that $B^i = D$. Kannan and Lipton further show that (B, D) is a yes instance of the matrix power problem if and only if $B^i = q(B)$ for some nonnegative integer i, where $q(x) \in \mathbb{Q}[x]$ is a polynomial that depends on B and D and its coefficients can be computed in polynomial time. Here the degree of $q(x)$ is one less than the degree of the minimal polynomial of B. The rest of the algorithm in [7] focuses on checking if there is an $i \in \mathbb{Z}^+$ satisfying $B^i = q(B)$. Assume that we have computed the polynomial $q(x)$, and let α be an eigenvalue of B. Now, if there exists $i \in \mathbb{Z}^+$ such that $B^i = q(B)$ then $\alpha^i = q(\alpha)$. The algorithm in [7] uses this fact repeatedly while considering different cases: when $q(x)$ has a root that is not a root of unity, or when all roots of $q(x)$ are roots of unity with multiplicity 1, or the case when all the roots of $q(x)$ are roots of unity but with at least one root of multiplicity greater than 1. Kannan and Lipton design their algorithm based on this case analysis.

In this paper, we broadly follow the Kannan-Lipton algorithm [7], but we need to differently analyze the complexity of the main steps involved in it. This forces

X. Hu and J. Wang (Eds.): COCOON 2008, LNCS 5092, pp. 160–169, 2008.

us to modify several subroutines in the algorithm. Since these steps basically require linear algebraic computation over $\mathbb{Q}$, we obtain an upper bound in the GapL hierarchy. Some of the steps involve checking if a set of vectors are linearly independent over $\mathbb{Q}$, computing the determinant of a matrix over $\mathbb{Q}$, computing the inverse of a matrix, computing powers and the minimal polynomial of a rational matrix etc. We crucially use earlier work [1],[5],[6] classifying the complexity of various linear-algebraic problems using logspace counting classes. Among the new observations, we show that testing if all roots of a univariate polynomial over $\mathbb{Q}$ are complex roots of unity is in $AC^0(GapL)$. Furthermore, if all roots of a polynomial are complex roots of unity we can factorize $p(x)$ into its irreducible factors in $AC^0(GapL)$.

Finally, we show that the orbit problem is hard for $C_=L$ under logspace many-one reductions.

2   Basic Results

In this section we recall basic definitions, notation, and results.

Definition 1. *A complex number θ is a nth root of unity if $\theta^n - 1 = 0$. Furthermore, θ is a primitive n^{th} root of unity if θ is a nth root of unity and $\theta^m - 1 \neq 0$ for all integers $0 < m < n$.*

Clearly, an nth root of unity is of the form $e^{(2\pi\sqrt{-1})j/n}$ for $0 \leq j \leq (n-1)$. Also, $e^{(2\pi\sqrt{-1})j/n}$ is a primitive n^{th} root of unity if and only if $\gcd(j,n) = 1$. We denote $\sqrt{-1}$ by ι. Let $\varphi(j)$ denote the Euler totient function: the number of positive integers less than and relatively prime to j.

Definition 2. *Let $\theta_1, \ldots, \theta_{\varphi(j)}$ be primitive j^{th} roots of unity. Then, the jth cyclotomic polynomial is defined as $C_j(x) = \prod_{i=1}^{\varphi(j)} (x - \theta_i)$.*

It is well-known that $C_j(x)$ is irreducible over $\mathbb{Q}$ and hence it must divide any polynomial $h(x) \in \mathbb{Q}[x]$ that has as root one of the primitive n^{th} roots of unity.

Proposition 1. *Let $h(x) \in \mathbb{Q}[x]$. If $h(\theta) = 0$ for a primitive n^{th} root of unity θ then $h(\theta') = 0$ for every other primitive n^{th} root of unity θ'.*

We assume that each rational entry of an input matrix $A \in \mathbb{Q}^{n \times m}$ is given in terms of its numerator and denominator. Also, we will assume that an algorithm computing $\det(A)$ for a rational matrix $A \in \mathbb{Q}^{n \times n}$ will output two integers p and q such that $\det(A) = p/q$. Furthermore, we will *not* require that p and q be relatively prime, that is $\gcd(p,q)$ need not be 1. This assumption is necessary because computing the GCD of two integers is not known to be in NC. This representation of rationals does not affect our algorithm so long as the size in binary of the two integers p and q is bounded by a polynomial in the size of the input. We will make a similar assumption for other computations involving rational inputs. We now recall the following results concerning rational matrices. These are usually stated for integer matrices.

Lemma 1. *Let $A \in \mathbb{Q}^{n \times m}$ be the given input rational matrix. Then,*

1. *[2],[3],[8],[9],[10] When $n = m$, computing the determinant of A denoted by $\det(A)$, computing the $(i,j)^{th}$ entry of A^{-1}, and computing the $(i,j)^{th}$ entry of A^l for a given positive integer l are complete for GapL under logspace many-one reductions.*

2. *[1] Checking if the set of column vectors of A are linearly dependent is complete for $C_{=}L$ under logspace many-one reductions.*

3. *[1] Let $\mathbf{b} \in \mathbb{Q}^n$ be an n-dimensional rational vector. Then, determining if the system of linear equations $A\mathbf{x} = \mathbf{b}$ has a rational vector $\mathbf{x}$ as a solution is complete for $L^{C_{=}L}$ under logspace truth-table reductions.*

4. *Computing a maximal set of linearly independent columns from A is in $FL^{C_{=}L}$.*

5. *[5] Given $B \in \mathbb{Q}^{n \times n}$, we can compute the coefficients of the minimal polynomial of B in $AC^0(GapL)$.*

Proof. Let $A \in \mathbb{Q}^{n \times m}$ be the given input rational matrix. Let $A_{ij} = p_{ij}/q_{ij}$, where $1 \le i \le n$ and $1 \le j \le m$. Also, we can assume the size of each p_{ij} and q_{ij} is at most $\max(m,n)$. Let q be the product of all the denominators of the entries in A. It follows from [4] that, for any positive integer n, we can compute the ith bit of the product of n integers, each of size n, using an NC^1 circuit and therefore we can compute q which is a product of nm integers in NC^1 as well. Let us consider the matrix (qA), obtained by multiplying each entry of A by q. Clearly (qA) is an integer matrix and $A = (qA)/q$. In problems involving an additional vector $\mathbf{b}$, we multiply q with the denominators of the entries occurring in $\mathbf{b}$ to reduce the problem to the case when the inputs are integer matrices. In all these cases, the size of q as well as entries of (qA) and $(q\mathbf{b})$ are bounded by a polynomial in the size of the input, where $1 \le i \le n$ and $1 \le j \le m$. Thus we can compute the ith bit of any entry of these matrices in logspace. The results stated above then follow by applying known complexity bounds (proven in the references appearing in the Theorem statement) on linear algebraic problems involving integer matrices to (qA), and $(q\mathbf{b})$. ∎

The next lemma shows that a solution to a feasible system of linear equations over rationals can be computed in the GapL hierarchy. The proof is omitted due to lack of space.

Lemma 2. *Let $A \in \mathbb{Q}^{m \times n}$ and $\mathbf{b} \in \mathbb{Q}^n$. If the system of linear equation $A\mathbf{x} = \mathbf{b}$ is feasible, then a solution to it can be computed in $AC^0(GapL)$.*

We also need the following result of [6] computing the GCD of two polynomials over $\mathbb{Q}$.

Lemma 3. *[6] Given polynomials $f(x)$, $g(x) \in \mathbb{Q}[x]$ as input their GCD can be computed in the complexity class PL, which is contained in L^{GapL}.*

3 Kannan-Lipton Algorithm

We next recall the definition of the GapL hierarchy from [2]: the first level $GapLH_1$ is defined to be GapL. For $i \geq 1$, $GapLH_{i+1}$ is defined as all functions $f : \Sigma^* \to \mathbb{Z}$, such that for some logspace-bounded non-deterministic oracle Turing machine M with a function $g \in GapLH_i$ as oracle, we have $f(x) = acc_M(x) - rej_M(x)$. We denote the GapL hierarchy by GapLH. Here the oracle access is according to the Ruzzo-Simon-Tompa restriction [2].

We say that a language $L \in AC^0(GapL)$ if there exists a logspace uniform AC^0 oracle circuit family $\{C_n\}_{n \geq 1}$ with oracle gates computing a function $g \in GapL$, such that on any input x of length n, we have $C_n(x) = 1$ if and only if $x \in L$. It is shown in [2] that $GapLH = AC^0(GapL)$.

We now proceed to show that the orbit problem is in GapLH, and hence in $AC^0(GapL)$. We first describe the main steps in Kannan-Lipton algorithm [7] for the orbit problem. They first reduce the orbit problem to the *Matrix Power problem*: Given $B, D \in \mathbb{Q}^{n \times n}$ does there exists a non-negative integer i such that $B^i = D$.

We now describe the reduction. Let $(A, \mathbf{x}, \mathbf{y})$ be an instance of the orbit problem. Let $V \subseteq \mathbb{Q}^n$ denote the subspace spanned by $\{\mathbf{x}, A\mathbf{x}, A^2\mathbf{x}, \cdots, A^{n-1}\mathbf{x}\}$. Clearly V is k-dimensional for the largest k such that $\{\mathbf{x}, A\mathbf{x}, A^2\mathbf{x} \cdots, A^{k-1}\mathbf{x}\}$ are linearly independent, and a basis for V is this set $\{\mathbf{x}, A\mathbf{x}, A^2\mathbf{x} \cdots, A^{k-1}\mathbf{x}\}$. We can compute this basis in $AC^0(GapL)$: with an L^{GapL} computation we can first compute $A^j\mathbf{x}$ for $1 \leq j \leq n - 1$. This machine's output is taken as input by another L^{GapL} computation that will find the largest k such that $\{\mathbf{x}, A\mathbf{x}, A^2\mathbf{x} \cdots, A^{k-1}\mathbf{x}\}$ is linearly independent.

The subspace V is A-invariant. That is, $x \in V$ if and only if $A^i\mathbf{x} \in V$ for each $i \geq 0$. Consequently, $(A, \mathbf{x}, \mathbf{y})$ is a 'yes' instance for the orbit problem only if $\mathbf{y} \in V$. We can check if $\mathbf{y} \in V$ in L^{GapL}. If $\mathbf{y} \notin V$ then the reduction outputs the pair (O_n, I_n) of the matrix power problem, where O_n is the $n \times n$ zero matrix and I_n is the identity matrix. Therefore, in the sequel we can assume that $\dim(V) = k$ and $\mathbf{y} \in V$. Let

$$A^k\mathbf{x} = \sum_{j=0}^{k-1} \alpha_j A^j\mathbf{x}, \quad \mathbf{x} = \sum_{j=0}^{k-1} \beta_j A^j\mathbf{x}, \quad \mathbf{y} = \sum_{j=0}^{k-1} \gamma_j A^j\mathbf{x}.$$

Notice that $\beta_0 = 1$ and $\beta_j = 0$ for all $j > 0$. We can compute the scalars $\alpha_j, \beta_j, \gamma_j$ in L^{GapL} by solving each of the above three systems of linear equations using Cramér's rule.

The $k \times k$ matrix for the linear transformation A from V to V has $e_{j+1}, 1 \leq j \leq k - 1$ as its first $k - 1$ columns and $(\alpha_0, \cdots, \alpha_{k-1})^T$ as the last column.[1] Call this matrix A'. Likewise, let $\mathbf{x}' = (\beta_0, \cdots, \beta_{k-1})^T$ and $\mathbf{y}' = (\gamma_0, \cdots, \gamma_{k-1})^T$. Clearly, $(A', \mathbf{x}', \mathbf{y}')$ is a yes instance of the orbit problem if and only if $(A, \mathbf{x}, \mathbf{y})$ is a yes instance. This is because $A', \mathbf{x}', \mathbf{y}'$ are essentially A, $\mathbf{x}$, and $\mathbf{y}$ expressed

---

[1] Here the vectors e_{j+1} denote the standard basis vectors of $\mathbb{R}^k$.

using the basis $\mathbf{x}, A\mathbf{x}, \cdots, A^{k-1}\mathbf{x}$ of V. Now, let C denote the $k \times k$ invertible matrix $[\mathbf{x}'|A'\mathbf{x}'|\cdots|A'^{k-1}\mathbf{x}']$. Similarly, let C' denote the $k \times k$ matrix $[\mathbf{y}'|A'\mathbf{y}'|\cdots|A'^{k-1}\mathbf{y}']$. Then, there exists an $i \geq 0$ such that $A'^i\mathbf{x}' = \mathbf{y}'$ if and only if $A'^iC = C'$, which we can rewrite as $A'^i = C'C^{-1}$ as C is invertible. Thus, $(A', C'C^{-1})$ is the instance of the matrix power problem to which we have reduced $(A, \mathbf{x}, \mathbf{y})$. We formally state this as a lemma.

Lemma 4. *The orbit problem can be reduced to the matrix power problem in* $AC^0(GapL)$.

Proof. The correctness of the reduction follows from the above argument. To see that it is computable in $AC^0(GapL)$, we note that a set of L^{GapL} computations need to be carried out that involves a nesting of at most two levels of GapL queries.

We now turn to the matrix power problem. Let $B, D \in \mathbb{Q}^{n \times n}$ be an input instance. Following [7] we further reduce it to a more tractable problem.

Lemma 5. *Given* $B, D \in \mathbb{Q}^{n \times n}$, *we can compute in* $AC^0(GapL)$ *a polynomial* $q(x) \in \mathbb{Q}[x]$ *of degree at most* $n - 1$ *such that there exists a non-negative integer* i *satisfying* $B^i = D$ *if and only if* $B^i = q(B)$.

Proof. Let $p(x)$ be the minimal polynomial of B (computable in $AC^0(GapL)$ [5]). We have $p(B) = 0$ and $\deg(p(x)) = r \leq n$. If there exists an $i \geq 0$ such that $B^i = D$, then we show that there is a polynomial $q(x)$ of degree at most $n - 1$ such that $D = q(B)$. We divide x^i by $p(x)$ and take the remainder as the polynomial $q(x)$. Thus, $q(x) \equiv x^i (\mathrm{mod}\ p(x))$, and $\deg(q(x)) \leq (\deg(p(x)) - 1) \leq (n - 1)$. Therefore, (B, D) is a yes instance of the matrix power problem only if such a polynomial $q(x)$ exists. We can test this and compute the coefficients of $q(x)$ by solving the following system of n^2 linear equations over n variables: $\sum_{j=0}^{(r-1)} q_j B^j = D$ where the unknowns are the coefficients q_j of the polynomial $q(x)$. Given B and D as input, an L^{GapL} computation will first compute B^j for $1 \leq j \leq n - 1$ and pass it as input to another L^{GapL} computation to check the feasibility of the above system and find a solution $q(x)$ using Lemma 2. Thus, the polynomial $q(x)$ can be computed in $AC^0(GapL)$. Clearly, $B^i = q(B)$ if and only if $B^i = D$.

As mentioned previously, the overall reduction from the orbit problem involves composing computations, each of which is in some constant level of the GapL hierarchy. Since we will do only a constant number of such compositions the overall computation is still in a constant level of the GapL hierarchy. Continuing with the proof, as a consequence of Lemma 4 and Lemma 5, we obtain the following.

Corollary 1. *Given an instance* $A \in \mathbb{Q}^{n \times n}$ *and* $\mathbf{x}, \mathbf{y} \in \mathbb{Q}^n$ *of the orbit problem, for some* $m \leq n$ *we can compute a matrix* $B \in \mathbb{Q}^{m \times m}$ *and a polynomial* $q(x) \in \mathbb{Q}[x]$ *of degree at most* $(m - 1)$ *in* $AC^0(GapL)$, *such that* $A^i\mathbf{x} = \mathbf{y}$ *for some* $i \geq 0$ *if and only if* $B^i = q(B)$.

The following easy lemma is a useful property for the next step.

Lemma 6. *Suppose $p(x) \in \mathbb{Q}[x]$ is the minimal polynomial of matrix $B \in \mathbb{Q}^{n \times n}$. For any two polynomials $r(x), q(x) \in \mathbb{Q}[x]$ we have $r(B) = q(B)$ if and only if $r(x) = q(x) (\bmod\ p(x))$.*

In particular, it follows that $B^i = q(B)$ for some $i \geq 0$ if and only if $x^i = q(x) (\bmod\ p(x))$. As a consequence of Corollary 1 and Lemma 6, it suffices to solve in $AC^0(GapL)$ the problem of checking if $x^i = q(x) (\bmod\ p(x))$ for some $i \geq 0$, where $p(x)$ is the minimal polynomial of the matrix B. Given polynomials $p, q \in \mathbb{Q}[x]$, where p is a monic, the goal is to test in $AC^0(GapL)$ if $x^i = q(x) (\bmod\ p(x))$ for some $i \geq 0$. Following [7], we need to handle different cases depending on the roots of $p(x)$. A crucial property is a bound from algebraic number theory [7, Theorem 3]. For a polynomial $f \in \mathbb{Q}[x]$ let $|f|$ denote the ℓ_2 norm of the vector of its coefficients.

Theorem 1. *[7, Theorem 3] There is a polynomial P such that for any algebraic number $\alpha \in \mathbb{C}$ that is not a root of unity and any polynomial $q(x) \in \mathbb{Q}[x]$, if $\alpha^i = q(\alpha)$ for $i \in \mathbb{Z}^+$ then $i \leq P(\deg(f_\alpha), \log(|f_\alpha|), \log(|q|))$, where $f_\alpha \in \mathbb{Q}[x]$ is the minimal polynomial of α.*

Thus, if the given polynomial $p(x)$ has a root α that is not a root of unity then, by Theorem 1, we can test if there is an i such that $x^i = q(x) (\bmod\ p(x))$ by trying the polynomially many values of i in the range $i \leq P(\deg(f_\alpha), \log(|f_\alpha|), \log(|q|))$. Since f_α is an irreducible factor of $p(x)$, we know that $|f_\alpha|$ is polynomially bounded by $|p|$. Hence the range of values for i is indeed polynomially bounded by the input size. Indeed, since this test involves only division of polynomials, using the result of [4, Corollary 6.5] it can be carried out in logspace.

We now consider the case when all the roots of $p(x)$ are complex roots of unity. We use key properties of the cyclotomic polynomial $C_j(x)$. First we show that $C_j(x)$ can be computed in $AC^0(GapL)$ by an algorithm that takes j in unary as input.

Lemma 7. *Given 1^j as input the jth cyclotomic polynomial $C_j(x)$ can be computed in $AC^0(GapL)$.*

Proof. The jth cyclotomic polynomial $C_j(x) = \prod_{r=1}^{\varphi(j)}(x - \omega_r)$ where the ω_r are the $\varphi(j)$ different primitive j^{th} roots of unity and $C_j(x)$ is an irreducible factor of $x^j - 1$.

We first define the polynomial $t_j(x) = \prod_{i=1}^{j-1}(x^i - 1)$. The polynomial t_j is of degree $j(j-1)/2$ with rational coefficients. We can also compute the coefficients of this polynomial in logspace by substituting a large power of 2 for the indeterminate x and extracting the bits of the coefficients from the resulting value. Furthermore, it is clear that $b_j(x) = \gcd(t_j(x), x^j - 1)$ contains as roots precisely all non-primitive j^{th} roots of unity. Therefore, it follows that $C_j(x)$ is the quotient obtained on dividing $x^j - 1$ by $b_j(x)$. Given the coefficients of $t_j(x)$ we can apply Lemma 3 to compute $\gcd(t_j(x), x^j - 1)$ in $AC^0(GapL)$. Therefore, the overall computation is clearly in $AC^0(GapL)$.

We can easily show that testing if all roots of $p(x)$ are complex roots of unity is in $AC^0(GapL)$.

Lemma 8. *Given* $p(x) \in \mathbb{Q}[x]$ *as input we can test in* $\mathrm{AC}^0(\mathrm{GapL})$ *if all roots of* $p(x)$ *are complex roots of unity, and if so we can factorize* $p(x)$ *into its irreducible factors in* $\mathrm{AC}^0(\mathrm{GapL})$.

Proof. Let $\deg(p(x)) = d$. We first compute $C_j(x), 1 \le j \le d$ using Lemma 7. Next, since division of polynomials with rational coefficients can be carried out in logspace using [4, Corollary 6.5], we can find the highest power of $C_j(x)$ that divides $p(x)$ in logspace. Putting it together will give us all the irreducible factors of $p(x)$, with multiplicity, from the set $C_j(x), 1 \le j \le d$.

After applying Lemma 8 we will know whether $p(x)$ has a root that is not a root of unity (in which case we can use the easy logspace algorithm based on Theorem 1). Thus, we now consider only the case when $p(x) = \prod_{j=1}^{d} C_j(x)^{k_j}$, where $k_j \ge 0$. An easy and useful lemma is the following.

Lemma 9. *Let* $q(x)$ *be an arbitrary polynomial and let* $C_j(x)$ *be the* j^{th} *cyclotomic polynomial. The congruence* $x^{\ell} \equiv q(x) \pmod{C_j(x)}$ *holds for some nonnegative integer* ℓ *if and only if it holds for some* unique ℓ *in the range* $0 \le \ell \le (j-1)$.

Proof. Since $C_j(x)$ divides $x^j - 1$, it follows that $x^{\ell} \equiv q(x) \pmod{C_j(x)}$ implies $x^{\ell \ (mod \ j)} \equiv q(x) \pmod{C_j(x)}$.

Using the above result we first handle the case when $k_j \in \{0,1\}$ in $p(x) = \prod_{j=1}^{d} C_j(x)^{k_j}$.

Lemma 10. *If* $p(x) = \prod_{j=1}^{d} C_j(x)^{k_j}$ *for* $k_j \in \{0,1\}$, *then the problem of testing for a given polynomial* $q(x) \in \mathbb{Q}[x]$ *if* $x^i \equiv q(x) \pmod{p(x)}$ *for some positive integer* i, *is in* $\mathrm{AC}^0(\mathrm{GapL})$.

Proof. By the Chinese remainder theorem, it suffices to check if there is a positive integer i such that $x^i \equiv q(x) \pmod{C_j(x)}$ for every C_j such that $k_j = 1$. By Lemma 9 there is an $i \ge 0$ such that $x^i \equiv q(x) \pmod{C_j(x)}$ if and only if there is an $i_j \in \{0, 1, \cdots, j - 1\}$ such that $x^{i_j} \equiv q(x) \pmod{C_j(x)}$. Notice that such an i_j, if it exists, has to be *unique*. If for some C_j such that $k_j = 1$ no such i_j exists we reject the input. Otherwise, we would have computed i_j for each C_j with $k_j = 1$. We only need to check if there exists a positive integer i such that

$$i \equiv i_j \pmod{j} \tag{1}$$

for all j such that $k_j = 1$. We cannot directly apply the chinese remainder theorem to check this congruence as the different j's need not be relatively prime. However, since each such j is bounded by d, it follows that j has logarithmically many bits. Hence we can compute the prime factorization for each j such that $k_j = 1$ in deterministic logspace. Let $p_1, p_2, \cdots, p_k$ denote the set of all prime factors of any $j \le d$. Clearly, each p_i is logarithmic in size and k is also logarithmic in the input size. Then we can rewrite the congruences in Equation 1 above as

$$i \equiv i_j \pmod{p_{\ell}^{r_{j,\ell}}}, \tag{2}$$

where $1 \le \ell \le k$ and j such that $k_j = 1$ and $j = \prod p_\ell^{r_{j,\ell}}$. Now, for each prime p_ℓ above we club together all congruences of the type $i \equiv i_j \pmod{p_\ell^{r_{j,\ell}}}$ for all the j's. Let j' be a value of j for which $r_{j',\ell}$ is maximum. Then, a necessary condition that Equation 2 has a solution for i is that $i_j = i_{j'} \pmod{p_\ell^{r_{j,\ell}}}$ for all j which we can check in logspace. Having checked this condition we can replace all the congruences in Equation 2 by the single congruence $i \equiv i_{j'} \pmod{p_\ell^{r_{j',\ell}}}$. Thus, for each p_ℓ we will have a single congruence and we can *now* invoke the chinese remainder theorem to check in logspace if there is a solution for Equation 1.

It now remains to handle the case when for some j, the exponent k_j of $C_j(x)$ is at least 2 in the factorization of $p(x)$.

Lemma 11. *Given $q(x) \in \mathbb{Q}[x]$ and a cyclotomic polynomial $C_j(x)$, we can compute in deterministic logspace a set $S_{q(x),j}$ of positive integers such that $|S_{q(x),j}|$ is polynomially bounded in $\log|q|$ and j, with the property that $x^i \equiv q(x)(\bmod\ C_j(x)^2)$ can have solutions only for $i \in S_{q(x),j}$.*

Proof. Suppose $x^i \equiv q(x)(\bmod\ C_j(x)^2)$. Then we have $x^i - q(x) = r(x)C_j(x)^2$. Taking the formal derivative on both sides we obtain $ix^{i-1} - q'(x) = 2C_j(x)r(x) + r'(x)C_j(x)^2$, implying that $ix^{i-1} - q'(x) \equiv 0 \pmod{C_j(x)}$, where $q'(x)$ and $r'(x)$ are the derivatives of $q(x)$ and $r(x)$ respectively. Let P_ℓ denote the polynomial $x^\ell \pmod{C_j(x)}$ for $0 \le \ell \le j - 1$. Notice that each P_ℓ is of degree at most $\varphi(j) - 1$. Furthermore, let $q'_1(x) = q'(x) \pmod{C_j(x)}$. Thus, i is a candidate solution only if for some ℓ we have $iP_\ell = q'_1(x)$. We define the set $S_{q(x),j} = \{s \mid s = \frac{q'_1(x)}{P_\ell}$ for some $\ell\}$. Clearly, $|S_{q(x),j}| \le j$ and can be computed in deterministic logspace. $\square$

We obtain the following corollary which limits the search space for the index i to such a set $S_{q(x),j}$.

Corollary 2. *Suppose $p(x) = \prod_{j=1}^{d} C_j(x)^{k_j}$ such that $k_{j'} \ge 2$ for some j'. Then $x^i \equiv q(x) \pmod{p(x)}$ for some i if and only if $x^i \equiv q(x) \pmod{p(x)}$ for some $i \in S_{q(x),j'}$.*

The rest of the algorithm is as follows: we need to check if there is an $i \in S_{q(x),j'}$ such that for each $k_j > 0$ we have $x^i \equiv q(x) \pmod{C_j(x)^{k_j}}$. Such an i is a solution. Notice that we cannot directly check this by division because $i \in S_{q(x),j'}$ may be an integer that is polynomially many bits long. Thus we need to devise a different test for checking if $x^i \equiv q(x) \pmod{C_j(x)^{k_j}}$ for a given i. This is described in our final lemma that will also complete the upper bound description.

Lemma 12. *Given as input a polynomial $q(x) \in \mathbb{Q}[x]$, and integer i (encoded in binary), a cyclotomic polynomial $C_j(x)$ and an integer k, where k and j are encoded in unary, we can test in deterministic logspace if $x^i \equiv q(x) \pmod{C_j(x)^k}$.*

Proof. Let ω denote a primitive j^{th} root of unity. Since $C_j(x)$ is irreducible it follows that $C_j(x)^k$ divides $x^i - q(x)$ if and only if $(x - \omega)^k$ divides $x^i - q(x)$. That means ω is a root of multiplicity k for $f(x) = x^i - q(x)$. Equivalently, we need to check if ω is a root of the ℓth formal derivative $f^{(\ell)}(x)$ of the polynomial $f(x)$ for each $0 \leq \ell \leq k-1$. Notice that $f^{(\ell)}(x)$ assumes the form $i(i-1)\cdots(i-\ell)x^{i-\ell} - q^{(\ell)}(x)$. Computing the coefficient $i(i-1)\cdots(i-\ell)$ is iterated integer multiplication that can be done in deterministic logspace. Furthermore, the ℓth derivative of the polynomial can be done term by term, which will also involve a similar iterated integer multiplication for each term and it can be done in deterministic logspace. Now, checking if ω is a root of $f^{(\ell)}(x)$ is equivalent to checking if $C_j(x)$ divides $f^{(\ell)}(x)$, again by the irreducibility of $C_j(x)$. But $f^{(\ell)}(x)$ has the nice form $i(i-1)\cdots(i-\ell)x^{i-\ell} - q^{(\ell)}(x)$ which is easy to divide by $C_j(x)$ as we can replace the exponent $i - \ell$ in the first term by $(i - \ell) \pmod{j}$. This completes the proof.

We now show that the orbit problem is hard for $C_{=}L$ under logspace many-one reductions.

Theorem 2. *Orbit problem is hard for $C_{=}L$ under logspace many-one reductions.*

Proof. Given a directed graph $G = (V, E)$, and vertices $u, v \in V$, the problem of checking is there is a directed path from u to v is NL-complete. In fact, this problem remains NL-complete for input graphs that are layered, directed, and acyclic with u as its unique source node and v its unique sink node, where u is the unique node in the first layer and v is the unique node in the last layer. By a layered digraph we mean for each edge $(s, t) \in E$ in the graph if s is in layer i then t is in layer $(i + 1)$. The counting version of this problem: namely, counting the number of directed u-v paths is #L complete under logspace many-one reductions. Furthermore, verifying if the number of directed u-v paths is a given nonnegative integer m is $C_{=}L$-complete under logspace many-one reductions. Therefore, it suffices to show a logspace many-one reduction from this problem to the orbit problem.

Let A be the adjacency matrix of an input digraph G as described above with vertex set $V = \{1, 2, \cdots, n\}$. Let 1 be its unique source node and let its sink node be n. We want to check if the number of paths from 1 to n is m. Suppose G is a layered digraph with $\ell + 1$ layers. All directed paths from 1 to n are of length ℓ, assuming there is a directed path from 1 to n in G. Clearly, A is an $n \times n$ matrix with 0-1 entries with rows and columns indexed by V. For $k \in \mathbb{Z}^+$, the $(i, j)^{th}$ entry of A^k is the number of walks from vertex i to vertex j in G. Since G is acyclic, all walks are directed paths. Define vectors $\mathbf{x} = (0, \ldots, 0, 1)^T \in \mathbb{Q}^{n \times 1}$, and $\mathbf{y} = (m, 0, \ldots, 0)^T \in \mathbb{Q}^{n \times 1}$. Since n is the unique node on the $(\ell+1)^{st}$ layer, the number of directed paths in G from 1 to n is m if and only if $A^\ell \mathbf{x} = \mathbf{y}$. More precisely, there is some $i \in \mathbb{Z}^+$ such that $A^i \mathbf{x} = \mathbf{y}$ if and only if there are exactly m directed paths in G from 1 to n.

4   Concluding Remarks

The main open question here is to close the gap between the $AC^0(\mathrm{GapL})$ upper bound and the $C_{=}L$ hardness result for the orbit problem. Some other interesting questions also arise. In Lemma 8 we have shown that factoring univariate polynomials whose roots are all complex roots of unity is in $AC^0(GapL)$. Using the LLL algorithm, all univariate polynomials over $\mathbb{Q}$ can be factored in polynomial time. To the best of our knowledge, there is no known P-hardness result for polynomial factorization. It would be interesting to either obtain a better complexity upper bound or show P-hardness.

Acknowledgments. We thank the anonymous referees for their thoughtful comments.

References

1. Allender, E., Beals, R., Ogihara, M.: The Complexity of Matrix Rank and Feasible Systems of Linear Equations. Computational Complexity 8(2), 99–126 (1999)
2. Allender, E., Ogihara, M.: Relationships among PL, #L and the Determinant. RAIRO - Theoretical Informatics and Applications 30, 1–21 (1996)
3. Damm, C.: DET=$L^{\#L}$. Informatik-Preprint 8, Fachbereich Informatik der Humboldt-Universitat zu, Berlin (1991)
4. Hesse, W., Allender, E., Barrington, D.A.M.: Uniform Constant-Depth Threshold Circuits for Division and Iterated Multiplication. Journal of Computer and System Sciences 65(4), 695–716 (2002)
5. Hoang, T.M., Thierauf, T.: The Complexity of the Characteristic and the Minimal Polynomial. Theoretical Computer Science 295(1-3), 205–222 (2003)
6. Hoang, T.M., Thierauf, T.: The Complexity of the Inertia and Some Closure Properties of Gapl. In: Proceedings of 20th IEEE Conference on Computational Complexity, pp. 28–37 (2005)
7. Kannan, R., Lipton, R.: Polynomial-Time Algorithm for the Orbit Problem. Journal of the ACM 33(4), 808–821 (1986)
8. Toda, S.: Counting Problems Computationally Equivalent to Computing the Determinant. Technical report 91-07, Department of Computer Science, University of Electro-Communications, Tokyo (1991)
9. Valiant, L.G.: Why is Boolean Complexity Theory Difficult? In: Proceedings of the London Mathematical Society symposium on Boolean function complexity, pp. 84–94. Cambridge University Press, New York (1992)
10. Vinay, V.: Counting Auxiliary Pushdown Automata and Semi-Unbounded Arithmetic Circuits. In: Proceedings of 6th Structure in Complexity Theory Conference (CCC 1991), pp. 270–284 (1991)

Quantum Separation of Local Search
and Fixed Point Computation

Xi Chen[1,*], Xiaoming Sun[2,**], and Shang-Hua Teng[3,***]

[1] Institute for Advanced Study
csxichen@gmail.com
[2] Tsinghua University
xiaomings@tsinghua.edu.cn
[3] Boston University
steng@cs.bu.edu

Abstract. We give a lower bound of $\Omega(n^{(d-1)/2})$ on the quantum query complexity for finding a fixed point of a discrete Brouwer function over grid $[n]^d$. Our lower bound is nearly tight, as Grover Search can be used to find a fixed point with $O(n^{d/2})$ quantum queries. Our result establishes a nearly tight bound for the computation of d-dimensional approximate Brouwer fixed points defined by Scarf and by Hirsch, Papadimitriou, and Vavasis. It can be extended to the quantum model for Sperner's Lemma in any dimensions: The quantum query complexity of finding a panchromatic cell in a Sperner coloring of a triangulation of a d-dimensional simplex with n^d cells is $\Omega(n^{(d-1)/2})$. For $d = 2$, this result improves the bound of $\Omega(n^{1/4})$ of Friedl, Ivanyos, Santha, and Verhoeven.

More significantly, our result provides a quantum separation of local search and fixed point computation over $[n]^d$, for $d \geq 4$. Aaronson's local search algorithm for grid $[n]^d$, using Aldous Sampling and Grover Search, makes $O(n^{d/3})$ quantum queries. Thus, the quantum query model over $[n]^d$ for $d \geq 4$ strictly separates these two fundamental search problems.

1 Introduction

In this paper, we give a nearly tight bound on the quantum query complexity of fixed point computation over grid $[n]^d = \{1, 2, ..., n\}^d$. Our result demonstrates a strict separation of fixed point computation and local search in the quantum query model, resolving an open question posed in [7]. We also solve the problem left open in [7] about both the randomized and quantum query complexity of discrete fixed point computation over hypercube $\{0, 1\}^n$.

Motivation

In various applications, we often need not only to decide whether solutions satisfying certain properties exist, but also to find a desirable solution. This family

* Supported by NSF Grant DMS-0635607.
** Supported by the National Natural Science Foundation of China Grant 60553001, 60603005, 60621062, and the National Basic Research Program of China Grant 2007CB807900, 2007CB807901.
*** Supported by NSF grants CCR-0635102 and ITR CCR-0325630.

X. Hu and J. Wang (Eds.): COCOON 2008, LNCS 5092, pp. 170–179, 2008.
© Springer-Verlag Berlin Heidelberg 2008

of computational problems is usually referred to as the *search problem*. Three fundamental types of search problems are global optimization, local search, and fixed point computation (FPC). In a global optimization problem, we are given an objective function h over a domain D and are asked to find a solution $\mathbf{x} \in D$ such that $h(\mathbf{x}) \leq h(\mathbf{y})$, for all $\mathbf{y} \in D$. In local search, we are given a function h over a domain D and a neighborhood function $N : D \to 2^D$. We are asked to find a solution $\mathbf{x} \in D$ such that $h(\mathbf{x}) \leq h(\mathbf{y})$, for all $\mathbf{y} \in N(\mathbf{x})$.

FPC arises in geometry, topology, game theory, and mathematical economics. Brouwer proved that every continuous map f from a 3D simplex S to itself has a fixed point, i.e., $\mathbf{x} \in S$ such that $f(\mathbf{x}) = \mathbf{x}$. Applying Brouwer's theorem, Nash established that every finite, n-player game has an equilibrium point [14]. Arrow and Debreu [5] then extended the equilibrium theory to exchange markets.

FPC is somewhat related with optimization. One can reduce FPC to root finding: $\mathbf{x} \in S$ is a fixed point of f, if $f(\mathbf{x}) - \mathbf{x} = 0$, or $\|f(\mathbf{x}) - \mathbf{x}\| = 0$. Every global optimum of $g(\mathbf{x}) = \|f(\mathbf{x}) - \mathbf{x}\|$ is a fixed point of f. One can also view a local optimum of h as a fixed point. For every $\mathbf{x} \in D$, let $f_h(\mathbf{x})$ be a point in $N(\mathbf{x})$ that minimizes $h(\mathbf{x})$. Then, $\mathbf{x}$ is a fixed point of f_h iff $\mathbf{x}$ is a local optimum of h. Of course, this reduction from local search to FPC is less formal than the reduction from FPC to global optimization because the function f_h may not satisfy the "continuity" condition required by the Fixed Point Theorems. The following are two fundamental complexity questions about these search problems:

- Is global optimization strictly harder than fixed point computation?
- Is a fixed point harder to find than a local optimum?

To address these questions in the framework traditionally considered in Theoretical Computer Science, one may want to consider global optimization, local search, and FPC over discrete domains. For optimization problems, it is somewhat easier to define the discrete or combinatorial analog of continuous optimization, by considering discrete input domains, such as the hypercube $\{0, 1\}^n$ or grid $[n]^d$: Given a function h over $D = \{0, 1\}^n$ or $[n]^d$, find a global or local optimum of h. In local search one may consider $N(\mathbf{x})$ to be the direct neighbors of $\mathbf{x}$ in $\{0, 1\}^n$ or $[n]^d$.

The discrete FPC is less straightforward and some inaccuracy must be introduced to ensure the existence of a solution with finite description [17,18,15,12,9]. One idea is to consider approximate fixed points as suggested by Scarf [17] over a finite discretization of the convex domain, where a vertex $\mathbf{x}$ in the discretization is an approximate fixed point of a continuous map f if $\|f(\mathbf{x}) - \mathbf{x}\| \leq \epsilon$ for a given $\epsilon > 0$. Another idea is to use the direction-preserving functions (see Section 2) as introduced by Iimura, Murota, and Tamura [13] over $[n]^d$. One can also use Sperner's definition of discrete fixed points. Sperner's famous lemma states: Suppose that Ω is a d-dimensional simplex with vertices $v_1, v_2, ..., v_{d+1}$, and that $\mathcal{S}$ is a simplicial decomposition of Ω. Suppose Π assigns to each vertex of $\mathcal{S}$ a color from $\{1, 2, ..., d + 1\}$ such that, for every vertex v of $\mathcal{S}$, $\Pi(v) \neq i$ if the ith component of the barycentric coordinates of v (the convex combination of $v_1, v_2, ..., v_{d+1}$ to express v), is 0. Sperner's Lemma asserts that there exists a cell in $\mathcal{S}$ that contains all the $d + 1$ colors. This fully-colored simplex cell is often referred to as a *Sperner simplex* of $(\mathcal{S}, \Pi)$. Now consider a Brouwer map f with Lipschitz constant L over the simplex Ω. Suppose further that the diameter of each simplex cell in $\mathcal{S}$ is at most ϵ/L. Then, one can define a color assignment Π_f such that each fully-colored simplex in $(\mathcal{S}, \Pi_f)$ must have a vertex $\mathbf{v}$ satisfying $\|f(\mathbf{v}) - \mathbf{v}\| \leq \Theta(\epsilon)$. Thus, a fully-colored cell of $(\mathcal{S}, \Pi_f)$ can be viewed as an approximate, discrete fixed point of f. The Hirsch, Papadimitriou, and Vavasis model [12] extends Sperner's Lemma from the simplex to the hypergrid $[n]^d$.

If the function h for optimization or the map f for FPC is given concisely by a boolean circuit, then these three problems are search problems in complexity classes **FNP**, **PLS**, and **PPAD**, respectively. Other than **PPAD** $\subseteq$ **FNP** and **PLS** $\subseteq$ **FNP**, the precise relations between these classes remain unclear.

In a recent paper, Chen and Teng demonstrated that the randomized query model over $[n]^d$ strictly separates these three search problems [7]:

Global optimization is harder than FPC and FPC is harder than local search.

In particular, they proved that given a black-box, discrete Brouwer function f from $[n]^d$ to $[n]^d$, the randomized query complexity for finding an $\mathbf{x}$ such that $f(\mathbf{x}) = \mathbf{x}$ is $\Theta(n^{d-1})$. The separation statement then follows from two earlier results: A folklore theorem states that the randomized query complexity for finding a global optimum of a black-box function h from $[n]^d$ to $\mathbb{R}$ is $\Theta(n^d)$; Aldous [2] showed that the randomized query complexity for finding a local optimum of a black-box function h from $[n]^d$ to $\mathbb{R}$ is $O(n^{d/2})$.

They further conjectured that FPC is also strictly harder than local search in the quantum query model over $[n]^d$. In particular, they conjectured that the quantum query complexity of FPC over $[n]^d$ is $\Theta(n^{d/2})$. If this conjecture is true, then just like in its randomization counterpart, FPC is harder than local search under the quantum query model in two or higher dimensions.

Our Contributions
We prove a nearly tight bound of $\Omega(n^{(d-1)/2})$ on the quantum query complexity of FPC over grid $[n]^d$ — Grover Search solves FPC with $O(n^{d/2})$ quantum queries. Our result gives a nearly tight bound for the computation of d-dimensional approximate Brouwer fixed points as defined by Scarf and by Hirsch, Papadimitriou, and Vavasis [12]. It can be extended to the quantum model for Sperner's Lemma in any dimensions: The quantum query complexity of finding a panchromatic cell in a Sperner coloring of a uniform triangulation of a d-dimensional simplex with n^d cells is $\Omega(n^{(d-1)/2})$. For $d = 2$, this result improves the bound of $\Omega(n^{1/4})$ obtained by Friedl, Ivanyos, Santha, and Verhoeven [10].

Our result provides a quantum separation between local search and FPC over $[n]^d$, for $d \geq 4$. Aaronson's local search algorithm over $[n]^d$ makes $O(n^{d/3})$ quantum queries [1]. Thus, the quantum query model over $[n]^d$ strictly separates these two fundamental search problems when $d \geq 4$.

We use the quantum adversary argument of Ambainis [4,20] in the lower bound proof. We hide a distribution of random, directed paths in the host grid graph (over $[n]^d$) with a known starting vertex, and ask the algorithm to find the ending vertex of the path. This "path hiding" approach was used in [1,21,19] for deriving quantum lower bounds of local search. It was also used in [10] for deriving the $\Omega(n^{1/4})$ lower bound of the two-dimensional Sperner's problem. The main difference between our work and previous works is that the paths used in previous works have some monotonicity properties: there is an increasing (or decreasing) value along the path. Given any two vertices on the path, without querying other vertices one can decide which vertex appears earlier on the path. Such a monotonicity property makes it easier to derive good bound on the collision probability needed for a lower bound on local search, but limits the length of the path, making it impossible to derive tight lower bounds for FPC.

Instead of using "monotone" paths, we improve the path construction technique of Chen and Teng [7] in their randomized query lower bound for FPC, and make it work for the quantum adversary argument. We also find an interesting connection between the two discrete domains — hypergrid $[n]^d$ and hypercube $\{0,1\}^n$, which allows us to resolve a question left open in [7] on both the randomized and quantum query complexity of FPC over $\{0,1\}^n$. We show that they are $\Omega(2^{n(1-\epsilon)})$ and $\Omega(2^{n(1-\epsilon)/2})$ respectively, for all $\epsilon > 0$.

It remains open whether the quantum query complexity of fixed point computation over $[n]^d$ is indeed $\Theta(n^{d/2})$ or a better algorithm with quantum query complexity $\Theta(n^{(d-1)/2})$ exists.

Related Work

Hirsch *et al* [12] introduced the first query model for discrete FPC over $[n]^d$. They proved a tight $\Theta(n)$ deterministic bound for $[n]^2$, and an $\Omega(n^{d-2})$ deterministic lower bound in general. Chen and Deng [8] improved their bound to $\Theta(n^{d-1})$. Friedl, Ivanyos, Santha, and Verhoeven considered the 2-dimensional Sperner's problem [10]. They proved an $\Omega(n^{1/2})$ bound for its randomized query complexity and an $\Omega(n^{1/4})$ bound for its quantum query complexity.

Aaronson [1] was the first to introduce the quantum query complexity of local search over $[n]^d$ (and also $\{0,1\}^n$). He gave an upper bound $O(n^{d/3})$, and a lower bound $\Omega(n^{d/4-1/2}/\sqrt{\log n})$. Santha and Szegedy [16] proved a lower bound $\Omega(n^{1/4})$ for $d = 2$. Zhang [21] then obtained a lower bound of $\Omega(n^{d/3})$ (which is tight up to a log factor), for $d \geq 5$. In the same paper, he also obtained a nearly tight quantum bound for $\{0,1\}^n$. For $d = 2$ and $d = 3$, Sun and Yao [19] eventually gave an almost optimal lower bound.

2   Definition of Problems

We start with some notations. We let $\mathbb{Z}_n^d$ denote set $\{1, 2, ..., n\}^d$, and G_n^d denote the natural *directed* graph over $\mathbb{Z}_n^d$: edge $(\mathbf{u}, \mathbf{v}) \in G_n^d$ if there exists $i \in [d]$ such that $|u_i - v_i| = 1$ and $u_j = v_j$ for all other $j \in [d]$. We let H^n denote the following *directed* graph over hypercube $\{0,1\}^n$: edge $(\mathbf{u}, \mathbf{v}) \in H^n$ if there exists $i \in [n]$ such that $|u_i - v_i| = 1$ and $u_j = v_j$ for all other $j \in [n]$.

We use K_n to denote the complete *directed* graph of size n: the vertex set of K_n is $\{1, 2, ..., n\}$, and for all $1 \leq i \neq j \leq n$, (i, j) is an edge in K_n. We use K_n^d to denote the *Cartesian product* of d complete graphs: $K_n^d = K_n \square K_n \square ... \square K_n$.

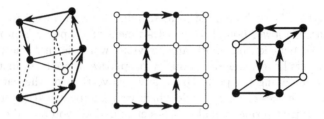

Fig. 1. END-OF-PATH problems over K_3^2, G_4^2 and H^3, respectively

More exactly, the vertex set of K_n^d is $\{1, 2, ..., n\}^d$; $(\mathbf{u}, \mathbf{v})$ is a directed edge of K_n^d if there exists $i \in [d]$ such that $u_i \neq v_i$ and $u_j = v_j$ for all other $j \in [d]$.

Let G be a directed graph and P be a *simple* directed path in G. We say $P = \mathbf{v}^1 \mathbf{v}^2 ... \mathbf{v}^k$, where $k \geq 1$, is simple if for all $1 \leq i \neq j \leq k$, $\mathbf{v}^i \neq \mathbf{v}^j$. Then P naturally induces a map $\mathcal{F}_P$ from the edge set of G to $\{0, 1\}$: for all $(\mathbf{u}, \mathbf{v}) \in G$, $\mathcal{F}_P(\mathbf{u}, \mathbf{v}) = 1$ if $(\mathbf{u}, \mathbf{v}) \in P$; and $\mathcal{F}_P(\mathbf{u}, \mathbf{v}) = 0$, otherwise. We let $\text{END}(P)$ denote the ending vertex of path P. Finally, we let $\mathbb{E}^d = \{\pm \mathbf{e}_1, \pm \mathbf{e}_2, ..., \pm \mathbf{e}_d\}$ denote the set of *principle unit-vectors* in d-dimensions. Let $\| \cdot \|$ denote $\| \cdot \|_\infty$.

Discrete Brouwer Fixed-Points

A function $f : \mathbb{Z}_n^d \to \{\mathbf{0}\} \cup \mathbb{E}^d$ is *bounded* if $f(\mathbf{x}) + \mathbf{x} \in \mathbb{Z}_n^d$ for all $\mathbf{x} \in \mathbb{Z}_n^d$; $\mathbf{v} \in \mathbb{Z}_n^d$ is a *zero point* of f if $f(\mathbf{v}) = \mathbf{0}$. Clearly, if $F(\mathbf{x}) = \mathbf{x} + f(\mathbf{x})$, then $\mathbf{v}$ is a fixed point of F if and only if $\mathbf{v}$ is a zero point of f.

A function f from set $S \subset \mathbb{Z}^d$ to $\{\mathbf{0}\} \cup \mathbb{E}^d$ is *direction-preserving* if $\| f(\mathbf{u}) - f(\mathbf{v}) \| \leq 1$ for all pairs $\mathbf{u}, \mathbf{v} \in S$ such that $\| \mathbf{u} - \mathbf{v} \| \leq 1$.

Following the discrete fixed-point theorem of [13], we have: for every function $f : \mathbb{Z}_n^d \to \{\mathbf{0}\} \cup \mathbb{E}^d$, if f is both bounded and direction-preserving, then it has at least one zero point. We refer to a bounded and direction-preserving function f over $\mathbb{Z}_n^d$ as a *Brouwer function* over $\mathbb{Z}_n^d$. In the query model, one can only access f by asking queries of the form: "What is $f(\mathbf{v})$?" for a point $\mathbf{v} \in \mathbb{Z}_n^d$.

The problem ZP^d that we will study is: *Given a Brouwer function* $f : \mathbb{Z}_n^d \to \{\mathbf{0}\} \cup \mathbb{E}^d$ *in the query model, find a zero point of* f. We use $\mathsf{QQ}_{\mathsf{ZP}}^d(n)$ to denote the *quantum query complexity* of problem ZP^d. A description of the quantum query model can be found in [6] and [3]. The main result of the paper is

Theorem 1 (Main). *For all $d \geq 2$ and large enough n, $\mathsf{QQ}_{\mathsf{ZP}}^d(n) = \Omega(n^{\frac{d-1}{2}})$.*

The End-of-Path Problems over Graphs K_n^d, G_n^d and H^n

To prove Theorem 1, we introduce the following d-dimensional problem KP^d (the END-OF-PATH problem over K_n^d): its input is a binary string of length $|K_n^d|$ (that is, the number of edges in K_n^d), which encodes the map $\mathcal{F}_P$ of a *simple directed* path P in K_n^d; P is known to start at $\mathbf{1} = (1, ..., 1) \in K_n^d$; and we need to find its ending vertex $\text{END}(P)$. We let $\mathsf{QQ}_{\mathsf{KP}}^d(n)$ denote the quantum query complexity of problem KP^d.

Similarly, for $d \geq 2$, we define the END-OF-PATH problem GP^d over G_n^d, and use $\mathsf{QQ}_{\mathsf{GP}}^d(n)$ to denote its quantum query complexity. The following reduction from GP^d to ZP^d was given in [7]: from any input $\mathcal{F}_P$ of GP^d, where P is a simple path in G_n^d (starting at $\mathbf{1}$), one can build a Brouwer function f over $\mathbb{Z}_{24n+7}^d$ such that: *1)* f has exactly one zero point $\mathbf{v}^*$. Once it is found, the ending vertex of P can be located immediately; *2)* For any $\mathbf{v} \in \mathbb{Z}_{24n+7}^d$, the value of f at $\mathbf{v}$ only depends on (at most) $4d$ bits of $\mathcal{F}_P$. By using Lemma 1 of [16], we have

Lemma 1. *For all $d \geq 2$, $\mathsf{QQ}_{\mathsf{GP}}^d(n) \leq O(d) \cdot \mathsf{QQ}_{\mathsf{ZP}}^d(24n+7)$.*

To give a lower bound for $\mathsf{QQ}_{\mathsf{GP}}^d(n)$, we reduce KP^d to GP^{d+1}, and prove the following lemma. The proof can be found in the full version.

Lemma 2. *For all $d \geq 1$, $\mathsf{QQ}_{\mathsf{KP}}^d(n) \leq O(d\sqrt{dn}) \cdot \mathsf{QQ}_{\mathsf{GP}}^{d+1}(4dn+1)$.*

Finally, in Section 3, we prove an almost-tight lower bound for problem KP^d.

Theorem 2. *For all $d \geq 1$ and large enough n, $\mathsf{QQ}_{\mathsf{KP}}^d(2n+3) = \Omega\big((n/2^{11})^{\frac{d+1}{2}}\big)$.*

As a result, Theorem 1 follows directly from Lemma 1, 2 and Theorem 2.

Besides, as a by-product, our work also implies an almost-tight lower bound for the END-OF-PATH problem HP over H^n: its input is a binary string of length $|H^n|$ which encodes the map $\mathcal{F}_P$ of a *simple directed* path P in H^n; P is known to start at $\mathbf{0} = (0, 0, ..., 0) \in \{0,1\}^n$; and we need to find $\mathrm{END}(P)$. Let $\mathsf{QQ}_{\mathsf{HP}}(n)$ denote its quantum query complexity, then in the full version, we show that

Lemma 3. *For all $d \geq 2$ and $n \geq 1$, $\mathsf{QQ}_{\mathsf{GP}}^d(2^n) \leq O(1) \cdot \mathsf{QQ}_{\mathsf{HP}}(dn)$.*

Corollary 1. *For all $\epsilon > 0$ and large enough n, $\mathsf{QQ}_{\mathsf{HP}}(n) = \Omega(2^{n(1-\epsilon)/2})$.*

3   An Almost-Tight Lower Bound for KP^d

Using Grover's search [11], we get the following upper bound for $\mathsf{QQ}_{\mathsf{KP}}^d$:

Lemma 4. *For all $d \geq 1$, $\mathsf{QQ}_{\mathsf{KP}}^d(n) = O(\sqrt{d \cdot n^{d+1}})$.*

To prove a matching lower bound, we need the following theorem from [4,20]:

Theorem 3. *Let $f : S \to \{0,1\}^{n_1}$ be a partial function, where $S \subset \{0,1\}^{n_2}$. Let $w : S \times S \to \{0,1\}$ be a map satisfying the following condition: $w(\mathbf{x}, \mathbf{y}) = w(\mathbf{y}, \mathbf{x})$ for all $\mathbf{x}, \mathbf{y} \in S$ and $w(\mathbf{x}, \mathbf{y}) = 0$ whenever $f(\mathbf{x}) = f(\mathbf{y})$. Then the quantum query complexity $\mathsf{QQ}(f)$ of f satisfies*

$$\mathsf{QQ}(f) = \Omega\left(\min_{\substack{\mathbf{x}, \mathbf{y}, i : x_i \neq y_i \\ w(\mathbf{x}, \mathbf{y})=1}} \sqrt{\frac{1}{\theta(\mathbf{x}, i)\theta(\mathbf{y}, i)}}\right), \text{ where } \theta(\mathbf{x}, i) = \frac{\sum_{\mathbf{y}' \in S, \, y_i' \neq x_i} w(\mathbf{x}, \mathbf{y}')}{\sum_{\mathbf{y}' \in S} w(\mathbf{x}, \mathbf{y}')}$$

for all $\mathbf{x} \in S$ and $i \in [n_2]$.

The sketch of the proof is as follows. First, we build a set of *hard paths* $\mathcal{S}_m^d$ in graph K_{2m+3}^d for $d \geq 1$ and $m \geq 2$. These paths induce a collection of binary strings $\{\mathcal{F}_P, P \in \mathcal{S}_m^d\}$ of length $|K_{2m+3}^d|$, which plays the role of S in Theorem 3. Naturally, the f in Theorem 3 maps each string $\mathcal{F}_P$ to $\mathrm{END}(P)$. Then, given any

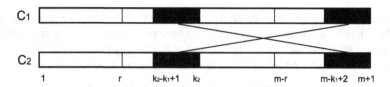

Fig. 2. Connectors C_1 and C_2, where C_1 can be r-transformed to C_2 with (k_1, k_2)

small enough $\beta > 0$, we define a relation $R^d_{m,\beta}$ over set $\mathcal{S}^d_m \times \mathcal{S}^d_m$. It induces a relation over $\{\mathcal{F}_P, P \in \mathcal{S}^d_m\} \times \{\mathcal{F}_P, P \in \mathcal{S}^d_m\}$, which satisfies all the conditions for w in Theorem 3. Finally, we analyze θ and use Theorem 3 to obtain a lower bound for $\mathsf{QQ}^d_{\mathsf{KP}}(2m+3)$.

We let $\mathbf{v} = (v_1, ..., v_d)$ denote a vertex in K^d_m, where $v_i \in [m]$ for all $i \in [d]$. For $d \geq 2$, we let D_d denote the map from $\mathbb{Z}^d$ to $\mathbb{Z}^{d-1}$: $D_d(\mathbf{v}) = (v_1, ..., v_{d-1})$. Let $\mathbf{v} \in \mathbb{Z}^{d-1}$ and $t \in \mathbb{Z}$, we use $(\mathbf{v}, t)$ to denote the vertex $\mathbf{u} \in \mathbb{Z}^d$ with $u_d = t$ and $u_i = v_i$ for all $i \in [d-1]$.

Construction of the Hard Paths

For all $d \geq 1$ and $m \geq 2$, we now construct, inductively, a set of simple paths $\mathcal{S}^d_m$ over K^d_{2m+3}. All the paths in $\mathcal{S}^d_m$ start with $\mathbf{1} = (1, 1, ..., 1) \in \mathbb{Z}^d$.

Definition 1 (m-connector). *Every permutation $\pi : [m+1] \to [m+1]$ with $\pi(1) = 1$ defines a sequence of $2m + 3$ integers: $C = 1 \circ 2\pi(1) \circ (2\pi(1) + 1) \circ 2\pi(2) \circ (2\pi(2) + 1) \circ ... \circ 2\pi(m+1) \circ (2\pi(m+1) + 1)$. Such a sequence C is called an m-connector. For each $i \in [m+1]$, we also use $C(i)$ to denote $2\pi(i)$, so $C = 1 \circ C(1) \circ (C(1) + 1) \circ ... \circ C(m+1) \circ (C(m+1) + 1)$. We let $\mathcal{C}_m$ denote the set of all m-connectors.*

The construction of $\mathcal{S}^d_m$ when $d = 1$ is straight-forward: $P \in \mathcal{S}^1_m$ if there is a $C \in \mathcal{C}_m$ such that P has the following edges: $(1, C(1)), (C(1), C(1) + 1), (C(1) + 1, C(2))...(C(m+1), C(m+1) + 1)$. We also say P is generated by C.

For the case when $d > 1$, we assume $\mathcal{S}^{d-1}_m$ has already been constructed. A path P is in $\mathcal{S}^d_m$ if it can be generated by a $(2m+4)$-tuple $(C, P_1, ..., P_{2m+3})$, where $C \in \mathcal{C}_m$ and $P_i \in \mathcal{S}^{d-1}_m$. The $2m + 3$ paths $P_1, P_2, ..., P_{2m+3}$ in the tuple must satisfy the following condition: $\mathrm{END}(P_1) = \mathrm{END}(P_{C(1)})$ and $\mathrm{END}(P_{C(i)+1}) = \mathrm{END}(P_{C(i+1)})$ for all $i \in [m]$.

Path P is a simple path in graph K^d_{2m+3} containing the following edges:

1. For all $i \in \{1, 3, ..., 2m+1, 2m+3\}$ and $\mathbf{u}, \mathbf{v} \in K^d_m$ with $u_d = v_d = i$, edge $(\mathbf{u}, \mathbf{v}) \in P$ if and only if $(D_d(\mathbf{u}), D_d(\mathbf{v})) \in P_i$;
2. For all $i \in \{2, 4, ..., 2m, 2m+2\}$ and vertices $\mathbf{u}, \mathbf{v} \in K^d_m$ with $u_d = v_d = i$, edge $(\mathbf{u}, \mathbf{v}) \in P$ if and only if $(D_d(\mathbf{v}), D_d(\mathbf{u})) \in P_i$;
3. For all $i \in [m+1]$, $((1, 1, ..., 1, C(i)), (1, 1, ..., 1, C(i) + 1)) \in P$;
4. Let $\mathbf{v} = \mathrm{END}(P_1)$, then $((\mathbf{v}, 1), (\mathbf{v}, C(1))) \in P$;
5. For all $i \in [m]$, let $\mathbf{v} = \mathrm{END}(P_{C(i)+1})$, then $((\mathbf{v}, C(i) + 1), (\mathbf{v}, C(i+1))) \in P$.

The Relation $R^d_{m,\beta}$ over $\mathcal{S}^d_m \times \mathcal{S}^d_m$

Let $r \in \mathbb{Z}^+$ be an integer such that $2r + 1 \leq m$.

Let C_1 and C_2 be two m-connectors. For integers $k_1 \in [r]$ and $r + 1 \le k_2 \le m - r$, we say C_1 can be r-*transformed* to C_2 with parameters (k_1, k_2) if: for all $i \in [k_1]$, $C_2(k_2 - k_1 + i) = C_1(m - k_1 + 1 + i)$ and $C_2(m - k_1 + 1 + i) = C_1(k_2 - k_1 + i)$; for all other indices $j \in [m + 1]$, $C_2(j) = C_1(j)$.

See Figure 2 for an example. Clearly, if connector C_1 can be r-transformed to C_2, then C_2 can also be r-transformed to C_1 with the same parameters.

Now, given a triple $\tau = (m, d, \beta)$ such that $\beta \in (0, 32^{-d}]$ and $m\beta \in \mathbb{Z}^+$, we inductively define a symmetric relation $R^d_{m,\beta} = R_\tau$ from set $S^d_m \times S^d_m$ to $\{0, 1\}$. We will use it as the function w in the quantum adversary argument (see Theorem 3). Before presenting details of the construction, we introduce the following useful notations: For path $P \in S^d_m$ and vertex $\mathbf{v} \in K^d_{2m+3}$, we let

- $S_\tau[P] = \{P' \in S^d_m, R_\tau(P, P') = 1\}$ and $N_\tau[P] = |S_\tau[P]|$;
- $S_\tau[P, \mathbf{v}] = \{P' \in S_\tau[P], \text{END}(P') = \mathbf{v}\}$ and $N_\tau[P, \mathbf{v}] = |S_\tau[P, \mathbf{v}]|$;
- $V_\tau[P] = \{\mathbf{v} \in K^d_{2m+3}, N_\tau[P, \mathbf{v}] > 0\}$.

When defining $R^d_{m,\beta} = R_\tau$, we will also prove that

$$|V_\tau[P]| = ((1 - 2\beta)m)^d \text{ and } V_\tau[P] \subset \{5, 7, ..., 2m + 3\}^d, \text{ for all } P \in S^d_m. \quad (1)$$

For the case when $d = 1$, assume P and P' are two paths in S^1_m which are generated by m-connectors C and C' in $\mathcal{C}_m$, respectively. Then, $R_\tau(P, P') = 1$ iff there exist integers k_1 and k_2 such that C can be (βm)-transformed to C' with parameters (k_1, k_2). Clearly, this definition of $R^1_{m,\beta}$ implies that $V_\tau[P] = \{C(r + 1) + 1, C(r + 2) + 1, ..., C(m - r) + 1\} \subset \{5, 7, ..., 2m + 3\}$, where $r = \beta m$, and thus, $|V_\tau[P]| = (1 - 2\beta)m$.

For the case when $d > 1$, we use $\bar{\tau}$ to denote the triple $(m, d - 1, \beta)$. By the inductive hypothesis, we assume relation $R_{\bar{\tau}}$ over $S^{d-1}_m \times S^{d-1}_m$ has already been defined, since $\beta \in (0, 32^{-d}] \subset (0, 32^{-(d-1)}]$. Furthermore, by Eq.(1), we have

$$|V_{\bar{\tau}}[P]| = ((1 - 2\beta)m)^{d-1}, \ V_{\bar{\tau}}[P] \subset \{5, 7, ..., 2m + 3\}^{d-1}, \text{ for all } P \in S^{d-1}_m. \quad (2)$$

Let P and P' be two paths in S^d_m. Assume they are generated by $(C, P_1, ..., P_{2m+3})$ and $(C', P'_1, ..., P'_{2m+3})$, respectively. We set $R_\tau(P, P') = 1$ if the following conditions are satisfied:

- Letting $r = \beta m$, there exist integers $k_1 \in [r]$ and $r + 1 \le k_2 \le m - r$ such that C can be r-transformed to C' with parameters (k_1, k_2);
- We let r_1, r_2, r_3, r_4 denote $C(k_2 - k_1) + 1$, $C(m + 1) + 1$, $C(m - k_1 + 1) + 1$ and $C(k_2) + 1$, respectively. We let l_1, l_2, l_3 denote $C(m - k_1 + 2)$, $C(k_2 + 1)$ and $C(k_2 - k_1 + 1)$, respectively;
- For each $i \in [3]$, there exists $\mathbf{v} \in V_{\bar{\tau}}[P_{r_i}] \cap V_{\bar{\tau}}[P_{l_i}]$ such that $P'_{r_i} \in S_{\bar{\tau}}[P_{r_i}, \mathbf{v}]$ and $P'_{l_i} \in S_{\bar{\tau}}[P_{l_i}, \mathbf{v}]$;
- $P'_{r_4} \in S_{\bar{\tau}}[P_{r_4}]$; For all other $j \in [2m + 3]$, $P'_j = P_j$.

Proof (Proof of Eq.(1)). We prove $V_\tau[P] = \bigcup_{t=C(k)+1, \ r+1 \le k \le m-r} V_{\bar{\tau}}[P_t] \times \{t\}$.

To prove this, it suffices to show that, for every $\mathbf{v} \in K^d_{2m+3}$ such that $v_d = C(k) + 1$ with $r + 1 \le k \le m - r$ and $(v_1, ..., v_{d-1}) \in V_{\bar{\tau}}[P_{v_d}]$, $N_\tau[P, \mathbf{v}] > 0$.

For all $i, j : 1 \leq i \neq j \leq m + 1$, we let

$$M_{i,j} = \sum_{\mathbf{v} \in V_{\bar{\tau}}[P_{C(i)+1}] \cap V_{\bar{\tau}}[P_{C(j)}]} N_{\bar{\tau}}[P_{C(i)+1}, \mathbf{v}] \cdot N_{\bar{\tau}}[P_{C(j)}, \mathbf{v}].$$

From Eq.(2), we have $|V_{\bar{\tau}}[P_{C(i)+1}] \cap V_{\bar{\tau}}[P_{C(j)}]| \geq 2((1 - 2\beta)m)^{d-1} - m^{d-1} > 0$, since $\beta < 32^{-d}$. Therefore, $M_{i,j} > 0$ for all i, j.

On the other hand, we can write $N_{\tau}[P, \mathbf{v}]$ as

$$\sum_{k=1}^{r} M_{k-k_1, m-k_1+2} \, M_{m+1, k+1} \, M_{m-k_1+1, k-k_1+1} \, N_{\bar{\tau}}[P_{v_d}, (v_1, ..., v_{d-1})]. \quad (3)$$

As a result, $N_{\tau}[P, \mathbf{v}] > 0$ since we assumed that $(v_1, ..., v_{d-1}) \in V_{\bar{\tau}}[P_{v_d}]$.

We now prove the following important lemma about relation $R_{m,\beta}^d = R_{\tau}$:

Lemma 5. *For $d \geq 1$ and $\beta \in (0, 32^{-d}]$ such that $r = \beta m \in \mathbb{Z}^{+}$, we have*

$$\frac{1}{\mu_d(\beta)} \leq \frac{N_{\tau}[P, \mathbf{v}]}{N_{\tau}[P', \mathbf{v}']} \leq \mu_d(\beta), \text{ for all } P, P' \in \mathcal{S}_m^d, \, \mathbf{v} \in V_{\tau}[P] \text{ and } \mathbf{v}' \in V_{\tau}[P'],$$

where $\mu_d(\beta)$ is defined inductively as follows: $\mu_1(\beta) = 1$; for $d \geq 2$,

$$\mu_d(\beta) = (\mu_{d-1}(\beta))^7 / (2(1 - 2\beta)^{d-1} - 1)^3 .$$

Proof. The case when $d = 1$ is trivial. For $d > 1$, suppose P is generated by $(C, P_1, ..., P_{2m+3})$ and P' is generated by $(C', P_1', ..., P_{2m+3}')$. We similarly define $M_{i,j}$ and $M_{i,j}'$ for P and P', respectively, then by using the inductive hypothesis and Eq.(1), we have $M_{i,j} / M_{i',j'} \leq (\mu_{d-1}(\beta))^2 / (2(1 - 2\beta)^{d-1} - 1)$. The lemma then follows directly by applying this inequality to every item in Eq.(3).

Using induction, one can prove that for $\beta \in (0, 32^{-d}]$, $\mu_d(\beta) \leq e^{32^{d-1}\beta}$.

Proof of the Lower Bound

For $d \geq 1$ and $\beta \in (0, 32^{-d}]$, we prove that, when m is large enough (and satisfies $\beta m \in \mathbb{Z}^{+}$), relation $R_{\tau} = R_{m,\beta}^d$ can serve as the function w in Theorem 3, and give us a lower bound for $QQ^d(2m + 3)$.

Let $(\mathbf{v}^1, \mathbf{v}^2)$ be a directed edge in K_{2m+3}^d, $\mathbf{v} \in K_{2m+3}^d$, and path $P \in \mathcal{S}_m^d$. We introduce the following notations:

- $\mathcal{S}_{\tau}[P, (\mathbf{v}^1, \mathbf{v}^2)] = \{P' \in \mathcal{S}_{\tau}[P], \, \mathcal{F}_P(\mathbf{v}^1, \mathbf{v}^2) \neq \mathcal{F}_{P'}(\mathbf{v}^1, \mathbf{v}^2)\}$;
- $\mathcal{S}_{\tau}[P, \mathbf{v}, (\mathbf{v}^1, \mathbf{v}^2)] = \mathcal{S}_{\tau}[P, \mathbf{v}] \cap \mathcal{S}_{\tau}[P, (\mathbf{v}^1, \mathbf{v}^2)]$;
- $N_{\tau}[P, (\mathbf{v}^1, \mathbf{v}^2)] = |\mathcal{S}_{\tau}[P, (\mathbf{v}^1, \mathbf{v}^2)]|, \, N_{\tau}[P, \mathbf{v}, (\mathbf{v}^1, \mathbf{v}^2)] = |\mathcal{S}_{\tau}[P, \mathbf{v}, (\mathbf{v}^1, \mathbf{v}^2)]|$

We also let

$$\theta_{\tau}(P, (\mathbf{v}^1, \mathbf{v}^2)) = \frac{N_{\tau}[P, (\mathbf{v}^1, \mathbf{v}^2)]}{N_{\tau}[P]} = \frac{\sum_{\mathbf{v} \in V_{\tau}[P]} N_{\tau}[P, \mathbf{v}, (\mathbf{v}^1, \mathbf{v}^2)]}{N_{\tau}[P]}.$$

Theorem 2 follows as a corollary of Theorem 3 and Lemma 6 below. The proof of Lemma 6 can be found in the full version.

Lemma 6. *Let $d \geq 1$ and $\beta \in (0, 32^{-d}]$. There exists a constant $L_{d,\beta}$ such that for all $m \geq L_{d,\beta}$ with $\beta m \in \mathbb{Z}^{+}$, if $P, P' \in \mathcal{S}_m^d$ satisfy $R_{m,\beta}^d(P, P') = R_{\tau}(P, P') = 1$ and $\mathcal{F}_P(\mathbf{v}^1, \mathbf{v}^2) \neq \mathcal{F}_{P'}(\mathbf{v}^1, \mathbf{v}^2)$ for some edge $(\mathbf{v}^1, \mathbf{v}^2) \in K_{2m+3}^d$, then*

$$\theta_{\tau}(P, (\mathbf{v}^1, \mathbf{v}^2)) \cdot \theta_{\tau}(P', (\mathbf{v}^1, \mathbf{v}^2)) \leq (2^{11}/m)^{d+1} .$$

References

1. Aaronson, S.: Lower Bounds for Local Search by Quantum Arguments. In: Proc. of the 36th STOC, pp. 465–474 (2004)
2. Aldous, D.: Minimization Algorithms and Random Walk on the d-Cube. Annals of Probability 11(2), 403–413 (1983)
3. Ambainis, A.: Quantum Lower Bounds by Quantum Arguments. In: Proc. of the 32nd FOCS, pp. 636–643 (2000)
4. Ambainis, A.: Polynomial Degree vs. Quantum Query Complexity. In: Proc. of the 44th FOCS, pp. 230–239 (2003)
5. Arrow, K., Debreu, G.: Existence of an Equilibrium for a Competitive Economy. Econometrica 22(3), 265–290 (1954)
6. Beals, R., Buhrman, H., Cleve, R., Mosca, M., de Wolf, R.: Quantum Lower Bounds by Polynomials. Journal of ACM 48(4), 778–797 (2001)
7. Chen, X., Teng, S.H.: Paths Beyond Local Search: A Tight Bound for Randomized Fixed-Point Computation. In: Proc. of the 48th FOCS, pp. 124–134 (2007)
8. Chen, X., Deng, X.: On Algorithms for Discrete and Approximate Brouwer Fixed Points. In: Proc. of the 37th STOC, pp. 323–330 (2005)
9. Deng, X., Papadimitriou, C., Safra, S.: On the Complexity of Price Equilibria. Journal of Computer and System Sciences 67(2), 311–324 (2003)
10. Friedl, K., Ivanyos, G., Santha, M., Verhoeven, F.: On the Black-Box Complexity of Sperner's Lemma. In: Liśkiewicz, M., Reischuk, R. (eds.) FCT 2005. LNCS, vol. 3623, pp. 245–257. Springer, Heidelberg (2005)
11. Grover, L.: A Fast Quantum Mechanical Algorithm for Database Search. In: Proc. of the 28th STOC, pp. 212–219 (1996)
12. Hirsch, M., Papadimitriou, C., Vavasis, S.: Exponential Lower Bounds for Finding Brouwer Fixed Points. Journal of Complexity 5, 379–416 (1989)
13. Iimura, T., Murota, K., Tamura, A.: Discrete Fixed Point Theorem Reconsidered. Journal of Mathematical Economics 41, 1030–1036 (2005)
14. Nash, J.: Equilibrium Point in n-Person Games. Proceedings of the National Academy of the USA 36(1), 48–49 (1950)
15. Papadimitriou, C.: On Inefficient Proofs of Existence and Complexity Classes. In: Proc. of the 4th Czechoslovakian Symposium on Combinatorics (1991)
16. Santha, M., Szegedy, M.: Quantum and Classical Query Complexities of Local Search are Polynomially Telated. In: Proc. of the 36th STOC, pp. 494–501 (2004)
17. Scarf, H.: The Approximation of Fixed Points of a Continuous Mapping. SIAM Journal on Applied Mathematics 15, 997–1007 (1967)
18. Scarf, H.: On the Computation of Equilibrium Prices. In: Fellner, W. (ed.) Ten Economic Studies in the Tradition of Irving Fisher. John Wiley & Sons, Chichester (1967)
19. Sun, X., Yao, A.C.: On the Quantum Query Complexity of Local Search in Two and Three dimensions. In: Proc. of the 47th FOCS, pp. 429–438 (2006)
20. Zhang, S.: On the Power of Ambainis's Lower Bounds. Theoretical Computer Science 339(2-3), 241–256 (2005)
21. Zhang, S.: New Upper and Lower Bounds for Randomized and Quantum Local Search. In: Proc. of the 38th STOC, pp. 634–643 (2006)

Multi-party Quantum Communication Complexity with Routed Messages*

Seiichiro Tani[1], Masaki Nakanishi[2], and Shigeru Yamashita[2]

[1] NTT Communication Science Laboratories, NTT Corporation
tani@theory.brl.ntt.co.jp
[2] Graduate School of Information Science,
Nara Institute of Science and Technology
{m-naka,ger}@is.naist.jp

Abstract. This paper describes a general quantum lower bounding technique for the communication complexity of a function that depends on the inputs given to two parties connected via paths, which may be shared with other parties, on a network of any topology. The technique can also be employed to obtain a lower-bound of the quantum communication complexity of some functions that depend on the inputs distributed over all parties on the network. As a typical application, we apply our technique to the *distinctness* problem of deciding whether there are at least two parties with identical inputs, on a k-party ring; almost matching upper bounds are also given.

1 Introduction

Studying *communication complexity* has been one of the central issues in computer science since its introduction by Yao [18]. Not only it is interesting in its own right, but it also has many applications such as analyzing VLSI circuit designs, data structures and networks (see the book [12] for more details).

In the simplest case where there are two parties connected to each other by a communication channel, the two parties, Alice and Bob, get inputs $x \in \{0,1\}^n$ and $y \in \{0,1\}^n$, respectively, and cooperatively compute $f(x,y) : \{0,1\}^n \times \{0,1\}^n \to \{0,1\}$ by exchanging messages. For example, Alice first performs local computation depending on her input and sends a message to Bob. He then does some local computation depending on his input and the received message, and sends a message back to Alice. This message exchange is repeated until Alice or Bob outputs the value of f. For any protocol $\mathcal{P}$ that computes f, the cost of $\mathcal{P}$ is the number of communication bits on the worst-case input (x,y). The communication complexity of f, $D(f)$, is the minimum cost of $\mathcal{P}$, over all protocols $\mathcal{P}$ that compute f. Protocol $\mathcal{P}$ may be randomized, i.e., Alice and Bob can access random strings r_A and r_B, respectively, in addition to the inputs they receive. The communication complexity of a randomized protocol that computes

---

* This work was supported in part by Grant-in-Aid for Scientific Research (KAKENHI): (16092218), (18700011), (19700010) and (19700019).

X. Hu and J. Wang (Eds.): COCOON 2008, LNCS 5092, pp. 180–190, 2008.

f is the number of communication bits in the worst-case over all inputs and all random strings. The communication complexity $R_\epsilon(f)$ of f for error probability ϵ is the minimum communication complexity over all randomized protocols that compute f with error probability at most ϵ for every input. If ϵ is bounded by a certain constant that is less than $1/2$, we call it *bounded error*. Without loss of generality, ϵ is assumed to be $1/3$ in the bounded error setting unless it is explicitly set to a different value. There is another randomized setting: a randomized protocol that never outputs an incorrect answer, but may give up with probability at most ϵ. We call such a protocol a Las Vegas protocol or a zero-error protocol. The communication complexity of f in the zero-error setting is denoted by $R_{0,\epsilon}(f)$. Furthermore, there is another way of giving random strings to Alice and Bob: they are allowed to access public coins (*or* a common random string). Formally, the output of protocol $\mathcal{P}$ depends on the inputs and common random string r. The public-coin versions of $R_\epsilon(f)$ and $R_{0,\epsilon}(f)$ are denoted by $R_\epsilon^{pub}(f)$ and $R_{0,\epsilon}^{pub}(f)$, respectively.

Quantum communication complexity, introduced by Yao [19], is the quantum counterpart of (classical) communication complexity. Parties are allowed to perform quantum computation and send/receive quantum bits (*or* qubits). The communication complexities, $Q_E(f)$, $Q_\epsilon(f)$ and $Q_{0,\epsilon}(f)$ are defined as the quantum counterparts of $D(f)$, $R_\epsilon(f)$ and $R_{0,\epsilon}(f)$, respectively. In particular, the quantum counter part of deterministic computation (protocol, algorithm, etc.) is called exact computation (protocol, algorithm, etc.); it runs in bounded time and always outputs the correct answer.

It is known that there are functions for which non-constant gaps exist between quantum and classical communication complexity. For exact computation, Buhrman et al. [4] proved that for a certain promise version of the equality function EQ'_n, $Q_E(EQ'_n) = O(\log n)$ while $D(EQ'_n) \in \Omega(n)$ [6]. In the bounded-error case, Raz [15] showed a promise problem that has an exponential gap between quantum and classical settings, i.e., $Q_{1/3}(f) = O(\log n)$ and $R_{1/3}(f) = \Omega(n^{1/4}/\log n)$. As for total functions, the largest known gap is quadratic: $Q_{1/3}(\mathsf{DISJ}_n) = \Theta(\sqrt{n})$ [1,16] and $R_{1/3}(\mathsf{DISJ}_n) = \Theta(n)$ [9], where DISJ_n is the $2n$-bit disjoint function, i.e., $\bigwedge_{i=1}^{n}(\overline{x_i y_i})$. Exponential gaps have been demonstrated for restricted or other models; examples include the bounded-error one-way communication model [7] and the bounded-error simultaneous message-passing model [5].

As mentioned above, there have been a lot of researches on the standard two-party communication model for quantum communication complexity. On the other hand, unlike the classical case, there is almost no research that considers more general (and more natural when we consider the Internet) model, i.e., distributed quantum computing over multiple parties on a network whose underlying graph is not necessarily complete. In this setting, a certain pair of parties may have to communicate with each other via some other parties; it seems difficult to directly apply known techniques for the standard two-party communication model.

Our Contribution. We first show how the quantum communication complexity $Q_{1/3}(f)$ of $f(x, y)$ in the standard two party case characterizes a lower bound of the number of communication qubits needed to compute $f(x, y)$ when x and y are given as input to two parties on any network G consisting of many parties. More strictly, the lower bound $Q_{1/3}^G(f)$ is characterized by the length s of the shortest path between the two parties on G and the maximum cut size w of G as well as the input size and $Q_{1/3}(f)$:

$$Q_{1/3}^G(f) = \Omega(s(Q_{1/3}(f) - \log\min\{n, s\})/\log w).$$

Our approach is to extend the classical deterministic lower bound technique in [17] to the quantum case. To do so, we introduce a new notion "*quantum protocol with classical public coins*," and then use the idea of the classical lower bound technique in a careful combination with the quantum version of the *public-to-private randomness conversion technique*. We also prove similar results in the zero-error setting.

We then apply the lower bound technique to lower-bound the quantum communication complexity of computing a fundamental problem, the distinctness problem, on a k-party ring with bounded error probability: the problem is deciding whether there are at least two parties who get identical input in $\{1, \ldots, L\}$ on a k-party ring, for which we derive lower bound $\Omega(k(\sqrt{k} + \log\log L))$. We also give two quantum protocols for the problem. The first algorithm gives almost the matching upper bound: $O(k(\sqrt{k}\log k + \log\log L))$. The second algorithm gives optimal bound $O(k\sqrt{k})$ in the case of $L = O(k)$. As far as we know, this is the first non-trivial result of almost tight bounds of multi-party quantum communication complexity on a network whose underlying graph is not complete.

2   Basic Tools

Converting Public Coins into Private Coins. In what follows, we assume that communication is quantum, but parties share no prior-entanglement. If a quantum protocol allows parties to access an arbitrary number of classical public coins, it is called a *quantum protocol with classical public coins*. $Q_\epsilon^{pub}(f)$ is defined as the minimum communication complexity over all *quantum protocols with classical public coins* that compute f with error probability at most ϵ.

As in the classical case [14], we would like to be able to replace many public coins with a small number of communication bits in the case of quantum protocols with classical public coins. Although it looks very similar to the classical case (also mentioned in [10]), the proof needs to be modified to handle quantum errors.

Lemma 1. *Let $f : \{0, 1\}^n \times \{0, 1\}^n \to \{0, 1\}$ be a Boolean function. For every positive real δ and ϵ ($\delta + \epsilon < 1/2$), any ϵ-error quantum protocol with classical public coins can be transformed into an $(\epsilon + \delta)$-error quantum protocol without classical public coins by using additional $\lceil \log n + 2\log 1/\delta \rceil$-bit communication.*

Proof. Suppose that we have any ϵ-error quantum protocol with classical public coins, $\mathcal{P}$, that computes f, and assume that $\mathcal{P}$ chooses a random string according to probability distribution Π over all possible random strings. Let $P(x, y, r)$ be the event that $\mathcal{P}$ is given input (x, y) and chooses particular string r as the random string. The error probability of $\mathcal{P}$ under event $P(x, y, r)$, i.e., the probability that the output of $\mathcal{P}$ under $P(x, y, r)$ is not equal to $f(x, y)$, is denoted by $\mathbf{Er}[P(x, y, r)]$.

We will show that there exist $t = \lceil n/\delta^2 \rceil$ strings $r_1, \ldots, r_t$ such that, for every input (x, y), the expected value of $\mathbf{Er}[P(x, y, r)]$ for random r chosen uniformly from the t strings is at most $\epsilon + \delta$. Therefore, if Alice randomly chooses one of the t strings and sends the $\log t$ bits specifying the chosen string to Bob, then they can compute f with error probability at most $\epsilon + \delta$. The lemma follows.

Choose t strings $r_1, \ldots, r_t$ according to probability distribution Π of common random strings. Since $0 \le \mathbf{Er}[P(x, y, r_i)] \le 1$, we can show by the Hoeffding inequality for fixed input (x, y) that

$$\mathbf{Pr}_{r_1, \ldots, r_t} \left[\left(\frac{1}{t} \sum_{i=1}^{t} \mathbf{Er}[P(x, y, r_i)] - \epsilon \right) > \delta \right] \le 2 e^{-2\delta^2 t}.$$

If we set t to $\lceil n/\delta^2 \rceil$, $2 e^{-2\delta^2 t}$ is smaller than 2^{-2n}. Therefore, the probability that, for some input (x, y), $\frac{1}{t} \sum_{i=1}^{t} \mathbf{Er}[P(x, y, r_i)] > \epsilon + \delta$ is smaller than $2^{-2n} \cdot 2^{2n} = 1$. This implies that there exist $r_1, \ldots, r_t$ such that for every input (x, y), $\frac{1}{t} \sum_{i=1}^{t} \mathbf{Er}[P(x, y, r_i)] \le \epsilon + \delta$. □

This lemma can be easily generalized to the case of k parties, in which every party i gets $x_i \in \{0, 1\}^n$ as input and they have to compute function f depending on x_i's.

Lemma 2. *Let $f : \{0, 1\}^{nk} \to \{0, 1\}$ be a Boolean function. For every positive real δ and ϵ ($\delta + \epsilon < 1/2$) , any ϵ-error quantum protocol with classical public coins that computes f on k parties can be transformed into an $(\epsilon + \delta)$-error quantum protocol without classical public coins, by using additional communication to broadcast a $\lceil \log(kn) + 2 \log 1/\delta \rceil$-bit message.*

Proof. Follow the same argument with $t = \lceil kn/(2\delta^2) \rceil$. □

In the case of a ring, the additional communication is just $k \lceil \log(kn) + 2 \log 1/\delta \rceil$-bits, since broadcasting involves passing the message around the ring. For zero-error quantum protocols, we can obtain similar results by considering the deviation of the average give-up probability over t random strings from the average give-up probability over all random strings.

3   General Lower Bound

Now we describe our key theorem which lower-bounds the total quantum communication complexity over all links of a network of any topology by using the ordinary quantum communication complexity of the two party case.

Theorem 1. *Suppose that n-bit strings x and y are given to two parties P_A and P_B, respectively, on network G of any topology, where s is the length of the shortest path between P_A and P_B and w is the size of the maximum cut of the underlying graph that separates P_A and P_B. Let $Q_\epsilon^G(f)$ be the total quantum communication complexity over all links in G of computing a Boolean function $f(x,y)$ with error probability at most ϵ ($0 \leq \epsilon < 1/2$). It follows that $Q_\epsilon^G(f)$ is at least $\frac{s}{\lceil \log w \rceil}$ times*

$$\max\{\delta_1(Q_{\epsilon+\delta_1/2+\delta_2}(f) - \lceil \log(n/\delta_2^2) \rceil), \delta_3(Q_{\epsilon+\delta_3/2}(f) - \lceil \log s \rceil)\},$$

where $0 < \delta_1, \delta_2, \delta_3 < 1$ such that $\epsilon + \delta_1/2 + \delta_2$ and $\epsilon + \delta_3/2$ are smaller than $1/2$, and $Q_\epsilon(f)$ denotes the quantum communication complexity of $f(x,y)$ for error probability at most ϵ in the ordinary two party case (where the two parties are directly connected by a quantum communication link).

Proof. The proof is similar to that given for the classical deterministic setting in [17], but we need to make some modification to it in order to handle bounded error setting. We first partition network G into $(s+1)$ layers as in Figure 1 so that

1. every layer is a disjoint subset of the set of all nodes in G,
2. the first layer has a unique member P_A,
3. the $(s+1)$st layer has a unique member P_B,
4. no edge jumps over a layer, i.e., there is no edge between a node in the ith layer and a node in the $(i+j)$th layer for any $1 \leq i < s$ and $i+1 < i+j \leq s+1$.

Note that we can do this by using breadth first traversal of G starting at P_A, since s is the length of the shortest path between P_A and P_B.

Let $\mathcal{P}$ be the best protocol between P_A and P_B on network G that computes $f(x,y)$ with error probability at most ϵ. We then construct a quantum protocol with classical public coins between two parties that are directly connected to each other, by simulating protocol $\mathcal{P}$ as follows: if the value of the public coins is $i \in \{1, \ldots, s\}$, the two parties, P_A and P_B, divide network G at the boundary of the i-th and the $(i+1)$-st layers into two parts, and they then simulate the left and the right parts, respectively. Let q_i be the number of qubits communicated between the i-th and the $(i+1)$-st layers during the execution of protocol $\mathcal{P}$, and let w_i be the number of links across the boundary of the i-th and the $(i+1)$-st layers. It follows that the number of qubits that P_A and P_B need to communicate in the above simulation is at most $q_i \lceil \log w_i \rceil$, since at most $\lceil \log w_i \rceil$ bits are needed, when simulating each message exchanged between the i-th and the $(i+1)$-st layers, to specify on which link among w_i links the message is sent. The obtained protocol between P_A and P_B computes f with error probability at most ϵ and with expected communication complexity $(1/s \sum_i q_i \lceil \log w_i \rceil)$. To guarantee the worst case communication complexity, we modify this protocol so that if the amount of communication exceeds $1/\delta_1$ times $(1/s \sum_i q_i \lceil \log w_i \rceil)$ for $0 < \delta_1 < 1$, it stops and randomly outputs 0 or 1. The probability of this event is at most δ_1 by Markov's inequality. On the condition that this event occurs, the probability of outputting the wrong value is exactly $1/2$. Thus, the modified

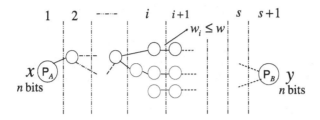

Fig. 1. Network G that is partitioned into $(s+1)$-layers

protocol has error probability at most $\epsilon + \delta_1/2$. Hence, we have obtained an $(\epsilon+\delta_1/2)$-error quantum protocol with classical public coins, whose complexity is $1/(s\delta_1)\sum_i q_i \lceil\log w_i\rceil \leq \lceil\log w\rceil/(s\delta_1)\cdot\sum_i q_i$. The last equality is due to $w_i \leq w$. This implies that $Q^{pub}_{\epsilon+\delta_1/2}(f) \leq \lceil\log w\rceil/(s\delta_1)\cdot\sum_i q_i$. Amount $\sum_i q_i$ is the total number of qubits communicated during execution of protocol $\mathcal{P}$, $Q^G_\epsilon(f)$, which is lower-bounded by $s\delta_1 Q^{pub}_{\epsilon+\delta_1/2}(f)/\lceil\log w\rceil$. By applying Lemma 1 to $Q^{pub}_{\epsilon+\delta_1/2}(f)$, we have $Q^G_\epsilon(f) \geq s\delta_1(Q_{\epsilon+\delta_1/2+\delta_2}(f) - \lceil\log n + 2\log 1/\delta_2\rceil)/\lceil\log w\rceil$.

There is another way of deciding the boundary between P_A's part and P_B's part: P_A randomly chooses one layer-boundary out of the s layer-boundaries and informs P_B of the chosen layer-boundary by a $\lceil\log s\rceil$-bit message. By an argument similar to the one stated above,

$$Q_{\epsilon+\delta_3/2}(f) \leq 1/(s\delta_3)\sum_i q_i \lceil\log w_i\rceil + \lceil\log s\rceil \leq \lceil\log w\rceil/(s\delta_3)\cdot\sum_i q_i + \lceil\log s\rceil.$$

This implies that $Q^G_\epsilon(f)$ is lower-bounded by $s\delta_3(Q_{\epsilon+\delta_3/2}(f) - \lceil\log s\rceil)/\lceil\log w\rceil$. □

If we set $\epsilon, \delta_1, \delta_2, \delta_3$ to constants such that $\epsilon + \delta_1/2 + \delta_2$ and $\epsilon + \delta_3$ are at most some constant less than $1/2$, then $Q_\epsilon(f)$, $Q_{\epsilon+\delta_1/2+\delta_2}(f)$ and $Q_{\epsilon+\delta_3/2}(f)$ differ by at most constant multiplicative factors.

Corollary 1. *Suppose that f, G, s and w are defined as above. Then, for constant $0 < \epsilon < 1/2$, $Q^G_\epsilon(f) = \Omega(s(Q_\epsilon(f) - \log\min\{n,s\})/\log w)$.*

If function f is derived from some symmetric function g, we can obtain a more concrete lower bound by using the lower bound results in [16].

Corollary 2. *Suppose that G, s and w are defined as above, $f(x,y)$ $(x,y \in \{0,1\}^n)$ is of the form $f(x,y) = g(|x \wedge y|)$ for any predicate $g : \{0,\ldots,n\} \to \{0,1\}$. If $l_0 \in \{0,1,\ldots,\lfloor n/2\rfloor\}$ and $l_1 \in \{0,1,\ldots,\lfloor n/2\rfloor\}$ are the smallest integers such that $g(h)$ is constant for $h \in \{l_0(g),\ldots,n-l_1(g)\}$. Then, the total quantum communication complexity over all links of computing $f(x,y)$ in the bounded error setting is $\Omega(s(\sqrt{nl_0(g)}+l_1(g) - \log\min\{n,s\})/\log w)$.*

For zero-error quantum protocols, we have a similar but slightly different result: $\delta_1/2$ and $\delta_3/2$ are replaced by δ_1 and δ_3, since the protocol must give up if the amount of communication exceeds $1/\delta_1$ $(1/\delta_3)$ times $1/s\sum_i q_i\lceil\log w_i\rceil$ in order to preserve the zero-error property.

4   Application: Almost Tight Bound of Distinctness on a Ring

This section applies the lower bound theorem of the previous section to a fundamental distributed computing problem, the distinctness problem, which emerges when checking whether the priorities of processors are totally ordered. The distinctness problem was first introduced by Tiwari [17] and is defined as follows.

Definition 1 (DISTINCT$_{k,L}^{G}$). *Let k parties be placed on a network whose underlying graph is G. Let each party P_i $(0 \leq i \leq k - 1)$ have an integer $x_i \in \{0, \ldots, L - 1\}$ $(k \leq L)$. The goal is to decide whether any two parties have different values, i.e., $i \neq j \longrightarrow x_i \neq x_j$. At termination, each party knows a one-bit result.*

The main theorem of this section gives almost tight bounds of the bounded-error quantum communication complexity for the distinctness problem on a ring-shaped network.

Theorem 2. *The quantum communication complexity of* DISTINCT$_{k,L}^{ring}$ *for $L = k + \Omega(k)$ in the bounded error setting is summarized as follows:*

- *if $L \leq k(\log k)^2$, $O(k\sqrt{L})$ and $\Omega(k\sqrt{k}) \subseteq \Omega(k\frac{\sqrt{L}}{\log k})$.*
- *if $L > k(\log k)^2$, $O(k(\sqrt{k}\log k + \log\log L))$ and $\Omega(k(\sqrt{k} + \log\log L))$*

The theorem implies that our bounds are almost tight up to a log factor. In particular, they are optimal $\Theta(k\sqrt{k})$ up to a constant factor for $L \in O(k)$. The theorem is directly obtained from the lemmas in the next subsections. Hereafter, we deal with only bounded-error computation.

4.1   Lower Bound

To get a lower bound, we will prove a lower bound of the quantum communication complexity for a certain distributed computing problem by applying Corollary 1, and then reduce the problem to DISTINCT$_{k,L}^{ring}$.

Lemma 3. *The quantum communication complexity of* DISTINCT$_{k,L}^{ring}$ *is $\Omega(k(\sqrt{k} + \log\log L))$, for $L = k + \Omega(k)$.*

Proof. We will reduce the following problem DISJ$_{k-2\lceil ck\rceil+2,\lceil ck\rceil}^{ring}$ to DISTINCT$_{k,L}^{ring}$: when party P_A is diametrically opposite P_C on a $(k - 2(\lceil ck\rceil - 1))$-party ring for any constant c $(\leq 1/4)$ (we assume here $2|(k - 2\lceil ck\rceil)$ for simplicity, but this assumption is not essential), and $\lceil ck\rceil$-bit strings x and y are given to P_A and P_C, respectively, the goal is to compute function DISJ$_{\lceil ck\rceil}(x,y) = \bigwedge_{i=1}^{\lceil ck\rceil} \overline{x_i y_i}$. Problem DISJ$_{k-2\lceil ck\rceil+2,\lceil ck\rceil}^{ring}$ has the total communication complexity over all links of $\Omega(k\sqrt{k})$ by Corollary 2 with $n = \lceil ck\rceil$, $w = 2$, $s = (k - 2(\lceil ck\rceil - 1))/2 = O(k)$, $l_0(g) = 1$, and $l_1(g) = 0$.

Now we describe the reduction. We will reduce $\mathsf{DISJ}^{\mathrm{ring}}_{k-2\lceil ck\rceil+2,\lceil ck\rceil}$ to $\mathsf{DISTINCT}^{\mathrm{ring}}_{k,L}$ for any $L \geq k + \lceil ck\rceil$. We first partition the k-party ring of $\mathsf{DISTINCT}^{\mathrm{ring}}_{k,L}$ into four segments A, B, C and D of size $\lceil ck\rceil$, $(k - 2\lceil ck\rceil)/2$, $\lceil ck\rceil$ and $(k - 2\lceil ck\rceil)/2$, respectively, where segment A is diametrically opposite C. Let $I_1 = \{0, 1, \ldots, \lceil ck\rceil - 1\}$, $I_2 = \{\lceil ck\rceil, \ldots, 3\lceil ck\rceil - 1\}$ and $I_3 = \{3\lceil ck\rceil, \ldots, L-1\}$. Next we construct an instance of $\mathsf{DISTINCT}^{\mathrm{ring}}_{k,L}$ from any instance of $\mathsf{DISJ}^{\mathrm{ring}}_{k-2\lceil ck\rceil+2,\lceil ck\rceil}$ as follows: (1) the ith party of A (C) has $(i-1) \in I_1$ as input if the ith bit of input to P_A (resp. P_C) of $\mathsf{DISJ}^{\mathrm{ring}}_{k-2\lceil ck\rceil+2,\lceil ck\rceil}$ is 1 for $i = 1, \ldots, \lceil ck\rceil$, otherwise every party in A and C is given any distinct value in I_2, (2) every party in B and D is given any distinct value in I_3. It is not hard to see that $\mathsf{DISTINCT}^{\mathrm{ring}}_{k,L}$ is true if and only if there is no i such that the ith party of A has the same input as the ith party of C. Thus, $\mathsf{DISJ}^{\mathrm{ring}}_{k-2\lceil ck\rceil+2,ck}$ can be solved if P_A and P_C simulate segments A and C, respectively, and run any protocol solving $\mathsf{DISTINCT}^{\mathrm{ring}}_{k,L}$ against the above instance. By setting c to an arbitrary small positive constant, the lemma holds for all $L = k + \Omega(k)$.

In the case where $L = 2^{k^{\omega(1)}}$, we will reduce the following problem $\mathsf{EQ}^{\mathrm{ring}}_{k,\lceil \log L\rceil-1}$ to $\mathsf{DISTINCT}^{\mathrm{ring}}_{k,L}$: when party P_A is diametrically opposite P_C on a k-party ring (we assume again $2|k$ for simplicity, but this assumption is not essential), and $(\lceil \log L\rceil - 1)$-bit strings x and y are given to P_A and P_C, respectively, the goal is to decide whether x equals y, i.e., to compute $\mathsf{EQ}_{\lceil \log L\rceil-1} = \bigwedge_{i=1}^{\lceil \log L\rceil-1}(\overline{x_i \oplus y_i})$. We apply Corollary 1 to $\mathsf{EQ}^{\mathrm{ring}}_{k,\lceil \log L\rceil-1}$ with $n = \lceil \log L\rceil - 1$, $w = 2$, $s = O(k)$, and $Q_{1/3}(\mathsf{EQ}_{\lceil \log L\rceil-1}) = \Omega(\log \log L)$ [11]. $\mathsf{EQ}^{\mathrm{ring}}_{k,\lceil \log L\rceil-1}$ has the total communication complexity over all links of $\Omega(k(\log \log L - \log \min\{O(k), O(\log L)\})) = \Omega(k \log \log L)$ when $L = 2^{k^{\omega(1)}}$. Thus, $\mathsf{DISTINCT}^{\mathrm{ring}}_{k,L}$ has quantum communication complexity $\Omega(k \log \log L)$.

For the reduction, we construct an instance of $\mathsf{DISTINCT}^{\mathrm{ring}}_{k,L}$ from any instance of $\mathsf{EQ}^{\mathrm{ring}}_{k,\lceil \log L\rceil-1}$ as follows. We partition the k-party ring of $\mathsf{DISTINCT}^{\mathrm{ring}}_{k,L}$ into four segments A, B, C and D of size 1, $(k - 2)/2$, 1 and $(k - 2)/2$, respectively, where segment A is diametrically opposite C. We then set the most significant bit (MSB) of the input of $\mathsf{DISTINCT}^{\mathrm{ring}}_{k,L}$ given to every party in segments A and C to 1, while we set the MSBs of the inputs to the other parties to 0. The remaining $(\lceil \log L\rceil - 1)$ bits of the input to the party in segment A (segment C) are set to the input values of $\mathsf{EQ}^{\mathrm{ring}}_{k,\lceil \log L\rceil-1}$ given to P_A (resp. P_C). The remaining $(\lceil \log L\rceil - 1)$ bits of the input to the other parties are set to distinct values (this is possible due to $L = 2^{k^{\omega(1)}}$). □

4.2 Upper Bounds

To show the optimality of our lower bound, we give almost matching upper bounds.

Lemma 4. *The quantum communication complexity of* $\mathsf{DISTINCT}^{ring}_{k,L}$ *is* $O(k(\sqrt{k} + \log \log L))$.

Proof. We consider the following search problem: is there any party P_i such that there exists party P_j ($j \neq i$) that has the same input as party P_i? Given an oracle that, for input i, answers 1 if there is a party $P_j (\neq P_i)$ having the same input as party P_i and otherwise answers 0, we can solve the search problem with $O(\sqrt{k})$ queries to the oracle by Grover's quantum search algorithm in [8]. Let party P_0 be distinguished, and P_0 executes the search algorithm on behalf of all the parties. The oracle is simulated in a distributed way by the k parties as follows.

The simulation for input i consists of two phases. The purpose of the first phase is for P_0 to get the information of x_i. If $i \neq 1$, party P_0 first prepares a $(\lceil \log k \rceil + \lceil \log L \rceil)$-qubit message $|i\rangle|0^{\lceil \log L \rceil}\rangle$; otherwise P_0 prepares message $|i\rangle|x_1\rangle$. Party P_0 then sends it to adjacent party P_1. Every party P_j ($1 \leq j \leq k - 1$) except P_i simply passes the received message to adjacent party $P_{j+1 \pmod k}$; party P_i changes message $|i\rangle|0^{\lceil \log L \rceil}\rangle$ to $|i\rangle|x_i\rangle$, and then sends it to adjacent party $P_{i+1 \pmod k}$. The purpose of the second phase is to check whether string x_i is identical to one of the $k - 1$ strings $\{x_1, x_2, \ldots, x_k\} \setminus \{x_i\}$. If $i \neq 1$, party P_0 prepares $(\lceil \log L \rceil + \lceil \log k \rceil)$-qubit message $|x_i\rangle|0^{\lceil \log k \rceil}\rangle$; otherwise it prepares message $|x_i\rangle|0^{\lceil \log k \rceil - 1}1\rangle$. Notice that the second register is used to count the number of parties that have values identical to x_i. Party P_0 then sends it to adjacent party P_1. For $1 \leq j \leq k - 1$, P_j just passes the received message to adjacent party $P_{j+1 \pmod k}$ if $x_j \neq x_i$; otherwise party P_j increments the counter, i.e., the contents of the last $\lceil \log k \rceil$ qubits, in the received message, and sends the updated message to adjacent party $P_{j+1 \pmod k}$. When the message arrives at P_0, the counter has value of at least two if and only if there are at least two parties that have values identical to x_i. Party P_0 then sets the content of a fresh qubit to 1 if the value of the counter is at least two; otherwise, P_0 sets it to 0. The content of the qubit is the answer of the oracle. Finally, every computation (except the last step) and communication performed in the first and second phases is inverted to disentangle all work qubits including the message qubits.

The first and the second phases including their inversions have the communication complexity of $O(k \log(kL))$ in total, implying that one query needs $O(k \log(kL))$.

By combining Grover's search algorithm with this distributed oracle, $O(k\sqrt{k} \log(kL))$-qubit communication is sufficient to find a party P_i such that there exists a party P_j ($j \neq i$) that has the same input as party P_i. If such a party is found, the answer to $\mathsf{DISTINCT}_{k,L}^{\mathrm{ring}}$ is false; otherwise the answer is true. (To inform every party of the answer, a one-bit message needs to be sent around the ring, which does not change the order of complexity.) This complexity is tight up to a log multiplicative factor when L is polynomial in k. For larger L, however, it is not tight. In what follows, we will show more efficient algorithm for larger L by adding a preprocess before running the above algorithm.

The idea is to map $\lceil \log L \rceil$-bit inputs to $3 \log k$-bit strings by using universal hashing (see, e.g., [13]) and classical public randomness, and then apply the above algorithm with the $3 \log k$-bit strings as inputs.

Suppose that every party shares classical public coins. By using the public coins, every party selects a common hash function $f : \{0, \ldots, L-1\} \to \{0, \ldots, k^3 - 1\}$ from the family of $O(L^2)$ hash functions. Every party sets its new input to the $(3 \log k)$-bit string, and runs the $O(k\sqrt{k} \log(kL))$ algorithm for the new input, yielding complexity $O(k\sqrt{k} \log k)$. By Lemma 2 with input size $k\lceil \log L \rceil$, $O(k \log(k \log L))$-bit classical communication is sufficient to realize public coins. Thus, the total communication complexity is $O(k \log(k \log L) + k\sqrt{k} \log k) = O(k(\sqrt{k} \log k + \log \log L))$.

The correctness of this algorithm is proved as follows. When there are a pair of parties that share a common value, the step of applying Grover's search obviously finds one of the parties with bounded error. If there are no such pair, the probability that there exist a pair of parties that share a common value of the hash function is at most $1/k^3 \times k(k-1)/2 \leq 1/(2k)$. Thus, even in this case, the algorithm also guarantees bounded error. $\qquad\square$

In the case of $L < k(\log k)^2$, we can obtain a better bound, which is optimal for $L = O(k)$. To do this, we consider the following search problem: is there any element $x \in \{0, \ldots, L-1\}$ such that at least two parties have x as their inputs. We employ the optimal protocol due to Aaronson and Ambainis [1] for the two-party disjointness function.

Lemma 5. *The quantum communication complexity of* $\mathsf{DISTINCT}_{k,L}^{ring}$ *is* $O(k\sqrt{L})$.

References

1. Aaronson, S., Ambainis, A.: Quantum Search of Spatial Regions. Theory of Computing 1, 47–79 (2005)
2. Boyer, M., Brassard, G., Høyer, P., Tapp, A.: Tight Bounds on Quantum Searching. Fortschritte Der Physik 46(4-5), 493–505 (1998)
3. Brassard, G., Høyer, P., Mosca, M., Tapp, A.: Quantum Amplitude Amplification and Estimation. In: Jr Lomonaco, S.J., Brandt, H.E. (eds.) Quantum Computation and Quantum Information: A Millennium Volume, AMS Contemporary Math. Series, vol.305, pp. 53–74. AMS (2003)
4. Buhrman, H., Cleve, R., Wigderson, A.: Quantum vs. Classical Communication and Computation. In: Proc. of 30th Annual ACM Symposium on Theory of Computing (STOC 1998), pp. 63–68. ACM, New York (1998)
5. Buhrman, H., Cleve, R., Watrous, J., de Wolf, R.: Quantum Fingerprinting. Phys. Rev. Lett. 87(16), 167902 (2001)
6. Frankl, P., Rödl, V.: Forbidden Intersections. Trans. Amer. Math. Soc. 300(1), 259–286 (1987)
7. Gavinsky, D., Kempe, J., Kerenidis, I., Raz, R., de Wolf, R.: Exponential Separations for One-Way Quantum Communication Complexity, with Applications to Cryptography. In: Proc. of 39th Annual ACM Symposium on Theory of Computing, pp. 516–525. ACM, New York (2007)
8. Grover, L.K.: A Fast Quantum Mechanical Algorithm for Database Search. In: Proc. of 28th Annual ACM Symposium on Theory of Computing (STOC 1996), pp. 212–219. ACM, New York (1996)

9. Kalyanasundaram, B., Schnitger, G.: The Probabilistic Communication Complexity of Set Intersection. SIAM Journal on Discrete Mathematics 5(4), 545–557 (1992)
10. Klauck, H.: On Quantum and Approximate Privacy. Theory of Computing Systems 37, 221–246 (2004)
11. Kremer, I.: Quantum Communication. Master's thesis, Computer Science Department, The Hebrew University (1995)
12. Kushilevitz, E., Nisan, N.: Communication Complexity. Cambridge University Press, Cambridge (1997)
13. Leiserson, C.E., Rivest, R.L., Stein, C., Cormen, T.H.: Introduction to Algorithms. MIT Press, Cambridge (2001)
14. Newman, I.: Private vs. Common Random Bits in Communication Complexity. Information Processing Letters 39, 67–71 (1991)
15. Raz, R.: Exponential Separation of Quantum and Classical Communication Complexity. In: Proc. of 31st Annual ACM Symposium on Theory of Computing (STOC 1999), pp. 358–367. ACM, New York (1999)
16. Razborov, A.A.: Quantum Communication Complexity of Symmetric Predicates. Izvestiya Mathematics 67(1), 145–159 (2003)
17. Tiwari, P.: Lower Bounds on Communication Complexity in Distributed Computer Networks. Journal of the ACM 34(4), 921–938 (1987)
18. Yao, A.C.-C.: Some Complexity Questions Related to Distributed Domputing. In: Proc. of 11th Annual ACM Symposium on Theory of Computing, pp. 209–213 (1979)
19. Yao, A.C.-C.: Quantum Circuit Complexity. In: 34th Annual IEEE Symposium on Foundations of Computer Science (FOCS 1993), pp. 352–361. IEEE, New York (1993)

Monotone DNF Formula That Has a Minimal or Maximal Number of Satisfying Assignments

Takayuki Sato[1], Kazuyuki Amano[2,*], Eiji Takimoto[3], and Akira Maruoka[4]

[1] Dept. of Information Engineering, Sendai National College of Technology
Chuo 4-16-1, Ayashi, Aoba, Sendai 989-3128, Japan
taka@info.sendai-ct.ac.jp
[2] Department of Computer Science, Gunma University
Tenjin 1-5-1, Kiryu, Gunma 376-8515, Japan
amano@cs.gunma-u.ac.jp
[3] Graduate School of Information Sciences, Tohoku University
Aoba 6-6-05, Aramaki, Sendai 980-8579, Japan
t2@ecei.tohoku.ac.jp
[4] Dept. of Information Technology and Electronics, Ishinomaki Senshu University
amaruoka@isenshu-u.ac.jp

Abstract. We consider the following extremal problem: Given three natural numbers n, m and l, what is the monotone DNF formula that has a minimal or maximal number of satisfying assignments over all monotone DNF formulas on n variables with m terms each of length l? We first show that the solution to the minimization problem can be obtained by the Kruskal-Katona theorem developed in extremal set theory. We also give a simple procedure that outputs an optimal formula for the more general problem that allows the lengths of terms to be mixed. We then show that the solution to the maximization problem can be obtained using the result of Bollobás on the number of complete subgraphs when $l = 2$ and the pair (n, m) satisfies a certain condition. Moreover, we give the complete solution to the problem for the case $l = 2$ and $m \leq n$, which cannot be solved by direct application of Bollobás's result. For example, when $n = m$, an optimal formula is represented by a graph consisting of $\lfloor n/3 \rfloor - 1$ copies of C_3 and one $C_{3+(n \bmod 3)}$, where C_k denotes a cycle of length k.

1 Introduction and Overview

A *monotone* disjunctive normal form (DNF) formula is a DNF formula in which there are no negated variables. Monotone DNF formulas are one of the most widely used representations of monotone functions. In this paper, we consider an extremal problem that is closely related to the problem of counting the satisfying assignments of a monotone DNF formula, which is known to be $\sharp P$-complete even if we restrict the terms to two variables each [12],[13].

The problem we tackle in the paper is as follows: Given three natural numbers n, m and l, what is the monotone DNF formula that has a minimal or maximal

---

* Corresponding author.

X. Hu and J. Wang (Eds.): COCOON 2008, LNCS 5092, pp. 191–203, 2008.

number of satisfying assignments over all monotone DNF formulas on n variables with m terms each of length l?

The problem seems to be very fundamental and relates to several important problems that have been widely investigated in set theory and graph theory. For example, if l is set to 2, a monotone DNF formula F on n variables with m terms is naturally represented by a graph G_F with n vertices and m edges by corresponding each variable to a vertex and each term to an edge. Then the number of *unsatisfying* assignments x with k ones is equal to the number of k-complete subgraphs in $\overline{G_F}$. The problem of the number of complete subgraphs in a graph with a certain number of edges has attracted the attention of many researchers for more than a half century (see e.g., [2],[4],[5],[11]). In addition, the solution to our problem gives upper and lower bounds on the number of satisfying assignments of a given monotone DNF formula which would help to design an algorithm for a related counting problem.

The main aim of this paper is to investigate the structure of formulas with such an extremal property. We discuss the problem of minimization in Section 3 and then discuss the problem of maximization in Section 4.

For the minimization problem, we show that an optimal DNF formula can be easily determined using the Kruskal-Katona theorem [8],[9] developed in extremal set theory. Intuitively, an optimal DNF formula is obtained by OR-ing terms selected in a lexicographic manner. We also give the solution to a more general problem. A length vector of a DNF formula F is defined by $S = (s_1, \ldots, s_n)$ where s_i is the number of terms in F that contains exactly i variables. We give a simple procedure that outputs a monotone DNF formula that minimizes the number of satisfying assignments over all monotone DNF formulas with a given length vector.

For the maximization problem, we use the result of Bollobás on the number of complete subgraphs contained in a graph [1]. When $l = 2$ and a pair (n, m) satisfies a certain condition, we can derive the structure of a formula that has the maximal number of satisfying assignments among all monotone 2-DNF formulas on n variables with m terms. The optimal DNF formula is represented by a graph consisting of q disjoint cliques of size as equal as possible for some natural number q (Corollary 6). Next we investigate the problem for "sparse" cases, in which the number of terms is not larger than the number of variables, which cannot be solved by a direct application of Bollobás's result. We show that if $l = 2$ and $n = m$, an optimal DNF formula is represented by a graph consisting of $\lfloor n/3 \rfloor - 1$ copies of C_3 and one $C_{3+(n \bmod 3)}$, where C_k denotes a cycle of length k (Theorem 7). We also give the complete solution to the maximization problem for the case $l = 2$ and $m < n$ (Theorem 12). These are the main technical contributions of this paper.

The outline of the proof of these results is the following. We first introduce a number of local transformation rules on formulas which preserves the number of terms and does not decrease the number of satisfying assignments in a formula. Then we will show that by applying these rules whenever possible, every monotone DNF formula will "converge" to a desired formula. Further extensions of our technique will hopefully solve the more challenging problem of "dense" cases.

2   Preliminaries

Let F be a monotone DNF formula on n variables (i.e., a DNF in which there are no negated variables). For a term t in F, $Var(t) \subseteq \{1, \ldots, n\}$ denotes the set of indices of variables that appear in t. Throughout this paper, we assume that a monotone DNF formula contains no redundant terms. That is, there are no two terms t_i and t_j such that $Var(t_i) \subseteq Var(t_j)$ or $Var(t_j) \subseteq Var(t_i)$. A monotone l-DNF is a monotone DNF such that each term of the formula contains exactly l variables. More generally, for a vector of nonnegative integers $S = (s_1, s_2, \ldots, s_n)$, a monotone DNF formula F is said to be S-DNF if F contains exactly s_l terms of length l for every $l = 1, \ldots, n$. Note that there are no monotone S-DNF formulas for some such vectors, e.g., $S = (n + 1, 0, \ldots, 0)$.

Let $\mathsf{SAT}(F) \subseteq \{0, 1\}^n$ denote the set of satisfying assignments for F, i.e.,

$$\mathsf{SAT}(F) = \{x \in \{0, 1\}^n \mid F(x) = 1\}.$$

For $x = (x_1, \ldots, x_n) \in \{0, 1\}^n$, let $|x|$ denote the number of 1's in x, i.e., $|x| = \sum_{i=1}^{n} x_i$. For $k = 0, \ldots, n$, let $\mathsf{SAT}_k(F)$ be the set of satisfying assignments for F that has k ones, i.e.,

$$\mathsf{SAT}_k(F) = \{x \in \{0, 1\}^n \mid F(x) = 1 \text{ and } |x| = k\}.$$

For a set S, $\sharp S$ or $|S|$ denotes the size of S.

3   Minimization

In this section, we consider the problem of determining a monotone S-DNF formula F that minimizes $\sharp\mathsf{SAT}(F)$ over all monotone S-DNF formulas for a given vector $S = (s_1, \ldots, s_n)$, and solve it by giving a procedure that outputs an optimal S-DNF formula. The construction depends heavily on a combinatorial result due to Kruskal and Katona [8],[9].

To begin, we consider a special case of the problem in which the length of each term is identical, say l, i.e., $s_i = 0$ for every $i \neq l$. In the following, the set $\{1, \ldots, n\}$ is denoted by $[n]$. For $X = [n]$, the set of all r-subsets of X is denoted by

$$X^{(r)} = \{A \subseteq X \mid |A| = r\}.$$

For an r-set $\mathcal{A} \subseteq X^{(r)}$, the *upper shadow* of $\mathcal{A}$ is defined as

$$\partial_u(\mathcal{A}) = \{B \in X^{(r+1)} \mid B \supset A \text{ for some } A \in \mathcal{A}\}.$$

For $x \in \{0, 1\}^n$, let A_x be the set of indices $i \in [n]$ with $x_i = 1$. If we define

$$\mathcal{A}_r = \{A_x \mid x \in \mathsf{SAT}_r(F)\},$$

then $\partial_u(\mathcal{A}_r)$ has a one-to-one correspondence to $\mathsf{SAT}_{r+1}(F)$, and $\partial_u(\partial_u(\mathcal{A}_r))$ has a one-to-one correspondence to $\mathsf{SAT}_{r+2}(F)$. Note that, if F is a monotone l-DNF formula with m terms, $\sharp\mathsf{SAT}_r(F) = 0$ for every $r = 0, \ldots, l-1$ and $\sharp\mathsf{SAT}_l(F) = m$.

Hence the problem of determining a monotone l-DNF formula with m terms that minimizes $\sharp \mathrm{SAT}(F)$ is equivalent to the problem of determining an l-set $\mathcal{A} \subseteq X^{(l)}$ with $|\mathcal{A}| = m$ that minimizes

$$\partial_u(\mathcal{A}) + \partial_u(\partial_u(\mathcal{A})) + \cdots + \underbrace{\partial_u(\partial_u(\cdots(\partial_u(\mathcal{A})\cdots))}_{n-l}.$$

We will now show that this problem is easily solved by applying the Kruskal-Katona theorem ([8],[9]) from extremal set theory.

We consider two natural orders on $X^{(r)}$, namely, the *lexicographic order* (or *lex order*) and the *colexicographic order* (or *colex order*). In what follows, we write $A, B \in X^{(r)}$ as $A = \{a_1, a_2, \ldots, a_r\}$, $B = \{b_1, b_2, \ldots, b_r\}$ with $a_1 < a_2 < \cdots < a_r$ and $b_1 < b_2 < \cdots < b_r$. In the lex order $A < B$ iff for some $1 \le k \le r$, $a_k < b_k$ and $a_i = b_i$ for $i = 1, \ldots, k-1$. In the colex order $A < B$ iff for some $1 \le k \le r$, $a_k < b_k$ and $a_i = b_i$ for $i = k+1, \ldots, r$. Equivalently, $A < B$ in the colex order if $A \neq B$ and for $s = \max\{t \mid a_t \neq b_t\}$ we have $a_s < b_s$. Note that $\{2,3,6,7\} < \{2,4,5,7\}$ in the lex order and $\{2,4,5,7\} < \{2,3,6,7\}$ in the colex order.

The lex ordering of $[5]^{(3)}$ is

$$\{1,2,3\}, \{1,2,4\}, \{1,2,5\}, \{1,3,4\}, \{1,3,5\},$$
$$\{1,4,5\}, \{2,3,4\}, \{2,3,5\}, \{2,4,5\}, \{3,4,5\},$$

and the colex ordering is

$$\{1,2,3\}, \{1,2,4\}, \{1,3,4\}, \{2,3,4\}, \{1,2,5\},$$
$$\{1,3,5\}, \{2,3,5\}, \{1,4,5\}, \{2,4,5\}, \{3,4,5\}.$$

It should be noted that the lex ordering is obtained from the colex ordering by mapping $\{1, 2, \ldots, n\}$ to $\{n, \ldots, 2, 1\}$ and then listing in reverse order, and vice versa.

The following theorem, which is an easy corollary of the Kruskal-Katona theorem, has the most convenient form for our purpose.

Theorem 1. [3, Theorem 5] *Let $1 \le r \le n-1$, $\mathcal{A} \subseteq X^{(r)}$ and let $\mathcal{B}$ be the set of the last $|\mathcal{A}|$ elements of $X^{(r)}$ in the colex order. Then, $|\partial_u(\mathcal{A})| \ge |\partial_u(\mathcal{B})|$.* □

The following corollary is immediate.

Corollary 2. *Let $1 \le r \le n-1$, $\mathcal{A} \subseteq X^{(r)}$ and let $\mathcal{B}$ be the set of the first $|\mathcal{A}|$ elements of $X^{(r)}$ in the lex order. Then, $|\partial_u(\mathcal{A})| \ge |\partial_u(\mathcal{B})|$.* □

The above corollary says that the size of the upper shadow of $\mathcal{A}$ is minimized when we set $\mathcal{A}$ to the lexicographically ordered first m-elements of $X^{(r)}$. Since the upper shadow of the set $\mathcal{A}$ of the first m elements of $X^{(r)}$ is the set of the first $|\partial_u(\mathcal{A})|$ elements of $X^{(r+1)}$, we get the following theorem by applying Corollary 2 inductively.

Theorem 3. *Let F be an m term monotone l-DNF formula $F = t_1 \vee \cdots \vee t_m$ on n variables where $Var(t_1), \ldots, Var(t_m)$ are the first m elements of $[n]^{(l)}$ in the* *lex order. Then for every m term monotone l-DNF formula G on n variables,* $\sharp SAT(G) \geq \sharp SAT(F)$. □

Now we return to the general problem. For a family of sets $\mathcal{A} \subseteq 2^X$ and for $0 \leq i \leq n$, we define $\partial^{(i)}(\mathcal{A})$ by

$$\partial^{(i)}(\mathcal{A}) = \{B \in X^{(i)} \mid B \supseteq A \text{ for some } A \in \mathcal{A}\}.$$

By the same arguments used above, we obtain the following theorem.

Theorem 4. *Let $S = (s_1, \ldots, s_n)$ be a vector of non-negative integers. Let $F = t_1 \vee \cdots \vee t_{s_1+\cdots+s_n}$ be a monotone S-DNF formula on n variables obtained by the following procedure:*

1. Let $\mathcal{A} = \emptyset$.
2. For $i = 1, \ldots, n$
    Add the first s_i elements of $([n]^{(i)} \setminus \partial^{(i)}(\mathcal{A}))$ in the lex order to $\mathcal{A}$.
3. Output the OR of terms t for $Var(t) \in \mathcal{A}$.

Then for every monotone S-DNF formula G on n variables, $\sharp SAT(G) \geq \sharp SAT(F)$. □

4 Maximization

In this section, we investigate the opposite problem: the determination of a monotone DNF formula that maximizes the number of satisfying assignments. To begin, we restrict ourselves to monotone 2-DNF formulas. For example, if $n = 6$, a simple examination shows that a monotone 2-DNF formula with m terms that has a maximal number of satisfying assignments is $F = x_1x_2 \vee x_3x_4 \vee x_5x_6$ when $m = 3$, and is $F = x_1x_2 \vee x_1x_3 \vee x_2x_3 \vee x_4x_5 \vee x_4x_6 \vee x_5x_6$ when $m = 6$.

A monotone 2-DNF formula is naturally represented by a graph. For a monotone 2-DNF formula over $\{x_1, \ldots, x_n\}$, let $G_F = (V, E)$ be the graph defined by (i) $V = \{x_1, \ldots, x_n\}$ and (ii) $(x_i, x_j) \in E$ if and only if the term x_ix_j appears in F. An *independent set* of a graph G is a subset S of the vertices such that no two vertices in S represent an edge of G. For $0 \leq r \leq n$, we denote by $\sharp IS_r(G)$ the number of independent sets of size r. The total number of independent sets of G, i.e., $\sum_{r=0}^{n} \sharp IS_r(G)$ is denoted by $\sharp IS(G)$.

It is easy to verify that there is a one-to-one correspondence between an unsatisfying assignment x of F with $|x| = r$ and an independent set of G_F of size r. This implies that $\sharp SAT_r(F) = {}_nC_r - \sharp IS_r(G_F)$ and $\sharp SAT(F) = 2^n - \sharp IS(G_F)$. Hence the problem of determining a monotone 2-DNF formula that has the maximal number of satisfying assignments among all 2-DNF formulas on n variables with m terms is equivalent to the problem of determining a graph that has the minimal number of independent sets among all graphs with n vertices and m edges.

To tackle the problem we use the results on the minimal number of complete subgraphs of a graph of given order and size, which have been widely investigated in the literature of *extremal graph theory*.

For two natural numbers n and q, let $\mathrm{CLQ}_q(n)$ be the graph on n vertices that consists of q disjoint complete graphs of sizes $\lfloor n/q \rfloor, \lfloor (n+1)/q \rfloor, \ldots, \lfloor (n+q-1)/q \rfloor$ and let $e(n,q)$ be the number of edges in $\mathrm{CLQ}_q(n)$. Note that $\mathrm{CLQ}_q(n)$ is the union of q cliques of size as equal as possible. For example, $\mathrm{CLQ}_2(10)$ is the union of two copies of K_5 and $\mathrm{CLQ}_3(10)$ is the union of two copies of K_3 and one K_4, where K_r denotes the complete graph of size r. In these examples $e(10,2) = 20$ and $e(10,3) = 12$.

For $2 < r \le n$, let $\sharp\mathrm{IS}_r(n,m)$ be the minimal number of independent sets of size r that must be contained in every graph with n vertices and at most m edges, i.e.,

$$\sharp\mathrm{IS}_r(n,m) = \min\{\sharp\mathrm{IS}_r(G) \mid G = (V,E) \text{ s.t. } |V| = n, |E| \le m\}.$$

The following theorem, due to Bollobás, says that the graph $\mathrm{CLQ}_q(n)$ minimizes the number of independent sets in a graph with $e(n,q)$ edges.

Theorem 5. [1] (or see [2, Theorem 1.7, pp. 298]) *For every $q \le n$ and for every r, $\sharp\mathrm{IS}_r(n, e(n,q)) \ge \sharp\mathrm{IS}_r(CLQ_q(n))$.*   □

Since the above theorem holds for every fixed r, by summing over r, we immediately get the following corollary which says that the graph $\mathrm{CLQ}_q(n)$ gives the solution to our problem if $m = e(n,q)$.

Corollary 6. *For every $q \le n$, $\sharp\mathrm{IS}(n, e(n,q)) \ge \sharp\mathrm{IS}(CLQ_q(n))$.*   □

For example, if $n = 10$, the above corollary gives an explicit construction of the optimal 2-DNF formulas for $m = 1, 2, 3, 4, 5, 8, 12$ and 20 (see Fig. 1). Unfortunately, Corollary 6 gives no information about the structures of optimal formulas for "intermediate" values of m. Considerable effort has been devoted to determine the value of $\sharp\mathrm{IS}_r(n,m)$ for such values of m, and especially for $r = 3$ (see e.g., [6]).

It should be remarked that the general problem, i.e., the one allowing terms of mixed length, can be investigated by considering hypergraphs instead of graphs. However, the problem for hypergraphs is considered to be extremely difficult (see e.g. [4] for more background on such "Turán-type" problems for hypergraphs).

We now turn to the case $n = m$ as the starting point for tackling the maximization problems that cannot be solved by a direct application of Corollary 6. The main result is the following.

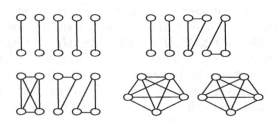

Fig. 1. The graphs representing optimal 2-DNF formulas for $n = 10$ and $m = 5, 8, 12$ and 20. The cases for $m < 5$ are graphs with m independent edges and $n - 2m$ isolated vertices.

Theorem 7. *Let F be a monotone 2-DNF formula with n variables and n terms represented by a graph G such that*

(i) *G is the union of k disjoint copies of C_3 if $n = 3k$,*
(ii) *G is the union of $k - 1$ disjoint copies of C_3 and a C_4 if $n = 3k + 1$, and*
(iii) *G is the union of $k - 1$ disjoint copies of C_3 and a C_5 if $n = 3k + 2$.*

Then for every monotone 2-DNF H with n variables and n terms, $\sharp\mathsf{SAT}(H) \le \sharp\mathsf{SAT}(F)$. Equivalently, for every graph $G' = (V', E')$ with $|V'| = |E'| = n$, $\sharp\mathrm{IS}(G') \ge \sharp\mathrm{IS}(G)$.

The following figure shows the graphs representing the optimal DNF formulas shown in Theorem 7.

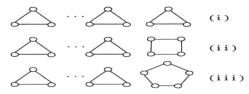

For the case $n = 3k$, Theorem 7 is already proved by Corollary 6 because $e(n, n/3) = n$. For the case $n = 3k + 1$, Theorem 7 can easily be derived from a more general version of Bollobás's theorem.

Theorem 8. [2, Theorem 1.7, pp. 298]   *For every $q \le n$ and for every $l \in [e(n, q), e(n, q - 1)]$,*

$$\sharp\mathrm{IS}_r(n, l) \ge \sharp\mathrm{IS}_r(CLQ_{q-1}(n)) + \frac{\sharp\mathrm{IS}_r(CLQ_q(n)) - \sharp\mathrm{IS}_r(CLQ_{q-1}(n))}{e(n, q - 1) - e(n, q)}(e(n, q - 1) - l).$$

In other words, $\sharp\mathrm{IS}_r(n, l)$ is bounded below by the linear function that passes through the two points $(e(n, q), \sharp\mathrm{IS}_r(CLQ_q(n)))$ and $(e(n, q - 1), \sharp\mathrm{IS}_r(CLQ_{q-1}(n)))$. □

A corollary follows immediately from Theorem 8 by letting $q = \lceil n/3 \rceil$.

Corollary 9. *All of the following are true:*

(i) *If $n = 3k$, then for every $l \in [0, 6]$ and for every $G = (V, E)$ with $|V| = n$ and $|E| = n + l$, $\sharp\mathrm{IS}(G) \ge (256 - \frac{131}{6}l)4^{k-4}$.*
(ii) *If $n = 3k + 1$, then for every $l \in [-2, 2]$ and for every $G = (V, E)$ with $|V| = n$ and $|E| = n + l$, $\sharp\mathrm{IS}(G) \ge (7 - l)4^{k-1}$.*
(iii) *If $n = 3k + 2$, then for every $l \in [-1, 4]$ and for every $G = (V, E)$ with $|V| = n$ and $|E| = n + l$, $\sharp\mathrm{IS}(G) \ge (48 - \frac{23}{5}(l + 1))4^{k-2}$.* □

Statement (ii) of the above corollary immediately implies the result for $n = 3k + 1$ in Theorem 7.

Proof. (Part (ii) of Theorem 7.)   Let G be the graph defined in (ii) of Theorem 7. It is easy to show that $\sharp\mathrm{IS}(G) = \sharp\mathrm{IS}(C_3)^{k-1} \cdot \sharp\mathrm{IS}(C_4) = 7 \cdot 4^{k-1}$. Then statement (ii) of Corollary 9 with $l = 0$ implies the result for $n = 3k + 1$ in Theorem 7. □

The remaining case, $n = 3k + 2$, is the most difficult of the three cases. There is a gap between the number of independent sets of the graph G described in part (iii) of Theorem 7, that is $11 \cdot 4^{k-1} = 44 \cdot 4^{k-2}$, and the lower bound guaranteed by Corollary 9, that is $43.4 \cdot 4^{k-2}$. In the rest of this section, we improve the lower bound to show the optimality of G.

Here we introduce two famous integer sequences known as the Fibonacci numbers and the Lucas numbers. The *Fibonacci numbers* $\mathrm{Fib}(0), \mathrm{Fib}(1), \mathrm{Fib}(2), \ldots$ are produced by the recursive definition $\mathrm{Fib}(0) = 0$, $\mathrm{Fib}(1) = 1$ and $\mathrm{Fib}(n) = \mathrm{Fib}(n-1) + \mathrm{Fib}(n-2)$ for $n \geq 2$. The *Lucas numbers* $\mathrm{Luc}(0), \mathrm{Luc}(1), \mathrm{Luc}(2), \ldots$ are produced by the recursive definition $\mathrm{Luc}(0) = 2$, $\mathrm{Luc}(1) = 1$ and $\mathrm{Luc}(n) = \mathrm{Luc}(n-1) + \mathrm{Luc}(n-2)$ for $n \geq 2$. For example, $\mathrm{Fib}(5) = 5$ and $\mathrm{Luc}(5) = 11$. Note that $\mathrm{Luc}(n) = \mathrm{Fib}(n-1) + \mathrm{Fib}(n+1)$ for every $n \geq 1$. Let P_k denote a path consisting of k vertices and $k - 1$ edges, i.e., $P_k = (V, E)$ is a graph with $V = \{v_1, \ldots, v_k\}$ and $E = \{v_1 v_2, \ldots, v_{k-1} v_k\}$. Note that P_1 is an isolated vertex. The following fact is easy to verify (or see e.g., [10, Examples 10, 11, pp. 144–145]).

Fact 10. *For every k, $\sharp \mathrm{IS}(P_k) = \mathrm{Fib}(k+2)$ and $\sharp \mathrm{IS}(C_k) = \mathrm{Luc}(k)$.*   □

The following fact is useful for later analysis. Recall that a number of independent sets of a graph G with components $G_1, G_2, \ldots, G_p$ is given by the product $\sharp \mathrm{IS}(G_1) \cdot \sharp \mathrm{IS}(G_2) \cdots \sharp \mathrm{IS}(G_p)$. Here a *component* of a graph G is a maximal connected subgraph of G.

Fact 11. *All of the following are true:*

(i) *For every $k \geq 4$, $\sharp \mathrm{IS}(P_k) \geq \sharp \mathrm{IS}(C_3) \cdot \sharp \mathrm{IS}(P_{k-3})$,*
(ii) *For every $k \geq 6$, $\sharp \mathrm{IS}(C_k) \geq \sharp \mathrm{IS}(C_3) \cdot \sharp \mathrm{IS}(C_{k-3})$,*
(iii) *$\sharp \mathrm{IS}(C_4) \cdot \sharp \mathrm{IS}(C_4) \geq \sharp \mathrm{IS}(C_3) \cdot \sharp \mathrm{IS}(C_5)$,*
(iv) *$\sharp \mathrm{IS}(C_4) \cdot \sharp \mathrm{IS}(C_5) \geq \sharp \mathrm{IS}(C_3) \cdot \sharp \mathrm{IS}(C_3) \cdot \sharp \mathrm{IS}(C_3)$, and*
(v) *$\sharp \mathrm{IS}(C_5) \cdot \sharp \mathrm{IS}(C_5) \geq \sharp \mathrm{IS}(C_3) \cdot \sharp \mathrm{IS}(C_3) \cdot \sharp \mathrm{IS}(C_4)$.*

Intuitively, parts (ii) to (v) of Fact 11 say that long cycles can be transformed into shorter cycles without increasing the number of independent sets in a graph (See Fig. 2).

Proof. (i) By Fact 10, we have

$$\sharp \mathrm{IS}(P_k) = \mathrm{Fib}(k+2)$$
$$\geq 4\mathrm{Fib}(k-1) = \sharp \mathrm{IS}(C_3) \cdot \sharp \mathrm{IS}(P_{k-3}).$$

The last inequality follows from $\mathrm{Fib}(k-1) \geq (1/2)\mathrm{Fib}(k)$ for every k. This completes the proof of (i).

Statement (ii) is proved in a similar way. Statements (iii), (iv) and (v) are obvious since $\sharp \mathrm{IS}(C_3) = 4$, $\sharp \mathrm{IS}(C_4) = 7$ and $\sharp \mathrm{IS}(C_5) = 11$.   □

The next lemma says that every tree can be transformed into a path of the same size without increasing the number of independent sets in a graph.

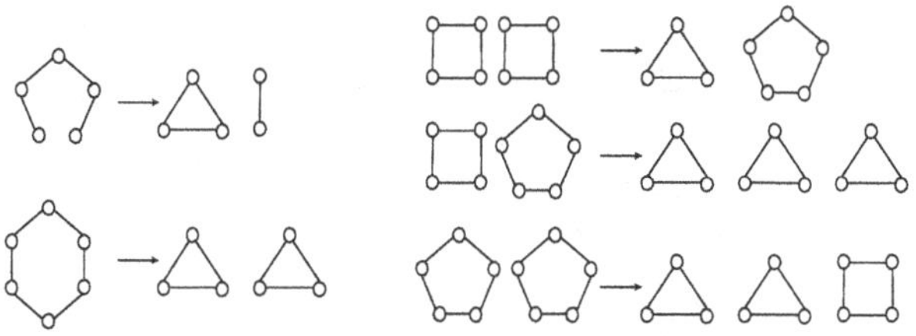

Fig. 2. The transformation rules of graphs described in Fact 11. Each rule preserves the total number of edges without increasing the number of independent sets.

Lemma 1. *For every tree T with k vertices, $\sharp\mathrm{IS}(T) \geq \sharp\mathrm{IS}(P_{k-1})$.*

Proof. We prove the lemma by induction on k. For $k \leq 3$, the lemma is obvious.

Let $T = (V, E)$ be a tree with $k \geq 4$ vertices. Let a be an arbitrary leaf of T and b be the adjacent vertex of a. Let $G_0 = (V_0, E_0)$ and $G_1 = (V_1, E_1)$ be two induced subgraphs of T such that $V_0 = V - \{a\}$ and $V_1 = V - \{a, b\}$. It is obvious that

$$\sharp\mathrm{IS}(T) = \sharp\mathrm{IS}(G_0) + \sharp\mathrm{IS}(G_1). \tag{1}$$

Since T is a tree, G_0 is a tree with $k - 1$ vertices and G_1 is a set of trees, i.e., a forest, which has $k - 2$ vertices. By adding suitable edges to the forest G_1 if necessary, we can obtain a tree G' with $k - 2$ vertices. The induction hypothesis implies that $\sharp\mathrm{IS}(G_0) \geq \sharp\mathrm{IS}(P_{k-2})$ and $\sharp\mathrm{IS}(G_1) \geq \sharp\mathrm{IS}(G') \geq \sharp\mathrm{IS}(P_{k-3})$. By Eq. (1) and Fact 10, we have

$$
\begin{aligned}
\sharp\mathrm{IS}(T) &= \sharp\mathrm{IS}(G_0) + \sharp\mathrm{IS}(G_1)\\
&\geq \sharp\mathrm{IS}(P_{k-2}) + \sharp\mathrm{IS}(P_{k-3})\\
&= \mathrm{Fib}(k) + \mathrm{Fib}(k-1) = \mathrm{Fib}(k+1) = \sharp\mathrm{IS}(P_{k-1}).
\end{aligned}
$$

This completes the induction step and, thus, the proof of the lemma.   □

The next lemma says that a cycle C_k contains a minimal number of independent sets among all *connected* graphs with k vertices and k edges. The proof of the following lemma is similar to the proof of the previous lemma and we omit it here for reasons of limited space.

Lemma 2. *For every connected graph $G = (V, E)$ with $|V| = |E| = k$, $\sharp\mathrm{IS}(G) \geq \sharp\mathrm{IS}(C_k)$.*

For a connected graph that contains one more edge, we have the following bound.

Lemma 3. *Let $G = (V, E)$ be a connected graph with $|V| = k$ and $|E| = k + 1$. Then $\sharp\mathrm{IS}(G) \geq \frac{3}{4}\sharp\mathrm{IS}(C_k)$.*

Proof. Let $G = (V, E)$ be a connected graph with k vertices and $k + 1$ edges. Let $(a, b) \in E$ be an arbitrary edge such that $G' = (V, E - \{(a, b)\})$ is a connected graph with $|V| = |E| = k$. Note that any edge on some cycle in G satisfies this condition.

For $M \subseteq \{a, b\}$, let s_M be the number of independent sets S of G' such that $S \cap \{a, b\} = M$. It is easy to check that $s_\emptyset \geq s_{\{a\}} \geq s_{\{a,b\}}$ and $s_\emptyset \geq s_{\{b\}} \geq s_{\{a,b\}}$. Hence we have

$$\sharp \mathrm{IS}(G) = s_\emptyset + s_{\{a\}} + s_{\{b\}}$$
$$\geq \frac{3}{4}(s_\emptyset + s_{\{a\}} + s_{\{b\}} + s_{\{a,b\}}) \geq \frac{3}{4}\sharp\mathrm{IS}(G') \geq \frac{3}{4}\sharp\mathrm{IS}(C_k)$$

where the last inequality follows from Lemma 2. $\qquad\square$

A graph $G = (V, E)$ with $|V| = n$ and $|E| = m$ is said to be *optimal* for (n, m) if G contains a minimal number of independent sets over all graphs with n vertices and m edges. For a connected graph $G = (V, E)$, the *value* of G is defined to be $|E| - |V|$. Note that for every connected graph G, the value of G is at least -1.

If G is the union of $k - 1$ disjoint copies of C_3 and a C_5, then $\sharp\mathrm{IS}(G) = 11 \cdot 4^{k-1}$. Thus every graph G with $n(= 3k + 2)$ vertices and n edges such that $\sharp\mathrm{IS}(G) > 11 \cdot 4^{k-1}$ is not optimal for $(3k + 2, 3k + 2)$. Thus, by Corollary 9 we can eliminate some possibilities for the structure of optimal graphs.

Lemma 4. *Each of the following graphs is not optimal for* $(n, m) = (3k+2, 3k+2)$.

(i) *A graph that contains an isolated vertex.*
(ii) *A graph that contains two vertex-disjoint copies of P_2.*
(iii) *A graph that contains a path P_k for some $k \geq 3$.*

Proof. (i) Let $G = (V, E)$ be a graph with $|V| = |E| = 3k + 2$ that contains an isolated vertex a. Then $\sharp\mathrm{IS}(G) = 2 \cdot \sharp\mathrm{IS}(G')$ where G' is a graph with $3k + 1$ vertices and $3k+2$ edges obtained from G by removing a. By applying statement (ii) of Corollary 9 with $l = 1$, we have $\sharp\mathrm{IS}(G') \geq 6 \cdot 4^{k-1}$. This implies $\sharp\mathrm{IS}(G) \geq 12 \cdot 4^{k-1}$, which completes the proof.

(ii) Let $G = (V, E)$ be a graph with $|V| = |E| = 3k + 2$ that is the union of two copies of P_2 and G' where G' has $3k - 2$ vertices and $3k$ edges. By applying statement (ii) of Corollary 9 with $l = 1$, we have $\sharp\mathrm{IS}(G) = 9 \cdot \sharp\mathrm{IS}(G') \geq 9 \cdot (6 \cdot 4^{k-2}) > 11 \cdot 4^{k-1}$, which completes the proof.

(iii) The proof is analogous to the previous proof. Let $G = (V, E)$ be a graph that contains a path P_k and G' be a graph with $n - k$ vertices obtained from G by removing P_k. For $k = 3$, we have $\sharp\mathrm{IS}(G) = 5 \cdot \sharp\mathrm{IS}(G') \geq 5 \cdot (38.8 \cdot 4^{k-3}) > 11 \cdot 4^{k-1}$. Here we use statement (iii) of Corollary 9 with $l = 1$. For $k = 4$, we have $\sharp\mathrm{IS}(G) = 8 \cdot \sharp\mathrm{IS}(G') \geq 8 \cdot (6 \cdot 4^{k-2}) > 11 \cdot 4^{k-1}$. For $k = 5$, we have $\sharp\mathrm{IS}(G) = 13 \cdot \sharp\mathrm{IS}(G') \geq 13 \cdot (256 - \frac{131}{6} \cdot 4^{k-5}) > 11 \cdot 4^{k-1}$. For $k \geq 6$, by applying part (i) of Fact 11 recursively, the problem is reduced to the problem for the case $k = 3, 4$ or 5. This completes the proof of the lemma. $\qquad\square$

We need one more lemma on the transformation of a graph with two components.

Lemma 5. *Let G be a graph with n vertices consisting of two components, P_2 and G', where G' is a connected graph with $n - 2$ vertices whose value is 1. Then*

(i) $\sharp\mathrm{IS}(G) \geq \sharp\mathrm{IS}(C_3)^k$ *if* $n = 3k$,
(ii) $\sharp\mathrm{IS}(G) \geq \sharp\mathrm{IS}(C_3)^{k-1} \cdot \sharp\mathrm{IS}(C_4)$ *if* $n = 3k + 1$, *and*
(iii) $\sharp\mathrm{IS}(G) \geq \sharp\mathrm{IS}(C_3)^{k-1} \cdot \sharp\mathrm{IS}(C_5)$ *if* $n = 3k + 2$.

Proof. We have already proved parts (i) and (ii) because the total number of edges in G is equal to the number of vertices in G.

To prove part (iii), we first consider the case $k = 4$ for which Lemma 3 implies that

$$\sharp\mathrm{IS}(G) = \sharp\mathrm{IS}(P_2) \cdot \sharp\mathrm{IS}(G') \geq 3 \cdot (3/4)\sharp\mathrm{IS}(C_{12})$$
$$= \frac{9 \cdot 322}{4} = 724.5 > 704 = 4^3 \cdot 11 = \sharp\mathrm{IS}(C_3)^3 \cdot \sharp\mathrm{IS}(C_5).$$

The cases for $k > 4$ can be proved by using the relation $\sharp\mathrm{IS}(C_{l+3}) \geq 4 \cdot \sharp\mathrm{IS}(C_l)$. For $k = 2$ and $k = 3$, the statement is proved by the facts that $\sharp\mathrm{IS}(G') \geq 15$ for every G' with 6 vertices and 7 edges, and $\sharp\mathrm{IS}(G') \geq 60$ for every G' with 9 vertices and 10 edges. Both of them can be verified by a simple case analysis. $\square$

We can now proceed to the proof of the optimality of case (iii) in Theorem 7. Note that the optimal graph for $(3k + 2, 3k + 2)$ is not unique. For example, a cycle C_5 in a graph G can be replaced by a graph $D = (V', E')$ with $V' = \{x_1, \ldots, x_5\}$ and $E' = \{x_1x_2, x_1x_3, x_2x_3, x_1x_4, x_4x_5\}$ without changing the number of independent sets because $\sharp\mathrm{IS}(C_5) = \sharp\mathrm{IS}(D) = 11$.

Proof. (Part (iii) of Theorem 7.)   Let $G = (V, E)$ be an arbitrary graph with $|V| = |E| = 3k + 2$ and let $G_1, G_2, \ldots, G_p$ denote the components of G. In what follows, we show that the desired graph, which is the union of $k - 1$ disjoint copies of C_3 and a C_5, can be obtained by applying a series of transformations which do not increase the number of independent sets in a graph.

First, we claim that if there are two components whose value is -1 then a graph is not optimal. Suppose that G_i and G_j have the value -1. Let G' be a graph obtained from G by changing G_i and G_j to two disjoint paths without changing the size of each component. Lemma 1 guarantees that this transformation does not increase the number of independent sets. However, Lemma 4 says that such a graph is not optimal. This completes the proof of the claim.

By a similar argument based on Lemma 4, if G contains a component C of value -1, then C must be P_2. In such a case, there is only one component D whose value is 1 and all other components have the value 0. By Lemma 5, we can modify the graph G to a graph G' such that all components of G' have the value 0 without increasing the number of independent sets. Now, by applying Lemma 2, we can transform G into a graph in which each component is a cycle. Finally, by applying parts (ii) to (v) of Fact 11 recursively, we eventually obtain the desired graph in which the number of independent sets is not larger than the original graph. $\square$

We can also give the complete solution to the more general problem in which the number of terms m is less than the number of variables n.

Theorem 12. *Suppose that $m < n$. Let F be a monotone 2-DNF formula with n variables and m terms represented by a graph G such that*

(i) G is the union of m vertex-disjoint edges and $n - 2m$ isolated vertices, if $m \leq n/2$,

(ii) G is the union of $(2m - n)/3$ copies of C_3 and $n - m$ copies of P_2, if $n/2 < m < n$ and $2m - n \equiv 0 \pmod{3}$,

(iii) G is the union of $(2m - n - 1)/3$ copies of C_3, $n - m - 1$ copies of P_2 and a P_3, if $n/2 < m < n$ and $2m - n \equiv 1 \pmod{3}$,

(iv) G is the union of $(2m - n + 1)/3$ copies of C_3, $n - m - 1$ copies of P_2 and an isolated vertex, if $n/2 < m < n$ and $2m - n \equiv 2 \pmod{3}$.

Then for every monotone 2-DNF H with n variables and m terms, $\sharp SAT(H) \leq \sharp SAT(F)$. Equivalently, for every graph $G' = (V', E')$ with $|V'| = n$ and $|E'| = m$, $\sharp IS(G') \geq \sharp IS(G)$.

The following figure shows the graphs representing the optimal 2-DNF formulas on n variables with $m(< n)$ terms given in Theorem 12.

The proof of Theorem 12 is analogous to the proof of Theorem 7 and is omitted because of limited space.

5   Future Work

We have provided an algorithm for deriving a DNF formula which minimizes the number of satisfiable assignments, but the computational effort was not considered. The design of an algorithm which derives the result more efficiently is thus a task for future work.

Acknowledgment

The second author would like to thank Tatsuie Tsukiji for helpful discussions.

References

1. Bollobás, B.: Relations between Sets of Complete Subgraphs. In: Proc. 5th British Comb. Conf., pp. 79–84 (1975)
2. Bollobás, B.: Extremal Graph Theory. Academic Press, New York (1978)
3. Bollobás, B.: Combinatorics. Cambridge University Press, Cambridge (1986)
4. Chung, F., Graham, R.: Erdős on Graphs. Wellesley, Massachusetts (1998)
5. Erdős, P.: On the Number of Complete Subgraphs Contained in Certain Graphs. Caspois Pest. Mat. 94, 290–296 (1969)

6. Fisher, D.: Lower Bounds on the Number of Triangles in a Graph. J. Graph Theory 13(4), 505–512 (1989)
7. Greenhill, C.: The Complexity of Counting Colorings and Independent Sets in Sparse Graphs and Hypergraphs. Computational Complexity 9, 52–73 (2000)
8. Kruskal, J.B.: The Number of Simplices in a Complex. In: Bellman, R. (ed.) Mathematical Optimization Techniques, pp. 251–278. University of California Press (1963)
9. Katona, G.O.H.: A Theorem on Finite Sets. In: Erdős, P., Katona, G. (eds.) Theory of Graphs, pp. 187–207. Akadémiai Kiadó and Academic Press (1968)
10. Rosen, K.H.: Handbook of Discrete and Combinatorial Mathematics. CRC Press, New York (2000)
11. Turán, P.: On an Extremal Problem in Graph Theory (in Hungarian). Mat. Fiz. Lapok 48, 436–452 (1941)
12. Vadhan, S.P.: The Complexity of Counting in Sparse, Regular, and Planar Graphs. SIAM J. Comput. 31(2), 398–427 (2001)
13. Valiant, L.G.: The Complexity of Computing the Permanent. Theoret. Comput. Sci. 8, 189–201 (1979)

Approximating Alternative Solutions

Michael Krüger and Harald Hempel

Institut für Informatik, Friedrich-Schiller-Universität Jena,
D-07743 Jena, Germany
{krueger,hempel}@minet.un-jena.de

Abstract. We study the approximability of alternative solutions for
NP-problems. In particular, we show that approximating the second
best solution is in many cases, such as MAXCUT, MAXSAT, MINIMUM
STEINER TREE, and others, substantially easier than approximating a
first solution. We prove that our polynomial-time approximation scheme
for the second best solution of MINIMUM STEINER TREE is optimal. In
contrast we also argue that for the problems MINIMUM INDEPENDENT
DOMINATING SET and MINIMUM TRAVELING SALESPERSON PROBLEM a
given optimal solution does not simplify finding a second best solution.

1 Introduction

The general notion of alternative solution problems stems from the in practice
often occurring desire to not just have one solution to a given problem but
several of similar quality, so the user and not the algorithm can have the final
say about which one to use. So in the framework of NP-complete problems a
natural question to study is if the knowledge of a solution to a given instance
does make the computation of an alternative solution easier or not.

In this paper we show that in many cases the approximability of a alternative solution is a lot easier then approximating a first solution. This is in some
sense surprising since when it comes to exactly computing an alternative solution, it has been shown that in general this is as hard, namely NP-hard, as
computing a first solution [19],[2],[18],[21],[11],[5],[10] for various types of puzzles [19],[18],[21],[11],[5] and other combinatorial problems [10]. Relatedly, it is
known that approximating a (*kth*) alternative maximum a posteriori assignment
for belief networks is as hard as approximating a first such assignment [2]. A similar result holds for approximatimg the second best longest cycle in a Hamiltonian
cubic graph [4].

We show that NP-complete problems differ widely when it comes to approximating an alternative solution. While for some NP-complete problems such as
MINIMUM VERTEX COVER, MAXIMUM INDEPENDENT SET, MAXIMUM CLIQUE,
MINIMUM SET COVER, MINIMUM DOMINATING SET the approximation of an
alternative optimal solution is straight forward and efficient—they allow a constant absolute error 1 approximation of an alternative optimal solution by adding
or deleting a single vertex—the approximation of alternative solutions is by no
means trivial in general and depends heavily on the problem itself. In fact, we

X. Hu and J. Wang (Eds.): COCOON 2008, LNCS 5092, pp. 204–214, 2008.

show on one hand that for MinIndDomSet and MinTSP even the knowledge of an optimal solution does not simplify the approximability of an alternative solution, i.e., a second solution is as hard to approximate as the first solution. On the other hand, we prove that for problems such as MaxCut, MaxSat, MinMaxMatching, Minimum Steiner Tree, and Min-Δ-TSP the presence of an optimal solution helps in approximating alternative ones. For instance, the NP-complete problem MaxCut that, given a graph, asks for a partition of the vertex set into two parts V_1 and V_2 that minimizes the number of edges between V_1 and V_2 is APX-complete [15] and thus admits no polynomial-time approximation scheme for computing such a partition. Yet we show that there exists a polynomial-time approximation scheme for approximating an alternative partition.

The following table gives an overview of our results for the treated alternative solution problems (AS-problems), that will be stated and proved in Section 3. Note that we assume that P $\neq$ NP and thus, FPTAS $\subsetneq$ PTAS $\subsetneq$ APX $\subsetneq$ NPO. Also, *abs error* $\leq k$ in the table below means that there exists a polynomial-time absolute error k approximation algorithm, that is, there exists an efficient algorithm providing a feasible solution y for each input x whose cost differs from the cost of an optimal solution for x by at most k.

NPO-Problem A	Known approximab. of A	New approximab. of AS-A
MaxCut	APX-complete	ptas
MaxSat	APX-complete	ptas
MinMaxMatching	absolute error ≤ 2	absolute error ≤ 1
MinST	APX-complete	ptas but no fptas
Min-Δ-TSP	APX-complete	ptas
MinIndDomSet	NPO PB-complete	NPO PB-complete
MinTSP	NPO-complete	NPO-complete

2   Preliminaries

We assume the reader to be familiar with the basic definitions and notations of approximability theory [1].

Definition 1. *An* NP-*optimization problem A is a tuple $(I, sol, cost, goal)$ such that the following holds:*

1. *The set of instances I can be decided in polynomial time.*
2. *For every $x \in I$ the set of feasible solutions $sol(x)$ for x can be decided in polynomial time and there exists a polynomial p_A such that $(\forall x)(\forall y \in sol(x))[|y| \leq p_A(|x|)]$.*
3. *Let $cost(x,y) \in \mathbb{N}$ denote the costs of solution y for the instance x. The function cost, also called the objective function, can be computed in polynomial time.*
4. *$goal \in \{min, max\}$.*

The class NPO *is the class of all* NP-*optimization problems.*

Similar to NPO-problems we now define alternative solution approximation problems as follows.[1]

Definition 2. *Let $A = (I, sol, cost, goal)$ be an NPO-problem. The problem of approximating an alternative solution for A,AS-A, is defined as*

Problem Description (AS-A)

Instance: *A pair (x, y_{opt}), where $x \in I$ and $y_{opt} \in sol(x)$ is an optimal solution for x (with respect to cost and goal).*
Feasible Solution: *A string y from $sol(x) \setminus \{y_{opt}\}$.*
Costs: *The cost $cost(x, y)$ of y with respect to x.*
Goal: *goal.*

3   Approximability Results

In this section we state and prove our approximability results for the AS-versions of several NPO-problems.

3.1   Maximum Cut, Maximum Satisfiability, Minimum Metric Traveling Salesperson Problem, and Minimum Maximal Matching

Problem Description (MaxCut)

Instance: A graph $G = (V, E)$.
Feasible Solution: A partition of V into disjoint sets V_1 and V_2.
Costs: The cardinality of the cut, i.e., the number of edges with one end point in V_1 and one endpoint in V_2.
Goal: max.

There exists a constant factor approximation algorithm with an approximation ratio of 1.14 for MAXCUT [6]. Since MAXCUT is APX-complete [15], MAXCUT does not admit a ptas. In contrast, there exists a ptas for AS-MAXCUT.

Theorem 1. AS-MAXCUT *is in* PTAS.

Proof. Let $G = (V, E)$ be a graph with a maximum cut V_1, V_2 with cardinality $opt(G)$.

The algorithm for AS-MAXCUT either moves one vertex from V_1 to V_2 or one vertex from V_2 to V_1 and selects this vertex such that the new cut is as large as

---

[1] Note that AS-A is formally speaking not an NPO-problem since checking if a string y_{opt} is truly an optimal solution for x can in general not be done in polynomial-time, in other words the set of instances I is not in P unless P = NP. One might view AS-A as a sort of promise problem, where the promise is the optimality of the given solution. However, since we are interested in approximating alternative solutions under the assumption that an optimal solution is given we take the freedom to nevertheless use concepts such as ptas and fptas that are formally defined only for NPO-problems but could be easily modified for our purposes.

possible. This can easily by done, by testing the size of the modified cuts for all possible choices to put a vertex $v \in V$ from V_1 to V_2 or from V_2 to V_1.

For the analysis, let k_v be the number of edges between V_1 and V_2 including v, for all vertices $v \in V$. Furthermore, for $v \in V_i$, $i \in \{1,2\}$, let ℓ_v denote the number of edges from v to some vertex in V_i. It is not hard to see that moving a vertex v from V_1 to V_2 or from V_2 to V_1 results in a cut of size $opt(G) - (k_v - \ell_v)$.

Furthermore, it holds that

$$\sum_{v \in V} k_v = 2 \cdot opt(G)$$

and thus, $min\{k_v : v \in V\} \leq 2 \cdot opt(G)/|V|$. It follows that $min\{(k_v - \ell_v) : v \in V\} \leq 2 \cdot opt(G)/|V|$ and hence, the above algorithm provides a solution of size at least $(1 - 2/|V|) \cdot opt(G)$, a $(1 + 2/(|V| - 2))$-approximation. This is a $(1 + \varepsilon)$-approximation for $|V| \geq 2/\varepsilon + 2$. Thus, the algorithm is a ptas. $\square$

A similar proof can be used to show the same result for MAXIMUM DIRECTED CUT which is also APX-complete [15]. Note furthermore, that applying similar local modifications leads to ptas' for the APX-complete problems MAXSAT [15] and MINIMUM METRIC TRAVELING SALESPERSON PROBLEM [16]. Similarly we can show that for the APX-complete problem MINIMUM MAXIMAL MATCHING [20] that has an approximation algorithm with a guaranteed absolute error of at most 2 this bound drops to 1 in the alternative solution version. Due to space restrictions we omit the details of the proofs which can be found in [10].

3.2   Minimum Independent Dominating Set

Problem Description (MinIndDomSet)

Instance: A graph $G = (V, E)$.
Feasible Solution: An independent dominating set (IDS) for G, i.e., a subset $V' \subseteq V$ such that for all $u \in V \setminus V'$ there exists a $v \in V'$ for which $\{u, v\} \in E$, and such that no two vertices in V' are joined by an edge from E.
Costs: The cardinality of the IDS, i.e., $|V'|$.
Goal: min.

The problem MININDDOMSET was shown to be not approximable within $|V|^{1-\varepsilon}$ for any $\varepsilon > 0$ [7]. Furthermore, MININDDOMSET is known to be NPO PB-complete [8]. It turns out that AS-MININDDOMSET is also not approximable within $|V|^{1-\varepsilon}$ and NPO PB-complete.

Theorem 2. *AS-MININDDOMSET is not approximable within* $|V|^{1-\varepsilon}$.

Proof. Assume to the contrary that A is a $|V|^{1-\varepsilon}$-approximation algorithm for AS-MININDDOMSET for some $\varepsilon > 0$. Using this assumption we will give a $|V|^{1-\varepsilon}$-approximation algorithm for MININDDOMSET.

So let $G = (V, E)$ be an instance for MININDDOMSET. We add a new vertex w to V and the edges $\{\{w, v\} : v \in V\}$ to E. Thus, $\{w\}$ is a minimal IDS of the resulting graph G'. It is not hard to see, that w is no member of any alternative independent dominating set. Furthermore, each IDS of G is also an IDS of G' and conversely, each IDS $V' \neq \{w\}$ of G' is also an IDS of G. Now, we apply algorithm A to $(G', \{w\})$, which produces a $|V|^{1-\varepsilon}$-approximation V' for the second smallest IDS of G'. As seen above, this is also a $|V|^{1-\varepsilon}$-approximation for the minimal IDS of G, a contradiction.

Hence, the assumption is wrong which implies the theorem.   ❑

Note that the above idea represents a linear reduction from MinIndDomSet to AS-MinIndDomSet and thus, AS-MinIndDomSet is also NPO PB-complete (also see [8]).

3.3 Minimum Steiner Tree

Problem Description (MinST)

Instance: A complete graph $G = (V, E)$, edge weights $s : E \to \mathbb{N}$, and a subset $S \subseteq V$ of required vertices.

Feasible Solution: A Steiner tree T, i.e., a subtree of G that includes all the vertices in S.

Costs: The sum of the weights of the edges in the subtree T, i.e., $cost(T) =$
$$\sum_{e \in T} s(e).$$
Goal: min.

To the best of our knowledge the current best approximation algorithm for MinST is a 1.55-approximation [17]. Furthermore, MinST is known to be APX-complete [3], even when restricted to edge weights from $\{1, 2\}$. Thus, there is no ptas for MinST. In contrast, we now show that AS-MinST admits a ptas.

Theorem 3. *AS-MinST is in PTAS.*

Proof. We give a ptas for AS-MinST. Let $\varepsilon > 1$ be a rational number. We give an ε-approximation algorithm with runtime p_ε for some polynomial p_ε. So let $((G, s, S), T_{opt})$ be an instance for AS-MinST, where $G = (V, E)$, $|V| = n$, and T_{opt} is an optimal Steiner tree for the MinST-instance (G, s, S). Let $E(T)$ denote the edges of a graph T, e.g., $E(T_{opt})$ are the edges of T_{opt}. We assume that the triangle inequality holds for s, since we can replace $s(\{v_1, v_3\})$ by $s(\{v_1, v_2\}) + s(\{v_2, v_3\})$ if $s(\{v_1, v_3\}) > s(\{v_1, v_2\}) + s(\{v_2, v_3\})$. In this case using the edge $\{v_1, v_3\}$ represents using $\{v_1, v_2\}$ and $\{v_2, v_3\}$. Furthermore, we can assume that there are no weightless edges in G since otherwise, the respective vertices can be united to a new vertex.

Let $T_{secbest}$ denote an (unknown) second best Steiner tree of (G, s, S), i.e., an optimal solution for $((G, s, S), T_{opt})$ as AS-MinST-instance. Let $E^*_{T_{secbest}}$ denote the edges from $T_{secbest} \setminus T_{opt}$, that share a vertex with an edge from T_{opt}, that is,

$$E^*_{T_{secbest}} = \{e \in E(T_{secbest}) \setminus E(T_{opt}) : (\exists e' \in E(T_{opt}))[|e \cap e'| = 1]\}.$$

Our algorithm consists of two different procedures called algorithm A and algorithm B. Depending on the number of edges in $E^*_{T_{secbest}}$, algorithm A or algorithm B outputs a sufficiently good approximation of $T_{secbest}$. Since $E^*_{T_{secbest}}$ is unknown, both algorithms are applied consecutively, both results are compared, and the better Steiner tree is output.

First, assume that $|E^*_{T_{secbest}}| \geq k := \lceil 1/(\varepsilon - 1) \rceil$. In this case the following easy algorithm A provides a sufficiently good approximation. Algorithm A adds the edge, that realizes the minimum $min\{s(e) : e \in E \setminus E(T_{opt}) \wedge (\exists e' \in E(T_{opt}))[|e \cap e'| = 1]\}$. In other words, A adds the smallest possible edge e, such that the resulting graph $T_{opt} + e$ is still connected. If adding e leads to a cycle, A removes the cycle edge with the biggest weight (of course, e is not removed). Note that all edges from $E^*_{T_{secbest}}$ have weight greater or equal the weight of e. Thus, $cost(T_{secbest}) \geq ks(e)$ and it holds

$$cost(T_{opt} + e) = cost(T_{opt}) + s(e)$$
$$\leq cost(T_{secbest}) + \frac{1}{k}cost(T_{secbest})$$
$$\leq \varepsilon cost(T_{secbest}).$$

Thus, algorithm A is an ε-approximation if $|E^*_{T_{secbest}}| \geq k$. Obviously, algorithm A has a polynomial-runtime p_A.

If $|E^*_{T_{secbest}}| < k$, algorithm A must not be good enough. In this case algorithm B exactly computes $T_{secbest}$ or another second best solution of (G, s, S), which is obviously an ε-approximation. If (G, s, S) has several second best Steiner trees we use the notion $T_{secbest}$ very flexibly, that is, $T_{secbest}$ always denotes a second best Steiner tree that still fulfils all assumptions made w.l.o.g.

Algorithm B performs a smart exhaustive search over all Steiner trees T with $|E^*_T| < k$. In a first step B guesses a candidate E^* for $E^*_{T_{secbest}}$ (see also Figure 1). Since, $|E| \leq n^2$, there are at most $n^{2\ell}$ possibilities for $|E^*| = \ell$. Since

$$\sum_{\ell=1}^{k-1}(n^2)^\ell = \frac{(n^2)^\ell - 1}{n^2 - 1} \leq n^{2k},$$

n^{2k} is an upper bound for the number of possibilities in this step. Note that guessing means here, that all possibilities are consecutively generated and handed over to the next step of the algorithm.

Assume in the following that $E^*_{T_{secbest}}$ is guessed in the first step. The sought-after Steiner tree $T_{secbest}$ decomposes into $T_{secbest} \cap T_{opt}$ and $T_{secbest} \setminus T_{opt}$. In the second step, we guess all possibilities for $T_{secbest} \setminus T_{opt}$ using the fact that $E^*_{T_{secbest}}$ is the set of edges, where $T_{secbest}$ "leaves" T_{opt}. We know that (a) $T_{secbest} \setminus T_{opt}$ is a forest, where (b) the endpoints of the edges $E^*_{T_{secbest}}$ that are in T_{opt} are exactly the leaves of this forest, since (a) $T_{secbest}$ is a tree and (b) if there were further leaves in $T_{secbest} \setminus T_{opt}$, these leaves could be cut from $T_{secbest}$ to gain a better Steiner tree[2]. Furthermore, since the triangle inequality holds for s and since $T_{secbest} \setminus T_{opt}$ is an optimal forest (edge disjoint with T_{opt}) having these leaves, we can assume that this forest contains no vertices of degree 2. Because

---

[2] Note that this is wrong in case of $T_{secbest}$ is T_{opt} plus an edge of small weight. However, in this situation algorithm A outputs $T_{secbest}$.

of that and since $|E^*_{T_{secbest}}| \leq k - 1$, this forest contains at most $k - 2$ further vertices. After guessing a set of at most $k - 2$ vertices ($\leq n^{k-2}$ possibilities) the number of possibilities to form a forest from the given leaves and inner vertices is a number only depending on k, say $g(k)$. So for this step we have $\leq g(k)n^{k-2}$ possibilities (see also Figure 1).

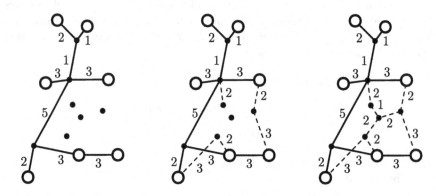

Fig. 1. An illustration of the first and second step of algorithm B. (left) The optimal tree T_{opt}, (middle) the optimal tree T_{opt} plus the edges $E^*_{T_{secbest}}$ guessed in the first step, (right) the optimal tree T_{opt} plus the forest $T_{secbest} \setminus T_{opt}$ computed in the third step.

Assume that $T_{secbest} \setminus T_{opt}$ was guessed in the second step and add it to T_{opt}. The resulting graph has several cycles. The number of cycles is bounded by $(k - 1)^2$, since T_{opt} and $T_{secbest} \setminus T_{opt}$ contain no cycles and there is at most one new cycle per path from a leave of $T_{secbest} \setminus T_{opt}$ to another. The number of such pathes is bounded by $(k - 1)^2$. Now, the algorithm B has to find out, which is the best possible way to remove edges from T_{opt} such that the resulting graph is a tree and thus, $T_{secbest}$. This is done by a simple recursive subroutine C as follows. The routine C picks one cycle and consecutively works through all possibilities to delete one edge from the cycle and restarts C on the respective modified graph (see also Figure 2). After destroying all cycles algorithm B checks whether further edges (and possibly vertices) can be deleted without isolating a vertex from S. If this is possible, it is done as often as possible (see also Figure 2). During this process, the algorithm stores the recent best solution and finally deletes the according edges (and possibly vertices) from $T_{secbest} \cup T_{opt}$, resulting in $T_{secbest}$. An upper bound on the number of tested possibilities in this third step is $n^{2(k-1)}$ since, n is an upper bound for the number of edges in each cycle. Note that this subroutine C is not very clever and the estimate of the number of possibilities very rough. Both can surely be improved.

Let us take stock. Assume that $|E^*_{T_{secbest}}| < k$. Algorithm B guesses all $\leq n^{2k}$ possibilities for E^* with $|E^*| < k$ and thus also guesses $E^*_{T_{secbest}}$ in the first step. In the second step all possible forests connecting the chosen edges E^* are guessed. Here, for the correctly guessed $E^*_{T_{secbest}}$, we among others obtain $T_{secbest} \setminus T_{opt}$. In

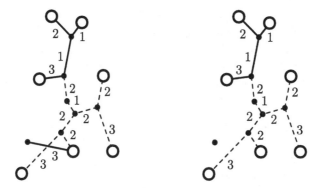

Fig. 2. An illustration of the third step of algorithm B. (left) T_{opt} plus $T_{secbest} \setminus T_{opt}$ after subroutine C destroyed all cycles, (right) The second best solution $T_{secbest}$ obtained after deleting the useless edge.

the second step there are $\leq g(k)n^{k-2}$ possibilities for some function g. The third step deals with finding the best possibility to remove T_{opt}-edges from $T_{secbest} \cup T_{opt}$ such that the resulting graph is a tree. Starting with the correctly guessed $T_{secbest} \cup T_{opt}$ this third step outputs $T_{secbest}$ after at most $n^{2(k-1)}$ tested possibilities. So, algorithm B tests overall $n^{2k} \cdot g(k)n^{k-2} \cdot n^{2(k-1)} \leq g(k)n^{5k}$ possibilities which needs $\leq g(k)n^{5k} \cdot p(n)$ time for some polynomial p independent from k, since all but the guessing steps can be managed in polynomial time in n. In case of $|E^*_{T_{secbest}}| < k$, algorithm B finds and outputs $T_{secbest}$. Observe, that the running-time of algorithm B is a polynomial in n depending only from k, that is, from ε.

So executing algorithm A and algorithm B can be done in time p_ε for some polynomial p_ε and the better result is always an ε-approximation for $T_{secbest}$. $\square$

Furthermore, we can show, that a ptas is the best possible approximation algorithm for AS-MINST, i.e., there is no fptas for AS-MINST. Due to space restrictions we omit the proof which is based on a fptas-preserving reduction from MINST to AS-MINST. It can be found in [10].

Theorem 4. *There is no* fptas *for* AS-MINST.

3.4   Minimum Traveling Salesperson Problem

Problem Description (MinTSP)

Instance: A complete undirected graph $G = (V, E)$ on n vertices and a weight function $d : E \to \mathbb{N}$.

Feasible Solution: A Hamiltonian cycle in G.

Costs: The weight of the cycle, i.e., the sum of the edge weights of the cycle.

Goal: min.

It is known that MINTSP is NPO-complete [12]. For AS-MINTSP, it turns out that a given optimal solution does not help in finding the second best solution, that is, AS-MINTSP is also NPO-complete.

Theorem 5. AS-MINTSP *is* NPO-*complete.*

Proof. For the proof we combine the argument for the NPO-completeness of MINTSP from [12], where Hamiltonian cycles play an important role, with an idea to generate an additional Hamiltonian cycle in a graph [13].

First, we give a sketch of the proof of the NPO-completeness of MINTSP from [12], that is, we give an idea of the $\leq_{ptas}$-reduction[3] from the NPO-complete problem WEIGHTED 3SAT (see also [12]) to MINTSP.

Problem Description (Weighted 3SAT)

Instance: A Boolean formula F in 3CNF over a variable set X and a weight function $w : X \rightarrow \mathbb{N}$.

Feasible Solution: All satisfying assignments α. The assignment $\alpha' : x \mapsto true$, $x \in X$, is always a feasible solution even if $F(\alpha') = false$.

Costs: The sum of the weights of the variables set to be true, i.e., $cost(\alpha)$ is
$$cost(\alpha) = \sum_{\alpha(x)=1} w(x) \text{ if } F(\alpha) = 1 \text{ and } cost(\alpha') = 1 + \sum_{x \in X} w(x) \text{ if } F(\alpha') = 0.$$

Goal: min.

This reduction is based on a well-known $\leq_m^p$-reduction from 3SAT to HAMILTONIAN CYCLE from [14]. In this construction, a Boolean formula F over X is mapped to a graph G_F such that F is satisfiable if and only if G_F is Hamiltonian. Moreover, for each variable from $x \in X$, G_F includes an edge e_x, such that each Hamiltonian cycle C of G_F canonically corresponds to the satisfying assignment α_C, defined via $\alpha_C(x) = 1$ if $e_x \in C$ and $\alpha_C(x) = 0$, otherwise. Conversely, for each satisfying assignment α of F, G_F has a Hamiltonian cycle C_α with $\alpha_{C_\alpha} = \alpha$.

Based on this construction, the reduction WEIGHTED 3SAT $\leq_{ptas}$ MINTSP can be described as follows. Let (F, w) be an instance of WEIGHTED 3SAT, and let $G_F = (V, E)$ be the above graph. Let $G'_F = (V, E')$ be the complete graph over V. The distance function $d : E' \rightarrow \mathbb{N}$ is defined as follows,

$$d(e) = \begin{cases} 1 + \sum_{x \in X} w(x), & \text{if } e \in E' \setminus E \\ w(x), & \text{if } e = e_x, \text{ for some } x \in X \\ 0, & \text{otherwise.} \end{cases}$$

So, all edges not in G_F have a very large penalty weight $1 + \sum_{x \in X} w(x)$ and a Hamiltonian cycle C_α corresponding to a satisfying assignment α of F has weight $\sum_{\alpha(x)=1} w(x) = cost(\alpha)$, which is smaller than the penalty weight. Thus, it is not hard to see, that the functions f, g, and c defined via

- $f(((F,w),\varepsilon)) = (G'_F, d)$,
- $g((((G'_F, d), C, \varepsilon))) = \alpha_C$ if $cost((G'_F, d), C) \leq \sum_{x \in X} w(x)$, and
  $g(((x, C, \varepsilon))) = \alpha'$, otherwise, and
- $c(\varepsilon) = \varepsilon$

actually realize the reduction WEIGHTED 3SAT $\leq_{ptas}$ MINTSP.

---

[3] A $\leq_{ptas}$-reduction is a ptas-preserving reduction between NPO-problems. Due to space restrictions we refer to [1] for the definition and basic properties.

We modify this reduction as follows. We make use of a construction given in [13], where, a graph G is mapped to a graph $\hat{G}$ with an additional Hamiltonian cycle C. In particular, $\hat{G}$ always has the Hamiltonian cycle C and it has an alternative Hamiltonian cycle for each Hamiltonian cycle of G. We apply this construction to the above graph G_F and call the resulting graph $\hat{G}_F = (\hat{V}_F, \hat{E}_F)$. Let C denote the additional cycle in $\hat{V}_F$.

Recall that Hamiltonian cycles of G_F correspond to satisfying assignments of F, and note that this correspondence is preserved by the construction of $\hat{G}_F$. In particular, there are still edges e_x, $x \in X$, in $\hat{G}_F$ such that each Hamiltonian cycle C', different from C, corresponds to a satisfying assignment $\alpha_{C'}$ of F as follows, $\alpha_{C'}(x) = 1$ if and only if $e_x \in C'$. It is important to note, that the cycle C contains none of the edges e_x, $x \in X$.

Now we can define the modified $\leq_{ptas}$-reduction. Let $\hat{G}'_F = (\hat{V}_F, \hat{E}'_F)$ be the complete graph over $\hat{V}_F$. Analogous to the case of WEIGHTED 3SAT $\leq_{ptas}$ MINTSP, the distance function $\hat{d} : \hat{E}'_F \to \mathbb{N}$ is defined as follows,

$$
\hat{d}(e) = \begin{cases} 1 + \sum_{x \in X} w(x), & \text{if } e \in \hat{E}'_F \setminus \hat{E}_F \\ w(x), & \text{if } e = e_x, \text{ for some } x \in X \\ 0, & \text{otherwise.} \end{cases}
$$

Again, edges not in $\hat{G}_F$ have large penalty weight. Since, C is a Hamiltonian cycle of $\hat{G}_F$ that uses non of the edges e_x, $x \in X$, it is the optimal solution and has weight zero. For each satisfying assignment α of F, $\hat{G}'_F$ has an alternative Hamiltonian cycle C'_α, having weight $\sum_{\alpha(x)=1} w(x) = cost(\alpha)$, which is again smaller than the penalty weight. Hence, the functions $\hat{f}$, $\hat{g}$, and $\hat{c}$ defined via

- $\hat{f}(((F, w), \varepsilon)) = ((\hat{G}'_F, d), C)$,
- $\hat{g}(((\hat{G}'_F, d), C), C', \varepsilon) = \alpha'_C$ if $cost(((\hat{G}'_F, d), C), C') \leq \sum_{x \in X} w(x)$, and

  $\hat{g}(((\hat{G}'_F, d), C), C', \varepsilon) = \alpha'$, otherwise, and
- $\hat{c}(\varepsilon) = \varepsilon$

realize the reduction WEIGHTED 3SAT $\leq_{ptas}$ AS−MINTSP.   ❑

A very similar construction [9] can be used to show the same approximability result for approximating the nth best tour given the best $n - 1$ tours.

Acknowledgments. The authors would like to thank Tobias Berg and the anonymous referees for their very helpful comments.

References

1. Ausiello, G., Crescenzi, P., Gambosi, G., Kann, V., Marchetti-Spaccamela, A., Protasi, M.: Complexity and Approximation. Springer, Berlin (1999)
2. Abdelbar, A.M., Hedetniemi, S.M.: Approximating MAPs for Belief Networks is NP-hard and other Theorems. Artificial Intelligence 102(1), 21–38 (1998)

3. Bern, M., Plassmann, P.: The Steiner Problem with Edge Lengths 1 and 2. Inform. Process. Lett. 32(4), 171–176 (1989)
4. Bazgan, C., Santha, M., Tuza, Z.: On the Approximation of Finding another Hamiltonian Cycle in Cubic Hamiltonian Graphs. Journal of Algorithms 31(1), 249–268 (1999)
5. de Bondt, M.: On the ASP-Completeness of Cryptarisms. Technical Report 0419, Department of Mathematics, Radboud University of Nijmegen (2004)
6. Goemans, M.X., Williamson, D.P.: Improved Approximation Algorithms for Maximum Cut and Satisfiability Problems Using Semidefinite Programming. J. Assoc. Comput. Mach. 42(6), 1115–1145 (1995)
7. Halldórsson, M.M.: Approximating the Minimum Maximal Independence Number. Inform. Process. Lett. 46(4), 169–172 (1993)
8. Kann, V.: Polynomially Bounded Minimization Problems that are Hard to Approximate. Nordic J. Comput. 1(3), 317–331 (1994); Selected papers of the 20th International Colloquium on Automata, Languages and Programming (ICALP 1993), Lund (1993)
9. Krüger, M.: Weitere Hamiltonkreise in Graphen und Inverse Hamiltonkreisprobleme (German). Master's thesis, Friedrich-Schiller-University Jena, Germany (2005)
10. Krüger, M.: On the Complexity of Alternative Solutions. PhD Thesis, Friedrich-Schiller-University Jena, Germany (2008)
11. McPhail, B.P.: The Complexity of Puzzles: NP-Completeness Results for Nurikabe and Minesweeper. Bachelor thesis, The Division of Mathematics and Natural Sciences, Reed College, Portland (2003)
12. Orponen, P., Manila, H.: On Approximation Preserving Reductions: Complete Problems and Robust Measures (1990)
13. Papadimitriou, C.H., Steiglitz, K.: Some Complexity Results for the Traveling Salesman Problem. In: Proc. of Eighth Annual ACM Symposium on Theory of Computing, Hershey, Pennsylvania,US, pp. 1–9. Assoc. Comput. Mach., New York (1976)
14. Papadimitriou, C.H., Steiglitz, K.: Combinatorial Optimization: Algorithms and Complexity. Prentice-Hall Inc., Englewood Cliffs (1982)
15. Papadimitriou, C.H., Yannakakis, M.: Optimization, Approximation, and Complexity Classes. J. Comput. System Sci. 43(3), 425–440 (1991)
16. Papadimitriou, C.H., Yannakakis, M.: The Traveling Salesman Problem with Distances One and Two. Math. Oper. Res. 18(1), 1–11 (1993)
17. Robins, G., Zelikovsky, A.: Improved Steiner Tree Approximation in Graphs. In: Proc. of SODA 2000, pp. 770–779 (2000)
18. Seta, T.: The Complexity of Puzzles, Cross Sum and Their Another Solution Problems (ASP). Senior Thesis, Department of Infomation Science, the Faculty of Science, the University of Tokyo (2002)
19. Ueda, N., Nagao, T.: NP-Completeness Results for Nonogram via Parsimonious Reductions. Technical report, Tokyo Institute of Technology (1996)
20. Yannakakis, M., Gavril, F.: Edge Dominating Sets in Graphs. SIAM J. Appl. Math. 38(3), 364–372 (1980)
21. Yato, T., Seta, T.: Complexity and Completeness of Finding Another Solution and its Application to Puzzles. In: Proceedings of the National Meeting of the Information Processing Society of Japan (IPSJ) (2002)

Dimensions of Points in Self-similar Fractals

Jack H. Lutz[1],[*],[**] and Elvira Mayordomo[2],[**],[***]

[1] Department of Computer Science, Iowa State University, Ames, IA 50011 USA
lutz@cs.iastate.edu
[2] Departamento de Informática e Ingeniería de Sistemas, María de Luna 1,
Universidad de Zaragoza, 50018 Zaragoza, Spain
elvira@unizar.es

Abstract. We use nontrivial connections between the theory of computing and the fine-scale geometry of Euclidean space to give a complete analysis of the dimensions of individual points in fractals that are computably self-similar.

1 Introduction

This paper analyzes the dimensions of points in the most widely known type of fractals, the self-similar fractals. Our analysis uses nontrivial connections between the theory of computing and the fine-scale geometry of Euclidean space. In order to explain our results, we briefly review self-similar fractals and the dimensions of points.

1.1 Self-similar Fractals

The class of self-similar fractals includes such famous objects as the Sierpinski triangle, the Cantor set, the von Koch curve, and the Menger sponge, along with many more exotic sets in Euclidean space [2],[11],[12],[14]. To be concrete, consider the Sierpinski triangle, which is constructed by the process illustrated in Figure 1. We start (at the left) with the equilateral triangle D whose vertices are the points $v_0 = (0,0)$, $v_1 = (1,0)$, and $v_2 = (\frac{1}{2}, \frac{\sqrt{3}}{2})$ in $\mathbb{R}^2$ (together with this triangle's interior). The construction is carried out by three functions $S_0, S_1, S_2 : \mathbb{R}^2 \to \mathbb{R}^2$ defined by $S_i(x) = v_i + \frac{1}{2}(x - v_i)$ for each $x \in \mathbb{R}^2$ and $i = 0, 1, 2$. Note that $|S_i(x) - S_i(y)| = \frac{1}{2}|x - y|$ always holds, i.e., each S_i is a *contracting similarity*

---

* Research supported in part by National Science Foundation Grants 0344187, 0652569, and 0728806. Part of this author's research was performed during a sabbatical at the University of Wisconsin and two visits at the University of Zaragoza.
** Research supported in part by Spanish Government MEC and the European Regional Development Fund (ERDF) under Project TIN 2005-08832-C03-02.
*** Part of this author's research was performed during a visit at Iowa State University, supported by Spanish Government (Secretaría de Estado de Universidades e Investigación del Ministerio de Educación y Ciencia) grant for research stays PR2007-0368.

X. Hu and J. Wang (Eds.): COCOON 2008, LNCS 5092, pp. 215–224, 2008.
© Springer-Verlag Berlin Heidelberg 2008

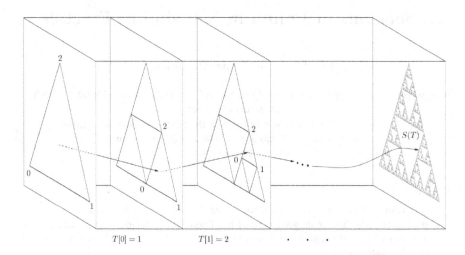

$T[0] = 1$ $T[1] = 2$ $\cdot$ $\cdot$ $\cdot$

Fig. 1. A sequence $T \in \{0, 1, 2\}^\infty$ codes a point $S(T)$ in the Sierpinski triangle

with *contraction ratio* $c_i = \frac{1}{2}$. Note also that each S_i maps the triangle D onto a similar subtriangle containing the vertex v_i.

We use the alphabet $\Sigma = \{0, 1, 2\}$ to specify the contracting similarities S_0, S_1, S_2. Each infinite sequence $T \in \Sigma^\infty$ over this alphabet *codes* a point $S(T)$ in the Sierpinski triangle via the following recursion. (See Figure 1.) We start at time $t = 0$ in the triangle $\Delta_0 = D$. At time $t + 1$, we move into the sub-triangle Δ_{t+1} of Δ_t given by the (appropriately rescaled) contracting similarity $S_{T[t]}$, where $T[t]$ is the tth symbol in T. The point $S(T)$ is then the unique point in $\mathbb{R}^2$ lying in all the triangles $\Delta_0, \Delta_1, \Delta_2, \ldots$. Finally, the Sierpinski triangle is the set $F(S) = \{S(T) \mid T \in \Sigma_k^\infty\}$ of all points coded in this fashion.

Self-similar fractals are defined by generalizing the above construction. We work in a Euclidean space $\mathbb{R}^n$. An *iterated function system (IFS)* is a list $S = (S_0, ..., S_{k-1})$ of two or more contracting similarities $S_i : \mathbb{R}^n \to \mathbb{R}^n$ that map an initial nonempty, closed set $D \subseteq \mathbb{R}^n$ into itself. Each S_i has a contraction ratio $c_i \in (0, 1)$. (The contraction ratios $c_0, \ldots, c_{k-1}$ need not be the same.) The alphabet $\Sigma_k = \{0, \ldots, k - 1\}$ is used to specify the contracting similarities in S, and each sequence $T \in \Sigma_k^\infty$ codes a point $S(T) \in \mathbb{R}^n$ in the now-obvious manner. The *attractor* of the IFS S is the set $F(S) = \{S(T) \mid T \in \Sigma_k^\infty\}$. In general, the sets $S_0(D), \ldots, S_{k-1}(D)$ may not be disjoint, so a point $x \in F(S)$ may have many *coding sequences*, i.e., many sequences T for which $S(T) = x$. A *self-similar fractal* is a set $F \subseteq \mathbb{R}^n$ that is the attractor of an IFS S that satisfies a technical *open set condition* (defined in section 2), which ensures that the sets $S_0(D), \ldots, S_{k-1}(D)$ are "nearly" disjoint.

The *similarity dimension* of a self-similar fractal F is the (unique) solution $\text{sdim}(F)$ of the equation

$$\sum_{i=0}^{k-1} c_i^{\text{sdim}(F)} = 1, \tag{1}$$

where $c_0, \ldots, c_{k-1}$ are the contraction ratios of any IFS S satisfying the open set condition and $F(S) = F$. A classical theorem of Moran [29] and Falconer [13] says that, for any self-similar fractal F,

$$\dim_H(F) = \mathrm{Dim}_P(F) = \mathrm{sdim}(F), \tag{2}$$

i.e., the Hausdorff and packing dimensions of F coincide with its similarity dimension. In addition to its theoretical interest, the Moran-Falconer theorem has the pragmatic consequence that the Hausdorff and packing dimensions of a self-similar fractal are easily computed from the contraction ratios by solving equation (1).

1.2   Dimensions of Points

The theory of computing has recently been used to provide a meaningful notion of the *dimensions of individual points* in Euclidean space [26],[1],[18],[27]. These dimensions are robust in that they have many equivalent characterizations. For the purposes of this paper, we define these dimensions in terms of Kolmogorov complexities of rational approximations in Euclidean space.

For each $x \in \mathbb{R}^n$ and $r \in \mathbb{N}$, we define the *Kolmogorov complexity* of x at precision r to be the natural number $\mathrm{K}_r(x) = \min\{\mathrm{K}(q) \mid q \in \mathbb{Q}^n$ and $|q - x| \le 2^{-r}\}$, where $\mathrm{K}(q)$ is the Kolmogorov complexity of the rational point q [25]. That is, $\mathrm{K}_r(x)$ is the minimum length of any program $\pi \in \{0,1\}^*$ for which $U(\pi)$ – the output of a fixed universal Turing machine on input π – is a rational approximation of x to within 2^{-r}. (Related notions of approximate Kolmogorov complexity have recently been considered by Vitanyi and Vereshchagin [33] and Fortnow, Lee and Vereshchagin [16].)

Definition 1.1. *Let $x \in \mathbb{R}^n$, let $X \subseteq \mathbb{R}^n$.*

1. *The dimension of the point x is* $\dim(x) = \liminf_{r \to \infty} \frac{\mathrm{K}_r(x)}{r}$.
2. *The strong dimension of the point x is* $\mathrm{Dim}(x) = \limsup_{r \to \infty} \frac{\mathrm{K}_r(x)}{r}$.
3. *The constructive dimension of the set X is* $\mathrm{cdim}(X) = \sup_{x \in X} \dim(x)$.
4. *The constructive strong dimension of the set X is* $\mathrm{cDim}(X) = \sup_{x \in X} \mathrm{Dim}(x)$.

Intuitively, $\dim(x)$ and $\mathrm{Dim}(x)$ are the lower and upper asymptotic *information densities* of the point $x \in \mathbb{R}^n$.

It is easy to see that $0 \le \dim(x) \le \mathrm{Dim}(x) \le n$ for all $x \in \mathbb{R}^n$. In fact, this is the only restriction that holds in general, i.e., for any two real numbers $0 \le \alpha \le \beta \le n$, there is a point $x \in \mathbb{R}^n$ with $\dim(x) = \alpha$ and $\mathrm{Dim}(x) = \beta$ [1]. Points x that are computable have $\dim(x) = \mathrm{Dim}(x) = 0$, while points x that are random (in the sense of Martin-Löf [28]) have $\dim(x) = \mathrm{Dim}(x) = n$.

The dimensions $\dim(x)$ and $\mathrm{Dim}(x)$ are well defined and robust, but are they *geometrically* meaningful? Prior work already indicates an affirmative answer. By Hitchcock's correspondence principle for constructive dimension ([21], extending a result of [32]), together with the absolute stability of constructive dimension [26], if $X \subseteq \mathbb{R}^n$ is any countable (not necessarily effective) union of computably closed, i.e., Π_1^0, sets, then

$$\dim_H(X) = \sup_{x \in X} \dim(x). \tag{3}$$

That is, the *classical* Hausdorff dimension [14] of any such set is completely determined by the dimensions of its individual points. Many, perhaps most, of the sets which arise in "standard" mathematical practice are unions of computably closed sets, so (3) constitutes strong *prima facie* evidence that the dimensions of individual points are indeed geometrically meaningful.

The full version of this paper shows that definitions 1.1 are equivalent to the original definitions of dimension and strong dimension [26],[1] and are thus constructive versions of the two most important classical fractal dimensions, namely Hausdorff dimension and packing dimension, respectively.

1.3   Our Results

Our main theorem concerns the dimensions of points in fractals that are *computably self-similar*, meaning that they are attractors of *computable* iterated function systems satisfying the open set condition. (We note that most self-similar fractals occurring in practice – including the four famous examples mentioned in section 1.1 – are, in fact, computably self-similar.) Our main theorem says that, if F is any fractal that is computably self-similar with the IFS S as witness, then, for every point $x \in F$ and every coding sequence T for x, the dimension and strong dimension of the point x are given by the dimension formulas

$$\dim(x) = \text{sdim}(F)\dim^{\pi_S}(T) \tag{4}$$

and

$$\text{Dim}(x) = \text{sdim}(F)\text{Dim}^{\pi_S}(T), \tag{5}$$

where $\dim^{\pi_S}(T)$ and $\text{Dim}^{\pi_S}(T)$ are the dimension and strong dimension of T *with respect to* the probability measure π_S on the alphabet Σ_k defined by

$$\pi_S(i) = c_i^{\text{sdim}(F)} \tag{6}$$

for all $i \in \Sigma_k$. (We define $\dim^{\pi_S}(T)$ and $\text{Dim}^{\pi_S}(T)$ in the next paragraph.) This theorem gives a complete analysis of the dimensions of points in computably self-similar fractals and the manner in which the dimensions of these points arise from the dimensions of their coding sequences.

In order to understand the right-hand sides of equations (4) and (5), we now define the dimensions $\dim^{\pi_S}(T)$ and $\text{Dim}^{\pi_S}(T)$.

Definition 1.2. *Let Σ be an alphabet with $2 \leq |\Sigma| < \infty$, and let π be a positive probability measure on Σ. Let $w \in \Sigma^*$ and $T \in \Sigma^\infty$.*

1. *The Shannon self-information of w with respect to π is*
   $\mathcal{I}_\pi(w) = \log \frac{1}{\pi(w)} = \sum_{i=0}^{|w|-1} \log \frac{1}{\pi(w[i])}$, *where the logarithm is base-2 [10].*

2. *The dimension of T with respect to π is*

$$\dim^\pi(T) = \liminf_{j \to \infty} \frac{\mathrm{K}(T[0..j-1])}{\mathcal{I}_\pi(T[0..j-1])}. \tag{7}$$

3. *The strong dimension of T with respect to π is*

$$\mathrm{Dim}^\pi(T) = \limsup_{j \to \infty} \frac{\mathrm{K}(T[0..j-1])}{\mathcal{I}_\pi(T[0..j-1])}. \tag{8}$$

The dimensions $\dim^\pi(T)$ and $\mathrm{Dim}^\pi(T)$ are measures of the *algorithmic information density* of T, but the "density" here is now an information-to-cost ratio. In this ratio, the "information" is algorithmic information, i.e., Kolmogorov complexity, and the "cost" is the Shannon self-information. To see why this makes sense, consider the case of interest in our main theorem. In this case, (6) says that the cost of a string $w \in \Sigma_k^*$ is $\mathcal{I}_\pi(w) = \mathrm{sdim}(F) \sum_{j=0}^{|w|-1} \log \frac{1}{c_{w[j]}}$, i.e., the sum of the costs of the symbols in w, where the cost of a symbol $i \in \Sigma_k$ is $\mathrm{sdim}(F) \log(1/c_i)$. These symbol costs are computational and realistic. A symbol i with high cost invokes a similarity S_i with a small contraction ratio c_i, thereby necessitating a high-precision computation.

The full version of this paper shows that definitions (7) and (8) are equivalent to "gale characterizations" of these dimensions, and hence that $\dim^\pi(T)$ is a constructive version of Billingsley dimension [3],[9].

Although our main theorem only applies directly to computably self-similar fractals, we use relativization to show that the Moran-Falconer theorem (2) for arbitrary self-similar fractals is an easy consequence of our main theorem. Hence, as is often the case, a theorem of computable analysis (i.e., the theoretical foundations of scientific computing [6]) has an immediate corollary in classical analysis.

The proof of our main theorem has some geometric and combinatorial similarities with the classical proofs of Moran [29] and Falconer [13], but the argument here is information-theoretic. As such, it gives a more clear understanding of the *computational* aspects of dimension in self-similar fractals, even in the classical case.

We briefly mention some other recent research on fractals in theoretical computer science. Braverman and Cook [5],[6] have used computability and complexity of various fractals to explore the relationships between the two main models of real computation. Rettinger and Weihrauch [31], Hertling [20], and Braverman and Yampolsky [7] have investigated computability and complexity properties of Mandelbrot and Julia sets. Gupta, Krauthgamer, and Lee [19] have used fractal geometry to prove a lower bounds on the distortions of certain embeddings of metric spaces. Most of the fractals involved in these papers are more exotic than the self-similar fractals that we investigate here. Cai and Hartmanis [8] and Fernau and Staiger [15] have investigated Hausdorff dimension in self-similar fractals and their coding spaces. This work is more closely related to the present paper, but the motivations and results are different. Our focus here is on a *pointwise* analysis of dimensions.

Some of the most difficult open problems in geometric measure theory involve establishing lower bounds on the fractal dimensions of various sets. Kolmogorov complexity has proven to be a powerful tool for lower-bound arguments, leading to the solution of many long-standing open problems in discrete mathematics [25]. There is thus reason to hope that our pointwise approach to fractal dimension, coupled with the introduction of Kolmogorov complexity techniques, will lead to progress in this classical area. In any case, our results extend computable analysis [30],[23],[34] in a new, geometric direction.

Preliminaries. For functions on Euclidean space, we use the computability notion formulated by Grzegorczyk [17] and Lacombe [24] in the 1950's and exposited in the monographs by Pour-El and Richards [30], Ko [23], and Weihrauch [34] and in the recent survey paper by Braverman and Cook [6]. For subsets of Euclidean space, we use the computability notion introduced by Brattka and Weihrauch [4] (see also [34],[6]).

2   More on Self-similar Fractals

This expository section reviews a fragment of the theory of self-similar fractals that is adequate for understanding our main theorem and its proof. Our treatment is self-contained, but of course far from complete. The interested reader is referred to any of the standard texts [2,11,12,14] for more extensive discussion.

Definition 2.1. *An iterated function system (IFS) is a finite sequence $S = (S_0, \ldots, S_{k-1})$ of two or more contracting similarities on a nonempty, closed set $D \subseteq \mathbb{R}^n$. We call D the domain of S, writing $D = \mathrm{dom}(S)$.*

We use the standard notation $\mathcal{K}(D)$ for the set of all nonempty compact (i.e., closed and bounded) subsets of a nonempty closed set $D \subseteq \mathbb{R}^n$. For each IFS S, we write $\mathcal{K}(S) = \mathcal{K}(\mathrm{dom}(S))$.

For each IFS $S = (S_0, \ldots, S_{k-1})$, we define the transformation $S : \mathcal{K}(S) \to \mathcal{K}(S)$ by $S(A) = \bigcup_{i=0}^{k-1} S_i(A)$ for all $A \in \mathcal{K}(S)$, where $S_i(A)$ is the image of A under the contracting similarity S_i.

Observation 2.2. *For each IFS S, there exists $A \in \mathcal{K}(S)$ such that $S(A) \subseteq A$.*

For each IFS $S = (S_0, \ldots, S_{k-1})$ and each set $A \in \mathcal{K}(S)$ satisfying $S(A) \subseteq A$, we define the function $S_A : \Sigma_k^* \to \mathcal{K}(S)$ by the recursion $S_A(\lambda) = A$; $S_A(iw) = S_i(S_A(w))$ for all $w \in \Sigma_k^*$ and $i \in \Sigma_k$.

If $c = \max\{c_0, \ldots, c_{k-1}\}$, where $c_0, \ldots, c_{k-1}$ are contraction ratios of $S_0, \ldots, S_{k-1}$, respectively, then routine inductions establish that, for all $w \in \Sigma_k^*$ and $i \in \Sigma_k$, $S_A(iw) \subseteq S_A(w)$ and $\mathrm{diam}(S_A(w)) \leq c^{|w|}\mathrm{diam}(A)$. Since $c \in (0,1)$, it follows that, for each sequence $T \in \Sigma_k^\infty$, there is a unique point $S_A(T) \in \mathbb{R}^n$ such that $\bigcap_{w \sqsubseteq T} S_A(w) = \{S_A(T)\}$. In this manner, we have defined a function $S_A : \Sigma_k^\infty \to \bar{\bar{\mathbb{R}}}^n$. The following observation shows that this function does not really depend on the choice of A.

Observation 2.3. *Let S be an IFS. If $A, B \in \mathcal{K}(S)$ satisfy $S(A) \subseteq A$ and $S(B) \subseteq B$, then $S_A = S_B$.*

For each IFS S, we define the *induced function* $S : \Sigma_k^\infty \to \mathbb{R}^n$ by setting $S = S_A$, where A is *any* element of $\mathcal{K}(S)$ satisfying $S(A) \subseteq A$. By Observations 2.2 and 2.3, this induced function S is well-defined.

We now have the machinery to define a rich collection of fractals in $\mathbb{R}^n$.

Definition 2.4. *The attractor (or invariant set) of an IFS $S = (S_0, \ldots, S_{k-1})$ is the set $F(S) = S(\Sigma_k^\infty)$, i.e., the range of the induced function $S : \Sigma_k^\infty \to \mathbb{R}^n$.*

It is well-known that the attractor $F(S)$ is the unique fixed point of the induced transformation $S : \mathcal{K}(S) \to \mathcal{K}(S)$, but we do not use this fact here.

For each $T \in \Sigma_k^\infty$, we call T a *coding sequence*, or an *S-code*, of the point $S(T) \in F(S)$.

In general, the attractor of an IFS $S = (S_0, \ldots, S_{k-1})$ is easiest to analyze when the sets $S_0(\text{dom}(S)), \ldots, S_{k-1}(\text{dom}(S))$ are "nearly disjoint". (Intuitively, this prevents each point $x \in F(S)$ from having "too many" coding sequences $T \in \Sigma_k^\infty$.) The following definition makes this notion precise.

Definition 2.5. *An IFS $S = (S_0, \ldots, S_{k-1})$ with domain D satisfies the* open set condition *if there exists a nonempty, bounded, open set $G \subseteq D$ such that $S_0(G), \ldots, S_{k-1}(G)$ are disjoint subsets of G.*

We now define the most widely known type of fractal.

Definition 2.6. *A* self-similar fractal *is a set $F \subseteq \mathbb{R}^n$ that is the attractor of an IFS that satisfies the open set condition.*

3   Pointwise Analysis of Dimensions

In this section we prove our main theorem, which gives a precise analysis of the dimensions of individual points in computably self-similar fractals. We first recall the known fact that such fractals are computable.

Definition 3.1. *An IFS $S = (S_0, \ldots, S_{k-1})$ is* computable *if $\text{dom}(S)$ is a computable set and the functions $S_0, \ldots, S_{k-1}$ are computable.*

Theorem 3.2. *(Kamo and Kawamura [22]). For every computable IFS S, the attractor $F(S)$ is a computable set.*

One consequence of Theorem 3.2 is the following.

Corollary 3.3. *For every computable IFS S, $\text{cdim}(F(S)) = \dim_{\text{H}}(F(S))$.*

We next present three lemmas that we use in the proof of our main theorem. The first is a well-known geometric fact (e.g., it is Lemma 9.2 in [14]).

Lemma 3.4. *Let $\mathcal{G}$ be a collection of disjoint open sets in $\mathbb{R}^n$, and let $r, a, b \in (0, \infty)$. If every element of $\mathcal{G}$ contains a ball of radius ar and is contained in a ball of radius br, then no ball of radius r meets more than $\left(\frac{1+2b}{a}\right)^n$ of the closures of the elements of $\mathcal{G}$.*

Our second lemma gives a computable means of assigning rational "hubs" to the various open sets arising from a computable IFS satisfying the open set condition.

Definition 3.5. *A hub function for an IFS $S = (S_0, \ldots, S_{k-1})$ satisfying the open set condition with G as witness is a function $h : \Sigma_k^* \to \mathbb{R}^n$ such that $h(w) \in S_G(w)$ for all $w \in \Sigma_k^*$. In this case, we call $h(w)$ the hub that h assigns to the set $S_G(w)$.*

Lemma 3.6. *If $S = (S_0, \ldots, S_{k-1})$ is a computable IFS satisfying the open set condition with G as witness, then there is an exactly computable, rational-valued hub function $h : \Sigma_k^* \to \mathbb{Q}^n$ for S and G.*

For $w \in \Sigma_k^*$, we use the abbreviation $\mathcal{I}_S(w) = \mathcal{I}_{\pi_S}(w)$, where π_S is the probability measure defined in section 1.3.

Our third lemma provides a decidable set of well-behaved "canonical prefixes" of sequences in Σ_k^∞.

Lemma 3.7. *Let $S = (S_0, \ldots, S_{k-1})$ be a computable IFS, and let $c_{\min}$ be the minimum of the contraction ratios of $S = (S_0, \ldots, S_{k-1})$. For any real number $\alpha > \operatorname{sdim}(S) \log \frac{1}{c_{\min}}$, there exists a decidable set $A \subseteq \mathbb{N} \times \Sigma_k^*$ such that, for each $r \in \mathbb{N}$, the set $A_r = \{w \in \Sigma_k^* \mid (r, w) \in A\}$ has the following three properties.*

(i) *No element of A_r is a proper prefix of any element of $A_{r'}$ for any $r' \leq r$.*
(ii) *Each sequence in Σ_k^∞ has a (unique) prefix in A_r.*
(iii) *For all $w \in A_r$, $r \cdot \operatorname{sdim}(S) < \mathcal{I}_S(w) < r \cdot \operatorname{sdim}(S) + \alpha$.*

Our main theorem concerns the following type of fractal.

Definition 3.8. *A computably self-similar fractal is a set $F \subseteq \mathbb{R}^n$ that is the attractor of an IFS that is computable and satisfies the open set condition.*

Most self-similar fractals occurring in the literature are, in fact, computably self-similar.

We now have the machinery to give a complete analysis of the dimensions of points in computably self-similar fractals.

Theorem 3.9. *(main theorem). If $F \subseteq \mathbb{R}^n$ is a computably self-similar fractal and S is an IFS testifying this fact, then, for all points $x \in F$ and all S-codes T of x, $\dim(x) = \operatorname{sdim}(F) \dim^{\pi_S}(T)$ and $\operatorname{Dim}(x) = \operatorname{sdim}(F) \operatorname{Dim}^{\pi_S}(T)$.*

The proof of Theorem 3.9 uses the preceding lemmas to assign the coding sequence T a "canonical prefix" w_r for each $r \in \mathbb{N}$. This assignment enables us to construct a data compression argument establishing the estimate

$$\frac{\mathrm{K}_r(x)}{r \cdot \operatorname{sdim}(F)} \approx \frac{\mathrm{K}(w_r)}{\mathcal{I}_S(w_r)}$$

Taking limits, Theorem 3.9 follows. Details appear in the full version of this paper.

Finally, we use relativization to derive the following well-known classical theorem from our main theorem.

Corollary 3.10. *(Moran [29], Falconer [13]). For every self-similar fractal $F \subseteq \mathbb{R}^n$, $\dim_H(F) = \mathrm{Dim}_P(F) = \mathrm{sdim}(F)$.*

Acknowledgments

The first author thanks Dan Mauldin for useful discussions. We thank Xiaoyang Gu for pointing out that $\dim_H^\nu$ is Billingsley dimension.

References

1. Athreya, K.B., Hitchcock, J.M., Lutz, J.H., Mayordomo, E.: Effective Strong Dimension in Algorithmic Information and Computational Complexity. SIAM Journal on Computing 37, 671–705 (2007)
2. Barnsley, M.F.: Fractals Everywhere. Morgan Kaufmann Pub., San Francisco (1993)
3. Billingsley, P.: Hausdorff Dimension in Probability Theory. Illinois J. Math 4, 187–209 (1960)
4. Brattka, V., Weihrauch, K.: Computability on Subsets of Euclidean Space i: Closed and Compact Subsets. Theoretical Computer Science 219, 65–93 (1999)
5. Braverman, M.: On the Complexity of Real Functions. In: Proceedings of the Forty-Sixth Annual IEEE Symposium on Foundations of Computer Science (FOCS 2005), pp. 155–164 (2005)
6. Braverman, M., Cook, S.: Computing over the Reals: Foundations for Scientific Computing. Notices of the AMS 53(3), 1024–1034 (2006)
7. Braverman, M., Yampolsky, M.: Constructing Non-computable Julia Sets. In: Proceedings of the Thirty-Ninth Annual ACM Symposium on Theory of Computing (STOC 2007), pp. 709–716 (2007)
8. Cai, J., Hartmanis, J.: On Hausdorff and Topological Dimensions of the Kolmogorov Complexity of the Real Line. Journal of Computer and Systems Sciences 49, 605–619 (1994)
9. Cajar, H.: Billingsley Dimension in Probability Spaces. Lecture Notes in Mathematics. Springer, Heidelberg (1982)
10. Cover, T.M., Thomas, J.A.: Elements of Information Theory. John Wiley & Sons, Inc., New York (1991)
11. Edgar, G.A.: Integral, Probability, and Fractal Measures. Springer, Heidelberg (1998)
12. Falconer, K.: The Geometry of Fractal Sets. Cambridge University Press, Cambridge (1985)
13. Falconer, K.: Dimensions and Measures of Quasi Self-similar Sets. Proc. Amer. Math. Soc. 106, 543–554 (1989)
14. Falconer, K.: Fractal Geometry: Mathematical Foundations and Applications. John Wiley & sons, Chichester (2003)

15. Fernau, H., Staiger, L.: Iterated Function Systems and Control Languages. Information and Computation 168, 125–143 (2001)
16. Fortnow, L., Lee, T., Vereshchagin, N.: Kolmogorov Complexity with Error. In: Durand, B., Thomas, W. (eds.) STACS 2006. LNCS, vol. 3884, pp. 137–148. Springer, Heidelberg (2006)
17. Grzegorczyk, A.: Computable Functionals. Fundamenta Mathematicae 42, 168–202 (1955)
18. Gu, X., Lutz, J.H., Mayordomo, E.: Points on Computable Curves. In: Proceedings of the Forty-Seventh Annual IEEE Symposium on Foundations of Computer Science (FOCS 2006), pp. 469–474 (2006)
19. Gupta, A., Krauthgamer, R., Lee, J.R.: Bounded Geometries, Fractals, and Low-Distortion Embeddings. In: Proceedings of the Forty-Fourth Annual IEEE Symposium on Foundations of Computer Science (FOCS 2003), pp. 534–543 (2003)
20. Hertling, P.: Is the Mandelbrot Set Computable? Mathematical Logic Quarterly 51, 5–18 (2005)
21. Hitchcock, J.M.: Correspondence Principles for Effective Dimensions. Theory of Computing Systems 38, 559–571 (2005)
22. Kamo, H., Kawamura, K.: Computability of Self-Similar Sets. Math. Log. Q. 45, 23–30 (1999)
23. Ko, K.: Complexity Theory of Real Functions. Birkhäuser, Boston (1991)
24. Lacombe, D.: Extension de la Notion de Fonction Recursive aux Fonctions d'une ow Plusiers Variables Reelles and other Notes. Comptes Rendus 240, 2478–2480; 241, 13–14; 151–153, 1250–1252 (1955)
25. Li, M., Vitányi, P.M.B.: An Introduction to Kolmogorov Complexity and its Applications. 2nd edn. Springer, Berlin (1997)
26. Lutz, J.H.: The Dimensions of Individual Strings and Sequences. Information and Computation 187, 49–79 (2003)
27. Lutz, J.H., Weihrauch, K.: Connectivity Properties of Dimension Level Sets. In: Proceedings of the Fourth International Conference on Computability and Complexity in Analysis (2007)
28. Martin, D.A.: Classes of Recursively Enumerable Sets and Degrees of Unsolvability. Zeitschrift für Mathematische Logik und Grundlagen der Mathematik 12, 295–310 (1966)
29. Moran, P.A.: Additive Functions of Intervals and Hausdorff Dimension. Proceedings of the Cambridge Philosophical Society 42, 5–23 (1946)
30. Pour-El, M.B., Richards, J.I.: Computability in Analysis and Physics. Springer, Heidelberg (1989)
31. Rettinger, R., Weihrauch, K.: The Computational Complexity of some Julia Sets. In: Proceedings of the Thirty-Fifth Annual ACM Symposium on Theory of Computing (SOTC 2003), pp. 177–185 (2003)
32. Staiger, L.: A Tight Upper Bound on Kolmogorov Complexity and Uniformly Optimal prediction. Theory of Computing Systems 31, 215–229 (1998)
33. Vereshchagin, K., Vitanyi, P.M.B.: Algorithmic Rate-Distortion Function. In: Proceedings IEEE Intn'l Symp. Information Theory (2006)
34. Weihrauch, K.: Computable Analysis. An Introduction. Springer, Heidelberg (2000)

Visual Cryptography on Graphs

Steve Lu[1,*], Daniel Manchala[2,**], and Rafail Ostrovsky[3,***]

[1] University of California, Los Angeles
stevelu@math.ucla.edu
[2] Xerox Corporation
daniel.manchala@xerox.com
[3] University of California, Los Angeles
rafail@cs.ucla.edu

Abstract. In this paper, we consider a new visual cryptography scheme that allows for sharing of *multiple* secret images on graphs: we are given an arbitrary graph (V, E) where every node and every edge are assigned an arbitrary image. Images on the vertices are "public" and images on the edges are "secret". The problem that we are considering is how to make a construction in which every vertex image is encoded and printed on a transparency, such that if two adjacent vertices' transparencies are overlapped, the secret image of their edge is revealed. We define the most stringent security guarantees for this problem (perfect secrecy) and show a general construction for all graphs where the cost (in terms of pixel expansion and contrast of the images) is dependent on the chromatic number of the cube of the underlying graph. For the case of bounded degree graphs, this gives us constant-factor pixel expansion and contrast.

1 Introduction

Secret sharing, introduced independently by Blakley[6] and Shamir[20], is a scheme for an authority to encode a secret into shares to be distributed to a set of n participants such that only qualified subsets of these participants may reconstruct the secret. It is also required that unqualified subsets learn nothing about the secret. In their works, both Blakley and Shamir describe a k-out-of-n threshold secret sharing scheme, where any subset of at least k participants may reconstruct the secret. A generalization of this is if we let P be the set of n participants, we may denote the collection of subsets of P that can reconstruct the secret by Γ, called an access structure. Note that Γ must be monotone increasing, i.e. if $A \in \Gamma$ and $A \subset B \subset P$ then $B \in \Gamma$. The study of secret

* Work done in part while visiting IPAM. Partially supported by Xerox Innovation Group Award, NSF grants 0430254, 0716835, 0716389, and NSF VIGRE grant DMS-0502315.
** Partially supported by the Xerox Innovation Group Award.
*** Work done in part while visiting IPAM. Partially supported by Xerox Innovation Group Award, IBM Faculty Award, NSF grants 0430254, 0716835, 0716389 and U.C. MICRO grant.

X. Hu and J. Wang (Eds.): COCOON 2008, LNCS 5092, pp. 225–234, 2008.
© Springer-Verlag Berlin Heidelberg 2008

sharing schemes has been generalized[5],[14] for arbitrary access structures Γ of qualified participants. Multi-secret sharing involves multiple secrets, with possibly different access structures, to be shared across participants. In this scenario, the authority can distribute shares in a way that certain qualified participants may recover certain secrets. These schemes[10],[7],[9],[11] perform better than trivially instantiating multiple single-secret sharing schemes.

Visual cryptography schemes (VCS), introduced by Naor and Shamir[19], involve a dealer encoding a *secret* (or *target*) *image* into shares to be distributed to n participants. These shares, when printed on transparencies, may be recombined simply by overlapping them. When a qualified subset of the participants overlap their transparencies, a human-recognizable facsimile of the secret image appears. The main benefit of such schemes is that the participants do not need to rely on machines to perform the reconstruction. In a generalization of this scheme, it is sometimes required that, in addition, each of the shares is an image that may be recognized by the humans. In this type of extension, each participant may have their own *source image* (that is known to the authority) and the share generated for each user by the authority must "look" like their source image (see Section 2 for definitions). If the shares are so generated in this fashion to match the source images, we call the scheme an Extended Visual Cryptography Scheme (EVCS). Indeed, many researchers have worked on EVCS, giving constructions and proving bounds for such schemes [2],[1],[3].

ORGANIZATION OF OUR RESULTS. Works of [2],[1],[3] focused on the case where there was only one secret image to be reconstructed. In this paper, we consider the natural generalization of this for multiple secret images. The problem our paper addresses is to extend previous constructions a group of n participants where each pair of participants may have a secret image they can reconstruct. We may treat this as a graph where each vertex represents a participant and each edge represents a secret image to be shared between the two of them. We refer to this model as a Graph-Based Extended Visual Cryptography Scheme (GEVCS). We propose a definition of security and correctness for GEVCS in Section 2. We summarize our main results in Section 3 and spend the rest of the paper on the proofs and constructions. We will show first that the definition is satisfiable by a naïve construction in Section 4, and then describe a better general construction for any graph in Section 5. Finally in Section 6, we employ our construction on bounded degree graphs to give a GEVCS with constant-factor pixel expansion and contrast.

COMPARISON TO PREVIOUS RESULTS. The previous results most relevant to our work are the multi-secret visual cryptography schemes proposed in [16],[15],[13], [22],[23]. The schemes described in Katoh-Imai [16], Iwamoto-Yamamoto [15] and Chen-Wu-Laih [13] are special restricted cases of the problem we are addressing. Wang et. al.[22] and Yi et. al.[23] propose a scheme for multiple visual secrets and general access structures. Using binary tree graphs for comparison, the [22],[23] schemes' pixel expansion would grow on the order of the number of nodes, while our main construction has a pixel expansion of no more than 25 for even arbitrarily many nodes. Because of these practical considerations we had in mind (i.e. much better results, and with constant pixel expansion), we chose to use the graph-based model instead of a general access structure for this paper.

Our work differs in comparison to previous results, such as [19],[2],[1],[3] in visual cryptography by handling multiple secret images. These results use graph-based access structures as an example, however our scheme handles the case of one secret image per edge as opposed to only one secret image per graph structure. On the other hand, there are constructions of (non-visual) secret sharing or multi-secret sharing on a graph-based access structure[21],[8],[9],[11],[12]. These are special types of access structures in which a graph $G = (V, E)$ is used to represent the sets of qualified participants. Each vertex is treated as a participant, and an edge between two participants indicates the two of them together may recover a secret. The constructions given in this paper involve graph decompositions, but our methods differ from these previous constructions as we need to simultaneously take into account the visual aspects on top of the multi-secret requirements. We will describe our novel decomposition in the following sections.

1.1 Background

PHYSICAL MODEL. The physical model of our scheme will use images printed on transparencies (as in [19]). Black pixels will be printed onto the transparency making these portions completely opaque, leaving the remaining portion completely transparent (we will refer to these as white pixels). Thus the transparency can be viewed as a Boolean matrix, where a 1 in the (i, j)th entry represents a black pixel at that location and a 0 represents a white pixel. When overlapping two transparencies, the result may be viewed as the Boolean OR operation performed entrywise on the two matrices. Because our constructions are all pixelwise operations, we henceforth treat all images as just a single black or white pixel.

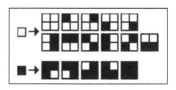

Fig. 1. Overlapping Operation   **Fig. 2.** Pixel Expansion

The operation of overlapping two transparencies is inherently a destructive operation, thus in order to retain information, we will introduce some redundancy. We sometimes refer to this process as *encoding* an image. In particular, we encode 1 pixel as m (usually chosen to be a perfect square) subpixels (known as the *pixel expansion*). If the original image was of size $p \times q$, then the encoded image will be of size $p\sqrt{m} \times q\sqrt{m}$. The natural visual interpretation is to say if there are more than some threshold d black subpixels then view it as black, otherwise view it as white (Figure 2). To accommodate the human eye, we may wish to preclude certain encodings that may look ambiguous (to the eye), so in addition we can have a contrast requirement, saying that only encodings with fewer than $d - \alpha$ black subpixels may be viewed as white (α is known as the absolute contrast, α/m the relative contrast).

EXTENDED VISUAL CRYPTOGRAPHY. We review the problem of extended visual cryptography for two participants and a dealer. Loosely speaking, the goal of the dealer is to take public images A_1 and A_2, a secret image B, and create secure encoded shares S_1 and S_2 such that S_i "looks like" A_i and the overlap of S_1 and S_2 "looks like" B. Formally, the setup is as follows: each participant has a public image, say "A_1" and "A_2", which are known as the two *source* images. There is a secret image, say "B", known as the *target* image, to be shared between them by a dealer. The dealer must then encode A_1 and A_2 into shares S_1 and S_2 (possibly under different encodings) by selecting the colors of the subpixels in a way so that when S_1 and S_2 are overlapped, the result is an encoding of B (possibly yet another encoding). In addition, like in a secret sharing scheme, there is a perfect secrecy requirement which should be satisfied. We will formalize and generalize these requirements for our purposes in the next section.

2 Our Definitions

In this section, we consider the problem of generating shares for n participants organized in a graph structure. Our interpretation of the graph is a way of denoting which pairs of participants will be overlapping their shares to obtain a secret image dealt between them. For example, a complete graph would mean any pair of participants may overlap their shares to get a secret image *for that pair*, resulting in a total of $\binom{n}{2}$ possible secret images. In this case, each vertex will have a source image A_i attached to it, and each edge will have a secret target image B_e attached to it. A (probabilistic) polynomial-time computable algorithm that takes these as input and produces image shares S_i (each of length m, the pixel expansion) that satisfy the properties defined below will be referred to as a *Graph-based Extended Visual Cryptography Scheme* or GEVCS.

2.1 Correctness and Security for GEVCS

When generating shares S_i (and overlapped shares T_e) given a graph $G = (V, E)$ with n vertices and r edges, we define the following contrast properties. Each S_i should have at least s_1^i 1's when encoding a 1 and at most s_0^i 1's when encoding a 0. We can similarly parameterize these thresholds for the T_e and obtain t_1^e and t_0^e. Define the *(absolute) contrast* to be $\alpha_i = s_1^i - s_0^i$, $\alpha_e = t_1^e - t_0^e$. Define the *relative contrast* to be $\frac{\alpha}{m}$, the ratio between the absolute contrast and the pixel expansion. In essence, it is the relative contrast that affects how clear the final images will appear to human eye. Visual cryptography schemes traditionally come with a guarantee of security by means of defining perfect secrecy. Usually, a set of forbidden players is not allowed to learn any information about the (one) secret image even under the possibility of collusion. In our scheme, participants share different secrets with different people, thus we need to take this into account when defining security. Thus we arrive at the following two definitions.

Definition 1. *We say a GEVCS satisfies the* contrast correctness *property with parameters* $(s_0^i, s_1^i, t_0^e, t_1^e)$ *if for every possible set of source images* A_i *and target images* B_e, *each share* S_i *that is generated is a valid encoding (under these parameters) of* A_i, *and overlapping any two of them along an edge* e *results in a valid encoding of* B_e.

Definition 2. *We say a GEVCS (a probabilistic polynomial-time algorithm named* $\mathcal{D}$*) is* secure *or* perfectly secret *for* G *if for any adversary* $\mathcal{A}$ *we have:*

$$Pr\left[(\{A_i\}, \{B_e\}_{e \neq e^*}, i^*, e^*) \leftarrow \mathcal{A}(G), B_{e^*} \xleftarrow{\text{R}} \{0,1\},\right.$$

$$\left.\{S_i\} \leftarrow \mathcal{D}(G, \{A_i\}, \{B_e\}), B \leftarrow \mathcal{A}(\{S_i\}_{i \neq i^*}); B = B_{e^*}\right] = \frac{1}{2}$$

In some of the constructions, the GEVCS will deal the shares by sampling from a collection of matrices. On a graph $G = (V, E)$ there will be collections $C_{\{b_e\}}^{\{a_i\}}$, one for each possible assignment of 0's and 1's to $\{a_i\}_{i \in V}, \{b_e\}_{e \in E}$. We will also make use of a so-called *basis matrix* – this $n \times m$ (n being the number of participants and m being the pixel expansion) matrix contains the m subpixels to be assigned to player i in row i. The collections will arise as all matrices obtained by permuting the columns of the basis matrices, thus we will have one basis matrix for each possible assignment of the vertex and edge source images. We will parameterize the basis matrices for a graph G and source images $\{a_i\}_{i \in V}$ and target images $\{b_e\}_{e \in E}$ by $S_{\{b_e\}}^{\{a_i\}}$. Our constructions will give an explicit algebraic formula to compute the basis matrix from given values of $\{a_i\}$ and $\{b_e\}$.

2.2   Graph Theoretic Terminology

A *star forest* is a graph where each connected component is a star. If we let $G = (V, E)$ be a graph, given a set of subgraphs $H_1, \ldots, H_k$ we say that they are a *graph (resp. star, star forest) cover* of G if every edge in E is contained in at least one H_i and each subgraph is a graph (resp. star, star forest). We let $N(v)$ denote the neighbors of a vertex v, or in other contexts, the neighborhood of v, i.e. the star centered at v with all its neighbors as points. Let $H_1, \ldots, H_k$ be subgraphs of $G = (V, E)$ such that each $H_i = (V_i, E_i)$, each $v \in V$ belongs to at most one H_i, and there are no edges in G between any two vertices that are not in the same subgraph, i.e. $\neg \exists i \neq j (v_i \in V_i \wedge v_j \in V_j \wedge (v_i, v_j) \in E)$. There can still be edges in G between vertices in the *same* H_i. In this case, we say that $H_1, \ldots, H_k$ form an *independent subgraph set* and $H = \bigcup_{i=1}^k H_i$ is an *independent subgraph* of G. In addition, if each H_k is a star, we say H is an *independent star forest subgraph* of G (see full version [18] for an example). The *cube* of G is a new graph $G^3 = (V, E')$ where $(v, w) \in E'$ if v and w are connected by a path of at most length 3. The *degree* of a graph is the maximum of the degrees of all its vertices and *(d-)bounded degree graphs* are those which have degree at most d.

3   Main Result

Our main result (which we will build up to in the remainder of the paper) can be stated as follows:

Theorem 1. *Let $G = (V, E)$ be a graph where no vertex has degree greater than d (i.e. a d-bounded degree graph) and let χ be the chromatic number of G^3. Then there exists a GEVCS on G with pixel expansion at most $m = \chi(5d + 1)$, and absolute contrast 2 for each source image on a vertex and 4 for each target image on an edge. Furthermore, we give an explicit construction for a GEVCS with pixel expansion at most $m = (d^3 + 1)(5d + 1)$.*

4   Warming Up: A Naïve Construction

For practical applications of GEVCSs, we wish to maximize the contrast and minimize the pixel expansion for the encoded images. We begin by exploring a naïve construction of a GEVCS for a complete graph that involves a pixel expansion of $m = 2n^2 - n$ with relative source contrast $\frac{1}{m}$ and relative target contrast $\frac{2}{m}$. Compare this to the optimal lower bounds in the recent work of Blundo et. al.[4]. They show a tight lower bound of $m \approx n^2/4$ with relative contrast $\frac{1}{m}$ for a $(2, n)$-VCS which has no source images (the shares are not required to look like anything) and only a single secret image to recover.

SATISFYING BOTH SECURITY AND CONTRAST FOR GENERAL GRAPHS. We present a construction of a GEVCS on any graph satisfying certain contrast parameters. This construction will turn out to be perfectly secret as well. For any complete graph $G = (V, E)$ of n vertices we give a construction with a pixel expansion of $m = 2n^2 - n$ and will determine the parameters $s_0^i, s_1^i, t_0^e, t_1^e$ after the construction. For each possible assignment of $\{a_i\}, \{b_e\}$ we will construct the basis matrix $S_{\{b_e\}}^{\{a_i\}}$ (we will write S for ease of reading). Each basis matrix will contain a so-called "source-contrast" block, U, meant to allow the source subpixels to pass the threshold for a black pixel, followed by n "target-contrast" blocks $T_1, \ldots, T_n$ meant to control the number of black subpixels of the target image. First define the $n \times n$ matrix U as $U_{ii} = a_i$ and $U_{ij} = 1$ for $i \neq j$. If a_i is black then row i will have one extra black subpixel. This will be used to differentiate a black pixel from a white pixel of the source image. The remaining matrices, T_i, will be used to control the darkness of the target image. We next define each T_i (which are $n \times 2(n-1)$ matrices) as follows. To form these matrices, start with an $(n - 1) \times (n - 1)$ matrix with diagonal entries $b_{(i,j)}$ (where $i \neq j$), then concatenate a matrix with diagonal entries $1 - b_{(i,j)}$, then insert a new row i which consists of $n - 1$ zeroes followed by $n - 1$ ones. Notice when any row j is overlapped with row i, the result will contain all 1's in the right half, and all zeroes except $b_{(i,j)}$ in the left half. On the other hand, when any row j is overlapped with row $k \neq i$, the result will contain exactly two 1's. Finally we let $S = U \| T_1 \| \cdots \| T_n$ (where $\|$ denotes horizontal matrix augmentation), a matrix with $2n^2 - n$ columns and n rows. We now count how many 1's there are in each row i corresponding to a white (resp. black) source pixel. There will be $n-1$ (resp. n) 1's in the U block, there will be a single 1 in each T_j block with $j \neq i$, and there will be $n-1$ 1's in the T_i block. Thus we can set $s_0^i = 3(n-1)$ and $s_1^i = 3(n-1)+1$. We may similarly count how many 1's there are in the overlap of two rows i and j with a white (resp. black) corresponding target pixel: n in the U block, n (resp. $n+2$) in the T_i and T_j blocks, and 2 in each T_k block for $k \neq i, j$. Thus we can set $t_0^e = 4n-4$ and $t_1^e = 4n-2$. Then our construction satisfies contrast for these parameters on a complete graph. To ensure security, we randomly permute the columns of the matrix S before setting share S_i as the ith row. We refer the reader to the full version ([18]) for the proof of security.

5   GEVCS for a General Graph

The idea is that we can view the act of overlapping a transparency with your neighbors as a local process so that we may seek to decompose our graphs into sufficiently independent local pieces and build GEVCSs for each piece, then patch them together in a meaningful way. Our construction idea is to construct GEVCS for building blocks, then somehow combine them to form any graph. We apply the naïve construction for a small subgraph, then describe how these subgraphs may be patched together. We now present two different ways the GEVCS described above may be patched together to form a GEVCS on a graph G. We take a graph cover of G and first show how to generate shares for an independent subgraph set in parallel. Then we will show how to take these sets of subgraphs and combine their shares sequentially. Below, we define two constructions and state two lemmas regarding the constructions. A sample construction and the proofs of the lemmas appear in the full version ([18]).

Definition 3 (Parallel Sharing on Independent Subgraphs). *Let H be an independent subgraph of G. We can write $H = \bigcup_{i=1}^{k} H_i$, where the H_i are the independent pieces (recall this means that each H_i has no edges connecting to an H_j). Obtain a GEVCS for each subgraph using the naïve construction above. Let m be the maximum pixel expansion over all the subgraphs. Construct a new distribution of share matrices for $H = \bigcup_{i=1}^{k} H_i$ by first sampling a share matrix from each GEVCS on H_j. The new matrix will have one row for each vertex in H, and because each vertex is uniquely contained in some H_j we may assign to it the corresponding row from the GEVCS on H_j (also, pad them with 0's at the end so that each row is of length m).*

Lemma 1. *By sampling the share matrix according to the distribution in definition 3, we obtain a secure GEVCS on H. The pixel expansion is equal to the maximum pixel expansion of the GEVCSs on the individual subgraphs and maintains the same contrast parameters for each vertex and edge. Also, this scheme satisfies the* smoothness *property.*

Definition 4 (Sequential Sharing on Dependent Subgraphs). *Consider a graph cover $K_1, \ldots, K_\ell$ of $G = (V, E)$ where each K_k is an independent subgraph. Use definition 3 on the K_i to obtain GEVCS schemes on each of these. For each K_i, first pad the shares in its GEVCS with rows $i \in V \setminus V_i$ by filling the these rows with 1's. Each of these matrices will have n rows, and we can then concatenate them horizontally. This completes the construction of a new distribution of shares on G.*

Lemma 2. *By sampling the share matrix according to the distribution in definition 4, we obtain a secure GEVCS on G. The pixel expansion is equal to the sum of the pixel expansions of the GEVCS on each of the subgraphs. In terms of absolute contrast, a source image has absolute contrast equal to the number of times its vertex appears in the covering, and a target image has absolute contrast equal to twice the number of times its edge appears in the covering.*

5.1   Construction of a GEVCS for a General Graph

Given a graph G with an independent subgraph cover $K_1, \ldots, K_\ell$ we can construct a GEVCS for G by applying the two previous constructions in Definitions 3 and 4. First construct a GEVCS for each component of K_i using the naïve GEVCS construction described above. Then combine the shares in parallel to obtain GEVCSs for each independent subgraph K_i. This will be followed by combining the shares sequentially by construction 4 to finally obtain a GEVCS on G. We give an explicit count of the final pixel expansion and contrast in the full version ([18]).

We make the observation that if one takes a coloring of G^3, one can make an independent star forest cover of G by taking K_i to be the union of all stars around centers of color i. If it uses χ colors, then there will be χ of the K_i's. This is explained in further detail in section 6. By combining the constructions above with the independent star forest cover in the following section, we obtain the main theorem:

Theorem 2. *Let $G = (V, E)$ be a graph where no vertex has degree greater than d (i.e. a d-bounded degree graph) and let χ be the chromatic number of G^3. Then there exists a GEVCS on G with pixel expansion at most $m = \chi(5d + 1)$, and absolute contrast 2 for each source image on a vertex and 4 for each target image on an edge. Furthermore, we give an explicit construction for a GEVCS with pixel expansion at most $m = (d^3 + 1)(5d + 1)$.*

6   Independent Star Forest Covers

In this section we describe how to find independent star forest covers for graphs to supply as input to our algorithm in the previous section. We describe the general construction of independent star forests for any graph G, and mention this construction leads to parameters depending only on the maximum degree of vertices of the graph. We begin by making the observation that given a k-coloring of G^3, we can decompose G as follows. Let $K_i = \bigcup_{v \text{ has color } i} N(v)$. Note this is an independent star forest cover as an edge between $N(v)$ and $N(w)$ implies there is a path of at most length 3 between v and w which translates to an edge (v, w) in G^3, hence they cannot be of the same color. We can find a coloring by remodeling the algorithm in Luby[17] into the algorithm in presented in Figure 3.

This algorithm will use at most $d^3 + 1$ colors (cf. [17] Section 7) if G is of degree d. This is because at each stage if the node itself is not colored then at

```
i ← 0
Construct G³ = (V, E)
while (V, E) is not empty do
    i ← i + 1
    Find a Maximal Independent Set S
    Color all the vertices in S by color i
    V ← V\S
end while
```

Fig. 3. Coloring G^3

least one neighbor is colored (by the property of a maximal independent set). Thus at the next stage, its degree will drop by at least 1, and since each vertex in G^3 has at most degree d^3, we arrive at the conclusion of at most $d^3 + 1$ colors. Combining this algorithm with the construction from the previous section gives rise to a construction of a GEVCS on any graph, and for d-bounded degree graphs a constant-factor (on the order of d^4) pixel expansion and contrast as stated in our main theorem. Unlike the naïve construction, this construction is independent of the number of participants. This concludes the construction of our GEVCS.

References

1. Ateniese, G., Blundo, C., De Santis, A., Stinson, D.R.: Constructions and Bounds for Visual Cryptography. In: auf der Heide, F.M., Monien, B. (eds.) ICALP 1996. LNCS, vol. 1099, pp. 416–428. Springer, Heidelberg (1996)
2. Ateniese, G., Blundo, C., De Santis, A., Stinson, D.R.: Visual Cryptography for General Access Structures. Electronic Colloquium on Computational Complexity 3(12) (1996)
3. Ateniese, G., Blundo, C., De Santis, A., Stinson, D.R.: Extended Capabilities for Visual Cryptography. heor. Comput. Sci. 250(1-2), 143–161 (2001)
4. Blundo, C., Cimato, S., De Santis, A.: De: Visual Cryptography Schemes with Optimal Pixel Expansion. Theor. Comput. Sci. 369(1–3), 169–182 (2006)
5. Benaloh, J.C., Leichter, J.: Generalized Secret Sharing and Monotone Functions. In: Goldwasser, S. (ed.) CRYPTO 1988. LNCS, vol. 403, pp. 27–35. Springer, Heidelberg (1990)
6. Blakley, G.: Safeguarding Cryptographic Keys. In: Proc. Am. Federation of Information Processing Soc., pp. 313–317 (1979)
7. Blundo, C., De Santis, A., Di Crescenzo, G., Gaggia, A.G., Vaccaro, U.: Multi-secret Sharing Schemes. In: Desmedt, Y.G. (ed.) CRYPTO 1994. LNCS, vol. 839, pp. 150–163. Springer, Heidelberg (1994)
8. Blundo, C., De Santis, A., Stinson, D.R., Vaccaro, U.: Graph Decompositions and Secret Sharing Schemes. J. Cryptology 8(1), 39–64 (1995)
9. Blundo, C., De Santis, A., De Simone, R., Vaccaro, U.: Tight Bounds on the Information Rate of Secret Sharing Schemes. Des. Codes Cryptography 11(2), 107–110 (1997)
10. Blundo, C., De Santis, A., Vaccaro, U.: Efficient Sharing of Many Secrets. In: Enjalbert, P., Wagner, K.W., Finkel, A. (eds.) STACS 1993. LNCS, vol. 665, pp. 692–703. Springer, Heidelberg (1993)

11. Di Crescenzo, G.: Sharing One Secret vs. Sharing Many Secrets. Theor. Comput. Sci. 295(1-3), 123–140 (2003)
12. Csirmaz, L.: Secret Sharing Schemes on Graphs. Cryptology ePrint Archive, Report 2005/059 (2005), http://eprint.iacr.org/
13. Chen, K.Y., Wu, W.P., Laih, C.S.: On the (2,2) Visual Multi-secret Sharing Schemes (2006)
14. Itoh, M., Saito, A., Nishizeki, T.: Secret Sharing Scheme Realizing General Access Structure. In: Proc. of IEEE Globecom, pp. 99–102 (1987)
15. Iwamoto, M., Yamamoto, H.: A Construction Method of Visual Secret Sharing Schemes for Plural Secret Images. IEICE Trans. on Fundamentals E86.A(10), 2577–2588 (2003)
16. Katoh, T., Imai, H.: An Extended Construction Method for Visual Secret Sharing Schemes. Electronics and Communications in Japan (Part III: Fundamental Electronic Science) 81(7), 55–63 (1998)
17. Luby, M.: A Simple Parallel Algorithm for the Maximal Independent Set Problem. SIAM J. Comput. 15(4), 1036–1055 (1986)
18. Manchala, D., Lu, S., Ostrovsky, R.: Visual Cryptography on Graphs. Cryptology ePrint Archive (2008), http://eprint.iacr.org/
19. Naor, M., Shamir, A.: Visual Cryptography. In: De Santis, A. (ed.) EUROCRYPT 1994. LNCS, vol. 950, pp. 1–12. Springer, Heidelberg (1995)
20. Shamir, A.: How to Share a Secret. Commun. ACM 22(11), 612–613 (1979)
21. Stinson, D.: Decomposition Constructions for Secret Sharing Schemes. IEEE Transactions on Information Theory 40(1), 118–125 (1994)
22. Wang, D., Yi, F., Li, X., Luo, P., Dai, Y.: On the Analysis and Generalization of Extended Visual Cryptography Schemes (2006)
23. Yi, F., Wang, D., Luo, P., Huang, L., Dai, Y.: Multi Secret Image Color Visual Cryptography Schemes for General Access Structures. Progress in Natural Science 16(4), 431–436 (2006)

Algebraic Cryptanalysis of CTRU Cryptosystem

Nitin Vats

Department of Computer Science and Automation,
Indian Institute of Science, Bangalore, India
nitinvatsa@gmail.com

Abstract. CTRU, a public key cryptosystem was proposed by Gaborit, Ohler and Sole. It is analogue of NTRU, the ring of integers replaced by the ring of polynomials $\mathbb{F}_2[T]$. It attracted attention as the attacks based on either LLL algorithm or the Chinese Remainder Theorem are avoided on it, which is most common on NTRU. In this paper we presents a polynomial-time algorithm that breaks CTRU for all recommended parameter choices that were derived to make CTRU secure against popov normal form attack. The paper shows if we ascertain the constraints for perfect decryption then either plaintext or private key can be achieved by polynomial time linear algebra attack.

1   Introduction

NTRU [8]was proposed in 1996 as a fast public key encryption system. Its security relies on the hardness of the short vector problem for some special lattice. It attracted considerable attention due to embarked application like hand held device and wireless system due to short key size and speed of encryption and decryption. CTRU [6] is a natural extension of NTRU in which role played by $\mathbb{Z}$ is replaced by the ring $\mathbb{F}_2[T]$ of polynomials in one variable T over the binary finite field $\mathbb{F}_2$.

It is considerable that the attacks based on either LLL algorithm or the Chinese Remainder Theorem are avoided on it, which is most common on NTRU. Till date many possible variants of NTRU have been proposed. It is interesting and important to study any possible variant of NTRU which is defined on ring $R = (\mathbb{F}_2[T])[X]/(X^n - 1)$. So far CTRU is the only attempt in this direction. Here we show that it is not beneficial to propose the variant of NTRU defined in the ring $(\mathbb{F}_2[T])[X]/(X^n - 1)$. Although it overcomes Lattice attack but totally insecure against more easy Linear algebra attacks.

The paper shows if we ascertain perfect decryption then either plaintext or private key can be achieved by polynomial time linear algebra attack.

CTRU simply replace the role played by $\mathbb{Z}$ in NTRU by $\mathbb{F}_2[T]$. The role of LLL algorithm is played by Popov form. It can be presumed that the hard lattice problem underlying NTRU become elementary linear algebra problem for CTRU Cryptosystem.

The paper is organized as follows. Section 1 gives some notation. Section 2 sketch briefly the CTRU Cryptographic system. Section 3 is security analysis and

X. Hu and J. Wang (Eds.): COCOON 2008, LNCS 5092, pp. 235–244, 2008.
© Springer-Verlag Berlin Heidelberg 2008

in Section 4 we do practical implementation of polynomial time linear algebra attack on CTRU for recommended parameters.

2   Notations

Here we use the same notation as in [6], more precisely Upper case letters are used for polynomials and Lower case letters for their corresponding degrees.

We work in the ring $R = D[X]/(X^n - 1)$ of truncated polynomials with polynomial coefficients. CTRU cryptosystem depends on an integer n and two irreducible polynomials P and Q of $D = \mathbb{F}_2[T]$. We assume that P and Q are polynomials of respective degrees s and m. Here s and m are relative prime to each other and follows the inequality $2 \leq s \leq m$.

Explicitly an element $L \in R$ can be written in the following form

$$L = L_0(T) + L_1(T)X + L_2(T)X^2 + \cdots + L_{n-1}(T)X^{n-1},$$

with $\{L_0, L_1, L_2, \ldots \ldots L_{n-1} \in \mathbb{F}_2[T]\}$, the coefficients $L_0, L_1, L_2, \ldots \ldots L_{n-1}$ are polynomials with maximum degree $l - 1$. We can also write L as a double sum in two variables ring $\mathbb{F}_2[T, X]/(X^n - 1)$. In this paper we speak of the coefficients of an element $L \in R$, We mean all the binaries L_{ij} as

$$L = \sum_{i=0}^{n-1} \left(\sum_{j=0}^{l-1} L_{ij} T^j \right) X^i = \sum_{0 \leq i < n, 0 \leq j < l} L_{ij} T^j X^i \quad \text{with} \quad L_{ij} \in \{0, 1\}$$

By $\deg(L)$, we denote the degree of L as a polynomial in T.

$$\deg(L) = \max \deg L_i(T), \quad \text{for} \quad 0 \leq i < n$$

For any integer $l \geq 0$, we define the set of polynomials of degree(in T), less than l.

$$R[l] = \{L \in R : \deg(L) < l\}$$

The set $R[l]$ will have 2^{ln} elements. CTRU cryptosystem depends on four integer parameters (n, d_f, d_g, d_r) and three sets (R_f, R_g, R_r) of polynomials.

Integer parameters must satisfy

$$\max\{d_f, d_g, d_r, s\} < m$$

Three sets (R_f, R_g, R_r) are defined as $R_f = R[d_f]$, $R_g = R[d_g]$, $R_r = R[d_r]$. P and Q are defined as two irreducible and relative prime polynomial with degrees s and m respectively. As s and m are relative prime, we notice that $\mathbb{F}_{2^s} \cap \mathbb{F}_{2^m} = \mathbb{F}_2$. Like in NTRU, independence of reduction $\mod(P)$ and $\mod(Q)$ is essential to avoid trivial attacks.

The quotient ring D_p and D_q of D by the ideals (P) and (Q) respectively are the finite fields $\mathbb{F}_{2^s}$ and $\mathbb{F}_{2^m}$. Elements in the quotient ring $\mathbb{F}_2[T]/Q$ has a unique element in $\mathbb{F}_2[T]$ of degree less than m.

Therefore the mapping

$$R_p \to \frac{R}{PR} \quad \text{and} \quad R_q \to \frac{R}{QR}$$

are one-to-one and onto. An element of R modulo P or Q will be chosen from ring R_p or R_q respectively.

3 The CTRU System

3.1 Generating Keys

We sketch the CTRU system, as developed in [1]. The private key contains two arbitrary polynomials F and G where $F \in R_f$ and $G \in R_g$.

Thus the secret key pair (F, G) is an element of space size $2^{nd_f} \times 2^{nd_g}$. Here F is required to be invertible in each of the quotient rings of R by the ideals (P) and (Q).

public key is defined as follows

$$F * H \equiv G \pmod{Q}, \quad \text{where } H \in R_q.$$

3.2 Encryption and Decryption

The encryption in CTRU is probabilistic as given same private key and same plain text will be encrypted differently at different time.

The plaintext is any polynomial $M \in R_p$. We also require a random polynomial as blinding value $R \in R_r$.

The ciphertext is the polynomial $E \in R_q$, described as below

$$E \equiv P * R * H + M \pmod{Q}.$$

Upon getting the encrypted message, decrypter compute $A \in R_q$ using his private key F

$$A \equiv F * E \pmod{Q}, \quad \text{where} \quad A \in R_q$$

In next step decrypter computes

$$B \equiv F_p^{-1} * A \pmod{P},$$

where F_p^{-1} is the inverse of F in the quotient ring R/PR.

The decrypter compute the polynomial B that is same as the message M.

4 Security Analysis

The authors of [6] have done security analysis against six kind of possible attacks. CTRU is analogue of NTRU. It is important to notice that CTRU enjoys the security against attacks based on LLL algorithm or Chinese Remainder Theorem which is biggest threat to original NTRU Cryptosystem. They have done cryptanalysis of CTRU through Popov normal form of matrices with polynomial entries. Here we present a new linear algebra based attack that is able to break the system completely. Our strategy is to find parameter constraints for private key and plaintext security provided valid decryption is guaranteed. Here is the analysis.

4.1 Decryption Criteria

In order to decrypt E, polynomial A is computed first

$$A \equiv F * E \pmod{Q}$$
$$A \equiv F(P * \mathrm{R} * H + M) \pmod{Q}$$
$$A \equiv (P * \mathrm{R} * G + M * F) \pmod{Q}$$

For valid decryption to work, A should be exactly equal to $P * \mathrm{R} * G + M * F$, not be congruent modulo Q as Next, we need to compute polynomial $B \in R_p$ on modulo P. It follows if and only if degree of polynomials $P * \mathrm{R} * G$ and $M * F$ is less than m. We can quantify it in terms of s, m, d_f, d_g and d_r if

$$\deg(P * \mathrm{R} * G) < \deg(Q)$$
$$s + (d_r - 1) + (d_g - 1) < m$$
$$s + d_r + d_g < m + 2$$
$$s + d_r + d_g \leq m + 1 \tag{1}$$
$$\text{and if} \quad \deg(M * F) < \deg Q$$
$$(s - 1) + (d_f - 1) < m$$
$$s + d_f < m + 2$$
$$s + d_f \leq m + 1 \tag{2}$$

It implies that valid decryption is guaranteed if and only if both conditions (1) and (2) satisfies. Collectively the decryption condition is

$$2s + d_r + d_f + d_g \leq 2m + 2 \tag{3}$$

4.2 Criteria for Private Key Security

Private key F satisfy the congruence

$$F * H \equiv G \pmod{Q}$$

This congruence can be written as equality

$$F * H = G + Q * U \quad \text{where} \quad U \in R \tag{4}$$

We treat the coefficients of variables as polynomials in the two variable polynomial ring $\mathbb{F}_2[T, X]$. We can compare the coefficient of monomials $T^i X^j$ appearing in equation (4). Here we estimate the number of equations and number of variables to achieve the private key. For this we need to know the maximum possible degree of U. We already know the degree bounds for rest of the elements F, H, G and Q so we can find maximum degree of U as

$$\deg(Q * U) = \deg(U) + m \quad \text{and} \quad \deg(U) = \deg(QU) - m$$
$$\begin{aligned} \deg(U) &= \deg(Q * U) - m \\ &= \deg(F * H - G) - m \\ &\leq \max\{\deg(F * H), \deg(G)\} - m \\ &\leq \max\{(d_f - 1) + (m - 1), (d_g - 1)\} - m \\ &= d_f - 2 \end{aligned}$$

We treat the coefficients of unknown variables $F \in R_f$, $G \in R_g$, and $U \in R[d_f - 1]$ as polynomials in the two variable polynomial ring $\mathbb{F}_2[T, X]$.

If we compare the coefficient of monomials $T^i X^j$ appearing in equation(4). We can calculate number of linear equations as follow

$$F * H = G + Q * U$$
$$\deg F * H \leq (d_f + m - 2)$$
$$\text{Number of linear equations} = n[(\deg F * H) + 1] = n(d_f + m - 1).$$

As F, G and U are unknown variables so number of unknown all together we have

$$\begin{aligned} \text{Number of variables} &= nd_f + nd_g + n(d_f - 1) \\ &= n(2d_f + d_g - 1) \end{aligned}$$

Attacker can achieve the private key in case the number of variables are less than or equal to the number of equations. In this case the system of equations will have unique solution. So private key will be secure if

$$\text{Number of variables} > \text{Number of equations}$$
$$\text{or} \quad n(2d_f + d_g - 1) > n(d_f + m - 1)$$
$$(d_f + d_g) > m$$
$$(d_f + d_g) \geq m + 1 \tag{5}$$

Alternatively we can find fast linear algebra attack against the private key with reduced number of equations and variables if we project the problem into proper vector space and exploit their properties precisely. Here is the analysis

Private key F satisfy the congruence

$$F * H \equiv G \pmod{Q}$$

It can be equivalently written as an equality

$$F * H \equiv G + Q * U \quad \text{where } U \in R$$

As $(G \in R_g) \subset R_q$, it follows that R_g has no nonzero element that will be devisable by Q. It implies that

$$R_g \cap (Q * R) = \{0\}$$

It is showed in [6] that popov normal form will be unable to break the system if

$$d_g < \frac{m}{2} \leq d_f$$

As degree of G is always less than degree of F and $H \in R_q$. It proves that

$$R_g \cap (R_f * H) = \{0\} \tag{6}$$

Result (6) implies that $Q * R$ can be parted into its components G and $F * H$, therefore to recover the private key F.

The Basis for the vector space R_g and R_f can be represented by the polynomials $X^i T^j$ with $0 \leq i < n$, $0 \leq j < d_g$ and $0 \leq i < n$, $0 \leq j < d_f$. So basis for the vector $R_f * H (\mathrm{mod} Q)$ will be formed by the polynomials

$$V_{ij} = X^i T^j * H \quad (\mathrm{mod} Q), \quad \text{with } 0 \leq i < n \text{ and } 0 \leq j < d_f$$

and basis for R_g is defined by polynomials

$$W_{ij} = X^i T^j \quad (\mathrm{mod} Q), \quad \text{with } 0 \leq i < n \text{ and } 0 \leq j < d_g$$

It is not required to take modulo Q of W_{ij} as $d_g < m$. So vector $F * H (\mathrm{mod} Q)$ and $G (\mathrm{mod} Q)$ can be represented by

$$\sum \alpha_{ij} V_{ij} \quad \text{and} \quad \sum \beta_{ij} W_{ij} \quad \text{with} \quad \alpha_{ij}, \beta_{ij} \in \mathbb{F}_2[T]$$

Attacker already knows integer d_f, d_g and polynomial Q. To find the private key F attacker need to know the coefficients $\alpha_{ij} \in \mathbb{F}_2[T]$. It can be achieved by the following analysis.

$$\text{As} \quad F * H \equiv G \quad (\mathrm{mod} Q)$$
$$\text{or} \quad \sum \alpha_{ij} V_{ij} = \sum \beta_{ij} W_{ij}$$

Due to the property (6) corresponding linear equations generated by comparing the coefficients of different $X^i T^j$'s. $(nm - nd_g)$ equation will be in only one variable α_{ij}. Attacker can equate them to zero due to property (6). So nd_f variable are defined in $(nm - nd_g)$ equations. These equations will have unique nonzero solution in case number of variables are less than or equal to number of equations. So constraints for private key security is

$$nm - nd_g < nd_f$$
$$m < d_f + d_g$$
$$m + 1 \leq d_f + d_g \tag{7}$$

This constraint is same as (5). In case we prove it to be false than attacker need to solve only $(nm - nd_g)$ equations in nd_f variables compare to $n(d_f + m - 1)$ linear equations in $n(2d_f + d_g - 1)$ variables.

4.3   Criteria for PlainText Security

Here we compute the bound on parameters to achieve plaintext M by considering linear equations generated by congruence

$$E \equiv P * \mathtt{R} * H + M \pmod{Q}$$

It can be written equivalently as an equality

$$E = P * \mathtt{R} * H + M + Q * K \qquad \text{where } K \in R \tag{8}$$

If we compare the coefficient of monomials $T^i X^j$ appearing in equation (8). We get number of linear equations equal to the degree of polynomial max of $(\deg(P * \mathtt{R} * H), \deg(M)$ and $\deg(Q * K))$.

Here we estimate the number of equations and number of variables to achieve the private key. For this we need to know the maximum possible degree of K as we already know the degree bounds for rest of the elements $H, \mathtt{R}, M, P$ and Q.

$$\begin{aligned}
\deg K &\leq \deg \mathtt{R} + \deg H + \deg P - \deg Q \\
&\leq (d_r - 1 + (m-1) + s - m) \\
&\leq (d_r + s - 2) \qquad \text{so} \quad K \in R[d_r + s - 1]
\end{aligned}$$

Number of linear equations $= \max\{\deg(P * \mathtt{R} * H), \deg(M), \deg(Q * K)\} + 1$

$$\deg(P * \mathtt{R} * H) = (s + (d_r - 1) + (m-1)) = (d_r + s + m - 2)$$
$$\text{and } \deg(Q * K) = ((d_r + s - 2) + m) = (d_r + s + m - 2)$$

As $\deg(M) < \deg(Q)$ so number of equations $= n\{(d_r + s + m - 2 + 1\}$
$$= n(d_r + s + m - 1)$$

as M, K and $\mathtt{R}$ are unknown variables so we can also calculate the number of variable all together as

$$\text{number of variables} = nd_r + ns + n(d_r + s - 1) = n(2d_r + 2s - 1)$$

Attacker can achieve the plaintext in case the number of variables are less than or equal to the number of equations. In this case the system of equations will have unique solution.

So plaintext will be secure if

$$\text{number of variables} > \text{number of equations}$$
$$\text{or} \quad n(2d_r + 2s - 1) > n(d_r + s + m - 1)$$
$$s + d_r > m$$
$$\text{or} \quad s + d_r \geq m + 1 \tag{9}$$

Alternatively, we can analyze the security of plaintext in different manner with reduced number of linear equations and variables if we project problem in proper vector space by exploiting more properties of variables. Consider the modular equation

$$E \equiv P * \mathtt{R} * H + M \pmod{Q}$$

Here two terms $P * R * H$ and M lies in the vector space $S = (P * R_r * H + Q * R) \cap R_q$ and R_p respectively.

Next, As M lies in vector space R_p that is quotient ring of R by ideal P. So there is no nonzero element in R_p that is divisible by P. It implies that

$$\{(P * R_r * H + Q * R) \cap R_q\} \cap R_p = \{0\}$$

As $(P * R_r * H + Q * R) \cap R_q$ has P as one of its multiplier. Here R belongs to vector space R_r. So the ciphertext E can be parted in terms of $(P * R * H)$ and M, both lies in different vector space S and R_p respectively as $S \cap R_p = \{0\}$.

Attacker knows R_r and S. Elementary linear algebra can be used to decompose E into its component parts and to recover plaintext M. basis of the vector space R_r is given by $X^i T^j$ with $0 \leq i < n$, $0 \leq j < d_r$ and for R_p with $0 \leq i < n$, $0 \leq j < s$. We can define the basis for the subspace S of R_q by

$$S_{ij} = P * X^i * T^j * H \pmod{Q}$$

with $0 \leq i < n$ and $0 \leq j < d_r$ due to uniqueness of direct sum decomposition, we can say that the element $(\sum \mu_{ij} * X^i T^j)$ of R_p will be equal to plaintext M Now we define coefficient $\delta_{ij} \in \mathbb{F}_2$. Consider the equation

$$\sum \mu_{ij} * X^i T^j = E - \sum \delta_{ij} * S_{ij} \quad \text{belongs to} \quad R_p$$

Here we notice that first if we compare the coefficients of all $X^i T^j$ for $0 < i \leq n, 0 < j \leq m$, then ns equations will be in $ns + nd_r$ unknowns and rest $nm - ns$ equations will be in nd_r unknowns. As ns dimensional subspace R_p is defined in the nm dimensional vector space R_q by $nm - ns$ linear independent known equations. CTRU will be secure against the plaintext attack if and only if the number of variables are more than number of equations

$$n(m - s) < nd_r$$
$$(m - s) < d_r$$
$$(m + 1) \leq s + d_r$$

Here also we get the same constraints as in condition(9). In case the condition is false then attacker need to solve less number of equations for less number of variables to recover the plain text.

5　Implementation of the Attack with Recommended Parameters

Ensuring valid decryption, Private key and plaintext security condition can be collectively written by following inequality derived by adding equation 1, 2, 5 and 9.

$$2s + d_r + d_g + d_f \leq 2m + 2 \leq s + d_f + d_g + d_r$$

It shows $s \leq 0$, means there can be no message which is secure provided correct decryption. Thats why equation 5 and 9 are false so we can use linear algebra to attack the system.

Authors of CTRU defined three levels of securities: moderate, high and very high according to the choice of parameters.

Consider moderate security parameter, which is defined as

$$n = 32, \; m = 32, \; s = 7$$

$$\text{Private} - \text{Key} : 1200 \text{ bits}, \quad \text{Public} - \text{Key} : 1024 \text{ bits}$$

$$\text{Key} - \text{Security} : (2^{200}), \; \text{message} - \text{Security} : (2^{200})$$

High security and very high security levels are CTRU-64 and CTRU-128 respectively.

Attacker can recover plaintext by solving a system of $n(d_r + s + m - 1)$ linear equations in $n(2d_r + 2s - 1)$ variables. Attacker can use the alternative method demonstrated in (4.3) to fasten the attack. Then he will have to solve $n(m - s)$ linear equations in nd_r unknowns.

If we consider moderate security parameters $n = 32, m = 32, s = 7$, then attacker has to solve a system of 800 $\mathbb{F}_2[T]$ linear equations in 384 variables with alternative method to recover corresponding plaintext. Attacker can use NTL or PARI-GP libraries to solve this system of equations. Pari took less than 2 minutes to solve it on a 2.13 GHZ Pentium-3 windows machine. For high and very high level security parameter we can fasten the attack by optimizing the code for this particular problem.

Here we have number of equations of $O(nm)$ with $O(nm)$ variables. We can use gaussian elimination method on $\mathbb{F}_2$ as the method can be performed on any field with complexity $O(nm)$ for $n \times n$ size matrix.

6   Conclusion

CTRU system is a natural variant of NTRU that enjoys the rigorous derivation of decryption, and avoid LLL and CRT based attacks. It has been claimed in [6] to be more secure than NTRU and parameters are recommended with consideration that it will be secure against popov normal form attack. However we have shown that CTRU is compleatly insecure to fulfill security criteria for valid decryption with given parameter choices. CTRU neither give any speed improvement over NTRU nor security against polynomial time linear algebra attacks.

Further research can be done in the direction which involve extension to non-commutative groups instead of using group algebra over $\mathbb{F}_2[T, X]/(X^n - 1)$.

It is recommended for the authors of CTRU to review the choice of parameters that may compromise between popov normal form attack and polynomial time linear algebra attacks.

References

1. Coglianese, M., Goi, B.M.: MaTRU : A New NTRU Based Cryptosystem. In: Maitra, S., Veni Madhavan, C.E., Venkatesan, R. (eds.) INDOCRYPT 2005. LNCS, vol. 3797, pp. 232–243. Springer, Heidelberg (2005)
2. Coppersmith, D.: Finding a Small Root of a Bivariate Integer Equation; Factoring with High Bits Known. In: Maurer, U.M. (ed.) EUROCRYPT 1996. LNCS, vol. 1070, pp. 178–189. Springer, Heidelberg (1996)
3. Coppersmith, D.: Small Solution to Polynomial Equations, and Low Exponent RSA Vulner-Abilities. Journal of Cryptology 10, 223–260 (1997)
4. Coppersmith, D., Shamir, A.: Lattice Attacks on NTRU. In: Fumy, W. (ed.) EUROCRYPT 1997. LNCS, vol. 1233, pp. 52–61. Springer, Heidelberg (1997)
5. Coppersmith, D.: Finding Small Solution to Small Degree Polynomials. In: Silverman, J.H. (ed.) CaLC 2001. LNCS, vol. 2146, pp. 20–31. Springer, Heidelberg (2001)
6. Gaborit, P., Ohler, J., Sole, P.: CTRU, a Polynomial Analogue of NTRU, INRIA. Rapport de recherche, N.4621 (November 2002), (ISSN 0249-6399), ftp://ftp.inria.fr/INRIA/publication/publi-pdf/RR/RR-4621.pdf
7. Hoffstein, J., Howgrave-Graham, N., Pipher, J., Silverman, J.H., Whyte, W.: NTRUSign: Digital Signatures Using the NTRU Lattice. In: Joye, M. (ed.) CT-RSA 2003. LNCS, vol. 2612, pp. 122–140. Springer, Heidelberg (2003)
8. Hoffstein, J., Pipher, J., Silverman, J.H.: NTRU : A Ring-Based Public Key Cryptosystem. In: Buhler, J.P. (ed.) ANTS 1998. LNCS, vol. 1423, pp. 267–288. Springer, Heidelberg (1998)
9. Hoffstein, J., Pipher, J., Silverman, J.H.: NSS: An NTRU Lattice-Based Signature Scheme. In: Pfitzmann, B. (ed.) EUROCRYPT 2001. LNCS, vol. 2045, pp. 211–228. Springer, Heidelberg (2001)
10. Hoffstein, J., Silverman, J.H.: Optimizations for NTRU. In: Public-Key Cryptography and computational Number Theory (2000)
11. Hoffstein, J., Silverman, J.H.: Random Small Hamming Weight Products with Applications to Cryptography. Discrete Applied Mathematics 130, 37–49 (2000)
12. Howgrave-Graham, N., Nguyen, P.Q., Pointcheval, D., Proos, J.: The Impact of Decrption Failures on the Security of NTRU Encryption. In: Boneh, D. (ed.) CRYPTO 2003. LNCS, vol. 2729, pp. 226–246. Springer, Heidelberg (2003)
13. Jaulmes, E., Joux, A.: A Chosen Ciphertext Attack on NTRU. In: Bellare, M. (ed.) CRYPTO 2000. LNCS, vol. 1880, pp. 20–35. Springer, Heidelberg (2000)
14. May, A., Silverman, J.H.: Dimension Reduction Methods for Convolution Modular Lattices. In: Silverman, J.H. (ed.) CaLC 2001. LNCS, vol. 2146, pp. 110–125. Springer, Heidelberg (2001)
15. McEliece, R.J.: A Public-Key Cryptosystem Based on Alzebraic Coding Theory. JPL DSN Progress report 42-44, 114–116 (1978)
16. Nguyen, P.Q., Pointcheval, D.: Analysis and Improvements of NTRU Encryption Paddings. In: Yung, M. (ed.) CRYPTO 2002. LNCS, vol. 2442, pp. 210–225. Springer, Heidelberg (2002)
17. Nguyen, P.Q., Stern, J.: The Two Faces of Lattice in Cryptology. In: Silverman, J.H. (ed.) CaLC 2001. LNCS, vol. 2146, pp. 148–180. Springer, Heidelberg (2001)
18. Rivest, R.L., Shamir, A., Adleman, L.: A Method for Obtaining Digital Signatures and Public Key Cryptosystem. Communications of the ACM 21, 120–126 (1978)
19. Schnorr, C.P.: A Hierarchy of Polynomial Time Lattice Basis Reduction Algorithms. Theoretical Computer Science 53, 201–224 (1987)

Detecting Community Structure by Network Vectorization[*]

Wei Ren[**], Guiying Yan[1], Guohui Lin[2], Caifeng Du[3], and Xiaofeng Han[4]

[1] Academy of Mathematics and Systems Science,
Chinese Academy of Science
renwei@amss.ac.cn, yangy@amt.ac.cn
[2] Department of Computing Science,
University of Alberta
[3] College of Mathematics and Computational Science,
China University of Petroleum
ducaif@amss.ac.cn
[4] College of Science, Shandong University of Science and Technology
hanxiaofeng@sdust.edu.cn

Abstract. With the growing number of available social and biological networks, the problem of detecting network community structure is becoming more and more important which acts as the first step to analyze these data. In this paper, we transform network data so that each node is represented by a vector, our method can handle directed and weighted networks. it also can detect networks which contain communities with different sizes and degree sequences. This paper reveals that network community can be formulated as a cluster problem.

1 Introduction

Many systems can be represented by networks where nodes denote entities and links denote existed relation between nodes, such systems may include the web networks [9], the biological networks [12], ecological web [14] and social organization networks [22]. Many interesting properties have also been identified in these networks such as small world[20] and power law distribution [1], one property that attracts much attention is the network community structure which is the phenomenon that nodes within the same community are more densely connected than those in different communities [10]. It is important in the sense that we can get a better understanding about the network structure.

Many earlier algorithms such as [10],[13],[23],[18] often design a similarity measure between each node pair and build a hierarchical tree based on this similarity. These algorithms give clear structure on how the nodes are organized and community structure is usually embedded in the hierarchy. However, they

---

[*] This work is supported by the NNSF (10531070) of China and Science Fund for Creative Research Group. The authors thank Dr. Martin Rosvall in Washington University for giving valuable comments on this paper.
[**] Corresponding author.

X. Hu and J. Wang (Eds.): COCOON 2008, LNCS 5092, pp. 245–254, 2008.

were ambiguous in the sense that there were no definite criterions on how to extract communities from this hierarchy. This, in fact, brings the fundamental problem: what is the definition of community structure? how to evaluate the goodness of a community structure detected?

The definition of strong and weak sense community structure is proposed [18] but the criterion of evaluating the goodness of community structure is not clear. Newman comes up with a modularity measure and shows that good community structure usually indicates a large modularity value. Also it is shown [8] that under some weak conditions, the communities detected by maximizing modularity should be the weak sense community structure. However, the definition of modularity has a scale problem and may have limit in resolving community structure, i.e, communities detected by maximizing modularity may be composed of two smaller modules. An information based algorithm is designed [19] and some asymmetric tests are devised in which the modularity based algorithm fail to perform well.

This comes again the fundamental question, is there any clear definition or description to characterize community structure? If there is, can this definition overcome the scale problem caused by modularity?

Intuitively, community should be the same meaning as node group or cluster. In data mining field, cluster usually refers to a set of closely located vectors, here the closeness depend on the distance defined, such as cosine product or euclidian distance. Now, what is the relationship between network community and vector cluster? If we could find some way to represent each node by a vector and these vectors maintain the node similarity, will the clusters among these vectors correspond to communities? This paper does vectorization using singular value decomposition and indeed verifies this view.

2   Method

Suppose that the network analyzed is unweighted and undirected, let the adjacent matrix be $\mathbf{A}$ with n nodes, $\mathbf{A}_{ij} = 1$ means that there is an edge from node i to node j. In this paper, we only consider that case in which community number c has already been known as a prior knowledge, so there is no need to do model selection. Note that our approach can handle weighted or directed networks.

This framework is composed of 3 parts: first, we obtain a similarity measure $\mathbf{B}$ between node pair based on adjacent matrix $\mathbf{A}$; second, do vectorization of the network using singular value decomposition on matrix $\mathbf{B}$ so that each node is represented by a vector; third, find clusters in these vectors using clustering algorithm, each cluster corresponds to a community in the original network. Detail of each part of the framework will be explained in the next few subsections.

2.1   Similarity Calculating

The calculating of the similarity matrix $\mathbf{B}$ aims at mining deep information from the network topology. This step serves as preprocessing step and can enhance the power in community detection. Any algorithms which defines similarity can be incorporated here, of course, different algorithms may impact the final results. To test the power of the framework, we adopt a very simple similarity measure explained below.

Researchers find that most real complex networks have a larger clustering coefficient than expected by chance [20], this means that a node pair may have many common neighbors which is a good indicator of the similarity between this node pair. Based on this fact, we set $\mathbf{B} = \mathbf{A} + \alpha \mathbf{A}^2$, this formula considers direct link and common neighbors with a scalar α balancing the weight from these two terms. Empirically, α is set to 0.3.

2.2   Vectorization Using Singular Value Decomposition

Given the similarity matrix $\mathbf{B}$, the goal of vectorization is to represent each node by a k dimensional vector while maintaining the similarity measure as much as possible. Suppose node i is represented by vector $\mathbf{n_i} = (n_{i1}, n_{i2}, ..., n_{ik})$.

According to Singular Value Decomposition, $\mathbf{B} = U_{n\times n}S_{n\times n}V_{n\times n}^T$ with the following properties:

- $U_{n\times n}^T U_{n\times n} = I_{n\times n}, V_{n\times n}^T V_{n\times n} = I_{n\times n}$
- $S_{n\times n}$ is a diagonal matrix, whose entry s_{ii} is the ith largest singular value of $\mathbf{B}$. Note that let r be the rank of $\mathbf{B}$, then $s_{ii} = 0$ for $i = r+1, ..., n$.
- $\mathbf{u_i}, \mathbf{v_i}$ which is the ith column of $U_{n\times n}$ and $V_{n\times n}$ are correlated by

$$\mathbf{u_i} = \mathbf{B}\mathbf{v_i}/s_{ii}, \mathbf{v_i} = \mathbf{B}^T\mathbf{u_i}/s_{ii}$$

- Let $U_{n\times k}, V_{n\times k}$ be the matrix consisting of the first $k(k \leq r)$ columns of $U_{n\times n}$ and $V_{n\times n}$, respectively, let S_{kk} be the $k \times k$ diagonal matrix whose ith diagonal entry is s_{ii}. Let $\mathbf{B_k} = U_{n\times k}S_k V_{n\times k}$ with $k \leq r$. In fact:

$$\mathbf{B_k} = \sum_{i=1}^{k} s_{ii}\mathbf{u_i}\mathbf{v_i}^T$$

$\mathbf{B_k}$ is the best approximation to $\mathbf{B}$ in term of matrix F norm.

Using the above properties, we can map each node i to be the $k-$dimensional vector

$$\mathbf{n_i} = (u_{i1}\sqrt{s_{11}}, u_{i2}\sqrt{s_{22}}, ..., u_{ik}\sqrt{s_{kk}})$$

This is exactly the same idea of **Latent Semantic Analysis** [5] widely used in text mining. How to determine k is really tricky, however, later in our experiments, we should that the results are robust to the choice of k.

2.3   Clustering Algorithms

The algorithms we employ are the hierarchical algorithm in [7] using average linkage scheme and the widely used k-means cluster method [15]. It should be clear that any clustering algorithm can be employed here.

Now let's consider the computational issue, it should be clear that the main computational burden is in the singular value decomposition phase. Singular value decomposition is equivalent to eigenvalue decomposition which already has highly efficient algorithms [3]. Take lanczos algorithm for example[3], when the matrix $\mathbf{B}$ is sparse(contains a lot of entries with value 0), computation of the top $k \ll n$ eigenvalues only costs a linear time $O(nk)$. Typically similarity matrix $\mathbf{B}$ of a complex networks is sparse, and fortunately the selection of k is insensitive (this will be shown in our experiment section), these two facts indicates our model is computational efficient.

3 Experiment

Empirically, the value of α in transforming the original adjacent matrix A to $B = A + \alpha(A \times A)$ is set to 0.3 in all our tests here, but it should be noted that this α does not have a significant impact on the results.

3.1 Zachary Club Network

The famous zachary club network is about acquaintance relationship between 34 members by [22]. The club splits to 2 parts due to an internal dispute so it naturally has community structure.

We use the hierarchical algorithm by Eisen in [7]. To prove the computational efficiency, k is set to only 2 and the result is shown in Fig. 1. The striking thing is that the original 2 community structure can be identified if each node is mapped into a 2 dimensional space, and the hierarchy structure in Fig. 1 shows that nodes in the same community are very near but further if they are in different communities. The more striking result is that for k from 2 to 34, our model can always detect the real community structure. This shows that results are insensitive to the choice of k.

3.2 Comparison with Other Algorithms

Newman in [17] has proposed a modularity measure $Q = \sum_{r=1}^{c} (\frac{l_{rr}}{l} - (\frac{d_r}{2l})^2)$ where l_{rr} is the number of links in community r, d_r be the total degree in community r, l is the total number of edges in the network and c be the number of communities detected. Many algorithms such as [21,6,16] try to maximize Q to detect the community structure. In fact, maximizing Q is to maximize the number of link within communities versus that are expected by chance. It is shown that there is a scale l in the definition of Q [8] and this may cause problem in networks whose communities vary in size and degree sequence.

Dolphin social network reported by Lusseau [14] provides a natural example where communities vary in size. In this network, two dolphins have a link with each other if they are observed together more often than expected by chance. The

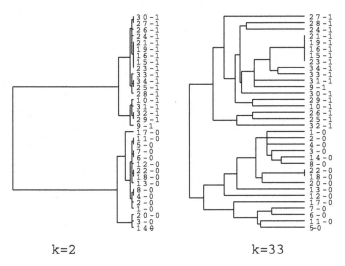

k=2   k=33

Fig. 1. Zachary club network: Hierarchy structure obtained when $k = 2, k = 33$. Each terminal in the hierarchy is labeled by node id and node community id. Our model can uncover the original community structure when k ranges from 2 to 34. Only the results for $k = 2$ and $k = 33$ are depicted.

original two subdivisions have different sizes, with one community 22 dolphins and the other 40. Setting $k = 2$, our model only misclassifies one node and gets exactly the same result as the GN algorithm [10] and the information based algorithm [19], however, the modularity based method get very different result, as depicted in Fig. 2.

Martin in [19] shows that the modularity based algorithms works well for networks whose communities roughly have the same size and degree sequence, but fails to get very good results when the communities differ in sizes and degree sequences. According to [4,19,10], we conduct the following three tests: symmetric, node asymmetric, link asymmetric. In the symmetric test, each network is composed of 4 communities with 32 nodes each, each node have an average degree of 16, k_{out} is the average number of edges linking to nodes in different communities. In the node asymmetric test, each network is composed of 2 communities with 96 and 32 nodes respectively. We set $k_{out} = 6, 7, 8$ in the symmetric and node asymmetric case, as k_{out} increases, it should be difficult to detect real community structure. In the link asymmetric test, 2 communities each with 64 nodes differ in their average degree sequence, nodes in one community have average 24 edges and in the other community have only 8 edges, we set $k_{out} = 2, 3, 4$. By setting different k ranging from 5 to 80 with an interval of 5, we employ the K-means clustering algorithm [15] to detect clusters or communities.

The result is represented in Tab. 1, note that the accuracy of our model is averaged over all k from 5 to 80. We also depict the result for different k in Fig. 3.

For the symmetric test, we model gets an average accuracy of 0.99,0.96,0.87 under $k_{out} = 6, 7, 8$ respectively which is as good as the currently best results in Tab. 1. It indeed gets better results than the information based algorithm [19]

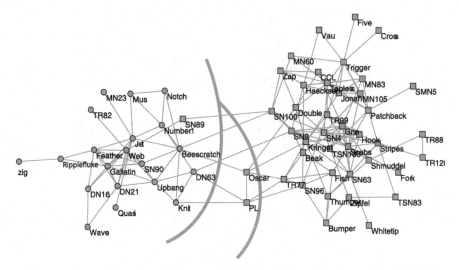

Fig. 2. Dolphin network community structure detected by our algorithm. Node shape denotes the real split and node color indicates the split by our model when $k = 2$. Only one node "SN89" is misclassified. The left line shows our split line, the right line is the split line by the modularity method [16].

in the node asymmetric case when $k_{out} = 8$ with an accuracy of at least 0.84 compared to the information based algorithm which is 0.82.

Another important fact from these tests is that the community structure detected is insensitive to the choice of k which can be clearly seen from Fig. 3, this makes it convenient to use our model.

What's more important is that, by transforming the network into vectors and employing ordinary clustering algorithm such as K-means, we get over the scale problem with modularity. This implies our approach, which mining these information from the transformed vectors, is reasonable.

Table 1. Results for three simulation tests: Symmetric, Node Asymmetric and Link Asymmetric. The accuracy of our model is averaged over all k from 5 to 80.

Test	k_{out}	Our model	Information[19]	Modularity[17]
Symmetric	6	0.99	0.99	0.99
	7	0.96	0.97	0.97
	8	0.87	0.87	0.89
Link asymmetric	2	1.00	1.00	1.00
	3	1.00	1.00	0.96
	4	0.99	1.00	0.74
Node asymmetric	6	0.99	0.99	0.85
	7	0.97*	0.96	0.80
	8	0.85*	0.82	0.74

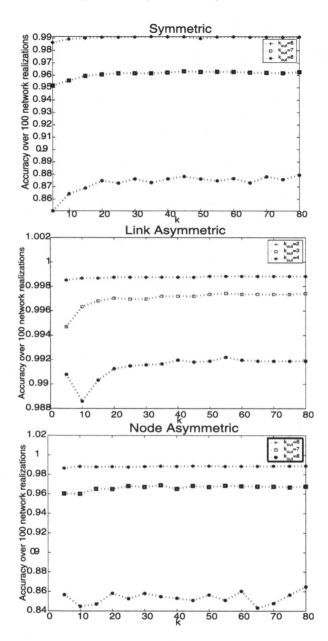

Fig. 3. Result on the benchmark test on 3 experiment under different k: symmetric, node asymmetric, link asymmetric. Each figure contains the accuracy obtained using different k under different k_{out}, by comparing with currently the best results in Tab. 1, it should concluded that our model can handle network whose communities vary in size and degree sequence. This shows that this vectorization model can overcome the drawback of modularity based algorithms. On the other hand, the results by our model are insensitive to the choice of k.

3.3 Power on Weighted Graphs

Our model is based on singular value decomposition and can naturally deal with weighted graphs. For unweighted graph with adjacent matrix A, we transform A to be $B = A + \alpha A^2$ and do singular value decomposition, however, for weighted graph with weighted adjacent matrix W, we directly apply singular value decomposition to W rather than $B = W + \alpha W^2$ and employ K-means clustering algorithm to detect clusters or communities.

To test our model on weighted networks, we follow the simulation test in [19]. This set of test is based on the above symmetric test when $k_{out} = 8$: we increase the weight of edges within a certain community to $w = 1.4, 1.6, 1.8, 2$, respectively, but keep the weight of edges running between communities unchanged(with weight 1). As weight w increases from 1.4 to 2, models should improve their power in detecting community structure. The results is represented in Tab. 2 and the results in [2] are also listed for comparison. As weight w becomes higher, both algorithms can enhance their detecting power, however, our algorithm generally outperforms the approach in [2]. Again, the community structure detected is insensitive to the choice of k, and our model gives better results on each weight $w = 1.4, 1.6, 1.8, 2.0$, which shows that our model is capable of handling weighted graph efficiently. What is more important is that, transforming the weighted network into vectors and treating the network community structure problem as an ordinary clustering algorithm is feasible and hence build the relationship between community structure and clustering algorithm.

Table 2. Benchmark test on weighted network designed by [2]. There are 4 communities each with 32 nodes in the network with $k_{out} = 8$. As w increase from 1.4 to 2, both methods improve their ability in detecting communities. However, our model get better results.

	$k = 20$	$k = 25$	$k = 30$	$k = 35$	Markov[2]
$w = 1.4$	0.95	0.95	0.95	0.95	0.89
$w = 1.6$	0.98	0.98	0.98	0.98	0.94
$w = 1.8$	0.99	0.99	0.99	0.99	0.97
$w = 2$	1.00	1.00	0.99	1.00	0.98

3.4 Detecting Communities in Directed Networks

To test the power of the model in detecting directed networks, we adopt the simulation test devised by [11]. Each network has 2 communities with 16 nodes each. First, this network is generated as an undirected network where every node pair is linked by a probability of p_r, obviously, this undirected network has no community structure. Next, we randomly shift the direction of each arc e_{ij} in the following way: if node i and j belong to the same community, this arc has the direction from i to j or j to i, each with probability of 0.5; if node i and j are not in the same community, then e_{ij} is given the direction from community 1 to community 2 with a probability of $p_b >= 0.5$ and the direction from community

Table 3. Results on directed graphs under different combination of p_r and p_b: As p_b increases, the community structure in the undirected graphs becomes more clear, this is reflected in our results

	$p_b = 0.6$	$p_b = 0.7$	$p_b = 0.8$	$p_b = 0.9$	$p_b = 1.0$
$p_r = 0.3$	0.66	0.69	0.74	0.77	0.82
$p_r = 0.5$	0.63	0.70	0.79	0.86	0.93
$p_r = 0.7$	0.62	0.71	0.85	0.93	0.98
$p_r = 0.9$	0.63	0.76	0.91	0.98	0.99
$p_r = 1.0$	0.63	0.76	0.92	0.98	1.00

2 to community 1 with a probability of $1 - p_b$. According to this scheme, only the direction of edges plays the key role in discerning these 2 communities. Larger p_b indicates better community structure and that using modularity algorithm by treating the directed network as an undirected network is not feasible.

Suppose that the adjacent matrix of this network is $D_{32 \times 32}$ with $D_{ij} = 1$ indicating that there is a directed edge from i to j, we use $\mathbf{B} = D + \alpha D \times D$ and perform singular value decomposition. Since in our previous analysis, the community structure detected is insensitive to the choice of k, so in this test, we set $k = 16$ and $\alpha = 0.3$ empirically. We let p_r and p_b ranging from 0.5 to 1.0 respectively, for each combination of p_r and p_b, 100 networks are generated and average accuracy is calculated. Tab. 3 shows that our model can effectively handle the undirected graph where only the direction of arcs matters. The result of our model also comply with intuition: as p_b increases, the modularity structure becomes clear.

4   Conclusion

The community structure problem attracts great attention from researchers and many specific algorithms have been proposed. It is generally believed the community structure detection has a very close relationship with clustering problem in data mining. We verify this perspective in this paper. Our model gives competitive results on widely known networks and can handle weighted and directed networks. Thus we can conclude that it indeed builds the bridge between the community detection problem and clustering problem.

References

1. Barabási, A.L., Albert, R.: Emergence of Scaling in Random Networks. Science 286, 509–512 (1999)
2. Alves, N.A.: Unveiling Community Structures in Weighted Networks. Phys. Rev. E 76, 036101 (2007)
3. Berry, M.W.: Large-Scale Sparse Singular Value Computations. The International Journal of Supercomputer Applications 6(1), 13–49 (1992)
4. Danon, L., Duch, J., Diaz-Guilera, A., Arenas, A.: Comparing Community Structure Identification (2005)

5. Deerwester, S., Dumais, S.T., Landauer, T.K., Furnas, G.W., Harshman, R.A.: Indexing by Latent Semantic Analysis. Journal of the Society for Information Science 41, 391–407 (1990)
6. Duch, J., Arenas, A.: Community Detection in Complex Network Using Extremal Optimization. Physical Review E 72, 027104 (2005)
7. Eisen, M.B., Spellman, P.T., Brown, P.O., Botstein, D.: Cluster Analysis and Display of Genome-Wide Expression Patterns. Proc. Natl. Acad. Sci. USA 95, 14863–14868 (1998)
8. Fortunato, S., Barthlemy, M.: Resolution Limit in Community Detection. Proc. Natl. Acad. Sci. USA 104(1), 36–41 (2007)
9. Freeman, L.C.: The Sociological Concept of "Group": An Empirical Test of Two Models. American Journal of Sociology 98, 152–166 (1992)
10. Girvan, M., Newman, M.E.J.: Community Structure in Social and Biological Networks. Proc. Natl. Acad. Sci. USA 99(2), 7821–7826 (2002)
11. Leicht, E.A., Newman, M.E.J.: Community Structure in Directed Networks. Phys. Rev. Lett. 100, 118703 (2008)
12. Hartwell, L.H., Hopfield, J.J., Leibler, S., Murray, A.W.: From molecular to modular cell biology. Nature 402, 6761 (1999)
13. Luo, F., Yang, Y., Chen, C.F., Chang, R., Zhou, J., Scheuermann, R.H.: Modular Organization of Protein Interaction Networks. Bioinformatics 23(2), 207–214 (2007)
14. Lusseau, D., Schneider, K., Boisseau, O.J., Haase, P., Slooten, E., Dawson, S.M.: The Bottlenose Dolphin Community of Doubtful Sound Features a Large Proportion of Long-Lasting Associations. Behavioral Ecology and Sociobiology 54, 396–405 (2003)
15. MacQueen, J.B.: Some Methods for Classification and Analysis of Multivariate Observations. In: Proceedings of 5-th Berkeley Symposium on Mathematical Statistics and Probability, vol. 1, pp. 281–297 (1967)
16. Newman, M.E.J.: Finding Community Structure in Networks Using the Eigenvectors of Matrices. Phys. Rev. E 74, 036104 (2006)
17. Newman, M.E.J., Girvan, M.: Finding and Evaluating Community Structure in Networks. Phys. Rev. E 69(2), 026113 (2004)
18. Radicchi, F., Castellano, C., Cecconi, F., Loreto, V., Parisi, D.: Defining and Identifying Communities in Networks. Proc. Natl. Acad. Sci. USA 101(9), 2658–2663 (2004)
19. Rosvall, M., Bergstrom, C.T.: An Information-Theoretic Framework for Resolving Community Structure in Complex Networks. Proc. Natl. Acad. Sci. USA 104(18), 7327–7331 (2007)
20. Watts, D.S.: Collective Dynamics of "Small-World" Networks. Nature 4 393(6684), 409–410 (1998)
21. White, S., Smyth, P.: A Spectral Clustering Approach to Finding Communities in Graphs. In: SIAM International Conference on Data Mining (2005)
22. Zachary, W.W.: An Information Flow Model for Conflict and Fission in Small Groups. Journal of Anthropological Research 33, 452–473 (1977)
23. Zhou, H.: Network Landscape from a Brownian Particle's Perspective. Phys. Rev. E 67, 041908 (2003)

Quasi-bicliques: Complexity and Binding Pairs

Xiaowen Liu[1,3], Jinyan Li[2], and Lusheng Wang[1,*]

[1] Department of Computer Science,
City University of Hong Kong, Kowloon, Hong Kong
lwang@cs.cityu.edu.hk
[2] School of Computer Engineering,
Nanyang Technological University, Singapore 639798
[3] Department of Computer Science,
University of Western Ontario, Canada

Abstract. Protein-protein interactions (PPIs) are one of the most important mechanisms in cellular processes. To model protein interaction sites, recent studies have suggested to find interacting protein group pairs from large PPI networks at the first step, and then to search conserved motifs within the protein groups to form interacting motif pairs. To consider noise effect and incompleteness of biological data, we propose to use *quasi-bicliques* for finding interacting protein group pairs. We investigate two new problems which arise from finding interacting protein group pairs: the maximum vertex quasi-biclique problem and the maximum balanced quasi-biclique problem. We prove that both problems are NP-hard. This is a surprising result as the widely known maximum vertex biclique problem is polynomial time solvable [16]. We then propose a heuristic algorithm which uses the greedy method to find the quasi-bicliques from PPI networks. Our experiment results on real data show that this algorithm has a better performance than a benchmark algorithm for identifying highly matched BLOCKS and PRINTS motifs.

1 Introduction

Proteins with interactions carry out most biological functions within living cells such as gene expression, enzymatic reactions, signal transduction, inter-cellular communications and immunoreactions. As the interactions are mediated by short sequence of residues among the long stretches of interacting sequences, these interacting residues or called interaction (binding) sites are at the central spot of proteome research. Although many imaging wet-lab techniques like X-ray crystallography, nuclear magnetic resonance spectroscopy, electron microscopy and mass spectrometry have been developed to determine protein interaction sites, the solved amount of protein interaction sites constitute only a tiny proportion among the whole population due to high cost and low throughput. Computational methods are still considered as the major approaches for the deep understanding of protein binding sites, especially for their subtle 3-dimensional structure properties that are not accessible by experimental methods.

---

* Corresponding author.

X. Hu and J. Wang (Eds.): COCOON 2008, LNCS 5092, pp. 255–264, 2008.
© Springer-Verlag Berlin Heidelberg 2008

The classical graph concept—maximal biclique subgraph (also known as maximal complete bipartite subgraph)—has been emerged recently for bioinformatics research closely related to topological structures of protein interaction networks and biomolecular binding sites. For example, Thomas *et al.* introduced complementary domains in [15], and they showed that the complementary domains can form near complete bipartite subgraphs in PPI networks. Morrison *et al.* proposed a lock-and-key model which is also based on the concept of maximal complete bipartite subgraphs [11]. Very recently, Andreopoulos *et al.* used clusters in PPI networks for identifying locally significant protein mediators [1]. Their idea is to cluster common-friend proteins, which are in fact complete-bipartite proteins, based on their similarity to their direct neighborhoods in PPI networks. Other computational methods studying bipartite structures of PPI networks focused on protein function prediction [4,8].

To identify motif pairs at protein interaction sites, Li *et al.* introduced a novel method with the core idea related to the concept of complete bipartite subgraphs from PPI networks [10]. The first step of the algorithm [10] finds large subnetworks with all-versus-all interactions (complete bipartite subgraphs) between a pair of protein groups. As the proteins within these protein groups have similar protein interactions and may share the same interaction sites, the second step of Li's algorithm is to compute conserved motifs (possible interaction sites) by multiple sequence alignments within each protein group. Thus, those conserved motifs can be paired with motifs identified from other protein groups to model protein interaction sites. One of the novel aspects of the algorithm [10] is that it combines two types of data: the PPI data and the associated sequence data for modeling binding motif pairs.

Each protein in the above PPI networks is represented by a vertex and every interaction between two proteins is represented by an edge. Discovering complete bipartite subgraphs in PPI networks can thus be formulated as the following biclique problem: Given a graph, the biclique problem is to find a subgraph which is bipartite and complete. The objective is to maximize the number of vertices or edges in the bipartite complete subgraph. We note that the maximum vertex biclique problem is polynomial time solvable [16]. This problem is also equivalent to the maximum independent set problem on bipartite graphs which is known to be solvable by a minimum cut algorithm. However, the maximum vertex balanced biclique problem is NP-hard [6]. The maximum edge biclique problem is proved to be NP-hard as well [12].

In this paper, we consider incompleteness of biological data, as the interaction data of PPI networks is usually not fully available. On the other hand, within an interacting protein group pair, some proteins in one group may only interact with a proportion of the proteins in the other group. Therefore, many subgraphs formed by interacting protein group pairs are not perfect bicliques—They are more often near complete bipartite subgraphs. Therefore, methods of finding bicliques may miss many useful interacting protein group pairs. To deal with this problem, we use quasi-bicliques instead of bicliques to find interacting protein group pairs. With the quasi-biclique, even though some interactions are missing in a protein interaction subnetwork, we can still find the two interacting protein groups. In this paper, we introduce and investigate two theoretical problems: the maximum vertex quasi-biclique problem and the maximum balanced quasi-biclique problem. We show that both problems are NP-hard. We also propose a heuristic algorithm for finding large quasi-bicliques in PPI networks.

2   Bicliques and Quasi-bicliques

Let $\mathcal{G} = (\mathcal{V}, \mathcal{E})$ be an undirected graph, where each vertex represents a protein and there is an edge connecting two vertices if the two proteins have an interaction. It is allowed that an edge $(u, v) \in \mathcal{E}$ and $u = v$, which is called a self-loop. Since $\mathcal{G}$ is an undirected graph, any edge $(u, v) \in \mathcal{E}$ implies $(v, u) \in \mathcal{E}$. For a selected edge (u, v) in $\mathcal{G}$, in order to find the two groups of proteins having the similar pairs of binding sites, we translate the graph $\mathcal{G} = (\mathcal{V}, \mathcal{E})$ into a bipartite graph. Let $X = \{x | (x, v) \in \mathcal{E}\}$, $Y_1 = \{y | (u, y) \in \mathcal{E} \& y \notin X\}$ and $Y_2 = \{w | (u, w) \in \mathcal{E} \& w \in X\}$. For a vertex $w \in Y_2$, w is incident to both u and v in $\mathcal{G}$, so both X and Y_2 contain w. We remain w in X and replace w in Y_2 with a new virtual vertex $\overline{w}$. After replacing all vertices w in Y_2 with $\overline{w}$, we get a new vertex set $\overline{Y_2}$. Let $Y = Y_1 \cup \overline{Y_2}$ and $E = \{(x, y) | (x, y) \in \mathcal{E} \& x \in X \& y \in Y_1\} \cup \{(x, \overline{w}) | (x, w) \in \mathcal{E} \& x \in X \& \overline{w} \in \overline{Y_2}\}$. In this way, we have a bipartite graph $G = (X \cup Y, E)$. A biclique in G corresponds to two subsets of vertices, say, subset A and subset B, in $\mathcal{G}$. In $\mathcal{G}$, every vertex in A is adjacent to all the vertices in B, and every vertex in B is adjacent to all the vertices in A. Moreover, $A \cap B$ may not be empty. In this case, for any vertex $w \in A \cap B$, $(w, w) \in \mathcal{E}$. This is the case, where the protein has self-loops. Self-loops are very common in practice. When self-loop appears, the protein contains two complementary motifs simultaneously.

In the following, we focus on the bipartite graph $G = (X \cup Y, E)$. For a vertex $x \in X$ and a vertex set $Y' \subseteq Y$, the degree of x in Y' is the number of vertices in Y' that are adjacent to x, denoted by $d(x, Y') = |\{y | y \in Y' \text{ and } (x, y) \in E\}|$. Similarly, for a vertex $y \in Y$ and $X' \subseteq X$, we use $d(y, X')$ to denote $|\{x | x \in X' \ (x, y) \in E\}|$. Now, we are ready to define the δ-quasi-biclique.

Definition 1. *For a bipartite graph $G = (X \cup Y, E)$ and a parameter $0 < \delta \leq \frac{1}{2}$, G is called a δ-quasi-biclique if for each $x \in X$, $d(x, Y) \geq (1 - \delta)|Y|$ and for each $y \in Y$, $d(y, X) \geq (1 - \delta)|X|$.*

Similarly, a δ-quasi-biclique in G corresponds to two subsets of vertices, say, subset A and subset B, in $\mathcal{G}$. In $\mathcal{G}$, every vertex in A is adjacent to at least $(1 - \delta)|B|$ vertices in B, and every vertex in B is adjacent to at least $(1 - \delta)|A|$ vertices in A. Moreover, according to the translation and the definition, $A \cap B$ may not be empty. Again, if a protein appears in both sides of a δ-quasi-biclique and there is an edge between the two corresponding vertices, the protein contains two complementary motifs simultaneously. In our experiments, we observe that about 22% of the δ-quasi-bicliques produced by our program contain self-loop proteins.

In many applications, due to various reasons, some edges in a biclique may missing and a biclique becomes a quasi-biclique. Thus, finding quasi-bicliques is more important in practice. The following theorem shows that large quasi-bicliques may not contain any large bicliques.

Theorem 1. *Let $G = (X \cup Y, E)$ be a random graph with $|X| = |Y| = n$, where for each pair of vertices $x \in X$ and $y \in Y$, (x, y) is chosen, randomly and independently, to be an edge in E with probability $\frac{2}{3}$. When $n \to \infty$, with high probability, G is a $\frac{1}{2}$-quasi-biclique, and G does not contain any biclique $G' = (X' \cup Y', E')$ with $|X'| \geq 2 \log n$ and $|Y'| \geq 2 \log n$.*

In the biological context, Theorem 1 indicates that some large interacting protein groups cannot be obtained simply by finding a maximal biclique. As large interacting protein groups are more useful, according to this theorem, we have to develop new computational algorithms to extract from PPI networks large interacting protein groups which form quasi-bicliques.

3   NP-Hardness

In this section, we study the following two problems: the maximum vertex quasi-biclique problem and the maximum balanced quasi-biclique problem.

3.1   The Maximum Vertex Quasi-biclique Problem

The maximum vertex quasi-biclique problem is defined as follows.

Definition 2. *Given a bipartite graph $G = (X \cup Y, E)$ and $0 < \delta \leq \frac{1}{2}$, the maximum vertex δ-quasi-biclique problem is to find $X' \subseteq X$ and $Y' \subseteq Y$ such that the $X' \cup Y'$ induced subgraph is a δ-quasi-biclique and $|X'| + |Y'|$ is maximized.*

The maximum vertex biclique problem, where $\delta = 0$, can be solved in polynomial time [16]. We can prove that the maximum vertex δ-quasi-biclique problem when $\delta > 0$ is NP-hard. The reduction is from $X3C$ (Exact Cover by 3-Sets), which is known to be NP-hard [9].

Exact Cover by 3-Sets
Instance: A finite set S of $3m$ elements, and a collection T of n triples (3-element subsets of S).
Objective: To determine whether T contains an exact cover of S, i.e. a subcollection $T' \subseteq T$ such that every element of S occurs in exactly one triple in T'.

Theorem 2. *For any constant integers $p > 0$ and $q > 0$ such that $0 < \frac{p}{q} \leq \frac{1}{2}$, the maximum vertex $\frac{p}{q}$-quasi-biclique problem is NP-hard.*

3.2   The Balanced Quasi-biclique Problem

A balanced quasi-biclique is a quasi-biclique in which the numbers of the vertices in both groups are similar. The maximum balanced quasi-biclique problem is defined as follows:

Definition 3. *Given a bipartite graph $G = (X \cup Y, E)$ and $0 < \delta \leq \frac{1}{2}$, the maximum balanced δ-quasi-biclique problem is to find $X' \subseteq X$ and $Y' \subseteq Y$ such that the $X' \cup Y'$ induced subgraph is a δ-quasi-biclique and $|X'| = |Y'|$ is maximized.*

We can also prove that the maximum balanced quasi-biclique problem is NP-hard.

Theorem 3. *For any constant integers $p > 0$ and $q > 0$ such that $0 < \frac{p}{q} \leq \frac{1}{2}$, the maximum balanced $\frac{p}{q}$-quasi-biclique problem is NP-hard.*

4   The Heuristic Algorithm

In practice, we need to find large quasi-bicliques in PPI networks. Here, we propose a heuristic algorithm to find large quasi-bicliques. Consider a PPI network $\mathcal{G} = (\mathcal{V}, \mathcal{E})$. Our heuristic algorithm has two steps. First, we construct the bipartite graph based on a pair of interacting proteins (u, v). Using the method described at the beginning of Section 2, we can get a bipartite graph $G = (X \cup Y, E)$ from $\mathcal{G} = (\mathcal{V}, \mathcal{E})$ and an edge (u, v).

In the algorithm, we have two parameters δ and τ, which control the quality and sizes of the quasi-bicliques. We can use a greedy method to get the seeds for finding large quasi-bicliques in G. At the beginning, we set $X' = \phi$ and $Y' = Y$. In each step, we find a vertex with the maximum degree in $X - X'$. The vertex is added into the biclique vertex set X', and we eliminate all vertices y in Y' such that $d(y, X') < (1 - \delta)|X'|$. We will continue this process until the size of Y' is less than τ. At each step, we get a seed for finding large quasi-bicliques.

The seeds may miss some possible vertices in the quasi-bicliques. We can extend the seeds to find larger quasi-bicliques. Let $X'' = X'$ and $Y'' = Y'$ be a pair of seed vertex sets. In the first step, we can find a vertex x in $X - X''$ with the largest degree $d(x, Y'')$ in $X - X''$. If $d(x, Y'') \geq (1 - \delta)|Y''|$, we add the vertex x to X''. In the second step, we can find a vertex y in $Y - Y''$ with the largest $d(y, X'')$ in $Y - Y''$. If $d(y, X'') \geq (1 - \delta)|X''|$, we add the vertex y to Y''. We repeat the above two steps until no vertex can be added. The whole algorithm is shown in Fig. 1. We can also exchange the two vertex sets X and Y to find more quasi-bicliques using the algorithm.

Let n be the number of vertices in the bipartite graph G. In the greedy algorithm, the time complexity of Step $3 - 5$ and Step 10 is $O(n)$, and the time complexity of Step $6 - 9$ is $O(n^2)$. So the time complexity of Step $3 - 10$ is $O(n^2)$. Step $3 - 10$ is repeated $O(n)$ times. Therefore, the time complexity of the whole algorithm is $O(n^3)$. To speed up the algorithm, we can do multiple vertex

---

The Greedy Algorithm

Input   A bipartite graph $(X \cup Y, E)$ and two parameters δ and τ.

Output   A set of δ-quasi-bicliques $(X' \cup Y', E')$ with $|X'| \geq \tau$ and $|Y'| \geq \tau$.

1.   Let $X' = \phi$ and $Y' = Y$.

2.   **while** $|Y'| \geq \tau$ and $X' \neq X$ **do**

3.   Find the vertex $x \in X - X'$ with the maximum degree $d(x, Y')$.

4.   Add x into X', $X' = X' \cup \{x\}$, and delete from Y' all vertices $y \in Y'$ such that $d(y, X') < (1 - \delta)|X'|$.

5. $X'' = X'$ and $Y'' = Y'$.

6.   **repeat**

7.   Find the vertex $x \in X - X''$ with the maximum degree $d(x, Y'')$. If $d(x, Y'') \geq (1 - \delta)|Y''|$, add x to X'', $X'' = X'' \cup \{x\}$.

8.   Find the vertex $y \in Y - Y''$ with the maximum degree $d(y, X'')$. If $d(y, X'') \geq (1 - \delta)|X''|$, add y to Y'', $Y'' = Y'' \cup \{y\}$.

9.   **until** no vertex is added in the step 7 and 8.

10.   **if** $|X''| \geq \tau$, $|Y''| \geq \tau$, for each $x \in X''$, $d(x, Y'') \geq (1 - \delta)|Y''|$, and for each $y \in Y''$, $d(y, X'') \geq (1 - \delta)|X''|$, output $(X'' \cup Y'')$ as a quasi-biclique.

Fig. 1. The greedy algorithm

addition in the algorithm. To do multiple vertex addition, we have an integer parameter $\alpha > 0$, and change the algorithm as follows. In Step 3, we select the best α vertices in $X - X'$ and add all the α vertices into X' in Step 4.

5   Experiments

We implemented the heuristic algorithm in JAVA. The software is called PPIExtend. As shown in the last step of the algorithm, some vertices in X'' may be adjacent to less than $(1 - \delta)|Y''|$ vertices in $|Y''|$, but the average degree of the vertices in X'' is no less than $(1 - \delta)|Y''|$. Similarly, some vertices in Y'' may be adjacent to less than $(1 - \delta)|X''|$ vertices in $|X''|$, but the average degree of the vertices in Y'' is no less than $(1 - \delta)|X''|$. In our experiments, these quasi-bicliques are still output to get more useful quasi-bicliques. Our algorithm consists of two steps: (i) find quasi-interacting protein group pairs, then (ii) use the sequence data of these protein groups to find conserved motifs. To evaluate the motif pairs found by our algorithm, we follow the two validation methods in [10]. First, we compare the single motifs with two block databases: BLOCKS [13] and PRINTS [3]. Second, we map our motif pairs into domain-domain interaction pairs in domain-domain interaction database iPfam [5]. We also study the overlapping between the protein group pairs found by PPIExtend and the protein group pairs found by FPClose* in [10]. Two interesting case studies on binding motif pairs are then followed.

5.1   Motif Mapping with BLOCKS, PRINTS, and iPfam

The protein interaction data of *Saccharomyces cerevisiace* (yeast) was downloaded from http://research.i2r.a-star.edu.sg/BindingMotifPairs/resources. The

Table 1. The mappings between the motifs and the two databases: BLOCKS and PRINTS. FPClose* uses BLOCKS 14.0 and PRINTS 37.0. Our PPIExtend method uses BLOCKS 14.3 and PRINTS 38.0. Each entry a/b means that the motifs are mapped to a blocks(domains) in all b blocks(domains) in the databases.

	BLOCKS		PRINTS		BOTH	
	blocks	domains	blocks	domains	blocks	domains
FPClose*	6408/24294	3128/4944	2174/11170	1093/1850	24.1%	62.1%
PPIExtend	9325/29767	4191/6149	2423/11435	1160/1900	28.5%	66.4%

data includes 10640 experimentally determined physical interactions of 4959 proteins in *Saccharomyces cerevisiace*. We set $\delta = 0.1$ and $\tau = 5$ and $\alpha = 5$. The greedy algorithm produced $59,124$ interacting protein group pairs. In all the protein group pairs, $13,266$ pairs (about 22%) contain self-loops, and a large number of protein group pairs (about 78%) do not contain self-loops. We then used PROTOMAT [13] to find the conserved motifs within the protein groups, as PROTOMAT is an algorithm that can find the multi-alignment of a group of sequences and can output the conserved motifs. By using the default parameters, PROTOMAT output $220,393$ motifs from our $59,124$ pairs of interacting protein groups.

The LAMA program [13] is a dynamic programming method that can find the optimal local alignment of two blocks where the Z-score is computed to evaluate the alignments. We make use of it to compare the block databases with our motifs. The default threshold of Z-score was used in the experiments. The mappings between the motifs and the two databases are also compared between our method and the FPClose* method [7] that was used in [10]. The comparison results are reported in Table 1. From this table, we can see that our method has more mappings to BLOCKS and PRINTS than FPClose* does. This indicates that the use of quasi-bicliques is effective to find more number of motif pairs at interaction sites.

The *i*Pfam database is built on top of the Pfam database [14] which stores the information of protein domain-domain interactions. To examine whether our binding motif pairs can match some pairs of interacting domains in *i*Pfam, we map our binding motif pairs through the integrated protein family database InterPro [2] which integrates a number of databases. We first map our motifs to domains in the BLOCKS and PRINTS databases. Then, we map the protein groups in BLOCKS and PRINTS to protein groups in InterPro. Finally, we map the protein groups in InterPro to domains in the Pfam database. In this way, our motif pairs can be well mapped to Pfam domain pairs. In fact, we strictly follow the procedure as suggested in [10] to map motif pairs to domain pairs.

In the experiments, we used Pfam 20.0 and *i*Pfam 20.0. We observed that the motif pairs found by our PPIExtend method can map to 81 distinct domain pairs in *i*Pfam. This is a much bigger number than 18 number of domain pairs that can be mapped from the motif pairs reported in [10]. This significant increase is mainly attributed to the use of quasi-bicliques because using quasi-bicliques can find many interacting protein group pairs with larger sizes. (See Theorem 1.)

Table 2. Left block l18493xB and right block r18493xA (output of PROTOMAT for protein group pair No.18493) aligning with the Bac_rhodopsin domain and the HAMP domain respectively. For brevity, only 5 sequences in each of the two blocks are shown. In the Bac_rhodopsin domain and HAMP domain, the capital letters are the amino acids with the highest frequency in each position. Pdb 1h2s_A and pdb 1h2s_B are chain A and chain B in protein complex 1h2s, respectively.

```
AC  l18493xB;                              AC  r18493xA;
    distance from previous block=(4,396)        distance from previous block=(7,177)
DE  none                                   DE  none
BL  IIK motif=[6,0,17] motomat=[1,1,-10]   BL  LLL motif=[6,0,17] motomat=[1,1,-10]
    width=20 seqs=7                            width=12 seqs=8
DIP:8095N  ( 206) VIGILIISYTKATCDMLAGK     DIP:7371N  (  10) LALIILYLSIPL
DIP:4973N  ( 536) MILILIAQFWVAIAPIGEGK     DIP:8128N  (  35) LSLRFLALIFDL
DIP:5150N  ( 417) LIKDEINNDKKDNADDKYIK     DIP:4176N  ( 106) LVLTSLSLTLLL
DIP:5371N  ( 384) IILALIVTILWFMLRGNTAK     DIP:7280N  (  11) LSLFLPPVAVFL
DIP:676N   ( 402) VIVAWIFFVVSFVTTSSVGK     DIP:5331N  ( 178) LSFFVLCGLARL
...                                        ...
pdb 1h2s_A ( 168) VILWAIYPFIWLLGPPGVA      pdb 1h2s_B (  61)      VSAILGLII
Bac_rhodopsin:    VVLWLAYPVVWLLGPEGIG      HAMP:                 IALLLALLL
```

In the 81 domain pairs, 48 pairs are domain-domain interactions on one protein (self-loops) and 33 pairs are domain-domain interactions on different proteins. Although the self-loops is a large portion of the pairs of domain interactions found by our method, we still found many other domain-domain interactions that are not self-loops.

We also examined the overlapping between the protein group pairs found by PPIExtend and the protein group pairs found by FPClose* in [10]. The overlapping criteria is that if one protein group pair G_1 contains more than 90% proteins of another protein group pair G_2, we say that G_1 covers G_2. We found that only 38 out of the 5,349 protein group pairs found by FPClose* can not be covered by our protein group pairs. However, there are 38,305 protein group pairs found by PPIExtend that cannot be covered by any protein group pairs found by FPClose*. This result demonstrates that our method not only can find much more number of bicliques and with larger size, also can find more binding motif pairs than the method presented in [10].

5.2 Case Studies

In this section, we present two binding motif pairs that can be mapped to interacting domain pairs. The first motif pair is derived from a protein group pair in which the left protein group contains 7 proteins and the right protein group contains 10 proteins. There are 66 interactions between the two groups of proteins. Using the hypergeometric probability model, the p-value of the protein group pair is less than 1.57×10^{-191}. PROTOMAT finds two left blocks and two right blocks in this protein group pair. The second left block contains 20 positions and the first right block contains 12 positions. By the mapping method, the positions $1 - 19$ of the second left block can be aligned with the positions $9 - 27$ of block

IPB001425B in BLOCKS, and the positions $4-12$ of the first right block can be aligned with the positions $1-9$ of block IPB003660A in BLOCKS. Block IPB001425B is in the Bac_rhodopsin domain, and block IPB003660A is in the HAMP domain, See Table 2 for more details. Our motif pair can map into the domain pair (PF00672, PF01036) in iPfam. iPfam shows that the HAMP domain interacts with the Bac_rhodopsin domain in protein complexes such as lh2s.

The second motif pair is derived from a protein group pair in which the left protein group contains 6 proteins and the right protein group contains 8 proteins. There are 43 interactions between the two groups of proteins. The p-value of the protein group pair is less than 1.09×10^{-122}. The motif pair can be mapped to the interacting PdxA domain pair (PF04166, PF04166) in iPfam. The domain pair has protein-protein interactions in protein complexes such as 1ps6.

6 Conclusion and Open Problem

We have proved that both the maximum vertex quasi-biclique problem and the maximum balanced quasi-biclique problem are NP-hard. However, the hardness of the maximum edge quasi-biclique problem is still an open problem. In this paper, we have shown the usefulness of the topology information of PPI networks for modeling the binding motifs at interaction sites. In future work, we will focus on how to integrate other information sources, such as protein functions and gene ontology localization information for identifying possible interaction sites.

Acknowledgments. We thank Dr. Haiquan Li for providing us the protein interaction data. We also thank Dr. Shmuel Pietrokovski for giving us the LAMA program. Lusheng Wang is fully supported by a grant from the Research Grants Council of the Hong Kong Special Administrative Region, China [Project No. CityU 121207].

References

1. Andreopoulos, B., An, A., Wang, X., Faloutsos, M., Schroeder, M.: Clustering by Common Friends Finds Locally Significant Proteins Mediating Modules. Bioinformatics 23(9), 1124–1131 (2007)
2. Apweiler, R., Attwood, T.K., Bairoch, A., Bateman, A., Birney, E., Biswas, M., Bucher, P., Cerutti, L., Corpet, F., Croning, M.D., Durbin, R., Falquet, L., Fleischmann, W., Gouzy, J., Hermjakob, H., Hulo, N., Jonassen, I., Kahn, D., Kanapin, A., Karavidopoulou, Y., Lopez, R., Marx, B., Mulder, N.J., Oinn, T.M., Pagni, M., Servant, F., Sigrist, C.J., Zdobnov, E.M.: The InterPro Database, an Integrated Documentation Resource for Protein Families, Domains and Functional Sites. Nucleic Acids Research 29(1), 37–40 (2001)
3. Attwood, T.K., Beck, M.E.: PRINTS-a Protein Motif Fingerprint Database. Protein Engineering, Design and Selection 7, 841–848 (1994)
4. Bu, D., Zhao, Y., Cai, L., Xue, H., Zhu, X., Lu, H., Zhang, J., Sun, S., Ling, L., Zhang, N., Li, G., Chen, R.: Topological Structure Analysis of the Protein-Protein Interaction Network in Budding Yeast. Nucleic Acids Research 31(9), 2443–2450 (2003)

5. Finn, R.D., Marshall, M., Bateman, A.: iPfam: Visualization of Protein-Protein Interactions in PDB at Domain and Amino Acid Resolutions. Bioinformatics 21(3), 410–412 (2005)
6. Garey, M.R., Johnson, D.S.: Computers and Intractability, A Guide to the Theory of NP-Completeness. Freeman, San Francisco (1979)
7. Grahne, G., Zhu, J.: Efficiently using Prefix-Trees in Mining Frequent Itemsets. In: Proceedings of the Workshop on Frequent Itemset Mining Implementations (FIMI) (2003)
8. Hishigaki, H., Nakai, K., Ono, T., Tanigami, A., Takagi, T.: Assessment of Prediction Sccuracy of Protein Gunction From Protein–Protein Interaction Data. Yeast 18(6), 523–531 (2001)
9. Karp, R.M.: Reducibility among Combinatorial Problems. In: Miller, R.E., Thatcher, J.W. (eds.) Complexity of Computer Computations, pp. 85–103 (1972)
10. Li, H., Li, J., Wang, L.: Discovering Motif Pairs at Interaction Sites from Protein Sequences on a Proteome-Wide Scale. Bioinformatics 22(8), 989–996 (2006)
11. Morrison, J.L., Breitling, R., Higham, D.J., Gilbert, D.R.: A Lock-and-Key Model for Protein-Protein Interactions. Bioinformatics 22(16), 2012–2019 (2006)
12. Peeters, R.: The Maximum Edge Biclique Problem is NP-Vomplete. Discrete Applied Mathematics 131(3), 651–654 (2003)
13. Pietrokovski, S.: Searching Databases of Conserved Sequence Regions by Aligning Protein Multiple-Alignments. Nucleic Acids Research 24, 3836–3845 (1996)
14. Sonnhammer, E.L.L., Eddy, S.R., Durbin, R.: Pfam: A Vomprehensive Database of Protein Domain Families Based on Seed Alignments. Proteins: Structure, Function and Genetics 28, 405–420 (1997)
15. Thomas, A., Cannings, R., Monk, N.A.M., Cannings, C.: On the Structure of Protein-Protein Interaction Networks. Biochemical Society Transactions 31(Pt 6), 1491–1496 (2003)
16. Yannakakis, M.: Node Deletion Problems on Bipartite Graphs. SIAM Journal on Computing 10, 310–327 (1981)

Complexity of a Collision-Aware String Partition Problem and Its Relation to Oligo Design for Gene Synthesis

Anne Condon[1], Ján Maňuch[2], and Chris Thachuk[1]

[1] University of British Columbia, Vancouver BC V6T 1Z4, Canada
{condon,cthachuk}@cs.ubc.ca
[2] Simon Fraser University, Burnaby BC V5A 1S7, Canada
jmanuch@sfu.ca

Abstract. Artificial synthesis of long genes and entire genomes is achieved by self-assembly of DNA oligo fragments - fragments which are short enough to be generated using a DNA synthesizer. Given a description of the duplex to be synthesized, a computational challenge is to select the short oligos so that, once synthesized, they will self-assemble without error. In this paper, we show that a natural abstraction of this problem, the *collision-aware string partition problem*, is NP-complete.

1 Introduction

There is extensive literature concerned with the study of string properties and algorithms for various string problems [8],[10],[7],[4]. Many well known string problems such as *longest common subsequence*, *hitting string*, and *bounded Post correspondence* have been shown to be NP-complete [2]. In this paper, we prove the hardness of another basic string problem, that of partitioning a string into unique substrings of bounded length, and demonstrate its relation to a problem in contemporary synthetic biology, that of synthesizing long strands of DNA.

A DNA *strand*, or *oligo*, is a string over the alphabet { A,C,G,T }. The *complement* O' of an oligo O, is determined from O by replacing each G with a C and vice versa, each T with an A and vice versa, and reversing the resulting string. Thus, the complement of the string CGCATAC is GTATGCG. Simplistically, a DNA *duplex* consists of a sense strand (top strand) S and an anti-sense strand (bottom strand) S', where S' is the complement of S.

Technology for synthesis of long DNA strands is enabling new advances in genomics and synthetic biology, such as production of novel or disease-resistant proteins [1]. Recently, this technology was used to artificially construct a complete bacterial genome larger than 500K bases [3]. Since DNA synthesis machines can be used to reliably produce only short DNA oligos, long DNA duplexes are typically synthesized via assembly of many short DNA oligo fragments [12].

To assemble correctly, the short DNA oligos must (a) be substrands of the sense and antisense strands of the given duplex, of length bounded by a given bound. They should (b) *cover* the duplex: if the oligos are ordered by distance

X. Hu and J. Wang (Eds.): COCOON 2008, LNCS 5092, pp. 265–275, 2008.
© Springer-Verlag Berlin Heidelberg 2008

from one end of the duplex, they should alternate between sense and antisense strands, with some overlap between successive oligos, thus enabling assembly via hybridization of complementary parts. Additionally, (c) no oligo should self-hybridize, and (d) no pair of oligos should *collide*, that is, hybridize to each other. If either (c) or (d) happens, proper assembly is foiled. Standard polynomial-time thermodynamically-drived nucleic acid secondary structure prediction algorithms can be used to test if an oligo or pair of oligos fail conditions (c) or (d). Since there is some flexibility in the length of the oligos, there are exponentially many ways to select the oligos. The *collision-aware oligo design for gene synthesis problem (CA-ODGS)* is: given a DNA duplex and length bound k for condition (a), determine whether there is a set of oligos that satisfies all conditions (a) through (d).

CA-ODGS has been widely studied in the literature[15],[11],[5],[1],[6]: for more details see the work of Villalobos *et al.* [15] and the references therein. The variant of the problem which removes the collision condition (d) can be solved in linear time [13]. However, no polynomial time algorithm is known for the general case when all four conditions must be satisfied. The importance of addressing the collision condition will increase as progress towards multiplexed gene synthesis and entire genome synthesis continues [14].

We conjecture that the CA-ODGS is NP-hard. To provide some evidence for this, we show NP-completeness of a simplified version of the problem, which abstracts away thermodynamic details, while retaining the key challenge of collision-aware partitioning. Informally, the variation asks whether a single string (as opposed to a duplex) can be partitioned into short substrings of bounded length, no two of which are identical. For example, consider partitioning the string *theimportantthingisnevertostopquestioning* into substrings having a maximum length of 3. One possible solution is shown in Figure 1 (top). Notice, however, that some partitions may produce substrings which are identical as is the case for the other partition shown in Figure 1 (bottom) where both the substrings *th* and *ing* appear twice. We note, however, that there are instances of the problem for which it is trivial to determine that no solution is possible, dependent on the string length, n, alphabet size, σ, and maximum substring length, k. For example, the string in Figure 1 cannot be partitioned into unique substrings when $k = 1$. In general, there is no solution when $n > \Sigma_{i=1}^{k} \sigma^i$. Likewise, the problem is solvable in constant time if both k and σ are constant: as above, if $n > \Sigma_{i=1}^{k} \sigma^i$ there is no solution, otherwise, n must also be constant and all possible partitions can be checked in constant time.

The paper is organized as follows. In Section 2, we formally introduce the collision-aware string partition (CA-SP) problem and motivate its similarity to the CA-ODGS problem, as well as point out differences between them. In Section 3, a polynomial time reduction from 3SAT(3) to CA-SP with an unbounded alphabet is given which implies the NP-completeness of the CA-SP problem. In Section 4, we show that the problem remains NP-complete for alphabet of size 4. In Section 5, we add complement-awareness to the string partitioning problem and show that the problem is still NP-complete.

Fig. 1. Two partitions are shown for the string *theimportantthingisnevertostopquestioning*. The substrings in both partitions have maximum length 3. The partition shown above the string is valid, in that no two substrings are identical; however, the partition shown below the string is invalid as there are two cases of identical substrings indicated with arrows.

2   The Collision-Aware String Partition Problem

Let Σ be a finite alphabet. A k-partition of a string $A \in \Sigma^*$ is a sequence $P = p_1, p_2, \ldots, p_l$, for some l, where each p_i is a string over Σ of length at most k and $A = p_1 p_2 \ldots p_l$. A sequence of strings $p_1, p_2, \ldots, p_l$ is *collision-free* if for all $i, j, 1 \le i \ne j \le l, p_i \ne p_j$. Given a string A and a partition P of A, we say that a substring a of A is *selected* if and only if a is an element of the sequence P.

Problem 1. Collision-Aware String Partition (CA-SP)

Instance: Finite alphabet Σ, a positive integer k, and a string A from Σ^*.
Question: Is there a collision-free, k-partition P of A?

CA-SP differs from CA-ODGS in several ways: in CA-SP, the goal is to partition a string, rather than a duplex; two strings are considered to collide if and only if they are equal, rather than if they bind stably; and the no-self-hybridization constraint (c) is not modeled at all. However, the task of designing a collision-free k-partition is quite similar to that of developing a collision-free oligo design. The design goal of CA-SP, that of avoiding identical substrings, is one of the design goals of CA-ODGS. To illustrate the importance of avoiding identical substrings in the CA-ODGS problem, consider the oligo design in Figure 2. The oligos labeled c and g are identical and on the same strand. If oligo g hybridized to oligo b, the gene construction could result in two distinct fragments (Figure 2 top). Those labeled e and h are also identical (read 5' to 3'), but on opposite strands. If h hybridized to d, one full fragment could result, however, it would contain a complementary inversion error (Figure 2 bottom). This type of error can be particularly troublesome, as the construct may need to be fully sequenced before the error is discovered.

3   Reducing 3SAT(3) to CA-SP with Unbounded Alphabet

We now describe a polynomial reduction from 3SAT(3) to CA-SP. Let ϕ be an instance of 3SAT(3), with set $C = \{c_1, \ldots, c_n\}$ of clauses, and set $X = x_1, \ldots, x_k$ of variables. We shall define an alphabet Σ and construct a string $A \in \Sigma^*$, such that A has a collision-free 2-partition if and only if ϕ is satisfiable. Let $|c_i|$ denote

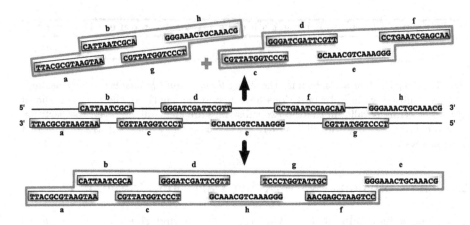

Fig. 2. An invalid design (middle) containing two pairs of identical oligos. The identical oligos c and g could hybridize out of order resulting in two distinct fragments (top). Identical oligos from opposite strands must also be avoided as entire regions could be assembled out of order (bottom).

the number of literals contained in the clause c_i and let $l_i^1, \ldots, l_i^{|c_i|}$ be the literals of clause c_i.

Let us start with the formal definition of 3SAT(3) problem shown to be NP-complete by Papadimitriou[9].

Problem 2. 3SAT(3)

Instance: A formula ϕ with a set C of clauses over a set X of variables in conjunctive normal form such that:

1. every clause contains two or three literals,
2. each variable occurs in exactly three clauses, once negated and twice positive.

Question: Is ϕ satisfiable?

We construct A to be the concatenation of three strings, $A = A'A''A'''$, with the following properties. First, the string A' encodes the clauses of ϕ, so that a collision-free 2-partition of A unambiguously selects a literal from each clause. Intuitively, the selected literals are intended to be a satisfying truth assignment for the variables of ϕ. Second, if A has a collision-free 2-partition, string A'' ensures that the selected literals are *consistent*, that is, no selected literal can be the negation of another. The constructions of A' and A'' rely on the fact that special short delimiting strings are *forbidden* — that is, cannot be selected in the 2-partition of $A'A''$. The string A''' is constructed to ensure that, in a collision-free 2-partition of A, the forbidden strings must be selected from A''', and thus cannot be selected from $A'A''$.

We use alphabet Σ and set of forbidden strings $\mathcal{F}$, defined as follows:

$$\Sigma = \{\xi_i; \ x_i \in X\} \cup \{\lambda_i^j; \ c_i \in C \wedge j \leq |c_i|\} \cup \{\boxplus, \square, \boxminus, \boxtimes\}$$
$$\mathcal{F} = \{\boxplus, \boxplus\boxplus, \square, \square\square, \boxminus, \boxminus\boxminus, \boxtimes, \boxtimes\boxtimes\}$$

Fig. 3. The construction of string $A = A'A''A'''$ for an instance ϕ of 3SAT(3), with $X = \{x_1, x_2, x_3, \ldots\}$ and $C = \{(\neg x_2, x_1, \neg x_3), (\neg x_1, x_2), (x_1, x_3, \neg x_4) \ldots\}$. Selections outlined with a box are required to avoid a collision. All potential partitions avoiding forbidden strings of clause strings, in A', and forced selection forbidden strings, in A''', are shown.

Note that the number of used symbols is linear in the size of the 3SAT(3) problem ϕ (at most $k + 3n + 4$).

Construction of A': For each clause $c_i \in C$, construct the *i-th clause string* to be $\lambda_i^1 \boxdot \lambda_i^2$ if $|c_i| = 2$, and $\lambda_i^1 \boxdot \lambda_i^2 \boxdot \lambda_i^3$ if $|c_i| = 3$. Next, construct the *i-th clause connector string* as $\lambda_i^{|c_i|} \boxplus \boxplus \lambda_{i+1}^1$. A' is the concatenation of the first clause string, the first clause connector, the second clause string, the second clause connector, and so on, up to the $|C|$-th clause string, and then terminated by $\boxplus \boxdot \boxdot \boxplus$. See Figure 3 for an example.

Lemma 1. *Given that no string from the forbidden set $\mathcal{F}$ is selected, exactly one literal symbol can be selected for each clause string in any collision-free 2-partition P' of A'. Additionally, the termination string $\boxplus \boxdot \boxdot \boxplus$ must be partitioned as $\boxplus \boxdot$, $\boxdot \boxplus$.*

Proof. Consider a collision-free 2-partition of A'. Since the strings $\boxplus$ and $\boxplus \boxplus$ are forbidden, it must be that the partition includes the substrings $\lambda_i^{|c_i|} \boxplus$ and $\boxplus \lambda_{i+1}^1$ of the ith clause connection string. Therefore, each clause string must be partitioned independently of the symbols in the adjoining clause connector strings.

Consider the clause string for clause c_i. Whether c_i has two or three literals, the forbidden substring $\boxdot$ cannot be selected alone. Therefore, each $\boxdot$ must be selected with an adjacent literal symbol. This leaves exactly one other literal symbol which must be selected.

Finally, neither of the termination symbols $\boxplus$ or $\boxdot$ can be selected alone in the string, since they are in the forbidden set, and similarly the string $\boxdot \boxdot$ cannot be selected since it too is in the forbidden set. Therefore, the termination string $\boxplus \boxdot \boxdot \boxplus$ must be partitioned as $\boxplus \boxdot$ and $\boxdot \boxplus$.

Construction of A'': We must now ensure that no literal of ϕ that is selected in A' is the negation of another selected literal. By definition of 3SAT(3), each variable appears exactly three times: twice positive and once negated. Let l_i^a and l_j^b be the two positive and l_k^c the negated occurrences of a variable x_v. Then construct the *enforcer string* for this variable as follows $\lambda_i^a \boxtimes \lambda_k^c \xi_v \lambda_k^c \xi_v \lambda_k^c \boxtimes \lambda_j^b$.

Now, construct A'' by concatenating all enforcer strings (in any order), separated by *enforcer connector strings*. The ith enforcer connector string is $\sigma \boxminus \boxminus \sigma'$, where σ and σ' are the last symbol of the ith enforcer string in A'' and the first symbol of the $i+1$st enforcer string in A'', respectively. Terminate the resulting string with $\boxdot \boxminus \boxminus \boxdot$. See Figure 3.

Lemma 2. *Given that no string from the forbidden set $\mathcal{F}$ is selected, any collision-free 2-partition of $A'A''$ must be consistent, and must partition the termination string of A'' into $\boxdot \boxminus$ and $\boxminus \boxdot$.*

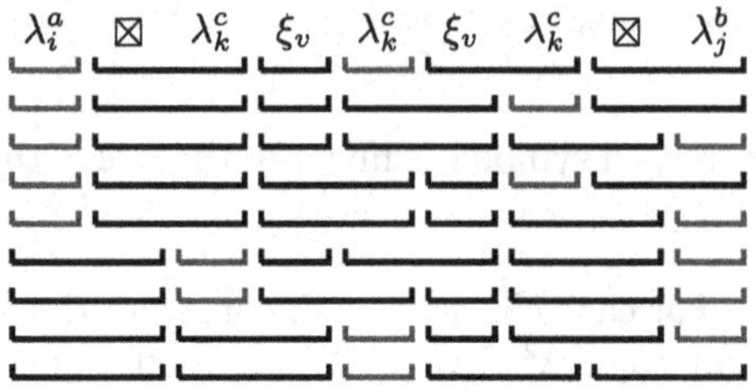

Fig. 4. All possible 2-partitions are shown for the enforcer string of a variable x_v having two positive literals l_i^a and l_j^b, and one negative literal l_k^c. In each partition, either λ_k^c is selected or both λ_i^a and λ_j^b are which guarantees that symbols for positive and negated literals of x_v cannot be simultaneously selected in A'.

Proof. Consider a collision-free 2-partition of $A'A''$. As was shown for clause strings in Lemma 1, the forbidden set forces that each enforcer connector string must be partitioned into two parts, each of size 2. Thus, each enforcer string must be partitioned independently.

Consider the enforcer string for variable x_v with positive literals $l_i^a = l_j^b = x_v$, and the negated literal $l_k^c = \neg x_v$. Figure 4 shows all 9 possible 2-partitions of the enforcer string. It follows that in each of them either λ_k^c is selected or both λ_i^a and λ_j^b are. In the first case, λ_k^c cannot be selected in $A'A''$ and thus satisfied literals are chosen consistently for x_v. In the second case, symbols for neither of the positive occurrences of x_v can be selected in $A'A''$. On the other hand, it is easy to see that there is 2-partition of the enforcer string compatible with any of four valid combinations of selecting symbols for literals in A'.

Finally, neither of the termination symbols $\boxdot$ or $\boxminus$ can be selected alone in the string, since they are in the forbidden set, and similarly the string $\boxminus \boxminus$ cannot be selected since it too is in the forbidden set. Therefore, the termination string $\boxdot \boxminus \boxminus \boxdot$ must be partitioned only as $\boxdot \boxminus$ and $\boxminus \boxdot$.

Construction of A''': To ensure that the forbidden substrings cannot be selected in A' or A'', we construct A''' as shown in Figure 3. A collision-free 2-partition of A''' selects every forbidden substring. Note that the connectors with a boxed outline in the figure must be selected, or a collision occurs. Then, for each sequence of forbidden symbols, there exists two possible partitions, both of which force selection of two forbidden substrings. Also, neither of the terminator strings at the end of A' or A'' appear in A'''. This completes the reduction. Notice that the reduction is linear as the size of the constructed string A is $|A| = |A'| + |A''| + |A'''| \leq 9n + 13k + 18$.

Theorem 1. *Collision-Aware String Partition (CA-SP) is NP-complete.*

Proof. It is easy to see that CA-SP is in NP: a nondeterministic algorithm need only guess a partition P where $|p_i| \leq k$ for all p_i in P and check in polynomial time that no two substrings in P are identical. Furthermore, it is clear that an arbitrary instance ϕ of 3SAT(3) can be reduced to an instance CA-SP, specified by a string A, in polynomial time and space by the reduction detailed above.

Now suppose there is a satisfying truth assignment for ϕ. Simply select one corresponding true literal per clause in A'. The construction of clause strings guarantees that a 2-partition of the rest of each clause string is possible, and this 2-partition can be extended to yield a 2-partition of A. Also, since a satisfying truth assignment for ϕ cannot assign truth values to opposite literals, then Lemma 2 guarantees that a valid partition of the enforcer strings are possible. Therefore, there exists a collision-free string partition of A.

Likewise, consider a collision-free string partition of A. Lemma 1 ensures that exactly one literal per clause is selected. Furthermore, Lemma 2 guarantees that if there is no collision, then no two selected variables in the clauses are negations of each other. Therefore, this must correspond to a satisfying truth assignment for ϕ (if none of the three literals of a variable is selected in the partition of A' then this variable can have arbitrary value in the truth assignment without affecting satisfiability of ϕ).

4   Reducing Unbounded CA-SP to CA-SP with 4 Letters

We now describe a reduction from the CA-SP problem with alphabet Σ of size N to the CA-SP problem with alphabet Ω of size 4. We can assume that $N \geq 2$. Let $K = \lceil \log_2 N \rceil$. Given a string A over Σ, we construct a string B over Ω such that A has a collision-free 2-partition if and only if B has a collision-free $2K$-partition.

We construct B to be a concatenation of three strings, $B = B'B''B'''$, with the following properties. First, the string B' is an encoding of A to the binary alphabet $\{0,1\} \subseteq \Omega$, where each letter in Σ is encoded as a distinct K-bit long string. If we would consider a special type of $2K$-partitions which only allow substrings of lengths K and $2K$ in the partition then, obviously, B' has such a collision-free partition if and only if A has a collision-free 2-partition. Let $\mathcal{F}_1$ be the set of all substrings of length $1, \ldots, K-1, K+1, \ldots, 2K-1$ of B'. We would like construct a string B'' over $\Omega = \{0, 1, \$, \cent\}$ such that in any valid $2K$-partition of B'', all elements of $\mathcal{F}_1$ are selected, in a similar manner as we forbid set $\mathcal{F}$ with the string A''' in the construction of the previous section. However, the set $\mathcal{F}$ was very simple and regular, and we needed a string of length only 18 to "forbid" it. In this case we need to construct another string B''' over alphabet

Ω which will forbid (select all elements of) the following large but regular set of strings $\mathcal{F}_2 = \cup_{i=1}^{2K} \{\$, \cent\}^i$, i.e., a set containing all strings over alphabet $\{\$, \cent\}$ of length at most $2K$. Then we will show that any collision-free $2K$-partition of $B''B'''$ selects all elements from $\mathcal{F}_1$. Consequently, any $2K$-partition of $B'B''B'''$ will partition B' into substrings of length K or $2K$, as described above, and the correctness of construction follows.

Construction of B': Consider any injective mapping $f : \Sigma \to \{0,1\}^K$. Clearly, such a mapping exists since $|\{0,1\}^K| = 2^K \geq N = |\Sigma|$. Let $B' = f(A)$. Let

$$\mathcal{F}_1 = \{w \in \cup_{\substack{1 \leq i \leq 2K-1 \\ i \neq K}} \{0,1\}^i; \ \exists u, v : \ B' = uwv\}$$

We have the following observation.

Observation 1. *Given that no string from the forbidden set $\mathcal{F}_1$ is selected, there is a collision-free 2-partition of A if and only if there is a collision-free $2K$-partition of B'.*

Construction of B'': Let $\{f_i\}_{i=0}^{t-1}$ be an enumeration of all elements in $\mathcal{F}_1$ such that $f_i \neq f_j$ for $i \neq j$, and $|f_i| \leq |f_j|$ for $i < j$. Let $\{s_i\}_{i=0}^{\lfloor t/2 \rfloor}$ be any enumeration of $\lfloor t/2 \rfloor + 1$ distinct elements of the set $\{\$, \cent\}^{2K-1} - \{\$^{2K-1}\}$. Consider the following substring of the whole sequence B: $0s_is_i0$. Assuming that no string from set $\mathcal{F}_2$ is selected, the only possible partition is

$$0s_i, s_i0.$$

Indeed, s_is_i is a string over $\{\$, \cent\}$ of length $4K-2$, and thus any further partitions of this substring would result in selection of an element from $\mathcal{F}_2$. Thus under the assumption that set $\mathcal{F}_2$ is forbidden, all substrings of type $0s_is_i0$ and $1s_is_i1$ can serve as the delimiters of the partition. We construct the string B'' as follows:

$$0s_0s_00f_01s_0s_01f_10s_1s_10f_2\ldots s_{\lfloor (t-1)/2 \rfloor}e_{t-1}f_{t-1}e_ts_{\lfloor t/2 \rfloor}, \text{ where } e_i = \begin{cases} 0 & \text{if } i \text{ is odd,} \\ 1 & \text{if } i \text{ is even.} \end{cases}$$

To verify that this construction is valid we need to check that the set $\{\$, \cent\}^{2K-1} - \{\$^{2K-1}\}$ contains enough elements, i.e., that $2^{2K-1} - 1 \geq \lfloor |\mathcal{F}_1|/2 \rfloor + 1$. In the worst case, the set $\mathcal{F}_1$ contains all binary strings of length $1, 2, \ldots, K-1, K+1, K+2, \ldots, 2K-1$. Thus the size of $\mathcal{F}_1$ is at most

$$2^1 + 2^2 + \cdots + 2^{K-1} + 2^{K+1} + \cdots + 2^{2K-1} = 2^{2K} - 2^K - 2.$$

And hence,

$$\left\lfloor \frac{|\mathcal{F}_1|}{2} \right\rfloor + 1 \leq 2^{2K-1} - 2^{K-1} \leq 2^{2K-1} - 1.$$

Lemma 3. *Given that no string from the set $\mathcal{F}_2$ is selected, the only collision-free $2K$-partition of B'' is the following one*

$$0s_0, s_00, f_0, 1s_0, s_01, f_1, 0s_1, s_10, f_2, \ldots, s_{\lfloor (t-1)/2 \rfloor}e_{t-1}, f_{t-1}, e_ts_{\lfloor t/2 \rfloor},$$

and hence in any $2K$-partition of $B'B''$, the strings in $\mathcal{F}_1$ cannot be selected in B'.

Proof. Since, the substrings $0s_is_i0$ and $1s_is_i1$ serve as delimiters in any partition, the substrings $f_0, \ldots, f_{t-1}$ have to be partitioned separately. It is enough to show that all these substrings are selected whole, i.e., they are not further partitioned into smaller substrings. Observe that set $\mathcal{F}_1$ has the following property: any partitioning of an element $f_i \in \mathcal{F}_1$ to smaller substrings will produce at least one substring contained in $\mathcal{F}_1$. Consider a simple backtracking algorithm which will try all possible partitions for f_0, f_1, etc. Obviously, f_0 is of length one, and thus the algorithm must partition f_0 as one piece. We will show by induction that this is the case in all steps of the algorithm. Assume the algorithm is considering substring f_i. Since any partition of f_i to smaller substrings contains a string $f_j \in \mathcal{F}_1$, which is shorter than f_i, f_j must have been already processed by the algorithm. By the induction hypothesis, it was selected, and thus cannot be used again when partitioning f_i. Thus f_i has to be selected as a whole piece. This completes the proof.

Construction of B''': Let B''' be the concatenation of all elements in $\mathcal{F}_2 = \cup_{i=1}^{2K}\{\$, \math{c}\}^i$ in an arbitrary order followed by $\$^{2K-1}0$.

Lemma 4. *Any $2K$-partition of B''' will select all elements of the $\mathcal{F}_2$ and the substring $\$^{2K-1}0$.*

Proof. Assume that a $2K$-partition of B''' does not select $w \in \mathcal{F}_2$ of length $j \geq 1$. Then the sum of the length of strings from $\mathcal{F}_2$ selected by this partition is at most $|B'''| - (2K + 1)$, and hence there is at least $2K$ letters of $B'''0^{-1}$ (B''' without the last letter) which are not covered by strings in $\mathcal{F}_2$. However, this is a contradiction as $B'''0^{-1} \in \{\$, \math{c}\}^*$ and $\mathcal{F}_2$ contains all possible strings over these two letters.

Assuming that a $2K$-partition of B''' selects all elements of $\mathcal{F}_2$, a similar argument shows that the last selected word is $\$^{2K-1}0$.

The NP-completeness of the CA-SP problem for a four-letter alphabet follows by Observation 1 and Lemmas 3 and 4.

Theorem 2. *Collision-Aware String Partition (CA-SP) is NP-complete for an alphabet of size 4.*

5   Adding Complement Awareness

In this section, we consider a string partition problem slightly closer to the CA-ODGS problem. We will require that not only the elements of the partitioned input string are distinct but also that one is not the complement of any other. We will call this problem the *collision-aware complement-aware string partition* (CACA-SP) problem and show that the proof from the previous section can be easily modified to yield the NP-completeness result for this problem as well.

Let $S = s_1 \ldots s_k$ be a DNA strand. The complement of S, denoted by S^c is obtained from S by replacing each A with a T and vice versa, and each C

with a G and vice versa, and reversing the resulting string. More formally, let $\Phi \colon \{A, C, G, T\} \to \{A, C, G, T\}$ be a mapping such that

$$\Phi(A) = T, \quad \Phi(T) = A, \quad \Phi(C) = G, \quad \text{and} \quad \Phi(G) = C,$$

and let S^R denote the mirror image of the sequence S. Then the complement of a strand S is $S^c = \Phi(S^R)$. Obviously, if the oligo design contains two oligos which are complements of each other, then these two oligos would hybridize together which would prevent them from interacting with the other oligos. Thus in a valid oligo design, no two oligos can be complements of each other. This is another necessary condition for the oligo design problem, which we now model in an extension to the collision-aware string partition problem.

Problem 3. Collision-Aware Complement-Aware String Partition (CACA-SP)

Instance: Finite alphabet Σ, a positive integer k, and a string A from Σ^*.
Question: Is there a collision-free, k-partition of A such that no two selected strings are complements of each other?

We will call such k-partitions, *valid k-partitions*. Note that some strings cannot be selected in any valid k-partition, in particular, self-complementary strings for which $w^c = w$. For instance, two copies of self-complementary string $AGCT^c = AGCT$ would form a DNA duplex.

Theorem 3. *Collision-Aware Complement-Aware String Partition (CACA-SP) problem is NP-complete for alphabet of size 4.*

Details omitted for space.

6 Conclusion

In this paper, we have shown that CA-SP, the collision-aware string partition problem, is NP-complete. It remains hard for a restricted alphabet of size 4, the alphabet size of DNA, and with the addition of complement awareness. This suggests that the collision-aware oligo design for gene synthesis problem (CA-ODGS) is also NP-complete, justifying the focus on heuristic algorithms for its solution.

An interesting theoretical problem is to determine whether the CA-SP problem remains NP-complete when we restrict the alphabet size to 3 or 2. An additional extension of our work would be to prove NP-completeness of an extension of CA-SP, pertaining to duplexes rather than single strands.

References

1. Cox, J.C., Lape, J., Sayed, M.A., Hellinga, H.W.: Protein Fabrication Automation. Protein Sci. 16(3), 379–390 (2007)
2. Garey, M., Johnson, D.: Computers and Intractability: A Guide to the Theory of NP-Completeness. Freeman Press, New York (1979)

3. Gibson, D.G., Benders, G.A., Andrews-Pfannkoch, C., Denisova, E.A., Baden-Tillson, H., Zaveri, J., Stockwell, T.B., Brownley, A., Thomas, D.W., Algire, M.A., Merryman, C., Young, L., Noskov, V.N., Glass, J.I., Venter, J.C., Hutchison, I., Clyde, A., Smith, H.O.: Complete Chemical Synthesis, Assembly, and Cloning of a Mycoplasma Genitalium Genome. Science 1151721 (2008)
4. Gusfield, D.: Algorithms on Strings, Trees, and Sequences. Cambridge Press (1997)
5. Hoover, D.M., Lubkowski, J.: DNAWorks: an Automated Method for Designing Oligonucleotides for PCR-based Gene Synthesis. Nuc. Acids Res. 30(10), 43 (2002)
6. Jayaraj, S., Reid, R., Santi, D.V.: GeMS: an Advanced Software Package for Designing Synthetic Genes. Nuc. Acids Res. 33(9), 3011–3016 (2005)
7. Lien, Y.E.: Periodic Properties of Strings. SIGACT News 7(1), 21–25 (1975)
8. Lothaire, M.: Combinatorics on Words. Encyclopedia of Mathematics and its Applications, vol. 17. Addison-Wesley, Reading (1983)
9. Papadimitriou, C.H.: Computational Complexity. Addison-Wesley, Reading (1994)
10. Post, E.L.: A Variant of a Recursively Unsolvable Problem. Bulletin of the American Mathematical Society 52, 264–268 (1946)
11. Rouillard, J.M., Lee, W., Truan, G., Gao, X., Zhou, X., Gulari, E.: Gene2Oligo: Oligonucleotide Design for in Vitro Gene Synthesis. Nuc. Acids Res. 32, 176–180 (2004) (Web Server issue)
12. Stemmer, W.P., Crameri, A., Ha, K.D., Brennan, T.M., Heyneker, H.L.: Single-Step Assembly of a Gene and Entire Plasmid from Large Numbers of Oligodeoxyribonucleotides. Gene 164(1), 49–53 (1995)
13. Thachuk, C., Condon, A.: On the Design of Oligos for Gene Synthesis. In: BIBE, pp. 123–130 (2007)
14. Tian, J., Gong, H., Sheng, N., Zhou, X., Gulari, E., Gao, X., Church, G.: Accurate Multiplex Gene Synthesis from Programmable DNA Microchips. Nature 432(7020), 1050–1054 (2004)
15. Villalobos, A., Ness, J.E., Gustafsson, C., Minshull, J., Govindarajan, S.: Gene Designer: a Synthetic Biology Tool for Constructing Artificial DNA Segments. BMC Bioinformatics 7, 285 (2006)

Genome Halving under DCJ Revisited

Julia Mixtacki

International NRW Graduate School in Bioinformatics and Genome Research,
Universität Bielefeld, Germany
julia.mixtacki@uni-bielefeld.de

Abstract. The Genome Halving Problem is the following: Given a rearranged duplicated genome, find a perfectly duplicated genome such that the rearrangement distance between these genomes is minimal with respect to a particular model of genome rearrangement. Recently, Warren and Sankoff studied this problem under the general DCJ model where the pre-duplicated genome contains both, linear and circular chromosomes. In this paper, we revisit the Genome Halving Problem for the DCJ distance and we propose a genome model such that constraints for linear genomes, as well as the ones for circular genomes are taken into account. Moreover, we correct an error in the original paper.

1 Introduction

Besides genome rearrangements, another important source for genome evolution is whole genome duplication. In the early 1970s, Susumu Ohno [13] came up with the hypothesis that whole genome duplication has occurred in mammalian evolution. Not without controversy, this question has been addressed several times in the last three decades, both in the biological and in the computational literature.

In fact, there is biological evidence for genome duplication among several eukaryotes. An outstanding example was the duplication in the yeast genome that was recently confirmed [12]. Even two rounds of duplication are found in vertebrates [5]. Duplication is a particularly common event in plants [1],[10] where most of the common crops have polyploid genomes.

The combinatorial problem, called the Genome Halving Problem, was first introduced in [8]: Assuming that a genome is duplicated and then rearranged over time, can we reconstruct an ancestral genome from the gene order that we observe today? The key to the solution of this question is the structure of the genome right after duplication: It must have been *perfect*, i.e. each chromosome has existed in two identical copies. Of course, there exist many perfectly duplicated genomes that could have been the ancestral genome. Therefore, we want to reconstruct one genome such that its *distance*, defined as the minimum number of rearrangements needed to transform it into the observed genome, is minimal.

Clearly, solutions to this problem depend on the underlying genome model and also on the rearrangement operations that are allowed. The most common genome rearrangement operations are *translocations, fusions, fissions, inversions*

X. Hu and J. Wang (Eds.): COCOON 2008, LNCS 5092, pp. 276–286, 2008.
© Springer-Verlag Berlin Heidelberg 2008

and *block interchanges*. It is remarkable that all these operations can be modelled by a single one, called *double cut and join* (DCJ) operation [15]. As shown in [4], the DCJ operation applies for genomes with a mixture of linear and circular chromosomes. In contrast, in the Hannenhalli-Pevzner (HP) theory [11] it is assumed that the genomes only consist of linear chromosomes and only translocations, fusions, fissions and inversions are considered.

El-Mabrouk and Sankoff [9] solved the Genome Halving Problem under the HP distance. Their algorithm for the reconstruction of doubled genomes is far from being trivial and is the final result of a whole series of papers [8],[7],[6]. In addition to the well-known *breakpoint graph*, they introduce further graphs, called *natural graph* and *signature graph*. Later, Alekseyev and Pevzner gave an alternative approach based on the notion of *contracted breakpoint graph* [2] and corrected in [3] an error in the El-Mabrouk-Sankoff analysis.

Very recently, Warren and Sankoff [14] studied the Genome Halving Problem under the more general DCJ model. This generalization yields a simplified problem since some of the complicated components of the breakpoint graph, such as *hurdles* and *knots*, can be ignored. Unfortunately, their solution still relies on the complex concepts introduced by El-Mabrouk and Sankoff. Indeed, as we will see in this paper, the problem can be solved by working directly on the natural graph.

In the following, we will revisit the Genome Halving Problem under the double cut and join operation where the ancestral genome may contain linear and circular chromosomes. Therefore, in our genome model, we take into account both, the constraints usually required for genomes with only linear chromosomes, as well as the ones for genomes with only circular chromosomes. Compared to the more general model studied in [14], these requirements on the ancestral genome increase the distance between the genomes. This yields a new proof and a simple algorithm for reconstructing an ancestral genome. Moreover, by our results, we will also correct an error in the Warren-Sankoff analysis.

The structure of this paper is as follows. We begin by formalizing the problem in the next section. Then, in Section 3, we study the effect of a DCJ operation on the natural graph. In Section 4 we present our distance formula and a linear-time algorithm to reconstruct an ancestral genome with the minimum number of DCJ operations. Finally, we will discuss the Warren-Sankoff formula in Section 5. The last section summarizes our results and addresses some open questions.

2   Problem Formulation

As usual, a *gene* is represented by a directed identifier where the direction is indicated by a *head* and a *tail*. These are called the *extremities* of the gene. The tail of a gene a is denoted by a^t, and its head is denoted by a^h.

An *adjacency* of two consecutive genes a and b can be of four different types:

$$\{a^h, b^t\}, \{a^h, b^h\}, \{a^t, b^t\}, \{a^t, b^h\}.$$

An extremity that is not adjacent to any other gene is called a *telomere*, represented by a singleton set $\{a^h\}$ or $\{a^t\}$.

Definition 1. *A duplicated genome A is a set of adjacencies and telomeres such that the head and the tail of every gene appears exactly twice.*

Thus, a duplicated genome has two identical copies of each gene that are called *paralogs* and we distinguish them by a subscript, called an *assignment of the paralogs*. For a gene a, we denote its copies by a_1 and a_2 and the *paralogous extremities* by a_1^t, a_2^t and a_1^h, a_2^h.

Example 1. Consider the following genome A_1 defined on the set of genes $\{a, b, c, d\}$:

$$\{\{d_2^h\}, \{d_2^t, a_2^t\}, \{a_2^h, d_1^h\}, \{d_1^t, c_2^h\}, \{c_2^t, b_2^t\}, \{b_2^h\}, \{b_1^h\}, \{b_1^t, c_1^t\}, \{c_1^h, a_1^t\}, \{a_1^h\}\}$$

A genome can be represented as a graph, called the *genome graph*, with vertices corresponding to the adjacencies and telomeres and edges joining the head and the tail of each paralogous extremity. Thus, we have:

$$\underset{d_2^h}{\bullet}\;\underset{d_2^t}{\bullet}\underset{a_2^t}{\bullet}\;\underset{a_2^h}{\bullet}\underset{d_1^h}{\bullet}\;\underset{d_1^t}{\bullet}\underset{c_2^h}{\bullet}\;\underset{c_2^t}{\bullet}\underset{b_2^t}{\bullet}\;\underset{b_2^h}{\bullet} \qquad\qquad \underset{b_1^h}{\bullet}\;\underset{b_1^t}{\bullet}\underset{c_1^t}{\bullet}\;\underset{c_1^h}{\bullet}\underset{a_1^t}{\bullet}\;\underset{a_1^h}{\bullet}$$

Suppose that the genome graph consists of N components C_1 to C_N. A *chromosome* is a set of adjacencies and telomeres that belong to the same component. Note that, by definition, each vertex in the genome graph has degree one or two, and thus the components of the genome graph are either *linear* or *circular*. We call a genome *linear* if all its chromosomes are linear. Similarly, a genome is *circular* if all its chromosomes are circular. For example, the above genome graph is a linear genome consisting of two linear chromosomes.

For paralogous extremities, we also use the following notation: if p is an extremity, then $\bar{p}$ is its corresponding paralogous extremity. By elevating this notation to sets of extremities, we can apply it to adjacencies and telomeres. For example, for an adjacency $x = \{a_1^h, b_2^t\}$, we have $\bar{x} = \{a_2^h, b_1^t\}$.

For a chromosome C, we define $\overline{C} = \{\bar{x} \mid x \text{ is an adjacency or telomere of } C\}$. This notation is useful to describe the different notions of a duplicated genome that can be found in the literature, for linear genomes in [9] and for circular genomes in [3]. By bringing this together for genomes with a mixture of linear and circular chromosomes, we have:

Definition 2. *A duplicated genome A consisting of chromosomes $C_1, \ldots, C_N$ is*

- *linear-perfectly duplicated, if for each linear chromosome C_i, we have $C_i = \overline{C_j}$ for some $j \in \{1, \ldots, N\}\backslash\{i\}$;*
- *circular-perfectly duplicated, if for each circular chromosome C_i, either we have $C_i = \overline{C_j}$ for some $j \in \{1, \ldots, N\}\backslash\{i\}$ or $C_i = C \cup \overline{C}$, where each adjacency of C_i occurs either in C or in $\overline{C}$, but not in both;*
- *perfectly duplicated, if it is linear- and circular-perfectly duplicated.*

Note that this definition does not depend on the assignment of the paralogs. Two examples of perfectly duplicated genomes are given in Fig. 1. From the

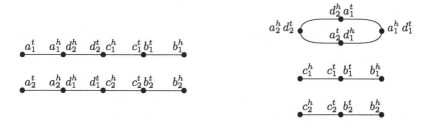

Fig. 1. Two perfectly duplicated genomes

right genome in that figure, we also see that the number of chromosomes of a
perfectly duplicated genome is not necessarily even.

Alternatively to the formulation on the level of chromosomes, a perfectly
duplicated genome can also be characterized locally, as stated by the next lemma.

Lemma 1. *A genome A is perfectly duplicated if and only if*

- *for each adjacency $\{u, v\}$ in A, also $\{\bar{u}, \bar{v}\}$ is in A and $u \neq \bar{v}$, and*
- *for each telomere $\{u\}$ in A, also $\{\bar{u}\}$ is in A.*

Now, let us consider rearrangement operations. Generally speaking, such an
operation applied to two adjacencies or telomeres of a genome disconnects the
incident edges of the genome graph, and reconnects them in one of the possible
other ways. More formally, given a graph with vertices of degree one (*external*
vertices) or degree two (*internal* vertices), we have:

Definition 3 ([4]). *The double cut and join (DCJ) operation acts on two vertices u and v of a graph with vertices of degree one or two in one of the following
three ways:*

*(a) If both $u = \{p, q\}$ and $v = \{r, s\}$ are internal vertices, these are replaced by
the two vertices $\{p, r\}$ and $\{s, q\}$ or by the two vertices $\{p, s\}$ and $\{q, r\}$.*
*(b) If $u = \{p, q\}$ is internal and $v = \{r\}$ is external, these are replaced by $\{p, r\}$
and $\{q\}$ or by $\{q, r\}$ and $\{p\}$.*
(c) If both $u = \{q\}$ and $v = \{r\}$ are external, these are replaced by $\{q, r\}$.

*In addition, as an inverse of case (c), a single internal vertex $\{q, r\}$ can be
replaced by two external vertices $\{q\}$ and $\{r\}$.*

Given two genomes A and B, the *DCJ distance* denoted by $d_{DCJ}(A, B)$ is the
minimum number of DCJ operations necessary to transform genome A into
genome B. Thus, we can formulate the following problem:

The Genome Halving Problem. Given a rearranged duplicated genome A,
find a perfectly duplicated genome B such that the DCJ distance between A
and B is minimal.

To solve this problem, we will construct another graph in the next section.
Again, the graph is defined on the adjacencies and telomeres of A, but this time
it represents the relation between paralogous extremities.

3   Natural Graphs

Let us consider a duplicated genome A with n genes each present in two copies. Assume that the two paralogs of every gene are assigned arbitrarily.

Definition 4. *The* natural graph $NG(A)$ *is a graph whose vertices are the adjacencies and telomeres of A and, for each extremity, the two paralogous extremities are connected by an edge, i.e. two vertices u and v are connected if $p \in u$ and $\bar{p} \in v$.*

Observe that the total number of edges in the graph equals two times the number of genes. The natural graph of genome A_1 from Example 1 is given in Fig. 2.

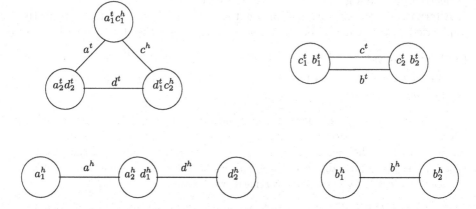

Fig. 2. Natural graph $N(A_1)$ of genome A_1 of Example 1

In a natural graph, by definition, every vertex has degree one or two. Thus, the natural graph consists only of cycles and paths.

Definition 5. *A cycle (or a path) with k edges, is a k-cycle (or k-path). If k is even, the cycle (or path) is called* even, *otherwise* odd.

Note that an adjacency $\{p, \bar{p}\}$ consisting of two paralogous extremities is a 1-cycle. The set of components of the natural graph can be partitioned into the following four disjoint subsets:

– EC := set of even cycles
– EP := set of even paths
– OC := set of odd cycles
– OP := set of odd paths

The following lemma is an immediate consequence of Lemma 1:

Lemma 2. *A genome A is perfectly duplicated if and only if all cycles in $NG(A)$ are 2-cycles and all paths in $NG(A)$ are 1-paths, i.e. $n = |EC| + |OP|/2$.*

4   Reconstructing an Ancestral Genome

In this section, we solve the genome halving problem by applying DCJ operations to the natural graph. This allows us to reconstruct a perfectly duplicated genome. We will first present our distance formula in Section 4.1 and then a linear time algorithm in Section 4.2.

4.1   Distance Formula

Consider a rearranged duplicated genome A. When a DCJ operation is applied to genome A, it acts on the adjacencies and telomeres of genome A. The same DCJ operation acts also on the natural graph $NG(A)$ since the adjacencies and telomeres of genome A are the vertices of this graph. Because the natural graph is a union of cycles and paths, all the properties of DCJ operations apply here as well, for instance: A DCJ operation can change the number of components only by one, as shown in [4]. Thus, we get a lower bound on the distance:

Lemma 3. *For a given genome A and any perfectly duplicated genome B over the same set of n genes, we have that*

$$d_{DCJ}(A, B) \geq n - \left(|EC| + \left\lfloor \frac{|OP|}{2} \right\rfloor \right).$$

In fact, there always exists a DCJ operation that increases either the number of even cycles or the number of odd paths. Thus, the distance decreases and the lower bound is strict as we see in the next theorem.

Theorem 1. *Let A be a rearranged duplicated genome with n genes each present in two copies, then the minimal distance between A and any perfectly duplicated genome B equals*

$$\min_{B} d_{DCJ}(A, B) = n - \left(|EC| + \left\lfloor \frac{|OP|}{2} \right\rfloor \right)$$

where $|EC|$ is the number of even cycles and $|OP|$ is the number of odd paths in the natural graph $NG(A)$.

Proof. We explain how to find a sequence of DCJ operations that achieves the lower bound of Lemma 3.

Let J, K, L and M be the total number of edges in all even cycles, even paths, odd cycles and odd paths, respectively. Note that the number of genes equals half of the total number of edges in $NG(A)$, i. e. $n = (J + K + L + M)/2$.

Consider a connected component G of $NG(A)$.

1. If G is an even j-cycle, we can create $\frac{j}{2}$ 2-cycles with $\frac{j}{2} - 1$ DCJ operations. Thus, for $|EC|$ even cycles with J edges in total, we need $\frac{J}{2} - |EC|$ DCJ operations to create $\frac{J}{2}$ 2-cycles.

2. If G is an even k-path, we can create $\frac{k}{2}$ 2-cycles with $\frac{k}{2}$ DCJ operations. Thus, for $|EP|$ even paths with K edges in total, we need $\frac{K}{2}$ DCJ operations to create $\frac{K}{2}$ 2-cycles.

3. If $|OP|$ is even, then $|OC|$ is also even.

   (a) If G is an odd l-cycle, we can create $\frac{l-1}{2}$ 2-cycles and one 1-cycle with $\frac{l-1}{2}$ DCJ operations. Thus, for $|OC|$ odd cycles with L edges in total, we need $\frac{L-|OC|}{2}$ DCJ operations to create $\frac{L-|OC|}{2}$ 2-cycles and $|OC|$ 1-cycles. We can choose two 1-cycles and create one 2-cycle. Since $|OC|$ is even, we can thus create $\frac{|OC|}{2}$ 2-cycles with $\frac{|OC|}{2}$ DCJ operations. Thus, in total we need $\frac{L-|OC|}{2} + \frac{|OC|}{2} = \frac{L}{2}$ DCJ operations.

   (b) If G is an odd m-path, we can create $\frac{m-1}{2}$ 2-cycles and one 1-path with $\frac{m-1}{2}$ DCJ operations. Thus, for $|OP|$ odd paths with M edges in total, we need $\frac{M-|OP|}{2}$ DCJ operations to create $\frac{M-|OP|}{2}$ 2-cycles and $|OP|$ 1-paths.

   Since L and M are even, summing up (a) and (b) gives us in total $\frac{L+M}{2} - \frac{|OP|}{2}$ DCJ operations.

4. If $|OP|$ is odd, then $|OC|$ is also odd.

   (a) If G is an odd l-cycle, we can create $\frac{l-1}{2}$ 2-cycles and one 1-cycle with $\frac{l-1}{2}$ DCJ operations. Thus, for $|OC|$ odd cycles with L edges in total, we need $\frac{L-|OC|}{2}$ DCJ operations to create $\frac{L-|OC|}{2}$ 2-cycles and $|OC|$ 1-cycles. We can choose two 1-cycles and create one 2-cycle. Since $|OC|$ is odd, there is one remaining 1-cycle that can be transformed into a 1-path by one extra DCJ operation. Thus, in total we need $\frac{L-|OC|}{2} + \frac{|OC|-1}{2} + 1 = \frac{L+1}{2}$ DCJ operations.

   (b) If G is an odd m-path, we can create $\frac{m-1}{2}$ 2-cycles and one 1-path with $\frac{m-1}{2}$ DCJ operations. Thus, for $|OP|$ odd paths with M edges in total, we need $\frac{M-|OP|}{2}$ DCJ operations to create $\frac{M-|OP|}{2}$ 2-cycles and $|OP|$ 1-paths.

   Since L and M are odd, summing up (a) and (b) gives us in total $\frac{L+1}{2} + \frac{M-|OP|}{2} = \frac{L+M}{2} - \frac{|OP|-1}{2}$ DCJ operations.

By bringing together the results, the distance formula follows. $\qquad\square$

4.2   Algorithm

In this section, we show how the distance computation as well as an algorithm for reconstructing an ancestral genome can be implemented in linear time. Based on the proof of Theorem 1, our strategy for reconstructing a perfectly duplicated genome is the following:

1. Construct the natural graph.
2. Maximize the number of even cycles and odd paths in the natural graph.
3. Reconstruct the perfectly duplicated genome from the resulting natural graph.

The natural graph can easily be constructed in $O(n)$ time and $O(n)$ space if we store the information about the adjacencies and the telomeres in two tables. The first table represents the vertices of the natural graph. Each of its entries contains one or two extremities, depending whether it represents an adjacency or a telomere. The edges can be obtained from the second table that stores for each paralogous extremity the index of the vertex that contains it. The two tables for genome A_1 of Example 1 are given in Tables 1 and 2. Thus, the natural graph $NG(A_1)$ has 10 vertices and 8 edges, for example one edge joining vertex 10 with vertex 3, another edge joining vertex 9 with vertex 2, and so on.

Table 1. Table storing the adjacencies and telomeres of genome A_1. Adjacencies have two entries, telomeres just one.

	1	2	3	4	5	6	7	8	9	10
first	d_2^h	d_2^t	a_2^h	d_1^t	c_2^t	b_2^h	b_1^h	b_1^t	c_1^h	a_1^h
second	$-$	a_2^t	d_1^h	c_2^h	b_2^t	$-$	$-$	c_1^t	a_1^t	$-$

Table 2. Table storing for each gene in A_1 the location of its head and its tail in Table 1

	a_1	a_2	b_1	b_2	c_1	c_2	d_1	d_2
head	10	3	7	6	9	4	3	1
tail	9	2	8	5	8	5	4	2

Using these tables, the connected components can be computed in linear time, and thus the distance as given by Theorem 1.

In order to reconstruct a perfectly duplicated genome, we maximize the number of even cycles and odd paths in the natural graph. This is done by Algorithm 1, following the idea used in the proof of Theorem 1. By marking each adjacency of Table 1, Algorithm 1 can be implemented in linear time. The adjacencies are processed in left-to-right order and each time an unmarked adjacency is detected, all adjacencies on its path or cycle are marked and transformed into 2-cycles and 1-paths by successively applying DCJ operations. Note that, by applying a DCJ operation, at most 4 entries in each of the two tables have to be updated. Eventually, all cycles are 2-cycles and all paths are 1-paths and a perfectly duplicated genome can be obtained as follows: By ignoring the assignment of the paralogs, each 2-cycle consists of two adjacencies of the form $\{u^x, v^y\}$, where $x, y \in \{t, h\}$, and each 1-path connects two telomeres of the form u^x, where $x \in \{t, h\}$. Thus, a perfectly duplicated genome can be reconstructed by replacing each 2-cycle by the adjacency $\{u^x, v^y\}$ and each 1-path by the telomere u^x. So, the overall running time of the algorithm for reconstructing a perfectly duplicated genome is linear.

5   A Note on the Warren-Sankoff Formula

In [14], Warren and Sankoff consider a more general genome model where the ancestral genome has to be neither circular-perfectly duplicated, nor linear-perfectly duplicated. Therefore, we will use the notion *general-perfectly dupli-cated* in order to distinguish it from our definition of a perfectly duplicated genome. More precisely, a genome is general-perfectly duplicated if and only if for each adjacency $\{u, v\}$ in A, also $\{\overline{u}, \overline{v}\}$ is in A, and for each telomere $\{u\}$ in A,

---

Algorithm 1. Reconstruction of a perfectly duplicated genome

1: Construct $NG(A)$, the natural graph of genome A
2: **while** there exists a k-path with $k > 1$ **do**
3:    Create a 2-cycle (and a $(k-2)$-path if $k > 2$)
4: **end while**
5: /* all remaining paths have length 1 */
6: **while** there exists a k-cycle with $k > 2$ **do**
7:    Create a 2-cycle and a $(k-2)$-cycle
8: **end while**
9: /* all remaining cycles have length 1 or 2 */
10: **while** there exists a 1-cycle **do**
11:    **if** there exists another 1-cycle **then**
12:       Create a 2-cycle
13:    **else**
14:       Create a 1-path
15:    **end if**
16: **end while**

---

also $\{\bar{u}\}$ is in A. Observe that, in contrast to our definition, a general-perfectly duplicated genome can have adjacencies of the type $\{u, \bar{u}\}$. For example, the following genome is general-perfectly duplicated, but not perfectly duplicated:

$$\bullet\!\!-\!\!\bullet\ b_1^h \quad b_1^t\ c_1^h \quad c_1^t\ c_2^t \quad c_2^h b_2^t \quad b_2^h \qquad\qquad d_2^t\, a_2^h \overset{\textstyle a_2^t\, a_1^t}{\underset{\textstyle d_2^h\, d_1^h}{\bigcirc}} a_1^h\, d_1^t$$

Now, let us denote by $d_{DCJ}^{general}(A, B)$ the minimum number of DCJ operations needed to transform a rearranged duplicated genome A into a general-perfectly duplicated genome B. By showing an upper and a lower bound, Warren and Sankoff finally claim that

$$\min_B d_{DCJ}^{general}(A, B) = n - \left(|EC| + |OP| + \left\lfloor \frac{|OC|}{2} \right\rfloor\right).$$

As a counterexample, consider a genome with just one gene a. Assume that the genome has two linear chromosomes, each consisting of one paralog a_1 and a_2. Note that the genome is general-perfectly duplicated and the natural graph has two paths of length one. Thus, the distance should be zero, but the above formula gives us

$$n - |OP| = 1 - 2 = -1.$$

Even though their distance formula is formulated in terms also defined in the natural graph, Warren and Sankoff follow a different approach. Therefore, instead of using their techniques, we will present in the following a correction of their result by modifying our algorithm.

As mentioned above, the difference is that a general-perfectly duplicated genome may have adjacencies that correspond to 1-cycles in the natural graph. Thus, we have:

Lemma 4. *A genome A is general-perfectly duplicated if and only if all cycles in $NG(A)$ are 2-cycles or 1-cycles, and all paths in $NG(A)$ are 1-paths.*

As a consequence of this lemma, we do not have to apply DCJ operations in order to get rid of 1-cycles in the natural graph as in our genome model. Since there are at most $\lceil |OC|/2 \rceil$ such DCJ operations, one can easily show that

$$\min_B d_{DCJ}(A, B) = \min_B d_{DCJ}^{general}(A, B) + \left\lceil \frac{|OC|}{2} \right\rceil.$$

By this fundamental relation, one can derive the distance formula for the general DCJ model studied by Warren and Sankoff in [14]:

Theorem 2. *Let A be a rearranged duplicated genome with $2n$ genes, then the minimal distance between A and any perfectly duplicated genome B equals*

$$\min_B d_{DCJ}^{general}(A, B) = n - (|EC| + \frac{|OP| + |OC|}{2})$$

where $|EC|$ is the number of even cycles, $|OC|$ the number of odd cycles and $|OP|$ the number of odd paths in the natural graph $NG(A)$.

It should be mentioned that an optimal algorithm for reconstructing a general-perfectly duplicated genome is obtained by just removing the last while-loop in our Algorithm 1.

6   Conclusion and Open Questions

In this paper, we solve the Genome Halving Problem for the DCJ distance under a general genome model with coexisting circular and linear chromosomes. Surprisingly, this can be done by working directly on the natural graph — all other graphs that are typically used in this context are bypassed. Moreover, our approach is also able to describe alternative genome models such as the one presented by Warren and Sankoff. Thus, our genome model represents a firm starting point for further studies and variants of the Genome Halving Problem.

One direction is to consider a more general set of rearrangement operations, the so-called *multi-break* rearrangements. By this generalization, a DCJ operation is equivalent to a 2-break operations and transpositions can be modelled by a 3-break operation instead of two DCJ operations as in our model. Therefore, the results of [3] can be extended to genomes with linear and circular chromosomes.

Finally, one can consider duplicated genomes with a higher multiplicity of each gene. This extension yields a natural graph with vertices of degree greater than two. It would have to be studied whether the DCJ operation can also be used on such a graph and how to partition the connected components.

Acknowledgments

The author would like to thank Jens Stoye and Robert Warren for helpful discussions. The anonymous reviewers gave valuable hints how to improve the paper.

References

1. Ahn, S., Tanksley, S.D.: Comparative Linkage Maps of Rice and Maize Genomes. Proc. Natl. Acad. Sci. 90(17), 7980–7984 (1993)
2. Alekseyev, M., Pevzner, P.: Whole Genome Duplications and Contracted Breakpoint Graphs. SIAM J. Comput. 36(6), 1748–1763 (2007)
3. Alekseyev, M., Pevzner, P.: Whole Genome Duplications, Multi-break Rearrangements, and Genome Halving Problem. In: Proceedings of SODA 2007, pp. 665–679 (2007)
4. Bergeron, A., Mixtacki, J., Stoye, J.: A Unifying View of Genome Rearrangements. In: Bücher, P., Moret, B.M.E. (eds.) WABI 2006. LNCS (LNBI), vol. 4175, pp. 163–173. Springer, Heidelberg (2006)
5. Dehal, P., Boore, J.L.: Two Rounds of Whole Genome Duplication in the Ancestral Vertebrate. PLoS Biology 3(10), 314 (2003)
6. El-Mabrouk, N.: Reconstructing an Ancestral Genome Using Minimum Segments Duplications and Reversals. J. Comput. Syst. Sci. 65(3), 442–464 (2002)
7. El-Mabrouk, N., Bryant, D., Sankoff, D.: Reconstructing the Pre-doubling Genome. In: Proceedings of RECOMB 1999, pp. 154–163 (1999)
8. El-Mabrouk, N., Nadeau, J., Sankoff, D.: Genome Halving. In: Farach-Colton, M. (ed.) CPM 1998. LNCS, vol. 1448, pp. 235–250. Springer, Heidelberg (1998)
9. El-Mabrouk, N., Sankoff, D.: The Reconstruction of Doubled Genomes. SIAM J. Comput. 32(3), 754–792 (2003)
10. Guyot, R., Keller, B.: Ancestral Genome Duplication in Rice. Genome 47, 610–614 (2004)
11. Hannenhalli, S., Pevzner, P.: Transforming Men into Mice (polynomial Algorithm for Genomic Distance Problem). In: Proceedings of FOCS 1995, pp. 581–592. IEEE Press, Los Alamitos (1995)
12. Kellis, M., Birren, B.W., Lander, E.S.: Proof and Evolutionary Analysis of Ancient Genome Duplication in the Yeast Saccharomyces Cerevisiae. Nature 428(6983), 617–624 (2004)
13. Ohno, S.: Ancient Linkage Group and Frozen Accidents. Nature 244, 259–262 (1973)
14. Warren, R., Sankoff, D.: Genome Halving with Double Cut and Join. In: Proceedings of APBC 2008, Series on Advances in Bioinformatics and Computational Biology, vol. 6 (2008)
15. Yancopoulos, S., Attie, O., Friedberg, R.: Efficient Sorting of Genomic Permutations by Translocation, Inversion and Block Interchange. Bioinformatics 21(16), 3340–3346 (2005)

Haplotype Inferring Via Galled-Tree Networks Is NP-Complete

Arvind Gupta*, Ján Maňuch**, Ladislav Stacho***, and Xiaohong Zhao†

School of Computing Science and Department of Mathematics
Simon Fraser University, Canada
{arvind,jmanuch,lstacho,xzhao2}@sfu.ca

Abstract. The problem of determining haplotypes from genotypes has gained considerable prominence in the research community since the beginning of the HapMap project. Here the focus is on determining the sets of SNP values of individual chromosomes (haplotypes), since such information better captures the genetic causes of diseases. One of the main algorithmic tools for haplotyping is based on the assumption that the evolutionary history for the original haplotypes satisfies perfect phylogeny. The algorithm can be applied only on individual blocks of chromosomes, in which it is assumed that recombinations either do not happen or happen with small frequencies. However, exact determination of blocks is usually not possible. It would be desirable to develop a method for haplotyping which can account for recombinations, and thus can be applied on multiblock sections of chromosomes. A natural candidate for such a method is haplotyping via phylogenetic networks or their simplified version: galled-tree networks, which were introduced by Wang, Zhang, Zhang ([25]) to model recombinations. However, even haplotyping via galled-tree networks appears hard, as the algorithms exist only for very special cases: the galled-tree network has either a single gall ([23]) or only small galls with two mutations each ([8]). Building on our previous results ([6]) we show that, in general, haplotyping via galled-tree networks is NP-complete, and thus indeed hard.

1 Introduction

With the completion of the Human Genome project, research has focused on the problem of determining variations in chromosomes among whole human population. This body of work is now encompassed in the international HapMap project [24],[3],[21]. Genetic variations, in particular SNPs (single nucleotide polymorphisms) are already playing a central role in determining the genetic causes of diseases and in designing individualized medicine [4],[5],[16],[20]. For complex diseases (those affected by more than a single gene), it is much more informative to have haplotype data (a set of SNP values on an individual chromosome)

---

* Research supported in part by NSERC grant.
** Research supported in part by MITACS.
*** Research supported in part by NSERC grant.
† The authors are listed alphabetically.

X. Hu and J. Wang (Eds.): COCOON 2008, LNCS 5092, pp. 287–298, 2008.

than the individual SNPs. However, experimental methods only allow for cost-effective determination of genotype information (the combined information of haplotypes for pairs of chromosomes) [18], and so the problem of computationally determining haplotypes from genotypes arises.

Various methods can be used to infer haplotypes from genotypes for population data. The first heuristic algorithm for computational haplotype inference was designed by Clark [2]. The exact version of Clark's problem was shown to be NP-hard [9]. Another approach, called pure-parsimony haplotyping, asking for a solution with the minimum number of distinct haplotypes, was shown to be NP-hard as well [11],[17]. Gusfield [10] developed the first exact polynomial algorithm based on the assumption of no recombinations happened during the evolutionary history of the haplotypes in consideration, which allowed him to make effective use of phylogenetic trees. This assumption was justified by experimental results that show many chromosomes are blocky with a strong correlation between sites on the same block [4],[20]. As such these experiments do not exclude recombinations within a block, models that allow for recombinations are needed.

The first attempt in haplotyping via models which allow a limited number of biological events that violate the perfect phylogeny model was taken in [23]. In this paper a polynomial algorithm for haplotyping via imperfect phylogenies with a single homoplasy was presented, as well as a practical algorithm for haplotyping via galled-tree networks with one recombination cycle (gall). Galled-tree networks are special instances of phylogenetic networks which in turn generalize phylogenetic trees by incorporating recombinations in the model [25]. There is always a phylogenetic network for any set of haplotypes, while finding such phylogenetic networks with the smallest number of recombinations is NP-hard [25],[1], and hence, haplotyping via phylogenetic networks is either easy and meaningless (any inferring is good) or intractable, depending on whether the minimum number of recombinations is required.

A galled-tree network is a special type of phylogenetic network in which recombination cycles do not intersect. Similar to phylogenetic trees, not every set of haplotypes admits a galled-tree network, however it can be decided in polynomial time whether it is the case [14]. In addition, if there is a galled-tree network, it is easy to find the one (reduced GTN) with the smallest number of recombinations, and no phylogenetic network for the same set of haplotypes has fever recombinations. In earlier work [7] we found a characterization of the existence of galled-tree networks. A similar characterization was independently discovered in [22]. Building on this characterization, we developed a polynomial algorithm for haplotype inference via galled-tree networks with galls having two mutations based on reduction of haplotyping problem to a hypergraph covering problem in [8]. It is very natural to ask whether the assumption on the number of galls or the size of galls can be dropped and still hope for a polynomial algorithm. In [6] we have reduced the haplotype inferring problem to a hypergraph covering problem for genotype matrices satisfying a combinatorial condition. Building on this work we show that the problem of inferring haplotypes via galled-tree networks is NP-complete by reduction from 3-SAT.

2   Definitions

2.1   Haplotype Inferring from Population Data

Single Nucleotide Polymorphisms (SNPs) are the most frequent form of human genetic variations. A set of SNP values (e.g., SNPs that sit on a gene) on a single chromosome is called a *haplotype*. SNPs usually take two values among all the human population. Therefore, haplotypes are commonly represented as sequences of 0 and 1, by fixing a mapping of $\{0, 1\}$ to two possible states in $\{A, C, G, T\}$ at each SNP position. A combined information from two haplotypes for a matching pair of chromosomes is called a *genotype*. Here, the information about which value comes from the first and which from the second copy of the chromosome is lost. Genotype sequence is usually represented as a sequence of $\{0, 1, 2\}$, where value 0 or 1 at certain position i represents the fact that both haplotypes have this value at i (*homozygous*), while value 2 means that the values on two haplotypes at position i differ (*heterozygous*). The *haplotype inference* problem, or simply *haplotyping*, asks for determining of haplotype sequences based on genotype sequences of a set of individuals:

Definition 1 (Haplotyping). *Given a genotype $n \times m$ matrix A with values $\{0, 1, 2\}$, we say that a haplotype $2n \times m$ matrix B with values in $\{0, 1\}$ is inferred from A if and only if for every SNP $c \in \{1, \ldots, m\}$,*

- *if $A(i, c) \in \{0, 1\}$, then $B(2i - 1, c) = B(2i, c) = A(i, c)$; and*
- *if $A(i, c) = 2$, then $B(2i - 1, c) \neq B(2i, c)$.*

Obviously, there is exponentially many ways in the number of 2's in a row how to infer two haplotypes from this row. Therefore, various types of parsimonious criteria are used to choose the most plausible inferring of the whole set of genomes, including maximum resolution problem of Clark, pure parsimony criteria, haplotyping via perfect phylogeny and several statistical methods, cf. [15] for an overview. In this paper, we are interested in haplotyping via galled-tree networks which allow for recombination events, defined in the next subsection.

2.2   Phylogenetic and Galled-Tree Networks

In phylogenetic trees, each vertex is labeled by a sequence of states of characters (e.g., SNPs) and is connected by a mutation edge to its parent along which one character changes its state. Phylogenetic networks introduced in [25] (sometimes called "recombination networks"), are an extension of phylogenetic trees in which a vertex can be connected by two recombination edges to two parents and the label sequence for this recombination vertex is formed by a recombination of sequences of its two parents.

Definition 2 (Phylogenetic network). *A phylogenetic network N on m characters is a directed acyclic graph containing exactly one vertex (the root) with no incoming edges, and each other vertex has either one incoming (mutation)* *edge or two incoming (mutation) edges. A vertex x with two incoming edges is called a* recombination vertex.

Each integer (character) from 1 to m is assigned to exactly one mutation edge in N and each mutation edge is assigned one character. Each vertex in N is labeled by a binary sequence of length m, starting with the root vertex which is labeled with the all-0 sequence. Since N is acyclic, all other vertices in N can be recursively labeled as follows:

- *For a non-recombination vertex v, let e be the mutation edge labeled c coming into v. The label of v is obtained from the label of v's parent by changing the value at position c from 0 to 1.*
- *Each recombination vertex x is associated with an integer $r_x \in \{2, \ldots, m\}$, called the* recombination point *for x. Label the two recombination edges coming to x, P and S, respectively. Let $P(x)$ $(S(x))$ be the sequence of the parent of x on the edge labeled P (S). Then the label of x is a recombination of labels of its parents: concatenation of the first $r_x - 1$ characters of $P(x)$ (prefix), followed by the last $m - r_x + 1$ characters of $S(x)$ (suffix).*

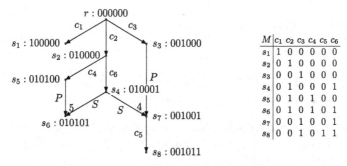

Fig. 1. A phylogenetic network for matrix M. In the network, each mutation edge is labeled by a character c_i, $i \in \{1, \ldots, 6\}$; recombination edges are labeled by P and S respectively; the integer label above each recombination vertex represents the recombination point.

In this paper, the sequence at the root of the phylogenetic network is always the all-0 sequence, and all results are relative to that assumption. More general phylogenetic networks with unknown root were studied in a recent paper by Gusfield [12]. A phylogenetic network for a given binary matrix M is illustrated in Figure 1.

Definition 3. *Given an $n \times m$ matrix A with values in $\{0, 1\}$, we say that a phylogenetic network N with m characters* explains A *if each sequence of A is a label of some vertex in N.*

Finding a phylogenetic network with the minimum number of recombination vertices for a given haplotype matrix is NP-hard [25],[1]. Hence, a more restricted version of phylogenetic networks was studied in several papers [13],[14],[25]. The restricted version can be defined as follows.

Definition 4 (Galled-tree network). *In a phylogenetic network N, let v be a vertex that has two paths out of it that meet at a recombination vertex x (v is the least common ancestor of the parents of x). The two paths together form a recombination cycle C. The vertex v is called the* coalescent *vertex. We say that C contains a character i, if i labels one of the mutation edges of C.*

A phylogenetic network is called a galled-tree *network if no two recombination cycles share an edge. A recombination cycle of a galled-tree network is sometimes referred to as a* gall.

Note that the example in Figure 1 is not a galled-tree network as two galls share an edge.

2.3 Inferring Haplotypes Via Galled-Tree Network and the Extended Hypergraph Covering Problem

In this section we will recall the characterization for the galled-tree network haplotyping (GTNH) problem using a hypergraph covering problem developed in[6]. This characterization works only for genotype matrices satisfying special combinatorial properties. To state the result of [6], we need the following definitions.

Definition 5. *Given a genotype matrix A, we say that A can be* explained *by a galled-tree network if there exists a haplotype matrix B inferred from A such that B can be explained by a galled-tree network.*

*Problem 1 (**Galled-Tree Network Haplotyping (GTNH) Problem**).* Given a genotype matrix A, decide if A can be explained by a galled-tree network.

Next, we will give the definitions of the combinatorial properties of genotype matrices used in [6].

Definition 6 (Simple genotype matrix). *We say that a genotype matrix is* simple *if every row contains either zero or three 2's.*

Definition 7. *Given a genotype matrix A, for every $x, y \in \{0, 1\}$, we say that a pair of columns c_1, c_2 induces $[x, y]$ in A, if A contains at least one of the pairs $[x, y]$, $[2, y]$ and $[x, 2]$ in columns c_1 and c_2. Similarly, a triple of columns c_1, c_2, c_3 induces $[x, y, z]$ in A, if A contains at least one of the triples $[x, y, z]$, $[2, y, z]$, $[x, 2, z]$ and $[x, y, 2]$ in columns c_1, c_2, c_3.*

Definition 8 (Weak diagonal (WD) property). *Given a genotype matrix A, we say that a pair of SNPs is* active *if it contains $[2, 2]$, or it induces all three pairs $[1, 1]$, $[0, 1]$ and $[1, 0]$. Further, we say that a pair c_1, c_2 is* weakly active *if either it is active, or if there is an SNP c_3 such that c_1, c_3 and c_2, c_3 are both active pairs. We say that A has the* weak diagonal *property if every weakly active pair of SNPs induces both $[0, 1]$ and $[1, 0]$.*

For purposes of this paper and clarity, we will consider the following (stripped) version of the extended genotype hypergraph.

Definition 9 (Extended genotype hypergraph (EGH)). *An extended genotype hypergraph is a hypergraph with hyperedges containing either two or three vertices, and a list of switches, which are ordered triples of vertices.*

Given a simple $n \times m$ genotype matrix A, the extended genotype hypergraph H_A *of A has the set of SNPs (columns) $\{1, \ldots, m\}$ as a vertex set. Hypergraph H_A contains only the following hyperedges:*

- *for every row r of A containing three 2's, say in columns c_1, c_2, c_3 there is a hyperedge $e_r = \{c_1, c_2, c_3\}$; and*
- *for every two columns c_1 and c_2 inducing $[0, 1]$, $[1, 0]$ and $[1, 1]$ in A, there is a hyperedge $\{c_1, c_2\}$ in H_A.*

Furthermore, for every triple of SNPs c_1, c_2, c_3 such that there are distinct hyperedges e and e' such that $c_1, c_2 \in e$ and $c_2, c_3 \in e'$ and the triple induces $[0, 1, 0]$ in A, H_A contains a switch $[c_1, c_2, c_3]$.

The following definition defines a graph covering of an extended genotype hypergraph.

Definition 10 (Covering of EGH). *Consider an extended genotype hypergraph H. We say that a graph G with the same vertex set as H covers H if G can be obtained as follows:*

- *for every 2-edge $\{c_1, c_2\}$ of H_A, add the edge (c_1, c_2) in G;*
- *for every 3-edge $\{c_1, c_2, c_3\}$ of H_A, add exactly one of the edges (c_1, c_2), (c_2, c_3) and (c_1, c_3) to G; and*

and for every switch $[c_1, c_2, c_3]$, at most one of the edges (c_1, c_2) and (c_2, c_3) is in G. A graph G that covers H is called a covering *of H.*

Finally, the EHC problem can be formulated as follows:

*Problem 2 (**Extended Hypergraph Covering (EHC) Problem**). Given an extended genotype hypergraph H, determine whether there is a covering G of H such that each connected component K of G is a path of length at most 3 satisfying the* ordered component property:

- *K is bipartite with partitions L and R such that all vertices in L are smaller than all vertices in R. Recall vertices of G and H are integers from 1 to m.*

The following characterization was shown in [6].

Theorem 1 ([6]). *Consider a simple genotype matrix with the WD property. Then A can be explained by a galled-tree network if and only if there exists a covering G of H_A such that every component of G is a path of length at most 3 and has the ordered-component property.*

Note that the WD property of the genotype matrix forces the components in the conflict graphs of inferred haplotype matrices to be small. In [6], it was also shown that the EHC problem is NP-complete, hence this characterization fails to provide a polynomial solution for the GTNH problem even for such special genotype matrices. On the other hand, since not every extended genotype hypergraph has a corresponding genotype matrix, in particular, the gadgets used to show NP-completeness of the EHC problem in [6] do not have a corresponding genotype matrix, this result does not imply that the GTNH problem is NP-complete. In the next section, we consider special instances of extended genotype hypergraphs for which, as we will see later, it is possible to construct a corresponding genotype matrix and show that the EHC problem for them remains NP-complete.

3   GTNH Problem Is NP-Complete

The proof of NP-completeness is done in two steps. First, we define special instances of extended genotype hypergraphs and show that the EHC problem for them is NP-complete. Then we show that it is possible to construct a genotype matrix for each such instance. The NP-completeness of the GTNH problem then follows by the characterization obtained in [6] (Theorem 1).

Definition 11 (Natural EGH). *We say that an extended genotype hypergraph H is natural if for any two hyperedges e, e' of H, $|e \cap e'| \leq 1$ and the list of switches contains all and only the following switches: for every vertex c of H with degree at least 3, and for every two hyperedges e_1 and e_2 containing c, and for every two vertices $c_1 \in e_1$ and $c_2 \in e_2$ such that c, c_1, c_2 are all distinct, there is a switch $[c_1, c, c_2]$.*

We will show that the extended genotype hypergraph covering problem for natural EGHs is NP-complete by reduction from 3-SAT. The proof follows the idea of the proof of NP-completeness of the EHC problem in [6]. However, the proof in [6], assumed that there are no switches in the EGH, which was the main reason why there was no corresponding genotype matrix for the constructed EGH. In the following proof, the gadgets had to be redesigned to take into account the existence of the switches.

Theorem 2. *The extended genotype hypergraph covering problem for natural extended genotype hypergraphs is NP-complete.*

Proof. The proof is done by a reduction from a special instance of the 3-SAT problem in which each clause contains two or three literals and every variable occurs in exactly three clauses — once positive and twice negated [19]. Let $f(x_1, x_2, \ldots, x_m) = C_1 \wedge \cdots \wedge C_k$ be such a formula in the conjunctive normal form, where $C_1, \ldots, C_k$ are the clauses. Let $p_1, \ldots, p_{3m}$ be the list of all occurrences of literals in f such that $p_{3i-2} = x_i$ and $p_{3i-1} = p_{3i} = \neg x_i$. Now every clause C_i can be written as

$$C_i = p_{s_{i,1}} \vee p_{s_{i,2}} \quad \text{or} \quad C_i = p_{s_{i,1}} \vee p_{s_{i,2}} \vee p_{s_{i,3}}$$

depending on whether C_i contains two or three literals.

Next, we construct a natural extended genotype hypergraph $H(f)$ for f which has a covering if and only if the formula f is satisfiable. The hypergraph $H(f)$ will be an edge-disjoint union of several gadgets, one for each clause and one for each variable. The only vertices in common among gadgets will be the literal vertices, in particular, each literal vertex will be shared between one clause gadget and one variable gadget. Furthermore, in each gadget we will mark every vertex either with a dot or a cross such that every literal vertex will be marked with a dot. This will guarantee that our marking will be consistent in whole $H(f)$. Using this marking we order the vertices of $H(f)$ such that every vertex marked with a dot precedes every vertex marked with a cross. This ordering implies that to verify the ordered component property of a covering it is enough to check that in such a covering every path of length two or three alternates between vertices with crosses and dots.

Now we define the gadgets; we start with the clause gadgets. Consider a covering c of $H(f)$. We say that a literal p_j has value 1 in this covering, if c restricted to the clause gadget containing p_j contains an edge incident to the vertex p_j. Note that this is well defined since for every literal vertex p_j there is a unique clause gadget containing it.

$$\begin{array}{cccc} \text{(a)} & \text{(b)} & \text{(c)} & \text{(d)} \end{array}$$

Fig. 2. The clause gadget for clause $C_i = p_{s_{i,1}} \lor p_{s_{i,2}}$ with two literals and all possible coverings. The coverings are depicted as solid edges joining particular pair of vertices. In (b), $p_{s_{i,1}}$ has value 1 and $p_{s_{i,2}}$ has value 0, in (c) $p_{s_{i,1}}$ has value 0 and $p_{s_{i,2}}$ has value 1, and in (d) both have value 1.

For every clause $C_i = p_{s_{i,1}} \lor p_{s_{i,2}}$ with two literals, we construct a gadget consisting of one 3-edge as depicted in Figure 2(a). Figure 2(b–d) shows all possible coverings of the gadget. Note that in each such covering, at least one of literals $p_{s_{i,1}}$ and $p_{s_{i,2}}$ has value 1.

For every clause $C_i = p_{s_{i,1}} \lor p_{s_{i,2}} \lor p_{s_{i,3}}$ with three literals, we construct a gadget consisting of one 2-edge and nine 3-edges as depicted in Figure 3(a). Figure 3(b–d) shows three possible coverings of the gadget, in which exactly one of the literals is set to 1. Note that in our proof, the restriction of a covering corresponding to a satisfiable assignment of the formula f to the gadget will be one of these three coverings. In any other covering of the gadget, an important property is that at least one of the literals has value 1. Indeed, assume that all three values are set to 0. Then in every covering of the gadget in which the condition for switches is satisfied, there is a path of length 4, cf. Figure 3(e), hence it is not a covering. This guarantees that in any covering of $H(f)$, in the corresponding assignment, every clause of f will be satisfied.

In the second part of the construction, for each variable x_i, we add a variable gadget which will guarantee that three occurrences of a variable x_i: p_{3i-2},

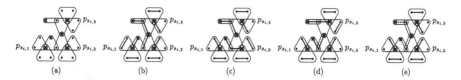

Fig. 3. (a) The part of hypergraph $\bar{H}(f)$ for clause $C_i = p_{s_{i,1}} \vee p_{s_{i,2}} \vee p_{s_{i,3}}$. (b–d) Three possible graph coverings, each representing one case how the clause can become satisfied. (e) The case when all three literals $p_{s_{i,1}}, p_{s_{i,2}}, p_{s_{i,3}}$ are set to 0 leads to a path of length 4 (dashed).

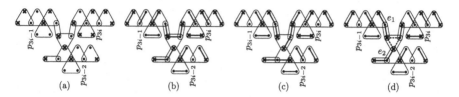

Fig. 4. (a) The part of hypergraph $H(f)$ verifying the values of three occurrences of a variable x_i. (b-c) Two possible coverings. In (b), we have depicted the unique covering if it is assumed that p_{3i-2} has value 1 (set by clause gadget). As can be seen, this forces values of p_{3i-1} and p_{3i} to 0 (in their clause gadgets). In (c), p_{3i-2} is forced to have value 0 (in its clause gadgets) and p_{3i-1} and p_{3i} can have arbitrary values. In (d), the case when p_{3i-2} and p_{3i-1} are set to 1 leads to a path of length 4 or 5 (depending on which of the dashed edges is chosen).

p_{3i-1}, p_{3i} must be assigned consistent values. That is if p_{3i-2} (positive occurrence) has value 1 then both p_{3i-1} and p_{3i} (negated occurrences) should have values 0, and if at least one of p_{3i-1} and p_{3i} has value 1 then p_{3i-2} should have value 0. This is achieved by a gadget consisting of three 2-edges and thirteen 3-edges depicted in Figure 4(a). Figures 4(b–c) show three possible coverings of the gadget. In these figures, a variable p_j has value 1 if no edge in the gadget joins p_j, which is in agreement with interpretation of values of p_i's in gadgets of the first part of construction.

Let us verify the claimed property of the gadget. Assume for instance that both p_{3i-2} and p_{3i-1} have value 1. No edge covering the variable gadget can be adjacent to any of these two vertices, otherwise the condition on switches (crossing from variable to clause gadgets) would be violated. Hence, the edges connecting the other two vertices of the 3-edges containing p_{3i-2} or p_{3i-1} have to be in the covering. Similarly, edges e_1, e_2 in Figure 4(e) have to be in the covering. Now, there is no edge to be selected to cover the 3-edge in the middle of the gadget, as the selection of any of the dashed edges would produce a path of length 4 or 5. The other cases can be proved using similar arguments.

Finally, we have to check that it is possible to find a covering of $H(f)$ which satisfies the conditions of the EHC problem (a solution to the EHC problem) if and only if f is satisfiable. First, consider a covering G that is the solution to the EHC problem for $H(f)$. For every clause C_i, at least one of $p_{s_{i,1}}, p_{s_{i,2}}$ (respectively, $p_{s_{i,1}}, p_{s_{i,2}}, p_{s_{i,3}}$) has value 1 in G. Let it be p_{q_i} (if there are several literals in C_i with value 1 in G, pick any of them). We will form a true assignment as follows. For every x_j, if there is $p_{q_i} = x_j$, set $x_j = 1$; if there is $p_{q_i} = \neg x_j$, set $x_j = 0$; otherwise set x_j to any value. As long as we guarantee that there are no i, i' such that $p_{q_i} = x_j$ and $p_{q_{i'}} = \neg x_j$, the above definition is correct and obviously is a true assignment for f. Assume that on the contrary, $p_{q_i} = x_j$ and $p_{q_{i'}} = \neg x_j$. Obviously, $p_{q_i} = p_{3j-2}$ and $p_{q_{i'}}$ is either p_{3j} or p_{3j-1}. Now, since p_{3j-2} has value 1 and one of p_{3j}, p_{3j-1} has value 1 in G, there is no valid covering of the variable gadget for x_j, a contradiction.

For the converse, consider a true assignment for f. For every clause $C_i = p_{s_{i,1}} \vee p_{s_{i,2}}$ with two literals, there is at least one literal in $q_i \in \{s_{i,1}, s_{i,2}\}$ with value 1 in this assignment. If it is $p_{s_{i,1}}$ (respectively, $p_{s_{i,2}}$), pick the covering of the clause gadget for C_i as depicted in Figure 2(b) (respectively, Figure 2(c)). Similarly, for every clause $C_i = p_{s_{i,1}} \vee p_{s_{i,2}} \vee p_{s_{i,3}}$ with three literals, pick the covering as depicted in Figures 3(b–d), depending on which literal has value 1 (if there are several pick one). For the variable gadgets we will select the coverings as follows. For every x_i, if value of x_i is 1, pick a hypergraph covering of the gadget for x_i depicted in Figure 4(b), and if value of $\neg x_i$ is 1, in Figure 4(d). Let G be the union of graphs that cover all gadgets. Now, we need to show that G satisfies the conditions of the EHC problem. The coverings of each gadgets satisfies these conditions. Since, the only common vertices among gadgets are literal vertices, which have degree 3, the switches with literal vertices in the middle forbid the connected components of the coverings of different gadgets to connect. Hence, G satisfies the conditions of the EHC problem as well.

The following lemma shows that for every natural EGH there is a corresponding simple genotype matrix with the WD property.[1]

Lemma 1. *For every natural EGH H, there is a simple genotype matrix A with the WD property such that $H_A = H$.*

The main result follows by Theorems 1 and 2, and Lemma 1.

Theorem 3. *The galled-tree network haplotyping problem is NP-complete. In addition, the problem remains NP-complete even if we require that each gall contains at most 4 mutation edges.*

4   Conclusions

We have shown that the GTNH problem is NP-hard in general. There are two very restricted instances for which the problem becomes tractable: the galled-tree networks with one gall and with simple galls, cf. [23] and [8], respectively. In the future work we would like to find less restricted instances solvable in polynomial time.

---

[1] Due to space limitation the proof of this lemma will appear in the full version.

References

1. Bordewich, M., Semple, C.: On the Computational Complexity of the Rooted Subtree Prune and Regraft Distance. Annals of Combinatorics 8, 409–423 (2004)
2. Clark, A.: Inference of Haplotypes from PCR-Amplified Samples of Dipoid Populations. Molecular Biology and Evolution 7, 111–122 (1990)
3. The International HapMap Consortium. A Haplotype Map of the Human Genome. Nature 437, 1299–1320 (2005)
4. Daly, M., Rioux, J., Schaffner, S., Hudson, T., Lander, E.: High-Resolution Haplotype Structure in the Human Genome. Nature Genetics 29(2), 229–232 (2001)
5. Gabriel, S., Schaffner, S., Nguyen, H., Moore, J., Roy, J., Blumenstiel, B., Higgins, J., DeFelice, M., Lochner, A., Faggart, M., Liu-Cordero, S., Rotimi, C., Adeyemo, A., Cooper, R., Ward, R., Lander, E., Daly, M., Altshuler, D.: The Structure of Haplotype Blocks in the Human Genome. Science 296 (2002)
6. Gupta, A., Manuch, J., Stacho, L., Zhao, X.: Haplotype Inferring via Galled-Tree Networks Using a Hypergraph Covering Problem for Special Genotype Matrices. Discr. Appl. Math. (to appear)
7. Gupta, A., Manuch, J., Stacho, L., Zhao, X.: Characterization of the Existence of Galled-Tree Networks. J. of Bioinform. and Comp. Biol. 4(6), 1309–1328 (2006)
8. Gupta, A., Manuch, J., Stacho, L., Zhao, X.: Algorithm for Haplotype Inferring via Galled-Tree Networks with Simple Galls (extended abstract). In: Istrail, S., Pevzner, P., Waterman, M. (eds.) ISBRA 2007. LNCS (LNBI), vol. 4463, pp. 121–132. Springer, Heidelberg (2007)
9. Gusfield, D.: Inference of Haplotypes from Samples of Diploid Populations: Complexity and Algorithms. J. Comp. Biology 8(3), 305–323 (2001)
10. Gusfield, D.: Haplotyping as Perfect Phylogeny: Conceptual Framework and Efficient Solutions. In: Proceedings of the Sixth Annual International Conference on Computational Biology (RECOMB 2002), pp. 166–175 (2002)
11. Gusfield, D.: Haplotype Inference by Pure Parsimony. In: Baeza-Yates, R., Chávez, E., Crochemore, M. (eds.) CPM 2003. LNCS, vol. 2676, pp. 144–155. Springer, Heidelberg (2003)
12. Gusfield, D.: Optimal, Efficient Reconstruction of Root-Unknown Phylogenetic Networks with Constrained and Structured Recombination. J. Comput. Syst. Sci. 70(3), 381–398 (2005)
13. Gusfield, D., Eddhu, S., Langley, C.: The Fine Structure of Galls in Phylogenetic Networks. INFORMS Journal on Computing 16(4), 459–469 (2004)
14. Gusfield, D., Eddhu, S., Langley, C.: Optimal, Efficient Reconstruction of Phylogenetic Networks with Constrained Recombination. Journal of Bioinformatics and Computational Biology 2(1), 173–213 (2004)
15. Gusfield, D., Orzack, S.H.: Handbook of Computational Molecular Biology, Chapter Haplotype Inference. CRC Computer and Information Science Series, p. 18C1C18C28. Chapman & Hall, Boca Raton (2005)
16. Helmuth, L.: Genome Research: Map of the Human Genome 3.0. Science 293(5530), 583–585 (2001)
17. Lancia, G., Pinotti, C., Rizzi, R.: Haplotyping Populations: Complexity and Aproximations. Dit-02-082, University of Trento (2002)
18. Mitra, R.D., Butty, V.L., Shendure, J., Williams, B.R., Housman, D.E., Church, G.M.: Digital Genotyping and Haplotyping with Polymerase Colonies. Proceedings of the Nationlal Academy of Sciences of the United States of America 100, 5926–5931 (2003)

19. Papadimitriou, C.H.: Computational Complexity. Addison-Wiesley Publishing Company, Inc. (1994)
20. Patil, N., Berno, A., Hinds, D., Barrett, W., Doshi, J., Hacker, C., Kautzer, C., Lee, D., Marjoribanks, C., McDonough, D., Nguyen, B., Norris, M., Sheehan, J., Shen, N., Stern, D., Stokowski, R., Thomas, D., Trulson, M., Vyas, K., Frazer, K., Fodor, S., Cox, D.: Blocks of Limited Haplotype Diversity Revealed by High-Resolution Scanning of Human Chromosome 21. Science 294(5547), 1719–1723 (2001)
21. Pennisi, E.: BREAKTHROUGH OF THE YEAR: Human Genetic Variation. Science 318(5858), 1842–1843 (2007)
22. Song, Y.S.: A Concise Necessary and Sufficient Condition for the Existence of a Galled-Tree. IEEE/ACM Transaction on Computational Biology and Bioinformatics 3(2), 186–191 (2006)
23. Song, Y.S., Wu, Y., Gusfield, D.: Algorithms for Imperfect Phylogeny Haplotyping (IPPH) with a Single Homoplasy or Recombination Event. In: Casadio, R., Myers, G. (eds.) WABI 2005. LNCS (LNBI), vol. 3692, pp. 152–164. Springer, Heidelberg (2005)
24. Thorisson, G., Smith, A., Krishnan, L., Stein, L.: The International HapMap Project Web Site. Genome Research 15, 1591–1593 (2005)
25. Wang, L., Zhang, K., Zhang, L.: Perfect Phylogenetic Networks with Recombination. Journal of Computational Biology 8(1), 69–78 (2001)

Adjacent Swaps on Strings

Bhadrachalam Chitturi[1], Hal Sudborough[1], Walter Voit[1], and Xuerong Feng[2]

[1] University of Texas at Dallas, Richardson, TX 75080, USA
{chalam,hal,wvoit}@utdallas.edu
[2] Valdosta State University, Valdosta, GA 31698, USA
xfeng@valdosta.edu

Abstract. Transforming strings by exchanging elements at bounded distance is applicable in fields like molecular biology, pattern recognition and music theory. A reversal of length two at position i is denoted by (i i+1). When it is applied to π, where $\pi = \pi_1, \pi_2, \pi_3, \ldots, \pi_i, \pi_{i+1}, \pi_n$, it transforms π to π', where $\pi' = \pi_1, \pi_2, \pi_3, \ldots, \pi_{i-1}, \pi_{i+1}, \pi_i, \pi_{i+1}, \ldots, \pi_n$. We call this operation an *adjacent swap*. We study the problem of computing the minimum number of adjacent swaps needed to transform one string of size n into another compatible string over an alphabet σ of size k, *i.e.* adjacent swap distance problem. $O(nlog_2n)$ time complexity algorithms are known for adjacent swap distance. We give an algorithm with $O(nk)$ time for both signed and unsigned versions of this problem where k is the number of symbols. We also give an algorithm with $O(nk)$ time for transforming signed strings with reversals of length up to 2, *i.e.* reversals of length 1 or 2.

1 Introduction

A gene is a code for a protein. A protein is made up of several *domains*. Each domain consists primarily of two types of *secondary structural elements*, namely, *helices* (α) and *strands* (β). Based on the type of secondary structural elements present and their arrangements, domains can be classified into four basic categories: (c1)α, (c2)β, (c3)$\alpha+\beta$, and (c4) α/β, where (c1) consists solely of helices, (c2) consists solely of strands, (c3) consists of both helices and strands where helices and strands are segregated within the domain, and (c4) consists of helices and strands that are intertwined. Thus, a gene (protein) is composed of a combination of 4 types of domains. The domains can be combined in a finite number of ways. Hence, genes can be categorized into finite structural categories.

Genes can also be categorized by function: (a) signaling, (b) metabolism, and (c) other functions. So, whether we use structural or functional categorization, a genome can be represented by a finite alphabet. The start of a gene is indicated by a start location. If an original genome (wild type) is read from left to right, the start location of each gene is at the left end. Some mutations cause the genome segment consisting of two adjacent genes to be involved in a reversal (of length 2). Such a mutation reverses the orientation of the two genes involved, motivating the study of signed adjacent swaps on strings. Likewise, in proteins,

X. Hu and J. Wang (Eds.): COCOON 2008, LNCS 5092, pp. 299–308, 2008.

the adjacent domains are swapped while retaining their orientation, mimicking a transposition of length 2. This motivates the study of adjacent swaps on unsigned strings. We solve the unsigned version of the problem, then also obtain a solution for the signed version. The final solution is applicable for the genes in a genome, whose count is in the order of 10^4.

An adjacent swap (j j+1) on a signed string $\alpha = \alpha_1, \ldots, \alpha_{j-1}, \alpha_j, \alpha_{j+1}, \ldots$ yields $\alpha_1, \alpha_2, \ldots, \alpha_{j-1}, -\alpha_{j+1}, -\alpha_j, \ldots$, where $-\alpha_j$ has the same symbol as α_j but the opposite sign. Where the sign denotes the orientation of the object that it represents. A positive sign indicates the normal orientation and the negative sign indicates the reverse orientation. We compute the minimum number of adjacent swaps or moves needed to transform one string of size n into another compatible string over an alphabet Σ of size k, i.e. the *adjacent swap distance*.

Hannenhalli and Pevzner [3] gave an $O(n^4)$ algorithm to optimally transform signed permutations by *unbounded* reversals. Several approximation algorithms are known for the unsigned version of this problem.

Sorting permutations of length n with reversals of size p was studied by Chen and Skiena [1]. They gave a complete characterization, for all n and p, of the number of equivalence classes of permutations of length n, under reversals of size p, for both permutations and circular permutations. They also gave an upper bound of $O(\frac{n^2}{p} + pn)$ and a lower bound of $\Omega(\frac{n^2}{p^2} + n)$ to sort all circular permutations of length n that can be sorted by reversals of length p.

In the literature, an adjacent swap has been called a mini-swap or a primitive-swap [4]. Toussaint [4] gives a simple method of calculating adjacent swap distance between compatible binary strings α, β in $O(n)$ time, denoted by $dswap(\alpha, \beta)$. If α and β both have p occurrences of 1, two new vectors of size p are created, say A and B, which contain the indices of 1's of α and β, respectively, in the correct order. Given such vectors, Toussaint [4] gives $dswap(\alpha, \beta) = \Sigma_{(1 \leq i \leq p)}(|A_i - B_i|)$.

Given a permutation π, two indices i and j $(i < j)$ have an inversion if $\pi_i > \pi_j$ (Cormen et al.[2]). The identity permutation of n symbols has no inversions. When an adjacent swap is performed over a pair of adjacent elements whose indices form an inversion, that inversion is eliminated. If the total number of inversions is t, then exactly t applications of (j j+1) eliminates all the inversions. Thus, the number of inversions in a permutation is equal to the minimum number of adjacent swaps required to sort that permutation. Cormen et al. [2] suggest a modification of mergesort to count the number of inversions in a permutation in $O(n log n)$ time. Due to a lack of space, some proofs are omitted.

2   Pairing Diagram

A sequence will be used to denote either a string or a permutation. Let the given source sequence be α and the destination sequence be β. The goal is to transform α into β with adjacent swaps. To achieve this, for each index of α its destination index is specified. This creates a one-to-one and onto mapping (bijection) from $(1, 2, \ldots, n)$ to $(1, 2, \ldots, n)$. If $\alpha = 3, 2, 1$ and $\beta = 2, 1, 3$ then the only possible mapping is $\{(1,3), (2,1), (3,2)\}$. If $\alpha = 1, 3, 2, 1$ and $\beta = 2, 1, 1, 3$, then there are two possible mappings $\{(1,2), (2,4), (3,1), (4,3)\}$ and $\{(1,3), (2,4), (3,1), (4,2)\}$, as the two 1's in α can be paired with the two 1's of β in either order. In general several mappings can lead to the same destination string.

Let $F = \{(i_1, j_1), (i_2, j_2), \ldots, (i_n, j_n)\}$ be a given mapping between a source sequence α and a destination sequence β. We construct a pairing diagram D_F as follows: (a) every element in α (and β) is a node, (b) the elements of α (top sequence) are arranged in order in a straight line starting with α_1 at the left end, and α_n at the right end, (c) the elements of β are arranged similarly, in order in a straight line, and (d) an edge is drawn between an element α_i and β_j, provided that (i, j) belongs to F. For any diagram D, let the intersection number of D, denoted by $I(D)$ be the number of intersections of edges of D.

Lemma 1. *For any mapping F, $I(D_F)$ adjacent swaps are necessary and sufficient to transform α into β when the transformation is done according to F.*

3   Adjacent Swaps on Strings

Let strings α and β of length n be defined on alphabet $\Sigma = \{0, 1, 2, \ldots, k-1\}$. Let $L_i(\alpha)$ and $L_i(\beta)$ $(0 \le i \le k-1)$ denote the vector of all indices with symbol i in strings α and β. Note that due to compatibility, for all i, $|L_i(\alpha)| = |L_i(\beta)|$. For example, if $\alpha = 0,1,2,3,0,1,2,3,0,1,2,3$, then $L_0(\alpha) = (1,5,9)$. Our algorithm calculates the number of moves necessary to transform α into β in $O(nk)$ time, which is linear for a fixed value of k.

The *optimum pairing* F^* of elements in α with the elements in β is defined by: for all i $(0 \le i \le k-1)$, if $L_i(\alpha) = (r1, r2, \ldots, rm)$ and $L_i(\beta) = (s1, s2, \ldots, sm)$, then add the pairs $(r1, s1), (r2, s2), \ldots, (rm, sm)$ to F^*. The optimum pairing diagram corresponds to the optimum pairing. We claim that among all mappings, the optimum pairing gives the minimum number of intersections. of vector Figure 1 shows the optimum pairing of the elements of α to the elements in β, where $n = 11$ and $k = 4$. The number of intersections of lines (14) in this diagram represents the number of adjacent swaps required to transform α into β.

Fig. 1. Optimum pairing for $k = 4$ and $n = 11$

A non-optimum pairing cannot yield the minimum number of intersections. Figure 2 is an illustration. Figure 2(a) shows a non-optimum pairing F. For example, F contains (2,5) instead of (2,2) and $I(F) = 6$. Figure 2(b) shows the optimum pairing F^* and $I(F^*) = 5$. In order to move the elements to their destination positions, F^* yields the following 5 moves: ((3 4), (2 3), (1 2), (4 5), (2 3)) and F yields the following 6 moves: ((3 4), (2 3), (1 2), (3 4), (2 3), (4 5)).

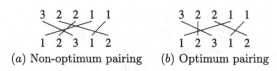

(a) Non-optimum pairing   (b) Optimum pairing

Fig. 2. Optimum and non-optimum pairing

Theorem 1. *The optimum pairing diagram for α and β yields the minimum number of intersections.*

Proof. **Basis:** When $n = 2$, if the elements are distinct then there is at most 1 intersection as dictated by the optimum pairing F^*.

Induction: Let the hypothesis hold for any two strings α_1 and β_1, where $|\alpha_1| = |\beta_1| = t$ $(1 \leq t \leq m)$. Now consider two strings α' and β' of length $m + 1$: If α' and β' end with the same symbol, the hypothesis trivially holds because the optimum pairing for the last elements of α' and β' produces no intersections and no additional adjacent swaps are necessary.

If α' and β' end with different symbols, without loss of generality assume α' ends with the symbol x and β' ends with symbol y $(x \neq y)$. Let for all j, $(m - p + 2 \leq j \leq m + 1)$, $\beta'_j \neq x$ and $\beta_{m-p+1} = x$. In that case optimum pairing will map α'_m to β_{m-p+1} and this forms the edge $(m+1, m-p)$ in D_{F^*}. Consider the p elements (nodes) $\beta'_{m-p+2...m+1}$, all and only the edges incident on these elements (nodes) and intersect with the edge $(m, m-p)$, creating p new intersections. The minimum increase in intersections is p, by the design of the optimum pairing. Let the string $\beta'_{1...m-p} \bullet \beta'_{m-p+1...m+1}$ be called β'', where $\bullet$ is the concatenation operator. Let the total number intersections in the optimum pairing F_1^* of $\alpha'_{1..m}$ and β'' be q. Then the total number of intersections in D_{F^*}, $I(D_{F^*}) = p+q$. Since, F_1^* is an optimum pairing for strings of length m, by the induction hypothesis q is the minimum possible value. By the design of the optimum pairing, the value of p is the minimum. Therefore, $I(D_{F^*})$ is the minimum possible value.

4 Algorithm to Transform Strings with Adjacent Swaps

For all i,j, the jth coordinate of $L_i(\alpha)$ is mapped to the jth coordinate of $L_i(\beta)$ to achieve the optimum pairing F^*. We use $2k$ queues to count the total number of edge intersections in $D(F^*)$, with just one scan across the strings α and β. For example, for $k = 3$ and $\Sigma = \{0, 1, 2\}$, there will be 6 queues: $q_{0\alpha}$, $q_{1\alpha}$, $q_{2\alpha}$, $q_{0\beta}$, $q_{1\beta}$, and $q_{2\beta}$. For each queue there will be 2 pointers. The tail pointer points to the most recent element added to the queue. The head pointer points to the next element to be removed from the queue (the oldest element in the queue). In addition, we maintain three global variables $Unmatched(\alpha)$, $Unmatched(\beta)$, and $intersection\_count$, which are initialized to 0. These variables track the total number of unmatched elements in queues of α and in the queues of β, and the number of intersections already discovered, respectively.

The elements of the strings are scanned in the following order α_1, β_1, α_2, $\beta_2, \ldots, \alpha_n$, β_n. If the current element from α (β) is a symbol x and $q_{x\beta}$ ($q_{x\alpha}$, respectively)is empty then we insert a vector of size $k - 1$ in the queue $q_{x\alpha}$ ($q_{x\beta}$, respectively). This vector contains the values of the tail pointers of all the other queues in α, $q_{y\alpha}$ ($x \neq y$). We do not need the tail pointer of the current symbol because there is no intersection of line segments joining the same symbols in optimum pairing. We call this a timestamp. It allows us to track when an element was added to its queue. In addition, $Unmatched(\alpha)$ ($Unmatched(\beta)$) is incremented by 1.

If the currently scanned element of α, say α_i, matches β_j (the element existing in a queue of β), then $j < i$ because we scanned α_i before β_i. In this case, α is the *current string* and β is the *alternate string*. If the currently scanned element of β, say β_j, matches α_i then $i \geq j$. In either case, the match represents an edge (i, j) in the $D(F^*)$. The edges that will eventually cross this edge should have one end to the left of (i, j) and the other to its right. We need the count of such edges which we denote by S_{ij}. We decrement $Unmatched(\beta)$ ($Unmatched(\alpha)$) by 1 because we made a match. Assume without loss of generality that a match is made when we scan a symbol β_j that matches α_i. The situation is symmetric if the currently scanned element is α_i and it is paired with β_j.

S_{ij} is composed of edges that originate from the elements that already exist in the queues and to the left of (i, j). These edges are incident on: (a) the current string and (b) the alternate string; so $S_{ij} = S_{ij}(current) + S_{ij}(alternate)$, where $S_{ij}(current)$ is the contribution of the current string to S_{ij} and $S_{ij}(alternate)$ is the contribution of the alternate string to S_{ij}.

The contribution of the current string to S_{ij} is straightforward to measure; if α is the current string then $S_{ij}(current) = Unmatched(\alpha)$. This is because the scanned element is the last element of the sequence so far, so all the unmatched elements on this string will eventually cross (i, j). The contribution of the alternate string to S_{ij} is measured by using the stored vector timestamps.

All the unpaired elements to the left of (i, j) will intersect the edge (i, j). Let the vector timestamp that was inserted for x be denoted by $inserted\_tail_x$, and let $inserted\_tail_{xy}$ denote the yth component of $inserted\_tail_x$, i.e. the value of the tail pointer of the queue representing element y when x is scanned (and hence $inserted\_tail_x$ is inserted into the appropriate queue). Let $Head_x$ and $Tail_x$ denote the values of the head pointer and the tail pointer, respectively, of the queue for element x. The total contribution of each queue (representing an element $\neq x$) of the alternate string to S_{ij} gives $S_{ij}(alternate)$.

We are interested in finding the unpaired elements that were introduced into the queues of the alternate string before β_j was introduced. The componenet $inserted\_tail_{\beta_{jy}}$ represents the value of the tail pointer of the queue for element y, at the time of insertion of vector $inserted\_tail_{\beta_j}$]. If the current head pointer of element y is greater than $inserted\_tail_{\beta_{jy}}$, then the elements of $\beta(\alpha)$ with value y that were scanned before β_j have already been paired and will not contribute to $S_{ij}(alternate)$. Otherwise, the pseudocode to compute S_{ij} is given below. Observe that intersection_count is a global variable and holds the running total for the total number of intersections in the pairing diagram.

$Total = 0;$
for all $p \neq x$ do
    if $(Head[p] > inserted\_tail_x][p])$ then continue;
    else Total = Total + $(Head[p]\text{-}inserted\_tail_x][p] + 1);$
end for;
$intersection\_count = intersection\_count + Total;$

After the count for the number of intersections is obtained, the vector times-tamps corresponding to the matched elements are removed. Let w be an element of β (or α) that contributed to S_{ij}. When a pairing occurs for w, the vector timestamps for β_j and α_i are no longer in the queues. This guarantees that each intersection is counted exactly once. As indicated, $S_{ij} = S_{ij}(current) + S_{ij}(alternate)$ is computed and is added to $intersection\_count$. If no match is made, we add the appropriate vector as described above, and continue to the next element given by the order $\alpha_1, \beta_1, \alpha_2, \beta_2, \ldots, \alpha_n, \beta_n$.

To illustrate, consider the example shown in Figure 1. We provide a description of an initial portion of our algorithm's execution on this example. The elements are scanned, as indicated, in the given zig-zag order. We show the new non-empty queue elements and the new non-zero variables:

(1) $\alpha_1 = A$ is scanned, $q_{A\alpha} = (0,0,0)$ and $Unmatched(\alpha) = 1$,
(2) $\beta_1 = T$ is scanned, $q_{T\beta} = (0,0,0)$ and $Unmatched(\beta) = 1$,
(3) $\alpha_2 = C$ is scanned, $q_{C\alpha} = (1,0,0)$ and $Unmatched(\alpha) = 2$ (Note: the first component of $q_{C\alpha}$ is 1 to denote that at the time it was inserted the queue $q_{A\alpha}$ had one unmatched element),
(4) $\beta_2 = T$ is scanned, $q_{T\beta} = \{(0,0,0),(0,0,0)\}$ and $Unmatched(\beta) = 2$,
(5) $\alpha_3 = G$ is scanned, $q_{G\alpha} = (1,1,0)$ and $Unmatched(\alpha) = 3$ (Note: the first two components of $q_{G\alpha}$ are 1 to denote that at the time it was inserted both of the queues $q_{A\alpha}$ and $q_{C\alpha}$ had one unmatched element).
(6) $\beta_3 = A$, and we make a match as $q_{A\alpha}$ is not empty, the vector $(0,0,0)$ is removed from $q_{A\alpha}$. As $Unmatched(\beta) = 2$, we add 2 to the $intersection\_count$ to denote two intersections from the β (current) string. As the vector $(0,0,0)$ from $q_{A\alpha}$ indicates the queues $q_{C\alpha}$, $q_{G\alpha}$, and $q_{T\alpha}$ were empty when it was added, it follows that no intersections come from the α (alternate) string. Both $Unmatched(\alpha)$ and $Unmatched(\beta)$ are then decremented by 1.
(7) $\alpha_4 = T$ is scanned, and we make a match as $q_{T\beta}$ is not empty, the vector $(0,0,0)$ is removed from $q_{T\beta}$. As $Unmatched(\alpha) = 3$, we add 3 to the global variable $intersection\_count$ to denote three intersections from the α (cur-rent) string. As the vector $(0,0,0)$ from $q_{T\beta}$ indicates the queues $q_{A\beta}$, $q_{C\beta}$, and $q_{G\beta}$ were empty when it was added, it follows that no intersections come from the β (alternate) string. Both $Unmatched(\alpha)$ and $Unmatched(\beta)$ are then decremented by 1.

The algorithm proceeds as indicated and terminates at β_{11}. Upon termination $intersection\_count$ holds the value of number of adjacent swaps necessary for the transformation. (In this example 14 swaps are necessary) .

When an element is inserted either an optimum pair is created or no optimum pairing is possible. In order to insert the vector timestamp for an element, $k-1$ queues are examined to get the tail pointer values, and a timestamp is inserted into the corresponding queue. Thus, the time consumed is $O(k)$. If no optimum pairing is possible, no further work is needed for this element. If an optimum pair exists, the current strings contribution, $Unmatched(current)$, is obtained in $O(1)$. We count the contribution of the alternate string by making $k-1$ comparisons in $O(k)$ time. Thus, the time consumed per element is $O(k)$. For all the $2n$ elements from α and β this yields $O(n*k)$. Likewise, when floating arrays are used, each element consumes $O(k)$ space, yielding $O(nk)$ space in total.

5   Transforming Signed Strings with Adjacent Swaps

In this section, we give a criteria for feasibility for signed string α to be transformed into a signed string β. If feasible, we give optimum algorithm for transformation by adjacent swaps. Every element of a signed string has a value (symbol) and a sign associated with it. Only the negative sign is explicitly shown. An adjacent swap (j j+1) on $\alpha = \alpha_1, \ldots, \alpha_{j-1}, \alpha_j, \alpha_{j+1}, \alpha_{j+2}, \ldots$ yields $\alpha_1, \alpha_2, \ldots, \alpha_{j-1}, -\alpha_{j+1}, -\alpha_j, \alpha_{j+2}, \ldots$, where $-\alpha_j$ has the same value (symbol) as α_j but the opposite sign. In this section, a *move* refers to an adjacent swap.

Feasibility
A necessary condition for two signed strings α and β to be compatible is that the corresponding unsigned strings are compatible with one another. However, this is not a sufficient condition. We note that an even number of moves are consumed by an element that is moved and returns to its original position. A direct consequence of this observation is that the *parity* (of the number of moves) between the elements is fixed. If an element is at an odd (even) distance from its destination position, it takes odd (even) number of moves for it to reach its destination.

Lemma 2. *A signed string α can be transformed into β iff there exists a pairing ψ of elements in α to the corresponding elements in β, where every pair in Ψ satisfies the following properties: (a) if the paired elements are at an even distance, then they have the same sign, or (b) if the paired elements are at an odd distance, then they have opposite signs.*

Compatibility theorem
Let α be the source string and β be the destination string over an alphabet $\Sigma = \{0, 1, \ldots, k-1\}$. For all $s \in \Sigma$, let $\alpha_{p,t}$ denote all the elements of α with index parity p and sign t, where $p \in \{e(even), o(odd)\}$ and $t \in \{-, +\}$. The indices start at 1. For example, if $\alpha = (0,-1,2,1)$ then $\alpha_{o,+} = (\alpha_1, \alpha_3)$, where the indices 1 and 3 have odd parity and the corresponding elements are all positive, and $\alpha_{e,-} = (\alpha_2)$. We define $g1(s, \alpha)$ as a vector containing all the elements from $\alpha_{e,-}(s)$ and $\alpha_{o,+}(s)$, in the order that they appear in α. We define $g2(s, \alpha)$ as a vector containing all the elements from $\alpha_{e,+}(s)$ and $\alpha_{o,-}(s)$

in the order that they appear in α. Every element of α is in either $g1(s,\alpha)$ or $g2(s,\alpha)$ for some s, and every element in $g1(s,\alpha)$ or $g2(s,\alpha)$ refers to a particular α_j for some j. Similar definitions hold for $g1(s,\beta)$ and $g2(s,\beta)$. In the previous example, $|g1(1,\alpha)| = 2$. $|g1(s,\beta)|$ and $|g2(s,\beta)|$ are similarly defined. If for a symbol s, $|g1(s,\alpha)| = |g1(s,\beta)|$ then α and β are G1 equivalent for s; if $|g2(s,\alpha)| = |g2(s,\beta)|$ then α and β are G2 equivalent for s. Theorem 2 states the necessary and sufficient condition to transform α into β. For example, if $\alpha = (0,-1,2,1,1)$, then $\alpha_{e,-}(1) = \alpha_2$, $\alpha_{o,+}(1) = \alpha_5$, $g1(1,\alpha) = (\alpha_2,\alpha_5)$, $\alpha_{e,+}(1) = \alpha_4$, $\alpha_{o,-}(1) = \{\}$, and $g2(1,\alpha) = (\alpha_4)$. Conversely, $g1(1,\alpha)_2 = \alpha_5$ and $g2(1,\alpha)_1 = (\alpha_4)$. Also, $|g1(1,\alpha)| = 2$ and $|g1(1,\alpha)| = 1$.

Theorem 2. *The strings α and β are compatible iff for all $s \in \Sigma$, α and β are G1 equivalent for s and G2 equivalent for s.*

Optimum algorithm
Let the original symbols in α and β be $\{0, 1, \ldots, j\}$. We will obtain a new problem involving α' and β' by mapping from α and β respectively, as per the following description. Let the vector V of length $2j$ be

$$V = (|g1(0,\alpha)|, |g2(0,\alpha)|, |g1(1,\alpha)|, |g2(1,\alpha)|, \ldots, |g1(j,\alpha)|, |g2(j,\alpha)|).$$

V has at least j non-zero entries and at most $2j$ non-zero entries. Let the pth non-zero entry of V be $gx(q,\alpha)$, where $x \in \{0,1\}$ and $q \leq p \leq 2q$. Replace each element corresponding to $gx(q,\alpha)$ with p (unsigned) to obtain α'. In order to obtain β', consider the vector

$$W = (|g1(0,\beta)|, |g2(0,\beta)|, |g1(1,\beta)|, |g2(1,\beta)|, \ldots, |g1(j,\beta)|, |g2(j,\beta)|)$$

and follow the preceding procedure (shown for α) for the string β. Observe that α' and β' are G1 and G2 equivalent for every symbol $s \in \{0, 1, .., j\}$. Run the unsigned optimal algorithm to transform α' into β' and note the adjacent swaps.

Claim. The adjacent swaps thus obtained give the optimal solution for transforming α into β.

6   Transformations of Signed Strings with Short Reversals

In this version of the problem, the operations that can be executed in any order are adjacent swaps and reversals of size one (a *1-flip*). We call them *short reversals* or *moves*. A 1-flip denoted by $s1(j)$, applied to $\alpha_1, \alpha_2, \ldots, \alpha_j, \alpha_{j+1}, ..\alpha_n$, yields $\alpha_1, \alpha_2, .. - \alpha_j, \alpha_{j+1}, ..\alpha_n$. Let the given strings be α and β. A necessary condition is that the corresponding unsigned versions of the strings are compatible. We note that the sign of an arbitrary element can be changed by a 1-flip.

We define *compatible distance* between a mapped pair of elements α_i and β_j, where $|j - i| = d$ as follows: (a) d is even and the signs of α_i and β_j are the same, or (b) d is odd and the signs of α_i and β_j are different. Contrarily, we define incompatible distance between a mapped pair of elements α_i and β_j where $|j - i| = d$ as follows: (a) d is odd and the signs of α_i and β_j are same, or (b) d is even and the signs of α_i and β_j are different. Let α_{i_1} and α_{i_2} be successive occurrences of a symbol ψ in α, for all k $(i_1 < k < i_2)$, $\alpha_k \neq \psi$. Let the pairing corresponding to α_{i_1} and α_{i_2} be $\{(i_1, j_1), (i_2, j_2)\}$. If both the pairs (i_1, j_1) and (i_2, j_2) are at incompatible distances then (i_1, i_2) is a *wrong sign couple*. For a given wrong sign couple (i_1, i_2) and its corresponding pairs $\{(i_1, j_1), (i_2, j_2)\}$, if there is no pair (p, q) for $(i_1 < p < i_2)$ and $(j_1 < q < j_2)$, then (i_1, i_2) is a *suboptimal couple*. Note that for a given wrong sign couple (i_1, i_2) and its corresponding pairs $\{(i_1, j_1), (i_2, j_2)\}$, if $|i_1 - i_2| = 1$ or $|j_1 - j_2| = 1$, then (i_1, i_2) is a suboptimal couple. Let the given suboptimal couple be (i_1, i_2) and its corresponding pairs be $\{(i_1, j_1), (i_2, j_2)\}$. If we apply *rectification* to (i_1, i_2), then the pairs $\{(i_1, j_1), (i_2, j_2)\}$ are replaced with $\{(i_1, j_2), (i_2, j_1)\}$.

Lemma 3. *Rectifications of a symbol ψ are non-overlapping.*

An algorithm that transforms α into β has to pair α_i with β_j for all values of i and j, where α_i and β_j have the same symbol. If all such pairs are at compatible distances, then no further work is necessary, else additional flip(s) are necessary. There are strings α and β for which there exists no pairing that has all pairs at a compatible distance. Consider $\alpha = 2,1,0,-2,-0$ and $\beta = 0,1,-2,0,-2$; all occurrences of 2 in α are at an incompatible distance from all occurrences of 2 in β. Therefore, apart from pairing elements at a compatible distance, an algorithm must be able to (a) pair elements at incompatible distances, (b) execute adjacent swaps to move the elements to their destination positions, and (c) change the signs of the elements to match the sign of the destination elements.

A block of a symbol ψ is a substring of contiguous occurrences of ψ that cannot be extended to either side. Note that we do not consider the sign associated with the symbol. If $\alpha = 1,0,2,-2,2,-2,1$ then $\alpha_{3..6}$ is a 2-block of size 4. A *run* of a symbol ψ is the maximal contiguous occurrences of ψ that have incorrect sign and every pair of succesive elements in a run form a suboptimal couple. Note that an element with a wrong sign that does not form a suboptimal couple constitutes a run of length 1. If $\alpha = 1,0,2,-2,2,-2,1$ and $\beta = -1,0,-2,2,-2,-2,1$ then α_1 is a "1" run of size 1 and $\alpha_{3..5}$ is a "2" run of size 3. Note that a ψ run is a substring of the corresponding ψ block. Let a run with length $k - j + 1$ of ψ in α have indices $(i_j, i_{j+1}, \ldots, i_k)$. Then, *left-continuous rectification* rectifies the pairs $\{(i_j, i_{j+1}), (i_{j+1}, i_{j+2}), \ldots, (i_{k-1}, i_k)\}$ when $k - j$ is odd, otherwise, rectifies the pairs $\{(i_j, i_{j+1}), (i_{j+1}, i_{j+2}), \ldots, (i_{k-2}, i_{k-1})\}$.

Lemma 4. *Let $r_{1..p}$ be the lengths of runs in a given string α. The minimum number of short reversals required to eliminate all the runs is:*

$$\sum_{1 \leq i \leq p} (\lfloor \frac{r_i}{2} \rfloor + r_i \bmod 2)$$

Lemma 4 shows that at most $\frac{(k-i)}{2}$ couples can be rectified. If k is even, then we rectify the pairs $\{(i_i, i_{i+1}), (i_{i+2}, i_{i+3}), \ldots, (i_{k-1}, i_k)\}$, else we rectify the pairs

$\{(i_i, i_{i+1}), (i_{i+2}, i_{i+3}), \ldots, (i_{k-2}, i_{k-1})\}$. In either case, we rectify $\frac{(k-i)}{2}$ couples. Thus, given a run of ψ of length $k - i$ with indices $(i_i, i_{i+1}, \ldots, i_k)$, the left-continuous rectification rectifies the maximal number of suboptimal couples.

Algorithm to find the number of short reversals

1. Disregard the signs and perform optimum pairing for α and β
2. For every symbol ψ, for every run (*length* > 0), perform left-continuous rectification.
3. Let the number of elements that still have the wrong signs be R.
4. Let the number of adjacent swaps required for optimum pairing be I(G) and let the number of suboptimal pairs that are rectified in step 2 be S^o. Then the total short reversals required are I(G) + S^o + R.

Claim. The above algorithm (A') consumes the optimal number of moves.

7   Conclusions

In this paper an algorithm to transform signed and unsigned strings of k symbols with adjacent swaps in $O(nk)$ time and $O(nk)$ space is presented. Also, an algorithm to transform signed strings with short reversals is presented with the same time and space complexity. This work is relevant for genomic mutations involving adjacent genes, when the genes are structurally or functionally classified into a small number of categories. Future work may focus on reductions of time and space. For example, it is unknown whether $O(n\sqrt{k})$ time (space) or $O(n \log k)$ time (space) is sufficient.

References

1. Chen, T., Skiena, S.S.: Sorting with Fixed-Length Reversals. Discrete Applied Mathematics 71, 269–295 (1996)
2. Cormen, T.H., Leiserson, C.E., Rivest, R.L., Stein, C.: Introduction to Algorithms. 2nd edn. McGraw-Hill Book Company, Cambridge (2002)
3. Hannenhalli, S., Pevzner, P.A.: Transforming Cabbage into Turnip: Polynomial Algorithm for Sorting Signed Permutations by Reversals. Journal of the ACM 46(1), 1–27 (1999)
4. Toussaint, G.: A Comparison of Rhythmic Similarity Measures. In: Proceedings of the Fifth International Conference on Music Information Retrieval, Barcelona, Spain (2004)

Efficient Algorithms for SNP Haplotype Block Selection Problems

Yaw-Ling Lin*

Dept. Computer Science and Information Engineering
Providence University, Taiwan
yllin@pu.edu.tw

Abstract. Global patterns of human DNA sequence variation (haplotypes) defined by common *single nucleotide polymorphisms* (SNPs) have important implications for identifying disease associations and human traits. Recent genetics research reveals that SNPs within certain haplotype blocks induce only a few distinct common haplotypes in the majority of the population. The existence of haplotype block structure has serious implications for association-based methods for the mapping of disease genes. Our ultimate goal is to select haplotype block designations that best capture the structure within the data.

Here in this paper we propose several efficient combinatorial algorithms related to selecting interesting haplotype blocks under different diversity functions that generalizes many previous results in the literatures. In particular, given an $m \times n$ haplotype matrix A, we show linear time algorithms for finding all interval diversities, farthest sites, and the longest block within A. For selecting the multiple long blocks with diversity constraint, we show that selecting k blocks with longest total length can be be found in $O(nk)$ time. We also propose linear time algorithms in calculating the all intra-longest-blocks and all intra-k-longest-blocks.

1 Introduction

Mutation in DNA is the principle factor that is responsible for the phenotypic differences among human beings, and SNPs (single nucleotide polymorphisms) are the most common mutations. An SNP is defined as a position in a chromosome where each one of two (or more) specific nucleotides are observed in at least 10% of the population [12]. The nucleotides involved in a SNP are called *alleles*. It has been observed that for almost all SNPs only two different alleles are present, in such case the SNP is said biallelic; otherwise the SNP is said multiallelic. In this paper, we will consider exclusively biallelic SNPs.

In diploid organisms, such as humans, each chromosome is made of two distinct copies and each copy is called a *haplotype*. There is currently a great deal of interest in how the recombination rate varies over the human genome. The

---

* This work is supported in part by the National Science Council, Taiwan, Grant NSC-96-2221-E-126-002.

X. Hu and J. Wang (Eds.): COCOON 2008, LNCS 5092, pp. 309–318, 2008.

primary reason for this is the excitement about using *linkage disequilibrium* (LD) to map and identify human disease genes. The blocks of LD are regions, typically less than 100 kb, in which LD decreases very little with distance between markers. Between these blocks, however, LD is observed to decay rapidly with physical distance. Concomitant with the blocks of LD, these studies also find low haplotype diversity within blocks. These results also show that the SNPs within each block induce only a few distinct common haplotypes in the majority of the population, even though the theoretical number of different haplotypes for a block containing n SNPs is exponential in n. The existence of haplotype block structure has serious implications for association-based methods for the mapping of disease genes. Conducting genome-wide association mapping studies could be much easier than it used to be based on single marker [3],[13],[12],[5],[4].

Terminology and Problem Definition

Abstractly, input to the haplotype blocking problem consists of m *haplotype vectors*. Each position in a vector is associated with a site of interest on the chromosome. Usually, the position in the haplotype vector has a value of 0 if it is the major allele or 1 if minor allele.

Let the *haplotype matrix* A be an $m \times n$ matrix of m observations over n markers (sites). We refer to the jth allele of observation i by A_{ij}. For simplicity, we first assume that $A_{ij} \in \{0,1\}$. A *block*, or marker interval, $[j,k] = \langle j, j+1, \ldots, k \rangle$ is defined by two marker indices $1 \le j \le k \le n$. A *segmentation* is a set of non-overlapping non-empty marker intervals. The data matrix limited to interval $[j,k]$ is denoted by $A(j,k)$; the values of the ith observation are denoted by $A(i,j,k)$, a binary string of length $k - j + 1$.

Given an interval $[j,k]$, a *diversity function*, $\delta : [j,k] \rightarrow \delta(j,k) \in \mathbb{R}$ is an evaluation function measuring the diversity of the submatrix $A(j,k)$. We say an interval $[j',k']$ is a *subinterval* of $[j,k]$, written $[j',k'] \subset [j,k]$, if $j \le j'$ *and* $k' \le k$. Note that δ-function is a *monotonic non-decreasing* function from $[1..n, 1..n]$ to the unit real interval $[0,1]$; that is, $0 \le \delta(j',k') \le \delta(j,k) \le 1$ whenever $[j',k'] \subset [j,k]$.

Given an input set of n haplotype vectors, a solution to the *Haplotype Block Selection (HBS)* problem is a segmentation of marker intervals, revealing these non-overlapped haplotype blocks of interest in the chromosome. Here in this paper we propose several efficient algorithms related to selecting interesting haplotype blocks under different evaluation (diversity) functions that generalizes many previous results in the literatures [12],[15],[5],[1].

Diversity Functions. Several operational definitions have been used to identify haplotype-block structures, including LD-based [5], recombination-based [10], information-complexity-based [1],[6] and diversity-based [12],[15] methods. The result of block partition and the meaning of each haplotype block may be different by using different measuring formula. For simplicity, haplotype samples can be converted into haplotype matrices by assigned major alleles to 0 and minor alleles to 1.

Definition 1 (haplotype block diversity). *Given an interval $[i, j]$ of a haplotype matrix A, a diversity function, $\delta : [i, j] \rightarrow \delta(i, j) \in \mathbb{R}$ is an evaluation function measuring the diversity of the submatrix $A(i, j)$.*

In terms of diversity functions, the block selection problem can be viewed as finding a segmentation of given haplotype matrix such that the diversities of chosen blocks satisfy certain value constraint. Following we examine several haplotype block diversity evaluation functions. Given an $m \times n$ haplotype matrix A, a block $S(i, j)$ $(i, j$ are the block boundaries) of matrix A is viewed as m haplotype strings; they are partitioned into groups by merging identical haplotype strings into the same group. The probability p_i of each haplotype pattern s_i, is defined accordingly such that $\sum p_i = 1$. As an example, Li [11] proposes a diversity formula defined by

$$\delta_D(S) = 1 - \sum_{s_i \in S} p_i^2. \tag{1}$$

Note that $\delta_D(S)$ is the probability that two haplotype strings chosen at random from S are different from each other. Other measurements of diversity can be obtained by choosing different diversity function; for example, to measure the information-complexity one can choose the information entropy (negative-log) function [1],[6]:

$$\delta_E(S) = - \sum_{s_i \in S} p_i \log p_i. \tag{2}$$

In the literatures [12],[15], Patil and Zhang et al. define a haplotype block as a region where at least 80% of observed haplotypes within a block must be common haplotype. As the same definition of common haplotype in the literatures, the coverage of common haplotype of the block can be formulated as a form of diversity:

$$\delta_C(S) = 1 - \frac{\sum\limits_{s_i \in C} p_i}{\sum\limits_{s_i \in U} p_i} = \frac{\sum\limits_{s_i \in M} \frac{1}{m}}{\sum\limits_{s_i \in U} p_i}. \tag{3}$$

Here U denotes the unambiguous haplotypes, C denotes the common haplotypes, and M denotes the singleton haplotypes. In other words, Patil et al. require that $\delta_C(S) \leq 20\%$.

Some studies [5],[15] propose the haplotype block definition based on LD measure D'; however, there is no consensus definition for it so far. Zhang and Jin [15] define a haplotype block as a region in which all pair-wise $|D'|$ values are not lower than a threshold α. Let S denote a haplotype interval $[i, j]$. We define the diversity as the complement of minimal $|D'|$ of S. By the definition, S is a haplotype block if its diversity is lower than $1 - \alpha$.

$$\delta_{L1}(S) = 1 - \min\{(|D'_{i'j'}|)|i \leq i' < j' \leq j\}. \tag{4}$$

Zhang et al. [15] also propose the other definition for haplotype block; they require at least α proportion of SNP pairs having strong LD (the pair-wise $|D'|$

greater than a threshold) in each block. Similarly, we can use the diversity to redefine the function. We define the diversity as the proportion of SNP pairs that do not have strong LD. Therefore, haplotype interval S is a feasible haplotype block if its diversity is smaller than a threshold. We can use the following diversity function to calculate the diversity of S. Here $N(i,j)$ denotes the number of SNP pairs that do not have strong LD in the interval $[i,j]$.

$$\delta_{L2}(S) = \frac{N(i,j)}{\binom{(j-i)+1}{2}} = \frac{N(i,j)}{\frac{1}{2}[(j-i)^2 + j - i]}. \tag{5}$$

Diversity measurement usually reflects the activity of recombination events occurred during the evolutionary process. Generally, haplotype blocks with low diversity indicates conserved regions of genome. By using appropriate diversity functions, the block selection problem can be viewed as finding a segmentation of given haplotype matrix such that the diversities of chosen blocks satisfy certain value constraint.

The rest of the paper is organized as the following. In Section 2, we propose an $O(mn+n^2)$ time, linear proportional to the input plus output size, algorithm for computing the ALL-INTERVAL-DIVERSITIES problem. We also show that all FARTHEST-SITES can be found totally in $O(mn)$ time, linear proportional to the input size. As a corollary, we show that the LONGEST-BLOCK can be found in $O(n)$ time by examining the $O(n)$-sized farthest-sites array; that is, the the LONGEST-BLOCK can be found in linear time. In Section 3, we show that the k-LONGEST-BLOCKS can be found in $O(nk)$. These blocks can be identified by examining the farthest-sites array and sing the technique of sparse dynamic programming. We also show that the ALL-INTRA-LONGEST-BLOCK problem can be found totally $O(n^2)$ time, which is a linear time algorithm proportional to the output size. The problem of ALL-INTRA-k-LONGEST-BLOCKS can be found totally in $O(n^2k)$ time, which is a linear time algorithm proportional to the output size.

2   Computing Diversities of All Blocks

Given an $m \times n$ haplotype matrix A, the goal here is to compute the all pairs block diversity values. That is, output the set $\{\delta(i,j) \mid 1 \le i \le j \le n\}$. Here we show that by using techniques of the suffix tree [7],[14], there exits an $O(mn+n^2)$ time, linear proportional to the input plus output size, algorithm for computing the ALL-INTERVAL-DIVERSITIES problem.

A naive algorithm would try to enumerate all different $O(n^2)$ blocks, each with data sized as much as $O(mn)$. While it takes at least linear time in computing the diversity of each submatrix, thus a totally $O(mn^3)$ operations would be needed.

To obtain the desired linear time algorithm for computing all diversities, we use the suffix tree data structure [7],[14]. It is well known that, given a string $s \in \sum^*$ of length n, the suffix tree of s, T_s, that consists of all suffixes of s as leaves can be constructed in linear time. Thus, given an $m \times n$ haplotype matrix A, our linear time algorithm begins by constructing m suffix trees of m rows

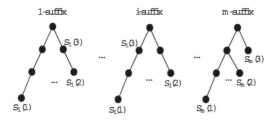

Fig. 1. The m suffix trees for m rows of a $m \times n$ haplotype matrix

of A, $\mathbb{T} = \{T_{A(1,1,n)}, T_{A(2,1,n)}, \ldots, T_{A(m,1,n)}\}$, in totally $O(mn)$ time. Figure 1 illustrates the m suffix trees of the given $m \times n$ haplotype matrix, where $S_i(j)$ denotes the jth suffix of row i.

After obtaining these m suffix trees, we can *merge* these m trees and obtain the *total suffix tree* T^* with mn leaves. Note that T^* can also be obtained by concatenating the original m row-strings of the haplotype matrix into one long string $u = A(1, 1, n)\sharp A(2, 1, n)\sharp \cdots \sharp A(m, 1, n)$ and constructing the corresponding suffix tree T_u, where '$\sharp$' denotes a symbol that does not appear in alphabets of the matrix. Note that the structure of T^* is equivalent to T_u. The next step of the algorithm is to construct n confluent subtrees, which is discussed in the following.

Confluent LCA Subtrees. The *lowest common ancestor* (LCA) between two nodes u and v in a tree is the furthest node from the root node that exists on both paths from the root to u and v. Harel and Tarjan [9] have shown that any n-node tree can be preprocessed in $O(n)$ time such that subsequently LCA queries can be answered in constant time.

Let T be a tree with leaf nodes L. Given $S \subset L$, let set $\Lambda(S) = \{\text{LCA}(x, y) \mid x \neq y \in S\}$ denote the collection of all (proper) lowest common ancestors defined over S.

Definition 2 (confluent subtree). *The* confluent subtree *of S in T, denoted by $T_{\uparrow S}$, is a subtree of T with leaves S and internal nodes $\Lambda(S)$. Further, $u \in \Lambda(S)$ is a parent of v in $T_{\uparrow S}$ if and only if u is the lowest ancestor of v in T comparing to any other node in $\Lambda(S)$.*

Our notation of confluent subtree is called *induced subtrees* in the literature [2]. Let T be a phylogenetic tree with leaf nodes L. A post-order (pre-order, or in-order) traversal of nodes of L within T defines a *tree ordering* of nodes on L. The following results is a generalization of the section 8 of [2].

Lemma 1. *Let T be an n-node phylogenetic tree with leaf nodes L. The following subsequent operation can be done efficiently after an $O(n)$ time preprocessing. Given a query set $A \subset L$, the confluent subtree $T_{\uparrow A}$ can be constructed in $O(|A|)$ time if A is given in sorted tree ordering; otherwise, $T_{\uparrow A}$ can be constructed in $O(|A| \log \log |A|)$ time.*

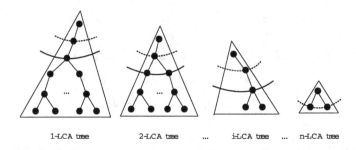

1-LCA tree   2-LCA tree   ...   i-LCA tree   ...   n-LCA tree

Fig. 2. The corresponding n confluent LCA trees for n columns in a haplotype matrix

By Lemma 1, in $O(mn)$ time, we can partition leaves of T^* into n sets $\{A_1, A_2, \ldots, A_n\}$, where A_i consists of all m suffixes started at column i in given haplotype matrix. These A_i's subsets can be used to construct the corresponding n confluent LCA subtrees, $T^*_{\uparrow A_i}$'s, as illustrated in Figure 2.

Each LCA tree $T^*_{\uparrow A_i}$ represent the changing structure of m rows of submatrix $A(i, n)$.

Event List. The idea is to build up the event-list for each LCA tree $T^*_{\uparrow A_i}$ so that all diversity values started at i, $\delta(i, \cdot)$'s, can be obtained in $O(m + n)$ time. To do so, we need to perform a BFS-like search on each LCA tree $T^*_{\uparrow A_i}$. This can be done by the bucket-sort technique. We first visit each node of all $T^*_{\uparrow A_i}$'s, and append each node into a global *depth*-list, ranked according to its distance to the root. After that, we pick up nodes in the depth-list by the increasing depth order, and append them into the final *event*-list.

Lemma 2. *The event-list of a given $m \times n$ haplotype matrix can be computed in $O(mn)$ time.*

By using the event-list, it is easily verified that diversities started at i, $\delta(i, \cdot)$'s, can be obtained by following the i-th entry of the event-list. Observe that there are at most m events along with each entry, and there are at most n diversities need to be calculated. Thus diversities $\delta(i, \cdot)$'s can be obtained in $O(m + n)$ time for each column i. It follows that ALL-INTERVAL-DIVERSITIES can be found in $O(mn + n^2)$ time, linear proportional to the input plus output size.

Theorem 1 (all-interval-diversities). *Given an $m \times n$ haplotype matrix A, $O(mn + n^2)$ time suffices to compute all interval diversities $\{\delta(i, j) \mid 1 \le i \le j \le n\}$.*

Recall that, in the FARTHEST-SITES problem, with a given real diversity upper limit D, for each marker site i we need to find its corresponding *farthest right marker* $j = R[i]$ so that $\delta(i, j) \le D$. By using the event-list and Lemma 2, it is easily verified that diversities started at i, $\delta(i, \cdot)$'s, can be examined by following the i-th entry of the event-list. Since there are at most m events along with each entry, the farthest site can thus be identified in $O(m)$ time for each column i. It follows that all farthest sites can be found in $O(mn)$ time, linear proportional to the input size.

Theorem 2 (farthest-sites). *Given an $m \times n$ haplotype matrix and a given real diversity upper limit D, for each marker site i we can find its corresponding farthest right marker $j = R[i]$ so that $\delta(i, j) \leq D < \delta(i, j+1)$, totally in $O(mn)$ time.*

Corollary 1 (longest-block). *Given an $m \times n$ haplotype matrix and a diversity upper limit D, the longest block with diversity not exceeding D can be found in $O(mn)$ time.*

3   Longest k Blocks with Diversity Constraints

We discuss a series of problem related to finding multiple blocks with diversity constraints in the section. Given a haplotype matrix A and a diversity upper limit D, let $S = \{B_1, B_2, \ldots, B_k\}$ be a segmentation of A with $\delta(B) \leq D$ for each $B \in S$. The *length* of S is the total length of all block in S; i.e., $\ell(S) = |B_1| + |B_2| + \cdots + |B_k|$. Our objective is to find a segmentation consists of k feasible blocks such that the total length $\ell(S)$ is maximized.

Given A and D, first we consider the most general form of the problem and define the block length evaluation function

$$f(k, i, j) = \max\{\ell(S) \mid S \text{ a feasible segmentation of } A(i, j) \text{ with } k \text{ blocks}\}$$

Note that the k-LONGEST-BLOCKS asks to find the value $f(k, 1, n)$. First of all, we prepare the *left farthest sites*, $L[j]$'s, for each site j of A; according to Theorem 2, the array can be identified in $O(mn)$ time. Here we show that the answer can be found in $O(nk)$ time after the preprocessing. The idea behind the dynamic programming formula is illustrated at Figure 3.

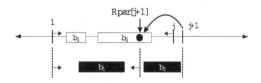

Fig. 3. The $O(kn)$ algorithm to compute $f(k, 1, n)$

It can be verified that

$$f(k, 1, j) = \max\{f(k, 1, j-1), f(k-1, 1, L[j]-1) + j - L[j] + 1\}$$

That is, the k-th block of the maximal segment S in $[1, j]$ either does not include site j; otherwise, the block $[L[j], j]$ must be the last block of S. Note that

$f(k, 1, j)$ can be determined in $O(1)$ time suppose $f(k-1, 1, \cdot)$'s and $f(k, 1, 1..(j-1))$'s being ready. It follows that $f(k, 1, \cdot)$'s can be calculated from $f(k-1, 1, \cdot)$'s, totally in $O(n)$ time. Thus a computation ordering from $f(1, 1, \cdot)$'s, $f(2, 1, \cdot)$'s, ..., to $f(k, 1, \cdot)$'s leads to the following result.

Theorem 3 (k-longest-blocks). *Given a haplotype matrix A and a diversity upper limit D, find a segmentation consists of k feasible blocks such that the total length is maximized can be done in $O(nk)$ time after a linear time preprocessing.*

Given A and D, the ALL-INTRA-LONGEST-BLOCK asks to find, for all constrained interval $[i, j] \subset [1, n]$, the corresponding *longest* block $B \subset [i, j]$ with $\delta(B) \leq D$. That is, the values $f(1, i, j)$'s. Similarly, we first prepare the farthest sites of A in $O(mn)$ time. Here we show that the answer can be found in $O(n^2)$ time after the preprocessing. Using similar idea as illustrated at Figure 3. It can be verified that

$$f(1, i, j) = \max\{f(1, i, j-1), (j - L[j] + 1)\}$$

That is, the longest block in $[i, j]$ either does not include site j; otherwise, the block $[L[j], j]$ must be the longest (feasible) block within $[i, j]$. Note that $f(1, i, j)$ can be determined in $O(1)$ time whenever $f(1, i, 1..(j-1))$'s being ready. It follows that $f(1, i, \cdot)$'s can be calculated totally in $O(n)$ time. Thus a computation ordering from $f(1, 1, \cdot)$'s, $f(1, 2, \cdot)$'s, ..., to $f(1, n, \cdot)$'s leads to the following result.

Theorem 4 (all-intra-longest-block). *Given a haplotype matrix A and a diversity upper limit D, find for all constrained interval $[i, j] \subset [1, n]$, the corresponding longest block $B \subset [i, j]$ with $\delta(B) \leq D$ can be done in $O(n^2)$ time after a linear time preprocessing.*

Note that the computation time is linear proportional to the output plus input size, and thus is an optimal algorithm in regarding of time efficiency or all constrained interval $[i, j] \subset [1, n]$, find the longest segmentation with k *feasible* blocks such that the total length is maximized. That is, output the set $S = \{B_1, B_2, \ldots, B_k\}$, with $\delta(B) \leq D$ for each $B \in S, B \subset [i, j]$, such that $|B_1| + |B_2| + \cdots + |B_k|$ is maximized.

The ALL-INTRA-k-LONGEST-BLOCKS asks to find, for all constrained interval $[i, j] \subset [1, n]$, the longest segmentation with k *feasible* blocks such that the total length is maximized. Again, the problem can be solved by first prepare the farthest sites of A in $O(mn)$ time. Here we show that the answer can be found in $O(n^2 k)$ time after the preprocessing. Using similar idea as illustrated at Figure 3. It can be verified that

$$f(k, i, j) = \max\{f(k, i, j-1), f(k-1, i, L[j]-1) + j - L[j] + 1\}$$

That is, the k-th block of the maximal segment S in $[i, j]$ either does not include site j; otherwise, the block $[L[j], j]$ must be the last block of S. Note that $f(k, i, j)$ can be determined in $O(1)$ time suppose $f(k-1, i, \cdot)$'s and $f(k, i, 1..(j-1))$'s being ready. It follows that $f(k, i, \cdot)$'s can be calculated from $f(k-1, i, \cdot)$'s, totally in $O(n)$ time. Thus $f(k, \cdot, \cdot)$'s can be calculated from $f(k-1, \cdot, \cdot)$'s totally in $O(n^2)$

time. Thus a computation ordering from $f(1, \cdot, \cdot)$'s, $f(2, \cdot, \cdot)$'s, ..., to $f(k, \cdot, \cdot)$'s leads to the following result.

Theorem 5 (all-intra-k-longest-blocks). *Given A, D, there exits an $O(kn^2)$-time algorithm to find, for all constrained interval $[i, j] \subset [1, n]$, the longest segmentation with k feasible blocks such that the total length is maximized.*

4   Experimental Results

In response to needs for analysis and observation of human genome variation, we establish web systems and apply the algorithms in this paper to provide several analysis tools for bioinformatics and genetics researchers. The web consists of a list of PHP and Perl CGI-scripts together with several C programs for selecting haplotype blocks and analyzing the haplotype diversity. We collect the haplotype data of human chromosome 21 from Patil et al. [12] and download haplotypes for all the autosomes from phase II data of HapMap [8] so that the bioinformatics researchers can use the data to evaluate the performance of the tools. Researchers can also input their own haplotype data by on-line inputting the haplotype sample or uploading the file of haplotype sample, or providing a hyperlink of other website's haplotype data.

The website provides tools to examine the diversity of haplotypes and partition haplotypes into blocks by using different diversity functions. Using the diversity visualization tool, researchers can observe the diversity of haplotypes in the form of the diagram of curves; the diversity function (1), δ_D, is used to calculate the diversities of intervals based on sliding widows manner. Our website also provides tools for researchers to partition haplotypes into blocks with constraints on diversity and tag SNP number; by using the tool, researchers can find the longest segmentation consists of non-overlapping blocks with limited number of tag SNPs.

Our web system is freely accessible at http://bioinfo.cs.pu.edu.tw/~hap/index.php. Some preliminary results, including the selection of different diversity functions as well as choosing meaningful diversity constraints can also be found in the web system.

5   Concluding Remarks

By using appropriate diversity functions, the block selection problem can be viewed as finding the segmentation of a given haplotype matrix such that the diversities of chosen blocks satisfy certain value constraint. In this paper, we propose several efficient combinatorial algorithms related to selecting interesting haplotype blocks under different diversity functions.

By using the powerful data structures including suffix trees [7],[14] and the confluent LCA subtrees [9],[2], we are able to show that the kernel combinatorial structure, *event-list*, can be constructed in time linear proportional to the input. We need to point out that these time-efficiency results of our algorithms can be applied in many different definitions of diversity functions; there we assume that the computation of the diversity of submatrix $A(j, k')$ can be obtained from a shorter prefix of $A(j, k)$, with $k < k'$, in time proportional to the size of changing events . It is not hard to verify that the diversity functions listed in Eq. (1), (2), (3), as well as many other different diversity functions, possess this property.

References

1. Anderson, E.C., Novembre, J.: Finding Haplotype Block Boundaries by Using the Minimum-Description-Length Principle. Am. J. of Human Genetics 73, 336–354 (2003)
2. Cole, R., Farach, M., Hariharan, R., Przytycka, T., Thorup, M.: An $O(n \log n)$ Algorithm for the Maximum Agreement Subtree Problem for Binary Trees. SIAM Journal on Computing 30(5), 1385–1404 (2002)
3. Daly, M., Rioux, J., Schafiner, S., Hudson, T., Lander, E.: Highresolution Haplotype Structure in the Human Genome. Nature Genetics 29, 229–232 (2001)
4. Dawson, E., Abecasis, G., et al.: A First-Generation Linkage Disequilibrium Map of Human Dhromosome 22. Nature 418, 544–548 (2002)
5. Gabriel, S.B., Schaffner, S.F., Nguyen, H., et al.: The Structure of Haplotype Blocks in the Human Genome. Science 296(5576), 2225–2229 (2002)
6. Greenspan, G., Geiger, D.: Model-Based Inference of Haplotype Block Variation. In: Seventh Annual International Conference on Computational Molecular Biology (2003)
7. Gusfield, D.: Algorithms on Strings, Trees and Sequences: Computer Science and Computational Biology. Cambridge University Press, Cambridge (1997)
8. International HapMap Project, http://www.hapmap.org/index.html.en
9. Harel, D., Tarjan, R.E.: Fast Algorithms for Finding Nearest Common Ancestors. SIAM Journal on Computing 13(2), 338–355 (1984)
10. Hudson, R.R., Kaplan, N.L.: Statistical Properties of the Number of Recombination Events in the History of a Sample of DNA Sequences. Genetics 111, 147–164 (1985)
11. Li, W.H., Graur, D.: Fundamentals of Molecular Evolution. Sinauer Associates, Inc. (1991)
12. Patil, N., Berno, A.J., Hinds, D.A., et al.: Blocks of Limited Haplotype Diversity Revealed by High Resolution Scanning of Human Chromosome 21. Science 294, 1719–1723 (2001)
13. Reich, D., Cargill, M., Lander, E., et al.: Linkage Disequilibrium in the Human Genome. Nature 411, 199–204 (2001)
14. Ukkonen, E.: On-Line Construction of Suffix Trees. Algorithmica 14(3), 249–260 (1995)
15. Zhang, K., Qin, Z., Chen, T., Liu, J.S., Waterman, M.S., Sun, F.: HapBlock: Haplotype Block Partitioning and Tag SNP Selection Software Using a Set of Dynamic Programming Algorithms. Bioinformatics 21(1), 131–134 (2005)

Sequence Alignment Algorithms for Run-Length-Encoded Strings

Guan Shieng Huang[1,*], Jia Jie Liu[2], and Yue Li Wang[1,**]

[1] Department of Computer Science and Information Engineering,
National Chi Nan University, Taiwan
shieng@ncnu.edu.tw, yuelwang@ncnu.edu.tw
[2] Department of Information Management, Shih Hsin University, Taiwan
jjliu@cc.shu.edu.tw

Abstract. A unified framework is applied to solving various sequence comparison problems for run-length encoded strings. All of these algorithms take $O(\min\{mn', m'n\})$ time and $O(\max\{m, n\})$ space, for two strings of lengths m and n, with m' and n' runs, respectively. We assume the linear-gap model and make no assumption on the scoring matrices, which maximizes the applicability of these algorithms. The trace (i.e., the way to align two strings) of an optimal solution can also be recovered within the same time and space bounds.

1 Introduction

We consider how to compare the similarity of two run-length-compressed strings in this paper. Let Σ be an alphabet with a constant number of symbols. A string like a^k for any $a \in \Sigma$ and $k \in \mathbb{N}$ is called a *run*. Strings can be compressed by runs. For example, $aaaabbbaa$ can be compressed to $(a, 4)(b, 3)(a, 2)$ as three runs in the run-length encoding. Let x and y be two strings over Σ with lengths m and n, respectively. Suppose x has m' runs and y has n' runs. In this paper, we show that the following problems can be solved in $O(m'n)$ time and $O(n)$ space under a unified framework: (A) the string edit distance problem; (B) the pairwise global alignment problem; and (C) the pairwise local alignment problem. We assume the linear-gap model, and the scoring matrices can be arbitrary.

Two special cases of (A) and (B) are the LCS (Longest-Common-Subsequence) metric and the Levenshtein metric [24,18]. For the LCS metric, Bunke and Csirik [8] first presented an $O(m'n + mn')$ time algorithm. It is further improved by Apostolico et al. [4] in time $O(m'n' \lg(m'n'))$, and by Mitchell [22] in time $O((m' + n' + d) \lg(m' + n' + d))$ where d is the number of matches of runs (in the worst case $d = O(m'n')$). Around 2002, three groups independently made a break through on this problem. Arbell et al. [5] solved the case for the Levenshtein metric, which is an edit distance problem with unit cost, in time $O(m'n + mn')$; Mäkinen et al. [21] solved the general edit distance problem,

---

* Research supported in part by NSC 95-2221-E-260-016-MY2.
** Research supported in part by NSC 96-2221-E-260-018.

X. Hu and J. Wang (Eds.): COCOON 2008, LNCS 5092, pp. 319–330, 2008.

under the assumption that the distance matrix is nonnegative and satisfies the triangle inequality, in time $O(m'n + mn')$; Crochemore et al. [11] solved the global alignment problem under the linear-gap model, and the latter can be easily converted to solving the general edit distance problem, also in $O(m'n + mn')$ time. Finally, Liu et al. [19,20] proposed an $O(\min\{m'n, mn'\})$-time algorithm for the Levenshtein and LCS metrics.

The paper of Crochemore et al. [11] originally solves sequence alignment problems under the Lempel-Ziv encoding. Strings being compressed by LZ78 have nice recursive structures. They employed *the Monge property* (which will be explained in Sect. 3) to solve the global and the local alignment problems in both $O(hn^2/\lg n)$ time and space where $0 < h \le 1$ is the entropy of input strings. The trace [24] can also be recovered in the same time and space bounds. This technique was also applied to run-length compressed strings for the global alignment problem. We remark that the approach of [11] has several advantages. First, its assumption is rather general. It assumes the linear-gap model with unrestricted scoring matrices. Second, this framework has great potential to solve other related problems such as the local alignment problem.

We combine techniques from [11] and [19], and improve several results of [21,11,19,20]. In addition, we propose an algorithm for the approximate matching under the run-length encoding. Let T be a long text, which is run-length-compressed into n' runs. Let P be a pattern with m characters. This problem asks one to find out all occurrences of P in T such that their distances are under some given threshold. Our algorithm takes $O(n'm)$ time and $O(m)$ space, which improves the previous $O(n'mm')$-time algorithm in [21]. Furthermore, we do not need the entries in a distance matrix and the threshold for approximate matching to be bounded by constants, which is implicitly assumed in the algorithm proposed by [21]. A comparison of related results is listed in Table 1.

Table 1. A comparison of related results

Problem	In This Paper	Previous Results
Edit distance/ Global alignment	$O(\min\{m'n, mn'\})$ time[a]	$O(m'n + mn')$ time [11][a] $O(\min\{m'n, mn'\})$ time[19,20][b]
Local alignment	$O(\min\{m'n, mn'\})$ time[a]	only for LZW compression [11][c]
Approximate matching	$O(n'm)$ time[a]	$O(n'mm')$ time [21][d]

[a] Linear gap cost with unrestricted scoring matrices.
[b] Limited to Levenshtein and LCS metrics.
[c] To the best of our knowledge.
[d] Assume the distance is a metric and the approximate threshold is a constant.

The organization of this paper is as follows. We explain the basic idea in Sect. 2 and give a quick review on related preliminaries in Sect. 3. Then subsequent sections are followed for each problem. Finally, a conclusion ends in Sect. 8.

2   The Idea

We integrate several important techniques from [11] and [19]. Let x be a run-length-compressed string which is put to the left side of the *edit graph* [12]. Let y be an uncompressed string which is put on the top. The edit graph is divided into several regions, and in each region, only the values on borders are computed. However, unlike [11,5,8,21] that partition the edit graph into blocks, we use strips. A strip R is a region defined by a run of x and the whole y (see Fig. 1). Let $O_R(j)$ be the jth cell in the last row of R. Let $I_R(i)$ be the ith cell in the last row of the region prior to R. We will show that, for each run of x, all values of O_R can be computed in $O(n)$ time based on values of I_R. Accordingly, the last row of the edit graph can be computed in overall $O(m'n)$ time.

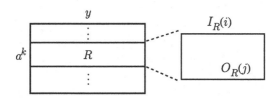

Fig. 1. A strip and its input/output borders

More specifically, let $DIST(i,j)$ be the cost of the optimal path starting from $I_R(i)$ and ending at $O_R(j)$ for $1 \leq i \leq j \leq n$. Define $OUT(i,j) = I_R(i) + DIST(i,j)$. Hirschberg [13] observed[1] that

$$O_R(j) = \min_{1 \leq i \leq j} OUT(i,j) \quad \text{for } 1 \leq j \leq n . \tag{1}$$

The matrix OUT defines a *Monge* matrix [2]. Therefore, all of the minima in (1) for all j can be determined in $O(n)$ time by applying the SMAWK algorithm [1]. $DIST$ and OUT matrices have also been used elsewhere, see [11,3,7,14,16,23,17]. Related preliminaries are given in the next section.

3   Preliminaries

Definition 1. *An $m \times n$ matrix $M = (c_{i,j})_{m \times n}$ is called* Monge *iff*

$$c_{i,j} + c_{i',j'} \leq c_{i,j'} + c_{i',j} \tag{2}$$

for all $1 \leq i \leq i' \leq m$ and $1 \leq j \leq j' \leq n$.

Symmetrically, we can define *an inverse Monge matrix* if (2) is replaced by $c_{i,j}+c_{i',j'} \geq c_{i,j'}+c_{i',j}$. Monge matrices have many nice properties and important applications. For surveys and recent applications, see [9,6,10].

Lemma 1 (Aggarwal and Park [2]). *All of the row minima and column minima in an $m \times n$ Monge or inverse Monge matrix can be determined in time*

---

[1] Hirschberg did not write in this form. It is borrowed from [11].

$O(m + n)$, *provided that each entry in the matrix can be accessed in time* $O(1)$. *(When there are many minima in a row or column, we can simply choose the first one.) Dually, all of the row and column maxima can also be found in the same time bound.*

The algorithm for Lemma 1, which is called the SMAWK algorithm, was first invented in [1] by applying the prune-and-search technique. Indeed, Aggarwal et al. in [1] defined a larger class called the *total monotonicity*. However, in real applications Monge or inverse Monge is sufficient for most cases. Extensions to higher dimensional arrays and on-line algorithms can be found in [2] and [6], respectively.

Lemma 2 (Aggarwal and Park [2]). *The matrices DIST and OUT used in (1) are Monge. When changing* min *to* max, *the corresponding DIST and OUT matrices are inverse Monge.*

In Appendix A, we give an informal proof of Lemma 2. As a consequence, if entries in $DIST$ can be accessed in $O(1)$ time, from Lemmas 1 and 2, all cells of O_R can be evaluated in $O(n)$ time for each strip R. In the following four sections, we will show how to define appropriate $O_R, I_R, DIST, OUT$ for various problems related to sequence alignment under this framework. The key point is on how to keep $DIST(i, j)$ accessible in $O(1)$ time.

4   Edit Distance with Unrestricted Scoring Matrices

Let $E(x, y)$ be the edit distance of strings x and y in the linear-gap model with scoring matrix δ. It accounts for the minimum number of weighted edit operations (i.e., the insertions, deletions, and substitutions) that transform x into y.

The traditional way to compute the edit distance is through the following recurrence relation:

$$E(ua, vb) = \min \{E(ua, v) + \delta(-, b), E(u, vb) + \delta(a, -), E(u, v) + \delta(a, b)\} \quad (3)$$

with base cases $E(ua, \epsilon) = E(u, \epsilon) + \delta(a, -)$, $E(\epsilon, vb) = E(\epsilon, v) + \delta(-, b)$, and $E(\epsilon, \epsilon) = 0$ where ϵ is the empty string and $-$ is the gap. We can replace (3) by

$$E(x'a^k, y[1..j]) = \min_{0 \le i \le j} \{E(x', y[1..i]) + E(a^k, y[(i + 1)..j])\} \quad (4)$$

when $x = x'a^k$ is run-length compressed and a^k is the last run of x. In fact, (4) is an instantiation of (1): set $I_R(i) = E(x', y[1..i])$ and $DIST(i, j) = E(a^k, y[(i + 1)..j])$. All we have to do is to show that $E(a^k, y[(i + 1)..j])$ can be evaluated in constant time, and consequently, the Monge paradigm described in Sects 2 and 3 ensures the $O(m'n)$ time and $O(n)$ space complexity.

For simplicity, let us first assume that the cost of an insertion or deletion is $d \ge 0$ and a substitution is $s \ge 0$. Then we have

Lemma 3. *Let* $a \in \Sigma$ *and* $z \in \Sigma^*$. *Let the length of* z *be* $|z|$ *and the number of occurrences of* a *in* z *be* $\sigma_a(z)$. *Then*

$$E(a^k, z) = d \max\{|z|, k\} - (d - s) \min\{|z|, k\} - s \min\{\sigma_a(z), k\} \quad \text{if } 0 \le s \le 2d \tag{5}$$

and

$$E(a^k, z) = d(|z| + k) - 2d \min\{\sigma_a(z), k\} \quad \text{if } s \ge 2d \ge 0 . \tag{6}$$

Proof. The edit distance of two strings is $s \times \alpha + d \times (\beta + \gamma)$ where α, β, and γ are the numbers of substitutions, insertions, and deletions, respectively. Since the string a^k contains identical letters, these numbers can be easily counted. Suppose $s \le 2d$; then $\gamma + \beta = \max\{|z|, k\} - \min\{|z|, k\}$ and $\alpha = \min\{|z|, k\} - \min\{\sigma_a(z), k\}$. Hence (5) follows. Suppose $s \ge 2d$; then any substitution can be replaced by a deletion followed by an insertion, and thus, $\alpha = 0$ and $\beta + \gamma = |z| + k - 2 \min\{\sigma_a(z), k\}$, the last term is the number of matches. Hence (6) follows.

We can spend $O(n)$ time to perform a linear scan on y in order to keep track of $\sigma_a(y[1..j])$ for $0 \le j \le n$, where $\sigma_a(y[1..0]) = \sigma_a(\epsilon) = 0$. Based on this preprocessing and the identity that $\sigma_a(y[(i+1)..j]) = \sigma_a(y[1..j]) - \sigma_a(y[1..i])$, Lemma 3 can be done in $O(1)$ time by instantiating z by $y[(i+1)..j]$.

Analysis of the time and space complexity. The preprocessing of $\sigma_a(y[1..j])$ for $0 \le j \le n$ takes $O(n)$ time and space, for each run in x. After this, each $DIST(i, j)$ and $OUT(i, j)$ can be accessed in $O(1)$ for $1 \le i \le j \le n$. Then applying the SMAWK algorithm on OUT matrix to find all of the column minima takes again $O(n)$ time and space. Note that the SMAWK algorithm need not evaluate all entries of the OUT matrix; it only scans $O(n)$ of them. Therefore, by using $O(n)$ time and space, we can advance the computation of the edit graph by a run of x. Hence the overall computation of $E(x, y)$ takes $O(m'n)$ time and $O(n)$ space. The time can be reduced to $O(\min\{m'n, mn'\})$ by choosing $E(x, y)$ or $E(y, x)$ according to which one costs less, and the corresponding space is $O(n)$ and $O(m)$, respectively.

As for a general scoring matrix, Lemma 3 can be extended smoothly when the size of the alphabet Σ is bounded. Let us reconsider the settings in Lemma 3, except now the weight for each edit operation is determined by δ. The following greedy algorithm can evaluate $E(a^k, y[(i+1)..j])$ for each i and j in $O(1)$ time. The evaluation of $E(a^k, z)$ inquires either to assign every a in a^k to a character in z or to delete itself. For those unmapped characters in z, insertions are applied. Define $\delta'(a, b) = \min\{\delta(a, b), \delta(a, -) + \delta(-, b)\}$ for all $b \in \Sigma$ (here a and b may be the same character). That is, if a substitution cannot be better than a deletion followed by an insertion, we never use it. Since the alphabet is bounded, without loss of generality, we can assume $\delta'(a, b_1) \le \delta'(a, b_2) \le \cdots \le \delta'(a, b_{|\Sigma|})$ for all $b_t \in \Sigma$. Hence this provides a priority to choose the mate for each a in a^k to z by $b_1 \prec b_2 \prec \cdots \prec b_{|\Sigma|} \prec -$. Let the occurrences of b_t in z be $\sigma_{b_t}(z)$. Also, set $\sigma_-(z)$ be k, which serves as the sentinel. Hence there exists u between 1 and $|\Sigma|$ such that $\sum_{1 \le t \le u} \sigma_{b_t}(z) \ge k$ but $\sum_{1 \le t \le u-1} \sigma_{b_t}(z) < k$ (here $b_{|\Sigma|+1} = -$). Let $A + B = \sigma_{b_u}(z)$ such that $A + \sum_{1 \le t \le u-1} \sigma_{b_t}(z) = k$. Then the value of $E(a^k, z)$ equals to

$$A \cdot \delta'(a, b_u) + B \cdot \delta'(-, b_u) + \sum_{1 \le t \le u-1} \delta'(a, b_t) \cdot \sigma_{b_t}(z) + \sum_{u+1 \le t \le |\Sigma|} \delta'(-, b_t) \cdot \sigma_{b_t}(z) \ ,$$

which generalizes Lemma 3 and can be computed in $O(u) = O(|\Sigma|) = O(1)$ time.

Theorem 1. *The edit distance problem in the linear-gap model with unrestricted scoring matrices can be solved in $O(\min\{mn', m'n\})$ time for run-length-encoded strings x and y with lengths m and n, being compressed into m' and n' runs, respectively.*

5 Global Alignment Algorithm

The algorithm for the global alignment problem is simply the dual of the one described in Sect. 4. In this case, (1) is modified into

$$O_R(j) = \max_{1 \le i \le j} OUT(i, j) \quad \text{for } 1 \le j \le n \ ,$$

and the inverse Monge property holds for the OUT matrix. Each entry in the OUT matrix can be accessed in $O(1)$ time by a similar greedy algorithm for the edit distance problem, as long as the alphabet is bounded. Then all of the column maxima, which are the values of O_R, can be found by the SMAWK algorithm in $O(n)$ time and space. This process can be continued, and finally the value of the optimal global alignment of x and y can be obtained in $O(m'n)$ time and $O(n)$ space. The trace of an optimal alignment can also be found in the same time and space bounds, by applying Hirschberg's technique [13].

Theorem 2. *The global alignment problem in the linear-gap model with unrestricted scoring matrices can be solved in $O(\min\{mn', m'n\})$ time for run-length-encoded strings x and y with lengths m and n, being compressed into m' and n' runs, respectively.*

6 Local Alignment Algorithm

In the local alignment problem, the goal is to identify a substring x' of x and a substring y' of y such that the global alignment score of x' and y' is maximized. Let $L(x, y)$ denote the score of the optimal local alignment for x and y ending at the ends of x and y. The traditional approach evaluates $L(x, y)$ through the following recurrence relation:

$$L(ua, vb) = \max\left\{0, L(ua, v) + \delta(-, b), L(u, vb) + \delta(a, -), L(u, v) + \delta(a, b)\right\} \quad (7)$$

with base cases $L(ua, \epsilon) = \max\{0, L(u, \epsilon) + \delta(a, -)\}$, $L(\epsilon, vb) = \max\{0, L(\epsilon, v) + \delta(-, b)\}$, and $L(\epsilon, \epsilon) = 0$. The score of the optimal local alignment for x and y can be obtained by finding the maximum over all cells in the induced alignment graph.

There are two kinds of local alignment paths in the alignment graph. The first kind categorizes paths fully contained in a strip. The second kind has paths that

y

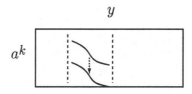

a^k

Fig. 2. An alignment inside a strip can always be moved down (or up) to touch the border in a local alignment graph

cross at least one border of a strip. Since the left side of a strip is always a whole run of x, paths of the first kind in fact can be *moved down or up* to touch the bottom or top of the strip (see Fig. 2 for an illustration). Hence it has at least one incarnation of the second kind, and thus can be ignored during the computation.

Paths of the second kind have a general pattern. Each path has three parts: S, H, E, which stand for segments of the path as follows (see Fig. 3):

S : a segment begins at some strip and ends at the bottom of this strip;
H : a segment begins at the top of a strip and ends at the bottom of some strip;
E : a segment begins at the top of a strip and ends within this strip.

Each part may be empty. The H part can be handled just like paths in the global alignment problem in a way propagating from one strip to the next. We can first ignore E part since their values cannot propagate to other strips, and after finishing values on borders of strips in the alignment graph, we can regain the contributions of E part into the alignment graph. As to incorporate S part into the alignment graph, we need to modify (4) into

$$L(x'a^k, y[1..j]) =$$
$$\max\left\{\max_{0\le i\le j}\left\{L(x', y[1..i]) + G(a^k, y[(i+1)..j])\right\}, L(a^k, y[1..j])\right\} \quad (8)$$

where $G(a^k, y[(i+1)..j]$ is the *global* alignment score that can be evaluated in $O(1)$ time. Note that (8) is different from (4) in the last term $L(a^k, y[1..j])$. Hence we create an array $S_R(j)$ for $0 \le j \le n$ at O_R in order to evaluate (8) where $S_R(j) = L(a^k, y[1..j])$. In the following paragraphs, we show how to calculate all entries of S_R in $O(n)$ time. Again, the trick is the inverse Monge property.

Let us reformulate the framework provided in Sect. 2 for the local alignment problem. The meanings of I_R and O_R are the same, but at this time they record the scores of optimal local alignments ending at the corresponding cells. In other words, $I_R(i) = L(x', y[1..i])$ and $O_R(j) = L(x'a^k, y[1..j])$. Leave $DIST(i, j) = G(a^k, y[(i+1)..j])$ and $OUT(i, j) = I_R(i) + DIST(i, j)$ unchanged. Then (8) can be rewritten into

$$O_R(j) = \max\left\{\max_{0\le i\le j} OUT(i, j), S_R(j)\right\}.$$

The computation of $\max_{0\le i\le j} OUT(i, j)$ for $0 \le j \le n$ is as before, and it can be finished in $O(n)$ time and space, and thus the H part follows.

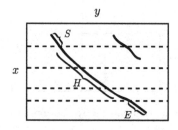

Fig. 3. Paths need to be considered in a local alignment graph

The calculations of S_R are closely related to $\max_{0 \leq i \leq j} DIST(i,j)$ for $0 \leq j \leq n$. Given j, let $i(j)$ be the first i that maximizes $DIST(i,j)$. It stands for the score of an optimal *local alignment* ending at $O_R(j)$ under the constraint that it must use exactly k copies of a's from a^k, a run of x. However, $S_R(j)$ stands for the optimal local alignment ending at $O_R(j)$ that uses *at most* k copies of a's. This difference can be resolved by applying this simple idea: set $\delta(a,-)$ to be zero; that is, the deletion of a's from a^k is free so that 'exactly' performs as well as 'at most'. This trick can be applied because we can always regroup all of the deletions of a's to its very beginning. As a consequence they can be truncated if deleting an a causes a negative score. By the above discussion, we have to consider two cases: (1) $\delta(a,-) < 0$ and (2) $\delta(a,-) \geq 0$. For the first case, let $\delta_{\mathrm{L}}(a,-) = 0$ and keep all other scores the same as δ. Let $DIST_{\mathrm{L}}(i,j)$ be the score of the optimal global alignment of a^k against $y[(i+1)..j]$ under δ_{L}. Then $DIST_{\mathrm{L}}$ is inverse Monge and the resulting $\max_i DIST_{\mathrm{L}}(i,j)$, which is the required $S_R(j)$, for $0 \leq j \leq n$, can be computed in O(n) time and space. The second case deals with $\delta(a,-) \geq 0$ (this could happen since δ is arbitrary). This case is much easier since $L(a^k, y[(i+1)..j])$ always uses all a^k and thus, $L(a^k, y[(i+1)..j])$ coincides with $DIST(i,j)$. Therefore $S_R(j) = \max_i DIST(i,j)$.

So far, we know how to handle S and H parts in the local alignment graph. In order to handle the E part, we need to create another array $E_R(i)$ for $0 \leq i \leq n$ at I_R such that $E_R(i) = L(a^k, y[(i+1)..n])$. This case can be processed similarly as the procedure for the S part.

The final optimal local alignment for $L(x,y)$ can be found by a linear scan over all entries in the output border O_R plus the corresponding contributions from the E part.

Analysis of the time and space complexity. It is easy to see that the time and space complexity is the same as the algorithm for the global alignment problem. The trace of an optimal local alignment can also be constructed in O($m'n$) time and O(n) space by Hirschberg's technique [13].

7   Approximate Matchings

Given a text T, a pattern P, a distance matrix δ, and a bound K, the problem of approximate matching is to find all occurrences of substrings T' of T such that the edit distance of T' and P under δ does not exceed the bound K. Its dual is to find out all occurrences of substrings whose global alignment scores against P are above the bound K.

Let T be run-length-encoded, with length n and n' runs. Let the length of P be m. We first propose an $O(mn')$ time and $O(m)$ space algorithm that can locate all end positions indexed by runs. The exact end locations of T' on T can further be reported as intervals on T in the same time and space bound. Consequently, all of the r occurrences can be reported with additional $O(r)$ time.

The idea is very similar to the algorithm in Sect. 6. We simply sketch the idea here. We put P on top of the edit graph and T to the left. Again, each optimal path can be broken into three parts: S, H, and E. However, in this time the prefixes and suffixes of P (which corresponds to y) cannot be truncated, but this facility is still available on T. The only difference is to set $S_R(j) = DIST_L(0, j)$ instead of $\min_i DIST_L(i, j)$. This prohibits the possibility to truncate the prefix of P. The E part is processed similarly, which prevents the removal of suffixes of P. If there exists a strip R and an index j such that $O_R(j) + E_{R'}(j)$ below the threshold K, where R' is the strip next to R, then we find an occurrence.

The next step is on how to locate all end positions of T' on T. Let T' be such an occurrence, and let R the the strip that contains the E part of T'. Let $I_R(i)$ be the starting position of E for T'. Since $I_R(i)$ records the distance of the optimal trace among all paths passing through it, its value cannot be worse than T'. Hence if we can find all segments, starting from $I_R(i)$ and ending to the right border of R, such that each of their edit distance against $P[(i+1)..m]$ plus the value of $I_R(i)$ is below K, then this is a valid occurrence. The greedy algorithm proposed in the end of Sect. 4 can be modified to fulfil this requirement. For each $I_R(i)$, the corresponding end locations of these segments (which are on the right border of R) then can be represented by an interval. (This is because we regroup letters in $P[(i+1)..m]$ into $|\Sigma|$ buckets, and the cost of mating or deleting one a to the same bucket costs the same, and the costs for buckets are sorted.) An additional check can also report approximate matches that occurred inside a strip.

Theorem 3. *Let T be a text with length n, being run-length-compressed into n' runs. Let P be a pattern with m characters. For any K, which serves as the threshold for matching, all of the occurrences T' of T such that the edit distance of T' and P is bounded above by K can be determined in $O(mn')$ time and $O(m)$ space. The threshold K and entries in the distance matrix δ can depend on m and n'.*

8   Conclusion

We improve previous $O(m'n + nm')$ time algorithms to $O(m'n)$, which makes one of the string truly compressed in the run-length encoding. However, another string is still left flat. An intriguing question is on (if it is possible) how to design an efficient algorithm for each problem whose time complexity only depends on the numbers of runs, as what were done in [4,22] for the LCS problem.

Another direction to extend the applicability of these algorithms is to introduce the *affine gap penalty*, which takes consecutive gaps as a whole unit and the costs for opening and extending a gap are different. In [15], Kim et al. proposed an $O(m'n + mn')$-time algorithm for the global alignment problem with affine gap penalty on run-length-encoded strings. It uses similar techniques as in [5,21]. However, Ledergerber et al. in [17] have pointed out that the matrix $OUT(i, j)$ described in Sect. 2 might not be (inverse) Monge. Therefore direct application of the model provided in Sect. 2 seems impossible. We note that it is possible to combine the results of [19] and [15] to get an $O(\min\{mn', m'n\})$-time algorithm for the affine gap penalty.

References

1. Aggarwal, A., Klawe, M.M., Moran, S., Shor, P., Wilher, R.: Geometric Applications of a Matrix-Searching Algorithm. Algorithmica 2(1), 195–208 (1987)
2. Aggarwal, A., Park, J.: Notes on Searching in Multidimensional Monotone Arrays. In: Proceedings of the 29th IEEE Symposium on Foundations of Computer Science (FOCS 1988), pp. 497–512 (1988)
3. Apostolico, A., Atallah, M.J., Larmore, L.L., Mcfaddin, S.: Efficient Parallel Algorithms for String Editing and Related Problems. SIAM Journal on Computing 19(5), 968–988 (1990)
4. Apostolico, A., Landau, G.M., Skiena, S.: Matching for Run-Length Encoded Strings. Journal of Complexity 15(1), 4–16 (1999)
5. Arbell, O., Landau, G.M., Mitchell, J.S.B.: Edit Distance of Run-Length Encoded Strings. Information Processing Letters 83(6), 307–314 (2002)
6. Bein, W.W., Golin, M.J., Larmore, L.L., Zhang, Y.: The Knuth-Yao Quadrangle-Inequality Speedup is a Consequence of Total-Monotonicity. In: Proceedings of the 7th annual ACM-SIAM Symposium on Discrete Algorithms (SODA 2006), pp. 31–40 (2006)
7. Benson, G.: A Space Efficient Algorithm for Finding the Best Nonoverlapping Alignment Score. Theoretical Computer Science 145(1–2), 357–369 (1995)
8. Bunke, H., Csirik, J.: An Improved Algorithm for Computing the Edit Distance of Run-Length Coded Strings. Information Processing Letters 54(2), 93–96 (1995)
9. Burkard, R.E., Klinz, B., Rudolf, R.: Perspectives of Monge Properties in Optimization. Discrete Applied Mathematics 70(2), 95–161 (1996)
10. Burkard, R.E.: Monge Properties, Discrete Convexity and Applications. European Journal of Operational Research 176(1), 1–14 (2007)
11. Crochemore, M., Landau, G.M., Ziv-Ukelson, M.: A Subquadratic Sequence Alignment Algorithm for Unrestricted Scoring Matrices. SIAM Journal on Computing 32(6), 1654–1673 (2003)
12. Gusfield, D.: Algorithms on Strings, Trees, and Sequences. Cambridge University Press, Cambridge (1997)
13. Hirschberg, D.S.: A Linear Space Algorithm for Computing Maximal Common Subsequences. Communications of the ACM 18(6), 341–343 (1975)
14. Kannan, S.K., Myers, E.W.: An Algorithm for Locating Nonoverlapping Regions of Maximum Alignment Score. SIAM Journal on Computing 25(3), 648–662 (1996)
15. Kim, J.W., Amir, A., Landau, G.M., Park, K.: Computing Similarity of Run-Length Encoded Strings with Affine Gap Penalty. In: Consens, M.P., Navarro, G. (eds.) SPIRE 2005. LNCS, vol. 3772, pp. 315–326. Springer, Heidelberg (2005)

16. Landau, G.M., Ziv-Ukelson, M.: On the Common Substring Alignment Problem. Journal of Algorithms 41(2), 338–359 (2001)
17. Ledergerber, C., Dessimoz, C.: Alignments with Non-overlapping Moves, Inversions and Tandem Duplications in $o(n^4)$ Time. Journal of Combinatorial Optimization (to appear, 2007)
18. Levenshtein, V.I.: Binary Codes Capable of Correcting, Deletions, Insertions and Reversals. Soviet Physics Doklady 10, 707–710 (1966)
19. Liu, J.J., Huang, G.S., Wang, Y.L., Lee, R.C.T.: Edit Distance for a Run-Length-Encoded String and an Uncompressed String. Information Processing Letters 105(1), 12–16 (2007)
20. Liu, J.J., Wang, Y.L., Lee, R.C.T.: Finding a Longest Common Subsequence Between a Run-Length-Encoded String and an Uncompressed String. Journal of Complexity (to appear, 2008)
21. Mäkinen, V., Navarro, G., Ukkonen, E.: Approximate Matching of Run-Length Compressed Strings. Algorithmica 35(4), 347–369 (2003)
22. Mitchell, J.: A Geometric Shortest Path Problem, with Application to Computing a Longest Common Subsequence in Run-Length Encoded Strings. Technical report, SUNY Stony Brook (1997)
23. Schmidt, J.P.: All Highest Scoring Paths in Weighted Grid Graphs and Their Application to Finding All Approximate Repeats in Strings. SIAM Journal on Computing 27(4), 972–992 (1998)
24. Wagner, R.A., Fischer, M.J.: The String-to-String Correction Problem. Journal of the ACM 21(1), 168–173 (1974)

A   Monge Property

The reason that OUT matrix in (1) is Monge (or inverse Monge) is as follows. As for minimization problems, such as the edit distance problem, we intend to find *shortest paths* in edit graphs. First, observe that $OUT(i,j)$'s are defined for $1 \leq i \leq j \leq n$, whose domain is not a square (see Fig. 4). Let us first consider $1 \leq i < i' \leq j < j' \leq n$. In Fig. 5, $DIST(i,j)$ and $DIST(i',j')$ are the lengths of shortest paths (drawn solid) from i to j and from i' to j', respectively. Similarly, the dashed lines are the shortest paths for i-j' and i'-j. Intuitively, $DIST(i,j) \leq DIST(i,c) + DIST(c,j)$ and $DIST(i',j') \leq DIST(i',c) + DIST(c,j')$, as long as lengths of paths are additive in the edit graph. This holds because $DIST(i,j)$ is the length of a shortest path connecting i and j, and $DIST(i,c) + DIST(c,j)$ represents the length of *another* path which may not be the shortest. We remark that this result does not rely on the condition of the triangle inequality for weights on edges; in fact, triangle inequality may not hold for paths on an edit graph. On the other hand, we have $DIST(i,j') + DIST(i',j) = DIST(i,c) + DIST(c,j') + DIST(i',c) + DIST(c,j)$. Combining them, we get

$$DIST(i,j) + DIST(i',j') \leq DIST(i,j') + DIST(i',j) \ ,$$

and thus

$$OUT(i,j) + OUT(i',j') \leq OUT(i,j') + OUT(i',j)$$

for $1 \leq i < i' \leq j < j' \leq n$. The above argument partially fulfils the Monge property. As for $1 \leq j < i \leq n$, this can be resolved by setting $OUT(i,j) = \infty$

and taking the convention that $r + \infty = \infty + r = \infty$ and $\infty + \infty = \infty$ for any real number r. Hence the matrix $OUT(i, j)$ for $1 \leq i, j \leq n$ in the minimization version is Monge.

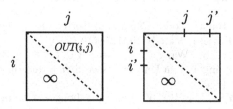

Fig. 4. Extension of the OUT matrix to the Monge matrix

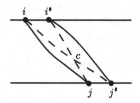

Fig. 5. A geometric interpretation of the quadrangle inequality

As for maximization problems, such as sequence alignment problems, we intend to find *longest paths* in the *alignment* graphs (note that an alignment graph is always acyclic). The argument in the above paragraph can also be applied by setting $OUT(i, j) = -\infty$ for $1 \leq j < i \leq n$ and taking the convention that $r + -\infty = -\infty + r = -\infty$ and $-\infty + -\infty = -\infty$ for any real number r. Finally, we conclude that the matrix $OUT(i, j)$ for $1 \leq i, j \leq n$ in the maximization version is inverse Monge.

A 2.25-Approximation Algorithm for Cut-and-Paste Sorting of Unsigned Circular Permutations*

Xiaowen Lou and Daming Zhu

School of Computer Science and Technology,
Shandong University, Jinan 250101, P.R. China
xwlou@mail.sdu.edu.cn, dmzhu@sdu.edu.cn

Abstract. We consider sorting unsigned circular permutations by cut-and-paste operations. For a circular permutation, a cut-and-paste operation can be a reversal, a transposition, or a transreversal. For the sorting of signed permutations, there are several approximation algorithms allowing various combinations of these operations. For the sorting of unsigned permutations, we only know a 3-approximation algorithm and an improved algorithm with ratio 2.8386+δ, both allowing reversals and transpositions. In this paper, by new observations on the breakpoint graph, we present a 2.25-approximation algorithm for cut-and-paste sorting of unsigned circular permutations.

1 Introduction

A genome can be represented by a permutation of integers, where each integer stands for a gene. Some genomes are represented by circular permutations, and others are represented by linear permutations [1],[6]. Data representing genomes can be signed or unsigned, depending on the sequencing method of the biological experiments. Both circular and linear permutations can be signed or unsigned.

Most typical rearrangement operations have the following features: cut a segment out of a genome, possibly reverse it, and then paste it back to the remaining genome. More precisely, a *reversal* cuts a segment out of the genome, reverses it, then pastes it back to exactly where it was cut; a *transposition* cuts a segment out of the genome, then pastes it back to a new position; a *transreversal* cuts a segment out of the genome, reverses it, then pastes it back to a new position. Therefore, each of the three operations can be viewed as a kind of *cut-and-paste* operation.

Rearrangement sorting on a permutation asks to find a shortest sequence of rearrangement operations that transform one permutation into another. The sorting results can be used to infer the evolutionary events between two genomes. Hannenhalli and Pevzner showed that reversal sorting can be solved in polynomial time for signed permutations [8]. To date, the optimal reversal sequence

---

* Supported by (1) National Nature Science Foundation of China, 60573024. (2) Chinese National 973 Plan, previous special, 2005cca04500.

X. Hu and J. Wang (Eds.): COCOON 2008, LNCS 5092, pp. 331–341, 2008.

can be computed in subquadratic time [10]. On the other hand, reversal sorting for unsigned permutations is proved to be NP-Hard [3]. The currently best performance ratio is 1.375 [2]. For the problem of transposition sorting, several 1.5-approximation algorithms are proposed [1],[6].

Each kind of reversal, transposition and transreversal events can happen during the evolution of a genome. Walter et al. gave a 2-approximation algorithm for sorting signed linear permutations by reversals and transpositions [11]. Gu et al. gave a 2-approximation algorithm for sorting signed linear permutations by transpositions and transreversals [5]. For unsigned linear permutations, Walter et al. gave a 3-approximation algorithm which allows reversals and transpositions [11]. And recently Rahman et al. improved the performance ratio for this problem to $2.8386+\delta$ for any $\delta > 0$ [9]. Hartman and Sharan gave a 1.5-approximation algorithm for sorting signed circular permutations by transpositions and transreversals [7].

Cranston et al. first studied the bounds for cut-and-paste sorting of permutations [4]. They showed that every unsigned linear permutation of n elements can be transformed to the identity permutation in at most $\lfloor 2n/3 \rfloor$ cut-and-paste operations [4]. But their algorithm does not give any approximation guarantee.

In this paper, we present a new algorithm to sort the unsigned circular permutations by cut-and-paste operations. By new observations on the breakpoint graph, we can averagely reduce more than $\frac{4}{3}$ breakpoints per single cut-and-paste operation. This leads to the performance ratio of 2.25. The time complexity is $O(n^2)$. Proofs for lemma 1, 3, 5, 8 are given in the full version of this paper.

2   Preliminaries

An *unsigned permutation* is a permutation of n unsigned integers $\{1, 2, \ldots, n\}$, denoted as $\pi = [g_1, \ldots, g_n]$. An *unsigned circular permutation* is an unsigned permutation where the first integer follows the last integer. We also use $\pi = [g_1, \ldots, g_n]$ to represent a circular permutation with $g_1 = 1$ as the first integer. To draw the figure of the circular permutation, we always write the elements of π counterclockwise to form a cycle. In the following, we identify $n+1$ with 1 in both indices and elements of a circular permutation.

2.1   What Is Cut-and-Paste

A *segment* $[g_i, \ldots, g_j]$ of π is a consecutive sequence in π from g_i to g_j. A *cut-and-paste* operation cuts a segment out of the permutation, possibly reverses it, then pastes it back to the remaining permutation. A cut-and-paste operation on a circular permutation can be a reversal, a transposition or a transreversal. A *reversal* $r(i, j)$ cuts the segment $[g_{i+1}, \ldots, g_j]$ out of π, reverses it, then pastes it back to exactly where it was cut, thus transforming π into $\sigma = [g_1, \ldots, g_i, g_j, \ldots, g_{i+1}, g_{j+1}, \ldots, g_n]$. A *transposition* $t(i, j, k)$ cuts the segment $[g_{i+1}, \ldots, g_j]$ out of π, pastes it to the position between g_k and g_{k+1}, thus transforming π into $\sigma = [g_1, \ldots, g_i, g_{j+1}, \ldots, g_k, g_{i+1}, \ldots, g_j, g_{k+1}, \ldots, g_n]$.

A *transreversal* $rt(i,j,k)$ cuts $[g_{i+1}, \ldots, g_j]$ out of π, reverses it, then pastes it to the position between g_k and g_{k+1}, thus transforming π into $\sigma = [g_1, \ldots, g_i, g_{j+1}, \ldots, g_k, g_j, \ldots, g_{i+1}, g_{k+1}, \ldots, g_n]$. The transformation from π into σ by operation ρ is denoted as $\pi \cdot \rho = \sigma$.

Let $\pi = [g_1, \ldots, g_n]$ and $\iota = [\iota_1, \ldots, \iota_n]$ be two unsigned circular permutations. The *cut-and-paste sorting problem* asks to find a sequence of cut-and-paste operations $\rho_1, \rho_2, \ldots, \rho_t$, such that $\pi \cdot \rho_1 \cdot \rho_2 \cdots \rho_t = \iota$ and t is minimized. We call t the *cut-and-paste distance* between π and ι, denoted as $d(\pi, \iota)$. A *sorting sequence* consists of all the rearrangement operations during the transformation from π to ι. By convention, it suffices to take ι as the identity permutation $[1, 2, \ldots, n]$.

2.2   The Breakpoint Graph of Unsigned Circular Permutation

Given two unsigned circular permutations π and ι as above. A pair of consecutive elements $[g_i, g_{i+1}]$ of π forms an *adjacency* if there is an integer k ($1 \leq k \leq n$) such that either $[g_i, g_{i+1}] = [\iota_k, \iota_{k+1}]$ or $[g_i, g_{i+1}] = [\iota_{k+1}, \iota_k]$. Otherwise, $[g_i, g_{i+1}]$ forms a *breakpoint*. The number of breakpoints in π is denoted as $b(\pi, \iota)$. A segment $[g_i, \ldots, g_j]$ of π is defined to be a *strip* if each $[g_k, g_{k+1}]$ is an adjacency for $i \leq k < j$, but either of $[g_{i-1}, g_i]$ and $[g_j, g_{j+1}]$ is a breakpoint. A strip of one element is a *singleton*.

Define the *breakpoint graph* $G(\pi, \iota)$ as follows: $G(\pi, \iota)$ has n vertices $g_1, g_2, \ldots, g_n$; set a *gray edge* between vertices g_i and g_j if g_i and g_j are consecutive elements of ι and $|i-j| \neq 1$; set a *black edge* between vertices g_i and g_{i+1} if they form a breakpoint. Clearly, $G(\pi, \iota)$ has the same number of black and gray edges, which equals to $b(\pi, \iota)$. If $\pi = \iota$, then $G(\pi, \iota)$ has no edge. Each vertex in the breakpoint graph has a degree from $\{0, 2, 4\}$, and each singleton has degree 4. A path in $G(\pi, \iota)$ is called *alternating* if every two consecutive edges on the path have distinct colors. An *alternating cycle* is a closed alternating path and has the same number of black and gray edges. The *length of a cycle* is the number of black edges in it. A cycle of length k is denoted as a k-cycle. In the following, we use g_i to denote an element of π or a vertex of $G(\pi, \iota)$ without making distinction, and a path (cycle) is always referred to as an alternating path (cycle).

It is clear that a cut-and-paste operation can reduce at most three breakpoints. Since $b(\pi, \iota) = 0$ for $\pi = \iota$, the following lower bound on cut-and-paste distance is obvious.

Theorem 1. *Given two unsigned permutations π and ι, $d(\pi, \iota) \geq b(\pi, \iota)/3$.*

3   A 2.25-Approximation Algorithm

3.1   Ties and Folds

Let $e_1 = (g_{i_1}, g_{i_2})$ and $e_2 = (g_{j_1}, g_{j_2})$ be two gray edges of $G(\pi, \iota)$. We say e_1 crosses e_2 if $i_1 < j_1 < i_2 < j_2$ or $j_1 < i_1 < j_2 < i_2$. We call a path consisting of three black edges and two gray edges a *tie*. The black edge in the middle of the path is called the *bottom* of the tie, the two black edges at the ends of the tie are called the *hands* of the tie, and the two gray edges adjacent to the bottom are called the *arms* of the tie. If a tie has one arm crossing the other, it is a *twisted tie*; otherwise, a *flat tie*. In the following, we represent a tie by the sequence of the six vertices on the tie, denoted by $t_0 = <u, i', i, j, j', v>$, where (u, i') and (j', v) are the hands, (i', i) and (j, j') are the arms, and (i, j) is the bottom of t_0.

By walking along a tie from one end to the other, each black edge of the tie can be assigned a direction of clockwise or counterclockwise. According to the relative directions of the bottom and the hands, we classify the ties into six types (Fig. 1). If a tie is twisted(flat) and its bottom has the same direction with both hands, it is defined to be *3-twisted(flat)* type. If a tie is twisted(flat) and its bottom has the same direction with only one hand, it is *2-twisted(flat)* type. If a tie is twisted(flat) and its bottom has the opposite direction with either of its two hands, it is *1-twisted(flat)* type. We also call a tie of k-twisted(flat) type a *k-twisted(flat)-tie* for $k \in \{1, 2, 3\}$. A tie divides the elements of the circular permutation into three disjoint segments. For simplicity, each segment is represented with two elements from the tie as its ending elements. The segment between two hands of the tie is called the *open segment*. Each of the other two segments is between the bottom and one hand of the tie, called the *shoulder segment*.

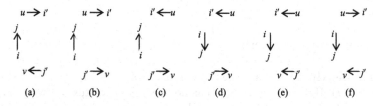

Fig. 1. Different types of a tie: (a) a 3-twisted-tie, (b) a 2-twisted-tie, (c) a 1-twisted-tie, (d) a 3-flat-tie, (e) a 2- flat-tie, and (f) a 1-flat-tie

Similarly, we call a path consisting of one gray edge and two black edges a *fold*, denoted by $f_0 = <p, m, m', s>$, where (p, m) and (m', s) are the black edges, (m, m') is the gray edge. When we walk along f_0, its two black edges can also be assigned their directions of clockwise or counterclockwise. If they have the same direction, call f_0 a *c-fold*; otherwise, *z-fold*. If f_0 is a c-fold, it divides the whole permutation into two disjoint segments: $[m, \ldots, m']$ and $[s, \ldots, p]$. For a black edge $e \notin \{(p, m), (m', s)\}$, we say e is *inside* (*or outside*) the c-fold f_0, if the segment $[s, \ldots, p]$ (or $[m, \ldots, m']$) contains e. We say a gray edge *crosses a fold*, if it crosses the gray edge of the fold.

We use *a k-y-move* to denote a cut-and-paste operation of type y which reduces k breakpoints for $0 \leq k \leq 3$ and $y \in \{r, t, rt\}$, where r, t and rt stand for reversal, transposition, and transreversal respectively. We also use a *k-move* to denote an operation which reduces k breakpoints; and use a *(i, j, k)-move* to denote three consecutive operations, where the first, the second and the third are i-move, j-move and k-move respectively. If the cut positions and the paste position of a cut-and-paste operation are black edges of the same tie (or fold), we call it a *move on the tie* (*or fold*).

A 2-cycle having one gray edge crossing the other can be removed by a 2-r-move. Thus we call such a 2-cycle a *good 2-cycle*; otherwise, a *bad 2-cycle*. We also classify the six kind of ties into two categories: the 3-twisted-ties, 2-twisted-ties and 1-flat-ties are called *good ties*, and the 1-twisted-ties, 2-flat-ties and 3-flat-ties are called *bad ties*, since the following lemma holds:

Lemma 1. *There exists a 2(3)-move on a good tie. Here 2(3) means either 2 or 3.*

Given a permutation π, if $\pi \neq \iota$ and $G(\pi, \iota)$ does not contain any good 2-cycles or good ties, we call π a *bad permutation*. For the initial permutation π and every intermediate permutation obtained after performing an operation, firstly we bypass its breakpoint graph and find a good 2-cycle or a good tie to remove if there exists one. We use *Search_good_ties*(π, ι) to denote such a greedy procedure. It will return a set S and a permutation π_1. Note that S is either an empty set or a set of 2(3)-moves, and π_1 is either ι or a bad permutation. If $\pi_1 = \iota$, the sorting of π is finished.

In fact, there may exist 2-moves for a bad permutation. For example, let $\pi = [1, 2, 13, 14, 10, 9, 12, 11, 7, 8, 15, 16, 5, 6, 3, 4]$. All the ties of $G(\pi, \iota)$ are bad. If we cut the segment [14, 10, 9], reverse it, and paste it between 8 and 15, we have a 2-move. This operation breaks one adjacency and two breakpoints, and creates three new adjacencies. In our algorithm, we do not perform such operations that break adjacencies.

3.2   Singleton Splitting

Given a bad permutation π_1, if π_1 contains singletons, we try to transform π_1 into a singleton-free permutation by splitting every singleton into two new elements. For every singleton g_i of π_1, we use a pair of consecutive elements $[g_i^-, g_i^+]$ to replace g_i in both π_1 and ι; thus get π_2 and ι' respectively. We call such a process transforming π_1 and ι into π_2 and ι' *singleton splitting*. If π_1 has s singletons, then both π_2 and ι' has $n' = n + s$ elements. As to the breakpoint graph, singleton splitting splits every singleton vertex g_i into two new vertices g_i^- and g_i^+; replaces the black edges (g_{i-1}, g_i) and (g_i, g_{i+1}) by (g_{i-1}, g_i^-) and (g_i^+, g_{i+1}); replaces the gray edges $(g_i - 1, g_i)$ and $(g_i, g_i + 1)$ by $(g_i - 1, g_i^-)$ and $(g_i^+, g_i + 1)$; thus transforms $G(\pi_1, \iota)$ into $G(\pi_2, \iota')$.

Since each singleton g_i in π_1 becomes an adjacency $[g_i^-, g_i^+]$ in π_2, obviously π_2 is a singleton-free permutation. Thus $G(\pi_2, \iota')$ must be uniquely decomposed into cycles. If π_1 is a bad permutation, π_2 is still a bad permutation. Moreover, π_2 and ι' have the same set of breakpoints as what π_1 and ι have. Any cut-and-paste operation we perform only cuts on breakpoints. Therefore, if we have got a sorting sequence for π_2, we can generate a sorting sequence for π_1 by merging g_i^- and g_i^+ into g_i, for every singleton g_i. Thus we have:

Lemma 2. *Every sorting sequence of π_2 mimics a sorting sequence of π_1 with the same number of cut-and-paste operations.*

3.3 Sorting the Singleton-Free Bad Permutations

In this section, we focus on how to find a feasible move for a singleton-free bad permutation. For convenience, we directly use permutation to represent a singleton-free permutation.

Lemma 3. *A gray edge must cross another gray edge in the breakpoint graph.*

Lemma 4. *If the breakpoint graph of a bad permutation has a bad 2-cycle, then a $(1,2(3))$-move is available.*

Proof. Let $c=<i,j,j',i'>$ be a bad 2-cycle with gray edges (i,i') and (j,j'). From lemma 3, there exists a gray edge crossing both (i,i') and (j,j'), let it be (m,m'), the gray edge of a fold $f_0=<p,m,m's>$. Note that f_0 shares no edges with c because there are no singletons. If f_0 is a z-fold, we can perform a reversal on (p,m) and (m',s). This is a 1-r-move and turns c into a good 2-cycle; thus a $(1,2)$-move is available. If f_0 is a c-fold, w.l.o.g, suppose that the black edge (i,j) is outside f_0. We can cut the segment $[p,\ldots,i',j',\ldots,s]$ and paste it reversely between i and j. This is a 1-rt-move and creates a 1-flat-tie $<p,i,i',j',j,s>$. From lemma 1, a $(1,2(3))$-move is available. □

Let t_0 be a bad tie, f_0 be a fold with gray edge $e=(m,m')$. We say f_0 *crosses* t_0, if e is not an arm of t_0 and e crosses at least one arm of t_0. If f_0 crosses only one arm of t_0, we call f_0 the *hand-fold* of t_0. If f_0 crosses both arms of t_0, we call f_0 the *bottom-fold* of t_0. A fold crossing a tie is either a hand-fold or a bottom-fold.

Lemma 5. *Both a 3-flat-tie and a 1-twisted-tie have an even number($\neq 0$) of hand-folds.*

From lemma 1 and 4, we can always remove a good tie or a 2-cycle by traversing the breakpoint graph if there is any. Thus lemma 6, 7, 9 will discuss each of the three bad ties in a bad permutation without any 2-cycles, call it *2-cycle-free*.

Lemma 6. *If there is a 1-twisted-tie in the breakpoint graph of a 2-cycle-free bad permutation, then a $(1,2(3))$-move is available.*

Proof. Let $t_0=<u,i',i,j,j',v>$ be a 1-twisted-tie. From lemma 5, t_0 has hand-folds. W.l.o.g., let $f_0=<p,m,m',s>$ be a hand-fold of t_0 and (m,m') crosses the arm (i,i').

Case 1. f_0 is a z-fold (Fig. 2(a)). We can perform a 1-r-move on f_0. This operation turns t_0 into a 2-twisted-tie; thus a $(1,2(3))$-move is available by lemma 1. For the special case of $(m',s)=(u,i')$ in Fig. 2(a), the 1-r-move on f_0 creates a good tie $<p,i',i,j,j',v>$, also a $(1,2(3))$-move.

Case 2. f_0 is a c-fold and (u,i') is inside f_0 (Fig. 2(b)). We can cut the segment $[s,\ldots,u,\ i',\ldots,p]$ and paste it reversely between j' and v. This is a 1-rt-move and creates a good tie $<u,i',i,j,j',p>$. Thus a $(1,2(3))$-move is available. Note that in Fig. 2(b), $(m',s)=(u,i')$ will not happen, since it will give rise to a good tie, a contradiction.

Case 3. f_0 is a c-fold and (u,i') is outside f_0 (Fig. 2(c)). We can cut the segment $[p,\ldots,j',v,\ldots,s]$ and paste it reversely between i' and u. This is a 1-rt-move and creates a good tie $<p,i',i,j,j',v>$. Thus a $(1,2(3))$-move is available.

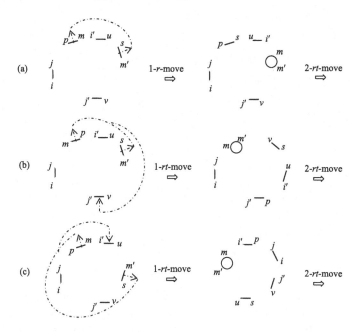

Fig. 2. Feasible $(1, 2(3))$-move for a 1-twisted-tie: The newly created adjacency (m, m') is denoted by a small circle

For the special case of $(s, m')=(j', v)$ in Fig. 2(c), the 1-rt-move on f_0 creates a good tie $<p, i', i, j, j', u>$, also a $(1, 2(3))$-move.   □

Lemma 7. *If there is a 2-flat-tie in the breakpoint graph of a 2-cycle-free bad permutation, then either a $(1, 2(3))$-move or a $(1, 1, 2(3))$-move is available.*

Proof. Let $t_0=<u, i', i, j, j', v>$ be a 2-flat-tie, where (v, j') is the hand having opposite direction with (u, i') and (i, j). From lemma 3, the arm (i, i') of t_0 must cross a gray edge, let it be (m, m'), the gray edge of a fold $f_0=<p, m, m', s>$.

Case 1. f_0 is t_0's hand-fold. We can always get a $(1, 2(3))$-move using the same way as in lemma 6 for the cases of f_0 being a z-fold or a c-fold with (u, i') outside it. If f_0 is a c-fold with (u, i') inside it, corresponding to the Case 2 of lemma 6, a 1-t-move on the black edges (p, m), (m', s) and (v, j') will create a good tie. Other details are omitted.

Case 2. f_0 is t_0's bottom-fold. There are three subcases according to the fold's type.

Case 2.1. f_0 is a z-fold (Fig. 3(a)). We can perform a reversal on f_0. This 1-r-move turns t_0 into a good tie. Thus a $(1, 2(3))$-move is available by lemma 1. For the special case of $(m', s)=(v, j')$ in Fig. 3(a), the first 1-r-move will create a good tie $<u, i', i, j, j', p>$; thus a $(1, 2(3))$-move is also available.

Case 2.2. f_0 is a c-fold and (i, j) is inside f_0 (Fig. 3(b)). We can cut the segment $[j, \ldots, s]$ and paste it reversely between v and j'. This is a 1-rt-move and creates a good tie $<p, m, m', i, i', u>$. Thus a $(1, 2(3))$-move is available.

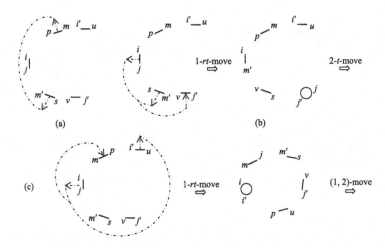

Fig. 3. Feasible $(1, 2(3))$-move or $(1, 1, 2(3))$-move for a 2-flat-tie

Case 2.3. f_0 is a c-fold and (i, j) is outside f_0 (Fig. 3(c)). We can cut the segment $[j, \ldots, v, j', \ldots, u]$ and paste it reversely between p and m. Since $(m', s) \neq (v, j')$ in a bad permutation, this is a 1-*rt*-move and creates a 1-twisted-tie $< s, m', m, j, j', v >$. From lemma 6, a $(1, 2(3))$-move follows this move; thus a $(1, 1, 2(3))$-move is available.   □

Lemma 8. *Let $t_0 = <u, i', i, j, j', v>$ be a 3-flat-tie in the breakpoint graph. If all the hand-folds of t_0 are c-folds crossing the arm (i', i), and (u, i') is outside all of them, then there exists a z-fold in the breakpoint graph. And such t_0 is called a bad 3-flat-tie.*

Lemma 9. *If there is a 3-flat-tie in the breakpoint graph of a 2-cycle-free bad permutation, then a $(1, 2(3))$-move or a $(0, 2(3), 2(3))$-move or a $(1, 1, 2(3))$-move is available.*

Proof. Let $t_0 = <u, i', i, j, j', v>$ be a 3-flat-tie. From lemma 5, t_0 has hand-folds. W.l.o.g., let $f_0 = <p, m, m', s>$ be a hand-fold of t_0 and (m, m') crosses the arm (i, i').

Case 1. f_0 is a z-fold (Fig. 4(a),(c)). From the definition of the gray edge, the shoulder segment $[j, \ldots, j']$ contains at least one black edge, let it be (q, n).

Case 1.1. f_0 takes the form as in Fig. 4(a), we can cut the segment $[n, \ldots, j', v, \ldots, m']$ and paste it reversely between p and m. This is a 1-*rt*-move and creates a good tie $<u, i', i, j, j', v>$ (Fig. 4(b)). Thus a $(1, 2(3))$-move is available.

Case 1.2. f_0 takes the form as in Fig. 4(c), we can cut the segment $[m, \ldots, i, j, \ldots, q]$ and paste it reversely between m' and s. This is a 1-*rt*-move and creates a good tie $<u, i', i, j, j', v>$ (Fig. 4(d)). Thus a $(1, 2(3))$-move is available. For the special case of $(s, m') = (j', v)$ in Fig. 4(c), the first 1-*rt*-move creates a good tie $<u, i', i, j, j', q>$.

Case 2. f_0 is a c-fold and (u, i') is inside f_0. Similarly as above, we can cut the segment $[m, \ldots, i, j, \ldots, q]$, and paste it between s and m'. This is a

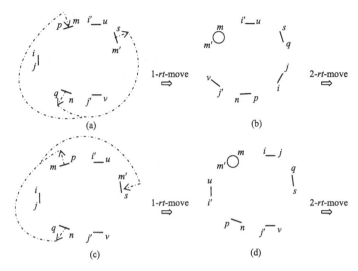

Fig. 4. Feasible $(1, 2(3))$-move for the 3-flat-tie in Case 1

1-t-move and turns t_0 into a good tie $< u, i', i, j, j', v >$. Thus a $(1, 2(3))$-move is available.

Case 3. f_0 is a c-fold and (u, i') is outside f_0. From lemma 5, t_0 has an even number of hand-folds. If there exists a z-fold crossing the arm (j, j') of t_0, we can have a $(1, 2(3))$-move as Case 1. If there exists a c-fold crossing the arm (j, j') and (j', v) is inside it, we can have a $(1, 2(3))$-move as Case 2. So we have only two subcases to discuss: arm (j, j') has a crossing c-fold with (j', v) outside it; or has no crossing hand-folds at all.

Case 3.1. There exists a c-fold $f_1 = <q, n, n', k>$ crossing the arm (j, j') of t_0 and (j', v) is outside f_1. We can cut the segment $[p, \ldots, i, j, \ldots, q]$ and paste it reversely between s and m'. This is a 0-rt-move which turns t_0 into a good tie and creates a new tie $t_1 = <k, n', n, m, m', p>$. If (n, n') crosses (m, m') (Fig. 5(a)), t_1 is a good tie of 3-twisted type (Fig. 5(b)). The 2-rt-move on t_0 turns t_1 into a good tie of 1-flat type (Fig. 5(c)). Thus a $(0, 2(3), 2(3))$-move is available. If (n, n') does not cross (m, m') (Fig. 5(d)), t_1 is a bad tie of 3-flat type (Fig. 5(e)). The 2-rt-move on t_0 turns t_1 into a good tie of 1-flat type (Fig. 5(f)). Thus a $(0, 2(3), 2(3))$-move is also available. Specially, in Fig. 5(a), if $(s, m') = (j', v)$, the first 0-rt-move turns t_0 into a good tie $<u, i', i, j, j', q>$; if $(u, i') = (n', k)$, the second 2-rt-move on t_0 creates a good tie $<v, n', n, m, m', p>$.

Case 3.2. t_0 is a bad 3-flat-tie. From lemma 8, a z-fold must exist, which implies the existence of 1-twisted-tie or 2-flat-tie in the breakpoint graph of a bad permutation. From lemma 6 and 7, either a $(1, 2(3))$-move or a $(1, 1, 2(3))$-move is available. $\qquad \square$

3.4 The Algorithm

Given the permutation π and ι, we compute a sequence of cut-and-paste operations that transform π into ι by gradually reducing $b(\pi, \iota)$. The algorithm

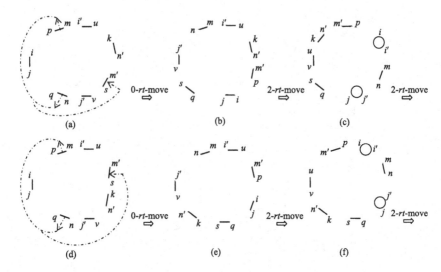

Fig. 5. Feasible $(0, 2(3), 2(3))$-move for the 3-flat-tie in Case 3.1

is given as *Circular_Sort*(π, ι). It only performs the following types of moves: a $2(3)$-move, a $(1, 2(3))$-move, a $(1, 1, 2(3))$-move, or a $(0, 2(3), 2(3))$-move. It takes $O(n)$ time to find a 2-cycle or a tie which is not a bad 3-flat-tie. Thus finding a feasible move as aforementioned and performing it takes $O(n)$ time. Each feasible move can averagely reduce at least $\frac{4}{3}$ breakpoints by every single operation. Since there are at most n breakpoints in $G(\pi, \iota)$, it only needs to perform $O(n)$ such feasible moves. Thus the time complexity is $O(n^2)$. And the performance ratio is described by the following theorem.

```
Algorithm Circular_Sort(π, ι)
1    (π₁, S) ← Search_good_ties(π, ι);
2    Transform π₁ and ι into π₂ and ι' by singleton splitting; S₁ ← ∅;
3    While π₂ ≠ ι'{
4        (π₂, S₂) ← Search_good_ties(π₂, ι); S₁ ← S₁ ∪ S₂;
5        If G(π₂, ι') contains a bad 2-cycle{
6            Perform a (1, 2(3))-move ρ; (Lemma 4)
7            π₂ ← π₂ · ρ; S₁ ← S₁ ∪ {ρ};
8        }//End If
9        Else{
10           Pick a tie t₀ that is not a bad 3-flat-tie;
11           If t₀ is a 1-twisted-tie, perform a (1, 2(3))-move ρ; (Lemma 6)
12           ElseIf t₀ is a 2-flat-tie, perform a (1, 2(3))-move or a (1, 1, 2(3))-move ρ; (Lemma 7)
13           Else perform a (1, 2(3))-move or (0, 2(3), 2(3))-move ρ; (Lemma 9)
14           π₂ ← π₂ · ρ; S₁ ← S₁ ∪ {ρ};
15       }//End Else
16   }//End While
17   Generate the sorting sequence S₃ for π₁ by mimicking the sequence S₁ for π₂; (Lemma 2)
18   Return S ← S ∪ S₃.
```

Theorem 2. *Let S be the set of ordered operations returned by algorithm Circular_Sort (π, ι), then $\frac{|S|}{d(\pi, \iota)} \leq 2.25$.*

Proof. Let π_1 be the permutation returned from *Search_good_ties*(π, ι) in step 1, π_2 be the permutation obtained from the singleton splitting in step 2. Obviously

$b(\pi_2, \iota') = b(\pi_1, \iota) \leq b(\pi, \iota)$. Therefore,

$$|S| \leq \frac{b(\pi, \iota) - b(\pi_1, \iota)}{2} + \frac{3b(\pi_2, \iota')}{4} = \frac{b(\pi, \iota) - b(\pi_2, \iota')}{2} + \frac{3b(\pi_2, \iota')}{4} \leq \frac{3b(\pi, \iota)}{4}$$

From theorem 1, $d(\pi, \iota) \geq \frac{b(\pi, \iota)}{3}$. This leads to $\frac{|S|}{d(\pi, \iota)} \leq \frac{9}{4} = 2.25$. □

References

1. Bafna, V., Pevzner, P.A.: Sorting by Transpositions. SIAM J. Discrete Math. 11, 272–289 (1998)
2. Berman, P., Hannenhalli, S., Karpinki, M.: 1.375-Approximation Algorithm for Sorting by Reversals. In: Möhring, R.H., Raman, R. (eds.) ESA 2002. LNCS, vol. 2461, pp. 200–210. Springer, Heidelberg (2002)
3. Caprara, A.: Sorting by Reversals is Difficult. In: Proc. of 1th Annual International Conference on Research in Computational Molecular Biology (RECOMB 1997), pp. 75–83 (1997)
4. Cranston, D.W., Sudborough, I.H., West, D.B.: Short Proofs for Cut-and-Paste Sorting of Permutations. Discrete Mathematics 307, 2866–2870 (2007)
5. Gu, Q.P., Peng, S.P., Sudborough, H.: A 2-Approximation Algorithm for Genome Rearrangements by Reversals and Transpositions. Theoretical Computer Science 210, 327–339 (1999)
6. Hartman, T., Shamir, R.: A Simpler 1.5-Approximation Algorithm for Sorting by Transpositions. In: Baeza-Yates, R., Chávez, E., Crochemore, M. (eds.) CPM 2003. LNCS, vol. 2676, pp. 156–169. Springer, Heidelberg (2003)
7. Hartman, T., Sharan, R.: A 1.5-Approximation Algorithm for Sorting by Transpositions and Transreversals. J. Computer and System Sciences 70, 300–320 (2005)
8. Hannenhalli, S., Pevzner, P.A.: Transforming Cabbage into Turnip (Polynomial Algorithm for Sorting Signed Permutations by Reversals). In: Proc. of 27th Annual ACM Symposium on Theory of Computing (STOC 1995), pp. 178–189 (1995)
9. Rahman, A., Shatabda, S., Hasan, M.: An Approximation Algorithm for Sorting by Reversals and Transpositions. J. Discrete Algorithms (in press) doi:10.1016/j.jda.2007.09.002
10. Tannier, E., Sagot, M.-F.: Sorting by Reversals in Subquadratic Time. In: Sahinalp, S.C., Muthukrishnan, S.M., Dogrusoz, U. (eds.) CPM 2004. LNCS, vol. 3109, pp. 1–13. Springer, Heidelberg (2004)
11. Walter, M.E.M.T., Dias, Z., Meidanis, J.: Reversal and Transposition Distance of Linear Chromosomes. In: Proc. of String Processing and Information Retrieval (SPIRE 1998), pp. 96–102 (1998)

A Practical Exact Algorithm for the Individual Haplotyping Problem MEC/GI[*]

Minzhu Xie[1,2], Jianxin Wang[1,**], and Jianer Chen[1,3]

[1] School of Information Science and Engineering,
Central South University, Changsha 410083, P.R. China
[2] College of Physics and Information Science,
Hunan Normal University, Changsha 410081, P.R. China
[3] Department of Computer Science,
Texas A&M University, College Station, TX 77843, USA
xieminzhu@hotmail.com,jxwang@mail.csu.edu.cn,chen@cs.tamu.edu
http://netlab.csu.edu.cn

Abstract. Given the genotype and the aligned single nucleotide polymorphism (SNP) fragments of an individual, *Minimum Error Correction with Genotype Information* (MEC/GI) is an important computational model to infer a pair of haplotypes compatible with the genotype by correcting minimum number of SNPs in the given SNP fragments. For the problem, there has been no practical exact algorithm. In DNA sequencing experiments, due to technical limits, the maximum length of a fragment sequenced directly is about 1kb. In consequence, the maximum number k of SNP sites that a fragment covers is usually small (usually smaller than 10). Based on the observation above, the current paper introduces a new parameterized dynamic programming algorithm of running time $O(mk2^k + m\log m + mk)$, where m is the number of fragments. The algorithm solves the MEC/GI problem efficiently even if the number of fragments and SNPs are large, and is practical in real biological applications.

1 Introduction

The differences in physical traits among human beings are genetically caused by DNA variations. *Single nucleotide polymorphism* (SNP) is believed to be the predominant form of human genetic variation, which is a single base mutation of a DNA sequence that occurs in at least 1% of the population [11]. In humans and other diploid organisms, chromosomes are paired up, and a *haplotype* describes the SNP sequence on a chromosome, while a *genotype* describes the conflated data of the SNP sequences on a pair of chromosomes.

---

[*] This research was supported in part by the National Natural Science Foundation of China under Grant Nos. 60433020 and 60773111, the National Basic Research 973 Program of China No.2008CB317107, the Program for New Century Excellent Talents in University No. NCET-05-0683, the Program for Changjiang Scholars and Innovative Research Team in University No. IRT0661, and the Scientific Research Fund of Hunan Provincial Education Department under Grant No.06C526.

[**] Corresponding author.

X. Hu and J. Wang (Eds.): COCOON 2008, LNCS 5092, pp. 342–351, 2008.
© Springer-Verlag Berlin Heidelberg 2008

Haplotypes provide greater power in disease association studies than individual SNPs [8]. However, Haplotyping has been time-consuming and expensive using biological techniques. Therefore, effective computational techniques have been attractive alternatives to determine haplotypes. The computational problems concerning haplotyping fall into two classes [13]: individual haplotyping and population haplotyping. Individual haplotyping, also called the haplotype assembly problem, is to reconstruct a pair of haplotypes of an individual according to some optimal criteria from his aligned SNP fragments [4]. While population haplotyping, also called the haplotype inference problem, is to infer the haplotypes of a population from their genotypes [2]. The current paper is focused on the individual haplotyping problem MEC/GI.

2   The Individual Haplotyping Problem MEC/GI

It is believed that SNPs have only two different alleles, and the major allele is denoted by '0' and the minor allele by '1'. Therefore, a haplotype can be represented as a string over a two-letter alphabet $\{0, 1\}$ instead of the four-letter alphabet$\{A, C, G, T\}$. And a genotype can be presented as a string over a three-letter alphabet $\{0, 1, 2\}$, where '0' (or '1') means that both chromosomes have the same allele '0' (or '1') at the same SNP site, and where '2' means that both chromosomes have different alleles, one is '0' and the other is '1'.

Let G and $\mathcal{H} = (H^1, H^2)$ be a genotype and a pair of haplotypes of length n, respectively. $\mathcal{H}$ is a *resolution* of G if the following condition holds: for each $1 \leq j \leq n$, if G_j='0' (or '1'), $H_j^1 = H_j^2 =$'0' (or '1'); if G_j='2', H_j^1='0', H_j^2='1' or H_j^1='1', H_j^2='0', where G_j, H_j^1 and H_j^2 denote the jth character of G, H^1 and H^2, respectively.

From DNA sequencing experiments, we can obtain short DNA fragments of a pair of chromosomes. The individual haplotyping problem takes the aligned DNA fragments as the input and considers only the values at the SNP sites of the fragments. Therefore, the aligned DNA fragments of an individual can be represented as an $m \times n$ SNP matrix M over the alphabet $\{0, 1, -\}$, where n columns represent a sequence of SNPs according to their order in the chromosome and m rows represent m aligned SNP fragments. In the matrix M, the ith row's value at the jth column is denoted by $M_{i,j}$. If the value of the ith fragment at the jth SNP site misses (i.e. there is a hole in the fragment) or the ith fragment doesn't cover the jth SNP site, then $M_{i,j}$ takes value '$-$' (value '$-$' is also called *empty value*).

The following are some definitions related to a SNP matrix M.

We say that the ith row *covers* the jth column if there are two indices k and r such that $k \leq j \leq r$, and both $M_{i,k}$ and $M_{i,r}$ are not empty.

The first and the last column that the ith row covers are denoted by $l(i)$ and $r(i)$ respectively.

If $M_{i,j} \neq$ '$-$', $M_{k,j} \neq$ '$-$' and $M_{i,j} \neq M_{k,j}$, then the ith and the kth rows of M are said to *conflict* at column j. If the ith and kth rows of M do not conflict at any column then they are *compatible*.

A SNP matrix M is *feasible* if its rows can be partitioned into two classes such that the rows in each class are all compatible.

Obviously, a set of aligned SNP fragments coming from a pair of chromosomes without error corresponds to a feasible SNP matrix, with each of the compatible classes in the matrix corresponding to a chromosome's haplotype. However, because of contaminants and read errors during DNA sequencing, the SNP matrix corresponding to the fragment data is generally not feasible. To make the SNP matrix feasible, there are many different ways, which result in different optimal models of the individual haplotype problem, and this paper is focused on the MEC/GI model [12].

> *Minimum error correction with genotype information* (MEC/GI): Given a $m \times n$ SNP matrix M and a genotype G of length n, correct a minimum number of elements ('0' into '1' and vice versa) of M so that the resulting matrix is feasible and G-compatible, i.e. the corrected SNP fragments can be partitioned into two classes of pairwise compatible fragments that determine a pair of haplotypes that is a resolution of G.

With the SNP matrix M and the genotype G, we use MEC/GI(M, G) to denote a solution to the MEC/GI problem (i.e. the minimum number of elements have to be flipped to make M feasible and G-compatible).

Zhang et al. [14] proved the MEC/GI problem, which they called the minimum conflict individual haplotyping (MCIH) problem, to be NP-hard. To solve the problem, Wang et al. [12] designed a genetic algorithm and a branch-and-bound algorithm, but the branch-and-bound algorithm cannot solve the problem in large size with time complexity $O(2^m)$, where m is the number of fragments, and the genetic algorithm cannot ensure the accuracy. Zhang et al. [14] designed a dynamic programming algorithm whose time complexity is $O(2^{2k}m)$, where k is the maximum length of SNP fragments. However Zhang et al.'s algorithm requires that for any two fragments i and j, if $l(i) \leq l(j)$, then $r(i) \leq r(j)$. To get rid of this constraint and improve the complexity, a new exact algorithm of time complexity $O(mk2^k + m\log m + mk)$ and space complexity $O(mk2^k)$ is proposed in the next section.

3   A New Practical Exact Algorithm for MEC/GI

Currently the main method to identify SNPs is DNA direct sequencing, and using the method SNP consortium has found 1.4 million SNPs [10]. The prevailing method to sequence a DNA fragment is the Sanger's technology [7], which cannot sequence any DNA fragment longer than 1200 bases. With the genome-wide SNPs distribution density of 1 SNP per 1 kb DNA sequence, the number of SNP sites in a fragment read is very small and usually smaller than 10 according to the available data in spite of the varying distribution density along a chromosome [1],[3]. In real applications, the number of SNP sites in a fragment read is generally between 3 and 8 [14]. Based on the fact above, we introduce the following parameterized condition.

Definition 1. *The k parameterized condition*: the number of SNP sites covered by a single fragment is bounded by k.

Accordingly, in a SNP matrix satisfying the k parameterized condition, each row covers at most k columns. For an $m \times n$ SNP matrix M, the parameter k can be obtained by scanning all rows of M.

In the following of the section, M is supposed to satisfy the k parameterized condition, and before introducing our parameterized algorithm in Subsection 3.2, we describe a preprocessing of the input SNP matrix M with the genotype G in Subsection 3.1.

3.1 Preprocessing

Suppose that the pair of haplotypes of an individual are H^1 and H^2. For each column j of M do:

Case 1. $G_j =$ '0': if (H^1, H^2) is a resolution of G, $H^1_j = H^2_j =$ '0'. Therefore, to make M compatible with G, for each row i, if $M_{i,j} =$ '1', $M_{i,j}$ must be flipped to '0'. Record the flipped elements, then delete column j from M and the jth character from G.

Case 2. $G_j =$ '1': if (H^1, H^2) is a resolution of G, $H^1_j = H^2_j =$ '1'. Therefore, to make M compatible with G, for each row i, if $M_{i,j} =$ '0', $M_{i,j}$ must be flipped to '1'. Record the flipped elements, then delete column j from M and the jth character from G.

Case 3. $G_j =$ '2': do nothing.

After the processing above, sort the rows in M in ascending order such that for any two rows i_1 and i_2, if $i_1 < i_2$, then $l(i_1) \leq l(i_2)$; scan M to obtain $l(i)$ and $r(i)$ for each row i;

It is obvious that after the preprocessing, the remaining characters of G are all '2'. If there is no remaining character, then the haplotype pair (H^1, H^2) has been determined and it is easy to solve the MEC/GI problem. For the case that there is at least one character left of G, we will propose a new algorithm to solve the MEC/GI problem.

In the following of this paper, the SNP matrix M is supposed to be beyond the preprocessing, and G contains only '2'. For briefness, MEC/GI(M, G) with G containing only '2' is denoted by MEC/GI(M).

3.2 A New Algorithm K-MEC/GI

For $x, y \in \{0, 1, -\}$, let

$$d(x, y) = \begin{cases} 1, & \text{if } x \neq -, y \neq - \text{ and } x \neq y \\ 0, & \text{otherwise.} \end{cases}$$

Given an $m \times n$ SNP matrix M and a pair of haplotypes $\mathcal{H} = (H^1, H^2)$, it is easy to see that in order to make the fragment corresponding to the ith row of M compatible with H^1 or H^2, the minimum number of SNPs of the fragment that should be flipped is $\min_{p=1..2}\left(\sum_{j=l(i)..r(i)} d(M_{i,j}, H_j^p)\right)$, and we denote it by $f_M(i, \mathcal{H})$, i.e.

$$f_M(i, \mathcal{H}) = \min_{p=1..2}\left(\sum_{j=l(i)..r(i)} d(M_{i,j}, H_j^p)\right).$$

Similarly, to make every row of M compatible with one haplotype of $\mathcal{H}$, the minimum number of elements of M that should be flipped is

$$D(M, \mathcal{H}) = \sum_{i=1..m} f_M(i, \mathcal{H}).$$

Obviously, for a $m \times n$ SNP matrix M and a genotype G containing only '2', the following equation holds:

$$\text{MEC/GI}(M) = \min_{\mathcal{H} \text{ is a solution of } G} (D(M, \mathcal{H})) \qquad (1)$$

To be a solution to a genotype G containing only '2', a pair of haplotypes $\mathcal{H} = (H^1, H^2)$ must be *heterozygous*, i.e. for any $1 \le j \le n$, H_j^1 and H_j^2 must be one of the following two cases: (1) $H_j^1 = \text{'0'}$, $H_j^2 = \text{'1'}$; (2) $H_j^1 = \text{'1'}$, $H_j^2 = \text{'0'}$. Therefore, based on Equation (1), we can solve the MEC/GI problem by enumerating all 2^n different heterozygous $\mathcal{H}$ in time complexity $O(mk2^n)$, which is impractical when n is large.

To reduce the time complexity, we must reduce the search space of $\mathcal{H}$. For briefness, in the following, haplotype pairs $\mathcal{H}$ or $\mathcal{L}$ are all heterozygous.

Let $r_m(i) = \max_{p=1..i} (r(p))$, and $len(i)$ denote $r_m(i) - l(i) + 1$. Let $H_{s..t}^1$ and $H_{s..t}^2$ denote the portion of H^1 and H^2 from column s to t, respectively, and $\mathcal{H}_{s..t}$ denote $(H_{s..t}^1, H_{s..t}^2)$.

Definition 2. Let i be a row of M, and $\mathcal{L} = (L^1, L^2)$ be a pair of haplotypes of length $len(i)$. $F(i, \mathcal{L})$ is defined as follows:

$$F(i, \mathcal{L}) = \min_{\mathcal{H} \text{ is heterozygous and } \mathcal{H}_{s..t} = \mathcal{L}} \left(\sum_{q=1..i} f_M(q, \mathcal{H})\right),$$

where $t = r_m(i)$ and $s = l(i)$.

Given a $m \times n$ SNP matrix M, it is easy to see the following equation holds:

$$\text{MEC/GI}(M) = \min_{\text{all possible heterozygous } \mathcal{L} \text{ of length } len(n)} (F(m, \mathcal{L})) \qquad (2)$$

For briefness, let $f_M'(i, \mathcal{L}) = \min_{p=1..2}\left(\sum_{j=l(i)..r(i)} d(M_{i,j}, L_{j-l(i)+1}^p)\right)$. It is easy to verify that

$$f_M(i, \mathcal{H}) = f_M'(i, \mathcal{L}), \quad \text{where } \mathcal{L} = \mathcal{H}_{l(i)..r_m(i)}. \qquad (3)$$

$f_M'(i, \mathcal{L})$ can be calculated by the function **CompF** (please see Figure 1), which takes time $O(len(i))$.

CompF$(i, \mathcal{L}, F, S)$ //$\mathcal{L}$ is coded by a binary number of $len(i)$ bits
//F denotes $f'_M(i, \mathcal{L})$, S keeps the corresponding elements flipped
1. $\mathcal{L}' = \mathcal{L}, F_1 = F_2 = 0, S_1 = S_2 = \varnothing$;
2.   **for** $j = 1..len(i)$ **do**   //obtain (L^1, L^2) from $\mathcal{L}$
        $c = \mathcal{L}' \bmod 2, \mathcal{L}' = \lfloor \mathcal{L}'/2 \rfloor$;
        if $c = 0$ **then** $L_j^1 = 0, L_j^2 = 1$;
        if $c = 1$ **then** $L_j^1 = 1, L_j^2 = 0$;
3.   **for** $l = 1..2$ **do**
        for $j = l(i)..r(i)$ **do**
            if $(M_{i,j} \neq -)$ and $(M_{i,j} \neq L_{j-l(i)+1}^l)$ **then**
                $F_l = F_l + 1, S_l = S_l \cup M_{i,j}$;
4.   **if** $F_1 \leq F_2$ **then** $F = F_1, S = S_1$;   **else** $F = F_2, S = S_2$;

Fig. 1. Function **CompF**

For a haplotype pair $\mathcal{L}$ of length $len(1)$, based on Definition 2 and Equation (3), we have the following equation:

$$F(1, \mathcal{L}) = f'_M(1, \mathcal{L}). \tag{4}$$

Once for every possible heterozygous $\mathcal{L}$ of length $len(i)$, $F(i, \mathcal{L})$ is known, we can calculate $F(i + 1, \mathcal{L}')$ for a given heterozygous $\mathcal{L}' = (L'^1, L'^2)$ of length $len(i + 1)$.
Let $s = l(i+1), t = r_m(i+1), s' = l(i+1) - l(i) + 1$, and $t' = r_m(i) - l(i+1) + 1$.
Based on Definition 2,

$$
\begin{aligned}
F(i + 1, \mathcal{L}') &= \min_{\mathcal{H}_{s..t} = \mathcal{L}'} \left(\sum_{q=1..i+1} f_M(q, \mathcal{H}) \right) \\
&= \min_{\mathcal{H}_{s..t} = \mathcal{L}'} \left(\sum_{q=1..i} f_M(q, \mathcal{H}) \right) + f'_M(i + 1, \mathcal{L}') \\
&= \min_{\mathcal{L}_{s'..len(i)} = \mathcal{L}'_{1..t'}} \left(\min_{\mathcal{H}_{l(i)..r_m(i)} = \mathcal{L}} \left(\sum_{q=1..i} f_M(q, \mathcal{H}) \right) \right) + f'_M(i + 1, \mathcal{L}')
\end{aligned}
$$

Therefore, the following equation holds:

$$F(i + 1, \mathcal{L}') = \min_{\mathcal{L}_{s'..len(i)} = \mathcal{L}'_{1..t'}} (F(i, \mathcal{L})) + f'_M(i + 1, \mathcal{L}') \tag{5}$$

Based on Equations (2)-(5), we have the following parameterized dynamic programming algorithm K-MEC/GI for the MEC/GI problem.

Theorem 1. *For an $m \times n$ SNP matrix M, if M satisfies the k parameterized condition, then the K-MEC/GI algorithm solves the MEC/GI problem correctly in time $O(mk2^k + m\log m + mk)$ and space $O(mk2^k)$.*

Due to space limit, the proof of Theorem 1 is omitted.

Algorithm K-MEC/GI
Input: an $m \times n$ SNP matrix M and a genotype G
Output: a solution to the MEC/GI problem for M
1. **preprocessing:**   see Subsection 3.1, keep the flipped elements at the deleted columns in S, let the number of rows and columns of M after the preprocessing be m' and n' ($m' \leq m$ and $n' \leq n$), respectively;
//r_m, len denote $r_m(i)$ and $len(i)$, respectively
2. $r_m = r(1)$,   len$= r_m(1) - l(1) + 1$, $i = 1$;
3. **for** $\mathcal{L} = 0..2^{len} - 1$ **do**
//$F[\mathcal{L}]$ denotes $F(i, \mathcal{L})$, i.e. the number of flipped element
//$S_E[\mathcal{L}]$ is the set of corresponding flipped elements
3.1.   call **CompF**$(1, \mathcal{L}, F[\mathcal{L}], S_E[\mathcal{L}])$ to calculate $F[\mathcal{L}], S_E[\mathcal{L}]$ using Eq. (4);
4. **while** $i < m'$ **do**   //recursion of Step 4 is based on Eq. (5)
//MAX denotes the maximum integer in the machine
4.1. $t' = r_m - l(i+1) + 1$;
//$F'[\mathcal{L}']$ denotes $\min_{\mathcal{L}'..len(i)=\mathcal{L}'_{1..t'}} (F(i, \mathcal{L}))$ of Eq. (5)
4.2.   **for** $\mathcal{L}' = 0..2^{t'} - 1$ **do** $F'[\mathcal{L}']$=MAX;
4.3.   **for** $\mathcal{L} = 0..2^{len} - 1$ **do**
4.3.1. $s' = l(i+1) - l(i) + 1$; $\mathcal{L}' = \mathcal{L}_{s'..len}$;
4.3.2.   **if** $F'[\mathcal{L}'] > F[\mathcal{L}]$ **then** $F'[\mathcal{L}'] = F[\mathcal{L}]$, $S_E'[\mathcal{L}'] = S_E[\mathcal{L}]$;
4.4. $i = i + 1$; $r_m = \max(r_m, r(i))$; len$= r_m - l(i) + 1$ //next row
4.5.   **for** $\mathcal{L} = 0..2^{len} - 1$ **do**
4.5.1.   **CompF**$(i, \mathcal{L}, \Delta F, \Delta S)$; $\mathcal{L}' = \mathcal{L}_{1..t'}$;
4.5.2. $F[\mathcal{L}] = \Delta F + F'[\mathcal{L}']$; $S_E[\mathcal{L}] = S_E'[\mathcal{L}'] \cup \Delta S$; //Eq. (5)
5. Let $\mathcal{H}$ be the $\mathcal{L}$ that minimizes $F[\mathcal{L}]$ over all $\mathcal{L} = 0..2^{len} - 1$;
6. MEC/GI$(M, G)=F[\mathcal{H}] + |S|$, and the corresponding flipped elements are in $S_E[\mathcal{H}] \cup S$.

Fig. 2. K-MEC Algorithm

4   Experimental Results

To solve the MEC/GI problem, Wang *et al.* [12] designed a branch-and-bound algorithm B-MEC/GI and a genetic algorithm G-MEC/GI. In this section, we will test the performance of our parameterized algorithm K-MEC/GI and Wang *et al.*'s algorithms. K-MEC/GI is implemented in C++, B-MEC/GI and G-MEC/GI come from Wang [12]. We ran our experiments on a Linux server (4 Intel Xeon 3.6GHz CPU and 4GByte RAM).

In the experiments, we use running time and reconstruction rate to test performance of an algorithm. Reconstruction rate [12] measures the similarity degree between the original haplotypes and the reconstructed haplotypes, and is defined as the ratio of the number of the SNP sites correctly reconstructed by an algorithm to the total number of the SNP sites of the haplotypes.

The haplotype data can be obtained by two methods [12]: the first is to get real haplotypes from public domain, and the second is to generate simulated haplotypes by computer. In our experiments, the real haplotypes were obtained from the file genotypes_chr1_CEU_r21_nr_fwd_phased.gz [1] , which was issued in July 2006 by the International HapMap Project [9]. The file contains 120 haplotypes on chromosome 1 of 60 individuals with each haplotype containing 193333 SNP sites. From the 60 individuals, select an individual at random. Then begining with a random SNP site, a pair of haplotypes of a given length can be obtained from the haplotypes of the selected individual. The simulated haplotypes can be generated as follows ([12],[6]). At first a haplotype h_1 of length n is generated at random, then another haplotype h_2 of the same length is generated by flipping every character of h_1 with the probability of d.

As to fragment data, to the best of our knowledge, real DNA fragments data in the public domain are not available, and references [12] and [6] used computer-generated simulated fragment data. After obtaining a pair of real or simulated haplotypes, in order to make the generated fragments have the same statistical features as the real data, a widely used shotgun assembly simulator Celsim ([5]) is invoked to generate m fragments whose lengths are between $lMin$ and $lMax$. At last the output fragments are processed to plant reading errors with probability e and empty values with probability p.

In our experiments, the parameters are set as follows: the probability d used in generating simulated haplotypes is 20%, fragment coverage rate $c = 10$, the minimal length of fragment $lMin = 3$, the maximal length of fragment $lMax = 7$ and empty value probability $p = 2\%$. The length of haplotype n and the reading error probability e are varied to compare the performance of the algorithms. Celsim takes a haplotype, the number m of fragments, the minimum fragment length $lMin$, and the maximum fragment length $lMax$ as input. Please refer to [6] and [5] for the details about how to generate artificial fragment data.

The experiment results are shown in Table 1 and Figure 3. In Table 1, the experiment results on real haplotype data are presented outside parentheses, and the results on simulated haplotype data are given inside parentheses. As exact algorihms, for a fixed SNP matrix M, both B-MEC/GI and K-MEC/GI select the same number of elements of M to correct. However, because which elements of M are to be selected may be not unique, the haplotype reconstruction rates of both algorithms are not completely the same, though there may be not obvious difference between them on the whole.

In Table 1, the experiment results, not only on real haplotype data but also on simulated haplotype data, show that when m, n, and the reading error probability e is small, these three algorithms have good performance in reconstructing the haplotypes. However, when m and n increase, the heuristic algorithm G-MEC/GI becomes less accurate than B-MEC/GI and K-MEC/GI, and the running time of B-MEC/GI increases sharply. When $n = 50$ and $m = 100$, B-MEC/GI can not work out with a solution in 4 days, but the running time of K-MEC/GI and G-MEC/GI is still small and less than 2 seconds.

With n increasing from 50 and 250, m increasing from 100 to 500 accordingly, and $e = 5\%$, the experiment results on real haplotype data are illustrated in Figure 3(a). When $n = 50$, $e = 5\%$, Figure 3(b) shows the experiment results

---

[1] From http://www.hapmap.org/downloads/phasing/2006-07_phaseII/phased/

Table 1. The comparative results of three algorithms for the MEC/GI problem

Parameters			Reconstruction rate (%)			Running time (s)		
n	m	e	K-MEC/GI	G-MEC/GI	B-MEC/GI	K-MEC/GI	G-MEC/GI	B-MEC/GI
		0.01	99.1 (99.1)	99.1 (99.1)	99.1 (99.1)	0.001 (0.001)	0.18 (0.19)	0.001 (0.001)
10	20	0.03	99.1 (99.1)	99.1 (99.1)	99.1 (99.1)	0.001 (0.001)	0.18 (0.19)	0.002 (0.002)
		0.05	99.0 (99.0)	98.7 (98.6)	99.0 (99.0)	0.001 (0.001)	0.19 (0.19)	0.011 (0.004)
		0.01	98.2 (98.3)	97.5 (97.6)	98.3 (98.3)	0.008 (0.007)	0.45 (0.44)	35.12 (34.99)
20	40	0.03	97.8 (97.5)	96.7 (96.7)	97.7 (97.4)	0.010 (0.010)	0.45 (0.44)	50.84 (51.96)
		0.05	97.2 (97.2)	96.5 (96.6)	97.3 (97.3)	0.015 (0.012)	0.47 (0.45)	83.03 (80.92)
		0.01	96.9 (97.0)	95.7 (95.4)	-	0.033 (0.040)	1.48 (1.49)	>96 hours
50	100	0.03	95.8 (96.6)	95.1 (95.2)	-	0.041 (0.042)	1.51 (1.48)	>96 hours
		0.05	95.5 (95.4)	94.8 (94.7)	-	0.047 (0.045)	1.58 (1.49)	>96 hours

The data not enclosed in parentheses are the experiment results coming from the experiments on the real haplotype data, and the data in parentheses show the ones on the simulated haplotypes data. All experiments are repeated 100 times except for B-MEC/GI with $n = 50$ and $m = 100$. Reconstruction rate and running time are over the repeated experiments with the same parameters.

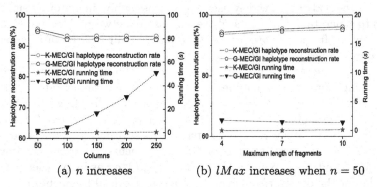

(a) n increases

(b) $lMax$ increases when $n = 50$

Fig. 3. The performance comparison of K-MEC/GI and G-MEC/GI

on real haplotype data with $lMax$ varying. In Figure 3, the comparison of re-construction rate is at the left Y axis, and the comparison of running time is at the right Y axis. Figure 3 shows again that K-MEC/GI is more accurate in hap-lotype reconstruction than G-MEC/GI, and illustrates that K-MEC/GI takes much less time than G-MEC/GI when n (or m) is large.

5   Conclusion

Haplotyping plays a more and more important role in some regions of genetics such as locating of genes, designing of drugs and forensic applications. Given the genotype and the aligned SNP fragments of an individual, MEC/GI is an important computational model to infer a pair of haplotypes compatible with the genotype by correcting minimum number of SNPs in the given SNP frag-ments. Based on the fact that the maximum length of fragments is small (usually smaller than 10 [1],[3]), the current paper introduced a new practical algorithm K-MEC/GI to solve the MEC/GI problem. With the fragments of maximum length k, the K-MEC/GI algorithm can solve the MEC/GI problem in time $O(mk2^k + mlogm + mk)$ and space $O(mk2^k)$. The experiment results show that our K-MEC/GI algorithm is more scalable than B-MEC/GI, and is more accurate in haplotype reconstruction than G-MEC/GI. Besides MEC/GI, there are many various computational models for the individual haplotyping problem, our parameterized method may deal with them with some modification.

Acknowledgments. We thank the anonymous referees for comments and help in improving the presentation of the earlier version of the paper.

References

1. Gabriel, S.B., Schaffner, S.F., Nguyen, H., et al.: The Structure of Haplotype Blocks in the Human Genome. Science 296(5576), 2225–2229 (2002)
2. Gusfield, D.: An Overview of Combinatorial Methods for Haplotype Inference. In: Istrail, S., Waterman, M.S., Clark, A. (eds.) DIMACS/RECOMB Satellite Workshop 2002. LNCS (LNBI), vol. 2983, pp. 9–25. Springer, Heidelberg (2004)
3. Hinds, D.A., Stuve, L.L., Nilsen, G.B., et al.: Whole-Genome Patterns of Common DNA Variation in Three Human Populations. Science 307(5712), 1072–1079 (2005)
4. Lancia, G., Bafna, V., Istrail, S., Lippert, R., Schwartz, R.: SNPs Problems, Complexity and Algorithms. In: Meyer auf der Heide, F. (ed.) ESA 2001. LNCS, vol. 2161, pp. 182–193. Springer, Heidelberg (2001)
5. Myers, G.: A Dataset Generator for Whole Genome Shotgun Sequencing. In: Lengauer, T., Schneider, R., Bork, P., et al. (eds.) Proc. ISMB 1999, pp. 202–210. AAAI Press, California (1999)
6. Panconesi, A., Sozio, M.: Fast Hare: a Fast Heuristic for Single Individual SNP Haplotype Reconstruction. In: Jonassen, I., Kim, J. (eds.) WABI 2004. LNCS (LNBI), vol. 3240, pp. 266–277. Springer, Heidelberg (2004)
7. Sanger, F., Nicklen, S., Coulson, A.R.: DNA Sequencing with Chain-Terminating Inhibitors. PNAS 74(12), 5463–5467 (1977)
8. Stephens, J.C., Schneider, J.A., Tanguay, D.A., et al.: Haplotype Variation and Linkage Disequilibrium in 313 Human Genes. Science 293(5529), 489–493 (2001)
9. The International HapMap Consortium: A Haplotype Map of the Human Genome. Nature 437(7063), 1299–1320 (2005)
10. The International SNP Map Working Group: A Map of Human Genome Sequence Variation Containing 1.42 Million Single Nucleotide Polymorphisms. Nature 409(6822), 928–933 (2001)
11. Wang, D.G., Fan, J.B., Siao, C.J., et al.: Large-Scale Identification, Mapping, and Genotyping of Single-Nucleotide Polymorphisms in the Human Genome. Science 280(5366), 1077–1082 (1998)
12. Wang, R.S., Wu, L.Y., Li, Z.P., Zhang, X.S.: Haplotype Reconstruction from SNP Fragments by Minimum Error Correction. Bioinformatics 21(10), 2456–2462 (2005)
13. Zhang, X.S., Wang, R.S., Wu, L.Y., Chen, L.: Models and Algorithms for Haplotyping Problem. Current Bioinformatics 1(1), 105–114 (2006)
14. Zhang, X., Wang, R., Wu, L., Zhang, W.: Minimum Conflict Individual Haplotyping from SNP Fragments and Related Genotype. Evolutionary Bioinformatics 2, 271–280 (2006)

Voronoi Diagram of Polygonal Chains under the Discrete Fréchet Distance[*]

Sergey Bereg[1], Kevin Buchin[2], Maike Buchin[2],
Marina Gavrilova[3], and Binhai Zhu[4]

[1] Department of Computer Science, University of Texas at Dallas,
Richardson, TX 75083, USA
besp@utdallas.edu
[2] Department of Information and Computing Sciences,
Universiteit Utrecht, The Netherlands
{buchin,maike}@cs.uu.nl
[3] Department of Computer Science, University of Calgary, Calgary,
Alberta T2N 1N4, Canada
marina@cpsc.ucalgary.ca
[4] Department of Computer Science, Montana State University,
Bozeman, MT 59717-3880, USA
bhz@cs.montana.edu

Abstract. Polygonal chains are fundamental objects in many applications like pattern recognition and protein structure alignment. A well-known measure to characterize the similarity of two polygonal chains is the (continuous/discrete) Fréchet distance. In this paper, for the first time, we consider the Voronoi diagram of polygonal chains in d-dimension under the discrete Fréchet distance. Given a set $\mathcal{C}$ of n polygonal chains in d-dimension, each with at most k vertices, we prove fundamental properties of such a Voronoi diagram $VD_F(\mathcal{C})$. Our main results are summarized as follows.

- The combinatorial complexity of $VD_F(\mathcal{C})$ is at most $O(n^{dk+\epsilon})$.
- The combinatorial complexity of $VD_F(\mathcal{C})$ is at least $\Omega(n^{dk})$ for dimension $d = 1, 2$; and $\Omega(n^{d(k-1)+2})$ for dimension $d > 2$.

1 Introduction

The Fréchet distance was first defined by Maurice Fréchet in 1906 [8]. While known as a famous distance measure in the field of mathematics (more specifically, abstract spaces), it was first applied in measuring the similarity of polygonal curves by Alt and Godau in 1992 [2]. In general, the Fréchet distance between 2D polygonal chains (polylines) can be computed in polynomial time [2],[3], even under translation or rotation (though the running time is much higher) [4],[15]. While computing (approximating) Fréchet distance for surfaces is NP-hard [9], it is polynomially solvable for restricted surfaces [5].

---

[*] The authors gratefully acknowledge the support of K.C. Wong Education Foundation, Hong Kong and NSERC of Canada.

X. Hu and J. Wang (Eds.): COCOON 2008, LNCS 5092, pp. 352–362, 2008.
© Springer-Verlag Berlin Heidelberg 2008

In 1994, Eiter and Mannila defined the *discrete Fréchet distance* between two polygonal chains A and B in d-dimension. This simplified distance is always realized by two vertices in A and B [7]. They showed that with dynamic programming the discrete Fréchet distance between polygonal chains A and B can be computed in $O(|A||B|)$ time. In [10], Indyk defined a similar discrete Fréchet distance in some metric space and showed how to compute approximate nearest neighbors using that distance.

Recently, Jiang, Xu and Zhu applied the discrete Fréchet distance in aligning the backbones of proteins (which are called the *protein structure-structure alignment* problem [11] and *protein local structure alignment* respectively [16]). In fact, in these applications the discrete Fréchet distance makes more sense as the backbone of a protein is simply a polygonal chain in 3D, with each vertex being the alpha-carbon atom of a residue.

On the other hand, a lot is still unknown regarding the discrete or continuous Fréchet distance. For instance, even though the Voronoi diagram has been studied for many objects and distance measures, it has not yet been studied for polygonal chains under the discrete (continuous) Fréchet distance. This problem is fundamental, it has potential applications, e.g., in protein structure alignment, especially with the ever increasing computational power. Imagine that we have some polylines $A_1, A_2, A_3...$ in space. If we can construct the Voronoi diagram for $A_1, A_2, A_3, ...$ in space, then given a new polyline B we can easily compute all the approximate alignments of B with the A_i's. The movement of B defines a subspace (each point in the subspace represents a copy of B) and if we sample this subspace evenly then all we need to do is to locate all these sample points in the Voronoi diagram for the A_i's.

Unfortunately, nothing is known about the Voronoi diagram under the discrete (continuous) Fréchet distance, even for the simplest case of line segments. In this paper, we will present the first set of such results by proving some fundamental properties for both the general case and some special case. We believe that these results will be essential for us to design efficient algorithms for computing/approximating the Voronoi diagram under the Fréchet distance.

2   Preliminaries

Given two polygonal chains A, B with $|A| = k$ and $|B| = l$ vertices respectively, we aim at measuring the similarity of A and B (possibly under translation and rotation) such that their distance is minimized under certain measure. Among the various distance measures, the Hausdorff distance is known to be better suited for matching two point sets than for matching two polygonal chains; the (continuous) Fréchet distance is a superior measure for matching two polygonal chains [2].

Let X be the Euclidean plane $\mathbb{R}^d$; let $d(a, b)$ denote the Euclidean distance between two points $a, b \in X$. The (continuous) Fréchet distance between two parametric curves $f : [0, 1] \to X$ and $g : [0, 1] \to X$ is

$$\delta_{\mathcal{F}}(f, g) = \inf_{\alpha, \beta} \max_{s \in [0,1]} d(f(\alpha(s)), g(\beta(s))),$$

where α and β range over all continuous non-decreasing real functions with $\alpha(0) = \beta(0) = 0$ and $\alpha(1) = \beta(1) = 1$.

Imagine that a person and a dog walk along two different paths while connected by a leash; they always move forward, though (possibly) at different paces. The minimum possible length of the leash is the Fréchet distance between the two paths. To compute the Fréchet distance between two polygonal curves A and B (in the Euclidean plane) of $|A|$ and $|B|$ vertices, respectively, Alt and Godau [3] presented an $O(|A||B|\log(|A||B|))$ time algorithm. We now define the discrete Fréchet distance following [7].

Definition 1. *Given a polygonal chain (polyline) in d-dimension $P = \langle p_1, \ldots, p_k \rangle$ of k vertices, a m-**walk** along P partitions the path into m disjoint non-empty subchains $\{\mathcal{P}_i\}_{i=1..m}$ such that $\mathcal{P}_i = \langle p_{k_{i-1}+1}, \ldots, p_{k_i} \rangle$ and $0 = k_0 < k_1 < \cdots < k_m = k$.*

*Given two polylines in d-dimension $A = \langle a_1, \ldots, a_k \rangle$ and $B = \langle b_1, \ldots, b_l \rangle$, a **paired walk** along A and B is a m-walk $\{\mathcal{A}_i\}_{i=1..m}$ along A and a m-walk $\{\mathcal{B}_i\}_{i=1..m}$ along B for some m, such that, for $1 \leq i \leq m$, $|\mathcal{A}_i| = 1$ or $|\mathcal{B}_i| = 1$ (that is, $\mathcal{A}_i$ or $\mathcal{B}_i$ contains exactly one vertex). The **cost** of a paired walk $W = \{(\mathcal{A}_i, \mathcal{B}_i)\}$ along two paths A and B is*

$$d_F^W(A, B) = \max_i \max_{(a,b) \in \mathcal{A}_i \times \mathcal{B}_i} d(a, b).$$

*The **discrete Fréchet distance** between two polylines A and B in d-dimension is*

$$d_F(A, B) = \min_W d_F^W(A, B).$$

*The paired walk that achieves the discrete Fréchet distance between two paths A and B is also called the **Fréchet alignment** of A and B.*

Consider the scenario in which the person walks along A and the dog along B. Intuitively, the definition of the paired walk is based on three cases:

1. $|\mathcal{B}_i| > |\mathcal{A}_i| = 1$: the person stays and the dog moves forward;
2. $|\mathcal{A}_i| > |\mathcal{B}_i| = 1$: the person moves forward and the dog stays;
3. $|\mathcal{A}_i| = |\mathcal{B}_i| = 1$: both the person and the dog move forward.

Eiter and Mannila presented a simple dynamic programming algorithm to compute $d_F(A, B)$ in $O(|A||B|) = O(kl)$ time [7]. Recently, Jiang *et al.* showed

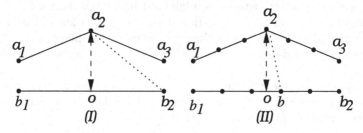

Fig. 1. The relationship between discrete and continuous Fréchet distances

that in 2D the minimum discrete Fréchet distance between A and B under translation can be computed in $O(k^3 l^3 \log(k+l))$ time, and under both translation and rotation it can be computed in $O(k^4 l^4 \log(k+l))$ time [11]. They are significantly faster than the corresponding bounds for the continuous Fréchet distance. In 2D, for the continuous Fréchet distance, under translation, the current fastest algorithm for computing the minimum (continuous) Fréchet distance between A, B takes $O((kl)^3 (k+l)^2 \log(k+l))$ time [4]; under both translation and rotation, the bound is $O((k+l)^{11} \log(k+l))$ time [15].

We comment that while the discrete Fréchet distance could be arbitrarily larger than the corresponding continuous Fréchet distance (e.g., in Fig. 1 (I), they are $d(a_2, b_2)$ and $d(a_2, o)$ respectively), by adding sample points on the polylines, one can easily obtain a close approximation of the continuous Fréchet distance using the discrete Fréchet distance (e.g., one can use $d(a_2, b)$ in Fig. 1 (II) to approximate $d(a_2, o)$). This fact has been pointed out in [7],[10]. Moreover, the discrete Fréchet distance is a more natural measure for matching the geometric shapes of biological sequences such as proteins.

In the remaining part of this paper, for the first time, we investigate the Voronoi diagram of a set of polygonal chains (polylines) in d-dimension. For the fundamentals regarding Voronoi diagram, the readers are referred to [12].

In our case, the Voronoi diagram of polygonal curves can be represented using the correspondence

polygonal curve with k vertices in $\mathbb{R}^d \leftrightarrow$ point in $\mathbb{R}^{dk}$

$$\langle (x_{11}, \ldots, x_{1d}), \ldots, (x_{k1}, \ldots, x_{kd}) \rangle \leftrightarrow (x_{11}, \ldots, x_{1d}, \ldots, x_{k1}, \ldots, x_{kd}).$$

In the following, we always use the parameters n, d, and k to denote n input curves in $\mathbb{R}^d$, each with at most k vertices. We give upper and lower bounds on the combinatorial complexity of the Voronoi diagram of the input curves, which is a partition of the space in $\mathbb{R}^{dk}$ into Voronoi regions associated with each input curve. By a Voronoi region we mean a set of curves with a common set of nearest neighbors under the discrete Fréchet distance in the given set of input curves.

3   The Combinatorial Upper Bound of $VD_F(\mathcal{C})$

In this section, we prove the combinatorial upper bound of $VD_F(\mathcal{C})$. We first show the case for $d = 2$, and then we sketch how to generalize the result to any fixed d-dimension.

Let $A_k = \langle a_1, a_2, \ldots, a_k \rangle$ and $B_l = \langle b_1, b_2, \ldots, b_l \rangle$ be two polygonal chains in the plane where $a_i = (x(a_i), y(a_i)), b_j = (x(b_j), y(b_j))$ and $k, l \geq 1$. We first have the following lemma, which is easy to prove.

Lemma 1. Let $A_2 = \langle a_1, a_2 \rangle$ and $B_2 = \langle b_1, b_2 \rangle$ be two line segments in the plane, then $d_F(A_2, B_2) = \max(d(a_1, b_1), d(a_2, b_2))$.

For general polylines, we can generalize the above lemma as follows. Notice that as $d_F()$ is a min-max-max measure and as we will be using the fact that the

Voronoi diagram is a minimization of distance functions, the following lemma is essential.

Lemma 2. *Let $A_k = \langle a_1, a_2, \ldots, a_k \rangle$ and $B_l = \langle b_1, b_2, \ldots, b_l \rangle$ be two polygonal chains in the plane where $k, l \geq 1$. The discrete Fréchet distance between A_k and B_l can be computed as $d_F(A_k, B_l) = \max\{d(a_i, b_1), i = 1, 2, \ldots, k\}$, if $l = 1$; $d_F(A_k, B_l) = \max\{d(a_1, b_j), j = 1, 2, \ldots, l\}$, if $k = 1$; and $d_F(A_k, B_l) = \max \{d(a_k, b_l), \min(d_F(A_{k-1}, B_{l-1}), d_F(A_k, B_{l-1}), d_F(A_{k-1}, B_l))\}$, if $k, l > 1$.*

Based on the above lemma, we investigate the combinatorial complexity of $VD_F(\mathcal{C})$, the Voronoi diagram of a set $\mathcal{C}$ of n planar polylines each with at most k vertices. Following [6],[14], a Voronoi diagram is a minimization of distance functions to the sites (in this case the polylines in $\mathcal{C}$). We first briefly review a result on the upper bound of lower envelopes in high dimensions by Sharir [14].

Let $\Sigma = \{\sigma_1, \ldots, \sigma_n\}$ be a collection of n $(d-1)$-dimensional algebraic surface patches in d-space. Let $\mathcal{A}(\Sigma)$ be the arrangement of Σ. The result in [14] holds upon the following three conditions.

(i) Each σ_i is monotone in the $x_1 x_2 \ldots x_{d-1}$-direction (i.e. any line parallel to x_d-axis intersects σ_i in at most one point). Moreover, each σ_i is a portion of a $(d-1)$-dimensional algebraic surface of constant maximum degree b.

(ii) The projection of σ_i in x_d-direction onto the hyperplane $x_d = 0$ is a semi-algebraic set defined in terms of a constant number of $(d-1)$-variate polynomials of constant maximum degree.

(iii) The surface patches σ_i are in *general position* meaning that the coefficients of the polynomials defining the surfaces and their boundaries are algebraically independent over the rationals.

Theorem 1. *[14] Let Σ be a collection of n $(d-1)$-dimensional algebraic surface patches in d-space, which satisfy the above conditions (i),(ii), and (iii). Then the number of vertices of $\mathcal{A}(\Sigma)$ that lie at the lower envelope (i.e., level one) is $O(n^{d-1+\epsilon})$, for any $\epsilon > 0$.*

We now show a general upper bound on the combinatorial complexity of $VD_F(\mathcal{C})$.

Lemma 3. *Let $B \in \mathbb{R}^{2l}$ be a polygonal chain of l vertices $b_1, \ldots, b_l$ in the plane where $b_i = (x(b_i), y(b_i))$. Let $f : \mathbb{R}^{2k} \to \mathbb{R}$ be the distance function defined as*

$$f(C) = d_F(C, B),$$

where $C = \langle c_1, \ldots, c_k \rangle \in \mathbb{R}^{2k}$ is a polyline with k vertices and $c_i = (x(c_i), y(c_i))$, $i = 1, \ldots, k$. The space $\mathbb{R}^{2k}$ can be partitioned into at most $(kl)!$ semi-algebraic sets $R_1, R_2, R_3, \ldots$ such that the function $f(C)$ with domain restricted to any R_i is algebraic. Thus, the function $f(C)$ satisfies the above conditions (i) and (ii).

We can prove the following theorem regarding the combinatorial upper bound for $VD_F(\mathcal{C})$.

Theorem 2. *Let $\mathcal{C}$ be a collection of n polygonal chains $C_1, \ldots, C_n$ each with at most k vertices in the plane. The combinatorial complexity of the Voronoi diagram $\mathrm{VD}_F(\mathcal{C})$ is $O(n^{2k+\epsilon})$, for any $\epsilon > 0$.*

For protein-related applications, the input are polygonal chains in 3D. So it makes sense to consider the cases when $d > 2$. We have

Theorem 3. *Let $\mathcal{C}$ be a collection of n polygonal chains $C_1, \ldots, C_n$ each with at most k vertices in $\mathbb{R}^d$. The combinatorial complexity of the Voronoi diagram $\mathrm{VD}_F(\mathcal{C})$ is $O(n^{dk+\epsilon})$, for any $\epsilon > 0$.*

4   The Combinatorial Lower Bounds of $VD_F(\mathcal{C})$

We now present a general lower bound for $VD_F(\mathcal{C})$. In fact we first show a result that even a slice of $VD_F(\mathcal{C})$ could contain a L_∞ Voronoi diagram in k dimensions, whose combinatorial complexity is $\Omega(n^{\lfloor \frac{k+1}{2} \rfloor})$. This result is somehow a 'folklore' as the relationship between discrete Fréchet distance and L_∞-distance is known to many researchers, e.g., in [10]. We treat this more like a warm-up of our lower bound constructions.

Schaudt and Drysdale proved that a L_∞ Voronoi diagram in k dimensions has combinatorial complexity of $\Omega(n^{\lfloor \frac{k+1}{2} \rfloor})$ [13]. Let $S = \{p_1, p_2, \ldots, p_n\}$ be a set of n points in $\mathbb{R}^k$ such that the L_∞ Voronoi diagram of S has complexity of $\Omega(n^{\lfloor \frac{k+1}{2} \rfloor})$. Let $M > 0$ be a real number such that the hypercube $[-M, M]^k$ contains S and all the Voronoi vertices of the L_∞ Voronoi diagram of S. We consider a k-dimensional flat F of $\mathbb{R}^{2k}$ defined as $F = \{(a_1, M, a_2, 2M, \ldots, a_k, kM) \mid a_1, \ldots, a_k \in \mathbb{R}\}$ and a projection $\pi : F \to \mathbb{R}^k$ defined as $\pi(b) = (b_1, b_3, \ldots, b_{2k-1})$, for $b = (b_1, b_2, b_3, \ldots, b_{2k-1}, b_{2k})$.

Let $\mathcal{C} = \{C_1, C_2, \ldots, C_n\}$, each C_i being a planar polygonal chain with k vertices. Let $C_i = \langle c_{i1}, c_{i2}, \ldots, c_{ik} \rangle$ and $c_{im} = (x(c_{im}), y(c_{im}))$, for $m = 1, 2, \ldots, k$. We set $c_{im} = (p_{im}, mM)$, for $1 \leq i \leq n, 1 \leq m \leq k$. Clearly, every $C_i \in F$. With C_i we associate a point $C_i' = \pi(C_i)$ in $\mathbb{R}^k$. We show that the intersection of F and a $VD_F(\mathcal{C})$ has complexity of $\Omega(n^{\lfloor \frac{k+1}{2} \rfloor})$.

Consider a point $T \in F$ such that $T' = \pi(T)$ is a L_∞ Voronoi vertex of S in $\mathbb{R}^k$. Then $T' \in [-M, M]^k$. At this point, the question is: what is the discrete Fréchet distance between T and a chain C_i? Note that $d_F^W(T, C_i) < M$ if and only if $W = W_0$ where $W_0 = \{(t_m, c_{im}) \mid m = 1, \ldots, k\}$. Therefore $d_F(T, C_i) = d_F^{W_0}(T, C_i) = \max\{|x(t_1) - x(c_{i1})|, |x(t_2) - x(c_{i2})|, \ldots, |x(t_j) - x(c_{ij})|, \ldots, |x(t_k) - x(c_{ik})|\}$. This is exactly the L_∞-distance between T' and C_i', or $d_F(T, C_i) = d_\infty(T', C_i')$. Then the slice of $VD_F(\mathcal{C})$ contains the L_∞ Voronoi diagram of S in k dimensions. We thus have the following theorem.

Theorem 4. *The combinatorial complexity of $\mathrm{VD}_F(\mathcal{C})$ for a set $\mathcal{C}$ of n planar polygonal chains with k vertices is $\Omega(n^{\lfloor \frac{k+1}{2} \rfloor})$; in fact even a k-dimensional slice of $\mathrm{VD}_F(\mathcal{C})$ can have a combinatorial complexity of $\Omega(n^{\lfloor \frac{k+1}{2} \rfloor})$.*

This lower bound on the combinatorial complexity is apparently not tight and we next show an improved lower bound construction which does not make use of the L_∞ Voronoi diagram. We summarize our result as follows.

Theorem 5. *For any d, k, n, there is a set of n polygonal curves in $\mathbb{R}^d$ with k vertices each whose Voronoi diagram under the discrete Fréchet distance, $\mathrm{VD}_F(\mathcal{C})$, has combinatorial complexity $\Omega(n^{dk})$ for $d = 1, 2$ and $k \in \mathbb{N}$ and complexity $\Omega(n^{d(k-1)+2})$ for $d > 2$ and $k \in \mathbb{N}$.*

We show the lower bounds in Theorem 5 first for dimension $d = 1$ (Lemma 4) and then for dimensions $d \geq 2$ (Lemma 5). For both lower bounds we construct a set S of n curves. Then we construct $g(n)$ query curves which all lie in different Voronoi regions of the Voronoi diagram of S. This implies that the Voronoi diagram has complexity $\Omega(g(n))$.

Lemma 4. *For all n and k, there is a set of n polygonal curves in $\mathbb{R}^1$ with k vertices each whose Voronoi diagram under the discrete Fréchet distance has at least $\lfloor \frac{n}{k} \rfloor^k$ Voronoi regions.*

Proof. We construct a set S of n curves with k vertices each for $n = m \cdot k$ with $m \in \mathbb{N}$. S will be a union of k sets $S_1, \ldots, S_k$ of m curves each. We show that the Voronoi diagram of S contains m^k Voronoi regions.

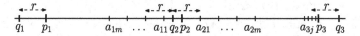

Fig. 2. Construction for $d = 1$ and $k = 3$

The construction for $k = 3$ is shown in Fig. 2. We place k points $p_1, \ldots, p_k$ with distance $2m$ between consecutive points on the real line. A curve in S has the form $\langle p_1, \ldots, p_{i-1}, p_i', p_{i+1}, p_k \rangle$ for some $i \in \{1, \ldots, k\}$ and point p_i' close to p_i. Our construction uses the following points, curves, and sets of curves. See Fig. 2 for an illustration for $k = 3$.

$$
\begin{aligned}
p_i &= (i-1)2m &&\text{for } i = 1, \ldots, k \\
a_{1j} &= p_2 - j, \quad a_{2j} = p_2 + j &&\text{for } j = 1, \ldots, m \\
a_{ij} &= p_i - j/(m+1) &&\text{for } i = 3, \ldots, k, \ j = 1, \ldots, m \\
S_{ij} &= \langle p_1, a_{ij}, p_2, \ldots, p_k \rangle &&\text{for } i = 1, 2, \ j = 1, \ldots, m \\
S_{ij} &= \langle p_1, \ldots, p_{i-1}, a_{ij}, p_{i+1}, \ldots, p_k \rangle &&\text{for } i = 3, \ldots, k, \ j = 1, \ldots, m \\
S_i &= \{S_{i1}, \ldots, S_{im}\} &&\text{for } i = 1, \ldots, k
\end{aligned}
$$

We claim that for all $1 \leq j_1, \ldots, j_k \leq m$ a query curve Q exists whose set of nearest neighbors in S under the discrete Fréchet distance denoted by $N_S(Q)$, is

$$N_S(Q) = \{S_{11}, \ldots, S_{1j_1}, \ldots, S_{k1}, \ldots, S_{kj_k}\}. \tag{1}$$

Since these are m^k different sets, this implies that there are at least m^k Voronoi regions.

The query curve Q will have k vertices $q_1, \ldots, q_k$ with q_i close to p_i for $i = 1, \ldots, k$. The discrete Fréchet distance of Q to any curve in S will be realized by a bijection mapping each p_i or p'_i to q_i. Because the p_i are placed at large pairwise distances, this is the best possible matching of the vertices for the discrete Fréchet distance.

Let $r = (a_{2j_2} - a_{1j_1})/2$ denote half the distance between a_{1j_1} and a_{2j_2}. We choose the first vertex of Q as $q_1 = -r$. The second vertex q_2 we choose as midpoint between a_{1j_1} and a_{2j_2}, i.e., $q_2 = (a_{1j_1} + a_{2j_2})/2$. Since $p_1 = 0$, the distance between p_1 and q_1 is r. Because all curves in S start at p_1, this is the smallest possible discrete Fréchet distance between Q and any curve in S. We now construct the remaining points of Q, such that the curves in $N_S(Q)$ are exactly those given in equation 1 and these have discrete Fréchet distance r to Q.

We have already constructed q_2 such that it has distance at most r to the points $a_{11}, \ldots, a_{1j_1}$ and $a_{21}, \ldots, a_{2j_2}$ (cf. Fig. 2). Now we choose the remaining points q_i as $q_i = a_{ij_i} + r$ for $i = 3, \ldots, k$. Then the point q_i has distance at most r to the points $p_i, a_{i1}, \ldots, a_{ij_i}$ for $i = 3, \ldots, k$. $\square$

Lemma 5. *For all n, k and for all $d \geq 2$, there is a set of n polygonal curves in $\mathbb{R}^d$ with k vertices each whose Voronoi diagram under the discrete Fréchet distance has at least $\lfloor \frac{n}{d(k-1)+2} \rfloor^{d(k-1)+2}$ Voronoi regions.*

Proof. We first give the construction for dimension $d = 2$ and then show how to generalize it for $d > 2$.

Construction for $d = 2$. We construct the set S as union of $2k = d(k-1) + 2$ sets $S_1, \ldots, S_{2k}$ of m curves each for $m \in \mathbb{N}$.

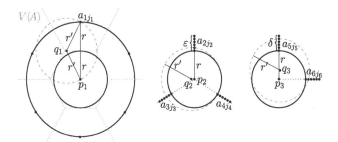

Fig. 3. Construction for $d = 2$ and $k = 3$

First, we place k points $p_1, \ldots, p_k$ at sufficient pairwise distance in $\mathbb{R}^2$, that is, at distance $4r$ for some distance $r > 0$. Let $a_{11}, \ldots, a_{1m}$ be points evenly distributed on the circle of radius $2r$ around p_1. Let a_{21}, a_{31}, and a_{41} be points evenly distributed on the circle with radius r around p_2. Let the points $a_{22}, \ldots, a_{2m}$ lie on the line through a_{21} and p_2 moved away from a_{21} by at most $\epsilon > 0$ as in Fig. 3. The distance ϵ is sufficiently small for our construction, namely $\epsilon < r\left(\frac{1}{\cos\left(\frac{\pi}{m}\right)} - 1\right)$ (assuming $m > 2$). Place the points $a_{32}, \ldots, a_{3m}$ and $a_{42}, \ldots, a_{4m}$ analogously.

For $i \geq 5$ the points a_{ij} are placed as follows. The points $a_{(i-1)j}$ and a_{ij} for $i = 2l$ are placed close to the point p_l. Then we place $a_{(i-1)1}$ and a_{i1} on the intersection of the coordinate axes originating in p_l with the circle of radius r around p_l. The points $a_{(i-1)j}, a_{ij}$ for $j \geq 2$ are placed on these axes, moved away from $a_{(i-1)1}, a_{i1}$ by at most $\delta > 0$. The distance δ is also sufficiently small for our construction, namely it is $\delta \leq (\sqrt{2} - 1)r$.

We can now define the curves in S. As in the construction for $d = 1$, the curves in S visit all but one of the points $p_1, \ldots, p_k$, and in the one point deviate slightly. We define

$$
\begin{aligned}
S_{1j} &= \langle a_{1j}, p_2, \ldots, p_k \rangle & &\text{for } j = 1, \ldots, m \\
S_{ij} &= \langle p_1, a_{ij}, p_2, \ldots, p_k \rangle & &\text{for } i = 2, 3, 4, \ j = 1, \ldots, m \\
S_{ij} &= \langle p_1, \ldots, p_{\lceil \frac{i}{2} \rceil - 1}, a_{ij}, p_{\lceil \frac{i}{2} \rceil}, \ldots, p_k \rangle & &\text{for } i = 5, \ldots, k, \ j = 1, \ldots, m \\
S_i &= \{ S_{i1}, \ldots, S_{im} \} & &\text{for } i = 1, \ldots, 2k
\end{aligned}
$$

Again we claim that for all $1 \leq j_1, \ldots, j_{2k} \leq m$ a query curve Q exists whose set of nearest neighbors in S is

$$
N_S(Q) = \{ S_{1j_1}, S_{21}, \ldots, S_{2j_2}, \ldots, S_{(2m)1}, \ldots, S_{(2m)j_{(2m)}} \}.
$$

This will imply that there are at least m^{2k} different Voronoi regions in the Voronoi diagram of S.

As second point q_2 of Q we choose the midpoint of the circle defined by the three points a_{2j_2}, a_{3j_3}, and a_{4j_4}. Let $r' > r$ be the radius of this circle. Note that $r' \leq r + \epsilon$ and that this circle contains the point p_2. Thus, the points $p_2, a_{21}, \ldots, a_{2j_2}, a_{31}, \ldots, a_{3j_3}$, and $a_{41}, \ldots, a_{4j_4}$ have distance at most r' to the point q_2.

As first point q_1 of Q we choose a point that has distance r' to both p_1 and a_{1j_1}, and a larger distance to all other a_{1j}. Consider the Voronoi diagram of the points $p, a_{11}, \ldots, a_{1m}$. Consider the edge between the cells of p and of a_{1j_1}. Because we chose ϵ sufficiently small, namely $\epsilon < r \left(\frac{1}{\cos\left(\frac{\pi}{m}\right)} - 1 \right)$, and because $r' \leq r + \epsilon$, there are two points in the interior of this edge with distance r' to p_1 and a_{1j_1}. We choose q_1 as one of these two points. Then the distance of q_1 to all a_{1j} for $j \neq j_1$ is larger than r'.

Now we choose the remaining points q_i of Q for $i = 3, \ldots, k$. Let $l = 2i, l' = 2i - 1$. There are two circles with radius r' that touch the points a_{lj_l} and $a_{l'j_l'}$. As q_i we choose the midpoint of the one circle that contains the point p_i. Then a point a_{lj} or $a_{l'j}$ has distance at most r' to q_i exactly if $l \leq j_l$ or $l' \leq j_l'$, respectively.

Construction for $d > 2$. The construction can be generalized to $d > 2$ giving a lower bound of $m^{d(k-1)+2}$ for $m \cdot d(k - 1) + 2$ curves.

The construction at p_1 remains the same. At p_2 we place $d + 1$ sets of points. Then one point from each set, i.e., $d + 1$ points, define a d-ball. At p_i for $i \geq 3$ we place d sets of points. Then one point from each set, i.e., d points, define a ball of the radius given by the choice of q_2. In total, this gives us $m^{d(k-1)+2}$ choices: m choices at p_1, m^{d+1} choices at p_2 and m^d choices each at $p_3, \ldots, p_k$. $\square$

5   Concluding Remarks

In this paper, for the first time, we study the Voronoi diagram of polygonal chains under the discrete Fréchet distance. We show combinatorial upper and lower bounds for such a Voronoi diagram. We conjecture that the upper bound is tight (up to ϵ) as we have shown here for dimension $d = 2$. For closing the gap, consider how the construction for $d = 2$ is generalized to $d > 2$ in the proof of Lemma 5. While the constructions at the vertices $p_2, \ldots, p_k$ are replaced by higher-dimensional analogs, the construction at p_1 stays the same as in the two-dimensional case. Improving the construction at p_1 might close the gap.

At this point, how to compute the diagram is still open. Our upper bound proof does not imply an algorithm as it is also an open problem to compute the lower envelope in high dimensions. If we simply want to compute low dimensional components, like vertices, in the envelope (or in our Voronoi diagram), then there is an $O(n^{kd+\epsilon})$ time algorithm [1]. However, due to the high combinatorial lower bound, to make the diagram useful, we probably need to develop efficient approximation algorithms to approximate such a Voronoi diagram for decent k (say $k = 20 \sim 30$) so that one can first use a $(k-1)$-link chain to approximate a general input polyline. The good news is that in many applications like protein structural alignment, n is not very large (typically ≤ 100). So designing such an (exact or approximate) algorithm is another major open problem along this line.

References

1. Agarwal, P., Aronov, B., Sharir, M.: Computing Envelopes in Four Dimensions with Applications. SIAM J. Comput. 26, 1714–1732 (1997)
2. Alt, H., Godau, M.: Measuring the Resemblance of Polygonal Curves. In: Proceedings of the 8th Annual Symposium on Computational Geometry (SoCG 1992), pp. 102–109 (1992)
3. Alt, H., Godau, M.: Computing the Fréchet Distance between Two Polygonal Curves. Intl. J. Computational Geometry and Applications 5, 75–91 (1995)
4. Alt, H., Knauer, C., Wenk, C.: Matching Polygonal Curves with Respect to the Fréchet Distance. In: Ferreira, A., Reichel, H. (eds.) STACS 2001. LNCS, vol. 2010, pp. 63–74. Springer, Heidelberg (2001)
5. Buchin, K., Buchin, M., Wenk, C.: Computing the Fréchet Distance between Simple Polygons in Polynomial Time. In: Proceedings of the 22nd Annual Symposium on Computational Geometry (SoCG 2006), pp. 80–87 (2006)
6. Edelsbrunner, H., Seidel, R.: Voronoi Diagrams and Arrangements. Discrete Comput. Geom. 1, 25–44 (1986)
7. Eiter, T., Mannila, H.: Computing Discrete Fréchet Distance. Tech. Report CD-TR 94/64, Information Systems Department, Technical University of Vienna (1994)
8. Fréchet, M.: Sur Quelques Points du Calcul Fonctionnel. Rendiconti del Circolo Mathematico di Palermo 22, 1–74 (1906)

9. Godau, M.: On the Complexity of Measuring the Similarity between Geometric Objects in Higher Dimensions. PhD thesis, Freie Universität, Berlin (1998)
10. Indyk, P.: Approximate Nearest Neighbor Algorithms for Fréchet Distance via Product Metrics. In: Proceedings of the 18th Annual Symposium on Computational Geometry (SoCG 2002), pp. 102–106 (2002)
11. Jiang, M., Xu, Y., Zhu, B.: Protein Structure-structure Alignment with Discrete Fréchet Distance. In: Proceedings of the 5th Asia-Pacific Bioinformatics Conference (APBC 2007), pp. 131–141 (2007)
12. Preparata, F.P., Shamos, M.I.: Computational Geometry: An Introduction. Springer, Heidelberg (1985)
13. Schaudt, B., Drysdale, S.: Higher Dimensional Delaunay Diagrams for Convex Distance Functions. In: Proceedings of the 4th Canadian Conference on Computational Geometry (CCCG 1992), pp. 274–279 (1992)
14. Sharir, M.: Almost Tight Upper Bounds for Lower Envelopes in Higher Dimensions. Discrete Comp. Geom. 12, 327–345 (1994)
15. Wenk, C.: Shape Matching in Higher Dimensions. PhD thesis, Freie Universität, Berlin (2002)
16. Zhu, B.: Protein Local Structure Alignment under the Discrete Fréchet Distance. J. Computational Biology 14(10), 1343–1351 (2007)

On Center Regions and Balls Containing Many Points*

Shakhar Smorodinsky[1], Marek Sulovský[2],**, and Uli Wagner[2],**

[1] Department of Mathematics, Ben-Gurion University, Be'er Sheva 84105, Israel
shakhar@cs.bgu.ac.il
[2] Institute of Theoretical Computer Science, ETH Zurich, 8092 Zurich, Switzerland
{smarek,uli}@inf.ethz.ch

Abstract. We study the disk containment problem introduced by Neumann-Lara and Urrutia and its generalization to higher dimensions. We relate the problem to centerpoints and lower centerpoints of point sets. Moreover, we show that for any set of n points in $\mathbb{R}^d$, there is a subset $A \subseteq S$ of size $\lfloor \frac{d+3}{2} \rfloor$ such that any ball containing A contains at least roughly $\frac{4}{5ed^3}n$ points of S. This improves previous bounds for which the constant was exponentially small in d. We also consider a generalization of the planar disk containment problem to families of pseudodisks.

1 Introduction

In 1988, Neumann-Lara and Urrutia [12] showed that for every set S of $n \geq 2$ points in the plane, there is a pair $\{a, b\} \subseteq S$ of *disk depth* at least $\frac{n-2}{60}$ with respect to S, i.e., every disk D containing $\{a, b\}$ contains at least $\frac{n-2}{60}$ points of S. The constant of 1/60 was repeatedly improved in a series of subsequent papers by Hayward, Rappaport and Wenger [9], by Bárány, Schmerl, Sidney and Urrutia [5], and by Hayward [8]. The best lower bound to date was obtained by Edelsbrunner, Hasan, Seidel, and Shen [7], who proved[1] that there is always a pair $\{a, b\} \subseteq S$ of disk depth at least

$$\frac{n-3}{2} - \sqrt{\frac{n^2 - 4n + 3}{12}} \approx \frac{1}{4.73}n$$

---

* This research was started at the Gremo Workshop on Open Problems 2007 organized by Emo Welzl in Illgau, Switzerland, 2-6 July.
** Research of Marek Sulovský and Uli Wagner was supported by the Swiss National Science Foundation (SNF Project 200021-116741).

[1] The proof is rather involved and combines an analysis of higher-order Vornoi diagrams with a lower bound of $3\binom{k+1}{2}$ for the number of so-called $(\leq k)$-*sets* of a point set, i.e., of subsets of size at most k that can be separated by a line, $0 \leq k < n/2$. The latter bound is tight for $k \leq n/3$, which is the range in which it is needed for the disk containment problem. The proof of that bound in [7] contains a lacuna; correct proofs were later furnished in [10] and [1] in the context of rectilinear crossing numbers of graphs. See also [3] for a particularly elegant proof and further improvements.

X. Hu and J. Wang (Eds.): COCOON 2008, LNCS 5092, pp. 363–373, 2008.
© Springer-Verlag Berlin Heidelberg 2008

w.r.t. S. On the other hand Hayward et al. [9] gave a construction of point sets for which no pair of points from the set has disk depth more than $\lceil n/4 \rceil - 1$. Thus, if we denote by c_2 the largest constant such that for every set S of n points in the plane, there is a pair $\{a, b\} \subseteq S$ of disk depth at least $c_2 \cdot n + o(n)$, then $1/4.73 \approx 1/2 - 1/\sqrt{12} \le c_2 \le 1/4$. The upper bound is conjectured to be tight[2] [9]. Hayward et al. [9] also considered a variant of the problem in which the point set is required to be in convex position. In this case, they proved that there is always a pair of disk depth at least $\lceil n/3 \rceil + 1$, and that this is tight in the worst case.

Bárány et. al. [5] considered a natural extension of the problem to higher dimensions (which was already raised in [12]). In dimension $d \ge 3$, pairs of points are no longer enough. In fact, a construction based on the moment curve in $\mathbb{R}^{d+1}$ shows that there are n-point sets in $\mathbb{R}^d$ for which every subset of size $\lfloor \frac{d+1}{2} \rfloor$ can be separated from the remaining points by a ball, i.e., has ball depth at most $\lfloor \frac{d+1}{2} \rfloor$. Bárány et al. show that this is the threshold: for every d, there exists a constant $c_d > 0$ such that for every set S of n points in $\mathbb{R}^d$, there is a subset $A \subseteq S$ of size at most $\lfloor \frac{d+3}{2} \rfloor$ and ball depth at least $c_d \cdot n$ w.r.t. S.

By abuse of notation, we denote the largest such constant by c_d as well. The proof of Bárány et al. yields a lower bound of

$$c_d \ge \frac{\lfloor \frac{d+3}{2} \rfloor}{\binom{d+3}{\lfloor \frac{d+3}{2} \rfloor}} \approx \frac{\sqrt{\pi} \left(\frac{d+3}{2} \right)^{3/2}}{2^{d+3}},$$

which is inversely exponential in d. In Section 3, we give an alternative proof which yields a lower bound that is inversely polynomial (cubic) in d, roughly $c_d \ge \frac{4}{5ed^3}$. In Section 2, we use the standard lifting map to the unit paraboloid in $\mathbb{R}^{d+1}$ to relate the ball containment problem in $\mathbb{R}^d$ to centerpoints and lower centerpoints in $\mathbb{R}^{d+1}$.

In Section 4, we consider an extension of the disk containment problem in a different direction and prove a dual version of the Neumann-Lara-Urrutia theorem for points and families of pseudo-disks in the plane.

2   Depth and the Region of (Lower) Centerpoints

To motivate the discussion in this section, we recall the paraboloidal lifting map. Let $U = \{x \in \mathbb{R}^{d+1} : x_{d+1} = x_1^2 + \ldots + x_d^2\}$. The lifting map takes a point $p = (p_1, \ldots, p_d) \in \mathbb{R}^d$ to the point $\hat{p} = (p_1, \ldots, p_d, p_1^2 + \ldots + p_d^2) \in U$. If B is a ball in $\mathbb{R}^d$ then there is a lower halfspace[3] $\hat{B}$ in $\mathbb{R}^{d+1}$ (open or closed according to

---

[2] Moreover, the error term is conjectured to be $O(1)$.

[3] Here, a halfspace is called *lower* (w.r.t. a fixed euclidean coordinate system) if it is of the form $H = \{x \in \mathbb{R}^d : x_d \le a_1 x_1 + \ldots + a_{d-1} x_{d-1}\}$; *upper* halfspaces are defined analogously. Furthermore, a hyperplane (or halfspace) is called *vertical* if it (or its bounding hyperplane) is of the form $\{x \in \mathbb{R}^d : a_1 x_1 + \ldots + a_{d-1} x_{d-1} = b\}$, not all a_i zero.

whether B is) such that a point $p \in \mathbb{R}^d$ lies in the interior of B, on the boundary of B, or outside of B, if and only if $\hat{p}$ lies in the interior of $\hat{B}$, on the boundary of $\hat{B}$, or in the upper open halfspace complementary to $\hat{B}$, respectively. We refer to [11, Section 5.7] for more details.

Thus, if S is a set of points in $\mathbb{R}^d$, then a set A has ball depth r w.r.t. S if and only if $\hat{A}$ has *lower halfspace depth* r w.r.t. $\hat{S}$. For the remainder of this section, we will abandon the lifting map and speak about point sets in $\mathbb{R}^d$. To unify our terminology and notation, we make the following general definition.

Let $\mathcal{R}$ be a family of subsets of $\mathbb{R}^d$, called *ranges*, and let $S \subseteq \mathbb{R}^d$, $|S| = n$. We define the $\mathcal{R}$-depth w.r.t. S of a set $A \subseteq \mathbb{R}^d$ (not necessarily a subset of S) as

$$\text{depth}_S^{\mathcal{R}}(A) := \min\{|P \cap R| : R \in \mathcal{R}, A \subseteq R\}.$$

We focus on the following families of ranges: $\mathcal{B}$, the family of all balls; $\mathcal{H}$, the family of all affine halfspaces; and the family $\mathcal{H}^-$, the family of all *lower* halfspaces. We assume that all points sets are in general position.[4]

For the remainder of this section, we restrict our attention to the case that all ranges are halfspaces, i.e., we assume[5] that $\mathcal{R} \subseteq \mathcal{H}$. For any integer $r \geq 0$, we define the r-*center* of S w.r.t. $\mathcal{R}$ as

$$C_r^{\mathcal{R}}(S) := \{x \in \mathbb{R}^d : \text{depth}_S^{\mathcal{R}}(x) \geq r\}.$$

Note that $\mathcal{R}' \subseteq \mathcal{R}$ implies $C_r^{\mathcal{R}'}(S) \supseteq C_r^{\mathcal{R}}(S)$ for all r and S. In the case $\mathcal{R} = \mathcal{H}$, it is known that $C_{\lceil \frac{n}{d+1} \rceil}^{\mathcal{H}}(S) \neq \emptyset$ for every n-point set S. The elements of this set are called *centerpoints* of S. In fact, it is easy to see that $C_r^{\mathcal{H}}(S)$ can be written as the intersection of all *closed* halfspaces that *miss* (i.e., whose complementary open halfspace contains) at most $r - 1$ points of S. If $(d + 1)(r - 1) < n$, i.e, $r \leq \lceil n/(d+1) \rceil$, then the intersection of any $d+1$ of these halfspaces is nonempty, and Helly's theorem implies $C_r^{\mathcal{H}}(S) \neq \emptyset$, see [11] for more details. In general this is best possible, i.e., there are examples of n-point sets with $C_r^{\mathcal{H}}(S) = \emptyset$ for $r > \lceil \frac{n}{d+1} \rceil$.

Agarwal et al. [2] showed that in the representation of $C_r^{\mathcal{H}}(S)$ as an intersection of closed halfspaces, it is sufficient to take only those halfspaces H that (1) miss *exactly* $r - 1$ points of S and (2) whose bounding hyperplane ∂H is spanned by d points of S. In particular, $C_r^{\mathcal{H}}$ is a convex polytope (it is obviously bounded).

We now turn to lower halfspaces. In this case, the center is still a convex polyhedron, but it is unbounded (if it is nonempty). We state this more precisely in the following lemma. We identify $\mathbb{R}^{d-1}$ with the hyperplane $\{x_d = 0\}$ and

---

[4] What exactly that means depends on the ranges under consideration: in the case of halfspaces, general position means that no $d + 1$ or fewer of the points are affinely independent. In the case of balls, we additionally require that no $d + 2$ of the points lie on a common $(d - 1)$-dimensional sphere.

[5] The reason is that for other ranges, the definiton of the center is usually not very meaningful since single points do not have large depth. For instance, in the case $\mathcal{R} = \mathcal{B}$ we have $C_1^{\mathcal{B}}(S) = S$ and $C_r^{\mathcal{B}}(S) = \emptyset$ for $r > 1$.

denote by π the orthogonal projection onto that hyperplane. Analogously to the proof of [2, Lemma 2.1], one can show the following:

Lemma 1. *The set $C_r^{\mathcal{H}^-}(S)$ is a convex polyhedron. It can be written as*

$$C_r^{\mathcal{H}^-}(S) = \left(\bigcap_{H \in \overline{\mathcal{H}}_{r-1}^+(S)} H \right) \cap \left(C_r^{\mathcal{H}}(\pi(S)) \times \mathbb{R} \right),$$

where $\overline{\mathcal{H}}_{r-1}^+(S)$ is the set of closed upper halfspaces whose bounding hyperplane is spanned by S and that miss exactly $r-1$ points of S.

Corollary 1. *If $r \leq \lceil n/d \rceil$ then $C_r^{\mathcal{H}^-}(S) \neq \emptyset$, and this is tight.*

Proof. If $r \leq \lceil n/d \rceil$, consider any point $(x_1, \ldots, x_d - 1) \in C^{\mathcal{H}}(\pi(S))$. By choosing a sufficiently large last coordinate x_d, we can lift it to a point $x = (x_1, \ldots, x_d)$ that lies above all hyperplanes spanned by S, and then $x \in C_r^{\mathcal{H}^-}(S)$. On the other hand, if $r > \lceil n/d \rceil$, we can first choose an n-point set $S' \subseteq \mathbb{R}^{d-1}$ with $C_r^{\mathcal{H}}(S') = \emptyset$. Then, by the preceding lemma, any lifting S' to $\mathbb{R}^d$, i.e., any $S \subseteq \mathbb{R}^d$ with $\pi(S) = S'$ will have $C_r^{\mathcal{H}^-}(S) = \emptyset$.

Next, we relate the center to the depth of larger (sub)sets. Let $\mathcal{R} \subseteq \mathcal{H}$. If a set A satisfies $\text{conv}(A) \cap C_r^{\mathcal{R}}(S) \neq \emptyset$ then $\text{depth}_S^{\mathcal{R}}(A) \geq r$ (in fact, for this implication we only need convexity of the ranges). What can we say about the converse? We need the following lemma.

Lemma 2. *Let $P \subseteq \mathbb{R}^d$ be a convex polyhedron and let $K \subseteq \mathbb{R}^d$ be a k-dimensional convex body disjoint from P. Then there exist $m = \min\{d, k+1\}$ facet-defining closed halfspaces $H_1, \ldots, H_m$ of P and a halfspace H containing K such that $H \cap H_1 \cap \ldots \cap H_m = \emptyset$.*

Proof. Since K and P are disjoint, both are closed, and K is compact, we can strictly separate them. That is, there is a hyperplane h such that P is contained in one of the corresponding open halfspaces, which we denote by h^+, and K is contained in the opposite open halfspace, h^-. Moreover, since the hyperplane h and P are polyhedral and disjoint, they have positive distance from each other. The points $q \in P$ for which $\text{dist}(q, h) = \text{dist}(P, h)$ form a face F of P of some dimension $d - r$, $1 \leq r \leq d$, and a suitable parrallel translate of h is a supporting hyperplane for P at F. Let $h_1, \ldots, h_r$, $r \leq d$, be the facet hyperplanes defining F, and let $H_1, \ldots, H_r$ be the corresponding closed halfspaces containing P. Then[6] $h^- \cap H_1 \cap \ldots \cap H_r = \emptyset$. A fortiori, $A \cap h^- \cap H_1 \cap \ldots \cap H_r = (A \cap h^-) \cap (A \cap H_1) \cap \ldots \cap (A \cap H_r) = \emptyset$, where A is the affine hull of K. Thus,

---

[6] To see this, assume that there exists some $x \in h^- \cap H_1 \cap \ldots \cap H_r$, and let q be a point in the relative interior of F. Since $x \in h^-$ and $q \in h^+$, it follows from convexity of $C := H_1 \cap \ldots \cap H_r$ that there is a point $y \in C \cap h$. Moreover, if H is a defining halfspace of the polyhedron P other than $H_1, \ldots, H_r$, then q lies in the interior of H. Thus, for sufficiently small $\varepsilon > 0$, $p := (1 - \varepsilon)q + \varepsilon y$ still lies in P, which is a contradiction, since it has smaller distance to h than q.

we have $r + 1$ convex sets in the k-dimensional space A that have an empty intersection. By Helly's theorem, there is a subfamily of at most $k + 1$ of these sets that have an empty intersection. Thus, by adding $h' \cap A$ to this subfamily and by relabeling the H_i, if either of these is necessary, we may assume that $A \cap h^- \cap H_1 \cap \ldots \cap H_m = \emptyset$, where $m = \min\{r, k + 1\}$. By further intersecting this empty intersection with K, we conclude that K is disjoint from the polyhedron $H_1 \cap \ldots \cap H_m$. Thus, we can separate them, i.e., find a halfspace H containing K such that $H \cap H_1 \cap \ldots \cap H_m = \emptyset$, as desired.

Corollary 2. *If* $\mathcal{R} = \mathcal{H}$ *or* $\mathcal{R} = \mathcal{H}^-$, *and if* K *is a convex body of dimension* k, *then* $K \cap C_r^{\mathcal{R}} = \emptyset$ *implies* $\mathrm{depth}_P^{\mathcal{R}}(K) \leq m(r - 1)$, *where* $m = \min\{d, k + 1\}$.

3   The Main Theorem

Theorem 1. *Let* $k = \lfloor \frac{d}{2} \rfloor + 1$ *and let*

$$\alpha = \alpha_d := \frac{1}{5ed(\lfloor \frac{d}{2} \rfloor + 1)(\lfloor \frac{d}{2} \rfloor + 3)}.$$

Then for every finite point set $S \subseteq \mathbb{R}^d$ *there exists a* k-*tuple* $A \subseteq S$ *such that*

$$\mathrm{depth}_S^{\mathcal{H}}(A) \geq \alpha n + O(1).$$

If we consider lower halfspaces instead of general halfspaces, the constant factor can be improved by a factor of 2. In the special case $d = 3$, a key step in our proof (tight lower bounds on the number of $(\leq k)$-sets can be refined to give a improved constants $1/(8e)$ (respectively, of $1/6e$, for points in convex position). We discuss these issues in more detail in Remark 1 below. We also note the following immediate consequence of the theorem, combined with Corollary 2:

Corollary 3. *Let* S *be a set of* n *points in general position in* $\mathbb{R}^d$. *Then there exists a subset* $A \subseteq S$ *of* $k_d = \lfloor \frac{d}{2} \rfloor + 1$ *elements such that* $\mathrm{conv}\, A \cap C_r^{\mathcal{H}}(S) \neq \emptyset$, *where* $r = \frac{\alpha_d}{k_d} n + O(1)$.

By the paraboloidal lifting map discussed in the beginning of Section 2, Theorem 1 also implies:

Corollary 4. *Let* S *be a set of* n *points in general position in* $\mathbb{R}^d$. *Then there is a subset* $A \subseteq S$ *of* $\lfloor \frac{d+3}{2} \rfloor$ $(= k_{d+1})$ *elements such that* $\mathrm{depth}_S^{\mathcal{B}}(A) \geq \alpha_{d+1} n + O(1)$.

In the special case $d = 2$, our proof yields a constant of $1/6e$ (instead of the currently best $1/4.73$).

Proof (Proof of Theorem 1). Let us assume that for some integer parameter ℓ to be specified in the course of the proof, every $A \in \binom{S}{k}$ satisfies $\mathrm{depth}_S^{\mathcal{H}}(A) \leq \ell$. We will show that $\ell \geq \alpha n + O(1)$, from which it follows that in particular there must be some A of $\mathrm{depth}_S^{\mathcal{H}}(A) = \ell - 1 = \alpha n + O(1)$, as promised.

Our assumption means that for every A there is a closed halfspace H which contains A and at most ℓ points of S in total. For any finite point set $X \subseteq \mathbb{R}^d$, let $a_{k,\le\ell}(X)$ (and $a_{k,\ell}(X)$) be the number of k-element subsets $A \subseteq X$ for which there is a halfspace containing A and at most ℓ (exactly ℓ) points of X. Then the above assumption implies

$$a_{k,\le\ell}(S) = \binom{n}{k}. \tag{1}$$

On the other hand, by the random sampling analysis of Clarkson and Shor [6], this number is bounded from above by

$$a_{k,\le\ell}(S) \le \mathbb{E}[a_{k,0}(X)]p^{-k}(1-p)^{-\ell}, \tag{2}$$

where X is a random subset of S obtained by choosing each point independently with probability p, for a parameter $0 < p < 1$ that will be specified later. Note that for any point set X,

$$a_k(X) := a_{k,0}(X)$$

is the number of k-sets of X, i.e., of k-element subsets $Y \in \binom{X}{k}$ such that Y and $X \setminus Y$ can be separated by a hyperplane.[7] This number is notoriously difficult to estimate in general. In a first step, we use the obvious estimate $a_k(X) \le a_{\le k}(X) := a_1(X) + \ldots + a_k(X)$. In our case, $k = \lfloor d/2 \rfloor + 1$ depends only on d, not on n, so we do not lose much in this step. The second step is to use worst-case estimates for $a_{\le k}(X)$. This was one of the first problems to which Clarkson and Shor applied their random sampling technique. Their bounds are tight up to a constant depending only on d, but for very small k, as in our case, this constant is quite large (exponential in d). Sharper bounds were proved by Wagner [14, Corollary 2.5]:

Lemma 3. *Let S be a set of n points in $\mathbb{R}^d$ and let $C_{n,d}$ denote a set of n points on the moment curve in $\mathbb{R}^d$. Then[8] for any integer $1 \le k \le n - 1$,*

$$a_{\le k}(S) \le 4 \cdot \sum_{i=1}^{d} a_{\le k}(C_{n,i}) \tag{3}$$

For points on the moment curve, the numbers a_k were computed exactly by Andrzejak and Welzl [4, Corollary 5.2]:

Lemma 4. *Let $C_{n,d}$ be a set of n points on the moment curve in $\mathbb{R}^d$. Then, for $1 \le k \le n - 1$, the number of k sets in $C_{n,d}$ equals*

$$a_k(C_{n,d}) = \sum_{s=0}^{d} \left[\binom{k-1}{\lfloor \frac{s-1}{2} \rfloor} \binom{n-k-1}{\lfloor \frac{s}{2} \rfloor} + \binom{k-1}{\lfloor \frac{s}{2} \rfloor} \binom{n-k-1}{\lfloor \frac{s-1}{2} \rfloor} \right] \tag{4}$$

---

[7] We remark that $a_{k,0}(X)$ is related, but in general not equal, to the number of $(k-1)$-dimensional faces of the convex hull $\mathrm{conv}(X)$.

[8] The results in [14] are phrased in terms of levels in arrangements of affine hyperplanes or halfspaces. By standard point-hyperplane duality, the k-sets of a set of n points in $\mathbb{R}^d$ correspond to the the d-dimensional cells at lower level k plus the d-dimensional cells at upper level k in an arrangement of n nonvertical affine hyperplanes in $\mathbb{R}^d$. The fact that we have to consider lower and upper levels separately leads to the factor of 4 in Lemma 3, instead of the factor of 2 in the quoted corollary.

We can regroup the terms with respect to the binomial coefficient with n and obtain an upper bound

$$a_k(C_{n,d}) \leq \sum_{t=0}^{\lfloor \frac{d}{2} \rfloor} \binom{n-k-1}{t} \cdot \left(\binom{k-1}{t-1} + 2\binom{k-1}{t} + \binom{k-1}{t+1} \right)$$

$$= \sum_{t=0}^{\lfloor \frac{d}{2} \rfloor} \binom{k+1}{t+1} \cdot \binom{n-k-1}{t} \leq \sum_{t=0}^{\lfloor \frac{d}{2} \rfloor} \binom{k+1}{k-t} \cdot \binom{n}{t}$$

We apply this with the particular $k = \lfloor d/2 \rfloor + 1$ of interest in our context. Recall that any subset of $C_{n,d}$ of size $i < k$ forms a face of the convex hull (the points on the moment curve are the vertices of a neighborly polytope), so formula (4) simplifies to $a_i(C_{n,d}) = \binom{n}{i}$. Thus,

$$a_{\leq k}(C_{n,d}) = \sum_{t=0}^{\lfloor \frac{d}{2} \rfloor} \left(1 + \binom{\lfloor \frac{d}{2} \rfloor + 2}{\lfloor \frac{d}{2} \rfloor + 1 - t} \right) \binom{n}{t} \leq \sum_{t=0}^{\lfloor \frac{d}{2} \rfloor} (\lfloor \tfrac{d}{2} \rfloor + 3)^{\lfloor \frac{d}{2} \rfloor - t + 1} \binom{n}{t}$$

Combining with Lemma 3 and the fact that $a_{\leq k}(C_{n,i}) \leq a_{\leq k}(C_{n,j})$ for $i \leq j$ [9], we conclude:

Lemma 5. *Let S be a set of n points in $\mathbb{R}^d$ and $k := \lfloor \frac{d}{2} \rfloor + 1$. Then*

$$a_{\leq k}(S) \leq 4d(\lfloor \tfrac{d}{2} \rfloor + 3) \sum_{t=0}^{\lfloor \frac{d}{2} \rfloor} (\lfloor \tfrac{d}{2} \rfloor + 3)^{\lfloor \frac{d}{2} \rfloor - t} \binom{n}{t}.$$

We now return to the random sampling setting. Note that the cardinality of the random subset X of S has the binomial distribution $\mathrm{Bi}(n, p)$. We need the following simple fact:

Lemma 6. *Let N be a binomially distributed random variable $N \sim \mathrm{Bi}(n, p)$ and let s be a positive integer. Then*

$$\mathbb{E}\left[\binom{N}{s} \right] = p^s \binom{n}{s}.$$

We apply this and Lemma 5 to the random subset X and obtain

$$\mathbb{E}[a_k(X)] \leq 4d(\lfloor \tfrac{d}{2} \rfloor + 3) p^{\lfloor d/2 \rfloor} \sum_{s=0}^{\lfloor \frac{d}{2} \rfloor} (\lfloor \tfrac{d}{2} \rfloor + 3)^s p^{-s} \binom{n}{\lfloor d/2 \rfloor - s}. \qquad (6)$$

---

[9] Indeed, every k-set of $C_{n,i}$ is lifted (by adding the missing $j - i$ coordinates) to a k-set of $C_{n,j}$ since we can extend the separating hyperplane to be vertical in those additional coordinates. This bound is very generous and a more careful and lenghtier calculation could replace the factor of d in the final result by some constant factor.

Substituting this into (2) yields (recall that $k = \lfloor d/2 \rfloor + 1$)

$$a_{k,\leq \ell}(S) \leq 4d(\lfloor \tfrac{d}{2} \rfloor + 3)p^{-1}(1-p)^{-\ell} \sum_{s=0}^{\lfloor \tfrac{d}{2} \rfloor}(\lfloor \tfrac{d}{2} \rfloor + 3])^s p^{-s} \binom{n}{\lfloor d/2 \rfloor - s}.$$

We set $p := \frac{1}{1+\ell}$ and obtain

$$a_{k,\leq \ell}(S) \leq 4ed(\lfloor \tfrac{d}{2} \rfloor + 3)(1+\ell) \sum_{s=0}^{\lfloor \tfrac{d}{2} \rfloor}(\lfloor \tfrac{d}{2} \rfloor + 3](1+\ell))^s \binom{n}{\lfloor d/2 \rfloor - s}.$$

Basic manipulations yield

$$a_{k,\leq \ell}(S) \leq 4ed(\lfloor \tfrac{d}{2} \rfloor + 3)(1+\ell) \sum_{s=0}^{\lfloor \tfrac{d}{2} \rfloor}(\lfloor \tfrac{d}{2} \rfloor \cdot \lfloor \tfrac{d}{2} + 3 \rfloor \cdot \tfrac{2\ell}{n})^s \frac{n^{\lfloor \tfrac{d}{2} \rfloor}}{(\lfloor d/2 \rfloor - s)! \lfloor \tfrac{d}{2} \rfloor^s}.$$

This can be further estimated by a geometric series

$$a_{k,\leq \ell}(S) \leq 4ed(\lfloor \tfrac{d}{2} \rfloor + 3)(1+\ell) \frac{n^{\lfloor \tfrac{d}{2} \rfloor}}{(\lfloor d/2 \rfloor)!} \sum_{s=0}^{\lfloor \tfrac{d}{2} \rfloor}(\lfloor \tfrac{d}{2} \rfloor \cdot \lfloor \tfrac{d}{2} + 3 \rfloor \cdot \tfrac{2\ell}{n})^s,$$

which sums up to

$$a_{k,\leq \ell}(S) \leq 5ed(\lfloor \tfrac{d}{2} \rfloor + 3)(1+\ell) \frac{n^{\lfloor \tfrac{d}{2} \rfloor}}{(\lfloor d/2 \rfloor)!} \tag{7}$$

under the assumption that $\lfloor \tfrac{d}{2} \rfloor \cdot \lfloor \tfrac{d}{2} + 3 \rfloor \cdot \tfrac{2\ell}{n} \leq \tfrac{1}{5}$. We combine the upper bound with (1) and obtain

$$\binom{n}{\lfloor \tfrac{d}{2} \rfloor + 1} \leq 5ed(\lfloor \tfrac{d}{2} \rfloor + 3)(1+\ell) \frac{n^{\lfloor \tfrac{d}{2} \rfloor}}{(\lfloor d/2 \rfloor)!}.$$

Futher simple manipulations give

$$\frac{1}{5ed(\lfloor \tfrac{d}{2} \rfloor + 1)(\lfloor \tfrac{d}{2} \rfloor + 3)} \cdot \frac{(n - \lfloor \tfrac{d}{2} \rfloor + 1)^{\lfloor \tfrac{d}{2} \rfloor + 1}}{n^{\lfloor \tfrac{d}{2} \rfloor}} \leq 1 + \ell,$$

which can be written as

$$\frac{n}{5ed(\lfloor \tfrac{d}{2} \rfloor + 1)(\lfloor \tfrac{d}{2} \rfloor + 3)} + O(1) \leq \ell.$$

This completes the proof of the theorem. The only thing that still has to be verified is validity of (7). But for sufficiently high n it is valid, since $2 < ed$.

Remark 1

1. In general, if we consider lower halfspaces instead of arbitrary ones, k-sets and $(\leq k)$-sets are replaced by lower k-sets and lower $(\leq k)$-sets, respectively. For these, the factor 4 in Lemma 3 can be replaced by a factor of 2, which

yields an improvement by a factor of 2 also for the final constant which we could denote by $\alpha_d^- = 2\alpha_d$.

2. If we consider arbitrary halfspaces by focus on the special case $d = 3$, exact upper bounds for the number e_k of so-called $(\leq k)$-*facets*[10] were proved by Welzl [15, Cor. 8]: for every set S of n points in $\mathbb{R}^3$, $e_{\leq k}(S) \leq 2\left[\binom{k+2}{2}n - 2\binom{k+3}{3}\right]$ (and this is attained for point sets in convex position). Together with the linear equation $a_k = (e_{k-1} + e_{k-2})/2 + 2$ valid for any point set in $\mathbb{R}^3$ (see [4, Remark 5]), this easily implies $a_2(S) \leq a_{\leq 2}(S) \leq 4n$ for the special $k = k_3 = 2$. This drastically simplifies the computations for the Clarkson-Shor estimate and yields $\binom{n}{2} = a_{2,\leq\ell}(S) \leq 4np(1-p)^{-\ell}$. Setting $p = 1/(1+\ell)$ as before then shows that there is a pair $A \in \binom{S}{2}$ of $\mathrm{depth}_S^{\mathcal{H}}(A) \geq \frac{n-1}{8e} - 2$.

3. If, in addition to $d = 3$, we further assume that the point set S is in convex position (which is the case, for instance, for points on the paraboloid U obtained by the lifting map), then a 2-set of S is actually an edge of the convex hull. Thus, $a_k(S) \leq 3n - 6$, which yields a further improvement in the constant from $1/8e$ to $1/6e$.

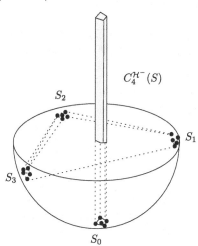

4. Consider the maximal value $r_d(n)$ such that in every point set S of size n in $\mathbb{R}^d$ there is a subset A of $k_d = \lfloor\frac{d}{2}\rfloor + 1$ elements satisfying $\mathrm{conv}\, A \cap C_{r_d}^{\mathcal{H}^-}(P) \neq \emptyset$. Corollary 3 shows that $r_d \geq \frac{1}{k_d}\alpha_d n + O(1)$. On the other hand, if $r > \lceil n/d \rceil$, then $C_r^{\mathcal{H}^-}(S)$ may be empty. There is a simple construction, which improves this trivial upper bound to $\lceil\frac{n}{d+1}\rceil$. In fact, it even shows that there is an n point set on *a paraboloid* such that for that $r = \lceil\frac{n}{d+1}\rceil + 1$, $\mathrm{conv}\, A \cap C_r^{\mathcal{H}^-}(S) = \emptyset$ for all $(d-1)$-element subsets. For simplicity, suppose that $n = t(d+1)$. Consider the vertex set of a regular $(d-1)$-simplex centered at the origin o in $\mathbb{R}^{d-1}$ and replace each vertex v_i by a sufficiently small cluster V_i of t

---

[10] A k-facet of a point set S in general position is a $(d-1)$-dimensional simplex σ such that one of the two open halfspaces determined by σ contains exactly k points of S.

points. Similarly, replace o by a small cluster of t points. Let $S_i = \hat{V}_i$ be the lifting of V_i to the paraboloid and let $S = S_0 \cup S_1 \cup \ldots \cup S_d$ (the lifted clusters $\hat{V}_i$ are located at the vertices of a d-simplex that is "pointing downward"). Consider $C := C_r^{\mathcal{H}^-}(S)$, with $r = \frac{n}{d+1} + 1$. Clearly, C is contained in the vertical cylinder conv$(V_0) \times \mathbb{R}$. Moreover, C lies above the lower convex hull of the "upper" clusters $S_1 \cup \ldots \cup S_d$. It is easy to see that conv $A \cap C = \emptyset$ for all $A \in \binom{S}{d-1}$.

5. The constants α_d are not tuned to optimality. In fact, the bound for lower dimensional $(\leq k)$-sets used to obtain Lemma 5 can be completely avoided by a longer precise calculation. This saves a factor of d at the cost of only slightly higher absolute constant and the bound is $\alpha_d = \frac{1}{\gamma_d 5.4e(\lfloor \frac{d}{2} \rfloor + 1)(\lfloor \frac{d}{2} \rfloor + 3.5)}$ where $\gamma_d := 1$ for d even and 2 for odd.

6. One might consider $k > k_d$ in Theorem 1. Our method gives improved values of the constant α for larger k. (Note that for $k = d$, one trivially gets a constant of $1/2$, by taking a d-tuple that that defines a halving facet of the point set.)

4   Pseudo-disks in the Plane

Definition 1. *A family D of simple closed Jordan regions is called a family of pseudo-disks, if for every pair $a, b \in D$ the boundaries of a and b intersect in at most two points.*

Theorem 2. *Let S be a set of points in $\mathbb{R}^2$ and let D be a family of pseudo-discs such that the boundary of each member of D passes through a pair of points in S and each pair out of the $\binom{n}{2}$ possible pairs has exactly one element of D whose boundary pass through it. Then there exists a member $d \in D$ that contain at least $\frac{1}{6e}n - 3$ points of S.*

Proof. This result can again be obtained by using the Clarkson-Shor method. Denote by $C_{\leq k}(P, D)$ the number of pseudocircles with at most k points in their interior. Sharir and Smorodinsky [13, Lemma 3.6] showed that for any such configuration of points and pseudodisks, $C_0(P, D) \leq 3n$. Then the Clarkson-Shor method gives $C_{\leq k}(P, D) \leq 3p^{-1}(1-p)^{-k}n$. Setting $p := \frac{1}{k+1}$ and applying basic estimates yields $C_{\leq k}(P, D) \leq 3e(k+1)n$. On the other hand, assuming that all the pseudodiscs have at most k points inside, we obtain $\binom{n}{2} \leq C_{\leq k}(P, D)$. Comparing these two numbers gives $\frac{n-1}{6e} - 1 \leq k$. Hence, choosing $k = \frac{n-1}{6e} - 2$ gives a desired contradiction.

References

1. Ábrego, B.M., Fernández-Merchant, S.: A Lower Bound for the Rectilinear Crossing Number. Graphs Comb. 21(3), 293–300 (2005)
2. Agarwal, P.K., Sharir, M., Welzl, E.: Algorithms for Center and Tverberg Points. In: Proceedings of the Twentieth Annual Symposium on Computational Geometry (SoCG), New York, pp. 61–67 (2004)

3. Aicholzer, O., Garcia, J., Orden, D., Ramos, P.: New Lower Bounds for the Number of ($\leq k$)-Edges and the Rectilinear Crossing Number of k_n. In: Proceedings of the 18th ACM-SIAM Symposium on Discrete Algorithms (2007)
4. Andrzejak, A., Welzl, E.: In Between k-Sets, j-Facets, and i-Faces: (i, j)-Partitions. Discrete Comput. Geom. 29(1), 105–131 (2003)
5. Bárány, I., Schmerl, J.H., Sidney, S.J., Urrutia, J.: A Combinatorial Result about Points and Balls in Euclidean Space. Discrete Comput. Geom. 4(3), 259–262 (1989)
6. Clarkson, K.L., Shor, P.W.: Applications of Random Sampling in Computational Geometry. II. Discrete Comput. Geom. 4(5), 387–421 (1989)
7. Edelsbrunner, H., Hasan, N., Seidel, R., Shen, X.J.: Circles Through Two Points that Always Enclose Many Points. Geom. Dedicata 32(1), 1–12 (1989)
8. Hayward, R.: A Note on the Circle Containment Problem. Discrete Comput. Geom. 4(3), 263–264 (1989)
9. Hayward, R., Rappaport, D., Wenger, R.: Some Extremal Results on Circles Containing Points. Discrete Comput. Geom. 4(3), 253–258 (1989)
10. Lovász, L., Vesztergombi, K., Wagner, U., Welzl, E.: Convex Quadrilaterals and k-Sets. In: Towards a theory of geometric graphs, Contemp. Math., vol. 342, pp. 139–148 (2004)
11. Matoušek, J.: Lectures on Discrete Geometry 212 of Graduate Texts in Mathematics. Springer, New York (2002)
12. Neumann-Lara, V., Urrutia, J.: A Combinatorial Result on Points and Circles on the Plane. Discrete Math 69(2), 173–178 (1988)
13. Smorodinsky, S., Sharir, M.: Selecting Points that Are Heavily Covered by Pseudo-Circles, Spheres or Rectangles. Combin. Probab. Comput. 13(3), 389–411 (2004)
14. Wagner, U.: On a Generalization of the Upper Bound Theorem. In: Proceedings of the 47 Annual Symposium on the Foundations of Computer Science (FOCS), pp. 635–645 (2006)
15. Welzl, E.: Entering and Leaving j-Facets. Discrete Comput. Geom. 25(3), 351–364 (2001)

On Unfolding 3D Lattice Polygons and 2D Orthogonal Trees

Sheung-Hung Poon

Department of Computer Science,
National Tsing Hua University, Hsinchu, Taiwan
spoon@cs.nthu.edu.tw

Abstract. We consider the problems of unfolding 3D lattice polygons embedded on the surface of some classes of lattice polyhedra, and of unfolding 2D orthogonal trees. During the unfolding process, all graph edges are preserved and no edge crossings are allowed. Let n be the number of edges of the given polygon or tree. We show that a lattice polygon embedded on an open lattice orthotube can be convexified in $O(n)$ moves and time, and a lattice polygon embedded on a lattice Tower of Hanoi, a lattice Manhattan Tower, or an orthogonally-convex lattice polyhedron can be convexified in $O(n^2)$ moves and time. The main technique in our algorithms is to fold up the lattice polygon from the end blocks of the given lattice polyhedron. On the other hand, we show that a 2-monotone orthogonal tree on the plane can be straightened in $O(n^2)$ moves and time. We hope that our results shed some light on solving the more general conjectures, which we proposed, that a 3D lattice polygon embedded on any lattice polyhedron can always be convexified, and any 2D orthogonal tree can always be straightened.

1 Introduction

Graph reconfiguration problems have wide applications in contexts including robotics, molecular conformation, animation, rigidity and knot theory. The motivation for reconfiguration problems of lattice graphs arises in applications in molecular biology [10],[11] and robotics.

A graph is *simple* if non-adjacent edges do not intersect. A *unit tree* (resp. *polygon*) is a tree (resp. polygon) containing only straight edges of unit length. An *orthogonal tree* (resp. *polygon*) is a tree (resp. polygon) containing only edges parallel to coordinate-axes. A *lattice tree* (resp. *polygon*) is a simple tree (resp. polygon) containing only edges from a square or cubic lattice. A *lattice polyhedron* is a three-dimensional solid whose surface is the union of lattice faces from a cubic lattice. A *k-monotone tree* along the x-axis is a tree for which every intersection with a line perpendicular to the x-axis is a set of at most k disjoint segments. In particular, a 1-monotone tree is also called a *monotone caterpillar*. In the whole paper, we treat a point as a degenerate line segment. We say that two edges e and e' *cross* each other if one of the edges penetrates into the interior of the other. We consider the problem about the reconfiguration of a

X. Hu and J. Wang (Eds.): COCOON 2008, LNCS 5092, pp. 374–384, 2008.

simple chain, polygon, or tree through a series of continuous motions such that the lengths and shapes of all graph edges are preserved and no edge crossings are allowed. A tree can be *straightened* if all its edges can be aligned along a common straight line such that each edge points "away" from a designated leaf node. In particular, a chain can be straightened if it can be stretched out to lie on a straight line. A polygon can be *convexified* if it can be reconfigured to a convex polygon. We say a chain or tree (resp. polygon) is *locked* if it cannot be straightened (resp. convexified). We consider one move in the reconfiguration as a continuous monotonic change for the joint angle at some vertex.

In four dimensions or higher, a tree (resp. polygon) can always be straightened (resp. convexified) [5]. In two dimensions, a chain (resp. polygon) can always be straightened (resp. convexified) [7],[14],[4]. However, there are some trees in two dimensions that can lock [2],[6],[12]. In three dimensions, even a 5-chain can lock [3]. Alt et al. [1] showed that deciding the reconfigurability for trees in two dimensions and for chains in three dimensions is PSPACE-complete. Due to the complexity of the problems in two and three dimensions, some special classes of trees and polygons have been considered. Poon [12],[13] showed that a unit tree of diameter 4 in 2D, a square lattice tree in 2D or 3D, and a square lattice polygon in 2D can be always straightened or convexified.

Definitions. A *pseudo-lattice tree* (resp. *polygon*) is a tree (resp. polygon) that contains only lattice edges, some of which may possibly coincide with each other. Suppose P is a pseudo-lattice tree or polygon. The *core* of P, denoted by $K(P)$, is the union of all lattice edges coincident to some edges in P. A lattice vertex (resp. edge) in $K(P)$ is called a *core vertex (resp. edge)* of P. A *spring* in P is a maximal connected zig-zag path of edges in P coincident to a common lattice edge. A spring with only one edge is called a *singleton*. A *dangling spring* is a spring with both its end vertices incident to a common core vertex. A pseudo-lattice polygon is called *nearly folded* if its core is a lattice face.

Outline. The rest of this paper is organized as follows. In Sections 2, 3, and 4, we show that a lattice polygon embedded on an open lattice orthotube, a lattice Tower of Hanoi, a lattice Manhattan Tower, or an orthogonally-convex lattice polyhedron can always be convexified. Finally, in Section 5, we show that a 2-monotone orthogonal tree can always be straightened.

2   Open Lattice Orthotube

An *open lattice orthotube* is a lattice polyhedron made out of lattice cubes that are glued face-to-face such that its face-to-face contact graph is a path. In an open lattice orthotube, the two blocks whose degrees in the face-to-face contact graph are one are called the *end blocks* of the given orthotube. In this section, we'll show that a lattice polygon P embedded on an open lattice orthotube Q can always be convexified.

Our algorithm proceeds by folding up polygon P from the end blocks of the orthotube Q successively. We fold up the part of P lying on an end block to form springs on its adjacent block using a constant number of moves so that the resulting pseudo-lattice polygon still has a simple core. Details are omitted in this abstract. We repeat this step until the final pseudo-lattice polygon becomes nearly-folded, which can then be convexified straightforwardly.

Theorem 1. *A lattice polygon embedded on an open lattice orthotube can be convexified in $O(n)$ moves and time.*

3   Lattice Towers

Let Z_k be the plane $z = k$ for $k \geq 0$. A Manhattan Tower [9] $\mathcal{Q}$ is an orthogonal polyhedron such that

(i) $\mathcal{Q}$ lies in the halfspace $z \geq 0$ and its intersection with Z_0 is a simply connected orthogonal polygon;

(ii) For $j > k \geq 0$, $\mathcal{Q} \cap Z_j \subset \mathcal{Q} \cap Z_k$: the cross section at a higher level is nested in that at a lower level.

A *Tower of Hanoi* $\mathcal{Q}$ is a Manhattan Tower such that its intersection with Z_k for $k \geq 0$ is either empty or a simply connected orthogonal polygon.

3.1   Lattice Tower of Hanoi

Given a lattice polygon P embedded on a lattice Tower of Hanoi $\mathcal{Q}$, we seek an algorithm that convexifies P. Our convexifying algorithm contains two phases. The first phase proceeds by pressing vertically downwards level by level from the highest level to the second lowest level. This phase is called the *level-pressing phase*. At the end of the phase, we obtain a lattice polyhedron of height one $\mathcal{Q}'$, and a pseudo-lattice polygon P' embedded on it. We then proceed to the second phase to collapse the end blocks of $\mathcal{Q}'$ one by one until P' is nearly folded, which can then be convexified straightforwardly. This phase is called the *end-block collapsing phase*. Below, we will describe both phases in detail.

Level-Pressing Phase. The overall picture of this phase is to press level by level vertically downwards from the highest level to the second lowest level. Let us first consider the details for pressing the highest level L down to the second highest level L' under the condition that L' is not the lowest level. Notice that between L and L', there are vertical lattice polygon edges connecting them, which we call *legs*. And we also call the end vertex of the leg at L' the *foot* of the leg. To press level L to level L', we press the maximal polygon paths on level L one by one onto the level L'. Each maximal polygon path α on level L has two legs connecting to level L'. We collapse the two legs together at the same time by rotating both legs down to level L' pivoting at their feet, respectively. During the motion, we make use of the vertically-shifted copies of the original path α between level L and L'. All the path vertices of α are pushed to slide on the continuously-shifted copy of the original α from one end of α to the other end of

α. The pressing step stops when the shifted copy of α reaches level L'. Note that at this stage, the two legs lie on level L', and each path vertex of α has moved one unit distance along the shifted copy of α. See the progress of the pushing step in operations (a) and (b) in Figure 1. After pressing α one level down, there might be some dangling springs formed at both ends of the pressed path α. Such dangling springs have to be collapsed since they might not even lie on the surface of the given polyhedron. A dangling spring can be collapsed by rotating it towards its adjacent spring around the common vertex incident to its adjacent spring until it is combined with its adjacent spring. If the combined spring is again a dangling spring, then we need to continue the collapsing operation until there are no more dangling springs near the both ends of the path α . See the operation (c) in Figure 1. We observe that pressing a maximal path to one level lower makes at least two vertical legs get collapsed , and takes at most $O(m)$ moves and time, where m is the length of the path. Thus pressing a complete level takes at most $O(n)$ moves and time.

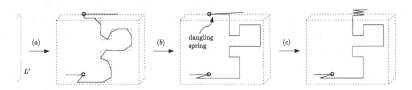

Fig. 1. $(a), (b)$ Pressing a maximal path from level L down to L'; (c) Collapsing dangling springs at the ends of the pressed path

We remark that after a complete level is pressed, the core on the current topmost level of the resulting pseudo-lattice polygon is not necessarily simple. This says that two maximal polygon paths may touch each other. However, as these paths do not cross each other, pressing such a maximal path is again not interfered by other parts of the polygon. As there can be at most $O(n)$ levels, the level-pressing phase takes $O(n^2)$ moves and time in total.

End-Block Collapsing Phase. After the level-pressing phase, we obtain a pseudo-lattice polygon P' embedded on a lattice polyhedron of height one Q'. At this stage, we proceed to the second phase of our algorithm by folding up P' from the end blocks of Q' successively. The end-block collapsing procedure is similar to what we did for orthotubes, but in a more complicated way. First we need to define some technical terms. A lattice face is called *external* if it is an xz- or yz-face and it lies on the surface of Q'; otherwise, it is called *internal* if it lies in the interior of Q'. We define an *x-block* (resp. *y-block*) as a block of width one along the y-axis (resp. x-axis) with two external yz-faces (resp. xz-faces) on its both ends. We call the span of an x-block (resp. y-block) along the x-axis (resp. y-axis) the *length* of the block. An *end block* of Q' is a block with three fully external xz- or yz-faces. In each collapsing step, our algorithm selects the shortest end block among all the end x- and y-blocks, and collapses it. The "shortest" ensures that either the non-fully external xz- or yz-face F of the end block is fully internal or F can be repaired to become internal. We then have the following lemma, whose proof is omitted.

Lemma 1. *Let P' be a non-nearly-folded pseudo-lattice polygon embedded on a lattice polyhedron of height one Q'. Then the shortest end block B of Q' can be collapsed in $O(n)$ moves and in $O(n + k)$ time, where k is the length of B. Moreover, the resulting pseudo-lattice polygon is either nearly folded or embedded on a lattice polyhedron of height one, and contains no dangling springs.*

In the light of Lemma 1, to fold up P', we have to collapse all lattice cells in Q', which takes time in the order of the size in Q'. Here we define the size of a polyhedron as the number of lattice cells it contains. If the size of Q' is large, this will make the algorithm inefficient. Hence we run a prepossessing step to reduce the size of Q' before we start to collapse the end blocks in Q'. First we select all the lattice cells from Q' containing at least one lattice edge from P'. We place these selected cells in S, initially empty. To include more cells to form a polyhedron, we scan the cells in S along the x-axis by sweeping a line ℓ parallel to the y-axis from left to right. Each time ℓ reaches a new lattice strip, we select all the lattice cells in Q' between the topmost and the bottommost cells in S intersecting ℓ, and include these selected cells in S. We sweep ℓ until all the cells in S are processed. Now, as P' is connected, it is not difficult to see that S is polyhedron and the size of S is $O(n^2)$, where n is the number of edges of P'. Now, we run the end-block collapsing phase on S, and the end-block collapsing phase takes $O(n^2)$ time in total due to Lemma 1. Hence we have the following theorem.

Theorem 2. *A lattice polygon embedded on a lattice Tower of Hanoi can be convexified in $O(n^2)$ moves and time.*

3.2   Lattice Manhattan Tower

Given a lattice polygon embedded on a lattice Manhattan Tower, we seek an algorithm that convexifies the lattice polygon. Our convexifying algorithm is the same as that for lattice Tower of Hanoi. The only difference is that when we press the highest level L to the second highest level L', we need to press several disconnected orthogonal polygonal regions on L instead of only one, as in the case of a lattice Tower of Hanoi. Thus we have the following theorem.

Theorem 3. *A lattice polygon embedded on a lattice Manhattan Tower can be convexified in $O(n^2)$ moves and time.*

4   Orthogonally-Convex Lattice Polyhedron

An *orthogonally-convex polygon* is an orthogonal polygon for which every intersection with a line perpendicular to a coordinate axis is either a segment or an empty set. An *orthogonally convex polyhedron* is an orthogonal polyhedron for which every intersection with a plane perpendicular to a coordinate axis is either an orthogonally-convex polygon or an empty set.

Consider a lattice polygon P embedded on an orthogonally-convex polyhedron Q. Our procedure to fold up P is again by pressing the levels down from top to bottom. Suppose we are pressing the highest level L down to the second highest level L' under the condition that L' is not the lowest level. In the level-pressing phase, we collapse the end y-blocks in level L from left and right to the middle. We fix a middle y-block Y in level L, where Y is connected to the level L'. We then collapse the end y-blocks towards Y. Details are omitted in this abstract. Finally, level L contains only the y-block Y, which can then be collapsed as an end block as the given polyhedron is orthogonally convex.

After the level-pressing phase, we obtain a lattice polyhedron of height one. We then apply the same end-block collapsing phase as in Section 3.1 to fold up the polygon. Hence, we obtain the following theorem.

Theorem 4. *A lattice polygon embedded on an orthogonally-convex lattice polyhedron can be convexified in $O(n^2)$ moves and time.*

5 Straightening 2-Monotone Orthogonal Trees

Given a 2-monotone orthogonal tree T along the x-axis in the plane, we seek an algorithm that straightens T. The skeleton of our algorithm runs by first dividing the whole tree into 2-monotone components and monotone caterpillars, and then straightening those components from right to left successively. We define a *maximal 2-monotone component U* of T to be a maximal subtree of T for which every intersection with a line perpendicular to the x-axis is a set of either two disjoint segments or an empty set except at the two end x-coordinates of U. We collect all the maximal 2-monotone components in a set $\mathcal{U}$, initially empty. After deleting these maximal 2-monotone components, the remaining parts of T form a set $\mathcal{K}$ of monotone caterpillars. Our algorithm runs iteratively to select the rightmost unstraightened component in $\mathcal{U}$ and $\mathcal{K}$ for straightening until the whole tree is straightened.

The rest of this section is organized as follows. In Subsection 5.1, we show that a monotone caterpillar can always be straightened. We present the algorithms to straighten a maximal 2-monotone component from its both ends, respectively, in Subsection 5.2. Finally, we present the whole algorithm to straighten T in Section 5.3.

5.1 Straightening a Monotone Orthogonal Caterpillar

Given a monotone orthogonal caterpillar K in the plane, we seek an algorithm that straightens K. We define the *backbone* of K to be a 1-monotone path from the vertex of K with the smallest x-coordinate to the vertex of K with the largest x-coordinate. The vertical segment attaching to a backbone vertex and connecting to a leaf vertex of K is called a *hair* of K. The straightening algorithm runs by successively straightening the rightmost unstraightened backbone edges of K together with the hairs attaching to them in the order from the right to the left so that finally all these edges point to the right horizontally. Thus we have the following lemma.

Lemma 2. *A monotone orthogonal caterpillar can be straightened in $O(n)$ moves and time, where n is the number of tree edges in the caterpillar.*

5.2   Straightening a Maximal 2-Monotone Component

We will describe in this subsection how we can straighten a single maximal 2-monotone component U in $\mathcal{U}$. First, we need the following property in Lemma 3 for a maximal 2-monotone component U, which says that U is basically an U-turn shape along the x-axis. Its proof is not difficult, and is omitted in this abstract. An *external segment* of T is a segment in the intersection of a vertical line with T such that it does not connect to T on both its left and right sides (see Figure 2). Otherwise, it is called *internal*.

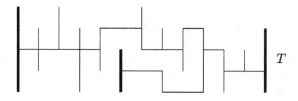

Fig. 2. The external segments of T are bold

Lemma 3. *Let U be a maximal 2-monotone component of 2-monotone orthogonal tree T in the plane; and let ℓ_l, ℓ_r be the vertical lines at the lowest and highest x-coordinates of U. Then one of $\ell_l \cap U$ and $\ell_r \cap U$ must contain exactly one segment, and the other must contain exactly two segments, at least one of which is an external segment of T.*

According to this property, we define some technical terms to name the different parts of U. We say that one of $\ell_l \cap U$ and $\ell_r \cap U$ containing exactly one segment is the *bridge* of U. The bridge divides U into two components, which we call *branches*. Observe that the two branches are monotone orthogonal caterpillars. Furthermore, the opening of U is defined as the vertical segment connecting the two segments of the non-bridge set of $\ell_l \cap U$ and $\ell_r \cap U$. We define the *backbone* of U to be the union of the bridge and the backbones of the two monotone caterpillar branches. Those vertical segments attaching to the backbone of U are called the *hairs* of U.

In this subsection, we will present the algorithms to straighten a maximal 2-monotone component from the opening and bridge ends, respectively. We need both of these two algorithms since the rightmost side of the unstraightened part of the tree may be the opening or the bridge sides of some maximal 2-monotone component. Now, suppose we are given a maximal 2-monotone component U of a 2-monotone orthogonal tree T.

Straightening from the opening end. Consider a maximal 2-monotone component U such that the opening is on its right end. Observe that both the upper and lower branches of U are monotone caterpillars. Thus our straightening algorithm follows the same strategy by straightening the rightmost unstraightened vertices of both caterpillars successively in parallel using the same routine in Section 5.1. Thus we have the following lemma.

Lemma 4. *A maximal 2-monotone component can be straightened in $O(n)$ moves and time from its opening end, where n is the number of tree edges in the component.*

Straightening from the bridge end. Consider a maximal 2-monotone component U in a 2-monotone orthogonal tree T such that the bridge is on its right end and the opening is on its left end. By Lemma 3, we know that the opening end of U on the left contains an external segment of T. We make use of that to pull the unstraightened parts of the two branches of U, which are close to the opening end, vertically apart while keeping the horizontal relative positions the same. And at the same time, we try to grow a convex balloon in U close to the bridge end on the right until finally the whole backbone of U becomes convex.

We now describe the intermediate balloon-growing step more precisely as follows. Suppose K_1, K_2 are the upper and lower branches of U. Let α be the convex path formed on the backbone of U at its bridge end in the unfolding process of our algorithm. Initially, α is set to be the bridge. At the beginning of each balloon-growing step, we assume $\alpha = q_0 q_1 \ldots q_m$ where the vertices are ordered in counter-clockwise order on the path. Let K_1' (resp. K_2') be the part of backbone of K_1 (resp. K_2) to the left of q_m (resp. q_0). Note that q_0 and q_m can have different x-coordinates. We also assume that all the vertical backbone edges of K_1' and K_2' lie to the left of the vertical supporting line of the leftmost vertex of q_0 and q_m; and q_m (resp. q_0) connects to a horizontal edge on K_1' (resp. K_2') on its left, say e_1 (resp. e_2). We will maintain this property in our balloon-growing process described below. Now, suppose e is the rightmost vertical backbone edge on K_1' and K_2'. Without loss of generality, we assume that e is on the upper branch K_1', and thus connects to e_1. A single balloon-growing step will try to include edges e and e_1 in the convex path α. We will describe how to do that as follows. Suppose that the downward extension of e intersects the horizontal edge of the backbone of K_2 at some point, say p. We consider two cases depending on the running direction of e: *Case 1:* e runs downward to get close to p—see the leftmost picture of Figure 3; *Case 2:* e runs upward to become farther from p—see the leftmost picture of Figure 4.

We first consider the unfolding procedure for *Case 1* as shown in Figure 3. During the unfolding motion, we keep the convex polygon $q_0 q_1 \ldots q_m$ rigid, and rotate it to the right pivoting at joint q_0. Moreover, the edge e_1 is always kept horizontal while the edge e is rotated to the right pivoting at u, and the part of K_1' to the left of u is moved directly upward. During the motion, we change the joint angles at u, v, q_0 and q_m. In the motion process, if edge $q_0 q_1$ becomes horizontal, then we will continue the motion by rotating the rigid polygon $q_1 q_2 \ldots q_m$ to the

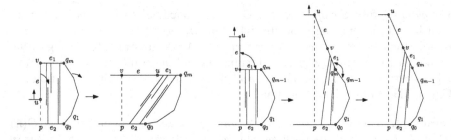

Fig. 3. The unfolding step for *Case 1*   **Fig. 4.** The unfolding step for *Case 2*

right pivoting at q_1, and other parts of the motion remain the same. We continue this operation until e becomes horizontal. In the above description, we did not give the details about the motion of the hairs, and will provide them now. At the initial configuration of U, for each downward hair h of K_1, suppose its downward extension intersects some backbone edge of K_2 at p_h. During the motion, the hair h will always point towards p_h. We do the symmetric thing for the upward hairs of K_2. Observe that during the motion, a new convex path $q_1 q_2 \ldots q_m v u$ is formed at the right end of U, and any pair of backbone edges of U are moving farther and farther away. Thus it is not difficult to see that these inner hairs will never interfere each other. For the outer hairs of K_1 and K_2, we always keep them vertical to their corresponding attaching backbone edge, and thus they will again not interfere with each other during the motion as α is convex.

Then we consider the unfolding procedure for *Case 2* as shown in Figure 4. During the first part of the unfolding motion, we keep the convex polygon $p q_0 q_1 \ldots q_m$ static, the whole part of K_1' to the left of u is moved directly upward, and the L-shaped path containing e and e_1 is stretched so that the angle between them increases until finally it becomes π and we stop. However, now the angle $\angle u q_m q_{m-1}$ may be concave. If this happens, then we continue to stretch the path $u q_m q_{m-1}$ by increasing the joint angle $\angle u q_m q_{m-1}$ at q_m until it becomes π. We continue this convexifying process until finally $q_0 q_1 \ldots q_m v u$ is convexified, which becomes the new current convex path in our algorithm. The motion of hairs is similar to what we have described for *Case 1*. This finishes the description of one single balloon-growing step.

By applying the balloon-growing steps recursively on the vertical edges of K_1 and K_2, we finally obtain convex backbone of U, say $q_0 q_1 \ldots q_m$ where the vertices are numbered in counter-clockwise order on the path. Due to our balloon-growing procedure as stated above, we know that both edges $q_0 q_1$ and $q_{m-1} q_m$ are horizontal. Now, we keep polygon $q_1 q_2 \ldots q_m$ rigid, and rotate it to the right pivoting at q_1 until $q_1 q_2$ becomes horizontal, then we proceed to rotate the rigid polygon $q_2 q_3 \ldots q_m$ to the right pivoting at q_2 until $q_2 q_3$ becomes horizontal. During these motions, the motion of hairs is similar to what we have described before. We continue this process until the whole backbone of U is straightened. At this stage, it is easy to see that all the hairs of U can be straightened by first positioning them vertically again and then applying Lemma 2. Hence, the whole component U is straightened. During each motion step, we either eliminate one concave corner on the backbone, or straighten some backbone vertex, and there can be $O(n)$ edges of U moved in such a step. As there are at most n concave corners and there are at most n unstraightened backbone vertices, at most $O(n)$ motion steps are needed. Thus U can be straightened in $O(n^2)$ moves and time.

Lemma 5. *A maximal 2-monotone component can be straightened in $O(n^2)$ moves and time from its bridge end, where n is the number of tree edges in the component.*

5.3   Straightening the Whole Tree

Our algorithm runs by successively straightening the rightmost unstraightened component in $\mathcal{U}$ and $\mathcal{K}$. A component is straightened from right to left by making use of Lemma 2, 4, or 5 Hence, we obtain the following theorem.

Theorem 5. *A 2-monotone orthogonal tree can be straightened in $O(n^2)$ moves and time, where n is the number of edges in the given tree.*

References

1. Alt, H., Knauer, C., Rote, G., Whitesides, S.: The Complexity of (Un)folding. In: Proc. of 19th ACM Symp. on Comput. Geom., pp. 164–170 (2003)
2. Biedl, T., Demaine, E., Demaine, M., Lazard, S., Lubiw, A., O'Rourke, J., Robbins, S., Streinu, I., Toussaint, G., Whitesides, S.: A Note on Reconfiguring Tree Linkages: Trees Can Lock. Disc. Appl. Math. 117(1-3), 293–297 (2002)
3. Biedl, T., Demaine, E., Demaine, M., Lazard, S., Lubiw, A., O'Rourke, J., Overmars, M., Robbins, S., Streinu, I., Toussaint, G., Whitesides, S.: Locked and Unlocked Polygonal Chains in Three Dimensions. Disc. & Comput. Geom. 26(3), 269–281 (2001)
4. Cantarella, J., Demaine, E.D., Iben, H., O'Brien, J.: An Energy-Driven Approach to Linkage Unfolding. In: Proc. of 20th Ann. ACM Symp. on Comput. Geom., pp. 134–143 (2004)
5. Cocan, R., O'Rourke, J.: Polygonal Chains Cannot Lock in 4D. Comput. Geom.: Theory & Appl. 20, 105–129 (2001)
6. Connelly, R., Demaine, E., Rote, G.: Infinitesimally Locked Self-Touching Linkages with Applications to Locked Trees. In: Physical Knots: Knotting, Linking, and Folding of Geometric Objects in R^3, American Mathematical Society, pp. 287–311 (2002)
7. Connelly, R., Demaine, E.D., Rote, G.: Straightening Polygonal Arcs and Convexifying Polygonal Cycles. Disc. & Comput. Geom. 30(2), 205–239 (2003)
8. Damian, M., Flatland, R., Meijer, H., O'Rourke, J.: Unfolding Well-Separated Orthotrees. In: Proc. of 15th Ann. Fall Workshop on Comput. Geom., pp. 23–25 (2005)
9. Damian, M., Flatland, R., O'Rourke, J.: Unfolding Manhattan Towers. In: Proc. 17th Canadian Conference on Computational Geometry, pp. 211–214 (2005)
10. Dill, K.A.: Dominant Forces in Protein Folding. Biochemistry 29(31), 7133–7155 (1990)
11. Hayes, B.: Prototeins. American Scientist 86, 216–221 (1998)

12. Poon, S.-H.: On Straightening Low-Diameter Unit Trees. In: Proc. of 13th Int. Symp. on Graph Drawing, pp. 519–521 (2005)
13. Poon, S.-H.: On Unfolding Lattice Polygons/Trees and Diameter-4 Trees. In: Chen, D.Z., Lee, D.T. (eds.) COCOON 2006. LNCS, vol. 4112, pp. 186–195. Springer, Heidelberg (2006)
14. Streinu, I.: A Combinatorial Approach for Planar Non-Colliding Robot Arm Motion Planning. In: Proc. of FOCS 2000, pp. 443–453 (2000)

New Algorithms for Online Rectangle Filling with k-Lookahead[*]

Haitao Wang, Amitabh Chaudhary, and Danny Z. Chen

Department of Computer Science and Engineering
University of Notre Dame, Notre Dame, IN 46556, USA
{hwang6,achaudha,dchen}@nd.edu

Abstract. We study the online rectangle filling problem which arises in channel aware scheduling of wireless networks, and present improved deterministic and first known randomized results for algorithms that are allowed a k-lookahead for $k \geq 2$. Our main result is a deterministic $\min\{1.848, 1 + 2/(k - 1)\}$-competitive online algorithm. The previous best-known solution for this problem has a competitive ratio of 2 for any $k \geq 2$. We also present a randomized online algorithm with a competitive ratio of $1 + 1/(k + 1)$. Our final result is a closely matching lower bound of $1 + 1/(\sqrt{k + 2} + \sqrt{k + 1})^2 > 1 + 1/(4(k + 2))$ on the competitive ratio of any randomized online algorithm against an oblivious adversary.

1 Introduction

We consider an online problem that arises in channel-aware scheduling in wireless networks: Given an online sequence of nonnegative real numbers $h(1), h(2), \ldots$, representing the maximum transmission capacities of a wireless channel at each time step, compute a sequence of transmission rates $u(1), u(2), \ldots$, for the wireless channel that satisfies two constraints: (1) at each time step t, $u(t) \leq h(t)$, and (2) changing the transmission rate will incur a penalty of one time step (this is required for the transmitter and receiver to coordinate and reset a new transmission rate); in other words, for any time step t, either $u(t) = u(t + 1)$ or at least one of $u(t)$ and $u(t + 1)$ is zero. The objective is to maximize the throughput, i.e., $\sum_{t=1,2,\ldots} u(t)$. The decision is made online with a lookahead of k time steps, i.e., at each time step t, we decide on $u(t)$ having seen the capacities $h(1), \ldots, h(t - 1)$ as well as the capacities $h(t), \ldots, h(t + k)$, but before we see $h(t + k + 1)$ and beyond.

In wireless networks [9], the channel conditions can change frequently, affecting the bit error rate and consequently the channel transmission capacity. The transmitter and receiver can monitor the channel capacity and change the transmission rate accordingly. To coordinate the changes in transmission rate, a change-over protocol is used, resulting in a temporary loss in the transmission of data. This is modeled by setting the transmission rate at one time step to zero. In addition, predicting the future channel conditions is expensive in terms

---

[*] This research was supported in part by NSF under Grant CCF-0515203.

X. Hu and J. Wang (Eds.): COCOON 2008, LNCS 5092, pp. 385–394, 2008.

of system resources such as time, bandwidth, and power [7]. This is modeled by regarding the channel capacities as an online sequence with a limited lookahead of k channel capacities. (See [1],[2],[4],[6],[7],[8],[10] for further details.)

The problem also has a clear geometric interpretation. The sequence of channel capacities corresponds to a sequence of "columns" of unit-width each such that the bases of all columns are on the x-axis and the height of column t is $h(t)$ (see Fig. 1). The transmission rate corresponds to a "used-height" in each column, from the base to a height $u(t) \leq h(t)$. The penalty for changing the transmission rate implies that in any solution, consecutive columns with nonzero used-height values have the same used-height and thus form a rectangle. The objective is, given an online sequence of columns with k-lookahead, to fill the region between the column "skyline" and the x-axis with rectangles in a manner that maximizes the total area covered by the rectangles, subject to the constraints that each rectangle lies on the x-axis and respects the skyline, i.e., the height of a rectangle is at most the height of the lowest column intersecting it, and that any two distinct rectangles are separated by at least one column with zero used-height. This is the *online rectangle filling problem with k-lookahead.*

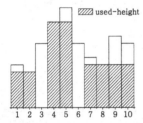
Fig. 1. A feasible solution

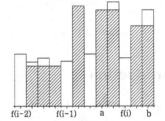
Fig. 2. Part of an SecondHalf solution

Previous Work. The online rectangle filling problem has been previously studied both for the specific 1-lookahead case and the general k-lookahead case. The first known algorithm with 1-lookahead was 4-competitive [2],[4]. (See [5] for the formal definitions of online algorithms, offline optimal algorithms, and competitive analysis.) The authors also showed that for any online algorithm with a finite lookahead, the competitive ratio is strictly larger than one. (It is easy to see that with no lookahead, the competitive ratio is unbounded.) Besides, the authors gave an offline optimal algorithm for an n-length sequence that takes $O(n^3)$ time. In a subsequent paper [3], the original 1-lookahead algorithm in [2],[4] was shown to be $(8/3 \approx 2.667)$-competitive and a lower bound of $8/5 = 1.6$ on the competitive ratio of any online algorithm with 1-lookahead was given. In the same paper [3], for a more general problem with a lookahead of k for any $k \geq 2$, an algorithm achieving a competitive ratio of 2 and a lower bound of $1 + 1/(k+1)$ were shown. The best known algorithm for the 1-lookahead case is 1.848-competitive [11]. The authors also proved a lower bound of 1.636 on the competitive ratio of any online algorithm with 1-lookahead, as well as giving an offline optimal algorithm with $O(n^2)$ time [11].

Our Contributions. In this paper, we study the online rectangle filling problem with k-lookahead for any $k \geq 2$. We first present a deterministic online algorithm with k-lookahead which attains a competitive ratio of $1 + 2/(k - 1)$ for $k \geq 4$ and 1.848 for $2 \leq k \leq 3$ (Section 2). This is the first algorithm for the problem with a competitive ratio that approaches 1 as k approaches $+\infty$, which significantly improves the 2-competitive algorithm in [3]. We also give the first randomized online algorithm for this problem with k-lookahead, attaining a competitive ratio of $1 + 1/(k + 1)$ against an oblivious adversary (Section 3). Further, we prove a lower bound of $1 + 1/(\sqrt{k + 2} + \sqrt{k + 1})^2$ on the competitive ratio against an oblivious adversary for any randomized online algorithm with k-lookahead (Section 4). Since it is larger than $1 + 1/(4(k + 2))$ and asymptotically equal to $1 + 1/(4(k + 1))$ as k is getting larger, the gap between the upper and lower randomized bounds is very small.

2   Online Rectangle Filling with k-Lookahead

Our online algorithm focuses on the second half of the lookahead window at every time step to make its online decisions. Let the online input sequence be $H = h(1), \ldots, h(t), \ldots$. At any time t, we need to choose $u(t)$ based on the knowledge of the past, i.e. $h(i)$ and $u(i)$ for $1 \leq i \leq t-1$, as well as the knowledge in the *lookahead window*, i.e. $h(i)$ for $t \leq i \leq t + k$. Denote the sequence of columns in the lookahead window by C. Further, let $\alpha = \lceil \frac{k+1}{2} \rceil$ and denote the set of the last α columns in C by $\overline{C}$. Given the entire input H, let OPT denote the offline optimal solution on H. For any consecutive subsequence X in H, let $OPT(X)$ denote the part of OPT that pertains to X. In contrast, let $OPT_r(X)$ denote the offline optimal solution if the input were *restricted* only to X, and there were no other columns before or after X. Clearly $OPT_r(X) \geq OPT(X)$. Assume $h(0)$ and $u(0)$ are both defined to be 0. The algorithm SecondHalf is described below and is initially invoked with $t = 1$.

Algorithm SecondHalf

1. Assume that the $u(i)$ values for all $i < t$ have been decided, and in particular $u(t - 1) = 0$. Compute the optimal solution $OPT_r(C)$ with the input restricted to C. Depending on whether there is a zero used-height column in this solution among the columns in $\overline{C}$, we have the following two strategies.
2. If all the α columns in $\overline{C}$ have nonzero used-heights (the used-heights are then all the same), then (i) set $u(t+k)$ to 0, (ii) compute the offline optimal solution for the sequence restricted to columns from t to $t + k - 1$, and set their used-heights accordingly, and (iii) invoke Step 1 with t set to $(t+k+1)$.
3. Else do: (i) find a column $t+i$ with the minimum height $h(t+i)$ among those in $\overline{C}$, (ii) set $u(t + i)$ to 0, (iii) compute the offline optimal solution for the sequence restricted to columns from t to $t + i - 1$ and set their used-heights accordingly, and (iv) invoke Step 1 with t set to $(t + i + 1)$.

It is clear that SecondHalf produces a feasible solution since whenever Step 1 is invoked with t, $u(t - 1)$ has been set to 0, so we are free to choose $u(t)$.

Theorem 1. *Algorithm* SecondHalf *attains a competitive ratio of at most* $1 + \frac{2}{k-1}$ *for any* $k \geq 2$.

2.1 Competitive Analysis of SecondHalf

We call the columns with used-height values set to 0 by SecondHalf as *zero columns*. Further, we call the zero columns created in Step 2 at $t + k$ and those created in Step 3 at $t + i$ as *partition columns*, and let the set of all partition columns be P. In particular, denote the set of all partition columns created in Step 2 (resp. Step 3) by P_I (resp. P_{II}). Note that every partition column is created after Step 1 of SecondHalf, and every time when Step 1 is executed, there are always a lookahead window C and its corresponding $\overline{C}$. Index all the partition columns in P with numbers from 1 to m, with $m = |P|$. Then the ith partition column is created right after the ith round execution of Step 1. We denote the lookahead window C and the corresponding $\overline{C}$ in the ith round execution of Step 1 by C_i and $\overline{C_i}$, respectively. Further, let $f(i)$ denote the time step of partition column i in the online input sequence H. Figure 2 shows a part of the online solution of SecondHalf including three partition columns $i-2$, $i-1$, and i, with a lookahead $k = 6$. In the figure, C_i and $\overline{C_i}$ consist of columns from $f(i-1) + 1$ to b and columns from $f(i) - 1$ to b, respectively. Both partition columns $i-1$ and i are in P_{II}. Note that for any i, C_i and C_{i+1} may overlap, but $\overline{C_i}$ and $\overline{C_{i+1}}$ cannot, as shown below.

Observation 1. *For every* i, $\overline{C_i}$ *has no column in common with* $\overline{C_{i+1}}$.

Since the partition column i is chosen from $\overline{C_i}$, we denote the columns in $\overline{C_i}$ to the left and right of the partition column i by L_i and R_i, respectively. In Fig. 2, L_i consists of column $f(i) - 1$ and R_i contains two columns $b - 1$ and b. It is clear that for every $i \in P_I$, $R_i = \emptyset$. For ease of presentation, define $R_0 = \emptyset$. Further, given a solution sequence S produced by SecondHalf, we say that the partition columns in P partition S into $m + 1$ *blocks*. Each block, except possibly the first one and the last one, contains the columns in S between two successive partition columns of P, and the partition columns do not belong to any block. Denote the block between partition columns $i-1$ and i by S_i, and the last block by S_{m+1}. The observation below follows directly from Observation 1.

Observation 2. *For every* $i \in P$, $R_{i-1} \cap L_i = \emptyset$ *and* $R_{i-1} \cup L_i \subseteq S_i$.

Given the online input sequence H, let ALG and OPT denote the solutions obtained by SecondHalf and the offline optimal algorithm, respectively. Of these solutions, let $ALG(i)$ and $OPT(i)$ denote the parts of ALG and OPT that pertain only to the columns in block S_i, respectively. Let $u'(t)$ denote the used-height of the column at time t in OPT and $\hat{u}(f(i))$ be the used-height of the partition column i in the solution $OPT_r(C_i)$, which is computed in the ith round execution of Step 1 of SecondHalf. Then we have: $ALG = \sum_{i=1}^{m+1} ALG(i)$ and $OPT = \sum_{i=1}^{m+1} OPT(i) + \sum_{i=1}^{m} u'(f(i))$.

Define the function $w(i)$ for every partition column $i \in P$: $w(i) = \hat{u}(f(i))$ if $i \in P_I$, and $w(i) = h(f(i))$ if $i \in P_{II}$. Then we have the following lemma.

Lemma 1. $ALG + \sum_{i \in P} w(i) \geq OPT$.

Proof. For each block S_i, consider the following three possible cases:

1. If $i \in P_I$, then $C_i = S_i \cup \{f(i)\}$. Since $OPT_r(C_i)$ is the optimal solution for C_i, $OPT_r(C_i) \geq OPT(i) + u'(f(i))$. Further, since SecondHalf ensures that $ALG(i) = OPT_r(S_i)$ is the optimal solution for the columns in S_i, and since the used-height of the $f(i)$th column in $OPT_r(C_i)$ is $\hat{u}(f(i))$, $ALG(i) + \hat{u}(f(i)) \geq OPT_r(C_i)$ holds. Thus, $ALG(i) + \hat{u}(f(i)) \geq OPT(i) + u'(f(i))$.
2. If $i \in P_{II}$, then since $ALG(i) = OPT_r(S_i)$ is the optimal solution for S_i and $h(f(i)) \geq u'(f(i))$, we have $ALG(i) + h(f(i)) \geq OPT(i) + u'(f(i))$.
3. If S_i is the last block ($i = m+1$), then similarly $ALG(m+1) \geq OPT(m+1)$.

Combining the three inequalities above, we have

$$\sum_{i=1}^{m+1} ALG(i) + \sum_{i \in P_I} \hat{u}(f(i)) + \sum_{i \in P_{II}} h(f(i)) \geq \sum_{i=1}^{m+1} OPT(i) + \sum_{i=1}^{m} u'(f(i))$$

which along with the definition of $w(i)$ leads to the lemma. $\square$

The next lemma identifies an important condition for Theorem 1.

Lemma 2. *If for every $i \in P$, $OPT_r(R_{i-1}) + OPT_r(L_i) \leq ALG(i)$ holds true, then* SecondHalf *attains a competitive ratio of at most $1 + \frac{1}{\alpha - 1}$.*

Proof. For each $i \in P_I$, since in $OPT_r(C_i)$ all columns in $\overline{C_i}$ have the same nonzero used-height $\hat{u}(f(i))$, a feasible solution for L_i has the same used-height $\hat{u}(f(i))$ for all columns. Further, since $i \in P_I$, $R_i = \emptyset$ and $|L_i| = \alpha - 1$. This implies $(\alpha - 1) \cdot \hat{u}(f(i)) \leq OPT_r(L_i) + OPT_r(R_i)$.

Similarly, for each $i \in P_{II}$, since $h(f(i))$ is the smallest among all columns in $\overline{C_i}$, and recalling that $|L_i| + |R_i| = \alpha - 1$, we have $(\alpha - 1) \cdot h(f(i)) \leq OPT_r(L_i) + OPT_r(R_i)$. By combining the above two inequalities, we get, for each $i \in P$,

$$(\alpha - 1) \cdot w(i) \leq OPT_r(L_i) + OPT_r(R_i) \tag{1}$$

Since $OPT_r(R_{i-1}) + OPT_r(L_i) \leq ALG(i)$ for any $i \in P$, we can obtain $(\alpha - 1) \cdot \sum_{i=1}^{m} w(i) \leq \sum_{i=1}^{m} (OPT_r(L_i) + OPT_r(R_i)) = \sum_{i=1}^{m} (OPT_r(R_{i-1}) + OPT_r(L_i)) + OPT_r(R_m) - OPT_r(R_0) \leq \sum_{i=1}^{m} ALG(i) + ALG(m+1) = ALG$.
Combining the above with Lemma 1 completes the proof. $\square$

Unfortunately, the condition in the above lemma is not always true. The following lemma specifies the situations in which the inequality can be satisfied. Refer to the full version of this paper [12] for the detailed proof.

Lemma 3. *For any $i \in P$, if $L_i = \emptyset$ or $|R_{i-1}| < k + 1 - \alpha$, then $OPT_r(R_{i-1}) + OPT_r(L_i) \leq ALG(i)$.*

Since $|R_{i-1}| \leq \alpha - 1$ for every $i \in P$, when k is odd we have $\alpha - 1 < k + 1 - \alpha$. Thus based on Lemmas 2 and 3, the competitive ratio is at most $1 + 1/(\alpha - 1)$. When k is even, however, $\alpha - 1 = k + 1 - \alpha$. It is not difficult to see that the only case where we cannot apply Lemma 3 is when $L_i \neq \emptyset$ and $|R_{i-1}| = \alpha - 1$. We refer to this situation as the *joined case* because there is no column between R_{i-1} and L_i (i.e. $S_i = R_{i-1} \cup L_i$). We also call the involved block S_i the *joined block*. In Fig. 2, S_i is a joined block since $k = 6$, $|R_{i-1}| = 3 = \alpha - 1 = k + 1 - \alpha$, and $L_i \neq \emptyset$. The joined case is handled by the next lemma.

Lemma 4. *When k is even, for any $i \in P$, if $|R_{i-1}| = \alpha - 1$ and $L_i \neq \emptyset$, then $w(i - 1) + w(i) \leq \frac{1}{\beta} \cdot (OPT_r(L_{i-1}) + ALG(i) + OPT_r(R_i))$, where $\beta = \frac{2(\alpha-1)^2}{2\alpha-1}$.*

Proof. Since $|R_{i-1}| = \alpha - 1$, $i - 1$ must be in P_{II} and $L_{i-1} = \emptyset$. Hence, $w(i-1) = h(f(i-1))$. Depending on $i \in P_{II}$ or $i \in P_I$, we analyze two possible cases.

Case 1: $i \in P_{II}$, which implies $w(i) = h(f(i))$. Since $i - 1 \in P_{II}$, $h(f(i-1))$ is no bigger than the minimum height of the columns in R_{i-1}. Similarly, since $i \in P_{II}$, $h(f(i))$ is no bigger than the minimum height of the columns in L_i and R_i. Let $h_1 = h(f(i-1))$ and $h_2 = h(f(i))$. We first assume $h_1 \geq h_2$ (the case with $h_1 < h_2$ will be discussed later). In the following, we consider two feasible solutions for the columns in S_i and the columns in R_i.

Feasible Solution SOL$_1$: Set the used-heights of all columns in R_{i-1} to h_1, of the column immediately after R_{i-1} to 0, of the remaining columns in L_i to h_2, and of the columns in R_i to h_2. This is a feasible solution as the height of any column in R_{i-1} is at least h_1 and the height of any column in L_i and R_i is at least h_2, and since there is a zero column between the columns with used-height h_1 and the columns with used-height h_2. Now, since $|R_{i-1}| = \alpha - 1$ and $|L_i| + |R_i| = \alpha - 1$, we get $SOL_1 = (\alpha - 1) \cdot h_1 + (\alpha - 2) \cdot h_2$.

Feasible Solution SOL$_2$: We set the used-heights of all columns in S_i and R_i to h_2. This leads to $SOL_2 = 2(\alpha - 1) \cdot h_2$.

Depending on the values of **SOL$_1$** and **SOL$_2$**, we have

1. If $SOL_1 \leq SOL_2$, then we have $(\alpha - 1) \cdot h_1 + (\alpha - 2) \cdot h_2 \leq 2(\alpha - 1) \cdot h_2$. This implies $\frac{h_1}{h_2} \leq \frac{\alpha}{\alpha-1}$ (we assume $h_2 > 0$ since otherwise we can always let h_2 be a relatively small value in the analysis). Furthermore, we get

$$\frac{SOL_2}{h_1 + h_2} = \frac{2(\alpha - 1)}{\frac{h_1}{h_2} + 1} \geq \frac{2(\alpha - 1)}{\frac{\alpha}{\alpha-1} + 1} = \frac{2(\alpha - 1)^2}{2\alpha - 1} = \beta . \qquad (2)$$

2. If $SOL_1 > SOL_2$, then we have $\frac{h_1}{h_2} > \frac{\alpha}{\alpha-1}$. This leads to

$$\frac{SOL_1}{h_1 + h_2} = \frac{(\alpha - 1) \cdot \frac{h_1}{h_2} + \alpha - 2}{\frac{h_1}{h_2} + 1} = \alpha - 1 - \frac{1}{\frac{h_1}{h_2} + 1} > \alpha - 1 - \frac{1}{\frac{\alpha}{\alpha-1} + 1} = \beta . \qquad (3)$$

Since $ALG(i) + OPT_r(R_i) \geq \max\{SOL_1, SOL_2\}$, by (2) and (3), we have $ALG(i) + OPT_r(R_i) \geq \beta \cdot (h_1 + h_2)$. A nearly identical argument leads to the same result when $h_1 < h_2$ (in SOL_1, set the used-height of all but the last column in R_{i-1} to h_1, of the last column to 0, and of the remaining columns to h_2; in SOL_2, set the used-height of all columns to h_1). Consequently,

$$h(f(i-1)) + h(f(i)) \leq \frac{1}{\beta} \cdot (ALG(i) + OPT_r(R_i)) \ . \qquad (4)$$

Case 2: $i \in P_I$, which implies $w(i) = \hat{u}(f(i))$. Here, again, we let $h_1 = h(f(i-1))$ and $h_2 = \hat{u}(f(i))$. Then by using reasoning very similar to the previous case (i.e., choosing the same SOL_1 and SOL_2 as in Case 1 and applying the same analysis), we have $h(f(i-1)) + \hat{u}(f(i)) \leq \frac{1}{\beta} \cdot ALG(i)$. Here, recall that $R_i = \emptyset$ and thus $OPT_r(R_i) = 0$. As a result, we get

$$h(f(i-1)) + \hat{u}(f(i)) \leq \frac{1}{\beta} \cdot (ALG(i) + OPT_r(R_i)) \ . \qquad (5)$$

The lemma follows from $L_{i-1} = \emptyset$ and the two inequalities in (4) and (5).   □

This gives the result for the joined case. To complete the proof for Theorem 1, we need to combine the joined case with the other general case (the case to which we can apply Lemma 3). The combination is done in the lemma below.

Lemma 5. *If k is even, then $\sum_{i \in P} w(i) \leq \frac{1}{\beta} \cdot ALG$.*

Proof. For every joined block S_i, call the partition column $i-1$ the *front* column and the partition column i the *rear* column, respectively. We claim that if S_i is joined, then neither S_{i-1} nor S_{i+1} can be joined. Since $L_i \neq \emptyset$, we know that $R_i < \alpha - 1$, and thus S_{i+1} is not joined. Similarly, since $|R_{i-1}| = \alpha - 1$, we have $L_{i-1} = \emptyset$, and thus S_{i-1} is not joined. This implies that any partition column can be involved in at most one joined block.

Let B denote the set of all partition columns in the joined cases and A denote the set of rear partition columns in the joined cases, i.e., $B = \{i \mid i \in A \text{ or } i+1 \in A\}$. Since $\frac{1}{\alpha-1} < \frac{1}{\beta}$, by Lemma 4 and (1),

$$\sum_{i \in P} w(i) = \sum_{i \in A} (w(i-1) + w(i)) + \sum_{i \in P \setminus B} w(i) \leq \frac{1}{\beta} \cdot \Big[\sum_{i \in A} ALG(i) +$$
$$\sum_{i \in A} (OPT_r(L_{i-1}) + OPT_r(R_i)) + \sum_{i \in P \setminus B} (OPT_r(L_i) + OPT_r(R_i)) \Big] \ ,$$

where $\sum_{i \in A} (OPT_r(L_{i-1}) + OPT_r(R_i)) + \sum_{i \in P \setminus B} (OPT_r(L_i) + OPT_r(R_i))$ can be equally transformed to $\sum_{i \in P \setminus A} (OPT_r(R_{i-1}) + OPT_r(L_i)) + OPT_r(R_m) - OPT_r(R_0)$ (refer to [12] for the detail of the transformation).

For any $i \in P \setminus A$, by Lemma 3, $OPT_r(R_{i-1}) + OPT_r(L_i) \leq ALG(i)$. Thus,

$$\sum_{i \in P} w(i) \leq \frac{1}{\beta} \cdot \Big[\sum_{i \in A} ALG(i) + \sum_{i \in P \setminus A} ALG(i) + ALG(m+1) \Big] = \frac{1}{\beta} \cdot ALG \ .$$

(Recall $OPT_r(R_m) \leq ALG(m+1)$ and $R_0 = \emptyset$.)   □

Based on Lemmas 1 and 5, when k is even, the competitive ratio of SecondHalf is at most $1 + 1/\beta$, which is less than $1 + 2/(k-1)$. Since when k is odd the competitive ratio is at most $1 + 1/(\alpha - 1) = 1 + 2/(k-1)$, Theorem 1 follows.

To complete the analysis for SecondHalf, we should discuss a little on our choice of the value for α. We find that if α is less than $\lceil \frac{k+1}{2} \rceil$, the whole analysis is still correct although there will be no joined case because $|R_{i-1}|$ is always less than $k + 1 - \alpha$ for any $k \geq 2$, but then the competitive ratio becomes larger. If α is larger than $\lceil \frac{k+1}{2} \rceil$, on the other hand, then the analysis will not be directly applicable since in that case, $R_{i-1} \cap L_i$ is not always $\emptyset$ and as a consequence, some properties we need are no longer true.

For $2 \leq k \leq 3$, the algorithm with 1-lookahead in [11] has a smaller competitive ratio of 1.848; but for $k \geq 4$, SecondHalf is better with a ratio of at most $1 + \frac{2}{k-1}$. Thus, the result below follows.

Theorem 2. *Our algorithm attains a competitive ratio of $1 + \frac{2}{k-1}$ for $k \geq 4$ and 1.848 for $2 \leq k \leq 3$.*

3   A Randomized Online Algorithm with k-Lookahead

In Algorithm SecondHalf, we choose zero columns by expecting the worst-case adversary input later on. In this section, we show that we can make such decisions randomly, by simply choosing a random initial offset, which leads to a highly competitive algorithm on average. Our algorithm begins at the first time step, with knowledge of heights $h(1), \ldots, h(k+1)$.

Algorithm RandomOffSet

1. Choose an integer l from $[1, 2, \ldots, k+2]$ uniformly at random. Set $u(l) = 0$. If $l > 1$, compute the optimal offline solution restricted to the columns 1 to $l - 1$ and set their used-heights accordingly. Move to the $(l + 1)$th column.
2. Suppose the current column is t. The used-heights of the columns 1 to $(t-1)$ have already been set; in particular, $u(t - 1)$ is 0. We know the heights of the columns t to $t + k$. Compute the offline optimal solution restricted to the columns t to $t + k$ and set their used-heights accordingly. Set $u(t+k+1) = 0$. Move to the $(t + k + 2)$th column and repeat this step.

It is clear that the above solution is feasible. The analysis of RandomOffSet is in our full paper [12], here we only give the theorem below.

Theorem 3. *For any $k > 0$, algorithm RandomOffSet attains a competitive ratio of $1 + 1/(k+1)$ against an oblivious adversary.*

4   A Lower Bound Against an Oblivious Adversary

Theorem 4. *For any randomized online algorithm for the rectangle filling problem with k-lookahead, the competitive ratio against an oblivious adversary is at least $1 + 1/(\sqrt{k+1} + \sqrt{k+2})^2$.*

We use Yao's principle [13] to obtain the above result. Denote the lower bound on the competitive ratio of any randomized algorithm against an oblivious adversary by $\overline{R}_{OBL}$. To apply Yao's principle, we need to specify a probability distribution $y(j)$ over input sequences σ_j and then use the relationship $\overline{R}_{OBL} \geq \min_i \frac{E_{y(j)}[OPT(\sigma_j)]}{E_{y(j)}[ALG_i(\sigma_j)]}$, in which OPT is the offline optimal algorithm, the ALG_is are all possible deterministic online algorithms, $ALG_i(\sigma_j)$ is the solution achieved by ALG_i on input σ_j, and the expectations are over the input distribution $y(j)$.

We use the following distribution on the input sequences: Let h_0 be such that $\frac{k+1}{k+2} < h_0 \leq 1$. (We will specify the exact value of h_0 later.) Consider the following two sequences σ_1 and σ_2, with $k+2$ columns each:

1. In σ_1, $h(t) = 1$ for $1 \leq t \leq k+1$, and $h(k+2) = h_0$.
2. In σ_2, $h(t) = 1$ for $1 \leq t \leq k+2$.

Let the probability of σ_1 be p and that of σ_2 be $1 - p$. (We will specify p later.)

Since $(k+1) < h_0(k+2)$, the offline optimal algorithm on σ_1 sets $u(t) = h_0$ for all t; thus $OPT(\sigma_1) = (k+2) \cdot h_0$. Similarly, $OPT(\sigma_2) = k+2$. Hence, the expected solution of the offline optimal algorithm is $E[OPT] = p \cdot OPT(\sigma_1) + (1 - p) \cdot OPT(\sigma_2) = p \cdot (k+2) \cdot h_0 + (1 - p) \cdot (k+2)$.

Partition all possible deterministic online algorithms into two groups: Those in the first group represented by ALG_1 set $u(1) \leq h_0$, and those in the second group represented by ALG_2 have $u(1) > h_0$. We now develop bounds on the solutions achieved by these two groups of algorithms. Due to the space limitation, we only list the results here: $ALG_1(\sigma_1) \leq (k+2) \cdot h_0$, $ALG_1(\sigma_2) \leq (k+2) \cdot h_0$, $ALG_2(\sigma_1) \leq u(1) \cdot (k+1)$, and $ALG_2(\sigma_2) \leq u(1) \cdot (k+2)$. Refer to [12] for the full explanation of the above bounds. Then we have $E[ALG_1] = p \cdot ALG_1(\sigma_1) + (1 - p) \cdot ALG_1(\sigma_2) \leq (k+2) \cdot h_0$ and similarly $E[ALG_2] \leq u(1) \cdot (k+2-p)$.

Let $\overline{R}_1$ (resp. $\overline{R}_2$) denote the ratio of the expected solutions of the offline optimal and of the algorithms represented by ALG_1 (resp. ALG_2). Thus

$$\overline{R}_1 = \frac{E[OPT]}{E[ALG_1]} \geq \frac{p \cdot (k+2) \cdot h_0 + (1 - p) \cdot (k+2)}{(k+2) \cdot h_0} = p + (1 - p) \cdot \frac{1}{h_0} ,$$

$$\overline{R}_2 = \frac{E[OPT]}{E[ALG_2]} \geq \frac{p \cdot (k+2) \cdot h_0 + (1 - p) \cdot (k+2)}{k+2-p} .$$

The inequality for $\overline{R}_2$ holds as $u(1) \leq 1$.

Based on Yao's principle, the lower bound $\overline{R}_{OBL} \geq \max_{p,h_0} \min\{\overline{R}_1, \overline{R}_2\}$, i.e.,

$$\overline{R}_{OBL} \geq \max_{0 \leq p \leq 1, \frac{k+1}{k+2} < h_0 \leq 1} \min\{p + (1 - p)\frac{1}{h_0}, \frac{p(k+2)h_0 + (1-p)(k+2)}{k+2-p}\} .$$

The specific values for h_0 and p that lead to Theorem 4 are obtained in the following lemma (the proof is in the full version of this paper [12]).

Lemma 6. *When $p = \sqrt{k+2}/(\sqrt{k+2} + \sqrt{k+1})$ and $h_0 = \sqrt{(k+1)/(k+2)}$, the value of the right hand of the above inequality is $1 + 1/(\sqrt{k+1} + \sqrt{k+2})^2$.*

References

1. Andrews, M., Zhang, L.: Scheduling over a Time-varying User-dependent Channel with Applications to High Speed Wireless Data. In: Proceedings of the 43rd IEEE Symposium on Foundations of Computer Science (FOCS), pp. 293–302 (2002)
2. Arora, A., Choi, H.: Channel Aware Scheduling in Wireless Networks. Technical Report 002, George Washington University (2006)
3. Arora, A., Jin, F., Choi, H.: Scheduling Resource Allocation with Timeslot Penalty for Changeover. Theoretical Computer Science 369(1-3), 323–337 (2006)
4. Arora, A., Jin, F., Sahin, G., Mahmoud, H., Choi, H.: Throughput Analysis in Wireless Networks with Multiple Users and Multiple Channels. Acta Informatica 43(3), 147–164 (2006)
5. Borodin, A., EI-Yaniv, R.: Online Computation and Competitive Analysis. Cambridge University Press, Cambridge (1998)
6. Borst, S.: User-level Performance of Channel-aware Scheduling Algorithms in Wireless Data Networks. In: Proceedings of IEEE INFOCOM 2003, pp. 321–331 (2003)
7. Catreux, S., Erceg, V., Gesbert, D., Heath, R.: Adaptive Modulation and MIMO Coding for Broadband Wireless Data Networks. IEEE Communications Magazine 40, 108–115 (2002)
8. Sahin, G., Jin, F., Arora, A., Choi, H.: Predictive Scheduling in Multi-carrier Wireless Networks with Link Adaptation. In: Proceedings of 60th Vehicular Technology Conference (VTC2004-Fall), vol. 7, pp. 5015–5020 (2004)
9. Stallings, W.: Wireless Communication & Networks. 1st edn. Prentice Hall, Englewood Cliffs (2001)
10. Tsibonis, V., Georgiadis, L., Tassiulas, L.: Exploiting Wireless Channel State Information for Throughput Maximization. In: Proceedings of IEEE INFOCOM 2003, pp. 301–310 (2003)
11. Wang, H., Chaudhary, A., Chen, D.Z.: Online Rectangle Filling. In: Kaklamanis, C., Skutella, M. (eds.) WAOA 2007. LNCS, vol. 4927, pp. 274–287. Springer, Heidelberg (2008)
12. Wang, H., Chaudhary, A., Chen, D.Z.: New Algorithms for Online Rectangle Filling with k-Lookahead. Manuscript (2008)
13. Yao, A.: Probabilistic Computations: Toward a Unified Measure of Complexity. In: Proceedings of the 18th IEEE Symposium on Foundations of Computer Science (FOCS), pp. 222–227 (1977)

Geometric Spanner of Objects under L_1 Distance*

Yongding Zhu[1], Jinhui Xu[1,**], Yang Yang[1],
Naoki Katoh[2], and Shin-ichi Tanigawa[2]

[1] Department of Computer Science and Engineering
State University of New York at Buffalo
Buffalo, NY 14260, USA
{yzhu3,jinhui,yyang6}@cse.buffalo.edu
[2] Department of Architecture and Architectural Systems
Kyoto University, Japan
{naoki,is.tanigawa}@archi.kyoto-u.ac.jp

Abstract. Geometric spanner is a fundamental structure in computational geometry and plays an important role in many geometric networks design applications. In this paper, we consider the following generalized geometric spanner problem under L_1 distance: Given a set of disjoint objects S, find a spanning network G with minimum size so that for any pair of points in different objects of S, there exists a path in G with length no more than t times their L_1 distance, where t is the stretch factor. Specifically, we focus on three types of objects: rectilinear segments, axis aligned rectangles, and rectilinear monotone polygons. By combining ideas of *t-weekly dominating set*, *walls*, *aligned pairs* and *interval cover*, we develop a 4-approximation algorithm (measured by the number of Steiner points) for each type of objects. Our algorithms run in near quadratic time, and can be easily implemented for practical applications.

1 Introduction

In this paper, we consider the following generalization of the classical geometric spanner problem: Given a set S of n disjoint objects in L_1^2 space (i.e., 2-dimensional space with L_1 norm) and a constant $t > 1$, construct a graph G for S of minimum size (i.e. the number of vertices and edges is minimized) so that for any pair of points $p_i \in o_i$ and $p_j \in o_j$, there exists a path $P(p_i, p_j)$ in G whose total length is at most $t \times d(p_i, p_j)$, where o_i and o_j are objects in S and $d(p_i, p_j)$ is the L_1 (or Manhanttan) distance between p_i and p_j. The path $P(p_i, p_j)$ consists of three parts, P_1, P_2 and P_3, where P_1 and P_3 are the portions of $P(p_i, p_j)$ inside o_i and o_j respectively. We assume that there implicitly

* The research of the first three authors was supported in part by National Science Foundation through CAREER award CCF-0546509 and grant IIS-0713489. The research of the last two authors was supported by the project New Horizons in Computing, Grant-in-Aid for Scientific Research on Priority Areas,MEXT Japan.
** Corresponding author.

X. Hu and J. Wang (Eds.): COCOON 2008, LNCS 5092, pp. 395–404, 2008.
© Springer-Verlag Berlin Heidelberg 2008

exists an edge (or path) between any pair of points inside each object $o \in S$. Thus, the objective of minimizing the size of G is equivalent to minimizing the total number of vertices, and edges between vertices in different objects. In this paper, we consider the cases where objects are disjoint rectilinear segments, axis aligned rectangles, and rectilinear monotone polygons in L_1^2 space.

Spanner is a fundamental structure in computational geometry and finds applications in many different areas. Extensive researches have been done on this structure and a number of interesting results have been obtained [11],[12],[13],[7], [2],[3],[4],[6],[8],[9],[10]. Almost all previous results consider the case in which the objects are points and seek to minimize the spanner's construction time, size, weight, maximum degree of vertex, diameter, or any combination of them.

A common approach for constructing geometric spanner is the use of Θ-graph [11],[12],[13],[7]. In [2], Arya *et al.* showed that a t-spanner with constant degree can be constructed in $O(n \log n)$ time. In [3],[4], they gave a randomized construction of a sparse t-spanner with expected spanner diameter $O(\log n)$. In [8],[9], Das *et al.* proposed an $O(n \log^2 n)$-time greedy algorithm for a t-spanner with $O(n)$ edges and $O(1)wt(MST)$ weight in 3-D space. Gudmundsson *et al.* showed in [10] that an $O(n)$ edges, and $O(1)wt(MST)$ weight t-spanner is possible to be constructed in $O(n \log n)$ time.

In graph settings, Chandar *et al.* [6] showed that for an arbitrary positive edge-weighted graph G and any $t > 1, \epsilon > 0$, a t-spanner of G with weight $O(n^{\frac{2+\epsilon}{t-1}})wt(MST)$ can be constructed in polynomial time. They also showed that $(\log^2 n)$-spanners of weight $O(1)wt(MST)$ can be constructed.

For geometric spanners of objects other than points, Asano *et al.* considered the problem of constructing a spanner graph for a set of axis-aligned rectangles using rectilinear bridges and under L_1 distance [5]. They showed that in general it is NP-hard to minimize the dilation, gave some polynomial time solutions for some special cases.

In [14], Yang *et al.* generalized the geometric spanner from points to segments and considered the problem of constructing a minimum-sized t-spanner for a set of disjoint segments in Euclidean space. They showed that a constant approximation can be obtained in $O(|Q| + n^2 \log n)$ time if the segments are relatively well separated, where Q is the set of vertices (called Steiner points) of G.

The problem considered in this paper is motivated by several applications in architecture, wireless mesh networks [14], and VLSI layout.

To solve the aforementioned problem, we further extend in this paper the concept of geometric spanner to polygons. Particularly, we consider three types of objects, rectilinear segments, axis-aligned rectangles, and rectilinear monotone polygons. We show that our framework for constructing geometric spanner of segments in [14] can be generalized to polygons and achieves much better performance ratios. Our approach builds the spanner in two steps. First, we identify a set of points, called *Steiner points*, from each object; Then a t-spanner is constructed for the Steiner points by applying some existing algorithms for point spanners such as the ones in [1]. Thus, our focus will be only on the first step. Furthermore, since most existing spanners are sparse graphs (i.e. consist of $O(n)$

edges), minimizing the size of the spanner for rectilinear polygons is equivalent (i.e., within a constant factor) to minimizing the total number of Steiner points. Our objective is hence to obtain a spanner with a minimum number of Steiner points.

Minimizing the number of Steiner points is in general quite challenging, mainly due to interference of Steiner points in different objects. To overcome this difficulty, we first generalize the concept of *weakly dominating set* in [14] to lower bound the number of Steiner points on one object. Then we show that under L_1 distance, weakly dominating set could be converted into *dominating set* by using several techniques such as *wall, aligned pairs*. Particularly we are able to find a set of strongly dominating set for each object, and show that the size of the strongly dominating set is a 4-approximation of the optimal solution. Our algorithm can be easily implemented and runs in near quadratic time.

Due to space limit, many details are omitted from this extended abstract.

2   Main Ideas

Let $S = \{O_1, O_2, ..., O_n\}$ be a set of n disjoint connected objects in L_1^2 space. A t-spanner G_S of S is a network which connects the objects in S and satisfies the following condition. For any two points p_i and p_j in objects $O_i \in S$ and $O_j \in S, i \neq j$, respectively, there exists a path (called spanner path) in G_S between p_i and p_j with length no more than $t|p_ip_j|$, where t is the *stretch factor* of the spanner and $|p_ip_j|$ is the L_1 distance between p_i and p_j. The spanner G_S consists of the objects, some sample points (called Steiner points) of the objects, and line segments (called *bridges*) connecting the Steiner points. We assume that there is an implicit path between p_i (or p_j) to any Steiner point in O_i (or O_j). Thus the spanner path between p_i and p_j includes an implicit path from p_i to some Steiner point $q_i \in O_i$ and and an implicit path from p_j to some Steiner point $q_j \in O_j$.

As mentioned in previous section, our main objective for the spanner G_S is to minimize its size. The size of G_S is the sum of the complexities of objects in S and the numbers of Steiner points and bridges. Since the total complexity of the objects is fixed, minimizing the size of G_S is equivalent to minimize the total number of Steiner points and bridges.

To simplify the optimization task, our main idea is to separate the procedure of minimizing the number of Steiner points from that of minimizing the number of bridges. In following sections, for each type of objects (i.e., rectilinear segments, axis aligned rectangles and rectilinear monotone polygons), we first compute a set Q of Steiner points with small size, and then construct a spanner G_Q for Q to minimize the number of bridges. The spanner G_Q together with the objects forms the spanner of S (i.e. G_S). Since most existing spanner algorithms for points yield spanners with linear number of edges, the difficulty of minimizing the size of G_S lies on minimizing the number of Steiner points (also called *dominating set minimization* problem).

To illustrate our main ideas on minimizing Steiner points, we first briefly discuss the framework for all three types of objects inherited from our algorithm for constructing segment spanners in [14]. We start with selecting Steiner points for a pair of objects.

Let O_1 and O_2 be two different objects in S and p_1 and p_2 be a pair of arbitrary points in O_1 and O_2 respectively. Let $q_1 \in O_1$ and $q_2 \in O_2$ be two Steiner points close enough to p_1 and p_2.

Definition 1 (t-Domination). *Steiner points q_1 and q_2 t-dominate p_1 and p_2 if the path $p_1 \rightarrow q_1 \rightarrow q_2 \rightarrow p_2$ is a t-spanner path for p_1 and p_2 (i.e., the length of the path is no more than $t \times |p_1p_2|$, where $|p_1p_2|$ is the length of the segment $\overline{p_1p_2}$). q_1 and q_2 are called the t-dominating pair of p_1 and p_2.*

From the definition, it is clear that the positions of q_1 and q_2 are constrained by p_1 and p_2. If we fix p_1, p_2, and one Steiner point q_1, then all possible positions of the other Steiner point q_2 form a (possibly empty) region denoted as $R(p_1, p_2, q_1)$ (which is a function of p_1, p_2 and q_1) in O_2. When q_1 moves in O_1, the region changes accordingly. Similarly, if we fix the two Steiner points q_2, q_1, together with p_2, all points in O_1 t-dominated by q_2 and q_1, with respect to p_2, also form an region $R(q_2, q_1, p_2)$ in O_1.

Since the spanner G_S needs to guarantee that there exists a spanner path (or equivalently a t-dominating pair of Steiner points) from p_2 to every point in O_1, from p_2's point of view, it expects q_2 to be in some position such that O_1 can be covered by a minimum number of q_1's. , i.e. the union of $R(q_2, q_1, p_2)$ covers O_1. Thus, to determine Steiner points in O_1, we need to (1) identify a minimum set of Steiner points to cover all points in O_1 and (2) find a way to deal with the influence of the Steiner points (e.g., q_2) in O_2 and other objects.

To overcome these two difficulties, we relax the constraints in the definition of t-domination.

Definition 2 (t-Weak Domination). *Steiner point q_1 t-weakly dominates p_1 and p_2 if q_1 and p_2 are the t-dominating pair of p_1 and p_2. q_1 t-weakly dominates p_1 if for any $p_2 \in O_2$, q_1 t-weakly dominates p_1 and p_2.*

In the above definition, we assume that q_2 can be placed at arbitrary position in O_2 (or equivalently every point in O_2 is a Steiner point), when placing Steiner points in O_1. With this relaxation, we only need to consider the relation between q_1 and p_1, p_2. More specifically, we only need to find a minimum number of points in O_1 so that every point p_1 in O_1 is t-weakly dominated by some selected Steiner point. We call such a set of points as a t-weakly dominating set of O_1. We will show in following sections how to select t-weakly dominating set for each object (i.e., overcoming difficulty (1)).

The concept of weakly dominating sets helps us to avoid the influence of Steiner points from other objects (i.e., difficulty (2)). However, t-weakly dominating sets alone do not guarantee the existence of t-dominating pair for each pair of points $p_1 \in O_1$ and $p_2 \in O_2$. To overcome this difficulty, in [14] we used the concept of *imaginary Steiner points*. In this paper, we show that using the concepts of *wall*, *aligned pairs*, and *interval covers*, we are able to convert t-weakly dominating sets into t-dominating (or t-strongly dominating) sets without introducing imaginary Steiner points.

For multiple objects, we first compute weak visibility graph for each object $O_i \in S$ and consider the Steiner-point-determination problem for O_i and each object weakly visible to O_i. The set of Steiner points in O_i computed from its weakly visible objects is called the *t-strongly dominating set* of O_i.

3 Constructing t-Spanner for Rectilinear Segments under L_1 Distance

In [14], an $O(1)$-approximation algorithm was designed for constructing a spanner of segments under L_2 distance. In this section, we consider a special case of the segment spanner problem in which the input is a set S of rectilinear segments, and the distance function is the L_1 norm (i.e., the Manhattan distance). We show that for this special case, a much better performance ratio (i.e., 4) can be achieved.

Let s_1 and s_2 be two rectilinear segments in S, and p_1 and p_2 be two arbitrary points on s_1 and s_2 respectively. Let $q_1 \in s_1$ and $q_2 \in s_2$ be two Steiner points of p_1 and p_2 respectively.

It is easy to see that when the two segments have different orientations (i.e., one horizontal and the other vertical, say s_1 is horizontal and s_2 is vertical), one Steiner point on each segment (i.e., the point closest to the other segment) is sufficient to t-dominate the corresponding segment. Thus we only focus on the case in which s_1 and s_2 have the same orientation. Without loss of generality, we assume that s_1 and s_2 are all horizontal segments.

Let q_1 be a Steiner point in s_1 t-weakly dominating p_1 and e_{1l} and e_{1r} be the two endpoints of the region $R(p_1, p_2, p_2)$ (i.e., the interval of all possible positions of the Steiner point q_1 when q_2 coincides with p_2). Then we have the following lemma.

Lemma 1. *Let s_1 and s_2 be defined as above. Then the two endpoints e_{1l} and e_{1r} of $R(p_1, p_2, p_2)$ locate on different sides of p_1 with either $|p_1 e_{1l}| = min\{|p_1 a_1|,$ $\frac{t-1}{2}|p_1 p_2| + \frac{t+1}{2}|p_1 p_2|_x\}$ and $|p_1 e_{1r}| = min\{|p_1 b_1|, \frac{t-1}{2}|p_1 p_2|\}$ or $|p_1 e_{1l}| = min\{ |p_1 a_1|, \frac{t-1}{2}|p_1 p_2|\}$ and $|p_1 e_{1r}| = min\{|p_1 b_1|, \frac{t-1}{2}|p_1 p_2| + \frac{t+1}{2}|p_1 p_2|_x\}$, where $|p_1 p_2|_x$ is the distance of p_1 and p_2 along the x-axis and a_1 and b_1 are the left and right endpoints of s_1 respectively.*

Lemma 2. *The minimum of $|p_1 e_{1l}|$ (or $|p_1 e_{1r}|$) is $min\{|p_1 a_1|, \frac{t-1}{2}|p_1 p_2|\}$ (or $min\{|p_1 b_1|, \frac{t-1}{2}|p_1 p_2|\}$), achieved either when e_{1l} coincides with a_1 (or e_{1r} coincides with b_1) or p_2 is at the endpoints of s_2, or $|p_1 p_2|$ is a constant that only depends on s_1 and s_2.*

Let m be the parameter of p_1 in its convex combination of the two endpoints of s_1, i.e. $p_1 = (1 - m)a_1 + mb_1$, for some $m \in [0, 1]$. Let $L_{1,2}(m)$ and $R_{1,2}(m)$ be the functions defining the positions of e_{1l} and e_{1r} (respectively) on s_1, i.e. $L_{1,2}(m) = m - |p_1 e_{1l}|/|a_1 b_1|$ and $R_{1,2}(m) = m + |p_1 e_{1r}|/|a_1 b_1|$.

Lemma 3. *$L_{1,2}(m)$ and $R_{1,2}(m)$ are piecewise linear functions of m.*

To efficiently compute a set of t-dominating set for each segment in S, we first introduce the concept of *wall*. Let s_1 and s_2 be two weakly visible segments in S, and p_1 and p_2 be their respective points. p_1 and p_2 are *horizontally (or vertically) visible pair* if p_1 and p_2 have the same y (or x) coordinate and the horizontal (or vertical) segment $\overline{p_1p_2}$ does not intersect the interior of any other segment in S. The union of all horizontally (or vertically) visible pairs forms one or more vertical (or horizontal) subsegments on each of s_1 and s_2. The corresponding subsegments on s_1 and s_2 have the same length and are called *wall* to each other. The set of such subsegments in each $s_i, i \in \{1,2\}$, is called the *wall portion* of s_i. We have the following lemmas about the positions of Steiner points.

Lemma 4. *Given a set of rectilinear segments S in L_1 space, to determine the position of the set Q of Steiner points, there exists a 2-approximation of Q (with respect to its size) with all the Steiner points located in the wall portions.*

Notice that given a set S of n rectilinear segments in L_1 space, the wall portions of each segment in S can be determined by using a trapezoid decomposition algorithm in $O(n \log n)$ time. For the simplicity of discussion, we also assume that there is a very large axis-aligned rectangle bounding all segments in S. The addition of the four edges of the rectangle into S could make each segment contain only wall portions, without changing the spanner problem itself (i.e., the Steiner points in S will still be a t-dominating set for S). Thus from now on, we assume that every segment in S only consists of wall portions.

Lemma 5. *Given a set of rectilinear segments S in L_1 space, the t-dominating set between two subsegments that are wall to each other can be computed optimally.*

Proof. Let $ss_1 \in s_1$ and $ss_2 \in s_2$ be two subsegments that are wall to each other. Notice that they have the same length. By Lemma 1 and Lemma 2, we know that $|p_1e_{1l}|$, $|p_1e_{1r}|$, $|p_2e_{2l}|$ and $|p_2e_{2r}|$ all have the same minimum value $\frac{t-1}{2}|p_1p_2|$. This implies the following two properties: i) (the figures of) $L_{1,2}(m)$ and $R_{1,2}(m)$ of these wall portions are straight line segments (in the coordinate system with m as the x-axis) and parallel to each other; ii) they form the same $B_{1,2}$ and $B_{2,1}$ bands (i.e., the region bounded by the $L_{1,2}(m)$ and $R_{1,2}(m)$ functions in the coordinate system; see [14] for more details). Property i) means that the t-weakly dominating points can therefore be determined by a horizontal interval cover in $B_{1,2}$ (see [14]). Property ii) means that the t-weakly dominating points are chosen as pairs on ss_1 and ss_2, i.e. if ss_1 and ss_2 are both horizontal, for each t-weakly dominating Steiner point q_1 on ss_1, there exists a t-weakly dominating Steiner point q_2 on ss_2 with the same x-coordinate as q_1. Together with the property of L_1 norm (i.e., $|p_1q_2| = |p_1q_1| + |q_1q_2|$), this implies that the minimum t-weakly dominating set computed using interval cover is also a minimum t-dominating set (i.e. no imaginary point is needed). □

The above lemma suggests an optimal way of determining Steiner points for a pair of wall portion. That is, compute an interval cover (using the band) for one wall portion and put vertically aligned Steiner points in the other wall portion. A Steiner point and its corresponding aligned Steiner point are called *aligned pair* to each other.

Below we consider how to determine Steiner point for a segment s_1. Without loss of generality, we assume s_1 is horizontal. For all the (sub)segments that are weakly visible to s_1, we only need to consider those that are wall to some subsegments of s_1 and those with different orientation (see Figure 1). Note that for any point $p_1 \in s_1$, its $L_1(m)$ and $R_1(m)$ may correspond to some points in s_1 but not in the same wall portion as p_1, where $L_1(m)$ and $R_1(m)$ are the functions defining the positions of e_{1l} and e_{1r} (of p_1), when considering all segments in $S \setminus \{s_1\}$. To simplify our discussion, we first assume that there is no vertical segment in S (later we will show how to handle the general case). From Lemmas 1 and 2, we know that $L_1(m) = \max(0, m - \frac{t-1}{2}d)$ and $R_1(m) = \min(1, m + \frac{t-1}{2}d)$, where d is the vertical distance from p_1 to the closest (sub-)segment which is a wall of the (sub-)segment containing p_1 (see Figure 1). Observe that $L_1(m)$ and $R_1(m)$ of s_1 are piecewise linear functions but may not always be continuous in the domain $[0, 1]$. As shown in Figure 2, we can apply the interval cover algorithm over the band of each segment $s_1 \in S$ to obtain the whole set Q of Steiner points for S. For every Steiner point putting on s_1, we also put its aligned pair on the corresponding wall portion of the other segment (i.e., the segment d distance away from s_1; see the algorithm below).

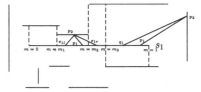
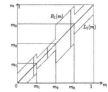

Fig. 1. Wall portions of segment s_1 **Fig. 2.** Band B_1 and an interval cover for segment s_1

In the general case in which vertical segments may present, let s_2 be such a vertical segment weakly visible to s_1. We denote d as the distance between s_1 and s_2 and l as the length of s_1 (see Figure 1). There are three cases to consider.

i) s_2 is to the right of s_1: It is easy to see that a Steiner point at the right endpoint of s_1, together with the Steiner points on s_2, t-dominate any pair of points in s_1 and s_2 respectively;

ii) s_2 is to the left of s_1: Similarly, we can put a Steiner point at the left endpoint of s_1;

iii) s_2 is between the two endpoints of s_1: Put a Steiner point at the closer endpoint of s_2 to s_1. Let $m' \in [0, 1]$ be the position of any point in s_2 in its parameterization. If $0 \leq m \leq m'$, $R_{1,2}(m) = 1$ and $L_{1,2}(m) = \max(0, m - \frac{t-1}{2}(m' - m + \frac{d}{l}))$; If $m' \leq m \leq 1$, $L_{1,2}(m) = 0$ and $R_{1,2}(m) = \min(1, m + \frac{t-1}{2}(m - m' + \frac{d}{l}))$.

Observe that the functions $L_{1,2}(m)$ and $R_{1,2}(m)$ in iii) is linear or piecewise linear. Thus in the general case, we consider the lower envelope $R_1(m)$ of all the

$R_{1,2}(m)$ functions and the upper envelope $L_1(m)$ of all the $L_{1,2}(m)$ functions. The band B_1 of s_1 is bounded by piecewise linear segments and still may not be continuous. Note that m' in case iii) may cause additional discontinuous points in $L_1(m)$ and $R_1(m)$. The discontinuous points may divide the band B_1 into multiple sub-bands with each bounded by continuously increasing and piecewise linear functions $L_1(m)$ and $R_1(m)$ (see Figure 3). This is the key property that we will use in the proof the main theorem (theorem 1).

Now we present the algorithm for finding Steiner points Q of S:

1. Parameterize each segment in S with the left endpoint or the upper endpoint be 0; Let $X = S$ and $Q = \Phi$;
2. For each segment s_i in S, construct the band B_i (i.e., the upper envelope of $L_{i,j}(m)$ and lower envelope of $R_{i,j}(m)$);
3. Take a segment s from X, and for each existing Steiner point in s, put a maximal horizontal interval in the band B of s; If there exists a segment in S having different orientation with s and located at the left (upward) of s, then add the left (upper) endpoint to Q; Similarly, if there exists a segment having different orientation and located at the right (downward) of s, then add the right (lower) endpoint to Q; Add the horizontal intervals corresponding to the new Steiner points into band B (see the green intervals in Figure 3);
4. In band B, check whether the set of intervals horizontally covers the interval $[0, 1]$. If there is a gap between some intervals, add Steiner points to s (using the interval cover algorithm [14]) until the gap is covered. Check whether there is any redundant Steiner point, and remove it, if any;
5. For each s' in X with wall portions that are wall to some subsegments of s, put aligned pairs in s' for Steiner points in s; Remove s from X;
6. Repeat steps 3-5 until X is empty, and return Q.

The following lemma is necessary for analyzing the algorithm.

Lemma 6. *Let s_1 and s_2 be two segments in S with the same orientation. Let q_1 and q_2 be an aligned pair in s_1 and s_2 respectively and p_1 and p_2 be two points in s_1 and s_2 respectively such that q_1 t-weakly dominates p_1 and q_2 t-weakly dominates p_2. Then q_1 and q_2 t-dominate p_1 and p_2. If s_1 and s_2 have different orientation with q_1 being its right endpoint of s_1, q_2 being a Steiner point in s_2, p_1 being any point in s_1 and p_2 being any point in s_2 t-weakly dominated by q_2 (with respect to s_1), then q_1 and q_2 t-dominate p_1 and p_2.*

The above lemma suggests that the solution obtained from the algorithm is a t-strongly dominating set for S. To analyze the quality of the solution (in terms of the size), we notice that for band B_1 of any segment s_1, $L_1(m)$ and $R_1(m)$ are not continuously increasing and the intervals covering B_1 may overlap. Thus the solution may not be optimal. The following theorem gives an estimation on the quality of solution.

Theorem 1. *Given a set of n rectilinear segments in L_1^2 space, a set of Steiner points with size no more than $4 \times |OPT|$ can be computed in $O(|Q| + n^2 \log n)$ time.*

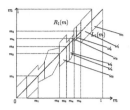

Fig. 3. Band and the proof of 4-approximation

Proof. As shown in Figure 3, the band B_1 can be divided into sub-bands by vertically cutting at the discontinuous points of $L_1(m)$ and $R_1(m)$. To prove Q is a 4-approximation of the optimal solution, it is sufficient to prove that for a consecutive set of sub-bands that require a Steiner point in the optimal solution, our algorithm provides at most 4 Steiner points. Let e_1 and e_2 be the two endpoints of such a consecutive set of sub-bands (called a strip) with $e_1 < e_2$. Let e_3 and e_4 be the leftmost and rightmost Steiner points in the strip (Note that e_3 and e_4 might be the same; For worst case analysis, we assume that they are different). Thus we have $e_1 \le e_3 < e_4 \le e_2$. We claim that there are at most two more Steiner Points between e_3 and e_4. To better understand this claim, consider the case in Figure 3, where the strip is between m_2 and m_5, and ic_2 is the corresponding interval of the only Steiner point in the strip in an optimal solution. In the worst case, our algorithm could have the leftmost Steiner point (i.e., e_3) at m_2 and the rightmost Steiner point (i.e., e_4) at m_5, with ic_4' and ic_1' as their corresponding intervals. The two intervals leave a gap. To fill the gap, our algorithm could generate at most two additional Steiner points at m' and m'' (i.e., ic_2' and ic_3'). To see this, we assume that there is another Steiner point β between m_2 and m_5. Without loss of generality, we assume $m' \le \beta \le m''$. By Step 4 of our algorithm, we know that β is a redundant Steiner point. This is because $ic_2' \cup ic_3' \supset ic_2$ and the interval that can be dominated by β is contained in $ic_2' \cup ic_3'$. Thus a contradiction. This shows that our algorithm generates at most 4 Steiner points for the strip. (Running time analysis is omitted.)   □

4   t-Spanner of Other Objects

Let $S = \{R_1, R_2, \cdots, R_n\}$ be a set of disjoint axis aligned rectangles, and $t > 1$ be the stretch factor. The follow theorem holds.

Theorem 2. *Given a set of n disjoint axis aligned rectangles in L_1^2 space, a set Q of Steiner points with size no more than $4 \times |OPT|$ can be computed in $O(|Q| + n^2 \log n)$ time.*

Let $S = \{P_1, P_2, \cdots, P_n\}$ be a set of disjoint rectilinear polygons in L_1^2 space with each polygon P_i ($i = 1, 2, ..., n$) being monotone in both the x and y directions. For this problem, we have the following theorem.

Theorem 3. *For a set S of n disjoint rectilinear monotone polygons in L_1^2 space, a set of t-strongly dominating Steiner points with size no more than $4 \times |OPT|$ can be computed in $O(|Q| + N^2 \log N)$ time, where N is the total number of vertices in S.*

References

1. Aronov, B., de Berg, M., Cheong, O., Gudmundsson, J., Haverkort, H., Smid, M., Vigneron, A.: Sparse Geometric Graphs with Small Dilation. In: Proceedings of the 12th Computing: The Australasian Theroy Symposium, vol. 51 (2006)
2. Arya, S., Das, G., Mount, D.M., Salowe, J.S., Smid, M.: Euclidean Spanners: Short, Thin, and Lanky. In: Proceedings of the Twenty-Seventh Annual ACM Symposium on Theory of Computing (STOC 1995), pp. 489–498 (1995)
3. Arya, S., Mount, D.M., Smid, M.: Dynamic Algorithms for Geometric Spanners of Small Diameter: Randomized Rolutions. Technical Report, Max-Planck-Institut für Informatik (1994)
4. Arya, S., Mount, D.M., Smid, M.: Randomized and Deterministic Algorithms for Geometric Spanners of Small Diameter. In: 35th IEEE Symposium on Foundtions of Computer Science, pp. 703–712 (1994)
5. Asano, T., de Berg, M., Cheong, O., Everett, H., Haverkort, H., Katoh, N., Wolff, A.: Optimal Spanners for Axis-Aligned Rectangles. Comput. Geom. Theory Appl. 30(1), 59–77 (2005)
6. Chandra, B., Das, G., Narasimhan, G., Soares, J.: New Spareness Results on Graph Spanners. In: Proceedings of the Eighth Annual Symposium on Computational Geometry, pp. 192–201 (1992)
7. Clarkson, K.L.: Approximation Algorithms for Shortest Path Motion Planning. In: Proceedings of the nineteenth annual ACM conference on Theory of computing, pp. 56–65 (1987)
8. Das, G., Heffernan, P., Narasimhan, G.: Optimally Sparse Spanners in 3-Dimensional Euclidean Space. In: Proceedings of the Ninth Annual Symposium on Computational Geometry, pp. 53–62 (1993)
9. Das, G., Narasimhan, G.: A Fast Algorithm for Constructing Sparse Euclidean Spanners. In: Proceedings of the Tenth Annual Symposium on Computational Geometry, pp. 132–139 (1994)
10. Gudmundsson, J., Levcopoulos, C., Narasimhan, G.: Fast Greedy Algorithms for Constructing Sparse Geometric Spanners. SIAM Journal on Computing 31(5), 1479–1500 (2002)
11. Keil, J.M.: Approximating the Complete Euclidean Graph. In: Proceedings of 1st Scandinavian Workshop on Algorithm Theory, pp. 208–213 (1988)
12. Keil, J.M., Gutwin, C.A.: Classes of Graphs which Approximate the Complete Euclidean Graph. Discrete and Computational Geometry 7, 13–28 (1992)
13. Rupper, J., Seidel, R.: Approximating the d-dimensional Complete Euclidean Graph. In: Proceedings of 3rd Canadian Conference on Computational Geometry, pp. 207–210 (1991)
14. Yang, Y., Zhu, Y., Xu, J., Katoh, N.: Geometric Spanner of Segments. In: Tokuyama, T. (ed.) ISAAC 2007. LNCS, vol. 4835, pp. 75–87. Springer, Heidelberg (2007)

Star-Shaped Drawings of Graphs with Fixed Embedding and Concave Corner Constraints*

Seok-Hee Hong[1] and Hiroshi Nagamochi[2]

[1] School of Information Technologies, University of Sydney
shhong@it.usyd.edu.au
[2] Department of Applied Mathematics and Physics, Kyoto University
nag@amp.i.kyoto-u.ac.jp

Abstract. A *star-shaped* drawing of a graph is a straight-line drawing such that each inner facial cycle is drawn as a star-shaped polygon, and the outer facial cycle is drawn as a convex polygon. In this paper, given a biconnected planar graph G with fixed plane embedding and a subset A of corners of G, we consider the problem of finding a star-shaped drawing D of G such that only corners in A are allowed to become concave corners in D. We first characterize a necessary and sufficient condition for a subset A of corners to admit such a star-shaped drawing D. Then we present a linear time algorithm for finding such a star-shaped drawing D. Our characterization includes Thomassen's classical characterization of biconnected plane graphs with a prescribed boundary that have convex drawings.

1 Introduction

Graph drawing has attracted much attention over the last twenty years due to its wide range of applications, such as VLSI design, software engineering and bioinformatics. Two or three dimensional drawings of graphs with a variety of aesthetics and edge representations have been extensively studied [2]. One of the most popular drawing conventions is the *straight-line drawing*, where all the edges of a graph are drawn as straight-line segments. Every planar graph is known to have a planar straight-line drawing [4]. A straight-line drawing is called a *convex drawing* if every facial cycle is drawn as a convex polygon. Note that not all planar graphs admit a convex drawing.

In general, the convex drawing problem has been well investigated. Tutte [13] showed that every *triconnected* plane graph with a given boundary drawn as a convex polygon admits a convex drawing using the polygonal boundary. However, not all biconnected planes graphs admit convex drawings. Thomassen [12] gave a necessary and sufficient condition for a *biconnected* plane graph with a prescribed convex boundary to have a convex drawing. Chiba et al. [1] presented a linear time algorithm for finding a convex drawing (if any) for a biconnected plane graph with a specified convex boundary. Miura et al. [11] gave a linear

---

* This is an extended abstract. For the full version, see [9]. This research was supported by a grant from the Australian Research Council, and the Scientific Grant-in-Aid from Ministry of Education, Culture, Sports, Science and Technology of Japan.

X. Hu and J. Wang (Eds.): COCOON 2008, LNCS 5092, pp. 405–414, 2008.
© Springer-Verlag Berlin Heidelberg 2008

time algorithm for finding a convex drawing with minimum outer apices for an *internally triconnected* plane graph. Hong and Nagamochi gave conditions for *hierarchical* plane graphs to admit a convex drawing [6], and for *c-planar clustered* graphs to admit a convex drawing in which every cluster is also drawn as a convex polygon [7].

However, not much attention has been paid to the problem of finding a convex drawing with a *non-convex boundary* or *non-convex faces*. Recently, Hong and Nagamochi [5] proved that every *triconnected* plane graph with a fixed star-shaped polygon boundary has an *inner-convex* drawing (a drawing in which every inner face is drawn as a convex polygon), if its kernel has a positive area. Note that this is an extension of the classical result by Tutte [13], since any convex polygon is a star-shaped polygon.

To draw biconnected graphs which do not admit convex drawings in a convex way as much as possible, it is natural to minimize the number of convex faces, concave vertices or concave corners in a drawing. However, Kant [10] already proved the NP-completeness of the problem of deciding whether a biconnected planar/plane graph can be drawn with at most k non-convex faces.

Recently, in our companion paper [8], we initiated a new notion of *star-shaped drawing* of a graph as a straight-line drawing such that each inner facial cycle is drawn as a star-shaped polygon, and the outer facial cycle is drawn as a convex polygon. Note that there is a biconnected plane graph which needs a concave corner in any of its straight-line drawings (including outer apices as concave corners). We proved that, given a biconnected *planar* graph G, a star-shaped drawing of G with the minimum number of concave corners can be found in linear time, where we are allowed to choose the plane embedding and the concave corners of G, based on the effective use of lower bounds [8].

In this paper, we deal with a start-shaped drawing of graphs with two given *constraints*: a fixed plane embedding constraints and a set of concave corner constraints.

Let G be a biconnected *plane* graph. We denote a *corner* λ around a vertex v by pair (v, f) of the vertex v and the facial cycle f whose interior contains the corner. Let $\Lambda(v)$ denote the set of all corners around a vertex v in G, and $\Lambda(G)$ denote the set of all corners in G. For a straight-line drawing D of a plane graph G, let $\Lambda^c(D)$ denote the set of concave corners in D. For a given subset $A \subseteq \Lambda(G)$ of corners of G, we consider the problem of whether G has a star-shaped drawing D^* such that $\Lambda^c(D^*) \subseteq A$. In this paper, we characterize when G admits such a drawing D^*, and give a linear time algorithm for testing the conditions in the characterization.

A corner (v, f^o) of a vertex v in the outer facial cycle f^o is an *outer corner* of f^o. We denote $\Lambda^o(f^o(G))$ the set of the outer corners of the outer facial cycle $f^o(G)$. We call a cycle C in G a *cut-cycle* if a cut-pair $\{u, v\} \subseteq V(C)$ separates the vertices outside C from those along C (including those inside C). A corner (v, f) of a vertex v in a cut-cycle C is an *outer corner* of C if v is not in the cut-pair of C, and f is one of the two facial cycles outside C that share the cut-pair of C. We denote by $\Lambda^o(C)$ the set of the outer corners of a cut-cycle (or the outer facial cycle) C, and $\Lambda^o(G)$ denote the set of all outer corners in G. For example, $C_1 = (u_6, u_9, u_2, u_{18}, u_8)$ of graph G in Fig. 1 is a cut-cycle, where $\Lambda^o(C_1) = \{(u_9, f_{12}), (u_{18}, f_9), (u_8, f_9)\}$ (see also Fig. 3(a)).

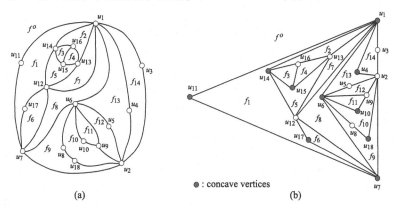

Fig. 1. (a) A biconnected plane graph G (b) A star-shaped drawing D of G

More formally, we prove the next result.

Theorem 1. *Let G a biconnected plane graph G, f^o be the outer facial cycle of G, and A be a subset of $\Lambda(G)$.*

(i) *There exists a straight-line drawing D of G such that $\Lambda^c(D) \subseteq A$ if and only if*

$$|A \cap \Lambda^o(f^o)| \geq 3 \text{ and } |A \cap \Lambda^o(C)| \geq 1 \text{ for all cut-cycle } C \text{ in } G. \qquad (1)$$

Testing whether A satisfies the condition or not can be done in linear time.

(ii) *Let A satisfy the condition in (i), and f^o be drawn as a convex polygon P with $\Lambda^c(P) = A \cap \Lambda^o(f^o)$. Then P can be extended to a star-shaped drawing D^* of G with $\Lambda^c(D^*) \subseteq A \cap \Lambda^o(G)$. Such a drawing D^* can be obtained in linear time.* □

From the theorem, we see that, for any straight-line drawing D, there exists a star-shaped drawing D^* of G with $\Lambda^c(D^*) \subseteq \Lambda^c(D) \cap \Lambda^o(G) \subseteq \Lambda^c(D)$. Theorem 1 also implies that any convex polygon P drawn for the boundary of a triconnected plane graph G can be extended to a convex drawing of G, since G has no cut-cycle. Moreover, Theorem 1 also contains as special cases Thomassen's characterization of biconnected plane graphs that have convex drawings [12], and the characterization of convex drawings with minimum outer apices by Miura et al. [11]. For details, see [9].

2   Preliminaries

Throughout the paper, a graph $G = (V, E)$ stands for a simple undirected graph. The set of vertices and set of edges of a graph G may be denoted by $V(G)$ and $E(G)$, resp.. The set of edges incident to a vertex $v \in V$ is denoted by $E(v)$. The degree of a vertex v in G is denoted by $d_G(v)$ (i.e., $d_G(v) = |E(v)|$). For a subset $X \subseteq E$ (resp., $X \subseteq V$), let $G - X$ denote the graph obtained from G by removing the edges in X (resp., the vertices in X together with the edges in $\cup_{v \in X} E(v)$).

2.1   Plane Graphs and Biconnected Plane Graphs

A graph G is called *planar* if its vertices and edges are drawn as points and curves in the plane so that no two curves intersect except for their end points, where no two vertices are drawn at the same point. In such a drawing, the plane is divided into several connected regions, each of which is called a *face*. A face is characterized by the cycle of G that surrounds the region. Such a cycle is called a *facial cycle*. A *plane embedding* of a planar graph G consists of an ordering of edges around each vertex and the outer face. A planar graph with a fixed plane embedding is called a *plane* graph. Let $f^o(G)$ denote the outer facial cycle of a plane graph G, and $V^o(G)$ denote $V(f^o(G))$. The set of faces of a plane graph G is denoted by $F(G)$. A vertex (resp., an edge) in the outer facial cycle is called an *outer vertex* (resp., an *outer edge*), while a vertex (resp., an edge) not in the outer facial cycle is called an *inner vertex* (resp., an *inner edge*).

Let $G = (V, E, F)$ be a biconnected plane graph. Let C be a cut-cycle and $\{u, v\} \in V(C)$ be the cut-pair that separates the vertices in $V(C) - \{u, v\}$ and inside C from the vertices outside C in G. We consider a subgraph H of G such that the boundary $f^o(H)$ is a cut-cycle in G, where we treat H as a plane graph under the same embedding of G. For such a plane graph H, we define the u, v-*boundary path* $f^o_{uv}(H)$ of H to be the path obtained by traversing the boundary $f^o(H)$ of H from u to v in the clockwise order. We denote $V(f^o_{uv}(H)) - \{u, v\}$ by $V^o_{uv}(H)$, and denote by $\Lambda^o_{uv}(H)$ the set of outer corners of $f^o_{uv}(H)$, i.e., $\Lambda^o_{uv}(H) = \Lambda^o(f^o(H)) \cap (\cup_{w \in V^o_{uv}(H)} \Lambda^o(w))$. For example, cut-cycle $C_1 = (u_6, u_9, u_2, u_{18}, u_8)$ in Fig. 1 has subgraph H_1 with edges $(u_6, u_9), (u_9, u_2), (u_9, u_{10}), (u_{10}, u_6), (u_2, u_{18}), (u_{18}, u_8)$, (u_8, u_6) such that $C_1 = f^o(H_1)$, where $\Lambda^o_{u_6, u_2}(C_1) = \{(u_9, f_{12})\}$ and $\Lambda^o_{u_2, u_6}(C_1) = \{(u_{18}, f_9), (u_8, f_9)\}$ hold.

For a cut-pair $\{u, v\}$ of a biconnected plane graph G, a u, v-*component* H is a connected subgraph of G that either consists of a single edge (u, v) or is a maximal subgraph such that $H - \{u, v\}$ remains connected. We may treat a u, v-component H of a plane graph G as a plane graph under the same embedding of G. In this case, the boundary $f^o(H)$ of H is a cut-cycle. For example, the subgraph H consisting of edges $(u_6, u_9), (u_9, u_2), (u_9, u_{10}), (u_{10}, u_6)$ is a u_6, u_2-component of graph G in Fig. 1. Note that a cut-cycle C is not necessarily the boundary $f^o(H)$ of some u, v-component H. For example, cut-cycle $C_1 = (u_6, u_9, u_2, u_{18}, u_8)$ in graph G has no such u, v-component H. A simple path with end vertices u and v of a graph G is called a u, v-*path*, and is called an *induced u, v-path* if every internal vertex (i.e., non end vertex) is of degree 2.

A biconnected plane graph G is called *internally triconnected* if, (i) for each inner vertex v with $d_G(v) \geq 3$, there exist three paths disjoint except for v, each connecting v and an outer vertex; and (ii) every cycle of G which has no outer edge has at least three vertices v with $d_G(v) \geq 3$.

Define the *contracted graph* G' of a biconnected plane graph G as the graph obtained from G by replacing each u, v-induced path Q in G with a single edge, where we replace Q with a path of length two if $(u, v) \in E$ and $E(Q) \subseteq E(f^o(G))$. Then, G is internally triconnected if and only if G' has no multiple edges and for every cut-pair $\{u, v\}$ in G', $u, v \in V^o(G')$ holds and each component in $G' - \{u, v\}$ contains an outer vertex. In an internally triconnected plane graph G, a cut-cycle C is *minimal* if there is no other cut-cycle C' with $\Lambda^o(C') \cap \Lambda^o(f^o(G)) \subset \Lambda^o(C) \cap \Lambda^o(f^o(G))$.

2.2 The SPQR Tree of a Biconnected Planar Graph

The SPQR tree of a biconnected planar graph G represents the decomposition of G into the triconnected components [3]. Each node ν in the SPQR tree is associated with a graph $\sigma(\nu) = (V_\nu, E_\nu)$ ($V_\nu \subseteq V$), called the *skeleton* of ν. In fact, we use a modified SPQR tree *without* Q-nodes. Thus, there are three types of nodes in the SPQR tree: S-node (the skeleton is a simple cycle), P-node (the skeleton consists of two vertices with at least 3 edges), and R-node (the skeleton is a triconnected simple graph). Fig. 2 shows the SPQR tree of the biconnected planar graph in Fig. 1.

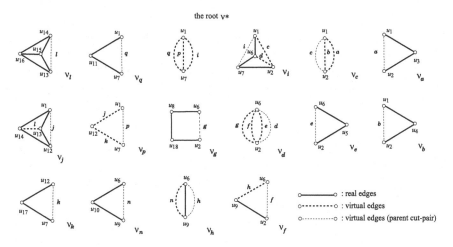

Fig. 2. The SPQR tree of the biconnected planar graph G in Fig. 1

We treat the SPQR tree as a rooted tree $\mathcal{T}$ by choosing a node ν^* as its root. For a node ν, let $Ch(\nu)$ denote the set of all children of ν and μ be the parent of ν. The graph $\sigma(\mu)$ has exactly one *virtual edge* e in common with $\sigma(\nu)$. The edge e is called the *parent virtual edge parent*(ν) of $\sigma(\nu)$, and the *child virtual edge* of $\sigma(\mu)$. We define a *parent cut-pair* of ν as the two end points of *parent*(ν). We denote the graph formed from $\sigma(\nu)$ by deleting its parent virtual edge as $\sigma^-(\nu) = (V_\nu, E_\nu^-)$, $(E_\nu^- = E_\nu - \{parent(\nu)\})$. Let $G^-(\nu)$ denote the subgraph of G which consists of the vertices and real edges in the graphs $\sigma^-(\mu)$ for all descendants μ of ν, including ν itself.

Consider the case where G is a plane graph. For a face $f \in F$ of G, we say that a node ν in the SPQR tree is incident to f if $\sigma(\nu)$ contains the face corresponding to f (note that there may exist more than one such node ν). We choose a node incident to the outer face $f^o(G)$ as the root ν^* of the SPQR tree. In particular, we choose a P- or S-node incident to $f^o(G)$ (if any) as the root ν^*, and choose an R-node incident to $f^o(G)$ only when there is no such P- or S-node. Hence, we can assume that if the root ν^* is an R-node then no outer edge in $\sigma(\nu^*)$ is a virtual edge. When G is a plane graph, we also treat graphs $\sigma^-(\nu)$ and $G^-(\nu)$ as plane graphs induced from the embedding of G. For a non-root node ν with $(u, v) = parent(\nu)$, two plane drawings for $\sigma(\nu)$ can be obtained from the plane graph $\sigma^-(\nu)$ by drawing the parent edge $e = (u, v)$ outside $\sigma^-(\nu)$; one has $f^o_{uv}(\sigma^-(\nu))$ plus e as its boundary, and the other has $f^o_{vu}(\sigma^-(\nu))$ plus e as its boundary, where we denote the former and latter plane graphs by $\sigma_{uv}(\nu)$ and $\sigma_{vu}(\nu)$, respectively.

If a subgraph H of G is the graph $G^-(\nu)$ for a node ν in the SPQR tree and $(u, v) = parent(\nu)$, then we denote $V^o_{uv}(H)$ by $V^o_{uv}(\nu)$, $\Lambda^o_{uv}(H)$ by $\Lambda^o_{uv}(\nu)$, $V^o_{uv}(\nu) \cup V^o_{vu}(\nu)$ by $V^o(\nu)$, and $\Lambda^o_{uv}(\nu) \cup \Lambda^o_{vu}(\nu)$ by $\Lambda^o(\nu)$, respectively. For the root ν^*, let $V^o(\nu^*) = V(f^o(G))$ and $\Lambda^o(\nu^*) = \Lambda^o(f^o(G))$.

2.3   Straight-Line Drawings, Convex Drawings and Star-Shaped Drawings

For two points p_1, p_2 in the plane, $[p_1, p_2]$ denotes the line segment with end points p_1 and p_2, and for three points p_1, p_2, p_3, $[p_1, p_2, p_3]$ denotes the triangle with three corners p_1, p_2, p_3. A *kernel* $K(P)$ of a polygon P is the set of all points from which all points in P are visible. The boundary of a kernel, if any, is a convex polygon. A polygon P is called *star-shaped* if $K(P) \neq \emptyset$.

A *straight-line drawing* D of a graph $G = (V, E)$ in the plane is an embedding of G in the two dimensional space $\Re^2$, such that each vertex $v \in V$ is drawn as a point $\tau_D(v) \in \Re^2$, and each edge $(u, v) \in E$ is drawn as a straight-line segment $[\tau_D(u), \tau_D(v)]$, where $\Re$ is the set of reals. Let D be a straight-line planar drawing of a biconnected plane graph G. A corner of G is called *concave* in D if its angle in D is greater than π. A vertex v in a straight-line drawing D is called *concave* if one of the corners around v is concave in D. For a straight-line drawing D of a biconnected plane graph G, let $\Lambda^c(D)$ denote the set of all concave corners in D.

A *star-shaped drawing* of a plane graph is a straight-line drawing such that each inner facial cycle is drawn as a star-shaped polygon and the outer facial cycle is drawn as a convex polygon. An outer vertex in a straight-line drawing of a plane graph is called an *apex* if it is concave in the drawing and its concave corner appears in the outer face. Fig. 1(b) shows a star-shaped drawing of the plane graph G in Fig. 1(a), where (u_1, f^o), (u_4, f_{13}), (u_6, f_8), (u_7, f^o), (u_{18}, f_9), (u_{10}, f_{10}), (u_{11}, f^o), (u_{14}, f_1), (u_{15}, f_5), and (u_{17}, f_1) are the concave corners and u_1, u_7, and u_{11} are the apices.

A straight-line drawing D of a plane graph $G = (V, E, F)$ is called a *convex drawing*, if every facial cycle is drawn as a convex polygon, equivalently if $\Lambda^c(D) = \emptyset$. We say that a drawing D of a graph G is *extended* from a drawing D' of a subgraph G' of G, if $\tau_D(v) = \tau_{D'}(v)$ for all $v \in V(G')$.

3   Proper Drawings and Configurations

This section investigates the structural property of the necessary condition (1) in Theorem 1.

Lemma 1. [9] *Let G be a biconnected plane graph. Condition (1) is necessary for a subset $A \subseteq \Lambda(G)$ to admit a straight-line drawing D of G such that $\Lambda^c(D) \subseteq A$.*

□

To show that (1) is also a sufficient condition, we define *proper drawings*, a restricted class of straight-line drawings of a given plane graph G. Let D be a straight-line drawing of G, and ν be a P-node of the SPQR tree of G, where (u, v) denotes *parent*(ν). Then we say that an edge e in the skeleton $\sigma(\nu)^-$ *intersects* with P-node ν in D if the drawing of $G^-(\mu)$ for the child node $\mu \in Ch(\nu)$ corresponding to the virtual edge e (or the drawing of the real edge e) intersects with line-segment $[u, v]$ (if we additionally draw $[u, v]$). We call such an edge the *central edge*.

We call a straight-line drawing D *proper* if for each P-node ν, at most one edge in $\sigma(\nu)^-$ intersects with ν. We call the set ψ of all central edges in a proper straight-line drawing D the *configuration* of D. Such a drawing D is also called ψ-*proper*. Fig. 3(b) illustrates a drawing of $G^-(\nu)$ in which two edges in $\sigma(\nu)^-$ intersect with ν, while Fig. 3(c) illustrates a drawing of $G^-(\nu)$ in which only one edge in $\sigma(\nu)^-$ intersects with ν. We will show that a configuration determines the structure of concave vertices in a proper straight-line drawing.

In the next section, we will prove that if (1) holds, then there exists a proper star-shaped drawing D of G such that $\Lambda^c(D) \subseteq A$. In this section, we show how to efficiently test whether a given subset $A \subseteq \Lambda(G)$ satisfies (1). For this, we investigate the structural property of (1) in terms of the SPQR tree of G.

For convenience, we fix a total ordering of all vertices in a given graph G, and we denote by $u < v$ if u is smaller than v in the ordering. For the parent cut-pair $\{u, v\}$ of a P-node ν, where we assume $u < v$, we number the child edges of ν as $e_1, e_2, \ldots, e_{k(\nu)}$ by traversing these edges from left to right, where $k(\nu)$ denotes the number of children of ν.

We are ready to formally define "central edge" and "configuration." For each P-node ν of T, we choose an edge e_{j^*} $(1 \leq j^* \leq k(\nu))$ of $\sigma^-(\nu)$, which we call the *central edge* of ν and denote it by $c(\nu)$. If ν has a real edge among its children, then we always choose the real edge as $c(\nu)$. Other virtual edge e_i in $\sigma^-(\nu)$ is called a *left edge* (resp., a *right edge*) if $i < j^*$ (resp., $i > j^*$). For each child $\mu_i \in Ch(\nu)$ corresponding to edge e_i, we call subgraph $G^-(\mu_i)$ with $i < j^*$ (resp., $i > j^*$) a *left component* (resp., a *right component*) of ν. If e_{j^*} is a real edge in $\sigma^-(\nu)$, then the edge is called the *central component* of ν; otherwise subgraph $G^-(\mu_{j^*})$ is called the *central component* of ν.

A *configuration* of a node ν is defined by a set ψ of central edges $c(\nu')$ for all descendants ν' of ν, and let $\Psi(\nu)$ denote the set of all configurations ψ of ν. Thus, for the root ν^*, a configuration $\psi \in \Psi(\nu^*)$ defines a central edge for each

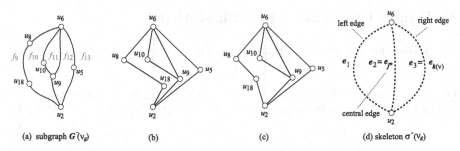

(a) subgraph $G^-(\nu_g)$ (b)   (c)   (d) skeleton $\sigma^-(\nu_g)$

Fig. 3. (a) Subgraph $G^-(\nu_g)$ of P-node ν_g in the SPQR tree of G (b) A straight-line drawing of $G^-(\nu_l)$ in which two edges $e_1, e_2 \in E_{\nu_l}$ intersect ν_l (c) A straight-line drawing of $G^-(\nu_l)$ in which only edge $e_2 \in E_{\nu_l}$ intersects ν_l (d) Skeleton $\sigma^-(\nu_g)$ of P-node ν_g

P-node in the SPQR tree of G. Now we characterize a subset $A \subseteq \Lambda^o(G)$ that satisfies (1) using the SPQR tree.

Definition 1. *For a biconnected plane graph G, and a configuration $\psi \in \Psi(\nu^*)$, let A be a subset of $\Lambda^o(G)$, and ν be a node in the SPQR-tree.*

(i) *Let ν be the root ν^*. Subset A is called* proper *with respect to ν^* if $|A \cap \Lambda^o(\nu^*)| \geq 3$;*

(ii) *Let ν be an R-node which is a child of an R-node or S-node, and $(u, v) = parent(\nu)$. Subset A is called* proper *with respect to ν if $A \cap \Lambda^o(\nu) \neq \emptyset$;*

(iii) *Let ν be a non-root P-node, $(u, v) = parent(\nu)$, and $E_\nu^- = \{e_1, e_2, \ldots, e_{k(\nu)}\}$, where $e_1, e_2, \ldots, e_{k(\nu)}$ appear from left to right and $e_{j^*} = c(\nu) \in \psi$. Let μ_j denote the child of ν corresponding to a virtual edge e_j. Subset A is called* proper *with respect to ν if it satisfies the following:*

$$A \cap \Lambda_{vu}^o(\mu_j) \neq \emptyset \ (1 \leq j \leq j^* - 1), \quad A \cap \Lambda_{uv}^o(\mu_j) \neq \emptyset \ (j^* + 1 \leq j \leq k(\nu)),$$

and if μ_{j^*} is not an induced path, then $A \cap \Lambda^o(\mu_{j^*}) \neq \emptyset$.

A subset $A \subseteq \Lambda^o(G)$ is called *proper* with respect to (ν^*, ψ) if A is proper with respect to all nodes ν in (i)-(ii) of Definition 1, and all P-nodes ν in (iii) of Definition 1.

Lemma 2. *Let G be a biconnected plane graph, ν^* be the root of the SPQR-tree, and A be a subset of $\Lambda^o(G)$. Then A satisfies (1) if and only if there is a configuration $\psi \in \Psi(G)$ such that A is proper with respect to (ν^*, ψ).* □

Lemma 3. *Let G be a biconnected plane graph, and A be a subset of $\Lambda^o(G)$. Then testing whether A satisfies (1) or not can be done in linear time.* □

We define *fringe corners* as those corners which are used as concave corners of any ψ-proper straight-line drawing. For a given configuration $\psi \in \Psi(\nu^*)$, Lemma 2 tells how to characterize "fringe corners." Given a node ν with a configuration $\psi \in \Psi(\nu^*)$, we define fringe corners in $G^-(\nu)$ as follows. Let μ be an arbitrary descendent of ν (including $\mu = \nu$).

Definition 2

(1) μ is an R-node that is not a child of a P-node: Let $C = f^o(G^-(\mu))$ of graph $G^-(\mu)$ if μ is not the root (where C is a cut-cycle of G); let $C = f^o(G)$ if μ is the root. Then any outer corner of the cut-cycle C is defined as a fringe corner of ν.

(2) μ is a P-node: Then a corner λ in graph $G^-(\mu)$ is called a fringe corner of ν if it satisfies one of the following: $\lambda \in \Lambda^o(H)$ for the central component H of μ; $\lambda \in \Lambda^o_{uv}(H)$ for a right component H of μ; and $\lambda \in \Lambda^o_{vu}(H)$ for a left component H of μ.

For a node ν and a configuration $\psi \in \Psi(\nu)$, we denote by $\Lambda^o(\nu, \psi)$ the set of all fringe corners of ν. Note that $\Lambda^o(\nu, \psi) \subseteq \Lambda^o(G)$ holds.

Lemma 4. *Let G be a biconnected plane graph, ν^* be the root of the SPQR-tree, and D be a straight-line drawing of G. Then there is a configuration $\psi \in \Psi(G)$ such that $A = \Lambda^c(D) \cap \Lambda^o(\nu^*, \psi)$ is proper with respect to (ν^*, ψ).* $\square$

For the proofs of Lemmas 2, 3 and 4, see [9].

Lemmas 1 and 3 imply the necessity and the time complexity in Theorem 1(i).

We now prove Theorem 1(ii), which also implies the sufficiency of Theorem 1(i).

4 Constructing Star-Shaped Drawings

Conversely, we can prove that, given a proper subset A, there exists a star-shaped drawing D such that $\Lambda^c(D) = A$.

Theorem 2. *Let G be a biconnected plane graph, ν^* be the root of the SPQR-tree of G, and $\psi \in \Psi(\nu^*)$ be a configuration of G. For any subset $A \subseteq \Lambda^o(\nu^*, \psi)$ proper with respect to (ν^*, ψ), a convex polygon P with $\Lambda^c(P) = A \cap \Lambda^o(f^o)$ drawn for $f^o(G)$ can be extended to a ψ-proper star-shaped drawing D such that $\Lambda^c(D) = A$. Such a drawing D can be constructed in linear time.* $\square$

Observe that Theorem 1(ii) follows from Theorem 2.

To prove Theorem 2, we design a divide-and-conquer algorithm that computes a star-shaped drawing for a subset $A \subseteq \Lambda^o(\nu^*, \psi)$ in Theorem 2 in a top-down manner along the SPQR tree $\mathcal{T}$. This is a modification of the algorithm for constructing a star-shaped drawing of a plane graph with a set of "concave vertices" [8]. A full description of the algorithm and the complexity analysis are given in the full version [9]. We here explain a brief sketch of the algorithm as follows. For the root ν^*, we first fix the boundary B_{ν^*} of G as a prescribed $|A \cap \Lambda^o(\nu^*)|$-gon P with $\Lambda^c(P) = A \cap \Lambda^o(\nu^*)$. We then start to process all nodes ν in $\mathcal{T}$ from the root to the leaves by repeatedly computing a drawing D_ν of skeleton $\sigma^-(\nu)$ (or $\sigma(\nu)$), where we first fix the boundary B_ν of $G^-(\nu)$ (or $G(\nu)$) and then extend B_ν to a convex drawing D_ν of the skeleton $\sigma^-(\nu)$ (or $\sigma(\nu)$). The line segments in D_ν for virtual edges will be replaced with new drawings D_μ of the nodes μ corresponding to the virtual edges, where a virtual edge in the boundary of the skeleton may be a sequence of line segments which forms part of the convex boundary B_ν.

References

1. Chiba, N., Yamanouchi, T., Nishizeki, T.: Linear Algorithms for Convex Drawings of Planar Graphs. In: Progress in Graph Theory, pp. 153–173. Academic Press, London (1984)
2. Di Battista, G., Eades, P., Tamassia, R., Tollis, I.G.: Graph Drawing: Algorithms for the Visualization of Graphs. Prentice-Hall, Englewood Cliffs (1998)
3. Di Battista, G., Tamassia, R.: On-line Planarity Testing. SIAM J. on Comput. 25(5), 956–997 (1996)
4. Fáry, I.: On Straight Line Representations of Planar Graphs. Acta Sci. Math. Szeged 11, 229–233 (1948)
5. Hong, S.-H., Nagamochi, H.: Convex Drawings of Graphs with Non-convex Boundary. In: Fomin, F.V. (ed.) WG 2006. LNCS, vol. 4271, pp. 113–124. Springer, Heidelberg (2006)
6. Hong, S.-H., Nagamochi, H.: Convex Drawings of Hierarchical Plane Graphs. In: Proc. of AWOCA 2006 (2006)
7. Hong, S.-H., Nagamochi, H.: Fully Convex Drawings of Clustered Planar Graphs. In: Proc. of WAAC 2007, pp. 32–39 (2007)
8. Hong, S.-H., Nagamochi, H.: Star-Shaped Drawings of Planar Graphs. In: Proc. of IWOCA (to appear, 2007)
9. Hong, S.-H., Nagamochi, H.: Star-Shaped Drawings of Graphs with Fixed Embedding and Concave Corner Constraints. Technical Report TR 2008-002, Department of Applied Mathematics and Physics, Kyoto University (2008)
10. Kant, G.: Algorithms for Drawing Planar Graphs. Ph.D. Dissertation, Department of Computer Science, University of Utrecht, Holland (1993)
11. Miura, K., Azuma, M., Nishizeki, T.: Convex Drawings of Plane Graphs of Minimum Outer Apices. Int. J. Found. Comput. Sci. 17, 1115–1128 (2006)
12. Thomassen, C.: Plane Representations of Graphs. In: Progress in Graph Theory, pp. 43–69. Academic Press, London (1984)
13. Tutte, W.T.: Convex Representations of Graphs. Proc. of London Math. Soc. 10(3), 304–320 (1960)

A New Characterization of P_6-Free Graphs*

Pim van 't Hof and Daniël Paulusma

Department of Computer Science, Durham University,
Science Laboratories, South Road, Durham DH1 3LE, England
{pim.vanthof,daniel.paulusma}@durham.ac.uk

Abstract. We study P_6-free graphs, i.e., graphs that do not contain an induced path on six vertices. Our main result is a new characterization of this graph class: a graph G is P_6-free if and only if each connected induced subgraph of G on more than one vertex contains a dominating induced cycle on six vertices or a dominating (not necessarily induced) complete bipartite subgraph. This characterization is minimal in the sense that there exists an infinite family of P_6-free graphs for which a smallest connected dominating subgraph is a (not induced) complete bipartite graph. Our characterization of P_6-free graphs strengthens results of Liu and Zhou, and of Liu, Peng and Zhao. Our proof has the extra advantage of being constructive: we present an algorithm that finds such a dominating subgraph of a connected P_6-free graph in polynomial time. This enables us to solve the HYPERGRAPH 2-COLORABILITY problem in polynomial time for the class of hypergraphs with P_6-free incidence graphs.

1 Introduction

All graphs in this paper are undirected, finite, and simple, i.e., without loops and multiple edges. Furthermore, unless specifically stated otherwise, all graphs are non-trivial, i.e., contain at least two vertices. For undefined terminology we refer to [4]. Let $G = (V, E)$ be a graph. For a subset $U \subseteq V$ we denote by $G[U]$ the subgraph of G induced by U. A subset $S \subseteq V$ is called a *clique* if $G[S]$ is a complete graph. A set $U \subseteq V$ *dominates* a set $U' \subseteq V$ if any vertex $v \in U'$ either lies in U or has a neighbor in U. We also say that U *dominates* $G[U']$. A subgraph H of G is a *dominating subgraph* of G if $V(H)$ dominates G. We write P_k, C_k, K_k to denote the path, cycle and complete graph on k vertices, respectively.

A graph G is called *H-free* for some graph H if G does not contain an induced subgraph isomorphic to H. For any family $\mathcal{F}$ of graphs, let $Forb(\mathcal{F})$ denote the class of graphs that are F-free for every $F \in \mathcal{F}$. We consider the class $Forb(\{P_t\})$ of graphs that do not contain an induced path on t vertices. Note that $Forb(\{P_2\})$ is the class of graphs without any edge and $Forb(\{P_3\})$ is the class of graphs all components of which are complete graphs.

The class of P_4-free graphs (or cographs) has been studied extensively (cf. [3]). The following characterization of $Forb(\{P_4, C_4\})$, i.e., the class of C_4-free cographs, is due to Wolk [12],[13] (see also Theorem 11.3.4 in [3]).

---

* This work has been supported by EPSRC (EP/D053633/1).

X. Hu and J. Wang (Eds.): COCOON 2008, LNCS 5092, pp. 415–424, 2008.
© Springer-Verlag Berlin Heidelberg 2008

Theorem 1 ([12],[13]). *A graph G is P_4-free and C_4-free if and only if each connected induced subgraph of G contains a dominating vertex.*

We can slightly modify this theorem to obtain a characterization of P_4-free graphs.

Theorem 2. *A graph G is P_4-free if and only if each connected induced subgraph of G contains a dominating induced C_4 or a dominating vertex.*

Since this theorem can be proven using similar (but much easier) arguments as in the proof our main result, its proof is omitted here.

The following characterization of P_5-free graphs is due to Liu and Zhou [7].

Theorem 3 ([7]). *A graph G is P_5-free if and only if each connected induced subgraph of G contains a dominating induced C_5 or a dominating clique.*

A graph G is called *triangle extended complete bipartite (TECB)* if it is a complete bipartite graph or if it can be obtained from a complete bipartite graph F by adding some extra vertices $w_1, \ldots, w_r$ and edges $w_i u, w_i v$ for $1 \le i \le r$ to exactly one edge uv of F (see Figure 1 for an example).

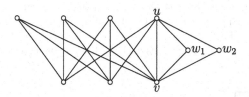

Fig. 1. An example of a TECB graph

The following characterization of P_6-free graphs is due to Liu, Peng and Zhao [8].

Theorem 4 ([8]). *A graph G is P_6-free if and only if each connected induced subgraph of G contains a dominating induced C_6 or a dominating (not necessarily induced) TECB graph.*

If we consider graphs that are not only P_6-free but also triangle-free, then we have one of the main results in [7].

Theorem 5 ([7]). *A triangle-free graph G is P_6-free if and only if each connected induced subgraph of G contains a dominating induced C_6 or a dominating (not necessarily induced) complete bipartite graph.*

A characterization of $Forb(\{P_t\})$ for $t \ge 7$ is given in [1]: $Forb(\{P_t\})$ is the class of graphs for which each connected induced subgraph has a dominating subgraph of diameter at most $t - 4$.

Our Results
Section 3 contains our main result.

Theorem 6. *A graph G is P_6-free if and only if each connected induced subgraph of G contains a dominating induced C_6 or a dominating (not necessarily induced) complete bipartite graph. Moreover, we can find such a dominating subgraph in polynomial time.*

This theorem strengthens Theorem 4 and Theorem 5 in two different ways. Firstly, Theorem 6 shows that we may omit the restriction "triangle-free" in Theorem 5 and that we may replace the class of TECB graphs by its proper subclass of complete bipartite graphs in in Theorem 4. Secondly, in contrast to the proofs of Theorem 4 and Theorem 5, the proof of Theorem 6 is constructive: we provide a (polynomial time) algorithm for finding the desired dominating subgraph. Note that we cannot use some brute force approach to obtain such a polynomial time algorithm, since a dominating complete bipartite graph might have arbitrarily large size.

In Section 3, we also show that the characterization in Theorem 6 is minimal in the sense that there exists an infinite family of P_6-free graphs for which a smallest connected dominating subgraph is a (not induced) complete bipartite graph. We mention that the algorithm used to prove Theorem 6 also works for an arbitrary (not necessarily P_6-free) graph G: in that case the algorithm either finds a dominating subgraph as described in Theorem 6 or finds an induced P_6 in G. We end Section 3 by characterizing the class of graphs for which each connected induced subgraph has a dominating induced C_6 or a dominating *induced* complete bipartite subgraph (again by giving a constructive proof). This class consists of graphs that, apart from P_6, have exactly one more forbidden induced subgraph. This generalizes a result in [2].

As an application of our main result, we consider the HYPERGRAPH 2-COLORABILITY problem in Section 4. This problem is NP-complete in general (cf. [5]). We prove that for the class of hypergraphs with P_6-free incidence graphs the problem becomes polynomially solvable. Moreover, we show that for any 2-colorable hypergraph H with a P_6-free incidence graph, we can find a 2-coloring of H in polynomial time. Section 5 contains the conclusions, discusses a number of related results in the literature and mentions open problems.

2   Preliminaries

We use the following terminology throughout the paper for a graph $G = (V, E)$. We say that an order $\pi = x_1, \ldots, x_{|V|}$ of V is *connected* if $G_i := G[\{x_1, \ldots, x_i\}]$ is connected for $i = 1, \ldots, |V|$. Let $w \in V$ and $D \subseteq V$. Then $N_G(w)$ denotes the set of neighbors of w in G. We write $N_D(w) := N_G(w) \cap D$ and $N_G(D) := \cup_{u \in D} N_G(u) \backslash D$. If no confusion is possible, we write $N(w)$ (respectively $N(D)$) instead of $N_G(w)$ (respectively $N_G(D)$). A vertex $v' \in V \backslash D$ is called a *D-private neighbor* (or simply private neighbor if no confusion is possible) of a vertex $v \in D$ if $N_D(v') = \{v\}$.

Let u, v be a pair of adjacent vertices in a dominating set D of a graph G such that $\{u, v\}$ dominates D. We call a dominating set $D' \subseteq D$ of G a *minimizer of D for uv* if $\{u, v\} \subseteq D'$ and each vertex of $D' \backslash \{u, v\}$ has a D'-private neighbor in G. We can obtain such a dominating set D' from D in polynomial time by repeatedly removing vertices without private neighbor from $D\backslash\{u,v\}$. This can be seen as follows. It is clear that D' dominates all vertices in $V(G)\backslash D$, since we only remove a vertex from $D\backslash\{u,v\}$ if all its neighbors outside D are dominated by remaining vertices in D. Moreover, since $\{u,v\}$ dominates D, all vertices removed from $D\backslash\{u,v\}$ are dominated by $\{u,v\}$. Note that the fact that u and v are adjacent means that the graph $G[D']$ is connected. We point out that D may have several minimizers for the same edge uv depending on the order in which its vertices are considered.

Example. Consider the graph G and its connected dominating set D in the left-hand side of Figure 2. All private neighbors are colored black. The set D' in the right-hand side is a minimizer of D for uv obtained by removing w_4 from D. Note that u does not have a D'-private neighbor but v does. Instead of removing w_4 we could also have chosen to remove w_2 first, since w_2 does not have a D-private neighbor. Let $D^1 := D\backslash\{w_2\}$. Since w_3 does not have a D^1-private neighbor, we can remove w_3 from D^1. The resulting set $D^2 := D^1\backslash\{w_3\}$ is a minimizer of D for uv in which every vertex of D^2 (including u) has a D^2-private neighbor.

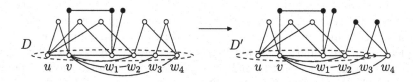

Fig. 2. A dominating set D and a minimizer D' of D for uv

3   Finding Connected Dominating Subgraphs in P_6-Free Graphs

Let G be a connected P_6-free graph. We say that D is a *type 1 dominating set* of G if D dominates G and $G[D]$ is an induced C_6. We say that D is a *type 2 dominating set* of G defined by $A(D)$ and $B(D)$ if D dominates G and $G[D]$ contains a spanning complete bipartite subgraph with partition classes $A(D)$ and $B(D)$.

Theorem 7. *If G is a connected P_6-free graph, then we can find a type 1 or type 2 dominating set of G in polynomial time.*

Proof. Let $G = (V, E)$ be a connected P_6-free graph with connected order $\pi = x_1, \ldots, x_{|V|}$. Recall that we write $G_i := G[\{x_1, \ldots, x_i\}]$, and note that G_i is connected and P_6-free for every i. For every $2 \leq i \leq n$ we want to find a type 1 or type 2 dominating set D_i of G_i. Let $D_2 := \{x_1, x_2\}$. Suppose $i \geq 3$. Assume D_{i-1} is a type 1 or type 2 dominating set of G_{i-1}. We show how we can use D_{i-1} to find D_i in polynomial time. Since the total number of iterations is $|V|$, we then find a desired dominating subgraph of $G_{|V|} = G$ in polynomial time.

We write $x := x_i$. If $x \in N(D_{i-1})$, then we set $D_i := D_{i-1}$. Suppose otherwise. Since π is connected, G_i contains a vertex y (not in D_{i-1}) adjacent to x.

Case 1. D_{i-1} is a type 1 dominating set of G_{i-1}.

We write $G[D_{i-1}] = c_1 c_2 c_3 c_4 c_5 c_6 c_1$. We claim that $D := N_{D_{i-1}}(y) \cup \{x, y\}$ dominates G_i, which means that we can choose $D_i := D$ as type 2 dominating set of G_i defined by $A(D_i) := \{y\}$ and $B(D_i) := \{x\} \cup N_{D_{i-1}}(y)$. Suppose otherwise. Then there exists a vertex $z \in V(G_i)$ not dominated by D. Since D_{i-1} dominates G_{i-1}, we may without loss of generality assume $yc_1 \in E(G_i)$.

Suppose $yc_4 \in E(G_i)$. Note that z is dominated by G_{i-1}. Without loss of generality, assume z is adjacent to c_2. Consequently, y is not adjacent to c_2. Since z is not adjacent to any neighbor of y and the path $zc_2c_1yc_4c_5$ cannot be induced in G_i, either z or y must be adjacent to c_5. If $zc_5 \in E(G_i)$, then $xyc_4c_5zc_2$ is an induced P_6 in G_i. Hence $zc_5 \notin E(G_i)$ and $yc_5 \in E(G_i)$. In case $zc_6 \in E(G_i)$ we obtain an induced path $xyc_5c_6zc_2$ on six vertices, and in case $zc_6 \notin E(G_i)$ we obtain an induced path $zc_2c_1c_6c_5c_4$. We conclude $yc_4 \notin E(G_i)$.

Suppose y is not adjacent to any vertex in $\{c_3, c_5\}$. Since G_i is P_6-free and $xyc_1c_2c_3c_4$ is a P_6 in G_i, y must be adjacent to c_2. But then $xyc_2c_3c_4c_5$ is an induced P_6 in G_i, a contradiction. Hence y is adjacent to at least one vertex in $\{c_3, c_5\}$, say $yc_5 \in E(G_i)$. By symmetry (using c_5, c_2 instead of c_1, c_4) we find $yc_2 \notin E(G_i)$.

Suppose z is adjacent to c_2. The path $zc_2c_1yc_5c_4$ on six vertices and the P_6-freeness of G_i imply $zc_4 \in E(G_i)$. But then $c_2zc_4c_5yx$ is an induced P_6. Hence $zc_2 \notin E(G_i)$. Also $zc_4 \notin E(G_i)$ as otherwise $zc_4c_5yc_1c_2$ would be an induced P_6, and $zc_3 \notin E(G_i)$ as otherwise $zc_3c_2c_1yx$ would be an induced P_6. Then z must be adjacent to c_6 yielding an induced path $zc_6c_1c_2c_3c_4$ on six vertices. Hence we may choose $D_i := D$.

Case 2. D_{i-1} is a type 2 dominating set of G_{i-1}.

Since D_{i-1} dominates G_{i-1}, we may assume that y is adjacent to some vertex $a \in A(D_{i-1})$. Let $b \in B(D_{i-1})$. Let D be a minimizer of $D_{i-1} \cup \{y\}$ for ab (note that $\{a, b\}$ dominates $D_{i-1} \cup \{y\}$). By definition, $G[D]$ dominates G_i. Also, $G[D]$ contains a spanning (not necessarily complete) bipartite graph with partition classes $A \subseteq A(D_{i-1})$, $B \subseteq B(D_{i-1}) \cup \{y\}$. Note that we have $y \in D$, because x is not adjacent to D_{i-1} and therefore is a D-private neighbor of y. Since y might not have any neighbors in B but does have a neighbor (vertex a) in A, we chose $y \in B$.

Claim 1. We may assume that $G[D]$ does not contain an induced P_4 starting in y and ending in some $r \in A$, as otherwise we are done (in polynomial time).

We prove Claim 1 as follows. Suppose $ypqr$ is an induced path in $G[D]$ with $r \in A$. Then $r \neq a$, so r has a D-private neighbor s. Since $xypqrs$ is a path on six vertices and $x \notin N(D_{i-1})$ holds, x must be adjacent to s. We first show that $D^1 := N_D(y) \cup \{x, y, q, r, s\}$ dominates G_i. See Figure 3 for an illustration of the graph $G[D^1]$.

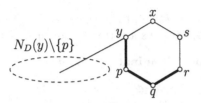

Fig. 3. The graph $G[D^1]$

Suppose D^1 does not dominate G. Then there exists a vertex $z \in N(D)\backslash N(D^1)$. Note that $G[(D\backslash\{y\}) \cup \{z\}]$ is connected because the edge ab makes $D\backslash\{y\}$ connected and $\{a, b\}$ dominates D. Let P be a shortest path in $G[(D\backslash\{y\})\cup\{z\}]$ from z to a vertex $p_1 \in N_D(y)$ (possibly $p_1 = p$). Since $z \notin N(D^1)$ and $p_1 \in D^1$, we have $|V(P)| \geq 3$. This means that $Pyxs$ is an induced path on at least six vertices, unless $r \in V(P)$ (since r is adjacent to s). However, if $r \in V(P)$ then the subpath $z\overrightarrow{P}r$ of P from z to r has at least three vertices (because $z \notin N(D^1)$). This means that $z\overrightarrow{P}rsxy$ contains an induced P_6. Hence D^1 dominates G_i.

To find a type 1 or type 2 dominating set D_i of G_i, we transform D^1 into D_i as follows. Suppose q has a D^1-private neighbor q'. Then $q'qpyxs$ is an induced P_6 in G_i, a contradiction. Hence q has no D^1-private neighbors and the set $D^2 := D^1\backslash\{q\}$ still dominates G_i. Similarly, r has no D^2-private neighbor r', since otherwise $r'rsxyp$ would be an induced P_6 in G_i. So the set $D^3 := D^2\backslash\{r\}$ also dominates G_i. Now suppose s does not have a D^3-private neighbor. Then the set $D^3\backslash\{s\}$ dominates G_i. In that case, we define a type 2 dominating set D_i of G_i by $A(D_i) := \{y\}$ and $B(D_i) := N_D(y) \cup \{x\}$. Assume that s has a D^3-private neighbor s' in G_i. Let $D^4 := D^3 \cup \{s'\}$.

Suppose $N_D(y)\backslash\{p\}$ contains a vertex p_2 that has a D^4-private neighbor p_2'. Then $p_2'p_2yxss'$ is an induced P_6, contradicting the P_6-freeness of G_i. Hence we can remove all vertices of $N_D(y)\backslash\{p\}$ from D^4, and the resulting set $D^5 := \{p, y, x, s, s'\}$ still dominates G_i. We claim that $D^6 := D^5 \cup \{q\}$ is a type 1 dominating set of G_i. Clearly, D^6 dominates G_i, since $D^5 \subseteq D^6$. Since $qpyxss'$ is a P_6 and $qpyxs$ is induced, q must be adjacent to s'. Hence D^6 is a type 1 dominating set of G_i, and we choose $D_i := D^6$. This proves Claim 1.

Let $A_1 := N_A(y)$ and $A_2 := A\backslash A_1$. Let $B_1 := N_B(y)$ and $B_2 := B\backslash(B_1 \cup \{y\})$. Since $a \in A_1$, we have $A_1 \neq \emptyset$. If $A_2 = \emptyset$, then we define a type 2 dominating set D_i of G_i by $A(D_i) := A$ and $B(D_i) := B$. Suppose $A_2 \neq \emptyset$. Note $|B| \geq 2$, because $\{b, y\} \subseteq B$. If $B_2 = \emptyset$, then we define D_i by $A(D_i) := A \cup \{y\}$ and $B(D_i) := B_1 = B\backslash\{y\}$. Suppose $B_2 \neq \emptyset$. If $G[A_1 \cup A_2]$ contains a spanning complete bipartite graph with partition classes A_1 and A_2, we define D_i by $A(D_i) := A_1$ and $B(D_i) := A_2 \cup B$. Hence we may assume that there exist two non-adjacent vertices $a_1 \in A_1$ and $a_2 \in A_2$. Let $b^* \in B_2$. Then $ya_1b^*a_2$ is an induced P_4 that contradicts Claim 1. This finishes the proof of Theorem 7. $\square$

We will now prove our main theorem.

Theorem 6. *A graph G is P_6-free if and only if each connected induced subgraph of G contains a dominating induced C_6 or a dominating (not necessarily induced)* *complete bipartite graph. Moreover, we can find such a dominating subgraph in polynomial time.*

Proof. Let G be a graph. Suppose G is not P_6-free. Then G contains an induced P_6 which contains neither a dominating induced C_6 nor a dominating complete bipartite graph. Suppose G is P_6-free. Let H be a connected induced subgraph of G. Then H is P_6-free as well. We apply Theorem 7 to H. □

The characterization in Theorem 6 is minimal due to the existence of the following family $\mathcal{F}$ of P_6-free graphs. For each $i \geq 2$, let $F_i \in \mathcal{F}$ be the graph obtained from a complete bipartite subgraph with partition classes $X_i = \{x_1, \ldots, x_i\}$ and $Y_i = \{y_1, \ldots, y_i\}$ by adding the edge $x_1 x_2$ as well as for each $h = 1, \ldots, i$ a new vertex x_h' only adjacent to x_h and a new vertex y_h' only adjacent to y_h (see Figure 4 for the graph F_3).

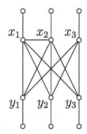

Fig. 4. The graph F_3

Note that each F_i is P_6-free and that the smallest connected dominating subgraph of F_i is $F_i[X_i \cup Y_i]$, which contains a spanning complete bipartite subgraph. Also note that none of the graphs F_i contain a dominating *induced* complete bipartite subgraph due to the edge $x_1 x_2$.

We conclude this section by characterizing the class of graphs for which each connected induced subgraph contains a dominating induced C_6 or a dominating *induced* complete bipartite subgraph. Again, these dominating induced subgraphs can be found in polynomial time. Let C_3^L denote the graph obtained from the cycle $c_1 c_2 c_3 c_1$ by adding three new vertices b_1, b_2, b_3 and three new edges $c_1 b_1, c_2 b_2, c_3 b_3$. Then we have the following result, the proof of which will be included in the journal version of this paper but is omitted here due to page restrictions.

Theorem 8. *A graph G is in $Forb(\{C_3^L, P_6\})$ if and only if each connected induced subgraph of G contains a dominating induced C_6 or a dominating induced complete bipartite graph. Moreover, we can find such a dominating subgraph in polynomial time.*

Bacsó, Michalak and Tuza [2] prove (non-constructively) that a graph G is in $Forb(\{C_3^L, C_6, P_6\})$ if and only if each connected induced subgraph of G contains a dominating induced complete bipartite graph. Note that Theorem 8 immediately implies this result.

4   The Hypergraph 2-Colorability Problem

A *hypergraph* is a pair (Q, S) consisting of a set $Q = \{q_1, \ldots, q_m\}$ and a set $S = \{S_1, \ldots, S_n\}$ of subsets of Q. With a hypergraph (Q, S) we associate its *incidence graph* I, which is a bipartite graph with partition classes Q and S, where for any $q \in Q, S \in S$ we have $qS \in E(I)$ if and only if $q \in S$. For any $S \in S$, we write $H - S := (Q, S \backslash S)$. A *2-coloring* of a hypergraph (Q, S) is a partition (Q_1, Q_2) of Q such that $Q_1 \cap S_j \neq \emptyset$ and $Q_2 \cap S_j \neq \emptyset$ for $1 \leq j \leq n$.

The HYPERGRAPH 2-COLORABILITY problem asks whether a given hypergraph has a 2-coloring. This is a well-known NP-complete problem (cf. [5]). Let $\mathcal{H}_6$ denote the class of hypergraphs with P_6-free incidence graphs.

Theorem 9. *The* HYPERGRAPH 2-COLORABILITY *problem restricted to* $\mathcal{H}_6$ *is polynomially solvable. Moreover, for any 2-colorable hypergraph* $H \in \mathcal{H}_6$ *we can find a 2-coloring of* H *in polynomial time.*

Proof. Let $H = (Q, S) \in \mathcal{H}_6$, and let I be the (P_6-free) incidence graph of H. We assume that I is connected, as otherwise we just proceed component-wise.

Claim 1. We may without loss of generality assume that S *does not contain two sets* S_i, S_j *with* $S_i \subseteq S_j$.

To prove Claim 1, suppose $S_i, S_j \in S$ with $S_i \subseteq S_j$. We can check in polynomial time whether such sets S_i, S_j exist. We show that H is 2-colorable if and only if $H - S_j$ is 2-colorable. Clearly, if H is 2-colorable then $H - S_j$ is 2-colorable. Suppose $H - S_j$ is 2-colorable. Let (Q_1, Q_2) be a 2-coloring of $H - S_j$. By definition, $S_i \cap Q_1 \neq \emptyset$ and $S_i \cap Q_2 \neq \emptyset$. Since $S_i \subseteq S_j$, we also have $S_j \cap Q_1 \neq \emptyset$ and $S_j \cap Q_2 \neq \emptyset$, so (Q_1, Q_2) is a 2-coloring of H. This proves Claim 1.

By Theorem 6, we can find a type 1 or type 2 dominating set D of I in polynomial time. Since I is bipartite, $G[D]$ is bipartite. Let A and B be the partition classes of $G[D]$. Since I is connected, we may without loss of generality assume $A \subseteq Q$ and $B \subseteq S$. Let $A' := Q \backslash A$ and $B' := S \backslash B$. We distinguish two cases.

Case 1. D is a type 1 dominating set of I.

We write $I[D] = q_1 S_1 q_2 S_2 q_3 S_3 q_1$, so $A = \{q_1, q_2, q_3\}$ and $B = \{S_1, S_2, S_3\}$. Suppose $A' = \emptyset$, so $Q = \{q_1, q_2, q_3\}$. Obviously, H has no 2-coloring. Suppose $A' \neq \emptyset$ and let $q' \in A'$. Since D dominates I, q' has a neighbor, say S_1, in B. If S_2 and S_3 both have no neighbors in A', then $q' S_1 q_2 S_2 q_3 S_3$ is an induced P_6 in I, a contradiction. Hence at least one of them, say S_2, has a neighbor in A'.

We claim that the partition (Q_1, Q_2) of Q with $Q_1 := A' \cup \{q_1\}$ and $Q_2 := \{q_2, q_3\}$ is a 2-coloring of H. We have to check that every $S \in S$ has a neighbor in both Q_1 and Q_2. Recall that S_1 has neighbors q_1 and q_2 and S_3 has neighbors q_1 and q_3, so S_1 and S_3 are OK. Since S_2 is adjacent to q_2 and has a neighbor in A', S_2 is also OK. It remains to check the vertices in B'. Let $S \in B'$. Since D dominates I and I is bipartite, S has at least one neighbor in A. Suppose S has exactly one neighbor, say q_1, in A. Then $S q_1 S_1 q_2 S_2 q_3$ is an induced P_6 in I, a contradiction. Hence S has at least two neighbors in A. The only problem occurs if S is adjacent to q_2 and q_3 but not to q_1. However, since S_2 is adjacent to q_2 and q_3, S must have a neighbor in A' due to Claim 1. Hence (Q_1, Q_2) is a 2-coloring of H.

Case 2. D is a type 2 dominating set of I.

Suppose $A' = \emptyset$. Then $|B| = 1$ as a result of Claim 1. Let $B = \{S\}$ and $q \in A$. Since S is adjacent to all vertices in A, we find $B' = \emptyset$ by Claim 1. Hence H has no 2-coloring if $|A| = 1$, and H has a 2-coloring $(\{q\}, A\backslash\{q\})$ if $|A| \geq 2$. Suppose $A' \neq \emptyset$. We claim that (A, A') is a 2-coloring of H. This can be seen as follows. By definition, each vertex in S is adjacent to a vertex in A. Suppose $|B| = 1$ and let $B = \{S\}$. Since S dominates Q and $A' \neq \emptyset$, S has at least one neighbor in A'. Suppose $|B| \geq 2$. Since every vertex in B is adjacent to all vertices in A, every vertex in S must have a neighbor in A' as a result of Claim 1. □

5 Conclusions

The key contributions of this paper are the following. We presented a new characterization of the class of P_6-free graphs, which strengthens results of Liu and Zhou [7] and Liu, Peng and Zhao [8]. We used an algorithmic technique to prove this characterization. Our main algorithm efficiently finds for any given connected P_6-free graph a dominating subgraph that is either an induced C_6 or a (not necessarily induced) complete bipartite graph. Besides these main results, we also showed that our characterization is "minimal" in the sense that there exists an infinite family of P_6-free graphs for which a smallest connected dominating subgraph is a (not induced) complete bipartite graph. We also characterized the class $Forb(\{C_3^L, P_6\})$ in terms of connected dominating subgraphs, thereby generalizing a result of Bacsó, Michalak and Tuza [2].

Our main algorithm can be useful to determine the computational complexity of decision problems restricted to the class of P_6-free graphs. To illustrate this, we applied this algorithm to prove that the HYPERGRAPH 2-COLORABILITY problem is polynomially solvable for the class of hypergraphs with P_6-free incidence graphs. Are there any other decision problems for which the algorithm is useful? In recent years, several authors studied the classical k-COLORABILITY problem for the class of P_ℓ-free graphs for various combinations of k and ℓ [6],[9],[11]. The 3-COLORABILITY problem is proven to be polynomially solvable for the class of P_6-free graphs [9]. Hoàng et al. [6] show that for all fixed $k \geq 3$ the k-COLORABILITY problem becomes polynomially solvable for the class of P_5-free graphs. They pose the question whether there exists a polynomial time algorithm to determine if a P_6-free graph can be 4-colored. We do not know yet if our main algorithm can be used for simplifying the proof of the result in [9] or for solving the open problem described above. We leave these questions for future research.

The next class to consider is the class of P_7-free graphs. Recall that a graph G is P_7-free if and only if each connected induced subgraph of G contains a dominating subgraph of diameter at most three [1]. Using an approach similar to the one described in this paper, it is possible to find such a dominating subgraph in polynomial time. However, a more important question is whether this characterization of P_7-free graphs can be narrowed down. Also determining the computational complexity of the HYPERGRAPH 2-COLORABILITY problem for the class of hypergraphs with P_7-free incidence graphs is still an open problem.

References

1. Bacsó, G., Tuza, Z.: Dominating Subgraphs of Small Diameter. Journal of Combinatorics. Information and System Sciences 22(1), 51–62 (1997)
2. Bacsó, G., Michalak, D., Tuza, Z.: Dominating Bipartite Subgraphs in Graphs. Discussiones Mathematicae Graph Theory 25, 85–94 (2005)
3. Brandstädt, A., Le, V.B., Spinrad, J.: Graph Classes: A Survey. In: SIAM Monographs on Discrete Mathematics and Applications, vol. 3. SIAM, Philadelphia (1999)
4. Diestel, R.: Graph Theory. 3rd edn. Springer, Heidelberg (2005)
5. Garey, M.R., Johnson, D.S.: Computers and Intractability. W.H. Freeman and Co., New York (1979)
6. Hoàng, C.T., Kamiński, M., Lozin, V.V., Sawada, J., Shu, X.: Deciding k-Colourability of P_5-Free Graphs in Polynomial Time (submitted, 2006), http://www.cis.uoguelph.ca/~sawada/pub.html
7. Liu, J., Zhou, H.: Dominating Subgraphs in Graphs with Some Forbidden Structures. Discrete Mathematics 135, 163–168 (1994)
8. Liu, J., Peng, Y., Zhao, C.: Characterization of P_6-Free Graphs. Discrete Applied Mathematics 155, 1038–1043 (2007)
9. Randerath, B., Schiermeyer, I.: 3-Colorability $\in$ for P_6-Free Graphs. Discrete Applied Mathematics 136, 299–313 (2004)
10. Seinsche, D.: On a Property of the Class of n-Colorable Graphs. Journal of Combinatorial Theory Series B 16, 191–193 (1974)
11. Sgall, J., Woeginger, G.J.: The Complexity of Coloring Graphs without Long Induced Paths. Acta Cybernetica 15(1), 107–117 (2001)
12. Wolk, E.S.: The Comparability Graph of a Tree. Proceedings of the American Mathematical Society 13, 789–795 (1962)
13. Wolk, E.S.: A Note on The Comparability Graph of a Tree. Proceedings of the American Mathematical Society 16, 17–20 (1965)

Maximum Connected Domatic Partition of Directed Path Graphs with Single Junction*

Masaya Mito and Satoshi Fujita

Department of Information Engineering
Graduate School of Engineering, Hiroshima University
{mito,fujita}@se.hiroshima-u.ac.jp

Abstract. In this paper, we consider the problem of finding a maximum connected domatic partition of a given graph. We propose a polynomial time algorithm for solving the problem for a subclass of directed path graphs which is known as a class of intersection graphs modeled by a set of directed paths on a directed tree. More specifically, we restrict the class of directed path graphs in such a way that the underlying directed tree has at most one node to have several incoming arcs.

1 Introduction

A connected dominating set (CDS, for short) for graph G is a dominating set which induces a connected subgraph of G [8],[13]. A **connected domatic partition** (CDP, for short) of G is a partition of the vertex set of G such that each subset in the partition is a CDS for G. In the literature, it is pointed out that CDS plays an important role in the resource allocation in computer networks, such as the message routing in wireless ad hoc networks [5],[14],[15], collective communication in sensor networks [3],[6],[9], and so on.

Let $d_c(G)$ denote the cardinality of a largest CDP of graph G [11],[12]. The problem of finding a CDP of maximum cardinality is known to be NP-hard for general graphs [6], and there have been derived several interesting results on the bound of value $d_c(G)$; e.g., it satisfies $d_c(G) \leq \kappa(G)$ unless G is a complete graph [16], where $\kappa(G)$ denotes the vertex connectivity of graph G, and it satisfies $d_c(G) \leq 4$ for any planar graph G [10]. It is also known that the above maximization problem can be solved in polynomial time for several classes of easy instances such as trees, cycles, and complete bipartite graphs. Unfortunately however, unlike ordinary domatic partition problem which has been investigated during these decades [1],[2], very few is known about the "connected" version of the partitioning problem.

In this paper, we first point out that the problem of finding a maximum CDP can be solved in linear time for the class of interval graphs. We then extend the result to a subclass of directed path graphs, which is an intersection graph modeled by a set of directed paths on a directed tree with a single "joining" node

---

* This work was partially supported by Kayamori Foundation of Information Science Advancement.

X. Hu and J. Wang (Eds.): COCOON 2008, LNCS 5092, pp. 425–433, 2008.

characterized by a vertex set called junction (a formal definition of the class of considered graphs will be given in Section 4). The basic idea of the proposed scheme is to focus on a set of *critical* vertices in a junction, and to carefully partition such vertex set by solving the *k-edge-coloring problem* for a corresponding bipartite multigraph. An extension of the greedy partitioning scheme used for the class of interval graphs could be effectively applied to complete the partitioning of the remaining vertices.

The remainder of this paper is organized as follows. Section 2 introduces necessary definitions. Section 3 describes a linear time algorithm for solving the partitioning problem for directed path graphs modeled by a directed tree with a single source. An extension of the algorithm to the case with a single junction is given in Section 4. Finally, we conclude the paper with future problems in Section 5.

2 Preliminaries

Let T be a directed tree consisting of node set W and arc set A. If there is a directed path from node x to node y in T, then we say that they satisfy relation $x < y$, and that x is an ancestor of y or y is a descendant of x. A node with no descendant is called a sink, and a node with no ancestor is called a source. Given directed path P in T, the first and the last nodes of the path are denoted as $\sigma(P)$ and $\tau(P)$, respectively. (For brevity, we will use a similar notation for the set of paths.)

Graph $G = (V, E)$ is called a **directed path graph** (DPG, for short) if it is an intersection graph modeled by a set of directed paths in a directed tree T, i.e., V corresponds to a set of directed paths in T and two vertices[1] in V are connected by an edge iff their corresponding paths share at least one node. In the following, we say that "G is modeled by T" if the meaning of the sentence is clear from the context. In addition, we identify a vertex v in G with its corresponding directed path in T. For example, we often say that vertex v "contains" node x if the corresponding path contains x. For any subset V' of V, we use symbol V' to denote a subgraph of T which is obtained by taking a union of paths corresponding to the vertices in V'. Moreover, for any subgraph T' of T and a vertex set $V' \subseteq V$, a subgraph of T, which is obtained by taking an intersection of T' and V', is denoted by $T' \cap V'$.

3 Tree with Single Source

In this section, we propose a simple algorithm to find an optimal CDP of DPG modeled by a tree with a single source. Note that this algorithm will be used as a subroutine in the next section.

3.1 Interval Graphs

To clarify the basic idea of the scheme, we first consider the case in which given G is modeled by a directed tree with a single source s and a single sink t; i.e.,

---

[1] Throughout this paper, we will use terms "vertex" and "edge" for graph G, and distinguish them from terms "node" and "arc" used for tree T.

when G is an interval graph. The proposed scheme partitions V into $\kappa(G)$ CDSs for G; i.e., it is optimal since $\kappa(G)$ is an upper bound on the cardinality of CDP. Let $\mathcal{C}$ be a set of $\kappa(G)$ subsets of vertices, each of which is intended to represent a CDS for G. The proposed algorithm proceeds as follows:

- Let $U \leftarrow V$ and initialize each element in $\mathcal{C}$ to $\emptyset$.
- Repeat the following operations until every element in $\mathcal{C}$ becomes a CDS for G.
  - Let c^* be an element in $\mathcal{C}$ such that $\tau(c^*) = \min_{c \in \mathcal{C}}\{\tau(c)\}$, where $\tau(c)$ denotes the last node of path c which is defined as $\min_{v \in U}\{\sigma(v)\}$ if c is an empty set.
  - Let v be a vertex in U such that $\sigma(v) \le \tau(c^*) < \tau(v)$. If U contains no such vertex, then output "failed" and terminate.
  - Move vertex v from U to c^*.
- Output $\mathcal{C}$ as a solution after adding the remaining vertices in U (if any) to an arbitrary element in $\mathcal{C}$, then terminate.

Proposition 1. *If G is a DPG modeled by a directed path, then the above algorithm outputs a CDP of cardinality $\kappa(G)$ in linear time.*

Proof. Since the time complexity is obvious by the description, in the following, we show that it always outputs a required partition. Let c be an element in $\mathcal{C}$ selected at the beginning of an iteration, and without loss of generality, let us assume that there is an arc a to have $\tau(c)$ as its predecessor (since otherwise, c has already been a CDS for G). Let α_a denote the number of paths containing arc a. By the description of the algorithm, if $\mathcal{C}$ contains an element (i.e., a path) which does not contain arc a, then any element (i.e., path) containing the arc can never be selected as the candidate. Thus, since $\kappa(G) = \min_{a \in A}\{\alpha_a\}$, the scheme associates $\kappa(G)$ vertices containing arc a to different elements in $\mathcal{C}$, in such a way that every element contains exactly one such vertex. A similar claim holds for every arc in the given T. Thus the proposition follows. $\square$

3.2 A Generalization

The above idea can be easily generalized to trees with a single source and several sinks. Let T be a directed tree with a single source s and ℓ sinks $t_1, t_2, \ldots, t_\ell$. Let P_j denote the path from s to t_j, and let $T_0, T_1, \ldots, T_{\ell-1}$ be subtrees of T defined as: $T_0 \overset{\text{def}}{=} \{s\}$, and $T_j \overset{\text{def}}{=} T_{j-1} \cup P_j$ for each $1 \le j \le \ell - 1$. Let $\hat{P}_j$ be the path from node $\tau(P_j \cap T_{j-1})$ to node t_j. Note that T_j can also be represented as $T_0 \cup (\bigcup_{i=1}^{j} \hat{P}_i)$.

A generalized algorithm for such tree T proceeds as follows:

- Let $U \leftarrow V$ and initialize each element in $\mathcal{C}$ to $\emptyset$.
- For $j = 1$ to ℓ sequentially, repeat the following operation until every element in $\mathcal{C}$ contains a vertex containing sink t_j:
  - Let c^* be an element in $\mathcal{C}$ such that $\tau(c^* \cap \hat{P}_j) = \min_{c \in \mathcal{C}}\{\tau(c \cap \hat{P}_j)\}$, where $\tau(c^* \cap \hat{P}_j)$ is defined as the first node of $\hat{P}_j$ if $c^* \cap \hat{P}_j = \emptyset$.

- Let v be a vertex in U such that $\sigma(v \cap \hat{P}_j) \leq \tau(c^* \cap \hat{P}_j) < \tau(v \cap \hat{P}_j)$. If U contains no such vertex, then output "failed" and terminate.
- Move the vertex v from U to c^*.
- Output C as a solution after adding the remaining vertices in U (if any) to an element in C, then terminate.

Proposition 2. *If G is a DPG modeled by a directed tree with a single source, the above algorithm outputs a CDP of size $\kappa(G)$ in linear time.*

Proof. The case of $\ell = 1$ is immediate from Proposition 1. To prove the claim for $\ell \geq 2$, it is enough to show that the following three conditions hold at the beginning of the jth iteration for each $2 \leq j \leq \ell$: 1) every element in C contains a vertex containing $\sigma(\hat{P}_j)$, 2) if there is an element in C which contains a vertex v containing an arc in $\hat{P}_j$, then the vertex v must be reachable from $\sigma(\hat{P}_j)$ in C, and 3) for any arc a in $\hat{P}_j$, each element in C contains at most one vertex containing a.

The first condition is immediate since every element in C is a CDS for a subgraph of G modeled by T_{j-1} at the beginning of the jth iteration. The second condition can be verified by considering a vertex v such that $\sigma(v \cap \hat{P}_j) > \sigma(\hat{P}_j)$, which is never being selected during the processing for sinks $t_1, \ldots, t_{j-1}$, since $T_{j-1} \cap v = \emptyset$. Finally, the third condition can be proved by using an argument similar to the proof of Proposition 1. Hence, the proposition follows. $\square$

4   Graph with Single Junction

4.1   Definitions

In the following, we let $k = \kappa(G)$, for brevity. In this section, we consider a class of trees which contains exactly one node to have more than one incoming arcs. Note that it is a generalization of the class of trees considered in the last section, since it allows the existence of several sources, as long as the number of "joins" is restricted to one.

Let J be the set of all vertices containing such a joining node w, which will be referred to as the **junction** with respect to w. Note that a junction induces a clique in the given graph, and that it partitions the given vertex set into three subsets as J, $V^+ \overset{\text{def}}{=} \{v \mid w < \sigma(v)\}$, and $V^- \overset{\text{def}}{=} V \setminus (J \cup V^+)$. For each node x in T, let $f(x)$ denote the number of paths in V containing both w and x. A **lower boundary** of a subtree of T centered at node w, is defined as follows:

Definition 1 (Lower Boundary). Let W^+ be the set of descendants x of w such that: 1) $f(x) \leq k$ and 2) a predecessor y of x satisfies $f(y) > k$. Given W^+, lower boundary W_L is defined as the set of nodes x satisfying either: 1) a successor of x is in W^+, or 2) x is a sink with $f(x) > k$.

A vertex in junction J is said to be **critical** if it corresponds to a path in the tree containing a node in the lower boundary W_L. Let J^* be the set of critical vertices. For each node x in T, let $f^*(x)$ denote the number of vertices in J^* containing x. With the above notions, an upper boundary of the subtree with respect to node w is defined as follows:

Definition 2 (Upper Boundary). If $f^*(w) = k$, upper boundary W_U and set W^- are both defined as $\{w\}$. Otherwise, W^- is the set of ancestors x of w such that: 1) $f^*(x) \leq k$ and 2) a successor y of x satisfies $f^*(y) > k$; and given set W^-, upper boundary W_U is defined as the set of nodes x satisfying either: 1) a predecessor of x is in W^-, or 2) x is a source with $f^*(x) > k$.

4.2   Basic Strategy for Partitioning Critical Vertices

The proposed scheme tries to find a partition $\mathcal{C}$ of J^* to satisfy the following two conditions:

- SHARE: For any $c \in \mathcal{C}$ and for any $x \in W^- \cup W^+$, $J^* \cap c$ contains at most one path containing node x.
- RESERVE: For any $c \in \mathcal{C}$ and for any $x \in W_U \cup W_L$, $J^* \cap c$ contains at least one path containing node x.

In this and the next subsection, we describe how to find such a partitioning in polynomial time, and in Subsection 4.4, we explain how those conditions are related with the overall partitioning of V. The basic idea for the partitioning of J^* to satisfy the above two conditions is to use k-edge-coloring of a bipartite multigraph reflecting the structure of critical vertices. More precisely, we consider a bipartite multigraph H with vertex set $V_H = W^- \cup W^+$ and edge set $E_H(\subset W^- \times W^+)$, where vertices x and y are connected by an edge in E_H iff J^* contains a path containing both x and y. Note that there may exist vertices in J^* which have no corresponding edge in E_H; e.g., a path terminating at the lower boundary does not contain a node in W^+ (handling of such "hidden" vertices will be discussed in the next subsection).

Let ϕ be a k-edge-coloring of H; i.e., ϕ is a function from E_H to $\{1, 2, \ldots, k\}$ such that any two adjacent edges are assigned different colors. Note that graph H is k-edge-colorable in $O(|E_H| \log k)$ time, since the maximum degree of H is at most k [4]. Given a k-edge-coloring ϕ of H, let us consider the following natural association of vertices in J^* to the elements in $\mathcal{C}$:

1) If an edge in E_H corresponding to vertex $v(\in J^*)$ is assigned color i by ϕ, then v is associated to the ith element in $\mathcal{C}$.
2) Otherwise, the vertex is associated to an element in $\mathcal{C}$ which contains no vertex sharing the same node in $W^- \cup W^+$ with v. Note that such element always exists in $\mathcal{C}$, since $f^*(x) \leq k$ for any $x \in W^- \cup W^+$.

The above assignment obviously satisfies condition SHARE. However, the second condition RESERVE cannot always be satisfied if we directly apply the procedure to the original H (due to the problem of "hidden" vertices). In the next subsection, we show that the second condition can always be satisfied if we conduct an appropriate preprocessing for the modification of graph H before conducting the above assignment procedure.

4.3 Preprocessing to Satisfy Condition RESERVE

In the preprocessing phase, bipartite multigraph H is modified for each node in W_L, according to the type of node defined as follows: A node of Type 1 is a sink; a node of Type 2 has a successor y with $f^*(y) \geq k$; and a node of Type 3 or 4 has a successor but none of them has a f^* value greater than or equal to k. The difference of the last two types is the summation of f^* values over all successors; i.e., in Type 3, the summation is bounded by k, but it is greater than k for Type 4. Recall that the objective of the preprocessing phase is to satisfy RESERVE after simply applying the above assignment procedure.

Type 1: Recall that a sink $x \in W_L$ is not contained in the vertex set of the original H. Let $S(\subseteq J^*)$ be a vertex set consisting of arbitrary k paths containing sink x. Add a new vertex corresponding to x to V_H, and connect it to k nodes in W^- via edges corresponding to the selected k paths.

Type 2: No modification is necessary if x has a successor y with $f^*(y) \geq k$. In fact, when $f^*(y) = k$, since it associates exactly one path containing y to c for any $c \in \mathcal{C}$, condition RESERVE obviously holds for its parent x. When $f^*(y) > k$, on the other hand, since x must have a descendant of the other type contained in W_L, the condition holds for x as long as it is satisfied for at least one of its descendants.

Type 3: Let S' be the set of successors of x, and $f^*(S') \overset{\text{def}}{=} \sum_{y \in S'} f^*(y)$. If $f^*(S') \leq k$, then contract vertices in $V_H \cap S'$ to a single vertex $\hat{x}$, i.e., $V_H \leftarrow (V_H \setminus S') \cup \{\hat{x}\}$, and connect it to nodes in W^- via corresponding edges. Then, after selecting arbitrary $f^*(x) - f^*(S')\ (> 0)$ paths terminating at node x from J^*, add edges corresponding to the selected paths to E_H. Note that this modification obviously satisfies RESERVE for node x, and does not violate SHARE for any $y \in S'$.

Type 4: Let $S' = \{y_1, y_2, \ldots, y_\ell\}$ be the set of successors of x. At first, for each $y_j \in V_H \cap S'$, add new vertex $\hat{y}_j$ to V_H, and connect it with y_j via $k - f^*(y_j)\ (> 0)$ parallel edges. Then, after adding another vertex p to V_H, connect it with vertices in $\{\hat{y}_1, \hat{y}_2, \ldots, \hat{y}_\ell\}$ in the following manner: Let i be an integer such that $\sum_{j=1}^{i-1} f^*(y_j) < k$ and $\sum_{j=1}^{i} f^*(y_j) \geq k$. Vertex p is connected with $\hat{y}_j$ via $f^*(y_j)$ parallel edges for $1 \leq j \leq i - 1$, and connected with vertex $\hat{y}_i$ via $k - \sum_{j=1}^{i} f^*(y_j)$ parallel edges. Note that in any k-edge-coloring of the resultant H, k edges incident on p are assigned distinct colors, which will be propagated to edges incident on S' via vertices in $\{\hat{y}_1, \hat{y}_2, \ldots, \hat{y}_\ell\}$. Thus, it satisfies condition RESERVE for node x.

A similar modification could be done for each node in W_U. Hence, a proof of the satisfaction of RESERVE will complete by proving the following lemma.

Lemma 1. *In the resultant graph H, each vertex in J^* corresponds to at most one edge in E_H.*

Proof. The claim apparently holds for the original H. An addition of edges corresponding to vertices in J^* takes place only when it examines node x of Type 1 or 3. In the former case, the path from w to x never contains a node in

W^+, and even if it contains a node in W_L, the modification of H for the node in W_L does not take place since it should be a node of Type 2. When x is a node of Type 3, on the other hand, every edge added by the modification corresponds to a path terminating at node x. Thus, it does not contain a node in W^+, and even if it contains a node in W_L, it does not cause a modification of H. A similar argument holds for the upper boundary W_U. Hence, the lemma follows.   □

4.4   Partition of the Remaining Vertices

Now, we have obtained a partition C of J^* satisfying conditions SHARE and RESERVE. Given such partition C, a greedy assignment scheme described in Section 3 correctly finds a (connected domatic) partition of $J^* \cup V^+$ in linear time, where $V^+ = \{v \mid w < \sigma(v)\}$. Thus, in the remaining of this section, we describe how to realize a correct partition of $V^- = V \setminus (J^* \cup V^+)$.

For each source s in T, let Z_s denote the set of sinks reachable from s without passing through node w; let T_s denote an out-tree which is obtained by taking a union of paths connecting from s to nodes in Z_s, and let P_s denote the unique path from s to w. The procedure for the partition of V^- proceeds as follows: First, for each source s, it assigns paths on P_s to C with an algorithm described below. After that, it assigns paths in T_s to C by using a greedy scheme for out-trees described in Section 3.2.

More concretely, the partition of paths on P_s proceeds as follows:

- Let $U \leftarrow V^-$, initialize each element in C to an empty set, and initialize D to the output of previous procedures; i.e., D is a (connected domatic) partition of $J^* \cup V^+$ of size k.
- Repeat the following operation until every element in C contains each arc on P_s.
  - Let c^* be an element in C such that $\tau(c^* \cap P_s) = \min_{c \in C}\{\tau(c \cap P_s)\}$, where $\tau(c^* \cap P_s)$ is defined as source s if $c = \emptyset$. Let d^* be an element in D such that $\sigma(d^* \cap P_s) = \min_{d \in D}\{\sigma(d \cap P_s)\}$.
  - Let v be a vertex in U such that $\sigma(v \cap P_s) \leq \tau(c^* \cap P_s) < \tau(v \cap P_s)$. If U contains no such vertex and $\tau(c^* \cap P_s) < \sigma(d^* \cap P_s)$, then output "failed" and terminate.
  - Move vertex v from U to c^* if U contains v. Otherwise, $c^* \leftarrow c^* \cup d^*$ and $D \leftarrow D \setminus \{d^*\}$.
- Output C as a solution, and terminate.

Proposition 3. *If D is initialized to the output of previous procedures, then the above procedure successfully outputs necessary partition in linear time.*

Proof. In the following, we will merely consider the correctness of scheme, since the time complexity is obvious by description. Let c be an element in C selected at the beginning of an iteration, and suppose that $\tau(c \cap P_s) < w$, without loss of generality. (Note that if $\tau(c \cap P_s) = w$, c has already contained every arc on P_s.) Let a be an arc on P_s starting from node $\tau(c \cap P_{s,w})$, and let α_a denote the number of vertices in G containing arc a. By the description of the algorithm, if there is an element in $\mathcal{C}$ which does not contain a, then any element in $\mathcal{C}$ containing arc a can never be selected as the candidate. In addition, since $\mathcal{D}$ is initialized to a partition of $J^* \cup V^+$ satisfying **SHARE** and **RESERVE** for nodes in $(W^- \cup W_U) \cap P_s$, it does not contain an element which contains more than one vertices containing arc a, or every element in $\mathcal{D}$ has contained arc a. Thus, since the vertex connectivity $k(= \kappa(G))$ is represented as $\min_{a \in A}\{\alpha_a\}$, element c can contain a vertex containing arc a. A similar claim holds for every arc on path P_s. Thus, the proposition follows. □

Finally, we can easily show that the greedy scheme given in Section 3.2 works well even for out-tree T_s. Thus, we have the following proposition.

Proposition 4. *If G is a DPG modeled by a directed tree containing at most one node to have several incoming arcs, then the connected domatic partition problem can be solved in polynomial time.*

5 Concluding Remarks

This paper proposed an algorithm for solving the connected domatic partition problem for a subclass of directed path trees. An important open problem is to examine if the proposed algorithm can be extended to general directed path graphs with more than one junctions. We have a positive conjecture for this problem, such that it could be solved by repeatedly applying the proposed assignment procedure for each junction (probably, it needs an appropriate ordering of such junctions). Since it can be easily shown that an undirected version of the problem is NP-hard using a reduction from the 3-edge-coloring problem [7], it would clarify a sharp boundary on the complexity of the connected domatic partition problem.

References

1. Bertossi, A.A.: On the Domatic Number of Interval Graphs. Information Processing Letters 28(6), 275–280 (1988)
2. Bonuccelli, M.A.: Dominating Sets and Domatic Number of Circular Arc Graphs. Discrete Applied Mathematics 12, 203–213 (1985)
3. Cardei, M., Du, D.-Z.: Improving Wireless Sensor Network Lifetime through Power Aware Organization. ACM Wireless Networks 11(3), 333–340 (2005)
4. Cole, R., Ost, K., Schirra, S.: Edge-Coloring Bipartite Multigraphs in $O(E \log D)$ Time. Combinatorica 21(1), 5–12 (2001)
5. Dai, F., Wu, J.: An Extended Localized Algorithm for Connected Dominating Set Formation in Ad Hoc Wireless Networks. IEEE Transactions on Parallel and Distributed Systems 53(10), 1343–1354 (2004)
6. Dong, Q.: Maximizing System Lifetime in Wireless Sensor Networks. In: Proc. of the 4th International Symposium on Information Processing in Sensor Networks, pp. 13–19 (2005)
7. Garey, M.R., Johnson, D.S.: Computers and Intractability: A Guide to the Theory of NP-Completeness. W.H. Freeman and Company, San Francisco (1979)

8. Guha, S., Khuller, S.: Approximation Algorithms for Connected Dominating Sets. In: Proc. European Symposium on Algorithms, pp. 179–193 (1996)
9. Ha, R.W., Ho, P.H., Shen, X., Zhang, J.: Sleep Scheduling for Wireless Sensor Networks via Network Flow Model. Computer Communications 29(13-14), 2469–2481 (2006)
10. Hartnell, B.L., Rall, D.F.: Connected Domatic Number in Planar Graphs. Czechoslovak Mathematical Journal 51(1), 173–179 (2001)
11. Haynes, T.W., Hedetniemi, S.T., Slater, P.J.: Fundamentals of Domination in Graphs. Marcel Dekker, New York (1998)
12. Haynes, T.W., Hedetniemi, S.T., Slater, P.J.: Domination in Graphs: Advanced Topics. Marcel Dekker, New York (1998)
13. Hedetniemi, S., Laskar, R.: Connected domination in Graphs. In: Graph Theory and Combinatorics, pp. 209–218. Academic Press, London (1984)
14. Wu, J., Li, H.: Domination and Its Applications in Ad Hoc Wireless Networks with Unidirectional Links. In: Proc. of International Conference on Parallel Processing, pp. 189–200 (2000)
15. Wu, J.: Extended Dominating-Set-Based Routing in Ad Hoc Wireless Networks with Unidirectional Links. IEEE Transactions on Parallel and Distributed Computing 22, 327–340 (2002)
16. Zelinka, B.: Connected Domatic Number of a Graph. Math. Slovaca 36, 387–392 (1986)

Efficient Algorithms for the k Smallest Cuts Enumeration*

Li-Pu Yeh and Biing-Feng Wang

Department of Computer Science, National Tsing Hua University
Hsinchu, Taiwan 30043
{dr928304,bfwang}@cs.nthu.edu.tw

Abstract. In this paper, we study the problems of enumerating cuts of a graph by non-decreasing weights. There are four problems, depending on whether the graph is directed or undirected, and on whether we consider all cuts of the graph or only s-t cuts for a given pair of vertices s, t. Efficient algorithms for these problems with $\tilde{O}(n^2 m)$ delay between two successive outputs have been known since 1992, due to Vazirani and Yannakakis. In this paper, improved algorithms are presented. The delays of the presented algorithms are $O(nm \log(n^2/m))$.

1 Introduction

Let $G = (V, E)$ be an edge-weighted, directed or undirected graph, where V is the vertex set and E is the edge set. Let $n = |V|$ and $m = |E|$. A cut is a partition of the vertex set V into two non-empty subsets. The *weight* of a cut (X, Y) is the total weight of the edges that go from X to Y. Let s, $t \in V$ be two vertices. An *s-t cut* is a cut (X, Y) such that $s \in X$ and $t \in Y$. The *minimum cut problem* is to find a cut of minimum weight and the *minimum s-t cut problem* is to find an s-t cut of minimum weight. Efficient algorithms for these two problems have numerous real-world applications such as generating traveling salesperson cutting planes, parallel computing, clustering, VLSI design, and network reliability [1], [3], [6], [8], [9], [17], [19]. The most fundamental tool for solving the minimum cut and minimum s-t cut problems is maximum flow computation. For the computation, Goldberg and Tarjan [6] had an $O(nm\log(n^2/m))$-time algorithm and King, Rao, and Tarjan [17] had an $O(nm\log_{m/n\log n} n)$-time algorithm. As a consequence of the well-known maximum-flow minimum-cut theorem [1], the minimum s-t cut problem can be solved in $\tilde{O}(nm)$ time. For the minimum cut problem, Hao and Orlin [9] had an $O(nm\log(n^2/m))$-time algorithm. For undirected graphs, better results for the minimum cut problem are known. Nagamochi and Ibaraki [19] gave an $O(nm + n^2\log n)$-time algorithm and Karger [15] gave an $O(m\log^3 n)$-time randomized algorithm.

In many important applications, such as the all terminal network reliability problem, the vertex packing problem, and the maximum closure problem, finding all minimum cuts or nearly minimum cuts might be more useful than finding a

---

* This research is supported by the National Science Council of the Republic of China under grants NSC-95-2213-E-007-029 and NSC-94-2752-E-007-082.

X. Hu and J. Wang (Eds.): COCOON 2008, LNCS 5092, pp. 434–443, 2008.

minimum cut [4], [5], [13], [14], [16], [21], [23]. Vazirani and Yannakakis [24] introduced the problems of enumerating cuts of a graph by non-decreasing weights. There are four problems, depending on whether the graph is directed or undirected, and on whether we consider all cuts of the graph or only s-t cuts for a given pair of vertices s, t. These enumeration problems have an application in studying the reliability and connectivity of networks [24]. For each of the problems, Vazirani and Yannakakis gave an efficient algorithm that requires at most $n - 1$ maximum flow computations between two successive outputs. Since a maximum flow computation can be done in $\tilde{O}(nm)$ time, the delays of their algorithms are $\tilde{O}(n^2m)$. In this paper, for each of the enumeration problems, an improved algorithm is presented. The delays of the presented algorithms are $O(nm\log(n^2/m))$. Our algorithms have the same schema as Vazirani and Yannakakis's algorithms. Our improvement is based on a delicate application of Hao and Orlin's minimum cut algorithm to their framework. Vazirani and Yannakakis's algorithms had been used as basic subroutines in the solutions of many problems. Our improvement directly reduces the running time of these solutions. For example, for the minimum k-cut problem, the upper bound is directly reduced by a factor of $\tilde{O}(n)$ for any $k \geq 3$.

Notation and preliminaries are given in the next section. In Section 3, we review Vazirani and Yannakakis's algorithm for the problem of enumerating all cuts of a directed graph. In Section 4, an improved algorithm is presented for the same problem. In Section 5, we show how to modify the algorithm in Section 4 so as to solve the other three problems. In Section 6, we describe existing algorithms whose running time can be reduced by our enumeration algorithms. Finally, in Section 7, we conclude this paper.

2   Preliminaries

Let $G = (V, E)$ be a directed or undirected graph, where V is the vertex set and E is the edge set. Let $n = |V|$ and $m = |E|$. Each edge $(i, j) \in E$ has a nonnegative real weight $w(i, j)$. A cut of G is a partition (X, Y) of the vertices into two non-empty subsets X and Y. For any cut (X, Y), we say that the vertices in X are on the *source side* and the vertices in Y are on the *sink side*. Let $C(G)$ be the set of cuts of G. If G is directed, the *weight* of a cut (X, Y) is the total weight of the edges going from X to Y; otherwise, it is the total weight of the edges having one end vertex in X and the other in Y. A *minimum cut* of G is a cut of minimum weight. For convenience, in this paper, we usually omit set braces around singletons, writing, for example, v instead of $\{v\}$.

Let S, T be two disjoint subsets of V. An S-T cut is a cut (X, Y) such that $S \subseteq X$ and $T \subseteq Y$. A *minimum S-T cut* is an S-T cut of minimum weight. The *partially specified cut set* with respect to (S, T) is defined as $P(S, T) = \{(X, Y)|(X, Y)$ is an S-T cut of $G\}$. Let $m(S, T)$ be a minimum cut in $P(S, T)$. By definition, $m(S, T)$ is just a minimum S-T cut of G. We have the following.

Lemma 1. **[1]** *If S, T are non-empty, a minimum S-T cut can be found in $\tilde{O}(nm)$ time.*

3   Vazirani and Yannakakis's Algorithm for Enumerating All Cuts in a Directed Graph

Number the vertices in V from 1 to n. Each cut (X, Y) of G is represented by an n-bit binary string $b_1 b_2...b_n$ as follows: $b_i = 0$ if and only if vertex $i \in X$. Consider a complete binary tree of height n. Its leaves are named in the standard way by binary strings of length n, internal nodes at depth $k \geq 1$ are represented by binary strings of length k, and root is represented by the empty string ε. Each node $v = b_1 b_2...b_k$ represents the partially specified cut set $P(S, T)$, where $S = \{i | b_i = 0, 1 \leq i \leq k\}$ and $T = \{i | b_i = 1, 1 \leq i \leq k\}$. For each internal node v, let $P(v)$ denote the partially specified cut set represented by v and let $m(v)$ denote the minimum cut in $P(v)$. Let $v = b_1 b_2...b_i$ be a node and $l = b_1 b_2...b_i b_{i+1}...b_n$ be a leaf in the subtree rooted at v. For $i < k \leq n$, we call $b_1 b_2...b_{k-1} \bar{b}_k$ an *immediate child* of the path from v to l. For example, given $v = 1$, $l = 1011$, the immediate children are 11, 100, and 1010. Clearly, the partially specified cut sets represented by the immediate children of the path from v to l form a partition of $P(v) - \{l\}$. Let x^k be the sequence of x repeated k times, where x is 0 or 1. Vazirani and Yannakakis's algorithm is as follows.

Algorithm 1. Enumeration_Vazirani_Yannakakis
Input: a directed graph $G = (V, E)$.
Output: all cuts of G in the order of non-decreasing weights
begin
    1. **for** $k \leftarrow 1$ **to** n **do** compute $m(0^k)$ and $m(1^k)$
    2. $\Pi \leftarrow \{(P(\varepsilon), m(\varepsilon))\}$ /* $P(\varepsilon) = C(G)$, $m(\varepsilon)$ is the minimum cut of G
    3. **while** $\Pi \neq \emptyset$ **do**
    4. **begin**
    5. $(P(v), m(v)) \leftarrow$ the element in Π with minimum $m(v)$
    6. $\Pi^* \leftarrow \{(P(u), m(u)) | u$ is an immediate child on the path from v to $m(v)\}$
    7. $\Pi \leftarrow \Pi - (P(v), m(v)) \cup \Pi^*$ /* delete $P(v)$ and insert a partition of
                                    $P(v) - m(v)$
    8. **output** $(m(v))$
    9. **end**
end

Lines 1 and 2 are initialization steps. The computation of $m(0^k)$ in Line 1 is done as follows. First, by using $n - 1$ maximum flow computations, we compute c_i as a minimum $\{1, 2, ..., i\}$-$\{i+1\}$ cut for $1 \leq i < n$. Then, we compute each $m(0^k)$ as the minimum cut in $\{c_k, c_{k+1}, ..., c_{n-1}\}$. The computation of $m(1^k)$ in Line 1 is done similarly. During the execution of the algorithm, the elements in Π are stored in a heap. Since $m(\varepsilon)$ can be computed as the minimum cut in $\{m(0), m(1)\}$, Line 2 requires $O(1)$ time. Therefore, the initialization steps require $2n - 2$ maximum flow computations. The delay between two successive outputs is analyzed as follows. The computation of $m(u)$ in Line 6 is the bottleneck. For each u, if $u = 0^k$ or $u = 1^k$ for some integer k, $m(u)$ was found in

Line 1; otherwise, it is computed by using a maximum flow computation. Since $|\Pi^*| \leq n - 1$, we have the following.

Theorem 1. [24] *The cuts of a directed graph can be enumerated in the order of non-decreasing weights with $\tilde{O}(n^2 m)$ time delay between two consecutive outputs.*

4 An Improved Algorithm for Enumerating All Cuts in a Directed Graph

4.1 The Algorithm

Let S, T be two disjoint subsets of V. By using a vertex v in $V - (S \cup T)$, we can partition the partially specified cut set $P(S, T)$ into two disjoint subsets $P(S \cup v, T)$ and $P(S, T \cup v)$. Let $U = (v_1, v_2, ..., v_{n-|S|-|T|})$ be a sequence of the vertices in $V - (S \cup T)$. The *extract-min partition* of $P(S, T)$ induced by U is a partition obtained as follows: First, partition $P(S, T)$ into two subsets by using the vertex v_1; then, recursively, partition the subset containing $m(S, T)$ by the sequence $(v_2, v_3, ..., v_{n-|S|-|T|})$. Note that in an extract-min partition of $P(S, T)$, the subset containing $m(S, T)$ is a singleton.

Select an arbitrary vertex s. We partition $C(G)$ into two subsets $P(s, \emptyset)$ and $P(\emptyset, s)$. Let $U = (v_1, v_2, ..., v_{n-1})$ be a sequence of the vertices in $V - s$. The *basic partition* of $P(s, \emptyset)$ induced by U is $\{P(\{s, v_1, ..., v_{i-1}\}, v_i) | 1 \leq i \leq n - 1\}$. Note that each subset in a basic partition of $P(s, \emptyset)$ contains non-empty source and sink sides. Similarly, define the *basic partition* of $P(\emptyset, s)$ induced by U as $\{P(v_i, \{s, v_1, ..., v_{i-1}\}) | 1 \leq i \leq n - 1\}$.

Our enumeration algorithm is as follows.

Algorithm 2. Enumeration_directed_Graph
Input: a directed graph $G = (V, E)$.
Output: all cuts of G in the order of non-decreasing weights
begin
   1. $\Pi \leftarrow$ Basic_Partition /* stores a partition of $C(G)$
                               and the minimum cuts of its subsets
   2. **while** $\Pi \neq \emptyset$ **do**
   3. **begin**
   4. $(P(S, T), m(S, T)) \leftarrow$ the element in Π with minimum $m(S, T)$
   5. $\Pi^* \leftarrow$ Extract_Min_Partition $(P(S, T), m(S, T))$
   6. $\Pi \leftarrow \Pi - (P(S, T), m(S, T)) \cup \Pi^*$
   7. **output** $(m(S, T))$
   8. **end**
end

Procedure Basic_Partition
begin
   1. $B_0 \leftarrow$ a basic partition of $P(s, \emptyset)$
   2. $B_1 \leftarrow$ a basic partition of $P(\emptyset, s)$

3. **for** each $P(S,T) \in B_0 \cup B_1$ **do** compute $m(S,T)$
4. **return** $(\{(P(S,T), m(S,T))|P(S,T) \in B_0 \cup B_1\})$
end

Procedure Extract_Min_Partition $(P(S,T), m(S,T))$
begin
1. $R \leftarrow$ an extract-min partition of $P(S,T)$
2. **for** each $P(S',T') \in R - \{m(S,T)\}$ **do** compute $m(S',T')$
3. **return** $(\{(P(S',T'), m(S',T'))|P(S',T') \in R - \{m(S,T)\}\})$
end

The detailed implementations of Basic_Partition and Extract_Min_Partition are described, respectively, in Sections 4.2 and 4.3.

4.2   Basic Partition

We only describe the computation of B_0 and the minimum cuts in its subsets. The computation of B_1 and the minimum cuts in its subsets is done similarly. A simple implementation is as follows: Select an arbitrary sequence U of the vertices in $V - s$, compute B_0 as the basic partition of $P(s, \emptyset)$ induced by U, and then compute the minimum cut of each subset by a maximum flow computation. Such an implementation needs $n - 1$ maximum flow computations and thus requires $\tilde{O}(n^2 m)$ time. In the following, an $O(nm\log(n^2/m))$-time implementation is presented.

The trick here is to select a specific sequence U. Hao and Orlin [9] had an efficient algorithm for computing a minimum cut of a directed graph. We determine the sequence U by making use of their algorithm. Given a directed graph $G = (V, E)$, Hao and Orlin's algorithm finds a minimum cut as follows. First, select an arbitrary vertex $s \in V$. Then, compute a minimum cut X_1 subject to the condition that s is on the source side. And then, compute a minimum cut X_2 subject to the condition that s is on the sink side. Clearly, the smaller one of X_1 and X_2 is a minimum cut. The cut X_2 is computed by firstly reversing each edge of G and then applying the same computation of X_1. The computation of X_1 is described below. First, set $S = \{s\}$ and $T = V - s$. Then, repeatedly, select a sink vertex $t \in T$, compute a minimum S-t cut, and then transfer t from T to S until T is empty. In total, $n - 1$ cuts are computed. Hao and Orlin showed that the $n - 1$ cuts can be computed in $O(nm\log(n^2/m))$ time if we select the sink vertex t in a careful way at each time. Finally, X_1 is computed as the smallest one of the $n - 1$ cuts. Let $(v_1, v_2, ..., v_{n-1})$ be the sequence of vertices in the order of their selection as sinks during the above computation of X_1. Then, the ith cut being computed is a minimum $\{s, v_1, ..., v_{i-1}\}$-v_i cut of G, $1 \le i \le n-1$. Therefore, we have the following.

Lemma 2. [9] *Given a directed graph* $G = (V, E)$ *and a vertex* $s \in V$, *we can determine a sequence* $(v_1, v_2, ..., v_{n-1})$ *of the vertices in* $V - s$ *and compute the cuts* $m(\{s, v_1, ..., v_{i-1}\}, v_i)$, $i = 1, 2, ..., n - 1$, *in* $O(nm\log(n^2/m))$ *time.*

According to Lemma 2, we implement the computation for B_0 as follows. First, by using Hao and Orlin's algorithm, we determine a sequence $U = (v_1, v_2, ..., v_{n-1})$ of the vertices in $V - s$ and compute the cuts $m(\{s, v_1, ..., v_{i-1}\}, v_i)$, $i = 1, 2, ...,$ $n - 1$. Then, we compute B_0 as the basic partition of $P(s, \emptyset)$ induced by U. The overall time complexity is $O(nm\log(n^2/m))$. Therefore, we have the following.

Lemma 3. *Basic_Partition requires $O(nm\log(n^2/m))$ time.*

4.3   Extract-Min Partition

Our implementation of Extract_Min_Partition consists of two phases. Let $P(S, T)$ be the given partially specified cut set. Let $m(S, T) = (S^*, T^*)$, $q = |S^* - S|$, and $r = |T^* - T|$. Phase 1 determines a sequence $(s_1, s_2, ..., s_q)$ of the vertices in $S^* - S$, partitions $P(S, T)$ into $q + 1$ subsets $P(S \cup \{s_1, s_2, ..., s_{i-1}\}, T \cup s_i)$, $i = 1, 2, ..., q + 1$, where $s_{q+1} = \emptyset$, and computes the minimum cut in each subset. After Phase 1, the minimum cut (S^*, T^*) is contained in the subset $P(S \cup \{s_1, s_2, ..., s_q\}, T) = P(S^*, T)$. Then, Phase 2 determines a sequence $(t_1, t_2, ..., t_r)$ of the vertices in $T^* - T$, further partitions the subset $P(S^*, T)$ into $r + 1$ subsets $P(S^* \cup t_i, T \cup \{t_1, t_2, ..., t_{i-1}\})$, $i = 1, 2, ..., r + 1$, where $t_{r+1} = \emptyset$, and computes the minimum cut in each subset.

We proceed to present the detailed implementation of Phase 1. For convenience, we assume that S contains only a single vertex s and T contains only a single vertex t. In case this is not true, we simply contract S and T, respectively, to create two new vertices. Our problem is the following: Given G, $P(s, t)$, and $m(s, t) = (S^*, T^*)$, determine a sequence $(s_1, s_2, ..., s_q)$ of the vertices in $S^* - s$ and compute the minimum cuts $m(S_i, \{t, s_i\})$, $i = 1, 2, ..., q$, where $S_i = \{s, s_1, ..., s_{i-1}\}$. Note that $m(S^*, T) = (S^*, T^*)$. We solve the above problem as follows. First, compute f as a maximum s-t flow in G. Next, obtain a graph G^- by removing T^* from G_f, where G_f is the residual graph of G induced by f. The vertex set of G^- is S^*. Then, by using Hao and Orlin's algorithm, determine a sequence $(s_1, s_2, ..., s_q)$ of the vertices in $S^* - \{s\}$ and compute a minimum S_i-s_i cut, denoted by (α_i, β_i), of G^- for $1 \leq i \leq q$. Finally, compute $m(S_i, \{t, s_i\}) = (\alpha_i, \beta_i \cup T^*)$ for $1 \leq i \leq q$. The overall time complexity is $O(nm\log(n^2/m))$. The correctness is ensured by the following lemma.

Lemma 4. *Let $x, y \in V$ be two vertices, (X^*, Y^*) be a minimum x-y cut of G, and f be a maximum x-y flow in G. Let G^- be the graph obtained by removing Y^* from the residual graph of G induced by f. Let X_1, $X_2 \subset X^*$ be two disjoint non-empty subsets such that $x \in X_1$, and let (α, β) be a minimum X_1-X_2 cut of G^-. Then, $(\alpha, \beta \cup Y^*)$ is a minimum X_1-$(X_2 \cup y)$ cut of G.*

Due to the page limit, the proof of Lemma 4 is omitted. Similarly, Phase 2 is implemented in $O(nm\log(n^2/m))$ time as follows. First, obtain a directed graph G^- from G_f by removing S^* and then reversing each edge. Next, by using Hao and Orlin's algorithm, determine a sequence $(t_1, t_2, ..., t_r)$ of the vertices in $T^* - \{t\}$ and compute a minimum T_i-t_i cut, denoted by (β_i, α_i), of G^- for $1 \leq i \leq r$, where $T_i = \{t, t_1, ..., t_{i-1}\}$. Finally, compute $m(S^* \cup t_i, T_i) = (\alpha_i \cup S^*, \beta_i)$ for $1 \leq i \leq r$. We obtain the following.

Lemma 5. *Extract_Min_Partition can be implemented in $O(nmlog(n^2/m))$ time.*

Consequently, we have the following.

Theorem 2. *The cuts of a directed graph can be enumerated in the order of non-decreasing weights with $O(nmlog(n^2/m))$ time delay between two consecutive outputs.*

5   Enumerating All Cuts of an Undirected Graph and All s-t Cuts of a Graph

We solve the problem of enumerating all cuts of an undirected graph G as follows. Since G is undirected, two cuts (X, Y) and (Y, X) are the same. To avoid encountering a cut twice, we firstly select an arbitrary vertex s and assume that s is always on the source side. That is, we only consider the cuts in $P(s, \emptyset)$. Next, G is transformed into a directed graph G' by replacing each undirected edge (i, j) with two directed edges (i, j) and (j, i), each having the same weight as the original edge. Then, we enumerate the cuts in $P(s, \emptyset)$ by applying Algorithm 2 to G' with the following slight modification: Basic_Partition only returns a basic partition of $P(s, \emptyset)$ and the minimum cuts in its subsets.

Theorem 3. *The cuts of an undirected graph can be enumerated in the order of non-decreasing weights with $O(nmlog(n^2/m))$ time delay.*

Enumerating all s-t cuts of a directed graph can be done by applying Algorithm 2 with the following modification: Basic_Partition only returns $(P(s, t), m(s, t))$.

Theorem 4. *The s-t cuts of a directed graph can be enumerated in the order of non-decreasing weights with $O(nmlog(n^2/m))$ time delay.*

As indicated in [24], the problem of enumerating all s-t cuts of an undirected graph can be treated as a special case of enumerating all s-t cuts of a directed graph by replacing each edge by two arcs with opposite directions. Therefore, we have the following.

Theorem 5. *The s-t cuts of an undirected graph can be enumerated in the order of non-decreasing weights with $O(nmlog(n^2/m))$ time delay.*

6   Applications

Let G be an undirected graph and $k \geq 2$ be an integer. A k-cut is a partition of the vertex set into k non-empty disjoint subsets. The *minimum k-cut problem* is to find a k-cut that minimizes the total weight of the edges whose endpoints are in different subsets. For any $A \subset V$, let $G[A]$ be the subgraph induced from G by A. If we can identify the first component V_1 of a minimum k-cut $(V_1, V_2, ..., V_k)$, then $(V_2, V_3, ..., V_k)$ can be computed by finding the minimum $(k - 1)$-cut on $G[V - V_1]$. We call a set D of 2-cuts a k-candidate set if it contains at least one such component V_1.

For $k = 3$, 4, 5, 6, the minimum k-cut problem has received much attention [2], [10], [11], [12], [18], [20], [22]. Previously, for $k = 3$, 4, 5, 6, the best upper bound was $\tilde{O}(n^k m)$. For $j \geq 1$, let $M(j)$ be the time required for computing the smallest j cuts. For $k = 3$ and 4, Nagamochi and Ibaraki [20] gave the following important result: a k-candidate set of size $O(n)$ can be computed in $M(O(n))$ time. Later, Nagamochi, Nishimura, and Ibaraki [22] extended this result to work for $k = 5$ and 6. By using Vazirani and Yannakakis's enumeration algorithm, $M(O(n)) = \tilde{O}(n^3 m)$. Given a k-candidate set D, a minimum k-cut can be found by simply applying a minimum $(k - 1)$-cut algorithm $|D|$ times. Therefore, for $k = 3$, 4, 5, 6, the minimum k-cut problem can be solved in $O(n^k m + n^{k-2} T_2)$ $= \tilde{O}(n^k m)$ time, where T_2 is the time required for finding a minimum 2-cut. By using our enumeration algorithm, $M(O(n)) = O(n^2 m \log(n^2/m))$ and thus the following is obtained.

Theorem 6. *The minimum k-cut problem can be solved in $O(n^{k-1} m \log(n^2/m))$ time for $k = 3$, 4, 5, 6.*

Goldschmidt and Hochbaum [7] showed that the minimum k-cut problem is NP-hard if k is part of the input and presented an $O(n^{k^2/2 - 3k/2 + 5} m \log(n^2/m))$-time algorithm. Very recently, based on the divide-and-conquer strategy, Kamidoi, Yoshida, and Nagamochi [11] successfully designed an efficient algorithm for the minimum k-cut problem. By using the minimum 5-cut and 6-cut algorithms in [22] as basic subroutines, the running time of their algorithm is $\tilde{O}(n^{g(k)+1} m)$ for any $k \geq 7$, where $g(k)$ is defined by the recurrence: $g(5) = 4$, $g(6) = 5$, and $g(k) = 2k - 5 + g(\lfloor (k + k^{1/2})/2 \rfloor + 1)$ for $k \geq 7$. By applying our result on the minimum $\{5, 6\}$-cut problem to the algorithm in [11], the following is obtained.

Theorem 7. *The minimum k-cut problem can be solved in $O(n^{g(k)} m \log(n^2/m))$ time for $k \geq 7$.*

An *ideal cut* of a directed acyclic graph is a cut (X, Y) such that there is no edges directed from Y to X. Vazirani and Yannakakis [24] showed that the problem of enumerating the ideal cuts of a directed acyclic graph by their weights can be reduced in $O(m)$ time to the problem of enumerating the s-t cuts of a directed graph. Therefore, we have the following.

Theorem 8. *The ideal cuts of a directed acyclic graph can be enumerated in the order of non-decreasing weights with $O(nm \log(n^2/m))$ time delay.*

Given a network of n vertices, each of whose m links is assumed to fail independently with some probability, the *all-terminal network reliability problem* is to determine the probability that the network becomes disconnected due to edge failures. This problem is NP-complete. Given an approximation ratio $\epsilon > 1$, Karger [14] had an approximation scheme that initially computes the smallest

$O(n^{2\alpha})$ cuts, where $\alpha = O(1 - \log \epsilon / \log n)$, and then determines a solution based on the cuts. By using Vazirani and Yannakakis's algorithm, the running time is $\tilde{O}(mn^{2+2\alpha} + (n/\epsilon)^{2^{O(-log n \, \epsilon)}})$. By using our result, the following is obtained.

Theorem 9. *The approximation scheme in* [14] *can be implemented in* $O(mn^{1+2\alpha} + (n/\epsilon)^{2^{O(-log n \, \epsilon)}})$ *time.*

7   Concluding Remarks

In this paper, improved algorithms were proposed for the problems of enumerating the cuts of a graph by their weights. The presented algorithms use the schema of Vazirani and Yannakakis's enumeration algorithms. To enumerate a cut, their algorithms require at most $n - 1$ maximum flow computations, and ours require one maximum flow computation and two invocations of Hao and Orlin's minimum cut algorithm. Hao and Orlin's algorithm is effective in practice and easy to implement [3]. Therefore, our improvement is significant from both the theoretical and practical points of view.

References

1. Ahuja, R.K., Magnanti, T.L., Orlin, J.B.: Network Flows: Theory, Algorithms, and Applications. Prentice-Hall, Englewood Cliffs (1993)
2. Burlet, M., Goldschmidt, O.: A New and Improved Algorithm for the 3-Cut Problem. Operations Research Letters 21, 225–227 (1997)
3. Chekuri, C.S., Goldberg, A.V., Karger, D.R., Levine, M.S., Stein, C.: Experimental Study of Minimum Cut Algorithms. In: Proceedings of the Eighteenth Annual ACM-SIAM Symposium on Discrete Algorithm, pp. 324–333 (1997)
4. Dinits, E.A., Karzanov, A.V., Lomonosov, M.V.: On the Structure of a Family of Minimal Weighted Cuts in a Graph. In: Fridman, A.A. (ed.) Studies in Discrete Optimization, Nauka, Moscow, pp. 290–306 (1976)
5. Fleischer, L.: Building Chain and Cactus Representations of All Minimum Cuts from Hao-Orlin in the Same Asymptotic Run Time. Journal of Algorithms 33, 51–72 (1999)
6. Goldberg, A.V., Tarjan, R.E.: A New Approach to the Maximum Flow Problem. Journal of the ACM 35, 921–940 (1988)
7. Goldschmidt, O., Hochbaum, D.S.: Polynomial Algorithm for the k-Cut Problem for Fixed k. Mathematics of Operation Research 19, 24–37 (1994)
8. Gomory, R.E., Hu, T.C.: Multi-terminal Network Flows. Journal of the Society for Industrial and Applied Mathematics 9, 551–570 (1961)
9. Hao, J., Orlin, J.B.: A Faster Algorithm for Finding the Minimum Cut in a Directed Graph. Journal of Algorithms 17, 424–446 (1994)
10. Kamidoi, Y., Wakabayashi, S., Yoshida, N.: A Divide-and-Conquer Approach to the Minimum k-Way Cut Problem. Algorithmica 32, 262–276 (2002)
11. Kamidoi, Y., Yoshida, N., Nagamochi, H.: A Deterministic Algorithm for Finding all Minimum k-Way Cuts. SIAM Journal on Computing 36, 1315–1327 (2006)
12. Kapoor, S.: On Minimum 3-Cuts and Approximating k-Cuts Using Cut Trees. In: Proceedings of the 5th Integer Programming and Combinatorial Optimization Conference, pp. 132–146 (1996)

13. Karger, D.R., Stein, C.: A New Approach to the Minimum Cut Problem. Journal of the ACM 43, 601–640 (1996)
14. Karger, D.R.: A Randomized Fully Polynomial Time Approximation Scheme for the All-Terminal Network Reliability Problem. SIAM Journal on Computing 29, 492–514 (1999)
15. Karger, D.R.: Minimum Cuts in Near-Linear Time. Journal of the ACM 47, 46–76 (2000)
16. Karzanov, A.V., Timofeev, E.A.: Efficient algorithms for Finding All Minimal Edge Cuts of a Nonoriented Graph. Cybernetics 22, 156–162 (1986)
17. King, V., Rao, S., Tarjan, R.E.: A Faster Deterministic Maximum Flow Algorithm. Journal of Algorithms 17, 447–474 (1994)
18. Levine, M.S.: Faster Randomized Algorithms for Computing Minimum 3, 4, 5, 6-way Cuts. In: Proceedings of the Eleventh ACM-SIAM Symposium on Discrete Algorithms, pp. 735–742 (2000)
19. Nagamochi, H., Ibaraki, T.: Computing the Edge-Connectivity of Multigraphs and Capacitated Graphs. SIAM Journal on Discrete Mathematics 5, 54–66 (1992)
20. Nagamochi, H., Ibaraki, T.: A Fast Algorithm for Computing Minimum 3-way and 4-way Cuts. Mathematical Programming 88, 507–520 (2000)
21. Nagamochi, H., Nishimura, K., Ibaraki, T.: Computing all Small Cuts in Undirected Networks. SIAM Journal on Discrete Mathematics 10, 469–481 (1997)
22. Nagamochi, H., Nishimura, K., Ibaraki, T.: A Faster Algorithm for Computing Minimum 5-Way and 6-Way Cuts in Graphs. Journal of Combinatorial Optimization 4, 151–169 (2000)
23. Picard, J.C., Queyrane, M.: On the Structure of All Minimum Cuts in a Network and Applications. Mathematical Programming Study 13, 8–16 (1980)
24. Vazirani, V., Yannakakis, M.: Suboptimal Cuts: Their Enumeration, Weight and Number. In: Proceedings of the 19th International Colloquium on Automata, Languages and Programming, pp. 366–377 (1992)

Covering Directed Graphs by In-Trees

Naoyuki Kamiyama* and Naoki Katoh**

Department of Architecture and Architectural Engineering, Kyoto University,
Kyotodaigaku-Katsura, Nishikyo-ku, Kyoto, 615-8540, Japan
{is.kamiyama,naoki}@archi.kyoto-u.ac.jp

Abstract. Given a directed graph $D = (V, A)$ with a set of d specified vertices $S = \{s_1, \ldots, s_d\} \subseteq V$ and a function $f \colon S \to \mathbb{Z}_+$ where $\mathbb{Z}_+$ denotes the set of non-negative integers, we consider the problem which asks whether there exist $\sum_{i=1}^{d} f(s_i)$ in-trees denoted by $T_{i,1}, T_{i,2}, \ldots, T_{i,f(s_i)}$ for every $i = 1, \ldots, d$ such that $T_{i,1}, \ldots, T_{i,f(s_i)}$ are rooted at s_i, each $T_{i,j}$ spans vertices from which s_i is reachable and the union of all arc sets of $T_{i,j}$ for $i = 1, \ldots, d$ and $j = 1, \ldots, f(s_i)$ covers A. In this paper, we prove that such set of in-trees covering A can be found by using an algorithm for the weighted matroid intersection problem in time bounded by a polynomial in $\sum_{i=1}^{d} f(s_i)$ and the size of D. Furthermore, for the case where D is acyclic, we present another characterization of the existence of in-trees covering A, and then we prove that in-trees covering A can be computed more efficiently than the general case by finding maximum matchings in a series of bipartite graphs.

1 Introduction

The problem for covering a graph by subgraphs with specified properties (for example, trees or paths) is very important from practical and theoretical viewpoints and have been extensively studied. For example, Nagamochi and Okada [10] studied the problem for covering a set of vertices of a given undirected tree by subtrees, and Arkin et al. [1] studied the problem for covering a set of vertices or edges of a given undirected graph by subtrees or paths. These results were motivated by vehicle routing problems. Moreover, Even et al. [2] studied the covering problem motivated by nurse station location problems.

This paper studies the problem for covering a directed graph by rooted trees which is motivated by the following evacuation planning problem. Given a directed graph which models a city, vertices model intersections and buildings, and arcs model roads connecting these intersections and buildings. People exist not only at vertices but also along arcs. Suppose we have to give several evacuation instructions for evacuating all people to some safety place. In order to avoid disorderly confusion, it is desirable that one evacuation instruction gives a single evacuation path for each person and these paths do not cross each other. Thus,

---

* Supported by JSPS Research Fellowships for Young Scientists.
** Supported by the project *New Horizons in Computing*, Grant-in-Aid for Scientific Research on Priority Areas, MEXT Japan.

X. Hu and J. Wang (Eds.): COCOON 2008, LNCS 5092, pp. 444–457, 2008.

we want each evacuation instruction to become an in-tree rooted at some safety place. Moreover, the number of instructions for each safety place is bounded in proportion to the size of each safety place.

The above evacuation planning problem is formulated as the following covering problem defined on a directed graph. We are given a directed graph $D = (V, A, S, f)$ which consists of a vertex set V, an arc set A, a set of d specified vertices $S = \{s_1, \ldots, s_d\} \subseteq V$ and a function $f: S \to \mathbb{Z}_+$ where $\mathbb{Z}_+$ denotes the set of non-negative integers. In the above evacuation planning problem, S corresponds to a set of safety places, and $f(s_i)$ represents the upper bound of the number of evacuation instructions for $s_i \in S$. For each $i = 1, \ldots, d$, we define $V_D^i \subseteq V$ as the set of vertices in V from which s_i is reachable in D, and we define an in-tree rooted at s_i which spans V_D^i as a (D, s_i)-in-tree. Here an in-tree is a subgraph T of D such that T has no cycle when the direction of an arc is ignored and all arcs in T is directed to a root. We define a set $\mathcal{T}$ of $\sum_{i=1}^{d} f(s_i)$ subgraphs of D as a D-proper set of in-trees if $\mathcal{T}$ contains exactly $f(s_i)$ (D, s_i)-in-trees for every $i = 1, \ldots, d$. If every two distinct in-trees of a D-proper set $\mathcal{T}$ of in-trees are arc-disjoint, we call $\mathcal{T}$ a D-proper set of arc-disjoint in-trees. Furthermore, if the union of arc sets of all in-trees of a D-proper set $\mathcal{T}$ of in-trees is equal to A, we say that $\mathcal{T}$ covers A.

Four in-trees illustrated in Figure 2 compose a D-proper set $\mathcal{T}$ of in-trees which covers the arc set of a directed graph $D = (V, A, S, f)$ illustrated in Figure 1(a) where $S = \{s_1, s_2, s_3\}$, $f(s_1) = 2$, $f(s_2) = 1$ and $f(s_3) = 1$. However, $\mathcal{T}$ is not a D-proper set of arc-disjoint in-trees.

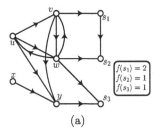
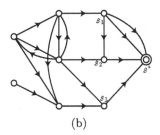

$$f(s_1) = 2$$
$$f(s_2) = 1$$
$$f(s_3) = 1$$

(a)   (b)

Fig. 1. (a) Directed graph D. (b) Transformed graph D^*.

We will study the problem for *covering directed graphs by in-trees* (in short CDGI), and we will present characterizations for a directed graph $D = (V, A, S, f)$ for which there exists a feasible solution of CDGI(D), and we will give a polynomial time algorithm for CDGI(D).

Problem: CDGI(D)
Input: a directed graph D;
Output: a D-proper set of in-trees which covers the arc set of D, if one exists.

A special class of the problem CDGI(D) in which S consists of a single vertex was considered by Vidyasankar [13] and Frank [3]. They showed the necessary

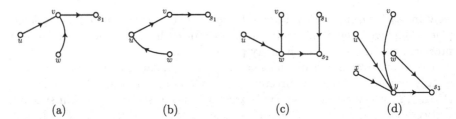

Fig. 2. (a) (D, s_1)-in-tree. (b) (D, s_1)-in-tree. (c) (D, s_2)-in-tree. (d) (D, s_3)-in-tree.

and sufficient condition in terms of linear inequalities that there exists a feasible solution of this problem. However, to the best of our knowledge, an algorithm for CDGI(D) was not presented.

Our results. We first show that CDGI(D) can be viewed as some type of the connectivity augmentation problem. After this, we will prove that this connectivity augmentation problem can be solved by using an algorithm for the weighted matroid intersection problem in time bounded by a polynomial in $\sum_{i=1}^{d} f(s_i)$ and the size of D (this generalizes the result by Frank [5]). Furthermore, for the case where D is acyclic, we show another characterization for D that there exists a feasible solution of CDGI(D). Moreover, we prove that in this case CDGI(D) can be solved more efficiently than the general case by finding maximum matchings in a series of bipartite graphs.

Outline. The rest of this paper is organized as follows. Section 2 gives necessary definitions and fundamental results. In Section 3, we give an algorithm for the problem CDGI. In Section 4, we consider the acyclic case.

2   Preliminaries

Let $D = (V, A, S, f)$ be a weakly connected directed graph which may have multiple arcs. Let $S = \{s_1, \ldots, s_d\}$. Since we can always cover by $|A|$ (D, s_i)-in-trees the arc set of the subgraph of D induced by V_D^i, we consider the problem by using at most $|A|$ (D, s_i)-in-trees. That is, without loss of generality, we assume that $f(s_i) \leq |A|$. For $B \subseteq A$, let $\partial^-(B)$ (resp. $\partial^+(B)$) be a set of tails (resp. heads) of arcs in B. For $e \in A$, we write $\partial^-(e)$ and $\partial^+(e)$ instead of $\partial^-(\{e\})$ and $\partial^+(\{e\})$, respectively. For $W \subseteq V$, we define $\delta_D(W) = \{e \in A: \partial^-(e) \in W, \partial^+(e) \notin W\}$. For $v \in V$, we write $\delta_D(v)$ instead of $\delta_D(\{v\})$. For two distinct vertices $u, v \in V$, we denote by $\lambda(u, v; D)$ the local arc-connectivity from u to v in D, i.e., $\lambda(u, v; D) = \min\{|\delta_D(W)|: u \in W, v \notin W, W \subseteq V\}$. We call a subgraph T of D *forest* if T has no cycle when we ignore the direction of arcs in T. If a forest T is connected, we call T *tree*. If every arc of an arc set B is parallel to some arc in A, we say that B is *parallel* to A. We denote a directed graph obtained by adding an arc set B to A by $D+B$, i.e., $D+B = (V, A \cup B, S, f)$. For $S' \subseteq S$, let $f(S') = \sum_{s_i \in S'} f(s_i)$. For $v \in V$, we denote by $R_D(v)$ a set of vertices in S which are reachable from v in D. For $W \subseteq V$, let $R_D(W) = \bigcup_{v \in W} R_D(v)$.

For an arc set B which is parallel to A, we clearly have for every $v \in V$

$$R_D(v) = R_{D+B}(v). \tag{1}$$

From (1), we have for every $i = 1, \ldots, d$

$$V_D^i = V_{D+B}^i. \tag{2}$$

We define D^* as a directed graph obtained from D by adding a new vertex s^* and connecting s_i to s^* with $f(s_i)$ parallel arcs for every $i = 1, \ldots, d$ (see Figure 1). We denote by A^* the arc set of D^*. From the definition of D^*,

$$|A^*| = \sum_{v \in V} |\delta_{D^*}(v)| = |A| + f(S). \tag{3}$$

We say that D is (S, f)-admissible if $|\delta_{D^*}(v)| \le f(R_D(v))$ holds for any $v \in V$.

2.1 Rooted Arc-Connectivity Augmentation by Reinforcing Arcs

Given a directed graph $D = (V, A, S, f)$, we call an arc set B with $A \cap B = \emptyset$ which is parallel to A a D^*-rooted connector if $\lambda(v, s^*; D^* + B) \ge f(R_D(v))$ holds for every $v \in V$. Notice that since a D^*-rooted connector B is parallel to A, B does not contain an arc which is parallel to an arc entering into s^* in D^*. Then, the problem rooted arc-connectivity augmentation by reinforcing arcs (in short RAA-RA) is formally defined as follows.

Problem: RAA-RA(D^*)

    Input: D^* of a directed graph D;
  Output: a minimum size D^*-rooted connector.

Notice that the problem RAA-RA(D^*) is not equivalent to the local arc-connectivity augmentation problem with minimum number of reinforcing arcs from $v \in V$ to $s_i \in R_D(v)$. For example, we consider D^* illustrated in Figure 3(a) of a directed graph $D = (V, A, S, f)$ where $S = \{s_1, s_2\}$, $f(s_1) = 2$ and $f(s_2) = 2$. The broken lines in Figure 3(b) represent a minimum D^*-rooted connector. For the problem that asks to increase the v-s_i local arc-connectivity for every $v \in V$ and $s_i \in R_D(v)$ to $f(s_i)$ by adding minimum parallel arcs to A (this problem is called the problem increasing arc-connectivity by reinforcing arcs in [7], in short IARA(D^*)), an optimal solution is a set of broken lines in Figure 3(c). While it is known [7] that IARA(D^*) is $\mathcal{NP}$-hard, it is known [5] that RAA-RA(D^*) in which S consists of a single element can be solved in time bounded by a polynomial in $f(S)$ and the size of D by using an algorithm for the weighted matroid intersection.

2.2 Matroids on Arc Sets of Directed Graphs

In this subsection, we define two matroids $M(D^*)$ and $U(D^*)$ on A^* for a directed graph $D = (V, A, S, f)$, which will be used in the subsequent discussion.

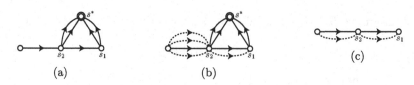

Fig. 3. (a) Input. (b) Optimal solution for RAA-RA. (c) Optimal solution for IARA.

We denote by $M = (E, \mathcal{I})$ a matroid on E whose collection of independent sets is $\mathcal{I}$. Introductory treatment of a matroid is given in [11].

For $i = 1, \ldots, d$ and $j = 1, \ldots, f(s_i)$, we define $M_{i,j}(D^*) = (A^*, \mathcal{I}_{i,j}(D^*))$ where $I \subseteq A^*$ belongs to $\mathcal{I}_{i,j}(D^*)$ if and only if both of a tail and a head of every arc in I are contained in $V_D^i \cup \{s^*\}$ and a directed graph $(V_D^i \cup \{s^*\}, I)$ is a forest. $M_{i,j}(D^*)$ is clearly a matroid (i.e. graphic matroid). Moreover, we denote the union of $M_{i,j}(D^*)$ for $i = 1, \ldots, d$ and $j = 1, \ldots, f(s_i)$ by $M(D^*) = (A^*, \mathcal{I}(D^*))$ in which $I \subseteq A^*$ belongs to $\mathcal{I}(D^*)$ if and only if I can be partitioned into $\{I_{i,1}, \ldots, I_{i,f(s_i)} : i = 1, \ldots, d\}$ such that each $I_{i,j}$ belongs to $\mathcal{I}_{i,j}(D^*)$. $M(D^*)$ is also a matroid (see Chapter 12.3 in [11]. This matroid is also called *matroid sum*). When $I \in \mathcal{I}(D^*)$ can be partitioned into $\{I_{i,1}, \ldots, I_{i,f(s_i)} : i = 1, \ldots, d\}$ such that a directed graph $(V_D^i \cup \{s^*\}, I_{i,j})$ is a tree for every $i = 1, \ldots, d$ and $j = 1, \ldots, f(s_i)$, we call I a *complete independent set of $M(D^*)$*.

Next we define another matroid. We define $U(D^*) = (A^*, \mathcal{J}(D^*))$ where $I \subseteq A^*$ belongs to $\mathcal{J}(D^*)$ if and only if I satisfies

$$|\delta_{D^*}(v) \cap I| \leq \begin{cases} f(R_D(v)), & \text{if } v \in V, \\ 0, & \text{if } v = s^*. \end{cases} \tag{4}$$

Since $U(D^*)$ is a direct sum of uniform matroids, $U(D^*)$ is also a matroid (see Exercise 7 of pp.16 and Example 1.2.7 in [11]). We call $I \in \mathcal{J}(D^*)$ a *complete independent set of $U(D)$* when (4) holds with equality.

For two matroids $M(D^*)$ and $U(D^*)$, we call an arc set $I \subseteq A^*$ *D^*-intersection* when $I \in \mathcal{I}(D^*) \cap \mathcal{J}(D^*)$. If a D^*-intersection I is a complete independent set of both $M(D^*)$ and $U(D^*)$, we call I *complete*.

When we are given a weight function $w \colon A^* \to \mathbb{R}_+$ where $\mathbb{R}_+$ denotes the set of non-negative reals, we define the weight of $I \subseteq A^*$ (denoted by $w(I)$) by the sum of weights of all arcs I. The *minimum weight complete intersection problem* (in short MWCI) is then defined as follows.

Problem: MWCI(D^*)
Input: D^* of a directed graph D and a weight function $w \colon A^* \to \mathbb{R}_+$;
Output: a minimum weight complete D^*-intersection, if one exists.

Lemma 1. *The problem MWCI(D^*) can be solved in $O(M|A^*|^6)$ time where $M = \sum_{v \in V} f(R_D(v))$.*

Proof. To prove the lemma, we use the following theorem concerning a matroid.

Theorem 1 ([9]). *Given a matroid $M = (E, \mathcal{I})$ which is a union of t ($\leq |E|$) matroids $M_1 = (E, \mathcal{I}_1), \ldots, M_t = (E, \mathcal{I}_t)$, we can test if a given set belongs to*

$\mathcal{I}$ in $O(|E|^3\gamma)$ time where γ is the time required to test if a given set belongs to $\mathcal{I}_1, \ldots, \mathcal{I}_t$.

Theorem 2 ([4]). *Given two matroids $M_1 = (E, \mathcal{I}_1)$ and $M_2 = (E, \mathcal{I}_2)$ with a weight function $w \colon E \to \mathbb{R}_+$ and a non-negative integer $k \in \mathbb{Z}_+$, we can find $I \in \mathcal{I}_1 \cap \mathcal{I}_2$ with $|I| = k$ whose weight is minimum among all $I' \in \mathcal{I}_1 \cap \mathcal{I}_2$ with $|I'| = k$ in $O(k|E|^3 + k|E|^2\gamma)$ time if one exists where γ is the time required to test if a given set belongs to both $\mathcal{I}_1$ and $\mathcal{I}_2$.*

Furthermore, we use the following lemma without the proof.

Lemma 2. *Given a directed graph $D = (V, A, S, f)$ and $I \subseteq A^*$, I is a complete D^*-intersection if and only if I is a D^*-intersection with $|I| = \sum_{v \in V} f(R_D(v))$.*

We consider the time required to test if a given set belongs to both $\mathcal{I}(D^*)$ and $\mathcal{J}(D^*)$. Since it is not difficult to see that we can test if a given set belongs to each $\mathcal{I}_{i,j}(D^*)$ in $O(|A^*|)$ time, we can test if a given set belongs to $\mathcal{I}(D^*)$ in $O(|A^*|^4)$ time from Theorem 1. Notice that $M(D^*)$ is the union of $f(S)$ matroids and $f(S) \leq |A^*|$ follows from (3). For $\mathcal{J}(D^*)$, the time complexity is clearly $O(|A^*|)$ time. From Lemma 2, a complete D^*-intersection is a D^*-intersection with $|I| = M$. Thus, the total time required for solving MWCI(D^*) is $O(M|A^*|^6)$ from Theorem 2. $\qquad\square$

2.3   Results from [8]

In this section, we introduce results concerning packing of in-trees given by Kamiyama et al. [8] which plays a crucial role in this paper.

Theorem 3 ([8]). *Given a directed graph $D = (V, A, S, f)$, the following three statements are equivalent: (i) For every $v \in V$, $\lambda(v, s^*; D^*) \geq f(R_D(v))$ holds. (ii) There exists a D-proper set of arc-disjoint in-trees. (iii) There exists a complete D^*-intersection.*

From Theorem 3, we obtain the following corollary.

Corollary 1. *Given a directed graph $D = (V, A, S, f)$ and an arc set B with $A \cap B = \emptyset$ which is parallel to A, the following three statements are equivalent: (i) B is a D^*-rooted connector. (ii) There exists a $(D + B)$-proper set of arc-disjoint in-trees. (iii) There exists a complete $(D + B)^*$-intersection.*

Proof. The equivalence of (ii) and (iii) follows from Theorem 3.
   (i)→(ii). Since B is parallel to A, we clearly have

$$(D + B)^* = D^* + B. \tag{5}$$

Since B is a D^*-rooted connector and from (5) and (1), we have for every $v \in V$

$$\lambda(v, s^*; (D + B)^*) = \lambda(v, s^*; D^* + B) \geq f(R_D(v)) = f(R_{D+B}(v)).$$

From this inequality and Theorem 3, this part follows.

(ii)→(i). Since there exists a $(D + B)$-proper set of arc-disjoint in-trees and from (5), Theorem 3 and (1), we have for every $v \in V$

$$\lambda(v, s^*; D^* + B) = \lambda(v, s^*; (D + B)^*) \geq f(R_{D+B}(v)) = f(R_D(v)).$$

This proves that B is a D^*-rooted connector.   □

Although the following theorem is not explicitly proved in [8], we can easily obtain it from the proof of Theorem 3 in [8].

Theorem 4 ([8]). *Given a directed graph $D = (V, A, S, f)$ which satisfies the condition of Theorem 3, we can find a D-proper set of arc-disjoint in-trees in $O(M^2|A^*|^2)$ time where $M = \sum_{v \in V} f(R_D(v))$.*

3   An Algorithm for Covering by In-Trees

Given a directed graph $D = (V, A, S, f)$, we present in this section an algorithm for CDGI(D). The time complexity of the proposed algorithm is bounded by a polynomial in $f(S)$ and the size of D. We first prove that CDGI(D) can be reduced to RAA-RA(D^*). After this, we show that RAA-RA(D^*) can be reduced to the problem MWCI.

3.1   Reduction from CDGI to RAA-RA

If $D = (V, A, S, f)$ is not (S, f)-admissible, i.e., $|\delta_{D^*}(v)| > f(R_D(v))$ for some $v \in V$, there exists no feasible solution of CDGI(D) since there can not be a D-proper set of in-trees that covers $\delta_D(v)$ from the definition of a D-proper set of in-trees. Thus, we assume in the subsequent discussion that D is (S, f)-admissible. The proof of the following proposition is omitted.

Proposition 1. *Given an (S, f)-admissible directed graph $D = (V, A, S, f)$, the size of a D^*-rooted connector is at least $\sum_{v \in V} f(R_D(v)) - (|A| + f(S))$.*

For an (S, f)-admissible directed graph $D = (V, A, S, f)$, we define opt_D by

$$\mathrm{opt}_D = \sum_{v \in V} f(R_D(v)) - (|A| + f(S)). \tag{6}$$

From Proposition 1, the size of a D^*-rooted connector is at least opt_D.

Lemma 3. *Given an (S, f)-admissible directed graph $D = (V, A, S, f)$, there exists a feasible solution of CDGI(D) if and only if there exists a D^*-rooted connector whose size is equal to opt_D.*

Proof. **Only if-part.** Suppose there exists a feasible solution of CDGI(D), i.e., there exists a D-proper set $\mathcal{T}$ of in-trees which covers A. For each $i = 1, \ldots, d$, we denote $f(s_i)$ (D, s_i)-in-trees of $\mathcal{T}$ by $T_{i,1}, \ldots, T_{i,f(s_i)}$. For each $e \in A$, let $P_e = \{(i, j): e$ is contained in $T_{i,j}\}$. Since $\mathcal{T}$ covers A, each $e \in A$ is contained in at least one in-tree of $\mathcal{T}$. Thus, $|P_e| \geq 1$ holds for every $e \in A$. Let an arc set

$B = \bigcup_{e \in A}\{|P_e| - 1 \text{ copies of } e\}$. We will prove that B is a D^*-rooted connector whose size is equal to opt_D.

We first prove $|B| = \text{opt}_D$. For this, we show that for every $v \in V$

$$\sum_{e \in \delta_D(v)}(|P_e| - 1) = f(R_D(v)) - |\delta_{D^*}(v)|. \tag{7}$$

Let us first consider $v \notin S$. For $s_i \in R_D(v)$, $T_{i,j}$ contains v since $T_{i,j}$ spans V_D^i and s_i is reachable from v. Hence, since $T_{i,j}$ is an in-tree and v is not a root of $T_{i,j}$ from $v \notin S$, $T_{i,j}$ contains exactly one arc $e \in \delta_D(v)$, i.e., (i, j) is contained in P_e for exactly one arc $e \in \delta_D(v)$. Thus, $\sum_{e \in \delta_D(v)}|P_e| = \sum_{s_i \in R_D(v)} f(s_i) = f(R_D(v))$. From this equation and since $|\delta_D(v)| = |\delta_{D^*}(v)|$ follows from $v \notin S$, (7) holds. In the case of $v \in S$, for $s_i \in R_D(v) \setminus \{v\}$, (i, j) is contained in P_e for exactly one arc $e \in \delta_D(v)$ as in the case of $v \notin S$. Thus, $\sum_{e \in \delta_D(v)}|P_e| = f(R_D(v)) - f(v)$. From this equation and $|\delta_{D^*}(v)| = |\delta_D(v)| + f(v)$,

$$\sum_{e \in \delta_D(v)}(|P_e| - 1) = f(R_D(v)) - f(v) - |\delta_D(v)| = f(R_D(v)) - |\delta_{D^*}(v)|.$$

This completes the proof of (7). Since B contains $|P_e| - 1$ copies of $e \in A$,

$$|B| = \sum_{v \in V}\sum_{e \in \delta_D(v)}(|P_e| - 1) = \sum_{v \in V}(f(R_D(v)) - |\delta_{D^*}(v)|) \quad \text{(from (7))}$$
$$= \text{opt}_D \quad \text{(from (3) and (6))}.$$

What remains is to prove that B is a D^*-rooted connector. From Corollary 1, it is sufficient to prove that there exists a $(D + B)$-proper set of arc-disjoint in-trees. For this, we will construct from $\mathcal{T}$ a set $\mathcal{T}'$ of arc-disjoint in-trees which consists of $T'_{i,1}, \ldots, T'_{i,f(s_i)}$ for $i = 1, \ldots, d$, and we prove that $\mathcal{T}'$ is a $(D + B)$-proper set of in-trees. Each $T'_{i,j}$ is constructed from $T_{i,j}$ as follows. When $e \in A$ is contained in more than one in-tree of $\mathcal{T}$, in order to construct $\mathcal{T}'$ from $\mathcal{T}$, we need to replace e of $T_{i,j}$ by an arc in B which is parallel to e for every $(i, j) \in P_e$ except one in-tree. For $(i_{\min}, j_{\min}) \in P_e$ which is lexicographically smallest in P_e, we allow $T'_{i_{\min}, j_{\min}}$ to use e, while for $(i, j) \in P_e \setminus (i_{\min}, j_{\min})$, we replace e of $T_{i,j}$ by an arc in B which is parallel to e so that for distinct $(i, j), (i', j') \in P_e \setminus (i_{\min}, j_{\min})$, the resulting $T'_{i,j}$ and $T'_{i',j'}$ contain distinct arcs which are parallel to e, respectively (see Figure 4).

We will do this operation for every $e \in A$. Let $\mathcal{T}'$ be the set of in-trees obtained by performing the above operation for every $e \in A$. Here we show that $\mathcal{T}'$ is a $(D + B)$-proper set of arc-disjoint in-trees. Since $T'_{i,j}$ and $T'_{i',j'}$ are arc-disjoint for $(i, j) \neq (i', j')$ from the way of constructing $\mathcal{T}'$, it is sufficient to prove that $T'_{i,j}$ is a $(D + B, s_i)$-in-tree. Since $T'_{i,j}$ is constructed by replacing arcs of $T_{i,j}$ by the corresponding parallel arc in B and $T_{i,j}$ is an in-tree rooted at s_i, $T'_{i,j}$ is also an in-tree rooted at s_i. Since $T_{i,j}$ spans V_D^i and from (2), $T'_{i,j}$ spans V_{D+B}^i. Hence, $T'_{i,j}$ is a $(D + B, s_i)$-in-tree. This completes the proof.

If-part. Let B be a D^*-rooted connector with $|B| = \text{opt}_D$. From Corollary 1, there exists a $(D + B)$-proper set $\mathcal{T}'$ of arc-disjoint in-trees. For each $i = 1, \ldots, d$, we denote $f(s_i)$ $(D + B, s_i)$-in-trees of $\mathcal{T}'$ by $T'_{i,1}, \ldots, T'_{i,f(s_i)}$. We will prove

Fig. 4. Illustration of the replacing operation. Let e be an arc in A, and let e', e'' be arcs in B. Assume that $P_e = \{(1,1), (1,2), (2,1)\}$. In this case, $T_{1,1}$, $T_{1,2}$ and $T_{2,1}$ contain e. Then, $T'_{1,1}$ contains e, $T'_{1,2}$ contains e', and $T'_{2,1}$ contains e''.

that we can construct from T' a D-proper set of in-trees covering A. We first construct from T' a set T of in-trees which consists of $T_{i,j}$ for $i = 1, \ldots, d$ and $j = 1, \ldots, f(s_i)$ by the following procedure **Replace**.

Procedure Replace: For each $i = 1, \ldots, d$ and $j = 1, \ldots, f(s_i)$, set $T_{i,j}$ to be a directed graph obtained from $T'_{i,j}$ by replacing every arc $e \in B$ which is contained in $T'_{i,j}$ by an arc in A which is parallel to e.

From now on, we prove that T is a D-proper set of in-trees which covers A. It is not difficult to prove that T is a D-proper set of in-trees from the definition of the procedure **Replace** in the same manner as the last part of the proof of the "only if-part". Thus, it is sufficient to prove that T covers A. For this, we first show that T' covers $A \cup B$. From $A \cap B = \emptyset$, $|B| = \mathrm{opt}_D$ and (6),

$$|A \cup B| = |A| + \mathrm{opt}_D = \sum_{v \in V} f(R_D(v)) - f(S). \tag{8}$$

Recall that each $v \in V$ is contained in $f(R_{D+B}(v))$ in-trees of T' from the definition of a $(D+B)$-proper set of in-trees. Thus, since in-trees of T' are arc-disjoint, it holds for each $v \in V$ that the number of arcs in $\delta_{D+B}(v)$ which are contained in in-trees of T' is equal to (i) $f(R_{D+B}(v))$ if $v \in V \setminus S$, or (ii) $f(R_{D+B}(v)) - f(v)$ if $v \in S$. Hence, the number of arcs in $A \cup B$ contained in in-trees of T' is equal to

$$\sum_{v \in V \setminus S} f(R_{D+B}(v)) + \sum_{v \in S}(f(R_{D+B}(v)) - f(v))$$
$$= \sum_{v \in V} f(R_{D+B}(v)) - f(S) = \sum_{v \in V} f(R_D(v)) - f(S) \quad \text{(from (1))}. \tag{9}$$

Since any arc of T' is in $A \cup B$ and the number of arcs in $A \cup B$ is equal to that of T' from (8) and (9), T' contains all arcs in $A \cup B$. Thus, T covers A from the definition of the procedure **Replace**.   □

As seen in the proof of the "if-part" of Lemma 3, if we can find a D^*-rooted connector B with $|B| = \mathrm{opt}_D$, we can compute a D-proper set of in-trees which covers A by using the procedure **Replace** from a $(D+B)$-proper set of arc-disjoint in-trees. Furthermore, we can construct a $(D+B)$-proper set of arc-disjoint in-trees by using the algorithm of Theorem 4. Since the optimal value of RAA-RA(D^*) is at least opt_D from Proposition 1, we can test if there exists a D^*-rooted connector whose size is equal to opt_D by solving RAA-RA(D^*). Assuming that we can solve RAA-RA(D^*), our algorithm for finding a D-proper set of in-trees which covers A called Algorithm **CR** can be illustrated as Algorithm 1 below.

From the Lemma 3, we can obtain the following lemma. The proof is omitted.

Algorithm 1. Algorithm CR

Input: a directed graph $D = (V, A, S, f)$
Output: a D-proper set of in-trees covering A, if one exists
 1: **if** D is not (S, f)-admissible **then**
 2:    Halt (there exists no D-proper set of in-trees covering A)
 3: **end if**
 4: Find an optimal solution B of RAA-RA(D^*)
 5: **if** $|B| > \mathrm{opt}_D$ **then**
 6:    Halt (there exists no D-proper set of in-trees covering A)
 7: **else**
 8:    Construct a $(D + B)$-proper set T' of arc-disjoint in-trees
 9:    Construct a set T of in-trees from T' by using the procedure **Replace**
10:    **return** T
11: **end if**

Lemma 4. *Given a directed graph $D = (V, A, f, S)$, Algorithm CR correctly finds a D-proper set of in-trees which covers A in $O(\gamma_1 + |V||A| + M^4)$ time if one exists where γ_1 is the time required to solve RAA-RA(D^*) and $M = \sum_{v \in V} f(R_D(v))$.*

3.2   Reduction from RAA-RA to MWCI

From the algorithm CR in Section 3.1, in order to present an algorithm for CDGI(D), what remains is to show how we solve RAA-RA(D^*). In this section, we will prove that we can test whether there exists a D^*-rooted connector whose size is equal to opt_D (i.e., Steps 4 and 5 in the algorithm CR) by reducing it to the problem MWCI. Our proof is based on the algorithm of [5] for RAA-RA(D^*) for $D = (V, A, S, f)$ with $|S| = 1$. We extend the idea of [5] to the general case by using Theorem 3. We define a directed graph D_+ obtained from an (S, f)-admissible directed graph $D = (V, A, S, f)$ by adding opt_D parallel arcs to every $e \in A$. Then, we will compute a D^*-rooted connector whose size is equal to opt_D by using an algorithm for MWCI(D_+^*) as described below. Since the number of arcs in a D^*-rooted connector whose size is equal to opt_D which are parallel to one arc in A is at most opt_D, it is enough to add opt_D parallel arcs to each arc of A in D_+ in order to find a D^*-rooted connector whose size is equal to opt_D.

We denote by A_+ and A_+^* the arc sets of D_+ and D_+^*, respectively. If $I \subseteq A_+^*$ is a complete D_+^*-intersection, since I is a complete independent set of $U(D_+^*)$ and from (4) and (1),

$$|I| = \sum_{v \in V} f(R_{D_+}(v)) = \sum_{v \in V} f(R_D(v)). \tag{10}$$

We define a weight function $w \colon A_+^* \to \mathbb{R}_+$ by

$$w(e) = \begin{cases} 0, & \text{if } e \in A^*, \\ 1, & \text{otherwise.} \end{cases} \tag{11}$$

The following lemma shows the relation between RAA-RA(D^*) and MWCI(D_+^*).

Lemma 5. *Given an (S, f)-admissible directed graph $D = (V, A, S, f)$ and a weight function $w\colon A_+^* \to \mathbb{R}_+$ defined by (11), there exists a D^*-rooted connector whose size is equal to opt_D if and only if there exists a complete D_+^*-intersection whose weight is equal to opt_D.*

To prove Lemma 5, we need the following lemmas. The proofs are omitted.

Lemma 6. *Given a directed graph $D = (V, A, S, f)$ and an arc set B with $A \cap B = \emptyset$ which is parallel to A, (i) if there is a complete D^*-intersection I, I is also a complete $(D + B)^*$-intersection, and (ii) if there is a complete $(D + B)^*$-intersection I such that $I \subseteq A^*$, I is also a complete D^*-intersection.*

Lemma 7. *Given D_+^* of an (S, f)-admissible directed graph $D = (V, A, S, f)$ and a weight function $w\colon A_+^* \to \mathbb{R}_+$ defined by (11), if there exists a complete D_+^*-intersection $I \subseteq A_+^*$, $w(I) \geq \mathsf{opt}_D$. Moreover, $w(I) = \mathsf{opt}_D$ if and only if $A^* \subseteq I$.*

Proof (Lemma 5). **Only if-part.** Assume that there exists a D^*-rooted connector whose size is equal to opt_D. Since D_+ has opt_D parallel arcs to every $e \in A$, there exists a D^*-rooted connector $B \subseteq A_+ \setminus A$ with $|B| = \mathsf{opt}_D$. Let us fix a D^*-rooted connector $B \subseteq A_+ \setminus A$ with $|B| = \mathsf{opt}_D$. From Corollary 1, there exists a complete $(D + B)^*$-intersection I. Thus, from (i) of Lemma 6, I is a complete D_+^*-intersection. Hence, what remains is to prove $w(I) = \mathsf{opt}_D$. Since the arc set of $(D + B)^*$ is equal to $A^* \cup B$ and I is a $(D + B)^*$-intersection, $I \subseteq A^* \cup B$ holds. Thus, since $w(A^* \cup B) = |B| = \mathsf{opt}_D$ follows from (11), $w(I) \leq w(A^* \cup B) = \mathsf{opt}_D$ holds. Hence, $w(I) = \mathsf{opt}_D$ follows from Lemma 7. This completes the proof.

If-part. Assume that there exists a complete D_+^*-intersection I with $w(I) = \mathsf{opt}_D$. Let B be $I \setminus A^*$, and we will prove that B is a D^*-rooted connector with $|B| = \mathsf{opt}_D$. We first prove B is a D^*-rooted connector by using (ii) of Lemma 6 and Corollary 1. We set B and D in Lemma 6 to be $A_+ \setminus (A \cup B)$ and $D + B$, respectively. Notice that $(D + B) + (A_+ \setminus (A \cup B)) = D_+$ follows from $B \subseteq A_+$ and $A_+ \setminus (A \cup B)$ is parallel to $A \cup B$. From $B = I \setminus A^*$, we have $I \subseteq A^* \cup B$. Thus, I is a complete $(D + B)^*$-intersection since I is a complete D_+^*-intersection and from (ii) of Lemma 6. Hence, from Corollary 1, B is a D^*-rooted connector.

What remains is to prove that $|B| = \mathsf{opt}_D$. From Lemma 7 and $w(I) = \mathsf{opt}_D$, $A^* \subseteq I$ holds. Thus, from $B = I \setminus A^*$ and (10), $|B| = |I \setminus A^*| = |I| - |A^*| = \sum_{v \in V} f(R_D(v)) - (|A| + f(S))$. This equation and (6) complete the proof. $\square$

As seen in the proof of the "if-part" of Lemma 5, if we can find a complete D_+^*-intersection I with $w(I) = \mathsf{opt}_D$, we can find a D^*-rooted connector B with $|B| = \mathsf{opt}_D$ by setting $B = I \setminus A^*$. Furthermore, we can obtain a complete D_+^*-intersection whose weight is equal to opt_D if one exists by using the algorithm for $\mathrm{MWCI}(D_+^*)$ since the optimal value of $\mathrm{MWCI}(D_+^*)$ is at least opt_D from Lemma 7. The formal description of the algorithm called Algorithm RM for finding a D^*-rooted connector whose size is equal to opt_D is illustrated in Algorithm 2.

The following lemma immediately follows from Lemma 5.

Algorithm 2. Algorithm RM

Input: D^* of an (S, f)-admissible directed graph $D = (V, A, S, f)$
Output: a D^*-rooted connector whose size is equal to opt_D, if one exists
 1: Find an optimal solution I for MWCI(D_+^*) with a weight function w defined by
    (11)
 2: **if** there exists no solution of MWCI(D_+^*) or $w(I) > \mathrm{opt}_D$ **then**
 3:     Halt (There exists no D^*-rooted connector whose size is equal to opt_D)
 4: **end if**
 5: **return** $I \setminus A^*$

Lemma 8. *Given D^* of an (S, f)-admissible directed graph $D = (V, A, f, S)$, Algorithm RM correctly finds a D^*-rooted connector whose size is equal to opt_D in $O(\gamma_2 + M|A|)$ time if one exists where γ_2 is the time required to solve MWCI(D_+^*) and $M = \sum_{v \in V} f(R_D(v))$.*

3.3   Algorithm for CDGI

We are ready to explain the formal description of our algorithm called Algorithm Covering for CDGI(D). Algorithm Covering is the same as Algorithm CR such that Steps 4, 5 and 6 are replaced by Algorithm RM.

Theorem 5. *Given a directed graph $D = (V, A, S, f)$, Algorithm Covering correctly finds a D-proper set of in-trees which covers A in $O(M^7|A|^6)$ time if one exits where $M = \sum_{v \in V} f(R_D(v))$.*

Proof. The correctness of the algorithm follows from Lemmas 4 and 8. We then consider the time complexity of this algorithm. From Lemmas 4 and 8, what remains is to analyze the time required to solve MWCI(D_+^*). If D is (S, f)-admissible, $|A^*| = \sum_{v \in V} |\delta_{D^*}(v)| \le \sum_{v \in V} f(R_D(v)) = M$. Thus, since D_+^* has opt_D parallel arcs of every $e \in A$, $|A_+^*| = |A^*| + \sum_{e \in A} \mathrm{opt}_D \le M + M|A|$. Hence we have $|A_+^*| = O(M|A|)$. Thus, from Lemma 1, we can solve MWCI(D^*) in $O(M^7|A|^6)$ time. From this discussion and Lemmas 4 and 8, we obtain the theorem. $\square$

4   Acyclic Case

In this section, we show that in the case where $D = (V, A, S, f)$ is acyclic, a D-proper set of in-trees covering A can be computed more efficiently than the general case. For this, we prove the following theorem.

Theorem 6. *Given an acyclic directed graph $D = (V, A, S, f)$, there exists a D-proper set of in-trees which covers A if and only if*

$$|B| \le f(R_D(\partial^+(B))) \text{ for every } v \in V \text{ and } B \subseteq \delta_D(v). \tag{12}$$

Proof (Sketch). For each $v \in V$, we define an undirected bipartite graph $G_v = (X_v \cup Y_v, E_v)$ which is necessary to prove the theorem. Let $X_v = \{x_e : e \in \delta_D(v)\}$

Fig. 5. (a) Input acyclic directed graph D. (b) Bipartite graph G_u for u in (a).

and $Y_v = \{y_{i,j} : s_i \in R_D(v), j = 1, \ldots, f(s_i)\}$. $x_e \in X_v$ and $y_{i,j} \in Y_v$ are connected by an edge in E_v if and only if s_i is reachable from $\partial^+(e)$ (see Figure 5).

It is well-known that (12) is equivalent to the necessary and sufficient condition that for any $v \in V$, there exists a matching in G_v which saturates vertices in X_v (e.g., Theorem 16.7 in Chapter 16 of [12]). Thus it is sufficient to prove that there exists a D-proper set of in-trees which covers A if and only if for any $v \in V$, there exists a matching in G_v which saturates vertices in X_v.   □

From Theorem 6, instead of the algorithm presented in Section 3, we can more efficiently find a D-proper set of in-trees covering A by finding a maximum matching in a bipartite graph $O(|V|)$ times. In regard to algorithms for finding a maximum matching in a bipartite graph, see e.g. [6].

Corollary 2. *Given an acyclic directed graph $D = (V, A, S, f)$, we can find a D-proper set of in-trees which covers A in $O(\mathsf{match}(M + |A|, M|A|))$ time if one exists where $\mathsf{match}(n, m)$ represents the time required to find a maximum matching in a bipartite graph with n vertices and m arcs and $M = \sum_{v \in V} f(R_D(v))$.*

Acknowledgement. We thank Prof. Tibor Jordán who informed us of the paper [5] and we are grateful to Shin-ichi Tanigawa for helpful comments.

References

1. Arkin, E.M., Hassin, R., Levin, A.: Approximations for minimum and min-max vehicle routing problems. J. Algorithms 59(1), 1–18 (2006)
2. Even, G., Garg, N., Könemann, J., Ravi, R., Sinha, A.: Min-max tree covers of graphs. Oper. Res. Lett. 32(4), 309–315 (2004)
3. Frank, A.: Covering branchings. Acta Scientiarum Mathematicarum [Szeged] 41, 77–81 (1979)
4. Frank, A.: A weighted matroid intersection algorithm. J. Algorithms 2(4), 328–336 (1981)
5. Frank, A.: Rooted k-connections in digraphs. Discrete Applied Mathematics (to appear)
6. Hopcroft, J.E., Karp, R.M.: An $n^{5/2}$ algorithm for maximum matchings in bipartite graphs. SIAM J. Comput. 2(4), 225–231 (1973)
7. Jordan, T.: Two NP-complete augmentation problems. Technical Report 8, Department of Mathematics and Computer Science, Odense University (1997)

8. Kamiyama, N., Katoh, N., Takizawa, A.: Arc-disjoint in-trees in directed graphs. In: Proc. the Nineteenth Annual ACM-SIAM Symposium on Discrete Algorithms (SODA 2008), pp. 518–526 (2008)
9. Knuth, D.: Matroid partitioning. Technical Report STAN-CS-73-342, Computer Science Department, Stanford University (1974)
10. Nagamochi, H., Okada, K.: Approximating the minmax rooted-tree cover in a tree. Inf. Process. Lett. 104(5), 173–178 (2007)
11. Oxley, J.G.: Matroid theory. Oxford University Press, Oxford (1992)
12. Schrijver, A.: Combinatorial Optimization: Polyhedra and Efficiency. Springer, Heidelberg (2003)
13. Vidyasankar, K.: Covering the edge set of a directed graph with trees. Discrete Mathematics 24, 79–85 (1978)

On Listing, Sampling, and Counting the Chordal Graphs with Edge Constraints

Shuji Kijima[1], Masashi Kiyomi[2], Yoshio Okamoto[3], and Takeaki Uno[4]

[1] Research Institute for Mathematical Sciences,
Kyoto University, Kyoto, 606-8502, Japan
kijima@kurims.kyoto-u.ac.jp
[2] School of Information Science, Japan Advanced Institute of Science and Technology,
1-1 Asahidai, Nomi, Ishikawa, 923-1292, Japan
mkiyomi@jaist.ac.jp
[3] Graduate School of Information Science and Engineering,
Tokyo Institute of Technology,
2-12-1-W8-88, Ookayama, Meguro-ku, Tokyo, 152-8552, Japan
okamoto@is.titech.ac.jp
[4] National Institute of Informatics,
2-1-2 Hitotsubashi, Chiyoda-ku, Tokyo, 101-8430, Japan
uno@nii.ac.jp

Abstract. We discuss the problems to list, sample, and count the chordal graphs with edge constraints. The objects we look at are chordal graphs sandwiched by a given pair of graphs where we assume at least one of the input pair is chordal. The setting is a natural generalization of chordal completions and deletions. For the listing problem, we give an efficient algorithm running in polynomial time per output with polynomial space. As for the sampling problem, we give two clues that seem to imply that a random sampling is not easy. The first clue is that we show #P-completeness results for counting problems. The second clue is that we give an instance for which a natural Markov chain suffers from an exponential mixing time. These results provide a unified viewpoint from algorithms theory to problems arising from various areas such as statistics, data mining, and numerical computation.

1 Introduction

A graph is *chordal* if it has no induced cycle of length more than three. The class of chordal graphs often appears as a tractable case of a lot of problems arising from various areas such as statistics, optimization, numerical computation, etc. In those areas, we often approximate a given graph by a chordal graph and then apply efficient algorithms for chordal graphs to the obtained graph. Evaluation criteria for chordal approximations depend on applications. For example, in the context of graphical modeling in statistics, a chordal approximation is desired to minimize AIC (Akaike's Information Criterion), BIC (Bayesian Information Criterion), MDL (Minimum Description Length), etc. [20],[26],[31]; in the context of numerical computation, a chordal approximation is desired to minimize the

X. Hu and J. Wang (Eds.): COCOON 2008, LNCS 5092, pp. 458–467, 2008.

number of added edges (a.k.a. the minimum fill-in problem) [22],[23],[29],[5]; in the context of discrete optimization, a chordal approximation is desired to minimize the size of a largest clique (a.k.a. the treewidth problem) [21],[14],[15],[4].

Since we are concerned with various sorts of criteria and often these computational problems are NP-hard, listing algorithms and random-sampling algorithms can be useful universal decision-support schemes. An exhaustive list found by an algorithm may provide an exact solution, whereas random samples may provide an approximative solution. Our goal is to provide efficient algorithms for listing problems and random-sampling problems of graphs, or to show the intractability of the problems.

As a chordal approximation, we consider the following two types of changes; either we just insert some edges or we just delete some edges to make a given graph G chordal. A result of the former operation is called a *chordal completion* of G, and a result of the latter operation is called a *chordal deletion* of G. As a computational problem, given a graph G we want to deal with all chordal completions of G or all chordal deletions of G.

In fact, we study a generalized problem of this kind. Namely, we are given two graphs $\overline{G}$ and $\underline{G}$ on the same vertex set such that $\underline{G}$ is contained in $\overline{G}$ and one of them is chordal, we want to deal with all chordal graphs that contain $\underline{G}$ and are contained in $\overline{G}$. When $\overline{G}$ is chordal, this problem generalizes the problem on chordal completions (since a complete graph is chordal), and when $\underline{G}$ is chordal, this problem generalizes the problem on chordal deletions (since an empty graph is chordal).

There are (at least) two reasons why we study this generalized problem. The first one is clear: this is more general. The second one comes from a more practical aspect. Since the number of chordal completions of a graph can be quite huge, it would be difficult and even impossible in most of the cases to run a listing algorithm to obtain the exhaustive list of the chordal completions. Also for random sampling, if the size of our sample space is quite large then the probability of needling a desired object will be pretty small. Indeed, as Wormald [32] showed, the number of chordal graphs with n vertices is asymptotically $\sum_{r=0}^{n} \binom{n}{r} 2^{r(n-r)}$, which is roughly $2^{n^2/4 + O(n \log n)}$. Thus, dealing with all chordal graphs is impractical, and a simple and natural way to narrow down the size of our list is to introduce a way to filter out some undesired candidates from the list, and a way to find a "suitable" chordal approximation in a more flexible manner when combined with several heuristics or local search strategies.

Results. We provide an efficient listing algorithm to enumerate all chordal graphs containing a given $\underline{G}$ and contained in a given $\overline{G}$ when $\overline{G}$ or $\underline{G}$ is chordal. The running time is polynomial in the input size per output, and the memory usage is bounded by a polynomial in the input size. The algorithm is based on a binary partition method, which is much simpler than the previous algorithms by Kiyomi and Uno [12] to list all chordal deletions and by Kiyomi, Kijima, and Uno [13] to list all chordal completions. Note also that these previous algorithms are not able to deal with our generalized problems.

As for the random sampling, we give two clues that seem to imply that a random sampling is not easy. The first clue is that counting the chordal graphs containing $\underline{G}$ and contained in $\overline{G}$ is #P-complete, even when $\underline{G}$ is chordal. The proof is done by a parsimonious reduction from the forest counting in a graph. To the best of our knowledge, this is the first result on #P-hardness for the graph sandwich problems. We also show that counting the chordal deletions is #P-complete by a Cook reduction from the forest counting. These results imply that a simple binary partition method does not yield a polynomial-time sampling algorithm. The second clue is the following. We apply the Markov chain Monte Carlo (MCMC) method to our problem. The MCMC is a promising approach for an efficient random sampling from a family of objects that is hard to count. We show that a simple and natural Markov chain suffers from slow mixing time; namely, we give an example for which the mixing time of the Markov chain is exponential. For detail of some proofs, we refer to the technical report version [11].

Related Work. Our generalized concept is actually a special case of the framework proposed by Golumbic, Kaplan, and Shamir [7] who studied the following *graph sandwich problem*: for a graph property Γ, we are given a pair of graphs $\overline{G}$ and $\underline{G}$ such that $\underline{G}$ is a subgraph of $\overline{G}$, and we are asked to decide if there exists a graph $G \in \Gamma$ that is a supergraph of $\underline{G}$ and at the same time a subgraph of $\overline{G}$. Golumbic, Kaplan, and Shamir [7] proved that graph sandwich problems are NP-complete for many graph properties, e.g., chordal graphs, perfect graphs, interval graphs, etc. As discussed above, for listing problems, Kiyomi and Uno [12] proposed algorithms to list the chordal deletions within constant time delay, and Kiyomi, Kijima, and Uno [13] proposed listing algorithms to list the chordal completions within a polynomial time delay. They are both based on the reverse search technique by Avis and Fukuda [1]. As for the counting problem, we are aware of the paper by Wormald [32] that gives an asymptotic number of the chordal graphs with n vertices. However, as far as graph sandwiches are concerned, neither algorithmic results nor hardness results seem to be known. There has been no result about random sampling of a chordal graph, as far as we see. For the related minimum chordal completion/deletion problems, both of which are well-known to be NP-hard [30],[19], there are some results on polynomial-time approximation, fixed parameter tractability, and exponential-time exact algorithms [18],[17],[4].

2   Preliminaries

All graphs in this paper are undirected and simple. For a graph G, we denote the set of vertices of G by $V(G)$ and the set of edges of G by $E(G)$. For a pair of graphs G and H on a common vertex set V, we write $G \subseteq H$ (and $G \subsetneqq H$) when $E(G) \subseteq E(H)$ (and $E(G) \subsetneqq E(H)$, respectively). For a graph $G = (V, E)$ and a pair of vertices $e = \{v_1, v_2\} \in (\binom{V}{2} \setminus E)$, we denote the graph $(V, E \cup \{e\})$ by $G + e$. Similarly, for a graph $G = (V, E)$ and an edge $e \in E$ we denote the graph $(V, E \setminus \{e\})$ by $G - e$. Given a pair of graphs $\overline{G}$ and $\underline{G}$ satisfying $\underline{G} \subsetneq \overline{G}$, we define the set $\Omega_C(\overline{G}, \underline{G})$ of chordal graphs sandwiched by $\overline{G}$ and $\underline{G}$ as

$$\Omega_C(\overline{G}, \underline{G}) \stackrel{\text{def.}}{=} \{G \mid G \text{ is chordal}, \ \underline{G} \subseteq G \subseteq \overline{G}\}. \tag{1}$$

A graph in $\Omega_C(\overline{G}, \underline{G})$ is called a *chordal sandwich* for the pair $\overline{G}$ and $\underline{G}$ while $\overline{G}$ and $\underline{G}$ are called the *ceiling graph* and the *floor graph* of $\Omega_C(\overline{G}, \underline{G})$, respectively. If $\overline{G}$ is a complete graph, then a chordal sandwich is called a *chordal completion* of $\underline{G}$. If $\underline{G}$ is an empty graph (i.e. has no edge), then a chordal sandwich is called a *chordal deletion* of $\overline{G}$.

Note that the graphs are "labeled" in $\Omega_C(\overline{G}, \underline{G})$, meaning that we distinguish $G \in \Omega_C(\overline{G}, \underline{G})$ from $G' \in \Omega_C(\overline{G}, \underline{G})$ when their edge sets are different even if they are isomorphic graphs.

We study the following three types of problems: given a pair of graphs $\overline{G}$ and $\underline{G}$ with $\underline{G} \subsetneq \overline{G}$

- output all graphs in $\Omega_C(\overline{G}, \underline{G})$ (listing);
- output the number $|\Omega_C(\overline{G}, \underline{G})|$ (counting);
- output one graph in $\Omega_C(\overline{G}, \underline{G})$ uniformly at random (sampling).

Golumbic, Kaplan, and Shamir [7] showed that, given a pair of graphs $\overline{G}$ and $\underline{G}$ satisfying $\underline{G} \subsetneq \overline{G}$, deciding whether $\Omega_C(\overline{G}, \underline{G})$ has an element is NP-complete. Therefore, three problems above are all intractable without any restriction. In this paper, we always assume that at least one of $\overline{G}$ and $\underline{G}$ is chordal. For later reference, we write this assumption as a condition.

Condition 1. *A pair of graphs $\overline{G}$ and $\underline{G}$ satisfies $\underline{G} \subsetneq \overline{G}$, and at least one of $\overline{G}$ and $\underline{G}$ is chordal.*

The following proposition is a key to some of our results.

Proposition 1. *Suppose a pair of chordal graphs $\overline{G} = (V, \overline{E})$ and $\underline{G} = (V, \underline{E})$ satisfies $\underline{G} \subseteq \overline{G}$, and let $k = |\overline{E} \setminus \underline{E}|$. Then there exists a sequence of chordal graphs $G_0, G_1, \ldots, G_k$ that satisfies $G_0 = \underline{G}$, $G_k = \overline{G}$, and $G_{i+1} = G_i + e_i$ with an appropriate edge $e_i \in \overline{E} \setminus \underline{E}$ for each $i \in \{0, \ldots, k-1\}$.*

Proof. We use the following result by Rose, Tarjan, and Lueker [23]: for a graph $G = (V, E)$ and a chordal graph $G' = (V, E \cup F)$ with $E \cap F = \emptyset$, the graph G' is a minimal chordal completion of G (i.e., $\Omega_C(G', G) = \{G'\}$) if and only if $G' - f$ is not chordal for each $f \in F$.

The proof is done by induction on k. If $k = 0$, then $\underline{G} = \overline{G}$ and we are done. Now assume that $k \geq 1$, and the proposition holds for all $k' < k$. In this case, $\overline{G}$ is not a minimal chordal completion of $\underline{G}$ since $\underline{G} \neq \overline{G}$ and $\underline{G}$ is actually a minimal chordal completion of itself. By the result of Rose, Tarjan, and Leuker above, there must exist an edge $f \in \overline{E} \setminus \underline{E}$ such that $\overline{G} - f$ is chordal. Then, letting $G_{k-1} = \overline{G} - f$ and $e_{k-1} = f$, we have $\overline{G} = G_k = G_{k-1} + e_{k-1}$. Further, by the induction hypothesis, there exists a sequence of chordal graphs $\underline{G} = G_0, G_1, \ldots, G_{k-1}$ such that $G_{i+1} = G_i + e_i$ for some $e_i \in (\overline{E} \setminus \{e_{k-1}\}) \setminus \underline{E}$. $\square$

Note that Proposition 1 implies that the set of chordal sandwiches forms a graded poset with respect to the inclusion relation of edge sets.

Procedure $A(\overline{G}, \underline{G})$ (when $\overline{G}$ is chordal) Procedure $B(\overline{G}, \underline{G})$ (when $\underline{G}$ is chordal)

1 begin	**1 begin**
2 find an edge $e \in \overline{E} \setminus \underline{E}$	2 find an edge $e \in \overline{E} \setminus \underline{E}$
such that $\overline{G} - e$ is chordal	such that $\underline{G} + e$ is chordal
3 If such e exists **do**	3 If such e exists **do**
4 output $\overline{G} - e$	4 output $\underline{G} + e$
5 call $A(\overline{G}, \underline{G} + e)$	5 call $B(\overline{G}, \underline{G} + e)$
6 call $A(\overline{G} - e, \underline{G})$	6 call $B(\overline{G} - e, \underline{G})$
7 **otherwise** halt	7 **otherwise** halt
8 end.	**8 end.**

Fig. 1. Procedures in the listing algorithms

3   Listing All Chordal Sandwiches

We give algorithms to list all chordal sandwiches in $\Omega_C(\overline{G}, \underline{G})$ for given $\overline{G}$ and $\underline{G}$ satisfying Condition 1.

First consider the case in which the ceiling graph $\overline{G}$ is chordal. Then, there exists an edge $e \in \overline{E} \setminus \underline{E}$ such that $\overline{G} - e$ is chordal if $\Omega_C(\overline{G}, \underline{G}) \setminus \{\overline{G}\} \neq \emptyset$, from Proposition 1. For the edge e, we consider a pair of sets $\Omega_C(\overline{G} - e, \underline{G})$ and $\Omega_C(\overline{G}, \underline{G} + e)$. Then, each graph of $\Omega_C(\overline{G}, \underline{G})$ without e is a member of $\Omega_C(\overline{G} - e, \underline{G})$, and each graph of $\Omega_C(\overline{G}, \underline{G})$ with e is a member of $\Omega_C(\overline{G}, \underline{G} + e)$, from the definition of a chordal sandwich. Thus, we obtain a binary partition of $\Omega_C(\overline{G}, \underline{G})$ as follows:

$$\Omega_C(\overline{G}, \underline{G}) = \Omega_C(\overline{G} - e, \underline{G}) \cup \Omega_C(\overline{G}, \underline{G} + e), \text{ and}$$
$$\Omega_C(\overline{G} - e, \underline{G}) \cap \Omega_C(\overline{G}, \underline{G} + e) = \emptyset.$$

Then, the ceiling graph $\overline{G}$ of $\Omega_C(\overline{G}, \underline{G} + e)$ is chordal, and the ceiling graph $\overline{G} - e$ of $\Omega_C(\overline{G} - e, \underline{G})$ is chordal again from the choice of e. We can repeat the binary partition recursively, until every set consists of a single element. More concretely, in our algorithm we first output $\overline{G}$ and call Procedure $A(\overline{G}, \underline{G})$ in Fig. 1.

Now we estimate the time complexity of our algorithm. Let $n = |V|$, $m = |\underline{E}|$ and $k = |\overline{E} \setminus \underline{E}|$. We can find an edge e in Step 2 in $O(k(n+m))$ time by a simple try-and-error approach with a linear-time algorithm to recognize a chordal graph by Rose, Tarjan, and Lueker [23] or by Tarjan and Yannakakis [25]. The try-and-error algorithm can be improved to $O(kn + n \log n)$ time by a dynamic data structure proposed by Ibarra [10].

The binary partition is valid in the sense that we always obtain a pair of non-empty sets in recursive calls. Thus, the accumulated number of recursive calls made by a call to $A(\overline{G}, \underline{G})$ is proportional to the number of outputs $|\Omega_C(\overline{G}, \underline{G})|$. Therefore, the total time complexity is $O(k(n + m) \cdot |\Omega_C(\overline{G}, \underline{G})|)$ or $O((kn + n \log n) \cdot |\Omega_C(\overline{G}, \underline{G})|)$, depending on the algorithm to find the edge e.

Consider next the case in which the floor graph $\underline{G}$ is chordal. In this case, we may obtain a similar algorithm. Procedure $B(\overline{G}, \underline{G})$ in Fig. 1 shows the concrete

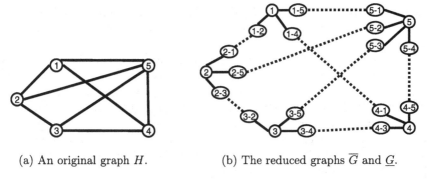

(a) An original graph H.   (b) The reduced graphs $\overline{G}$ and $\underline{G}$.

Fig. 2. An example of the transformation

algorithm. The time complexity can be estimated similarly, but in this case we can find an appropriate e faster, namely in $\mathrm{O}(k\log^2 n + n)$ [10].

4   Hardness of Counting the Chordal Sandwiches

Here, we show the #P-completeness of counting the chordal sandwiches by a parsimonious reduction. We also show the #P-completeness of counting the chordal deletions by a Cook reduction. These results imply that random sampling of chordal graphs is not easy, as indicated by many previous results about the relationship between the (approximate) counting and the random sampling (see e.g., [24]).

First we show the following theorem.

Theorem 1. *The computation of $|\Omega_C(\overline{G}, \underline{G})|$ is #P-complete, even when $\underline{G}$ is a connected chordal graph.*

We give a reduction from the problem to count the forests in a graph, which is known to be #P-complete [28]. Note that the reduction is parsimonious. Thus, if we have an approximation algorithm for the chordal sandwich counting, then we obtain an approximation algorithm for the forest counting with the same approximation ratio. Note that it is still a widely open problem whether or not a fully polynomial-time randomized approximation scheme exists for the forest counting problem.

Proof. The problem is clearly in #P. It is enough to show the #P-hardness.

First, we give a transformation of an instance (i.e., a graph) H of the forest counting problem into an instance (i.e., a pair of graphs) $\overline{G}$ and $\underline{G}$ of the chordal sandwich counting problem. To construct $\overline{G}$, we just replace every edge $\{u, v\} \in E(H)$ with a path of length three. Let $w_{u,v}$ and $w_{v,u}$ be new vertices of $\overline{G}$, which subdivide an edge $\{u, v\} \in E(H)$. Then, $|V(\overline{G})| = |V(H)| + 2|E(H)|$ and $|E(\overline{G})| = 3|E(H)|$ hold. To construct $\underline{G}$, we just remove every edge of the form $\{w_{u,v}, w_{v,u}\} \in E(\overline{G})$ from $\overline{G}$. Fig. 2 shows an example of the transformation. In Fig. 2 (b), the edges of $\underline{G}$ are drawn by solid lines, and the edges of $\overline{G}$ are drawn by solid lines and dashed lines. Note that $\underline{G}$ is a chordal graph consisting of n disjoint stars. Moreover, the girth (i.e., the length of a shortest cycle) of $\overline{G}$ is at least 9.

Next, we show that there exists one-to-one correspondence between the set of forests in H and $\Omega_{\mathrm{C}}(\overline{G}, \underline{G})$. For a forest $F = (V(H), E(F))$ in H, we define the corresponding graph $G \in \Omega_{\mathrm{C}}(\overline{G}, \underline{G})$ as $E(G) = E(\underline{G}) \cup \{\{w_{u,v}, w_{v,u}\} \in E(\overline{G}) \setminus E(\underline{G}) \mid \{u, v\} \in E(F)\}$. Then, G does not have any cycle, and G is chordal. Conversely, every graph in $\Omega_{\mathrm{C}}(\overline{G}, \underline{G})$ does not contain any cycle and is chordal since the girth of $\overline{G}$ is at least 9. Thus, for any $G \in \Omega_{\mathrm{C}}(\overline{G}, \underline{G})$, there exists a corresponding forest in H as the inverse of the above map. Hence, we obtain a bijection. Thus, we showed that the computation of $|\Omega_{\mathrm{C}}(\overline{G}, \underline{G})|$ is #P-hard even when $\underline{G}$ is chordal.

To obtain the full theorem we transform $\Omega_{\mathrm{C}}(\overline{G}, \underline{G})$ into $\Omega_{\mathrm{C}}(\overline{G}', \underline{G}')$ in which $\underline{G}'$ is connected and chordal. Let G be a graph. We transform G into $\Phi(G)$ defined as $V(\Phi(G)) \stackrel{\text{def.}}{=} V(G) \cup \{v_0\}$ and $E(\Phi(G)) \stackrel{\text{def.}}{=} E(G) \cup \{\{v_0, v\} \mid v \in V(G)\}$. Clearly $\Phi(G)$ is connected. Furthermore, G is chordal if and only if $\Phi(G)$ is chordal. Now, we define a pair of graphs $\overline{G}' \stackrel{\text{def.}}{=} \Phi(\overline{G})$ and $\underline{G}' \stackrel{\text{def.}}{=} \Phi(\underline{G})$ from the pair graphs $\overline{G}$ and $\underline{G}$. Then, $\underline{G}'$ is chordal when $\underline{G}$ is chordal, and $\Omega_{\mathrm{C}}(\overline{G}, \underline{G})$ and $\Omega_{\mathrm{C}}(\overline{G}', \underline{G}')$ are in one-to-one correspondence via Φ. Thus, we obtain the theorem.   □

Next, we discuss the hardness of counting the chordal deletions. The set of chordal deletions of G is described as the set of chordal sandwiches in $\Omega_{\mathrm{C}}(G, I_n)$, where I_n is an empty graph with n vertices and no edge.

Theorem 2. *The computation of $|\Omega_{\mathrm{C}}(G, I_n)|$ is #P-complete.*   □

Theorem 2 can be shown by a Cook-reduction from the forest counting problem in a graph. See [11] for detail. Note that the reduction does not preserve the approximation ratio.

5 A Simple Markov Chain and Its Slow-Mixing

Here, we consider a uniform sampling on $\Omega_{\mathrm{C}}(\overline{G}, \underline{G})$ satisfying Condition 1. We give a simple and natural Markov chain. Note that the following Markov chain can be easily modified into ones for non-uniform distributions by a Metropolis-Hastings method.

Let $\mathcal{M}$ be a Markov chain with a state space $\Omega_{\mathrm{C}}(\overline{G}, \underline{G})$ with Condition 1. A transition of $\mathcal{M}$ from a current state $G \in \Omega_{\mathrm{C}}(\overline{G}, \underline{G})$ to a next state G' is defined as follows; Choose an edge $e \in (\overline{E} \setminus \underline{E})$ uniformly at random. We consider the following three cases.

1. If $e \notin E(G)$ and $G + e$ is chordal, then set $H = G + e$.
2. If $e \in E(G)$ and $G - e$ is chordal, then set $H = G - e$.
3. Otherwise set $H = G$.

Let $G' = H$ with the probability $1/2$, otherwise let $G' = G$. Clearly $G' \in \Omega_{\mathrm{C}}(\overline{G}, \underline{G})$.

Table 1. The hardness of counting chordal sandwiches w.r.t. input pair

ceiling graph $\overline{G}$

		general	chordal	complete graph (cf. chordal completion)
floor graph $\underline{G}$	general	#P-complete	open	open
	chordal	#P-complete	open	open
	empty graph (cf. chordal deletion)	#P-complete	open	(cf. asymptotic analysis by Wormald [32])

The chain $\mathcal{M}$ is irreducible from the fact that $\Omega_C(\overline{G},\underline{G})$ on Condition 1 forms a graded poset. The chain $\mathcal{M}$ is clearly aperiodic, and hence $\mathcal{M}$ is ergodic. The unique stationary distribution of $\mathcal{M}$ is the uniform distribution on $\Omega_C(\overline{G},\underline{G})$, since the detailed balanced equation holds for any pair of $G \in \Omega_C(\overline{G},\underline{G})$ and $G' \in \Omega_C(\overline{G},\underline{G})$. From Proposition 1, the diameter of $\mathcal{M}$ is at most $2k$, where $k = |\overline{E} \setminus \underline{E}|$.

Now, we discuss the mixing time of the Markov chain. Let μ and ν be a pair of distributions on a common finite set Ω. The *total variation distance* $d_{\mathrm{TV}}(\mu,\nu)$ between μ and ν is defined by $d_{\mathrm{TV}}(\mu,\nu) \stackrel{\text{def.}}{=} \frac{1}{2}\sum_{x\in\Omega}|\mu(x)-\nu(x)|$. For an arbitrary positive ε, the *mixing time* $\tau(\varepsilon)$ of an ergodic Markov chain MC with a state space Ω is defined by $\tau(\varepsilon) \stackrel{\text{def.}}{=} \max_{x\in\Omega}\min\{t \mid \forall s \geq t,\ d_{\mathrm{TV}}(P_x^s,\pi) \leq \varepsilon\}$ where π is the unique stationary distribution of $\mathcal{M}$, and P_x^s denotes a distribution of $\mathcal{M}$ at time s starting from a state x.

Unfortunately, the Markov chain $\mathcal{M}$ is not rapidly mixing for some inputs. More precisely, we can show the following. See [11] for detail.

Proposition 2. *There exist infinitely many pairs of chordal graphs $\overline{G}$ and $\underline{G}$ satisfying $\underline{G} \subseteq \overline{G}$ for which the mixing time of $\mathcal{M}$ on $\Omega_C(\overline{G},\underline{G})$ is exponential in n, where n is the number of vertices of $\overline{G}$ (and $\underline{G}$).* $\square$

6   Concluding Remarks

We gave a simple and natural Markov chain for uniform sampling of $\Omega_C(\overline{G},\underline{G})$, and showed an example for which the mixing time of the chain is exponential, even when both of $\overline{G}$ and $\underline{G}$ are chordal. It is open if there is a rapidly mixing Markov chain. Our Markov chain uses the fact that $\Omega_C(\overline{G},\underline{G})$ for given $\overline{G}$ and $\underline{G}$ with Condition 1 forms a graded poset. However, it is known that the set of chordal sandwiches generally does not form a lattice even when both $\overline{G}$ and $\underline{G}$ are chordal. A future work would include a characterization of pairs $\overline{G}$ and $\underline{G}$ such that $\Omega_C(\overline{G},\underline{G})$ forms a lattice.

It is open that counting the chordal sandwiches is #P-hard when a given ceiling graph is restricted to be chordal (see Table 1). We conjecture that counting the chordal completions (i.e., when a given ceiling graph is complete) is #P-complete. We can consider listing, counting, and sampling the graph sandwiches for other graph classes, such as interval, proper interval, or perfect graphs. We can show that counting the interval sandwiches is #P-complete even when a given floor graph is connected and interval (see [11] for detail).

We gave an efficient algorithm to list chordal sandwiches. In our preliminary experiment with a simple implementation by Java on a standard PC (CPU: 3GHz, RAM: 3GB), the algorithm outputs about $24,000$ chordal graphs per second when $n = 10$, and about 400 chordal graphs per second when $n = 100$. We also implement the simple Markov chain by Java. In our preliminary experiment on a standard PC, about 100,000 transitions are executed per second when $n = 10$, and about 5,000 transitions are executed per second when $n = 100$.

References

1. Avis, D., Fukuda, K.: Reverse Search for Enumeration. Discrete Applied Mathematics 65, 21–46 (1996)
2. Dirac, G.A.: On Rigid Circuit Graphs. Abhandl. Math. Seminar Univ. Hamburg 25, 71–76 (1961)
3. Fulkerson, D.R., Gross, O.A.: Incidence Matrices and Interval Graphs. Pacific Journal of Mathematics 15, 835–855 (1965)
4. Fomin, F.V., Kratsch, D., Todinca, I.: Exact (Exponential) Algorithms for Treewidth and Minimum Fill-in. In: Díaz, J., Karhumäki, J., Lepistö, A., Sannella, D. (eds.) ICALP 2004. LNCS, vol. 3142, pp. 568–580. Springer, Heidelberg (2004)
5. Fukuda, M., Kojima, M., Murota, K., Nakata, K.: Exploiting Sparsity in Semidefinite Programming via Matrix Completion I: General Framework. SIAM Journal on Optimization 11, 647–674 (2000)
6. Golumbic, M.C.: Algorithmic Graph Theory and Perfect Graphs. Academic Press, New York (1980)
7. Golumbic, M.C., Kaplan, H., Shamir, R.: Graph Sandwich Problems. Journal of Algorithms 19, 449–473 (1995)
8. Heggernes, P.: Minimal Triangulations of Graphs: A Survey. Discrete Mathematics 306, 297–317 (2006)
9. Heggernes, P., Suchan, K., Todinca, I., Villanger, Y.: Characterizing Minimal Interval Vompletions: towards Better Understanding of Profile and Pathwidth. In: Thomas, W., Weil, P. (eds.) STACS 2007. LNCS, vol. 4393, pp. 236–247. Springer, Heidelberg (2007)
10. Ibarra, L.: Fully Dynamic Algorithms for Chordal Graphs. In: Proc. of SODA 1999, pp. 923–924 (1999)
11. Kijima, S., Kiyomi, M., Okamoto, Y., Uno, T.: On Counting, Sampling, and Listing of Chordal Graphs with Edge Constrains. RIMS-1610, Kyoto University (preprint, 2007), http://www.kurims.kyoto-u.ac.jp/preprint/file/RIMS1610.pdf
12. Kiyomi, M., Uno, T.: Generating Chordal Graphs Included in Given Graphs. IEICE Transactions on Information and Systems E89-D, 763–770 (2006)
13. Kiyomi, M., Kijima, S., Uno, T.: Listing Chordal Graphs and Interval Graphs. In: Fomin, F.V. (ed.) WG 2006. LNCS, vol. 4271, pp. 68–77. Springer, Heidelberg (2006)
14. Kloks, T., Bodlaender, H.L., Müller, H., Kratsch, D.: Computing Treewidth and Minimum Fill-in: All You Need are the Minimal Separators. In: Lengauer, T. (ed.) ESA 1993. LNCS, vol. 726, pp. 260–271. Springer, Heidelberg (1993)

15. Kloks, T., Bodlaender, H.L., Müller, H., Kratsch, D.: Erratum to the ESA 1993 proceedings. In: van Leeuwen, J. (ed.) ESA 1994. LNCS, vol. 855, p. 508. Springer, Heidelberg (1994)
16. Leckerkerker, C.G., Boland, J.C.: Representation of a Finite Graph by a Set of Intervals on the Real Line. Fundamenta Mathematicae 51, 45–64 (1962)
17. Marx, D.: Chordal Deletion is Fixed-Parameter Tractable. In: Fomin, F.V. (ed.) WG 2006. LNCS, vol. 4271, pp. 37–48. Springer, Heidelberg (2006)
18. Natanzon, A., Shamir, R., Sharan, R.: A Polynomial Approximation Algorithm for the Minimum Fill-in Problem. SIAM Journal on Computing 30, 1067–1079 (2000)
19. Natanzon, A., Shamir, R., Sharan, R.: Complexity Classification of Some Edge Modification Problems. Discrete Applied Mathematics 113, 109–128 (2001)
20. Pedersen, T., Bruce, R.F., Wiebe, J.: Sequential Model Selection for Word Sense Disambiguation. In: Proceedings of the Fifth Conference on Applied Natural Language Processing (ANLP 1997), pp. 388–395 (1997)
21. Robertson, N., Seymour, P.: Graph Minors II. Algorithmic Aspects of Tree-Width. Journal of Algorithms 7, 309–322 (1986)
22. Rose, D.J.: A Graph-Theoretic Study of the Numerical Solution of Sparse Positive Definite Systems of Linear Equations. In: Read, R.C. (ed.) Graph Theory and Computing, pp. 183–217. Academic Press, New York (1972)
23. Rose, D.J., Tarjan, R.E., Lueker, G.S.: Algorithmic Aspects of Vertex Elimination on Graphs. SIAM Journal on Computing 5, 266–283 (1976)
24. Sinclair, A.: Algorithms for Random Generation and Counting: A Markov Chain Approach. Birkhäuser, Boston (1993)
25. Tarjan, R.E., Yannakakis, M.: Simple Linear-Time Algorithms to Test Chordality of Graphs, Test Acyclicity of Hypergraphs, and Selectively Reduce Acyclic Hypergraphs. SIAM Journal on Computing 13, 566–579 (1984)
26. Takemura, A., Endo, Y.: Evaluation of Per-record Identification Risk and Swappability of Records in a Microdata Set via Decomposable Models. arXiv:math.ST/0603609
27. Valiant, V.G.: The Complexity of Computing the Permanent. Theoretical Computer Science 8, 189–201 (1979)
28. Valiant, V.G.: The Complexity of Enumeration and Reliability Problems. SIAM Journal on Computing 8, 410–421 (1979)
29. Yamashita, N.: Sparse Quasi-Newton Updates with Positive Definite Matrix Completion. Mathematical Programming (2007),
    http://www.springerlink.com/content/28271224t1nt3580/
30. Yannakakis, M.: Computing the Minimum Fill-in is NP-Complete. SIAM Journal on Algebraic and Discrete Methods 2, 77–79 (1981)
31. Whittaker, J.: Graphical Models in Applied Multivariate Statistics. Wiley, New York (1990)
32. Wormald, N.C.: Counting Labeled Chordal Graphs. Graphs and Combinatorics 1, 193–200 (1985)

Probe Ptolemaic Graphs

David B. Chandler[1], Maw-Shang Chang[2], Ton Kloks[*],
Van Bang Le[3], and Sheng-Lung Peng[4,**]

[1] Department of Mathematical Sciences
University of Delaware Newark, Delaware 19716, USA
davidbchandler@gmail.com
[2] Department of Computer Science and Information Engineering
National Chung Cheng University, Chiayi 62107, Taiwan
mschang@cs.ccu.edu.tw
[3] Institut für Informatik, Universität Rostock, 18051 Rostock, Germany
le@informatik.uni-rostock.de
[4] Department of Computer Science and Information Engineering
National Dong Hwa University, Hualien 97401, Taiwan
slpeng@mail.ndhu.edu.tw

Abstract. Given a class of graphs, $\mathcal{G}$, a graph G is a *probe graph of*
$\mathcal{G}$ if its vertices can be partitioned into two sets, $\mathbb{P}$ (the probes) and $\mathbb{N}$
(the nonprobes), where $\mathbb{N}$ is an independent set, such that G can be
embedded into a graph of $\mathcal{G}$ by adding edges between certain nonprobes.
In this paper we study the probe graphs of ptolemaic graphs when the
partition of vertices is unknown. We present some characterizations of
probe ptolemaic graphs and show that there exists a polynomial-time
recognition algorithm for probe ptolemaic graphs.

1 Introduction

In some applications, we want to determine the relation between every pair of
elements in a set. If the relation is of boolean type, it can be described by a simple
graph without self-loops. In some applications it is expensive to determine the
relation between every pair of elements. Therefore, the elements are partitioned
into *probes* and *nonprobes*. The relation between two elements is determined
whenever at least one of the two elements is a probe. In graph-theoretical terms,
we have a graph G whose vertices are partitioned into a set $\mathbb{P}$ of *probes* and a
set $\mathbb{N}$ of *nonprobes*. The set of nonprobes $\mathbb{N}$ is an independent set. We want to
know whether we can let G satisfy some property Π by adding edges between
certain nonprobes. Let $\mathcal{G}$ be the class of graphs satisfying the property Π.

A graph G is a *probe graph of $\mathcal{G}$* if its vertex set can be partitioned into a set $\mathbb{P}$
of *probes* and an *independent set* $\mathbb{N}$ of *nonprobes*, such that G can be embedded
into a graph of $\mathcal{G}$ by adding edges between certain nonprobes. If the partition of

---

[*] Currently, this author is visiting the School of Computing, University of Leeds, Leeds
LS2 9JT, UK.
[**] Corresponding author.

X. Hu and J. Wang (Eds.): COCOON 2008, LNCS 5092, pp. 468–477, 2008.
© Springer-Verlag Berlin Heidelberg 2008

the vertices of a graph G into a set of probes $\mathbb{P}$ and a set of nonprobes $\mathbb{N}$ is part of the input then we call G a *partitioned probe graph of $\mathcal{G}$* if G can be embedded into a graph of $\mathcal{G}$ by adding edges between certain vertices of $\mathbb{N}$. We denote a partitioned graph as $G = (\mathbb{P}+\mathbb{N}, E)$, and when this notation is used it is to be understood that $\mathbb{N}$ is an independent set. We will refer to the class of (partitioned) probe graphs of the class of $\mathcal{G}$ graphs as (partitioned) probe $\mathcal{G}$ graphs.

For the partitioned case, there are efficient algorithms for the recognition problem on some classes of graphs, *e.g.*, probe interval graphs [18],[22], probe permutation graphs [5], probe distance-hereditary graphs [6], probe comparability graphs [7], and so on. In graph theory, the unpartitioned case is more interesting. Only few graph classes have polynomial-time recognition algorithms for their probe versions in unpartitioned case. These classes are chordal [3], interval [8], cograph [9,21], split [9],[20], P_4-reducible [9], and P_4-sparse graphs [9].

In this paper, we study the unpartitioned case of probe ptolemaic graphs. The remaining of this paper is organized as follows. Some notation and preliminaries are given in Section 2. We then propose some characterizations of probe ptolemaic graphs in Section 3. In Section 4, we consider the recognition of a special class of probe ptolemaic graphs, called *probe chordal cographs*. In Section 5, we show that there exists a polynomial-time algorithm that checks whether a graph is a probe ptolemaic graph. Finally, we give conclusion in the last section.

2   Preliminaries

A graph G is a pair (V, E), where the elements of V are called the *vertices* of G and where E is a family of two-element subsets of V, called the *edges*. We use $V(G)$ and $E(G)$ to denote the vertex and edge sets of G, respectively. We write $n = |V|$ for the number of vertices and $m = |E|$ for the number of edges. We denote edges of a graph G as (x, y) (or xy) and we call x and y the end vertices of the edge. Unless stated otherwise, a graph is regarded as undirected. For a vertex x we write $N(x)$ for its set of neighbors in G, and for a subset $W \subseteq V$ we write $N(W) = \cup_{x \in W} N(x) - W$. For other conventions on graph-related notation we refer to any standard textbook. For graph classes not defined here we refer to [4].

For two sets A and B we write $A + B$ and $A - B$ instead of $A \cup B$ and $A \setminus B$ respectively. We write $A \subseteq B$ if A is a subset of B with possible equality and we write $A \subset B$ if A is a subset of B and $A \neq B$. For a set A and an element x we write $A - x$ instead of $A - \{x\}$ and $A + x$ instead of $A \cup \{x\}$.

For a graph $G = (V, E)$ and a subset $S \subseteq V$, we write $G[S]$ for the subgraph of G *induced* by S. For a subset $W \subseteq V$, we write $G - W$ for the graph $G[V - W]$, *i.e.*, the subgraph induced by $V - W$. For a vertex x we write $G - x$ rather than $G - \{x\}$.

For a partitioned probe graph $G = (\mathbb{P}+\mathbb{N}, E)$ of some graph class $\mathcal{G}$, an *embedding* of G is a graph of $\mathcal{G}$ obtained by adding edges between certain nonprobes, *i.e.*, vertices of $\mathbb{N}$.

Originally, ptolemaic graphs were defined as follows.

Definition 1 ([19]). *A connected graph is* ptolemaic *if for every four vertices x, y, u, and v:*

$$d(x,y)d(u,v) \leq d(x,u)d(y,v) + d(x,v)d(y,u)$$

A graph is ptolemaic *if every component is ptolemaic.*

Ptolemaic graphs can be characterized as those chordal graphs in which every 5-cycle has at least three chords, or, as those chordal graphs in which all chordless paths are shortest paths [4],[12],[27]. We will use the following characterization of ptolemaic graphs.

Theorem 1 ([17]). *A graph is ptolemaic if and only if it is distance hereditary and chordal.*

Recall that a graph is *chordal* if it has no induced cycle of length more than 3 and is *distance hereditary* if the distance between any two vertices remains the same in every connected induced subgraph. For the house, hole, domino, and gem, we refer to Fig. 1.

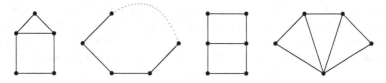

Fig. 1. A house, a hole, a domino, and a gem

Corollary 1. *Ptolemaic graphs are those chordal graphs which are gem-free.*

Definition 2 ([23]). *A* module *is a set M of vertices such that every vertex outside M is either adjacent to all vertices of M or to no vertex of M.*

A module which is also a clique is called a *clique module*. If $M = \varnothing$, $|M| = 1$, or M is the whole vertex set V, then M is called *trivial*.

One major characteristic, on which our recognition algorithms rely, is the following characterization of ptolemaic graphs. For similar characterizations, see also [24],[28].

Theorem 2 ([24],[28]). *A graph G is ptolemaic if and only if for every vertex x and for every component C of $G-N[x]$, $N(C)$ is a clique module in $G[C+N(C)]$.*

Proof. Since G is chordal, for every vertex x and every component C of $G-N[x]$, $N(C)$ is a clique. Assume some vertex $y \in C$ is adjacent to a vertex $\alpha \in N(C)$ and nonadjacent to a vertex $\beta \in N(C)$. Consider a β, y-path P with internal vertices in C. Then α is adjacent to every vertex of P, otherwise there is a chordless cycle of length at least 4 in G. Together with x this creates a gem.

Conversely, if G has a gem, then let x be one of the vertices of degree two. The neighborhood of x contains the universal vertex of the gem and the neighbor of x in the P_4 of the gem. The edge of the gem which does not intersect $N(x)$ is contained in some component C of $G - N[x]$. The other vertex of degree two contradicts the assumption that $N(C)$ is a clique module in the graph induced by $N[C]$.   □

It is well-known that chordal graphs can be characterized as those graphs in which every induced subgraph has a simplicial vertex [11],[14],[26].

Definition 3. *A vertex x is* simplicial *if its neighborhood $N(x)$ is a clique.*

We end this section with two other characterizations of ptolemaic graphs. Another way to characterize chordal graphs is by their minimal separators. A *minimal x, y-separator* for nonadjacent vertices x and y is a minimal set of vertices S such that x and y are in different components of $G - S$, the graph induced by $V - S$, where V is the set of vertices of G. A *minimal separator* is a set which is a minimal x, y-separator for some nonadjacent vertices x and y. A classic result of Dirac's says that a graph is chordal if and only if every minimal separator is a clique [11].

Theorem 3. *A connected graph is ptolemaic if and only if every connected induced subgraph either is a clique, or has a cutvertex, or has a separator which is a nontrivial clique-module.*

Proof. To see this, assume that G is ptolemaic and let H be a connected induced subgraph. Thus H is ptolemaic. If H is a clique or has a cutvertex, we are done. Otherwise, let x be a non-universal vertex. Let C be a component of $H - N[x]$ such that $N(C)$ is inclusion minimal. By Theorem 2 on the facing page, $S = N(C)$ is a clique module in $H[S + C]$. Since H is chordal, there exists a vertex $y \in C$ with S in its neighborhood. It follows, again by Theorem 2, that S is also a clique module in the component of $G - N[y]$ that contains x. Thus S is a separating clique module. To see the converse, notice that gems and chordless cycles of length at least 4 are not cliques, have no cutvertices, and have no nontrivial clique-modules. □

Theorem 4 ([17]). *A graph is G is ptolemaic if and only if for every pair of vertices x and y with $d(x, y) = 2$, $N(x) \cap N(y)$ is a clique separator in G.*

3   Characterizations of Probe Ptolemaic Graphs

A graph is *HHD-free* if none of its induced subgraphs is a hole, a house, or a domino. HHD-free graphs can be recognized in $O(n^3)$ time ([16]).

A *ptolemaic twin* in a partitioned graph $G = (\mathbb{P} + \mathbb{N}, E)$ is pair of vertices x, y such that

- $x, y \in \mathbb{P}$ and $N[x] = N[y]$, or
- $x, y \in \mathbb{N}$ and $N(x) = N(y)$, or
- $x \in \mathbb{P}, y \in \mathbb{N}$ and $N[x] \cap \mathbb{P} = N(y)$.

A *k-fan* is the graph consisting a k-path P_k and with a fully adjacent vertex. Thus a 4-fan is the gem. For a given graph $G = (V, E)$, a particular 2-CNF instance $F(G)$ is created as follows.

- The boolean variables are the vertices of G,
- for each edge ab of G, $(\bar{a} \vee \bar{b})$ is a clause, the *edge clause* for ab,
- for each $C_4 = abcd$ of G, $(a \vee b)$ and $(c \vee d)$ are two clauses, the *C_4 clauses* for that C_4,

- for each 4-fan with $P_4 = abcd$ of G, $(a \vee b)$, $(c \vee d)$ and $(\overline{a} \vee \overline{d})$ are three clauses, the 4-*fan clauses* for that 4-fan,
- for each 5-fan with $P_5 = abcde$ of G, $(\overline{b})$ and $(\overline{d})$ are two clauses, the 5-*fan clauses* for that 5-fan.

The formula $F(G)$ is the conjunction of all edge clauses, all C_4 clauses, all 4-fan clauses, and all 5-fan clauses. We will see that the recognition of probe ptolemaic graphs is 'in fact' a 2-SAT problem for 2-CNF instances like $F(G)$ defined above. Similar considerations have been made for other probe graph classes in [2],[20]; see also [25].

Finally, it is clear that a graph (partitioned or not) is probe ptolemaic if and only if each of its blocks (which can be detected in linear time) is probe ptolemaic. Moreover, probe ptolemaic graphs are HHD-free.

Theorem 5. *Let* $G = (\mathbb{P} + \mathbb{N}, E)$ *be a partitioned graph. Then the following statements are equivalent.*

(i) *G is a partitioned probe ptolemaic graph;*
(ii) *Each 2-connected induced subgraph of G has a ptolemaic twin;*
(iii) *G is HHD-free and $F(G)$ is satisfied by assigning nonprobe variables to* true *and probe variables to* false;
(iv) *G is HHD-free and $(F_1, \ldots, F_8)$-free; see Fig. 2.*

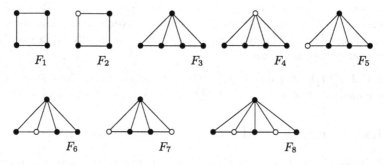

Fig. 2. Partitioned HHD-free graphs that are minimal but not probe ptolemaic; probes black, nonprobes white

Due to the limitation of space, we omit the proof of Theorem 5. For the unpartitioned case, by Theorem 5, we have the following theorem (for the same reason we omit the proof).

Theorem 6. *A graph G is a probe ptolemaic graph if and only if G is HHD-free and $F(G)$ is satisfiable.*

4   Probe Chordal Cographs

Assume G has a universal vertex, *i.e.*, a vertex x which is adjacent to all other vertices. Then, obviously, $G - x$ must be embedded into a chordal graph without an induced P_4, otherwise the embedding has a gem.

Definition 4 ([10],[13]). *A* cograph *is a graph without induced* P_4.

Chordal cographs can be characterized as follows. For a graph G let $\Upsilon = \Upsilon(G)$ be the set of *universal* vertices, *viz*, the set of vertices adjacent to all other vertices. Obviously, for any graph G, Υ induces a clique in G.

Proposition 1 ([29],[30]). *A connected graph is a chordal cograph if and only if it is complete or* $G - \Upsilon(G)$ *is a disconnected chordal cograph.*

Remark 1. Chordal cographs have the following tree-model: Let T be a rooted tree. Make any two vertices adjacent if they are contained in a common path from a leaf to the root.

From proposition 1 we easily obtain:

Lemma 1. *A partitioned graph* $G = (\mathbb{P} + \mathbb{N}, E)$ *can be embedded into a chordal cograph if and only if every connected induced subgraph has a nonempty set* Υ *of vertices that can be made universal by adding edges between nonprobes.*

The condition of Lemma 1 can be formulated in monadic second-order logic without edge-set quantifiers. Probe chordal-cographs have rankwidth at most 2. Therefore:

Corollary 2. *There exists an* $O(n^3)$ *algorithm which checks if a graph has an independent set* $\mathbb{N}$ *of nonprobes such that the partitioned graph* $G = (\mathbb{P} + \mathbb{N}, E)$ *can be embedded into a chordal cograph by adding some edges between nonprobes.*

For alternative algorithms see [6],[21].

5   Recognition of Probe Ptolemaic Graphs

In this section we show that there exists a polynomial-time recognition algorithm for the class of probe ptolemaic graphs. We assume that G is connected, otherwise we can test each component of G separately. By Section 4, we may assume that G has no universal vertices.

It is routine to check the following:

Proposition 2. *A probe ptolemaic graph has no induced house, hole, or domino. Furthermore, an induced gem has exactly two nonprobes, and they are at distance two in the* P_4 *of the gem.*

Lemma 2. *Consider two vertices* x *and* y *at distance 2 in* G. *Assume there exists an embedding of* G *in which* x *and* y *are probes. Let* H *be the graph obtained from* G *by adding the following edges:*

(i) *If* G *has an induced* C_4 *containing* x *and* y, *then the two nonprobes in the* C_4 *form an edge in* H.
(ii) *If* G *has a gem* Γ *with* x *and* y *at distance 2, then the two nonprobes in* Γ *form an edge in* H.

Then every gem or C_4 *in* H *induces also a gem or* C_4 *in* G.

Proof. By Theorem 4, the common neighborhood $S = N(x) \cap N(y)$ is a clique separator in any embedding. The claim follows from Proposition 2 and Theorem 2. $\square$

Lemma 3. *Assume G is ptolemaic. Let x be a non-universal vertex and let C_1 and C_2 be two components of $G - N[x]$. Then either $N(C_1) \subseteq N(C_2)$, $N(C_2) \subseteq N(C_1)$, or $N(C_1) \cap N(C_2) = \varnothing$.*

Proof. Write $S = N(x)$, and $S_i = N(C_i)$, $i = 1, 2$. Recall that S_1 and S_2 are cliques. If $N(C_1) = S$ or $N(C_2) = S$ then the claim is obvious, so assume that both $N(C_i)$ are proper subsets of S. Let $\gamma \in S_1 \cap S_2$, $\alpha \in S_1 - S_2$, and $\beta \in S_2 - S_1$. Let $c_i \in C_i$, $i = 1, 2$, be a vertex in C_i adjacent to all vertices of S_i. Since G is chordal these vertices exist. First assume that α and β are not adjacent. Then we find a gem with universal vertex γ and P_4; $[c_1, \alpha, x, \beta]$. If α and β are adjacent, we find a gem with universal γ and a P_4, $[c_1, \alpha, \beta, c_2]$ in its neighborhood. $\square$

Definition 5. *A block is a pair (x, C) where x is a vertex and C a component of $G - N[x]$.*

In our algorithm we first make a table of the blocks and order them according to increasing number of vertices in the component. For each block (x, C) we determine whether there exists an embedding of $G[x + N(C) + C]$ into a ptolemaic graph both in the case where x is a nonprobe and where x is a probe. When the type of x is fixed, the set A of vertices in $G[N[C] + x]$ which are at distance 2 from x in an embedding, is determined. We store in a table whether there exists an embedding of the block for both cases; with x of type 'probe' or 'nonprobe.' Furthermore, the table specifies if there exists such an embedding such that all vertices of A are nonprobes. We now describe how to obtain this information for each block.

Consider a block (x, C). Let $S = N(C)$. Suppose we want to check whether there exists an embedding of $G[x + S + C]$ with x as a probe. Notice that S is a clique in any embedding. Thus the partition of S is determined, except when it induces a clique in G in which case it can contain exactly one nonprobe.

We may add edges to destroy the induced 4-cycles and gems in $G[x + S + C]$ that contain x. This makes a clique of S and it makes S a module. Let $A \subseteq C$ be the set of vertices with a neighbor in S. If G has a nonedge between S and A then the partition into probes and nonprobes of $S + A$ is resolute. Otherwise, at most one of the two sets may contain nonprobes. Let $C_1, \ldots, C_t$ be the components of $C - A$.

Assume S has a probe u. Add edges to destroy C_4's that contain u and gems that contain u as a non-universal vertex. This makes a clique of each $S_i = N(C_i)$ and it makes S_i a module for $C_i + S_i$. Let $A_i \subseteq C_i$ be the set of vertices adjacent to S_i.

Now assume that some pair of minimal separators S_i and S_j overlap, i.e., $S_i \cap S_j \neq \varnothing$, $S_i - S_j \neq \varnothing$, and $S_j - S_i \neq \varnothing$. By Lemma 3, in any embedding all vertices of A_i are adjacent to $S_j - S_i$ or all vertices of A_j are adjacent to $S_j - S_i$. Then, say, all vertices of A_i must be nonprobes. Thus S_i contains only probes.

By table look-up, we check if there exists an embedding for the components of $C_i - A_i$ such that A_i consists of only nonprobes.

Assume that S has only nonprobes. Consider all gems and C_4's in $S + C$ incident with vertices of S, and add the edges between the nonprobes. By Lemma 2, this fixates A. Notice that the partition of A into probes and nonprobes is determined: the set of probes in A is exactly $N(\sigma) \cap A$ for any $\sigma \in S$. Let $C_1, \ldots, C_t$ be the components of $C - A$. Then S_i is a module for C_i, otherwise there would be a gem incident with a vertex of S. Let A_i be the set of vertices in C_i adjacent to S_i. By table look-up we check if there exist embedding for blocks (u, C_i) where u is a nonprobe in S. If there are overlapping separators S_i and S_j, we check if there are embeddings such that A_i or A_j consists only of nonprobes.

The case where x is a nonprobe, is similar to the description above.

Theorem 7. *There exists an efficient algorithm to check whether a given graph $G = (V, E)$ is probe ptolemaic.*

Proof. For the case where there exists a universal vertex, the algorithm is described in Section 4. Otherwise, choose a vertex x such that the largest component C of $G - N[x]$ has maximum cardinality. Let $S = N(C)$. By the choice of x, all vertices of $C_0 = V - (S + C)$ are adjacent to all vertices of S. Consider the case where S has at least one nonprobe. Then C_0 has only probes. By table look-up we can check if there is an embedding for the block (x, C).

Consider the case where S has only probes. If C_0 is not an independent set, it has at least one probe, say x. Again, we can check by table look-up if there is an embedding for (x, C). The algorithm in Section 4 checks if there is an embedding for $S + C_0$. If C_0 is an independent set of nonprobes, we can proceed as described prior to this theorem. □

Remark 2. There are $O(nm)$ different blocks in a graph. A rough estimate shows that each block can be processed in $O(n^2)$ time. This shows a time complexity of $O(n^3m)$ time as a rough upperbound. Notice that for the processing of a block only vertices are considered that are at distance at most 2 from the 'handle.' This shows that the algorithm can be implemented to run in $O(n^3)$ time. However, this needs a careful analysis and this is beyond the scope of today's abstract.

6   Concluding Remarks

Many probe graph classes have been investigated recently. However, many questions remain. To mention just one: Can probe perfect graphs be recognized in polynomial time?

References

1. Bandelt, H.J., Mulder, H.M.: Distance-Hereditary Graphs. Journal of Combinatorial Theory, Series B 41, 182–208 (1986)
2. Bayer, D., Le, V.B., de Ridder, H.N.: Probe Trivially Perfect Graphs and Probe Threshold Graphs (manuscript) (2006)

3. Berry, A., Golumbic, M.C., Lipshteyn, M.: Two Tricks to Triangulate Chordal Probe Graphs in Polynomial Time. In: Proceedings of 15th ACM–SIAM Symposium on Discrete Algorithms, pp. 962–969 (2004)
4. Brändstadt, A., Le, V.B., Spinrad, J.P.: Graph Classes: A Survey. In: SIAM Monographs on Discrete Mathematics and Applications, Philadelphia (1999)
5. Chandler, D.B., Chang, M.-S., Kloks, A.J.J., Liu, J., Peng, S.-L.: On Probe Permutation Graphs. In: Cai, J.-Y., Cooper, S.B., Li, A. (eds.) TAMC 2006. LNCS, vol. 3959, pp. 494–504. Springer, Heidelberg (2006)
6. Chandler, D.B., Chang, M.-S., Kloks, T., Liu, J., Peng, S.-L.: Recognition of Probe Cographs and Partitioned Probe Distance hereditary graphs. In: Cheng, S.-W., Poon, C.K. (eds.) AAIM 2006. LNCS, vol. 4041, pp. 267–278. Springer, Heidelberg (2006)
7. Chandler, D.B., Chang, M.-S., Kloks, T., Liu, J., Peng, S.-L.: Partitioned Probe Comparability Graphs. In: Fomin, F.V. (ed.) WG 2006. LNCS, vol. 4271, pp. 179–190. Springer, Heidelberg (2006)
8. Chang, G.J., Kloks, A.J.J., Liu, J., Peng, S.-L.: The PIGs Full Monty — a Floor Show of Minimal Separators. In: Diekert, V., Durand, B. (eds.) STACS 2005. LNCS, vol. 3404, pp. 521–532. Springer, Heidelberg (2005)
9. Chang, M.-S., Kloks, T., Kratsch, D., Liu, J., Peng, S.-L.: On the Recognition of Probe Graphs of Some Self-complementary Graph Classes. In: Wang, L. (ed.) COCOON 2005. LNCS, vol. 3595, pp. 808–817. Springer, Heidelberg (2005)
10. Corneil, D.G., Lerchs, H., Stewart, L.K.: Complement Reducible Graphs. Discrete Applied Mathematics 3, 163–174 (1981)
11. Dirac, G.A.: On Rigid Circuit Graphs. Abh. Math. Sem. Univ. Hamburg 25, 71–76 (1961)
12. Farber, M., Jamison, R.E.: Convexity in Graphs and Hypergraphs. SIAM J. Alg. Discrete Methods 7, 433–444 (1986)
13. Habib, M., Paul, C.: A Simple Linear Time Algorithm for Cograph Recognition. Discrete Applied Mathematics 145, 183–197 (2005)
14. Hajnal, A., Surányi, J.: Uber die Auflösung von Graphen in Vollständige Teilgraphen. Ann. Univ. Sci. Budapest, Eötvös Sect. Math 1, 113–121 (1958)
15. Hammer, P.L., Maffray, F.: Completely Separable Graphs. Discrete Applied Mathematics 27, 85–99 (1990)
16. Hoàng, C.T., Sritharan, R.: Finding Houses and Holes in Graphs. Theoretical Computer Science 259, 233–244 (2001)
17. Howorka, E.: A Characterization of Ptolemaic Graphs. Journal of Graph Theory 5, 323–331 (1981)
18. Johnson, J.L., Spinrad, J.: A Polynomial-Time Recognition Algorithm for Probe Interval Graphs. In: Proceedings of 12th ACM–SIAM Symposium on Discrete Algorithms (SODA 2001), pp. 447–486 (2001)
19. Kay, D.C., Chartrand, G.: A Characterization of Certain Ptolemaic Graphs. Canad. J. Math. 17, 342–346 (1965)
20. Le, V.B., de Ridder, H.N.: Probe Split Graphs. Discrete Mathematics and Theoretical Computer Science 9, 207–238 (2007)
21. Le, V.B., de Ridder, H.N.: Characterizations and Linear-Time Recognition for Probe Cographs. LNCS, vol. 4769, pp. 226–237 (2007)
22. McConnell, R.M., Spinrad, J.P.: Construction of Probe Interval Graphs. In: Proceedings 13th ACM–SIAM Symposium on Discrete Algorithms (SODA 2002), pp. 866–875 (2002)

23. Möhring, R.H., Radermacher, F.J.: Substitution Decomposition for Discrete Structures and Connections with Combinatorial Optimization. Ann. Discrete Math. 19, 257–356 (1984)
24. Nicolai, F.: Strukturelle und Algorithmische Aspekte Distanz-erblicher Graphen und Verwandter Klasses. Dissertation thesis, Gerhard-Mercator-Universität, Duisburg (1994)
25. de Ridder, H.N.: On Probe Graph Classes. PhD thesis, Universität Rostock, Germany (2007)
26. Rose, D.J., Tarjan, R.E., Lueker, G.S.: Algorithmic Aspects of Vertex Elimination on Graph. SIAM Journal of Computing 5, 266–283 (1976)
27. Soltan, V.P.: d-Convexity in Graphs. Soviet Math. Dokl. 28, 419–421 (1983)
28. Uehara, R., Uno, Y.: Laminar Structure of Ptolemaic and its Applications. In: Deng, X., Du, D.-Z. (eds.) ISAAC 2005. LNCS, vol. 3827, pp. 186–195. Springer, Heidelberg (2005)
29. Wolk, E.S.: The Comparability Graph of a Tree. Proceedings of the American Mathematical Society 13, 789–795 (1962)
30. Wolk, E.S.: A Note on The Comparability Graph of a Tree. Proceedings of the American Mathematical Society 16, 17–20 (1965)

Diagnosability of Two-Matching Composition Networks

Sun-Yuan Hsieh* and Chia-Wei Lee

Department of Computer Science and Information Engineering,
National Cheng Kung University,
No. 1, University Road, Tainan 70101, Taiwan
hsiehsy@mail.ncku.edu.tw, cwlee@csie.ncku.edu.tw

Abstract. Diagnosability is an important metric for measuring the reliability of multiprocessor systems. In this paper, we study the diagnosability of a class of networks, called Two-Matching Composition Networks (2-MCNs), each of which is constructed by connecting two graphs via two perfect matchings. By applying our result to multiprocessor systems, we also compute the diagnosability of folded hypercubes and augmented cubes, both of which belong to two-matching composition networks.

1 Introduction

Investigating the diagnosability of multiprocessor systems based on various strategies and models has been the focus of a great deal of research [2],[3],[4],[5], [6],[8], [9],[11],[12],[13],[14],[15],[16],[17],[18],[19],[20],[21],[22],[23]. Among the proposed models, two of which, namely the *PMC model* (Preparata, Metze, and Chien's model [17]) and the *MM model* (Maeng and Malek's model [16]), are well-known and widely used. In the PMC model, every node u is capable of testing whether another node v is faulty if there exists a communication link between them. The PMC model was also adopted in [3],[4],[11],[14]. In the MM model, on the other hand, every node tests each pair of its neighboring nodes by sending the same message to both of them and then checking if their responses agree. For this reason, the MM model is also called *the comparison diagnosis model*. Sengupta and Dahbura [18] further suggested a modification of the MM model, called the MM* model, in which any node has to test another two nodes if the former is adjacent to the later two. The MM* model was also adopted in [4],[12],[22].

In the MM* model, an n-dimensional hypercube has diagnosability n [22]; an n-dimensional enhanced hypercube has diagnosability $n+1$ [22]; an n-dimensional crossed cube has diagnosability n [12]; a k-ary n-dimensional butterfly graph has diagnosability $2k-2$ if $k \geq 3$ and $n \geq 3$ [1]. In this paper, we first propose a class of interconnection networks, called the *Two-Matching Composition Networks* (*2-MCNs* for short). Basically, a 2-MCN is constructed from

---

* Corresponding author.

X. Hu and J. Wang (Eds.): COCOON 2008, LNCS 5092, pp. 478–486, 2008.

the two given graphs, G_1 and G_2, with the same number of nodes by connecting the nodes of G_1 and G_2 via two perfect matchings. We call G_1 and G_2 the *components* of 2-MCN. Two well-known multiprocessor systems, such as folded hypercubes and augmented cubes, belong to 2-MCNs. Then, we study the diagnosability of 2-MCNs. Our main result is the following: suppose that the number of nodes in each component is at least $t + 3$ and both the order (which will be defined subsequently) and connectivity of each node in G_i is at least t, where $t \geq 3$. We prove that the diagnosability of 2-MCN constructed from G_1 and G_2 is $t + 2$ under the MM* model. In other words, the diagnosability of 2-MCNs is increased by two as compared with those of the components. As a result, it is possible to determine the diagnosability of many well-known and unknown, but potentially useful, multiprocessor systems. By applying our result, we show that the diagnosability of n-dimensional folded hypercubes and n-dimensional augmented cubes are $n + 1$ and $2n - 1$, respectively. The diagnosability of the above multiprocessor systems, as far as we know, have not yet been resolved until now.

The remainder of this paper is organized as follows: Section 2 introduces some definitions and notations. In Section 3, we demonstrate the diagnosability of 2-MCNs under the MM* model. In Section 4, we apply our result to folded hypercubes and augmented cubes. Finally, Section 5 contains a discussion and some concluding remarks.

2   Preliminaries

An *undirected graph* (*graph* for short) $G = (V, E)$ is comprised of a *node set* V and an *edge set* E, where V is a finite set and E is a subset of $\{(u, v)| (u, v)$ is an unordered pair of $V\}$. For an edge (u, v), we call u and v the *end-nodes* of the edge. A *subgraph* of $G = (V, E)$ is a graph (V', E') such that $V' \subseteq V$ and $E' \subseteq E$. Given $U \subseteq V(G)$, the *subgraph of G induced by U* is defined as $G[U] = (U, \{(u, v) \in E(G)| u, v \in U\})$, and let $\bar{U} = V(G) - U$. We also use $V(G)$ and $E(G)$ to denote the node set and edge set of G, respectively. A *multigraph* is like an undirected graph, but it can have multiple edges between its nodes as well as self-loops. A multiprocessor system can be modelled as an undirected graph G whose nodes represent processors and edges represent communication links.

The *node connectivity* (*connectivity* for short) of a graph G, denoted by $\kappa(G)$, is the minimum size of a node set S such that $G - S$ is disconnected or has only one node. A graph G is *t-connected* if $\kappa(G) \geq t$. A *node cover* of G is a subset $K \subseteq V(G)$ such that every edge of $E(G)$ has at least one end node in K. The node cover with the minimum cardinality is called a *minimum node cover*. Let V_1 and V_2 be two disjoint nonempty sets of $V(G)$. The *neighborhood set of V_1 in V_2* is defined as $N_G(V_1, V_2) = \{v \in V_2| \exists$ a node $u \in V_1$ such that $(u, v) \in E(G)\}$ and $N_G(V_1, \bar{V}_1)$ is reduced to $N_G(V_1)$. Similarly, if $u \in V(G)$ and $V' \subseteq V(G) - \{u\}$, then the *neighborhood set of u in V'* is defined as $N_G(u, V') = \{v \in V'| (u, v) \in E(G)\}$. When $V' = V(G) - u$, $N_G(u, V')$ is reduced to $N_G(u)$. We also call each node in $N_G(u)$ a *neighbor* of u. For convenience, the subscript G can be deleted from the above notations if no ambiguity occurs. The *degree* of a node u, denoted by $deg(u)$, in a simple graph G is the number of edges incident to u, i.e., $deg(u) = |N(u)|$. The *minimum degree* of a graph $G = (V, E)$, denoted by $\delta(G)$, is $\min_{u \in V}\{deg(u)\}$.

The MM model deals with the fault diagnosis by sending the same task from a node w to each pair of distinct neighbors, u and v, and then comparing their responses. In order to be consistent with the MM model, the following assumptions are made: (1) all faults are permanent; (2) a faulty processor produces incorrect output for each of its given task; (3) the outcome of a comparison performed by a fault processor is unreliable; and, (4) two faulty processors, when given the same inputs and task, do not produce the same output.

The *comparison scheme* of the system G can be modelled as a multigraph $M(V(G), L)$, where L is the labelled-edge set. Let $(u, v)_w$ denote an edge (u, v) labelled by w that represents a *comparison* in which nodes u and v are compared by w. A labelled-edge $(u, v)_w \in L$ represents that u and v are compared by w. The result of all comparisons in $M(V(G), L)$ is called the *syndrome* of the diagnosis. Briefly, a syndrome δ is a function from L to $\{0, 1\}$, and we use $\delta((u, v)_w)$ to represent the result of the comparison $(u, v)_w$ in δ. In the special case of the MM model, called the MM* model, if $u, v, w \in V(G)$ and $(u, w), (v, w) \in E(G)$, then $(u, v)_w$ must be in L. In other words, all possible comparisons of G must be in the comparison scheme of G. Hereafter, the results reported in this paper are for systems under the MM* model.

Let G be a system with its comparison scheme $M(V(G), L)$. Given a node $u \in V(G)$, we define the *order graph* of u, denoted by $G^u = (X^u, Y^u)$, as follows: (1) $X^u = \{v| (u, v) \in E(G) \text{ or } (u, v)_w \in L\}$, and (2) $Y^u = \{(v, w)| v, w \in X^u \text{ and } (u, v)_w \in L\}$. The *order* of u, denoted by $order_G(u)$, is defined as the cardinality of a minimum node cover of G^u. Given $V' \subseteq V(G)$, we define the *probing set* of V', denoted by $P(G, V')$, to be the set $\{u \in V'| (u, v)_w \in L \text{ and } v, w \in V(G) - V'\}$. It means that all nodes in V' can be compared with some node in $V(G) - V'$ by another node in $V(G) - V'$.

For a given syndrome δ, $S \subseteq V(G)$ is said to be *consistent* with δ if and only if the following conditions hold: (1) If $u \in S$ and $v, w \in V(G) - S$, $\delta((u, v)_w) = 1$, (2) If $u, v \in S$ and $w \in V(G) - S$, $\delta((u, v)_w) = 1$, and (3) If $u, v, w \in V(G) - S$, $\delta((u, v)_w) = 0$. The syndrome δ can be produced form the situation where all nodes in S are faulty and all nodes in $V(G) - S$ are fault-free. Because a faulty comparator w can lead to unreliable result, which means that the result by a faulty comparator may be 1 or 0 every time and thus a given faulty set S may produce different syndromes. Let $\delta(S)$ denote all syndromes which S is consistent with.

Two distinct subsets $S_1, S_2 \subseteq V(G)$ are said to be *indistinguishable* if $\delta(S_1) \cap \delta(S_2) \neq \emptyset$; otherwise, S_1 and S_2 are said to be *distinguishable*. The pair (S_1, S_2) is said to be *indistinguishable* (respectively, *distinguishable*) *pair* if S_1 and S_2 are indistinguishable (respectively, distinguishable). A system G is said to be *t-diagnosable* if all faulty nodes can be identified without replacement, provided that the number of faults presented does not exceed t and the *diagnosability* of G, denoted by $d(G)$, refers to the maximum number of faulty nodes that can be identified by the system. In other word, a system is t-diagnosable if and only if each pair of distinct sets $S_1, S_2 \subseteq V(G)$ with $|S_1|, |S_2| \leq t$ are distinguishable and the diagnosability of G is $\max\{t| G$ is t-diagnosable$\}$.

Lemma 1. [18] *For any $S_1, S_2 \subset V(G)$ and $S_1 \neq S_2$, (S_1, S_2) is a distinguishable pair if and only if at least one of the following conditional is satisfied:*

1. $\exists u, w \in V(G) - S_1 - S_2$ and $\exists v \in (S_1 - S_2) \bigcup (S_2 - S_1)$ such that $(u, v)_w \in L$,
2. $\exists u, v \in S_1 - S_2$ and $\exists w \in V(G) - S_1 - S_2$ such that $(u, v)_w \in L$, and
3. $\exists u, v \in S_2 - S_1$ and $\exists w \in V(G) - S_1 - S_2$ such that $(u, v)_w \in L$.

Lemma 2. [18] *A system G is t-diagnosable if and only if for each pair of sets $S_1, S_2 \subset V(G)$ such that $S_1 \neq S_2$ and $|S_1|, |S_2| \leq t$, at least one of the conditions of Lemma 1 is satisfied.*

Lemma 3. *Let G be a system with N nodes and $S_1, S_2 \subset V(G)$ be two distinct sets with $|S_1|, |S_2| \leq t$ and $|S_1 \bigcap S_2| = p$, where $0 \leq p \leq t - 1$. If $|P(G, S_1 \bigcup S_2)| > p$, then (S_1, S_2) is a distinguishable pair.*

Proof. Let $M(V(G), L)$ be the comparison scheme of G. If $|P(G, S_1 \bigcup S_2)| > p$, then there must exist a node $u \in P(G, S_1 \bigcup S_2)$ such that $u \notin S_1 \bigcap S_2$. This implies that $u \in (S_1 - S_2) \bigcup (S_2 - S_1)$. By the definition of the probing set, there exist two nodes $v, w \in V(G) - (S_1 \bigcup S_2)$ such that $(u, v)_w \in L$, which satisfies Condition 1 of Lemma 1. Therefore, (S_1, S_2) is a distinguishable pair.   □

Lemma 4. *Let $G = (V, E)$ be a connected graph. Suppose that each node in G have an order of at least t, where $t \geq 0$. If $U \subset V$ with $|U| \leq t$, then $P(G, U) = U$.*

Proof. Let u be a node in U. Since the order of u is at least t, the cardinality of the minimum node cover in the comparison graph $G^u = (X^u, Y^u)$ is also at least t. Clearly, $U - \{u\}$ is not a node cover of G^u because $|U - \{u\}| < t$. This implies that there exists an edge (w, v) in Y^u such that w and v are both in $\bar{U}$. Moreover, by the definition of the comparison graph, there exists a test $(u, v)_w$ or $(u, w)_v$. This implies that $u \in P(G, U)$. Therefore, $P(G, U) = U$.   □

A *matching* in a graph G is a set of non-loop edges with no shared end-nodes. The nodes incident to the edges of a matching M are *saturated* by M; the others are *unsaturated*. Two matchings are *distinct* if they have no common edge. A *perfect matching* in a graph is a matching that saturates every node. The class of *Two-Matching Composition Networks* is defined as follows.

Suppose that G_1 and G_2 are two graphs with the same number of nodes. Let PM be an arbitrary *perfect matching between $V(G_1)$ and $V(G_2)$*, i.e., PM is a set of edges connecting $V(G_1)$ and $V(G_2)$ in a one-to-one fashion, and let PM_2 be a set comprised of two arbitrary distinct perfect matchings between $V(G_1)$ and $V(G_2)$. Then, the resulting graph constructed from G_1 and G_2 by connecting $V(G_1)$ and $V(G_2)$ via PM_2 is called a *two-matching composition network*, denoted by $G(G_1, G_2; PM_2)$, where $V(G(G_1, G_2; PM_2)) = V(G_1) \bigcup V(G_2)$ and $E(G(G_1, G_2; PM_2) = E(G_1) \bigcup E(G_2) \bigcup PM_2$. Note that each node $v \in V(G_1)$ has exactly two neighbors in $V(G_2)$.

3   Diagnosability of 2-MCNs

In this section, we study the diagnosability of 2-MCNs under the MM* model. Some useful properties are shown as follows.

Lemma 5. *Let G_1 and G_2 be two graphs with the same number of nodes N, and let $t \geq 2$ be a positive integer. If both G_1 and G_2 are t-connected and $N \geq t+2$, then $G(G_1, G_2; PM_2)$ is $(t+2)$-connected.*

Proof. Let $G = G(G_1, G_2; PM_2)$. Suppose, by contradiction, that $\kappa(G) \leq t+1$. Then, there exists a set $U \subset V(G)$ with $|U| \leq t+1$ such that $G[G - U]$ is disconnected. Let $U_1 = U \bigcap V(G_1)$, $U_2 = U \bigcap V(G_2)$, $n_1 = |U_1|$, and $n_2 = |U_2|$. Without loss of generality, we assume that $n_1 \geq n_2$ and consider the following cases.

Case 1: $n_1 < t$ and $n_2 < t$. Since $\kappa(G_i) \geq t > n_1 \geq n_2$ for $i = 1, 2$, $G[G_1 - U_1]$ and $G[G_2 - U_2]$ are both connected. However, because $G[G - U]$ is disconnected, this implies that each edge $(u, v) \in PM_2$ must have at least one end-node u or v in U. Moreover, because each node in G_i for $i = 1, 2$ is incident to exactly two edges in PM_2, $N \leq |U| \leq t+1$. On the other hand, since $\kappa(G_i) \geq t$, the degree of each node u in G_i must be at least t, which implies that $|V(G_i)| = N \geq |\{u\} \bigcup N(u)| \geq t+1$. This will cause that $N = t+1$ which contradicts the assumption that $N \leq t+2$.

Case 2: $n_1 = t$ and $n_2 = 1$. By the definition of 2-MCNs and $n_2 = 1$, each node u in $G[G_1 - U_1]$ has at least one neighbor in $G[G_2 - U_2]$. Moreover, since G_2 is t-connected and $n_2 = 1 < t$, $G[G_2 - U_2]$ is connected. Then, $G[G - U]$ is connected, which is a contradiction.

Case 3: $n_1 = t+1$ and $n_2 = 0$. Since each node u in $G[G_1 - U_1]$ has two neighbors in G_2 and $n_2 = 0$, $G[G - U]$ is connected, which is a contradiction.

By combining the above cases, we complete the proof.   □

Let G_1 and G_2 be two networks with the same number of nodes N, and the order for each node v in G_i is at least t, where $i = 1, 2$. The following lemma shows that the order for each node v in $G(G_1, G_2; PM_2)$ is at least $t + 2$.

Lemma 6. *Let G_1 and G_2 be two networks with the same number of nodes N, and let $t \geq 2$ be a positive integer. If $order_{G_i}(v) \geq t$ for each node v in G_i, where $i = 1, 2$, then $order_G(v) \geq t+2$ for each node v in $G(G_1, G_2; PM_2)$.*

Now, we are ready to state and prove the following theorem about that the diagnosability of 2-MCNs under the MM* model.

Theorem 1. *Let $t \geq 3$ be a positive integer and let G_1 and G_2 be two graphs with the same number of nodes N, where $N \geq t + 3$. If each node in G_i has the order at least t and $\kappa(G_i) \geq t$ for $i = 1, 2$, 2-MCN $G(G_1, G_2; PM_2)$ is $(t+2)$-diagnosable.*

Proof. For convenience, let $G = G(G_1, G_2; PM_2)$. Let $S_1, S_2 \subset V(G)$ be two distinct sets with $|S_1|, |S_2| \leq t + 2$ and let $|S_1 \cap S_2| = p$, where $0 \leq p \leq t + 1$. Also, let $S_{G_i} = V(G_i) \cap (S_1 \cup S_2)$ for $i = 1, 2$ with $|S_{G_1}| = n_1$ and $|S_{G_2}| = n_2$. Clearly, $n_1 + n_2 = 2(t + 2) - p$. Without loss of generality, we assume that $n_1 \leq n_2$. We have the following cases.

Case 1: $n_1 \leq t$ and $n_2 \leq t$. By Lemma 4, $P(G_1, S_{G_1}) = S_{G_1}$ and $P(G_2, S_{G_2}) = S_{G_2}$. Then, $|P(G, S_1 \cup S_2)| \geq |P(G_1, S_{G_1})| + |P(G_2, S_{G_2})| = |S_{G_1}| + |S_{G_2}| = n_1 + n_2 = |S_1 \cup S_2| > |S_1 \cap S_2| = p$. Therefore, by Lemma 3, (S_1, S_2) is a distinguishable pair.

Case 2: $n_2 > t$. We have the following scenarios.

    Case 2.1: $n_1 < t$.

        Case 2.1.1: $0 \leq p \leq t$ or ($p = t + 1$ and $n_2 \geq t + 2$). By Lemma 4, $|P(G_1, S_{G_1})| = |S_{G_1}| = n_1$. Moreover, since each node in S_{G_2} is connected to exactly two nodes in G_1 (via PM_2) and $n_2 > n_1$, S_{G_2} contains a set R with $|R| \geq n_2 - n_1$ such that each node in R is connected to some node in $V(G_1) - S_{G_1}$. Let $u \in R$ be an arbitrary node and let w be a node in $V(G_1) - S_{G_1}$ connected to u. Since $\kappa(G_1) \geq t$ and $|S_{G_1}| = n_1 < t$, $G[V(G_1) - S_{G_1}]$ is connected, which implies that there is another node $v \in V(G_1) - S_{G_1}$ connected to w. Clearly, $(u, v)_w$ is a comparison. By combining the above arguments, we have that $|P(G, S_1 \cup S_2)| \geq |P(G_1, S_{G_1})| + (n_2 - n_1) = |S_{G_1}| + (n_2 - n_1) = n_2 > t \geq p$. Therefore, by Lemma 3, (S_1, S_2) is a distinguishable pair.

        Case 2.1.2: $p = t + 1 = n_2$. In this case, since $n_1 + n_2 = 2(t + 2) - p$, $n_1 \leq 2$. Since $n_1 + n_2 \leq t + 3$ and $p = t + 1$, there are at most two nodes in $(S_1 \cup S_2) - (S_1 \cap S_2)$, which leads to the following scenarios.

            Case 2.1.2.1: $|S_{G_1} - (S_1 \cap S_2)| \geq 1$. Let u be such a node in $S_{G_1} - (S_1 \cap S_2)$. Since $n_1 \leq 2$ and $|N_{G_1}(u)| \geq 3$ because $\kappa(G_1) \geq t \geq 3$, there exists a node $w \in V(G_1) - S_{G_1}$ connected to u. Moreover, because $n_1 \leq 2$ and $\kappa(G_1) \geq t \geq 3$, $G_1[V(G_1) - S_{G_1}]$ is connected, which implies that there is another node $v \in V(G_1) - S_{G_1}$ connected to w. Then, comparison $(u, v)_w$ satisfies Condition 1 of Lemma 1. Therefore, (S_1, S_2) is a distinguishable pair.

            Case 2.1.2.2: $|S_{G_1} - (S_1 \cap S_2)| = 0$ and $n_1 = 2$, which implies that there are exactly two nodes in $S_{G_2} - (S_1 \cap S_2)$. Let x and y be two nodes in $S_{G_2} - (S_1 \cap S_2)$. If x or y is connected to a node in $V(G_1) - S_{G_1}$, then the result that (S_1, S_2) is a distinguishable pair can be shown using a similar argument to that presented in Case 2.1.2.1. Otherwise, x and y are connected to two nodes in S_{G_1} via PM_2. Since $n_2 = |S_{G_2}| = t + 1$ and G_2 is t-connected for $t \geq 3$, $G[V(G_2) - (S_{G_2} - \{x,y\})]$ is connected. Hence, at least one node in $\{x,y\}$ must be connected to some node, say z, in $G[V(G_2) - S_{G_2}]$. Without loss of generality, we assume that y is connected to z. Note that z is connected to another node $z' \in V(G_1) - S_{G_1}$, where $(z,z') \in PM_2$. Then, comparison $(y,z')_z$ satisfies Condition 1 of Lemma 1. Therefore, (S_1, S_2) is a distinguishable pair.

Case 2.1.2.3: $|S_{G_1} - (S_1 \cap S_2)| = 0$ and $n_1 = 1$, which implies that there is a node in $S_{G_2} - (S_1 \cap S_2)$. Let x be a node in $S_{G_2} - (S_1 \cap S_2)$. Since $n_1 = 1$, there exists a node $w \in V(G_1) - G_{G_1}$ connected to x via PM_2. Then, the result that (S_1, S_2) is a distinguishable pair can be shown using a similar argument to that presented in Case 2.1.2.1.

Case 2.2: $n_1 \geq t$. In this case, $n_2 \in \{t+1, t+2, t+3, t+4\}$, which implies that $p \leq 3$. The proof of this case is similar to that used in Case 2.1.

By combining above cases, we complete the proof.   □

4   Application to Multiprocessor Systems

In this section, we apply the theorems presented in Sections 3 to two popular multiprocessor systems, n-dimensional folded hypercube and n-dimensional augmented cube. It is of course possible to apply them to many other potentially useful systems. Interested readers can refer to [10],[7] for the definitions of the above networks.

Since an n-dimensional hypercube is n-connected and the order of each node equals n [22] for $n \geq 3$, then by the definition of FQ_n and Theorem 1, the following lemma holds.

Lemma 7. FQ_n is $(n+1)$-diagnosable for $n \geq 4$.

Theorem 2. $d(FQ_n) = n+1$ for $n \geq 4$.

Proof. By Lemma 7, $d(FQ_n) \geq n+1$. Next, we show that $d(FQ_n) \leq n+1$. Suppose, by contradiction, that FQ_n is $(n+2)$-diagnosable. Let u be an arbitrary node in FQ_n, and let $S_1 = \{u\} \cup N(u)$ and $S_2 = N(u)$. Clearly, $S_1 \neq S_2$, $N(u) = S_1 \cap S_2$, and $|S_1|, |S_2| \leq n+2$. However, node u cannot be compared by any other node in $V(FQ_n) - S_1 - S_2$, which implies that none of the conditions in Lemma 1 is satisfied. This is contrary to that FQ_n is $(n+2)$-diagnosable, which leads to that $d(FQ_n) \leq n+1$. Hence, $d(FQ_n) = n+1$ for $n \geq 4$.   □

With the similar proof, we can show that the diagnosability of n-dimensional augmented cube AQ_n is $2n-1$, i.e., $d(AQ_n) = 2n-1$ for $n \geq 4$.

5   Concluding Remarks

Fault diagnosis of multiprocessor system has received a great deal of attention since Preparata *et al.* [17] introduced the concept of system-level diagnosis. In this paper, we have established sufficient conditions that can be utilized to determine the diagnosability of 2-MCNs. Furthermore, using the established theorem, we have successfully computed the diagnosability of folded hypercubes and augmented cubes, which is previously unknown in the literature.

References

1. Araki, T., Shibata, Y.: Diagnosability of Butterfly Networks under the Comparison Approach. IEICE Transactions on Fundamentals of Electronics Communications and Computer Sciences E85-A(5), 1152–1160 (2002)
2. Araki, T., Shibata, Y.: (t, k)-Diagnosable System: a Generalization of the PMC Models. IEEE Transactions on Computers 52(7), 971–975 (2003)
3. Armstrong, J.R., Gray, F.G.: Fault Diagnosis in a Boolean n Cube Array of Multiprocessors. IEEE Transactions on Computers 30(8), 587–590 (1981)
4. Chang, G.Y., Chang, G.J., Chen, G.H.: Diagnosabilities of Regular Networks. IEEE Transactions on Parallel and Distributed Systems 16(4), 314–323 (2005)
5. Chang, G.Y., Chen, G.H., Chang, G.J.: (t, k)-Diagnosis for Matching Composition Networks. IEEE Transactions on Computers 55(1), 88–92 (2006)
6. Chang, G.Y., Chen, G.H., Chang, G.J. (t, k)-Diagnosis for Matching Composition Networks under the MM* Model. IEEE Transactions on Computers 56(1), 73–79 (2007)
7. Choudum, S.A., Sunitha, V.: Augmented Cubes. Networks 40(2), 71–84 (2002)
8. Chwa, K.Y., Hakimi, L.: On Fault Identification in Diagnosable Systems. IEEE Transactions on Computers 30(6), 414–422 (1981)
9. Das, A., Thulasiraman, K., Agarwal, V.K.: Diagnosis of $t/(t+1)$-Diagnosable Systems. SIAM Journal on Computing 23(5), 895–905 (1994)
10. EI-Amawy, A., Latifi, S.: Properties and Performance of Folded Hypercubes. IEEE Transactions on Parallel and Distributed Systems 2(1), 31–42 (1991)
11. Fan, J.: Diagnosability of the Möbius Cubes. IEEE Transactions on Parallel and Distributed Systems 9(9), 923–928 (1998)
12. Fan, J.: Diagnosability of Crossed Cubes under the Comparison Diagnosis Model. IEEE Transactions on Parallel and Distributed Systems 13(7), 687–692 (2002)
13. Fan, J., Lin, X.: The t/k-Diagnosability of the BC Graphs. IEEE Transactions on Computers 54(2), 176–184 (2005)
14. Kavianpour, A., Kim, K.H.: Diagnosability of Hypercubes under the Pessimistic One-Step Diagnosis Strategy. IEEE Transactions on Computers 40(2), 232–237 (1991)
15. Lee, J.K., Butler, J.T.: A Characterization of t/s-Diagnosability and Sequential t-Diagnosability in Designs. IEEE Transactions on Computers 39(10), 1298–1304 (1990)
16. Maeng, J., Malek, M.: A Comparison Connection Assignment for Self-diagnosis of Multiprocessor Systems. In: Proceeding of the 11th International Symposium on Fault-Tolerant Computing, pp. 173–175 (1981)
17. Preparata, F.P., Metze, G., Chien, R.T.: On the Connection Assignment Problem of Diagnosable Systems. IEEE Transactions on Computers EC-16, 448–454 (1967)
18. Sengupta, A., Dahbura, A.: On Self-diagnosable Multiprocessor System: Diagnosis by the Comparison Approach. IEEE Transactions on Computers 41(11), 1386–1396 (1992)

19. Somani, A.K.: Sequential Fault Cccurrence and Reconfiguration in System Level Diagnosis. IEEE Transactions on Computers 39(12), 1472–1475 (1990)
20. Somani, A.K., Agarwal, V.K., Avis, D.: A Generalized Theory for System Level Diagnosis. IEEE Transactions on Computers 36(5), 538–546 (1987)
21. Somani, A.K., Peleg, O.: On Diagnosability of Large Fault Sets in Regular Topology-based Computer Systems. IEEE Transactions on Computers 45(8), 892–903 (1996)
22. Wang, D.: Diagnosability of Hypercubes and Enhanced Hypercubes under the Comparison Diagnosis Model. IEEE Transactions on Computers 48(12), 1369–1374 (1999)
23. Yang, C.L., Masson, G.M., Leonetti, R.A.: On Fault Isolation and Identification in t_1/t_1-Diagnosable Systems. IEEE Transactions on Computers 35(7), 639–643 (1986)

The Iterated Restricted Immediate Snapshot Model

Sergio Rajsbaum[1], Michel Raynal[2], and Corentin Travers[3]

[1] Instituto de Matemáticas, UNAM, D.F. 04510, Mexico
[2] IRISA, Campus de Beaulieu, 35042 Rennes Cedex, France
[3] Facultad de Informática, UPM, Madrid, Spain
rajsbaum@math.unam.mx, raynal@irisa.fr, ctravers@fi.upm.es

Abstract. In the *Iterated Immediate Snapshot* model (*IIS*) the memory consists of a sequence of one-shot *Immediate Snapshot* (*IS*) objects. Processes access the sequence of *IS* objects, one-by-one, asynchronously, in a *wait-free* manner; any number of processes can crash. Its interest lies in the elegant recursive structure of its runs, hence of the ease to analyze it round by round. In a very interesting way, Borowsky and Gafni have shown that the *IIS* model and the read/write model are equivalent for the wait-free solvability of decision tasks.

This paper extends the benefits of the *IIS* model to partially synchronous systems. Given a shared memory model enriched with a failure detector, what is an equivalent *IIS* model? The paper shows that an elegant way of capturing the power of a failure detector and other partially synchronous systems in the *IIS* model is by restricting appropriately its set of runs, giving rise to the *Iterated Restricted Immediate Snapshot* model (*IRIS*).

1 Introduction

A distributed model of computation consists of a set of n processes communicating through some medium (some form of message passing or shared memory), satisfying specific timing assumptions (process speeds and communication delays), and failure assumptions (their number and severity). A major obstacle in the development of a theory of distributed computing is the wide variety of models that can be defined – many of which represent real systems – with combinations of parameters in both the (a)synchrony and failure dimensions [4]. Thus, an important line of research is concerned with finding ways of unifying results, impossibility techniques, and algorithm design paradigms of different models.

An early approach towards this goal has been to derive direct simulations from one model to another, e.g., [3],[2],[6]. A more recent approach has been to devise models of a higher level of abstraction, where results about various more specific models can be derived (e.g., [12],[15]). Two main ideas are at the heart of the approach, which has been studied mainly for crash failures only, and is the topic of this paper.

X. Hu and J. Wang (Eds.): COCOON 2008, LNCS 5092, pp. 487–497, 2008.

Two bedrocks: wait-freedom and round-based execution It has been discovered [6],[16],[21] that the *wait-free* case is fundamental. In a system where any number of processes can crash, each process must complete the protocol in a finite number of its own steps, and "wait statements" to hear from another process are not useful. In a wait-free system it is easy to consider the *simplicial complex of global states* of the system after a finite number of steps, and various papers have analyzed topological invariants about the structure of such a complex, to derive impossibility results. Such invariants are based on the notion of *indistinguishability*, which has played a fundamental role in nearly every lower bound in distributed computing. Two global states are indistinguishable to a set of processes if they have the same local states in both. In the figure on the right, there is a complex with three triangles, each one is a *simplex* representing a global state; the corners of a simplex represent local states of processes in the global state. The center simplex and the rightmost simplex represent global states that are indistinguishable to p_1 and p_2, which is why the two triangles share an edge. Only p_3 can distinguish between the two global states.

Most attempts at unifying models of various degrees of asynchrony restrict attention to a subset of well-behaved, *round-based* executions. The approach in [7] goes beyond that and defines an *iterated* round-based model (*IIS*), where each communication object can be accessed only once by each process. These objects, called *Immediate Snapshot* objects [5], are accessed by the processes with a single operation denoted write_snap(), that writes the value provided by the invoking process and returns to it a snapshot [1] of its content. The sequence of *IS* objects are accessed asynchronously, and one after the other by each process. It is shown in [7] that the *IIS* model is equivalent (for bounded wait-free task solvability) to the usual read/write shared memory model.

Thus, the runs of the *IIS* model are not a subset of the runs of a standard (non-iterated) model as in other works, and the price that has to be payed is an ingenious simulation algorithm showing that the model is equivalent to a read/write shared memory model (w.r.t. wait-free task solvability). But the reward is a model that has an elegant recursive structure: the complex of global states after $i + 1$ rounds is obtained by replacing each simplex by a one round complex (see Figure 1). Indeed, the *IIS* model was the basis for the proof in [7] of the main characterization theorem of [16], and was instrumental for the results in [13].

Context and Goals of the Paper. The paper introduces the *IRIS model*, which consists of a subset of runs of the *IIS* model of [7], to obtain the benefits of the round by round and wait-freedom approaches in one model, where processes run wait-free but the executions represent those of a partially synchronous model. As an application, new, simple impossibility results for set agreement in several partially synchronous systems are derived.

In the construction of a distributed computing theory, a central question has been understanding how the degree of synchrony of a system affects its power to solve distributed tasks. The degree of synchrony has been expressed in various ways, typically either by specifying a bound t on the number of processes that can crash, as bounds on delays and process steps [11], or by a failure detector [8]. It has been shown multiple times that systems with more synchrony can solve more tasks. Previous works in this direction have mainly considered an asynchronous system enriched with a failure detector that can solve consensus. Some works have identified this type of synchrony in terms of fairness properties [22]. Other works have considered round-based models with no failure detectors [12]. Some other works [17] focused on performance issues mainly about consensus. Also, in some cases, the least amount of synchrony required to solve some task has been identified, within some paradigm. A notable example is the weakest failure detector to solve consensus [9] or k-set agreement [24]. Set agreement [10] represents a desired coordination degree to be achieved in the system, requiring processes to agree on at most k different values (consensus is 1-set agreement), and hence is natural to use it as a measure for the *synchrony degree* in the system. The fundamental result of the area is that k-set agreement is not solvable in a wait-free, i.e. fully asynchronous system even for $k = n - 1$ [6],[16],[21]. However, a clear view of what exactly "degree of synchrony" means is still lacking. For example, the same power as far as solving k-set agreement can be achieved in various ways, such as via different failure detectors [18] or t-resilience assumptions. A second goal for introducing the *IRIS* model, is to have a mean of precisely representing the degree of synchrony of a system, and this is achieved with the *IRIS* model by considering particular subsets of runs of the *IIS* model.

Capturing Partial Synchrony with a Failure Detector. A *failure detector* [8] is a distributed oracle that provides each process with hints on process failures. According to the type and the quality of the hints, several classes of failure detectors have been defined (e.g., [18],[24]).

As an example, this paper focuses on the family of *limited scope* accuracy failure detectors, denoted $\diamond\mathcal{S}_x$ [14],[23]. These capture the idea that a process may detect failures reliably on the same local-area network, but less reliably over a wide-area network. They are a generalization of the class denoted $\diamond\mathcal{S}$ that has been introduced in [8] ($\diamond\mathcal{S}_n$ is $\diamond\mathcal{S}$). Informally, a failure detector $\diamond\mathcal{S}_x$ ensures that there is a correct process that is eventually never erroneously suspected by any process in a cluster of x processes.

Results of the Paper. The paper starts by describing the read/write computation model enriched with a failure detector C of the class $\diamond\mathcal{S}_x$, and the *IIS* model, in Section 2. Then, in Section 3, it describes an *IRIS* model that precisely captures the synchrony provided by the asynchronous system equipped with C. To show that the synchrony is indeed captured, the paper presents two simulations in Section 4. The first is a simulation from the shared memory model with C to the *IRIS* model. The second shows how to extract C from the *IRIS* model, and then simulate the read/write model with C. From a technical point of view, this is the most difficult part of the paper. We had to develop a generalization of the wait-free simulation described in [7] that preserved consistency with the simulated failure detector.

The simulations prove Theorem 1: an agreement task is wait-free solvable in the read/write model enriched with C if and only if it is wait-free solvable in the corresponding *IRIS* model. Then, using a simple topological observation, it is easy to derive the lower bound of [14] for solving k-set agreement in a system enriched with C. In the approach presented in this paper, the technically difficult proofs are encapsulated in algorithmic reductions between the shared memory model and the *IRIS* model, while in the proof of [14] combinatorial topology techniques introduced in [15] are used to derive the topological properties of the runs of the system enriched with C directly[1].

2   Computation Model and Failure Detector Class

This section presents a quick overview of the background needed for the rest of the paper, more detailed descriptions can be found elsewhere, e.g., [4],[7],[8].

2.1   Shared Memory Model Enriched with a F.D. of the Class $\Diamond \mathcal{S}_x$

The paper considers a standard asynchronous system made up of n processes, $p_1, \ldots, p_n$, of which any of them can crash. A process is *correct in a run* if it takes an infinite number of steps. The shared memory is structured as an array $SM[1..n]$ of atomic registers, such that only p_i can write to $SM[i]$, and p_i can read any entry. Uppercase letters are used to denote shared registers. It is often useful to consider higher level abstractions constructed out of such registers, that are implementable on top of them, such as snapshots objects. In this case, a process can read the entire memory $SM[1..n]$ in a single atomic operation, denoted snapshot() [1].

A failure detector of the class $\Diamond \mathcal{S}_x$, where $1 \leq x \leq n$, provides each process p_i with a variable TRUSTED$_i$ that contains identities of processes that are believed to be currently alive. The process p_i can only read TRUSTED$_i$. When $j \in$ TRUSTED$_i$ we say "p_i trusts p_j". By definition, a crashed process trusts all processes. The failure detector class $\Diamond \mathcal{S}_x$ is defined by the following properties: (**Strong completeness**) There is a time after which every faulty process is never trusted by every correct process and, (**Limited scope eventual weak accuracy**) There is a set Q of x processes containing a correct process p_ℓ, and a (finite) time after which each process of Q trusts p_ℓ.

The following equivalent formulation of $\Diamond \mathcal{S}_x$ [18] is used in Section 4, assuming the local variable controlled by the failure detector is REPR$_i$: (**Limited eventual common representative**) There is a set Q of x processes containing a correct process p_ℓ, and a (finite) time after which, for any correct process p_i, we have $i \in Q \Rightarrow$ REPR$_i = \ell$ and $i \notin Q \Rightarrow$ REPR$_i = i$.

2.2   The Iterated Immediate Snapshot (*IIS*) Model

A *one-shot immediate snapshot* object *IS* is accessed with a a single operation denoted write_snap(). Intuitively, when a process p_i invokes write_snap(v) it is

---

[1] A companion technical report [19] extends the results presented here to other failure detector classes.

as if it instantaneously executes a write $IS[i] \leftarrow v$ operation followed by an IS.snapshot() operation.

The semantics of the write_snap() operation is characterized by the three following properties, where v_i is the value written by p_i and sm_i, the value (or *view*) it gets back from the operation, for each p_i invoking the operation. A view sm_i is a set of pairs (k, v_k), where v_k corresponds to the value in p_k's entry of the array. If $SM[k] = \bot$, the pair $(k, \bot)$ is not placed in sm_i. Moreover, we have $sm_i = \emptyset$, if the process p_i never invokes write_snap() on the corresponding object. The three properties are :(**Self-inclusion**) $\forall i : (i, v_i) \in sm_i$, (**Containment**) $\forall i, j : sm_i \subseteq sm_j \lor sm_j \subseteq sm_i$ and, (**Immediacy**) $\forall i, j : (i, v_i) \in sm_j \Rightarrow sm_i \subseteq sm_j$.

These properties are represented in the first image of Figure 1, for the case of three processes. The image represents a *simplicial complex,* i.e. a family of sets closed under containment; each set is called a *simplex,* and it represents the views of the processes after accessing the IS object. The *vertices* are the 0-simplexes, of size one; edges are 1-simplexes, of size two; triangles are of size three (and so on). Each vertex is associated with a process p_i, and is labeled with sm_i (the *view* p_i obtains from the object).

The highlighted 2-simplex in the figure represents a run where p_1 and p_3 access the object concurrently, both get the same views seeing each other, but not seeing p_2, which accesses the object later, and gets back a view with the 3 values written to the object. But p_2 can't tell the order in which p_1 and p_3 access the object; the other two runs are indistinguishable to p_2, where p_1 accesses the object before p_3 and hence gets back only its own value or the opposite. These two runs are represented by the corner 2-simplexes.

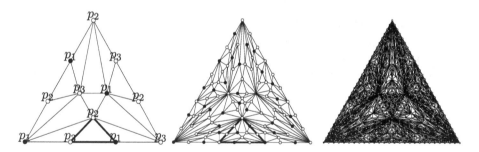

Fig. 1. One, two and three rounds in the *IIS* model

In the *iterated immediate snapshot model* (*IIS*) the shared memory is made up of an infinite number of one-shot immediate snapshot objects $IS[1], IS[2], \ldots$ These objects are accessed sequentially and asynchronously by each process. In Figure 1 one can see that the *IIS* complex is constructed recursively by replacing each simplex by the one round complex.

On the Meaning of Failures in the IIS Model. Consider a run where processes, p_1, p_2, p_3, execute an infinite number of rounds, but p_1 is scheduled before p_2, p_3 in every round. The triangles at the left-bottom corners of the complexes in

Figure 1 represent such a situation; p_1, at the corner, never hears from the two other processes. Of course, in the usual (non-iterated read/write shared memory) asynchronous model, two correct processes can always eventually communicate with each other. Thus, in the *IIS* model, the set of *correct processes* of a run, $Correct_{IIS}$, is defined as the set of processes that observe each other directly or indirectly infinitely often (a formal definition is given in [20]).

2.3   Tasks and Equivalence of the Two Models

An algorithm *solves a task* if each process starts with a private input value, and correct processes (according to the model) eventually decide on a private output value satisfying the task's specification. In an *agreement task,* the specification is such that, if a process decides v, it is valid for any other process to decide v (or some other function of v). The k-set agreement task is an agreement task, where processes start with input values of some domain of at least n values, and must decide on at most k of their input values.

It was proved in [7] that a task (with a finite number of inputs) is solvable wait-free in the read/write memory model if and only if it is solvable in the *IIS* model. As can be seen in Figure 1, the *IIS* complex of global states at any round is a subdivided simplex, and hence Sperner's Lemma implies that k-set agreement is not solvable in the *IIS* model if $k < n$. Thus, it is also unsolvable in the wait-free read/write memory model.

3   The *IRIS* Model

This section presents the *IRIS* model associated with a failure detector class C, denoted $IRIS(PR_C)$. It consists of a subset of runs of the *IIS* model, that satisfy a corresponding PR_C property. To distinguish the write-snapshot operation in the *IIS* model and its more constrained counterpart of the *IRIS* model, the former is denoted $R[r]$.write_snap(), while the latter is denoted $IS[r]$.WRITE_SNAPSHOT().

3.1   The Model $IRIS(PR_C)$ with $C = \diamond S_x$

Let sm_j^r be the view obtained by the process p_j when it returns from the $IS[r]$.WRITE_SNAPSHOT() invocation. As each process p_i is assumed to execute rounds forever, $sm_i^r = \emptyset$ means that p_i never executes the round r, and is consequently faulty. The property states that there is a set Q of x processes containing a process p_ℓ that does not crash, and a round r, such that at any round $r' \geq r$, each process $p_i \in Q \setminus \{\ell\}$ either has crashed ($sm_i^{r'} = \emptyset$) or obtains a view $sm_i^{r'}$ that contains strictly $sm_\ell^{r'}$. Formally, the property $PR_{\diamond S_x}$ is defined as follows:

$$PR_{\diamond S_x} \equiv \exists Q, \ell : |Q| = x \wedge \ell \in Q, \exists r :$$
$$\forall r' \geq r : (sm_\ell^{r'} \neq \emptyset) \wedge \left(i \in Q \setminus \{\ell\} \Rightarrow (sm_i^{r'} = \emptyset \vee sm_\ell^{r'} \subsetneq sm_i^{r'}) \right).$$

Figure 2 shows runs of the $IRIS(PR_{\diamond S_x})$ model for $x = 2$. The complex remains connected in this case and consequently consensus is unsolvable in that model.

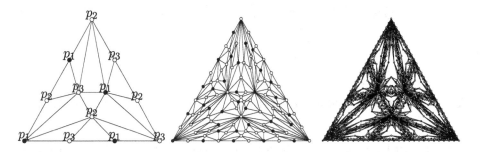

Fig. 2. One, two and three rounds in $IRIS(PR_{\diamond S_x})$ with $x = 2$ and $r = 2$

Theorem 1 (main). *An agreement task is solvable in the read/write model equipped with a failure detector of the class $\diamond S_x$ if and only if it is solvable in the $IRIS(PR_{\diamond S_x})$ model.*

We prove this theorem in Section 4 by providing a transformation from the read/write model enriched with $\diamond S_x$ to the $IRIS(PR_{\diamond S_x})$ model and the inverse transformation from the $IRIS(PR_{\diamond S_x})$ model to the read/write model with $\diamond S_x$.

3.2   The k-Set Agreement with $\diamond S_x$

The power of the $IRIS$ model becomes evident when we use it to prove the lower bound for k-set agreement in the shared memory model equipped with a failure detector of the class $\diamond S_x$.

Theorem 2. *In the read/write shared memory model, in which any number of processes may crash, there is no $\diamond S_x$-based algorithm that solves k-set agreement if $k < n - x + 1$.*

The proof consists of first observing that, if we partition the n processes in two sets: the low-order processes $L = \{p_1, \ldots, p_{n-x+1}\}$ and the high-order processes $H = \{p_{n-x+2}, \ldots, p_n\}$, and consider all IIS runs where the processes in H never take any steps, these runs trivially satisfy the $PR_{\diamond S_x}$ property. Therefore, as noticed at the end of Section 2.3, k-set agreement is unsolvable in the IIS model when $k < n - x + 1$, and hence unsolvable in our $IRIS(PR_{\diamond S_x})$ model. By Theorem 1 it is unsolvable in the read/write shared memory model equipped with a failure detector of the class $\diamond S_x$.[2]

4   Simulations

4.1   From the Read/Write Model with $\diamond S_x$ to $IRIS(PR_{\diamond S_x})$

This section presents a simulation of the $IRIS(PR_{\diamond S_x})$ model from the read/write model equipped with a failure detector $\diamond S_x$. The aim is to produce subsets of runs of the IIS model that satisfy the property $PR_{\diamond S_x}$. The algorithm is

---

[2] In the full paper, we show how to re-derive the more general result of [14] in the *IRIS* framework.

---

operation $IS[r]$.WRITE_SNAPSHOT$()(< i, v_i >)$:
(1)   **repeat** $m_i \leftarrow R[r]$.snapshot$()$; $rp_i \leftarrow$ REPR$_i$
(2)   **until** $((< rp_i, - > \in m_i) \lor rp_i = i))$ **end repeat**;
(3) $sm_i \leftarrow R[r]$.write_snap$(< i, v_i >)$;
(4)   **return** (sm_i).

Fig. 3. From the read/write model with $\Diamond S_x$ to the $IRIS(PR_{\Diamond S_x})$ model (code for p_i)

described in Figure 3. It uses the $\Diamond S_x$ version based on the representative variable REPR$_i$. Each round r is associated with an immediate snapshot object $R[r]$ that can in addition be read in snapshot. Sets returned by $R[r]$.snapshot$()$ and $R[r]$.write_snap$()$ are ordered by containment, and the operations can be consistently ordered. Objects $R[r]$ can be wait-free implemented from base read/write operations [1],[5].

At each round r, each process p_i repeatedly reads $R[r]$ until it observes its representative has already written or it discovers that it is its own representative ($rp_i = i$). This simple rule guarantees that processes that share the same representative p_ℓ eventually always return a view sm that contains the view returned by p_ℓ. Thus, the set of sequences of views produced by the algorithm satisfies the property $PR_{\Diamond S_x}$.

4.2 From $IRIS(PR_{\Diamond S_x})$ to the Read/Write Model Equipped with $\Diamond S_x$

We first show how to simulate the basic operations of an IIS model, namely write$()$ and snapshot$()$. This simulation works for any $IRIS(PR)$ model, as its runs are a subset of the IIS runs. Then a complete simulation that encompasses the failure detector $\Diamond S_x$ is given.

Simulating the write$()$ *and* snapshot$()$ *Operations.* The algorithm described in Figure 4 is based on the ideas of the simulation of [7]. Without loss of generality, we assume that (as in [7]) the kth value written by a process is k (consequently, a snapshot of the shared memory is a vector made up of n integers). To respect the semantics of the shared memory, vectors v returned as result of simulate(snapshot$()$) should be ordered and contain the integers written by the last simulate(write$()$) that precedes it.

As in [7], each process p_i maintains an estimate vector est_i of the current state of the simulated shared memory. When p_i starts simulating its k-th write$()$, it increments $est_i[i]$ to k to announce that it wants to write the shared memory (line 1). At each round r, p_i writes its estimate in $IS[r]$ and updates its estimate by taking the maximum component-wise, denoted $\max_{cw}$, of the estimates in the view sm_i it gets back (line 6). The main difference with [7] is the way processes compute valid snapshots of the shared memory. In [7], p_i returns a snapshot when all estimates in its view are the same. Here, for any round r, we define a valid snapshot as the maximum component-wise (denoted sm_min^r) of the estimates

```
init rᵢ ← 0; last_snapᵢ[1..n] ← [−1, ..., −1]; estᵢ[1..n] ← [0, ..., 0]; viewᵢ[1..] ← [∅, ..]
function simulate(op())                              % op ∈ {write(), snapshot()}
(1)   if op() = write() then estᵢ[i] ← estᵢ[i] + 1 endif; r_startᵢ ← rᵢ;
(2)   repeat rᵢ ← rᵢ + 1;
(3)     smᵢ ← IS[rᵢ].WRITE_SNAPSHOT(< i, estᵢ, viewᵢ[1..(rᵢ − 1)] >);
(4)     viewᵢ[rᵢ] ← { < i, {< j, estⱼ > such that < j, estⱼ, − >∈ smᵢ} > };
(5)     for each ρ ∈ {1, ..., rᵢ − 1} do
            viewᵢ[ρ] ← ⋃_{viewⱼ such that <j,−,viewⱼ>∈smᵢ} viewⱼ[ρ] endfor;
(6)     estᵢ ← max_cw{estⱼ such that < j, estⱼ, − >∈ smᵢ};
(7)     if (∃ρ > r_startᵢ | ∃ < −, smin >: ∀j ∈ smin :  < j, smin >∈ viewᵢ[ρ])
            % there is a smallest snapshot in viewᵢ[r_startᵢ + 1..rᵢ] known by pᵢ
(8)       then let ρ' be the greatest round ≤ rᵢ that satisfies predicate of line 7;
(9)         sminᵢ ← the smallest snapshot in viewᵢ[ρ'];
(10)        last_snapᵢ ← max_cw{estⱼ such that < j, estⱼ >∈ sminᵢ};
(11)        if last_snapᵢ[i] = estᵢ[i] then
(12)          if op = snapshot() then return (last_snapᵢ) else return() endif endif
(13)    endif endrepeat
```

Fig. 4. Simulation of the write() and snapshot() operations in $IRIS(PR_{\Diamond S_x})$ (p_i's code)

that appear in the smallest view (denoted $smin^r$) returned by $IS[r]$. Due to the fact that estimates are updated maximum component-wise, it follows from the containment property of views that $\forall r, r' : r < r' \Rightarrow sm\_min^r \leq sm\_min^{r'}$. As each snapshot returned is equal to sm_min^r for some r, it follows that any two snapshots of the shared memory are equal or one is greater than the other.

In order to determine smallest views, each process p_i maintains an array $view_i[1, ..]$ that aggregates p_i's knowledge of the views obtained by other processes. This array is updated at each round (lines 4-5) by taking into account the knowledge of other processes (that appear in sm_i).

Then, p_i tries to determine the last smallest view that it can know by observing the array $view_i$ (line 7). If there is a recent one (it is associated with a round greater than the round r_start_i at which p_i has started simulating its current operation), p_i keeps it in $smin_i$ (lines 8-9), and computes in $last\_snap_i$ the corresponding snapshot value of the shared memory (line 10). Finally, if p_i observes that its last operation announced (that is identified $est_i[i]$) appears in this vector, it returns $last\_snap_i$ (line 11). In the other cases, p_i starts a new iteration of the loop body.

From $IRIS(PR_{\Diamond S_x})$ to a Failure Detector of the Class $\Diamond S_x$. In a model equipped with a failure detector, each process can read at any time the output of the failure detector. We denote fd_query() this operation. A trivial algorithm that simulates $\Diamond S_x$-queries in the $IRIS(PR_{\Diamond S_x})$ is described in the figure on the right.

```
init rᵢ ← 0; TRUSTEDᵢ ← Π
function simulate(fd_query())
(1)   rᵢ ← rᵢ + 1; smᵢ ← IS[rᵢ].WRITE_SNAPSHOT(i);
(2)   TRUSTEDᵢ ← {j : j ∈ smᵢ}; return( TRUSTEDᵢ )
```

Simulation of fd_query() in $IRIS(PR_{\Diamond S_x})$

General Simulation. Given an algorithm $\mathcal{A}$ that solves a task T in the read/write model equipped with $\Diamond S_x$, we show how to solve T in the $IRIS(PR_{\Diamond S_x})$ model. Algorithm $\mathcal{A}$ performs local computation, write(), snapshot() and fd_query(). In the $IRIS(PR_{\Diamond S_x})$ model, processes run in parallel the algorithms described in Figures 4 and below in order to simulate these operations. More precisely, whatever the operation $op \in \{$write(), snapshot(), fd_query()$\}$ being simulated, each immediate snapshot object is used to update both the estimate of the shared memory and the output of the failure detector.

The simulations are proved correct in [20]. Theorem 1 then follows from the two simulations presented in Section 4.1 and Section 4.2.

References

1. Afek, Y., Attiya, H., Dolev, D., Gafni, E., Merritt, M., Shavit, N.: Atomic Snapshots of Shared Memory. J. ACM 40(4), 873–890 (1993)
2. Attiya, H., Bar-Noy, A., Dolev, D.: Sharing Memory Robustly in Message Passing Systems. J. ACM 42(1), 124–142 (1995)
3. Awerbuch, B.: Complexity of network synchronization. J. ACM 32, 804–823 (1985)
4. Attiya, H., Welch, J.: Distributed Computing: Fundamentals, Simulations, and Advanced Topics. Wiley, Chichester (2004)
5. Borowsky, E., Gafni, E.: Immediate Atomic Snapshots and Fast Renaming. In: Proc. of PODC 1993, pp. 41–51 (1993)
6. Borowsky, E., Gafni, E.: Generalized FLP Impossibility Results for t-Resilient Asynchronous Computations. In: Proc. 25th ACM STOC, pp. 91–100 (1993)
7. Borowsky, E., Gafni, E.: A Simple Algorithmically Reasoned Characterization of Wait-free Computations. In: Proc. 16th ACM PODC, pp. 189–198 (1997)
8. Chandra, T., Toueg, S.: Unreliable Failure Detectors for Reliable Distributed Systems. J. ACM 43(2), 225–267 (1996)
9. Chandra, T., Hadzilacos, V., Toueg, S.: The Weakest Failure Detector for Solving Consensus. J. ACM 43(4), 685–722 (1996)
10. Chaudhuri, S.: More Choices Allow More Faults: Set Consensus Problems in Totally Asynchronous Systems. Information and Computation 105, 132–158 (1993)
11. Dwork, C., Lynch, N., Stockmeyer, L.: Consensus in the Presence of Partial Synchrony. J. ACM 35(2), 288–323 (1988)
12. Gafni, E.: Round-by-round Fault Detectors: Unifying Synchrony and Asynchrony. In: Proc. 17th ACM Symp. on Principles of Distributed Computing, pp. 143–152 (1998)
13. Gafni, E., Rajsbaum, S., Herlihy, M.: Subconsensus Tasks: Renaming is Weaker than Set Agreement. In: Dolev, S. (ed.) DISC 2006. LNCS, vol. 4167, pp. 329–338. Springer, Heidelberg (2006)
14. Herlihy, M., Penso, L.D.: Tight Bounds for k-Set Agreement with Limited Scope Accuracy Failure Detectors. Distributed Computing 18(2), 157–166 (2005)
15. Herlihy, M.P., Rajsbaum, S., Tuttle, M.: Unifying Synchronous and Asynchronous Message-Passing Models. In: Proc. 17th ACM PODC, pp. 133–142 (1998)
16. Herlihy, M., Shavit, N.: The Topological Structure of Asynchronous Computability. J. ACM 46(6), 858–923 (1999)
17. Keidar, I., Shraer, A.: Timeliness, Failure-detectors, and Consensus Performance. In: Proc. 25th ACM PODC, pp. 169–178 (2006)

18. Mostefaoui, A., Rajsbaum, S., Raynal, M., Travers, C.: Irreducibility and Additivity of Set Agreement-oriented Failure Detector Classes. In: Proc. PODC 2006, pp. 153–162 (2006)
19. Rajsbaum, S., Raynal, M., Travers, C.: Failure Detectors as Schedulers. Tech Report # 1838, IRISA, Université de Rennes, France (2007)
20. Rajsbaum, S., Raynal, M., Travers, C.: The Iterated Restricted Immediate Snapshot Model. Tech Report # 1874, IRISA, Université de Rennes, France (2007)
21. Saks, M., Zaharoglou, F.: Wait-Free k-Set Agreement is Impossible: The Topology of Public Knowledge. SIAM Journal on Computing 29(5), 1449–1483 (2000)
22. Völzer, H.: On Conspiracies and Hyperfairness in Distributed Computing. In: Fraigniaud, P. (ed.) DISC 2005. LNCS, vol. 3724, pp. 33–47. Springer, Heidelberg (2005)
23. Yang, J., Neiger, G., Gafni, E.: Structured Derivations of Consensus Algorithms for Failure Detectors. In: Proc. 17th ACM PODC, pp. 297–308 (1998)
24. Zieliński, P.: Anti-Omega: the Weakest Failure Detector for Set Agreement. Tech Rep # 694, University of Cambridge (2007)

Finding Frequent Items in a Turnstile Data Stream

Regant Y.S. Hung, Kwok Fai Lai, and Hing Fung Ting*

Department of Computer Science,
The University of Hong Kong, Pokfulam, Hong Kong
{yshung,kflai,hfting}@cs.hku.hk

Abstract. Because of important applications such as denial-of-service attack detection, finding frequent items in data streams under different models has been studied extensively. Finding frequent items in a turnstile data stream is the most challenging because both insertions and deletions of items are allowed in the stream. In this paper, we propose a deterministic algorithm that solves the problem. Furthermore, we propose a randomized algorithm for the problem. Empirical results show that our randomized algorithm provides better results than existing randomized algorithms for the problem and our algorithm uses much smaller space, and supports faster query time and similar update time.

1   Introduction

This paper studies the problem of identifying frequent items in a turnstile data stream. Suppose that there are M distinct items $a_1, a_2, \ldots, a_M$. For each $1 \le i \le M$, the frequency $f(a_i)$ of item a_i is zero initially. A turnstile data stream is a sequence of updates (a_j, c_j) where a_j is an item and c_j is some integer that indicates the value of the update. After the update (a_j, c_j), the frequency $f(a_j)$ of a_j becomes $f(a_j) + c_j$ while the frequency of all other items remain unchanged. For example, suppose $M = 4$ and consider the stream of updates: $(a_4, 2), (a_2, 3), (a_1, 6), (a_2, -2), (a_4, 1)$. After the arrival of the last update $(a_4, 1)$, the frequencies of the items are: $f(a_1) = 6$, $f(a_2) = 1$, $f(a_3) = 0$ and $f(a_4) = 3$. If $c_j \ge 0$, we call (a_j, c_j) an increment update; otherwise, it is a decrement update. In our previous example, $(a_2, -2)$ is a decrement update while the rests are increment updates. The model described above is called turnstile model. As pointed out in [6], turnstile model is the hardest among the three popular models, namely the time-series model, the cash register model and the turnstile model, for data mining because it is the only model that allows arbitrary additions and subtractions of items (see [16] for more details on these three models). There are two parameters in our study, namely, ϵ the error bound and θ the threshold where $\theta > \epsilon$. We are interested in designing data structures that allow us to answer at any time the following query efficiently:

> Return a set Π of items such that (i) every item in Π has frequency no less than $(\theta - \epsilon)N$ and (ii) all items with frequencies no less than θN must be in Π.

---

* This research was supported in part by Hong Kong RGC Grant HKU-7163/07E.

X. Hu and J. Wang (Eds.): COCOON 2008, LNCS 5092, pp. 498–509, 2008.

Here, N is the sum of the frequencies of all the items, or equivalently, the sum of the values of all the updates in the whole stream. We call Π a set of frequent items. Let Frequent-(θ, ϵ) denote this problem. In our previous example, if $\theta = 0.6$ and $\epsilon = 0.1$, then only a_1 should be in Π since Π will not contain any item with frequency less than 5. As common to all data stream applications, our solution must satisfy the following requirements: process the stream in one-pass, use sublinear space and support fast update and query time. In this paper, we make the common, and practically reasonable assumption that no item will have frequency less than zero (see [16]).

Previous Works. There are many interesting algorithms for finding frequent items in a data stream under the cash register model in which only insertions are allowed in the stream. Manku and Motwani [14] gave two algorithms, namely, the sticky sampling and lossy counting for the problem. Sticky sampling is a randomized algorithm that finds the frequent items using $O(\frac{1}{\epsilon} \log \frac{1}{\theta\delta})$ space with success probability $1 - \delta$. The lossy counting finds the frequent items deterministically using $O(\frac{1}{\epsilon} \log \epsilon N)$ space where N is the number of items in the whole stream. Misra and Gries [15] designed a deterministic $O(\frac{1}{\epsilon})$-space algorithm that finds the set of frequent items, and this algorithm was rediscovered independently by Demaine et al. [5] and Karp et al. [12]. Arasu and Manku [2] gave an $O(\frac{1}{\epsilon} \log^2 \frac{1}{\epsilon})$-space algorithm for finding frequent items over sliding window (not the entire data stream). Lee and Ting [13] gave a better algorithm that reduces the space to $O(\frac{1}{\epsilon})$.

For turnstile data stream, Cormode and Muthukrishnan [4] gave a data structure called CM sketch, which is essentially a collection of hash tables, for estimating the frequency of an item. Based on CM sketch, they showed how to solve Frequent-(θ, ϵ) with probability $1 - \delta$ using $O(\frac{1}{\epsilon} \log M \log \frac{\log M}{\delta\theta})$ space where M is the number of distinct items in the data stream. The algorithm supports $O(\log M \log \frac{\log M}{\delta\theta})$ query and $O(\frac{1}{\theta} \log M)$ update time. Jin et al. [11] proposed two algorithms hCount and hCount*. The algorithm hCount is based on a data structure that is very similar to CM sketch. The algorithm hCount* extends hCount with some heuristic for improving the precision of the answer returned by hCount. The algorithm hCount* uses $O(\frac{1}{\epsilon} \log(\frac{M}{\delta}))$ space, and supports $O(M)$ query and $O(\log \frac{M}{\delta})$ update times.

There are also other interesting problems related to the turnstile model. Alon et al. [1] proposed an algorithm that estimates self-join sizes of a turnstile stream. Ganguly et al. [7], [8], [9] has studied counting distinct items in a turnstile data stream. Cormode et al. [3] designed Group-Count Sketch for tracking the wavelet representation of one-dimensional and multi-dimensional turnstile streams.

Our Contributions. In this paper, we propose a deterministic algorithm that solves the problem. Let $S_{-,t}$ be the sum of the values of all the decrement updates among the first t updates in the stream and $S_{+,t}$ be the sum of values in all the increment updates among the first t updates in the stream. Obviously $S_{+,t} \geq 0$. Note that $S_{-,t} \leq 0$ since every decrement update has value smaller than zero. Let r denote the maximum of the ratio between the absolute values of $S_{-,t}$ and

$S_{+,t}$, i.e. $r = \max_t \frac{|S_{-,t}|}{|S_{+,t}|}$. Since every item has frequency no less than zero, we have $r \leq 1$. If $r < 1$, our algorithm solves the problem using $O(\frac{1}{(1-r)\epsilon})$ space, update and query times. Note that it is not unusual to make the assumption that $\frac{1}{1-r}$ is a constant: Alon, Gibbons, Matias and Szegedy [1] proposed an algorithm for finding self-join sizes in a turnstile stream with the assumption that $r \leq 1/4$. Recently, Ganguly and Majumder proposed the first deterministic algorithm for finding frequent items in turnstile data stream [10]. They solved the problem using $O(\frac{1}{\epsilon^2\theta^2} \log \epsilon\theta M \log M)$ space. Compared with the work of Ganguly and Majumder [10], our algorithm does not need to make assumption that the universe is known in advance, and our algorithm uses smaller space asymptotically when $\frac{1}{1-r} = o(\frac{1}{\theta^2\epsilon} \log^2 M)$. We also derive a lower bound on the space complexity of any algorithm that solves the problem. Here, we consider the case that the algorithm solves the problem by storing a subset of items from the stream in the memory. Under this assumption, we prove that given a stream in which there are M distinct items and L updates, any deterministic algorithm that solves the problem correctly must use at least $\min(M, L, \frac{1}{(1-r_L)\theta})$ space where $r_L = \frac{|S_{-,L}|}{|S_{+,L}|}$. Note that θ is larger than ϵ and thus, $\frac{1}{(1-r_L)\theta} \leq \frac{1}{(1-r)\epsilon}$.

Furthermore, we propose a randomized algorithm that improves existing randomized algorithms for the problem. Note that all existing algorithms for the problem are based on CM sketch. There are two disadvantages of this approach. First, we need to compute from the CM sketch the estimated frequencies of a large number of items in order to find the set of frequent items. Second, in order to guarantee small error probability, we need to keep many hash tables in the sketch in order to reduce the error probability. We observe that we should not resort totally to probability in order to identify the set of frequent items, which is of size $O(1/\theta)$. After answering a query and deciding a set S of frequent items, the elements in this set should be the elements we check first for answering the next query. Thus, to have an accurate estimates of these $O(1/\theta)$ items, we can keep a counter for each of them to remember their frequencies so that when the next query comes, we can first check the counters of this set to have a quick and more accurate estimations to decide which of them should still be in the set. Certainly, when the frequency of an item in S decreases to a certain threshold, we should replace it by some more promising candidate. (Hence, to ensure we always have enough candidates, we will use much more than $1/\theta$ counters.)

To implement this idea, we introduce an auxiliary data structure called TIS, the top item set, that keeps track of the top items. Combined with TIS, we can reduce that number of hash tables substantially while maintaining similar precision of the answers. We have carried out many experiments to evaluate the performance of our approach, which we called CM sketch with TIS and compare it with CM Sketch and hCount*. Experiments show that our approach performs better than both CM sketch and hCount* even though our approach uses much less space. More precisely, with the use of real data [17], we tested the *precision* of the answers (i.e. proportion of frequent items returned among all the items returned) returned by various algorithms. It turns out that the precision of our approach is higher than that of CM sketch and hCount*. For *recall* (i.e. the ratio between (i) the number of items whose frequencies are no less than θN and are returned by the algorithm and (ii) the total number of items whose frequencies are no less than θN), we found that our approach gives false negative only once out of a few hundreds queries. The recall of the answer returned by our approach is similar with that of CM sketch, and better than hCount*. Since our approach answers the query by scanning the TIS (which is of size $O(\frac{1}{\theta})$) once only while CM sketch and hCount* need to estimate the frequency of all the M items, our approach answers query much faster than that of CM sketch and hCount*. We also demonstrate how to slightly modify our approach to ensure that there is no false negative in our answer.

2   Deterministic Algorithm for the Problem

Let S_+ and S_- denote the sum of values in all the increment updates and decrement updates in the whole stream, respectively. S_- is a negative number since every decrement update has a negative value. We give a lower bound below.

Theorem 1. *Any deterministic algorithm A that solves Frequent-(θ, ϵ) by storing a subset of items from the stream must store at least $\min(M, L, \frac{1}{\theta(1-r)})$ items where $r = |S_-|/|S_+|$, M and L denote the number of possible distinct items in the stream and the number of the updates in the stream, respectively.*

Proof. Suppose we have a data stream as follows: we have a sequence of increment updates at the beginning, and then followed with a sequence of decrement updates. More precisely, suppose A can store less than m items and we have the following stream: first of all, we have a sequence of m increment updates: $(a_1, x), (a_2, x), \ldots, (a_m, x)$, where x is some integer greater than or equal to one. Suppose A does not store the name of some item a_i for some $1 \leq i \leq m$ since A cannot store all the m distinct items in the stream. Then, the sequence of increment updates is followed by a sequence of decrement updates such that there is no decrement update for a_i and the sum of the values of all decrement updates S_- equals $x/\theta - S_+$. Note that $x = \theta(S_+ + S_-) = \theta N$ where N is the sum of all the updates in the stream. Thus, A fails to solve Frequent-(θ, ϵ) since A fails to return a_i that is one of the answers. Thus, A must store at least m items where $m = \frac{S_+}{x} = \frac{S_+}{\theta(S_+ + S_-)} = \frac{1}{\theta(1 + S_-/S_+)} = \frac{1}{\theta(1 - |S_-|/|S_+|)} = \frac{1}{\theta(1-r)}$ since $S_+ = mx$. Note that if $\frac{1}{\theta(1-r)} > \min(M, L)$, we can solve the problem by either keeping M counters for all possible items or recording all the updates in the stream. Thus, A must store at least $\min(M, L, \frac{1}{\theta(1-r)})$ items. $\qquad\square$

We now propose a deterministic algorithm for Frequent-(θ, ϵ). Recall that $S_{-,t}$ and $S_{+,t}$ denote the sum of values of all the decrement updates among the first t updates in the stream and the sum of values of all the increment updates among the first t updates in the stream respectively. Note that $|S_{+,t}| \geq |S_{-,t}|$ for any t in our model. We assume that we know the ratio $r = \max_t \frac{|S_{-,t}|}{|S_{+,t}|}$ and $r < 1$. If M, the number of distinct items in the stream, is not large, we can simply keep M counters, each for an item, such that we can record the exact frequency of every item. Then when there is a query for Frequent-(θ, ϵ), we can return all items whose frequencies are at least θN. If M is large, instead of keeping a counter for every item, we will keep a set S of at most $\frac{1}{(1-r)\epsilon}$ counters all the time. Each counter stores the estimated frequency of an item. Intuitively, if we can keep a counter for every frequent item, then we can solve Frequent-(θ, ϵ). We demonstrate how to achieve this by storing physically $\frac{1}{(1-r)\epsilon}$ counters.

Algorithm. Initially S is empty. We also keep a counter C_N that can keep track of the value of N accurately. C_N has value 0 initially. Whenever an update (a_k, v_k) arrives, we will first update C_N by adding v_k to it, and

If there is a counter kept for a_k in S: we will update the value of this counter by adding v_k to it. If this counter has value no larger than zero afterwards, we will remove it.

If there is no counter kept for a_k in S and $v_k > 0$: we will add one more counter for a_k with value v_k in S. Then, if there are $\frac{1}{(1-r)\epsilon}$ counters kept in S, we will have a batch decrement: the values of all the counters will be deducted by the minimum of the values of all the counters. Then we will remove the counters with value of zero. Thus, we will always keep no more than $\frac{1}{(1-r)\epsilon}$ counters all the time.

If there is no counter kept for a_k in S and $v_k \leq 0$: We do nothing.

Answer the query. For any item a, the estimated frequency $\hat{f}(a)$ of a is defined as the value of the counter of a kept in S. If no counter is kept for a, $\hat{f}(a) = 0$. When there is a query, we will check all counters in S and return the items whose estimated frequencies are no less than $(\theta - \epsilon)N$.

Theorem 2. *If $r = \max_t \frac{|S_{-,t}|}{|S_{+,t}|} < 1$, our algorithm solves Frequent-(θ, ϵ) with the use of $O(\frac{1}{(1-r)\epsilon})$ space, update and query times.*

Proof. Consider any item a. Let $f(a)$ denote the frequency of a. We can prove that $f(a) - \epsilon N \leq \hat{f}(a) \leq f(a)$. From this inequality, it is easy to verify that returning items with estimated frequencies no less than $(\theta - \epsilon)N$ solves the problem. We prove this inequality below.

Suppose we have a query for Frequent-(θ, ϵ) now and there are L updates in the stream. Let p_a be some integer such that once p_a-th update in the stream arrives, we keep a counter for a thereafter. In the other words, we have been keeping a counter for a since the arrival of p_a-th update. Let $f_{a,+}(i, j)$ denote the sum of the increment updates for a from the arrival of i-th update to the arrival of j-th update in the stream. Similarly, let $f_{a,-}(i, j)$ denote the sum of the decrement updates for a from the arrival of i-th update to the arrival of j-th update. Note that $f_{a,-}(i, j)$ is a non-positive number since each decrement update has value less than zero. Let $d(i, j) \geq 0$ denote the total amount deducted from a counter due to batch decrements during the period from the arrival of i-th update to the arrival of j-th update. It can be verified that

$$\hat{f}(a) = f_{a,+}(p_a, L) + f_{a,-}(p_a, L) - d(p_a, L). \tag{1}$$

We first prove that $\hat{f}(a) \leq f(a)$. In our model, we have $f_{a,+}(1, i) + f_{a,-}(1, i) \geq 0$ for any $i \geq 0$. Thus, $f_{a,+}(p_a, L) + f_{a,-}(p_a, L) \leq f(a)$ since $f_{a,+}(1, p_a - 1) + f_{a,-}(1, p_a - 1) \geq 0$ and $f_{a,+}(p_a, L) + f_{a,-}(p_a, L) + f_{a,+}(1, p_a - 1) + f_{a,-}(1, p_a - 1) = f(a)$. Together with $d(p_a, L) \geq 0$ and Equation 1, we conclude that $\hat{f}(a) \leq f(a)$. Since no counter is kept for a at the moment just before p_a-th update arrived in the stream, it can be proved by contradiction that $f_{a,+}(1, p_a - 1) + f_{a,-}(1, p_a - 1) - d(1, p_a - 1) \leq 0$. Together with Equation 1, we have $f(a) - d(1, L) \leq \hat{f}(a)$. Note that if $d(1, L) \leq \epsilon N$, then $\hat{f}(a) \geq f(a) - \epsilon N$.

We now prove that the amount deducted for a counter cumulatively from the beginning, $d(1, L)$, is no larger than ϵN. Since we will deduct each of the $\frac{1}{(1-r)\epsilon}$ counters when there is a batch decrement, we will deduct the product of $\frac{1}{(1-r)\epsilon}$ and $d(1, L)$ from all the counters cumulatively. Since this amount is no larger than $S_{+,L}$, we have $d(1, L) \leq \frac{S_{+,L}}{1/((1-r)\epsilon)} \leq \epsilon(S_{+,L} - S_{+,L} \cdot \frac{|S_{-,L}|}{|S_{+,L}|}) = \epsilon N$.

Our algorithm uses $O(\frac{1}{(1-r)\epsilon})$ space since we keep at most $\frac{1}{(1-r)\epsilon}$ counters at any time. When we update the data structure and answer the query, we need to access the whole data structure several times, that takes $O(\frac{1}{(1-r)\epsilon})$ time.   □

3   Randomized Algorithm

We will first describe the randomized algorithm that keeps a CM sketch with a deterministic data structure TIS. Then we show the experimental results. Finally, we give slight modifications to ensure that there is no false negative in our answer.

3.1   Algorithm

It can be observed that simply using CM sketch for finding frequent items has two problems. First, when we answer a query, we need to estimate the frequency of every item, that takes too long time. Second, we need to keep large number of rows in CM sketch to maintain acceptable error probability. To solve these problems, we propose an algorithm that maintains two data structures: a CM sketch [4], that is used to estimate the frequency of an item and a data structure called TIS, that maintains the estimated frequency of the items with relatively high frequencies. Whenever an update arrives, we will update the CM sketch and update the estimated frequency of the corresponding item in TIS (if it is not in TIS, we will add a counter for it). If we keep more than $\frac{1}{\theta} + C$ counters in TIS where C is some constant larger than $\frac{1}{\theta}$, we will remove the counter with the smallest estimated frequency. Conceptually, we maintain the set of $\frac{1}{\theta} + C$ items with highest frequencies. Note that the answer to Frequent-(θ, ϵ) contains at most $\frac{1}{\theta}$ items. Thus, if TIS can keep the top $\frac{1}{\theta}$ items, we can return all items whose estimated frequencies are no less than θN by just scanning only the items and its counters maintained in TIS.

Maintaining TIS solves the first problem since whenever there is a query, we can just check TIS to answer the query. The query time is greatly reduced from M, the number of distinct items, to $\frac{1}{\theta} + C$, the size of TIS. Surprisingly, empirical results show that our algorithm also solves the second problem: our approach (CM sketch with TIS) performs better than CM sketch that keeps more rows. A possible reason is: compared with CM sketch in which every counter will be shared by many items, TIS reserves the counters for maintaining estimated frequencies of the items with relatively high frequencies. This will improve the accuracy of our estimation on the frequencies of those items with relatively high frequencies, that includes the items with frequencies close to θN. Imagine that there are some items whose frequencies are $\theta N - 1$. If a data structure estimates their frequencies more accurately, it will not report them as frequent items and thus, the precision of answers given by this data structure will be improved. This may explain why TIS can reduce the error probability.

Unluckily, TIS has one disadvantage: when TIS probably does not contain all the frequent items, we will carry out a renew operation: empty TIS, estimate the frequencies of all M items and put the items with relatively high frequencies into TIS. This step is slow. Luckily, this seldom occurs: in a stream of tens of millions updates, less than three renew operations has been carried out. We will describe the CM sketch and TIS separately as below.

The CM Sketch Component

We maintain a 2-D array of counters with d rows and w columns: $count[1,1], \ldots, count[d,w]$ where each row is a hash table conceptually. We set $d = \lceil \ln \frac{M}{\delta} \rceil$ and $w = \lceil \frac{e}{\epsilon} \rceil$ where M denotes the number of distinct items appearing in the stream, δ denotes the failure probability that is specified by users and e denotes the base of natural logarithm. We also maintain d pairwise-independent hash functions $h_1 \cdots h_d : \{1 \cdots M\} \rightarrow \{1 \cdots w\}$. Every entry in the array has value of zero initially. When an item (a_i, c_i) arrives, for all $1 \leq j \leq d$, we will increment the counter $count[j, h_j(i)]$ by c_i, i.e. $count[j, h_j(i)] \leftarrow count[j, h_j(i)] + c_i$.

Note that in order to maintain TIS, we may need to get the estimated frequency of some item a_i from CM sketch. Let $f(a_i)$ denote the frequency of an item a_i. The estimated frequency $\hat{f}_{CM}(a_i)$ of CM sketch on $f(a_i)$ is defined as $\min_j count[j, h_j(i)]$. By having a trivial modification in the proof of Theorem 1 of [4], we have the following theorem.

Theorem 3. *In CM sketch, the estimate $\hat{f}_{CM}(a_i) = \min_j count[j, h_j(i)]$ for any item a_i has the following guarantees: $f(a_i) \leq \hat{f}_{CM}(a_i)$; and $\hat{f}_{CM}(a_i) \leq f(a_i) + \epsilon N$ with probability at least $1 - \frac{\delta}{M}$.*

The TIS Component

We define the notations as follows. Let TIS_{topK} be a subset of TIS. Let $\min(\text{TIS}_{\text{topK}})$ denote the minimum value of all the counters in TIS_{topK}. We have two phases: growing phase and maintenance phase. When TIS keeps less than $\frac{1}{\theta} + C$ counters, it will be in growing phase. Once the number of counters in TIS increases to $\frac{1}{\theta} + C$, we will have maintenance phase afterwards.

Growing Phase. Initially, TIS keeps no counter. In this phase, TIS keeps a counter for every item with frequency greater than zero: the value of the counter equals the frequency of the item. For example, when update (a_i, c_i) arrives such that $f(a_i)$ increases from zero to c_i, then TIS will keep one more counter having value c_i for a_i. In the growing phase, $\text{TIS}_{\text{topK}} = \text{TIS}$.

Maintenance Phase. Suppose after some update, the number of counters kept in TIS increases to $\frac{1}{\theta} + C$, we will have maintenance phase thereafter. In maintenance phase, when an update (a_i, c_i) arrives, we will first update the CM sketch as described. Then we will update TIS as follows:

(1) We check if a_i is in TIS. If it is in TIS, set its estimated frequency $\hat{f}_{\text{TIS}}(a_i)$ in TIS as $\min(\hat{f}_{CM}(a_i), \hat{f}_{\text{TIS}}(a_i) + c_i)$ (we set it as the minimum of the estimates from TIS and CM sketch since CM sketch never underestimates the frequency of an item by Theorem 3). Otherwise, we add TIS a counter with value $\hat{f}_{CM}(a_i)$ for a_i, and then remove the counter with smallest value. If there is a draw, remove one of them arbitrarily. Thus, we will maintain exactly $\frac{1}{\theta} + C$ counters in TIS in maintenance phase.

(2) Let j denote the minimum value $\min(\text{TIS}_{\text{topK}})$ of all the counters in TIS_{topK} at the moment just before (a_i, c_i) arrives. If the counter for a_i is originally in TIS_{topK}, the counter for a_i will be removed from TIS_{topK} when $\hat{f}_{\text{TIS}}(a_i) < j$. If the counter for a_i is originally not in TIS_{topK}, we will add it into TIS_{topK} when the counter for a_i is present in TIS and $\hat{f}_{\text{TIS}}(a_i) \geq j$.

(3) If TIS_{topK} contains less than $\frac{2}{\theta}$ counters (recall that $\frac{1}{\theta} + C > \frac{2}{\theta}$), we will renew TIS: We first empty TIS. Then we estimate the frequency of every item by CM sketch and put the set of $\frac{1}{\theta} + C$ items whose estimated frequencies are largest among all the M items into TIS and set $\text{TIS}_{\text{topK}} = \text{TIS}$.

Answer the query. When there is a query, we will check all counters in TIS_{topK} and return the items whose counters have values no less than θN where N is the sum of the frequencies of all the items. Note that we can maintain a counter to keep track the value of N.

3.2   Empirical Results

We have carried out many experiments for testing various aspects of the performance of our approach, such as accuracy (precision and recall) of the answers returned, the processing time and query time of the algorithms. Precision is defined as the ratio between the number of items whose frequencies are no less than θN and are returned by the algorithm to the total number of items returned by the algorithm. Recall is defined as the ratio between (i) the number of items whose frequencies are no less than θN and are returned by the algorithm and (ii) the total number of items whose frequencies are no less than θN. We have compared the performance of our approach (i.e. CM sketch with TIS) with that of CM sketch and hCount* [11]. Note that all approaches will keep a CM sketch. Let $CM[d, w]$ denote a CM sketch that has d hash tables and each hash table has w slots (i.e. there are d rows and w columns in the two-dimensional array).

We used two sets of real data for experiments. First, we used the packet traces from UCLA D-WARD project [17]. The packets are collected at the border router of Computer Science Department, UCLA. The algorithms will be used to find the set of frequent source IP addresses. We used $21,040,262$ packets for experiments. Another set of real data is a set of pages collected at Wikipedia. We have extracted $550,240$ pages from Wikipedia. Each word in the page is treated as an item. There are $484,561,588$ words in all the pages. The algorithms will be used to identify the frequent words appearing in the pages. In all our experiments, we choose $\epsilon = 0.001$ and $\delta = 0.05$. Recall that we have kept at most $\frac{1}{\theta} + C$ counters in TIS. We now set $C = 1000$. Note that the user specified threshold θ is set as 0.005 in all the following experiments unless specified explicitly. All experiments are carried out on a PC with an Intel Pentium 4 3.2GHz and main memory of 512MB.

We suppose the items come one by one, that forms a data stream. Note that there are both increment and decrement updates in turnstile model. As in [11], in order to simulate the turnstile model, we set up a sliding window of size W that covers the most recent W items and we need to find the frequent items in this sliding window. More precisely, we create a turnstile data stream for experiment as follows: we read first item, second item and so on when we create the turnstile data stream. Suppose we read the k-th item a now. If $k \leq W$, we will create an update $(a, 1)$ in the turnstile stream. If $k > W$, we will have two updates in the stream: insertion of k-th item a and deletion of $(k - W)$-th item a'. i.e. we create two updates in stream: $(a, 1)$ and $(a', -1)$. We carried out 10 sets of experiments with different values of W. In each set of experiment, we have queries after the arrival of a fixed number of updates. For each set of experiment, we take the average of the precisions/recalls of all the queries.

Tradeoff Between Space and Precision. The purpose of this experiment is to find how much space TIS can save while maintaining the same level of precision. In Figure 1(a), we have four CM sketches with different numbers of hash tables while the number of entries (i.e., counters) of each hash table is set to 601. Furthermore, for our approach, we keep CM[6, 601] with TIS. The results show that the precision of the answers returned by our approach is higher than that of all the four CM sketches, even with the one with 18 rows, i.e. CM[18, 601]. Note that CM[18, 601] stores $18 \times 601 = 10818$ counters while our approach uses only $(6 \times 601) + (1000 + 1/0.005) = 4806$ counters; thus our approach saves more than 6000 counters without sacrificing the precision. Note that this not only improves the space, but also improves the update time as the update time for CM sketch depends on how many rows it has. In Figure 1(b), we compare our approach with the CM[6, w] for different w. We find that our approach (CM[6, 601] with TIS) performs nearly the same as the CM[6, 1201]. As shown above, TIS has 4806 counters, while CM[6, 1201] uses 7206 counters; a significant saving of 2400 counters. Note that we have not compared our approach with hCount*. We will show that the precision of these two algorithms are very similar, though hCount* has a much worse recall (see Figure 2).

Precision and Recall. In Figure 2, we compare the precision and recall of the answers returned by our approach with that of CM sketch and hCount*. All three approaches keep CM[6, 601]. For precision, our approach performs a bit better than hCount* and much better than CM sketch. For recall, the result

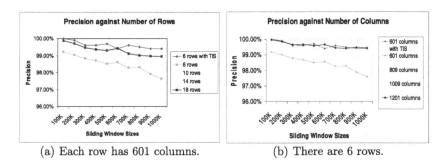

(a) Each row has 601 columns.   (b) There are 6 rows.

Fig. 1. Comparison between our approach and different sizes of CM sketch

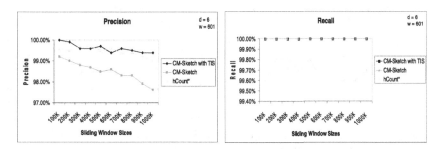

Fig. 2. Precision and recall of the answers returned by different algorithms

shows that hCount* gives many false negatives; in order to reduce false positives, hCount* has created false negatives. Although our approach does not guarantee that there is no false negative, we found that only one out of a few hundreds queries has a false negative. This performance is comparable with that of CM sketch, and is much better than that of hCount*.

Update Time and Query Time. Essentially, the update and query times of hCount* are no smaller than those of CM sketch. So we only compare our approach with CM sketch here. Note that the more the number of rows the CM sketch has, the slower is the query time for CM sketch. On the other hand, our approach will answer query quickly no matter how many rows we keep since we just check TIS for answering query. Together with the fact that CM sketch estimates the frequency of all distinct items where this step is very slow, our approach answers query much faster than CM sketch. Our experiments have verified this. Consider a CM sketch and our approach that keeps a CM sketch with the same size. Experimental results show that CM sketch answers query slowly, especially when there are many rows in the CM sketch. Our approach answers quickly (less than 0.1 ms) no matter how large the CM sketch we maintain. See Figure 3(a) for details. Figure 3(b) shows the overall running time for CM sketch and our approach with different number of queries made during the experiment. Here, overall running time includes the time for reading data and updating the data structure, and the query time. For small amount of query, the total running time of our

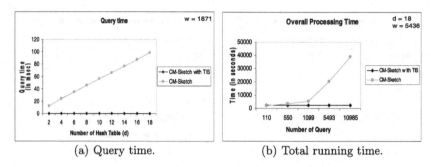

(a) Query time.   (b) Total running time.

Fig. 3. Running time analysis

approach is longer relatively because our approach needs some additional time for maintaining TIS. However, as the number of queries increases, query time will become dominant. The results show that if the number of queries is about 11000, the total running time of our approach is 2151 seconds while that of CM sketch is 38950 seconds. Note that there are very few renew operations carried out by our approach for most of the cases. In a stream of tens of millions updates in the turnstile stream, less than three renew operations were carried out in each set of experiments. This shows that renew operation seldom occurs. Moreover, the time for a renew operation is roughly the same as the query time for hCount*. Thus, renew operations would not affect the performance of our approach.

3.3 Improvement

Note that our heuristic does not guarantee no false negative. We modify our algorithm below such that we can guarantee there is no false negative in our answer: (1) recall that we will keep at most $\frac{1}{\theta} + C$ counters in TIS for some constant $C \geq \frac{1}{\theta}$. Note that we now set $C = \frac{2}{\theta}$; (2) originally, we will renew TIS whenever TIS_{topK} contains less than $\frac{2}{\theta}$ counters. We change the condition for renew operation: let $f_{\text{TIS}_{\text{topK}}}$ denote the minimum of the estimated frequency of all the counters in TIS_{topK} (i.e. $\min(\text{TIS}_{\text{topK}})$) if there are $\frac{3}{\theta}$ counters in TIS. And $f_{\text{TIS}_{\text{topK}}} = 0$ otherwise. Whenever $f_{\text{TIS}_{\text{topK}}} \geq \theta N$, we will renew TIS.

By the design of our data structure and by Theorem 3, it can be verified that

Lemma 1. *Our data structure* TIS_{topK} *always keeps a superset of items whose frequencies are greater than* $f_{\text{TIS}_{\text{topK}}}$.

Note that after renew operation, $f_{\text{TIS}_{\text{topK}}}$ may still be no less than θN since CM sketch will overestimate the frequency of an item. Although we can prove that this will happen with probability no larger than δ, we still need to handle this exceptional case by adding some special criteria, say, we renew TIS only when $f_{\text{TIS}_{\text{topK}}} \geq \theta N$ and the last renew operation has been carried out for more than t time units ago where t is some user-specified parameter.

When there is a query for frequent items, if $f_{\text{TIS}_{\text{topK}}} < \theta N$, we will scan TIS and return all items whose estimated frequencies are no less than θN. By Lemma 1, the answer will not have any false negative. If $f_{\text{TIS}_{\text{topK}}} \geq \theta N$, we will estimate the frequency of every item by CM sketch and return the items whose estimated frequencies are no less than θN. Thus, we have the following theorem.

Theorem 4. *Our algorithm returns all the items whose frequencies are no less than θN.*

References

1. Alon, N., Gibbons, P.B., Matias, Y., Szegedy, M.: Tracking Join and Self-join Sizes in Limited Storage. In: Symposium on Principles of Database Systems, pp. 10–20 (1999)
2. Arasu, A., Manku, G.S.: Approximate Counts and Quantiles over Sliding Windows. In: Symposium on Principles of Database Systems, pp. 286–296 (2004)
3. Cormode, G., Garofalakis, M., Sacharidis, D.: Fast Approximate Wavelet Tracking on Streams. In: International Conference on Extending Database Technology, pp. 4–22 (2006)
4. Cormode, G., Muthukrishnan, S.: An Improved Data Stream Summary: The Count-Min Sketch and its Applications. Journal of Algorithms 55(1), 58–75 (2005)
5. Demaine, E.D., López-Ortiz, A., Munro, J.I.: Frequency Estimation of Internet Packet Streams with Limited Space. In: European Symposium on Algorithms, pp. 348–360 (2002)
6. Gaber, M.M., Zaslavsky, A., Krishnaswamy, S.: Mining Data Streams: a Review. SIGMOD Record 34(2), 18–26 (2005)
7. Ganguly, S.: Counting Distinct Items over Update Streams. In: International Symposium on Algorithms and Computation, pp. 505–514 (2005)
8. Ganguly, S., Garofalakis, M.N., Kumar, A., Rastogi, R.: Join-Distinct Aggregate Estimation over Update Streams. In: Symposium on Principles of Database Systems, pp. 259–270 (2005)
9. Ganguly, S., Majumder, A.: Deterministic k-Set Structure. In: Symposium on Principles of Database Systems, pp. 280–289 (2006)
10. Ganguly, S., Majumder, A.: CR-precis: A Deterministic Summary Structure for Update Data Streams. In: IntErnational Symposium on Combinatorics, Algorithms, Probabilistic and Experimental Methodologies, pp. 48–59 (2007)
11. Jin, C., Qian, W., Sha, C., Yu, J.X., Zhou, A.: Dynamically Maintaining Frequent Items over a Data Stream. In: International Conference on Information and Knowledge Management, pp. 287–294 (2003)
12. Karp, R.M., Shenker, S., Papadimitriou, C.H.: A Simple Algorithm for Finding Frequent Elements in Streams and Bags. ACM Transactions on Database Systems 28(1), 51–55 (2003)
13. Lee, L.K., Ting, H.F.: A Simpler and More Efficient Deterministic Scheme for Finding Frequent Items over Sliding Windows. In: Symposium on Principles of Database Systems, pp. 290–297 (2006)
14. Manku, G.S., Motwani, R.: Approximate Frequency Counts over Data Streams. In: Very Large Data Bases Conference, pp. 346–357 (2002)
15. Misra, J., Gries, D.: Finding Repeated Elements. Science of Computer Programming 2, 143–152 (1982)
16. Muthukrishnan, S.: Data Streams: Algorithms and Applications. Now Publishers (2005)
17. Sanitized UCLA CSD Traffic Traces,
   http://www.lasr.cs.ucla.edu/ddos/traces/

A Linear Programming Duality Approach to Analyzing Strictly Nonblocking d-ary Multilog Networks under General Crosstalk Constraints

Hung Q. Ngo*, Yang Wang, and Anh Le

Computer Science and Engineering Department,
State University of New York at Buffalo,
201 Bell Hall, Amherst, NY 14260, USA

Abstract. In an optical crossconnect architecture, two routes sharing a common switching element suffer from crosstalk. Vaez and Lea [16] introduced a parameter c which is the maximum number of distinct switching elements a route can share with other routes in the crossconnect. We present a new method of analyzing strictly nonblocking d-ary multilog networks under this general crosstalk constraint using linear programming duality.

We improve known results on several fronts: (a) our sufficient conditions are better than known sufficient conditions for $\log_d(N, 0, m)$ to be nonblocking under the general crosstalk constraint, (b) our results are on d-ary multilog networks while known results are on binary networks, and (c) for several ranges of the parameter c, we give the first known necessary conditions for this problem which also match our sufficient conditions from the LP-duality approach.

1 Introduction

The d-ary multilog networks have been attractive for both electronic and photonic domains [10], [16], [5], [15], [13], [11], [8], because they have small depth ($O(\log N)$), absolute signal loss uniformity, and good fault tolerance. Henceforth, let $\log_d(N, 0, m)$ denote a d-ary multilog network with m vertically stacked inverse Banyan planes $BY^{-1}(n)$, as illustrated in Figure 1. Three levels of non-blockingness typically studied in the switching network literature are: rearrangeably nonblocking (RNB), wide-sense nonblocking (WSNB), and strictly nonblocking (SNB). The reader is referred to [6] for their precise definitions. This paper focuses on the SNB case.

In an optical crossconnect, no two routes are allowed to share a link, just as in the circuit switching case. However, if two routes share too many switching elements (SE), then crosstalk at those SEs degrades signal quality. Vaez and Lea introduced a parameter c which is the maximum number of distinct switching elements (SE) a route can share with other routes in the network. This is called

---

* Corresponding Author. The work of Hung Q. Ngo was supported in part by NSF CAREER Award CCF-0347565.

X. Hu and J. Wang (Eds.): COCOON 2008, LNCS 5092, pp. 510–520, 2008.

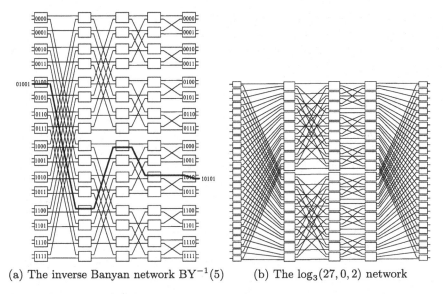

(a) The inverse Banyan network $BY^{-1}(5)$ (b) The $\log_3(27, 0, 2)$ network

Fig. 1. Illustration of the Multi-log Network

the *general crosstalk constraint*. The shared SEs are also called *crosstalk SEs* (CSE).

We shall refer to a switching network which is SNB under this crosstalk constraint a *c-SNB network*. For example, when $c = n$ a route can share any number of SEs with other routes (no shared link, still), because each route in a $\log_d(N, 0, m)$ network contains exactly n SEs. In this case only link-blocking takes effect, and it is commonly referred to as the *link-blocking case*, which is relevant to electronic switches [9], [3], [6], [1], [14], [7], [4], [5], [15], [11]. On the other hand, when $c = 0$ no CSE is allowed and we have the *node-blocking case* (or *crosstalk free*), relevant in optical crossconnect designs [10], [16], [13], [8], [2], [12], [17].

Prior to this paper, the only known results are sufficient conditions for the binary $\log_2(N, 0, m)$ network to be c-SNB [16]. Specifically, let $N = 2^n$ Vaez and Lea showed that, when $n = 2k + 1$

$$m \geq \begin{cases} 2^{k+1} + \left\lfloor \frac{2^{k+1}}{c+2} \right\rfloor - 1 & 0 < c \leq k \\ 2^{k+1} + \left\lfloor \frac{2^k}{c+2} \right\rfloor & k < c \leq 2k \end{cases} \tag{1}$$

is sufficient for $\log_2(N, 0, m)$ to be c-SNB. When, $n = 2k$ the sufficient condition is

$$m \geq \begin{cases} 2^k + 2^{k-1} + \left\lfloor \frac{2^k}{c+1} \right\rfloor - 1 & 0 < c \leq k \\ 2^k + 2^{k-1} - 1 & k < c \leq 2k - 1 \end{cases} \tag{2}$$

Our main contributions were outlined in the abstract. An important advantage of the LP-duality approach is the ease and brevity of sufficiency proofs. All we have to do is to check that a solution is indeed dual-feasible and the dual-objective value automatically gives us a sufficient condition. Earlier works on this problem relied on combinatorial arguments which are quite intricate and somewhat error-prone.

2   Preliminaries

For any positive integers l, d, let $[l]$ denote the set $\{1, \ldots, l\}$, $\mathbb{Z}_d$ denote the set $\{0, \ldots, d-1\}$ which can be thought of as d-ary "symbols," $\mathbb{Z}_d^l$ denote the set of all d-ary strings of length l, b^l denote the string with symbol $b \in \mathbb{Z}_d$ repeated l times (e.g., $3^4 = 3333$), $|\mathbf{s}|$ denote the length of any d-ary string $\mathbf{s}$ (e.g., $|31| = 2$), $\mathbf{s}_{i..j}$ denote the substring $s_i \cdots s_j$ of a string $\mathbf{s} = s_1 \ldots s_l \in \mathbb{Z}_d^l$. If $i > j$ then $\mathbf{s}_{i..j}$ is the empty string.

Let $N = d^n$. We consider the $\log_d(N, 0, m)$ network, which denotes the stacking of m copies of the d-ary inverse Banyan network $BY^{-1}(n)$ with N inputs and N outputs. Label the inputs and outputs of $BY^{-1}(n)$ with d-ary strings of length n. Specifically, each input $\mathbf{x} \in \mathbb{Z}_d^n$ and output $\mathbf{y} \in \mathbb{Z}_d^n$ have the form $\mathbf{x} = x_1 \cdots x_n$, $\mathbf{y} = y_1 \cdots y_n$, where $x_i, y_i \in \mathbb{Z}_d, \forall i \in [n]$. Also, label the $d \times d$ SEs in each of the n stages of $BY^{-1}(n)$ with d-ary strings of length $n-1$. An input $\mathbf{x}$ (resp. output $\mathbf{y}$) is connected to the SE labeled $\mathbf{x}_{1..n-1}$ in the first stage (resp. $\mathbf{y}_{1..n-1}$ in the last stage). In general, the unique path $R(\mathbf{x}, \mathbf{y})$ in $BY^{-1}(n)$ from an arbitrary input $\mathbf{x}$ to an arbitrary output $\mathbf{y}$ is exactly the following: input $\mathbf{x} = x_1 x_2 \ldots x_{n-1} x_n$ to stage-1 SE $x_1 x_2 \ldots x_{n-1}$ to stage-2 SE $y_1 x_2 \ldots x_{n-1}$ to stage-3 SE $y_1 y_2 \ldots x_{n-1}$, $\ldots$, to stage-n SE $y_1 y_2 \ldots y_{n-1}$, to output $\mathbf{y}$ $y_1 y_2 \ldots y_{n-1} y_n$.

Now, consider two unicast requests $(\mathbf{a}, \mathbf{b})$ and $(\mathbf{x}, \mathbf{y})$. From the observation above, the two routes $R(\mathbf{a}, \mathbf{b})$ and $R(\mathbf{x}, \mathbf{y})$ share a SE (also called a node) iff there is some $j \in [n]$ such that $\mathbf{b}_{1..j-1} = \mathbf{y}_{1..j-1}$ and $\mathbf{a}_{j..n-1} = \mathbf{x}_{j..n-1}$. In this case, the two paths intersect at a stage-j SE. It should be noted that two requests' paths may intersect at more than one SE. For any two d-ary strings $\mathbf{u}, \mathbf{v} \in \mathbb{Z}_d^l$, let $\mathrm{PRE}(\mathbf{u}, \mathbf{v})$ denote the *longest common prefix*, and $\mathrm{SUF}(\mathbf{u}, \mathbf{v})$ denote the *longest common suffix* of $\mathbf{u}$ and $\mathbf{v}$, respectively. The following propositions are immediate.

Proposition 1. *Let $(\mathbf{a}, \mathbf{b})$ and $(\mathbf{x}, \mathbf{y})$ be two unicast requests. Then their corresponding routes $R(\mathbf{a}, \mathbf{b})$ and $R(\mathbf{x}, \mathbf{y})$ in a $BY^{-1}(n)$-plane share at least a common SE iff*

$$|\mathrm{SUF}(\mathbf{a}_{1..n-1}, \mathbf{x}_{1..n-1})| + |\mathrm{PRE}(\mathbf{b}_{1..n-1}, \mathbf{y}_{1..n-1})| \geq n - 1. \qquad (3)$$

Moreover, the routes $R(\mathbf{a}, \mathbf{b})$ and $R(\mathbf{x}, \mathbf{y})$ intersect at exactly one SE iff

$$|\mathrm{SUF}(\mathbf{a}_{1..n-1}, \mathbf{x}_{1..n-1})| + |\mathrm{PRE}(\mathbf{b}_{1..n-1}, \mathbf{y}_{1..n-1})| = n - 1, \qquad (4)$$

in which case the common SE is an SE at stage $|\mathrm{PRE}(\mathbf{b}_{1..n-1}, \mathbf{y}_{1..n-1})| + 1$ of the $BY^{-1}(n)$-plane.

Proposition 2. *Let $(\mathbf{a}, \mathbf{b})$ and $(\mathbf{x}, \mathbf{y})$ be two unicast requests. Then their corresponding routes $R(\mathbf{a}, \mathbf{b})$ and $R(\mathbf{x}, \mathbf{y})$ in a $BY^{-1}(n)$-plane share at least a common link iff*

$$|\mathrm{SUF}(\mathbf{a}_{1..n-1}, \mathbf{x}_{1..n-1})| + |\mathrm{PRE}(\mathbf{b}_{1..n-1}, \mathbf{y}_{1..n-1})| \geq n. \qquad (5)$$

3   Sufficient Conditions

Let $(\mathbf{a}, \mathbf{b})$ be an arbitrary request. For each $i, j \in \{0, \ldots, n-1\}$, define

$$A_i := \{\mathbf{x} \in \mathbb{Z}_d^n - \{\mathbf{a}\} \mid \text{SUF}(\mathbf{x}_{1..n-1}, \mathbf{a}_{1..n-1}) = i\},$$

$$B_j := \{\mathbf{y} \in \mathbb{Z}_d^n - \{\mathbf{b}\} \mid \text{PRE}(\mathbf{y}_{1..n-1}, \mathbf{b}_{1..n-1}) = j\}.$$

Note that $|A_i| = |B_i| = d^{n-i} - d^{n-1-i}$, for all $0 \leq i \leq n-1$. Suppose the network $\log_d(N, 0, m)$ already had some routes established. Consider a $\text{BY}^{-1}(n)$ plane which blocks the new request $(\mathbf{a}, \mathbf{b})$. There can only be three cases for which this happens:

Case 1: There is a request $(\mathbf{x}, \mathbf{y})$ routed in the plane for which $R(\mathbf{x}, \mathbf{y})$ and $R(\mathbf{a}, \mathbf{b})$ share a link. By Proposition 2, it must be the case that $(\mathbf{x}, \mathbf{y}) \in A_i \times B_j$ for some $i + j \geq n$. We will refer to $(\mathbf{x}, \mathbf{y})$ as a *link-blocking request*.

Case 2: There is a request $(\mathbf{x}, \mathbf{y})$ whose route $R(\mathbf{x}, \mathbf{y})$ already intersects c other routes at c distinct SEs on the same plane, and adding $(\mathbf{a}, \mathbf{b})$ would introduce an additional intersecting SE to the route $R(\mathbf{x}, \mathbf{y})$. We will refer to these c other requests as *secondary requests* accompanying $(\mathbf{x}, \mathbf{y})$, and refer to $(\mathbf{x}, \mathbf{y})$ as a *node-blocking request of type* 1.

In particular, by Proposition 1 we must have $(\mathbf{x}, \mathbf{y}) \in A_i \times B_j$ for some $i + j = n - 1$. Note that the common SE between $R(\mathbf{a}, \mathbf{b})$ and $R(\mathbf{x}, \mathbf{y})$ is at stage $j + 1$. The routes for secondary requests must thus intersect $R(\mathbf{x}, \mathbf{y})$ at stages strictly less than $j + 1$ or strictly greater than $j + 1$. If a secondary request $(\mathbf{u}, \mathbf{v})$ has its route intersects $R(\mathbf{x}, \mathbf{y})$ at stage $1 \leq s < j+1$, it follows that $\text{PRE}(\mathbf{y}, \mathbf{v}) = s - 1 < j$, and $\text{SUF}(\mathbf{x}, \mathbf{u}) = n - 1 - (s-1) = n - s > n - 1 - j = i$. Hence, $(\mathbf{u}, \mathbf{v}) \in A_i \times B_{s-1}$. We will refer to such $(\mathbf{u}, \mathbf{v})$ as a *left secondary request*. If a secondary request $(\mathbf{u}, \mathbf{v})$ has its route intersects $R(\mathbf{x}, \mathbf{y})$ at stage $j + 1 < s \leq n$, it follows that $\text{PRE}(\mathbf{y}, \mathbf{v}) = s - 1 > j$, and $\text{SUF}(\mathbf{x}, \mathbf{u}) = n - 1 - (s - 1) = n - s < n - 1 - j = i$. Hence, $(\mathbf{u}, \mathbf{v}) \in A_{n-s} \times B_j$. We will refer to such $(\mathbf{u}, \mathbf{v})$ as a *right secondary request*.

To summarize, there are two types of secondary requests accompanying $(\mathbf{x}, \mathbf{y})$: the *left secondary requests* are the requests $(\mathbf{u}, \mathbf{v}) \in A_i \times B_{j'}$ for some $j' < j$, and the *right secondary requests* are the requests $(\mathbf{u}, \mathbf{v}) \in A_{i'} \times B_j$ for some $i' < i$. For each $i' < i$ there is at most one right secondary request in $A_{i'} \times B_j$. Similarly, for each $j' < j$ there is at most one left secondary request in $A_i \times B_{j'}$.

Case 3: There are $c + 1$ requests in the plane each of whose routes intersects $(\mathbf{a}, \mathbf{b})$ at exactly one SE. These will be called *node-blocking requests of type* 2. If $(\mathbf{x}, \mathbf{y})$ is such a request, then $(\mathbf{x}, \mathbf{y}) \in A_i \times B_j$ for some $i + j = n - 1$.

Theorem 3. *The number of blocking planes is the objective value of a feasible solution to the primal linear program as shown in Figure 2.*

Proof. Define the following variables. For each pair i, j such that $i + j \geq n$, let x_{ij} be the number of link-blocking requests in $A_i \times B_j$. For each pair i, j such that $i + j = n - 1$, let y_{ij} be the number of node-blocking requests of type 1, and z_{ij} be the number of node-blocking requests of type 2 in $A_i \times B_j$. For each pair i, j such that $i + j < n - 1$, let l_{ij} and r_{ij} be the number of left and right

---

The Primal LP

Maximize

$$\sum_{i+j \geq n} x_{ij} + \sum_{i+j=n-1} y_{ij} + \frac{1}{c+1} \sum_{i+j=n-1} z_{ij} \qquad (6)$$

Subject to

$$\sum_{j:\, i+j \geq n} x_{ij} + \sum_{j:\, i+j=n-1} (y_{ij} + z_{ij}) + \sum_{j:\, i+j<n-1} (l_{ij} + r_{ij}) \leq d^{n-i} - d^{n-1-i}, \;\forall i \quad (7)$$

$$\sum_{i:\, i+j \geq n} x_{ij} + \sum_{i:\, i+j=n-1} (y_{ij} + z_{ij}) + \sum_{i:\, i+j<n-1} (l_{ij} + r_{ij}) \leq d^{n-j} - d^{n-1-j}, \;\forall j \quad (8)$$

$$cy_{ij} - \left(\sum_{i'<i} r_{i'j} + \sum_{j'<j} l_{ij'} \right) = 0, \; i+j = n-1 \qquad (9)$$

$$l_{ij'} - y_{ij} \leq 0, \; i+j = n-1, \qquad (10)$$
$$j' < j$$

$$r_{i'j} - y_{ij} \leq 0, \; i+j = n-1, \qquad (11)$$
$$i' < i$$

$$x_{ij}, y_{ij}, z_{ij}, l_{ij}, r_{ij} \geq 0, \;\forall i,j \qquad (12)$$

Fig. 2. The Primal Linear Program

---

secondary requests in $A_i \times B_j$. The number of blocking planes is thus expressed in the objective function (6).

We next explain why the variables satisfy all the constraints. Recall that $|A_i| = d^{n-i} - d^{n-1-i}, \forall i$. Thus, the number of requests out of A_i is at most $d^{n-i} - d^{n-1-i}$, justifying constraint (7). Similarly, bounding the number of requests to B_j explains constraint (8). Constraint (9) expresses the fact that, for every node-blocking request of type 1 in $A_i \times B_j$ ($i+j = n-1$), there must be c accompanying secondary requests (left or right). Lastly, for each node-blocking request of type 1 in $A_i \times B_j$ ($i + j = n - 1$) constraint (10) says that for each $j' < j$ there is at most one left secondary request in $A_i \times B_{j'}$, and constraint (11) says that for each $i' < i$ there is at most one right secondary request in $A_{i'} \times B_j$. □

The dual linear program is given in Figure 3. The key idea is the following: due to weak duality every dual-feasible solution induces an objective value which is at least the number of blocking planes. As an illustration of the power of our method, let us reproduce a few known results. The examples should give the reader the correct insight without delving into too much technicality.

Example 1 ($c = 0$, the node-blocking case). When $c = 0$, the problem becomes the SNB problem in the node-blocking sense. It has been shown in [17] (which addressed the node-blocking problem in f-cast switches) that $m \geq d^{\lceil (n-1)/2 \rceil} + d^{n-\lceil (n-1)/2 \rceil} - 1$ is necessary and sufficient for $\log_d(N, 0, m)$ to be SNB. Let $i_0 = n - \lceil (n-1)/2 \rceil$ and $j_0 = \lceil (n-1)/2 \rceil$. Assign $u_i = 1$ for all i such that $i_0 \leq i \leq n-1$; and, $v_j = 1$ for all j such that $j_0 \leq j \leq n-1$. All other variables

---

The Dual LP

Minimize

$$\sum_{i=0}^{n-1}(d^{n-i} - d^{n-1-i})u_i + \sum_{j=0}^{n-1}(d^{n-j} - d^{n-1-j})v_j \tag{13}$$

Subject to

$$u_i + v_j \geq 1, \quad i+j \geq n \tag{14}$$

$$u_i + v_j + cw_{ij} - \sum_{i'<i} s_{i'j} - \sum_{j'<j} t_{ij'} \geq 1, \quad i+j = n-1 \tag{15}$$

$$u_i + v_j \geq \tfrac{1}{c+1}, \quad i+j = n-1 \tag{16}$$

$$u_i + v_j - w_{i,n-1-i} + t_{ij} \geq 0, \quad i+j < n-1 \tag{17}$$

$$u_i + v_j - w_{n-1-j,j} + s_{ij} \geq 0, \quad i+j < n-1 \tag{18}$$

$$u_i, v_j, s_{ij}, t_{ij} \geq 0, \quad \forall i,j \tag{19}$$

Fig. 3. The Dual Linear Program

---

are zeros. It is easy to check that this is a dual-feasible solution with objective value precisely $d^{\lceil (n-1)/2\rceil} + d^{n-\lceil (n-1)/2\rceil} - 2$. Thus, at most one more $BY^{-1}(n)$-plane is needed, for a total of $d^{\lceil (n-1)/2\rceil} + d^{n-\lceil (n-1)/2\rceil} - 1$ planes in the worst case, matching the known necessary and sufficient condition.

Example 2 ($c = n$, the link-blocking case). When $c = n$, the problem becomes the SNB problem in the link-blocking sense. It has been shown in [17,4] that $m \geq d^{\lceil n/2\rceil-1} + d^{\lfloor n/2\rfloor} - 1$ is necessary and sufficient for $\log_d(N,0,m)$ to be SNB. Since there can be no path with $n+1$ distinct SEs, the variables y_{ij} and z_{ij} are all zeros, so are the l_{ij} and r_{ij}. The primal LP becomes much simpler:

$$\max \quad \sum_{i+j\geq n} x_{ij}$$
$$\text{subject to} \sum_{j:\, i+j\geq n} x_{ij} \leq d^{n-i} - d^{n-1-i} \;\forall i$$
$$\sum_{i:\, i+j\geq n} x_{ij} \leq d^{n-j} - d^{n-1-j} \;\forall j$$
$$x_{ij} \geq 0 \qquad\qquad \forall i,j$$

The dual LP is:

$$\min \sum_i (d^{n-i} - d^{n-1-i})u_i + \sum_j (d^{n-j} - d^{n-1-j})v_j$$
$$\text{subject to} \qquad u_i + v_j \geq 1, \quad i+j \geq n, \qquad\qquad u_i, v_j \geq 0, \;\; \forall i,j$$

Let $i_0 = n - \lceil n/2\rceil + 1$ and $j_0 = \lfloor n/2\rfloor$. Assign $u_i = 1$ for all i such that $i_0 \leq i \leq n-1$; and, $v_j = 1$ for all j such that $j_0 \leq j \leq n-1$. All other variables are zeros. This solution is dual-feasible with objective value $d^{\lceil n/2\rceil-1} + d^{\lfloor n/2\rfloor} - 2$. Hence, $m \geq d^{\lceil n/2\rceil-1} + d^{\lfloor n/2\rfloor} - 1$ is sufficient for $\log_d(N,0,m)$ to be SNB in this case. Again, the sufficient condition obtained with our method matches the known necessary and sufficient condition.

We consider $1 \leq c \leq n - 1$ henceforth. First, a few technical lemmas are in order.

Lemma 4. *Suppose* $n = 2k + 1$ *and* $1 \leq c \leq 2k$. *For any integer* p *where* $0 \leq p \leq k$, *there exists a feasible solution to the dual LP with objective value*

$$2d^k - 2 + \frac{2d^k(d-1)}{c+2} \cdot f(c,d,p),$$

where

$$f(c,d,p) = \begin{cases} p + 1 - \dfrac{1 - \left(\frac{1}{d(c+1)}\right)^p}{d - \frac{1}{c+1}} & c = d - 1 \\[3mm] \dfrac{1 - \left(\frac{d}{c+1}\right)^{p+1}}{1 - \frac{d}{c+1}} - \dfrac{1 - \left(\frac{1}{d(c+1)}\right)^p}{d - \frac{1}{c+1}} & c \neq d - 1 \end{cases} \tag{20}$$

Proof. Consider the following assignment to the dual variables: $u_i = v_i = 1$ for $k + p + 1 \leq i \leq n - 1$, $u_i = v_i = 1 - \frac{1}{(c+2)(c+1)^{i-k-1}}$ for $k + 1 \leq i \leq k + p$, $u_i = v_i = \frac{1}{(c+2)(c+1)^{k-i}}$ for $k - p \leq i \leq k$, $w_{ij} = \min\{u_i, u_j\}$ for all $i + j = n - 1$, all other variables are 0. It is straightforward to verify that this solution is dual-feasible with the claimed objective value. $\qquad\square$

Lemma 5. *Suppose* $n = 2k + 1$, $k \geq 2$, *and* $c = k$. *There exists a feasible solution to the dual LP with objective value*

$$2d^k - 2 + \frac{2(d-1)d^{k-2}}{c+2}\left(d^2 - d + \frac{d^3 - 1}{c+2}\right)$$

Proof. Consider the following assignment to the dual variables: $u_i = v_i = 1$ when $k + 3 \leq i \leq n - 1$, $u_{k+2} = v_{k+2} = 1 - \frac{1}{(c+2)^2}$, $u_{k+1} = v_{k+1} = 1 - \frac{1}{(c+2)}$, $u_k = v_k = \frac{1}{(c+2)}$, $u_{k-1} = v_{k-1} = \frac{1}{(c+2)^2}$, $u_i = v_i = 0$ when $0 \leq i \leq k - 2$, $w_{kk} = \frac{1}{c+2}$, $w_{k+1,k-1} = w_{k-1,k+1} = \frac{2}{(c+2)^2}$, $w_{k+2,k-2} = w_{k-2,k+2} = \frac{1}{(c+2)^2}$, $w_{ij} = 0$ for all other i, j, $s_{i,k-2} = t_{k-2,i} = \frac{1}{(c+2)^2}$ when $0 \leq i \leq k - 2$, $s_{i,k-1} = t_{k-1,i} = \frac{1}{(c+2)^2}$ when $0 \leq i \leq k - 2$, $s_{ij} = t_{ij} = 0$ for all other i, j. $\qquad\square$

Lemma 6. *Suppose* $n = 2k + 1$ *and* $k + 1 \leq c \leq 2k$. *There exists a feasible solution to the dual LP with objective value*

$$2d^k - 2 + \frac{2(d-1)^2 d^{k-1}}{c+2}$$

Proof. Consider the following assignment to the dual variables: $u_i = v_i = 1$ when $k + 2 \leq i \leq n - 1$, $u_{k+1} = v_{k+1} = 1 - \frac{1}{(c+2)}$, $u_k = v_k = \frac{1}{(c+2)}$, $u_i = v_i = 0$ when $0 \leq i \leq k - 1$, $w_{kk} = \frac{1}{c+2}$, $w_{k+1,k-1} = w_{k-1,k+1} = \frac{1}{(c+2)}$, $w_{ij} = 0$ for all other i, j, $s_{i,k-1} = t_{k-1,i} = \frac{1}{(c+2)}$ when $0 \leq i \leq k-1$, $s_{ij} = t_{ij} = 0$ for all other i, j. $\qquad\square$

Lemma 7. *Suppose $n = 2k$. For any integer $p, 1 \leq p \leq k$, there exists a feasible solution to the dual LP with objective value*

$$2d^{k-1} - 2 + (d-1)d^{k-1} \cdot g(c,d,p),$$

where

$$g(c,d,p) = \begin{cases} p + 1 - \dfrac{1 - \left(\frac{1}{d(c+1)}\right)^{p-1}}{d(c+1) - 1} & c = d - 1 \\[4mm] \dfrac{1 - \left(\frac{d}{c+1}\right)^{p+1}}{1 - \frac{d}{c+1}} - \dfrac{1 - \left(\frac{1}{d(c+1)}\right)^{p-1}}{d(c+1) - 1} & c \neq d - 1. \end{cases} \tag{21}$$

Proof. Consider the following assignment to the dual variables: $u_i = v_i = 1$ when $k + p \leq i \leq n - 1$, $u_i = v_i = 1 - \frac{1}{2(c+1)^{i-k}}$ when $k + 1 \leq i \leq k + p - 1$, $u_i = v_i = \frac{1}{2(c+1)^{k-i}}$ when $k - p \leq i \leq k$, $u_i = v_i = 0$ when $0 \leq i \leq k - p - 1$, $w_{ij} = \min\{u_i, u_j\}$ when $i + j = n - 1$, $s_{ij} = t_{ij} = 0$ for all $i + j < n - 1$. □

Lemma 8. *Suppose $n = 2k$ and $k \leq c$. There exists a feasible solution to the dual LP with objective value $d^k + d^{k-1} - 2$.*

Proof. Consider the following assignment to the dual variables: $u_i = v_i = 1$ for $k+1 \leq i \leq n-1$, $u_k = v_k = \frac{1}{2}$, $u_i = v_i = 0$ for $0 \leq i \leq k-1$, $w_{k,k-1} = w_{k-1,k} = 1$, $w_{ij} = 0$ for all other i, j, $s_{i,k-1} = t_{k-1,i} = 1$ when $0 \leq i \leq k - 1$, $s_{ij} = t_{ij} = 0$ for all other i, j. □

Theorem 9. *Consider the case when $n = 2k+1$. Recall that $f(c,d,p)$ is defined in (20).*

(a) *If $1 \leq c \leq k - 1$, then $m \geq 2d^k - 1 + \left\lfloor \frac{2d^k(d-1)}{c+2} \cdot f(c,d,\lfloor \log_d(c)/2 \rfloor) \right\rfloor$ is sufficient for $\log_d(N,0,m)$ to be c-SNB. In particular, when $c \leq \min\{k - 1, d^2 - 1\}$ $m \geq 2d^k - 1 + \left\lfloor \frac{2d^k(d-1)}{c+2} \right\rfloor$ is sufficient.*

(b) *If $c = k$, then $m \geq 2d^k - 1 + \left\lfloor \frac{2(d-1)d^{k-2}}{c+2} \left(d^2 - d + \frac{d^3 - 1}{c+2} \right) \right\rfloor$ is sufficient for $\log_d(N,0,m)$ to be c-SNB.*

(c) *If $k + 1 \leq c \leq 2k$, then $m \geq 2d^k - 1 + \left\lfloor \frac{2(d-1)^2 d^{k-1}}{c+2} \right\rfloor$ is sufficient for $\log_d(N,0,m)$ to be c-SNB.*

Proof. To see (a), consider two cases. When $c = d - 1$, we apply Lemma 4 with $p = 0$. In this case, $f(d-1,d,p)$ is minimized at $p = 0$; moreover, $f(d-1,d,0) = 1$. When $c \neq d - 1$, we apply the same Lemma with $p = \lfloor \frac{1}{2} \log_d c \rfloor$. When $c \leq d^2 - 1$ the f's is minimized at $f(c,d,\lfloor \log_d(c)/2 \rfloor) = 1$. Parts (b) and (c) are straightforward from Lemmas 5 and 6. □

Similarly, with the help of Lemmas 7 and 8 we can show the following

Theorem 10. *Consider the case when $n = 2k$. Recall that $g(c,d,p)$ is defined in (21).*

(a) If $1 \leq c \leq k-1$, then $m \geq 2d^{k-1} - 1 + \lfloor (d-1)d^{k-1} \cdot g(c, d, \lceil \log_d(c)/2 \rceil) \rfloor$
is sufficient for $\log_d(N, 0, m)$ to be c-SNB. In particular, when $1 \leq c \leq$
$\min\{k-1, d^2\}$ $m \geq 2d^{k-1} - 1 + \left\lfloor \frac{(d-1)d^{k-1}}{c+1} \right\rfloor$ is sufficient.

(b) If $k \leq c \leq 2k$, then $m \geq d^k + d^{k-1} - 1$ is sufficient for $\log_d(N, 0, m)$ to be
c-SNB.

4   Necessary Conditions

We have seen in Examples 3 and 3 that our method gives sufficient conditions
which are also necessary when $c = 0$ (node-blocking case) and $c = n$ (link-
blocking case). A natural question is: "how good are our sufficient conditions in
Theorems 9 and 10 when $1 \leq c \leq n-1$?" In this section, we will show that our
sufficient conditions are also necessary for $1 \leq c \leq \min\{k-1, d^2-1\}$ when n
is odd, and necessary for $1 \leq c \leq \min\{k-1, d^2\}$ and when $c \geq n/2$ when n is
even.. This is the first time in the literature that any necessary and sufficient
conditions are derived for $1 \leq c \neq n$. It is an open question whether our sufficient
conditions are also necessary for other ranges of c.

Theorem 11

(a) When $n = 2k+1$ and $1 \leq c \leq \min\{k-1, d^2-1\}$, the sufficient condition
$m \geq 2d^k - 1 + \left\lfloor \frac{2d^k(d-1)}{c+2} \right\rfloor$ of Theorem 9 is also necessary for $\log_d(N, 0, m)$
to be c-SNB.

(b) When $n = 2k$ and $1 \leq c \leq \min\{k-1, d^2\}$ the sufficient condition $m \geq$
$2d^{k-1} - 1 + \left\lfloor \frac{(d-1)d^{k-1}}{c+1} \right\rfloor$ of Theorem 10 is also necessary for $\log_d(N, 0, m)$ to
be c-SNB.

(c) Moreover, when $n = 2k$ and $c \geq k$ the sufficient condition $m \geq d^k + d^{k-1} - 1$
of Theorem 10 is also necessary for $\log_d(N, 0, m)$ to be c-SNB.

Proof. Due to space limitation and for the sake of clarity, we will only prove
part (a) of this theorem when $d = 2$ and $c = 2$. Hopefully the reader will be able
to see the main line of thought. When $d = 2$ and $c = 2$, the sufficient condition
becomes $m \geq 2^{k+1} + 2^{k-1} - 1$. The main idea is to create a request $(\mathbf{a}, \mathbf{b})$ and
a network state compatible with this request in which there are $2^{k+1} + 2^{k-1} - 2$
blocking planes.

Let $\mathbf{a} = 0^n$ and $\mathbf{b} = 0^n$. The blocking planes are constructed as follows. For
each i where $k+2 \leq i \leq 2k = n-1$, each string $\mathbf{s} \in \mathbb{Z}_2^{2k-i-1}$ and each bit $b \in \mathbb{Z}_2$
create requests $(\mathbf{s}10^i b, 0^{2k+1-i}1\mathbf{s}1^{2i-2k-1}b)$, $(1^{2i-2k-1}\mathbf{s}10^{2k+1-i}b, 0^i 1\mathbf{s}b)$, and let
each of them be routed through a separate $BY^{-1}(n)$ plane. Since each of the
above requests link-blocks $(\mathbf{a}, \mathbf{b})$, all the above planes are blocking planes. The
number of blocking planes is $2 \left(\sum_{i=k+2}^{2k} 2^{2k-i} \right) = 2(2^{k-1} - 1) = 2^k - 2$.

Next, we will construct some more blocking planes which route node-blocking
requests of type 1. We certainly cannot use any of the inputs and outputs which
have already been used to create the blocking planes above. For each string $\mathbf{s} \in$

$\mathbb{Z}_2^{k-2}$ and each symbol $b \in \mathbb{Z}_2$, route the following three requests through a separate plane: $(s10^{k+1}b, 0^{k-1}1s00b)$, $(0^k s10b, 0^{k-1}1s01b)$, $(0^{k-1}s100b, 0^{k-1}1s10b)$. It is not difficult to check that each of the above planes are blocking-planes. Similarly, for each string $s \in \mathbb{Z}_2^{k-2}$ and each symbol $b \in \mathbb{Z}_2$, route the following three requests through a separate plane: $(00s10^{k-1}b, 0^{k+1}1sb)$, $(10s10^{k-1}b, 01s0^k b)$, $(01s10^{k-1}b, 001s0^{k-1}b)$. The total number of blocking planes created this way is $2 \cdot 2^{k-1} = 2^k$.

Finally, for each string $s \in \mathbb{Z}_2^{k-1}$, route the following three requests through a separate plane: $(s10^k 0, 0^k 1s0)$, $(s10^k 1, 10^{k-1}1s0)$, $(s10^{k-1}10, 0^k 1s1)$. The total number of blocking planes created this way is 2^{k-1}. We have not used any input nor output twice in creating a total of $2^k - 2 + 2^k + 2^{k-1} = 2^{k+1} + 2^{k-1} - 2$ blocking planes. The necessary condition is thus established. □

References

1. Chen, H.-B., Hwang, F.K.: On Multicast Rearrangeable 3-stage Clos Networks Without First-Stage Fan-Out. SIAM Journal on Discrete Mathematics 20, 287–290 (2006)
2. Chinni, V.R., Huang, T.C., Wai, P.-K.A., Menyuk, C.R., Simonis, G.J.: Crosstalk in a Lossy Directional Coupler Switch. J. Lightwave Technol. 13, 1530–1535 (1995)
3. Hwang, F.K., Lin, B.-C.: Wide-Sense Nonblocking Multicast $\log_2(n, m, p)$ Networks. IEEE Transactions on Communications 51, 1730–1735 (2003)
4. Hwang, F.K., Wang, Y., Tan, J.: Strictly Nonblocking f-cast $\log_d(n, m, p)$ Networks. IEEE Transactions on Communications 55, 981–986 (2007)
5. Hwang, F.: Choosing the Best $\log_2(n, m, p)$ Strictly Nonblocking Networks. IEEE Transactions on Communications 46, 454–455 (1998)
6. Hwang, F.K.: The Mathematical Theory of Nonblocking Switching Networks. World Scientific Publishing Co. Inc., River Edge (2004)
7. Jiang, X., Pattavina, A., Horiguchi, S.: Rearrangeable f-cast Multi-log $2n$ Networks. IEEE Transactions on Communications (to appear, 2007)
8. Jiang, X., Shen, H., Khandker, M.M.R., Horiguchi, S.: Blocking Behaviors of Crosstalk-Free Optical Banyan Networks on Vertical Stacking. IEEE/ACM Trans. Networking 11, 982–993 (2003)
9. Kabacinski, W., Danilewicz, G.: Wide-Sense and Strict-Sense Nonblocking Operation of Multicast Multi-$\log_2 n$ Switching Networks. IEEE Transactions on Communications 50, 1025–1036 (2002)
10. Lea, C.-T.: Muti-$\log_2 n$ Networks and Their Applications in High Speed Electronic and Photonic Switching Systems. IEEE Transactions on Communications 38, 1740–1749 (1990)
11. Lea, C.-T., Shyy, D.-J.: Tradeoff of Horizontal Decomposition Versus Vertical Stacking in Rearrangeable Nonblocking Networks. IEEE Transactions on Communications 39, 899–904 (1991)
12. Li, D.: Elimination of Crosstalk in Directional Coupler Switches. Optical Quantum Electron 25, 255–260 (1993)
13. Maier, G., Pattavina, A.: Design of Photonic Rearrangeable Networks with Zero First-Order Switching-Element-Crosstalk. IEEE Transactions on Communications 49, 1268–1279 (2001)
14. Pattavina, A., Tesei, G.: Non-blocking Conditions of Multicast Three-Stage Interconnection Networks. IEEE Transactions on Communications 46, 163–170 (2005)

15. Shyy, D.-J., Lea, C.-T.: $\log_2(n, m, p)$ Strictly Nonblocking Networks. IEEE Transactions on Communications 39, 1502–1510 (1991)
16. Vaez, M.M., Lea, C.-T.: Strictly Nonblocking Directional-Coupler-Based Switching Networks Under Crosstalk Constraint. IEEE Transactions on Communications 48, 316–323 (2000)
17. Wang, Y., Ngo, H.Q., Jiang, X.: Strictly Nonblocking f-cast d-ary Multilog Networks under Fanout and Crosstalk Constraints. In: Proceedings of the 2008 International Conference on Communications (ICC), Bejing, China (2008)

Optimal Tree Structures for Group Key Tree Management Considering Insertion and Deletion Cost[*]

Weiwei Wu[1], Minming Li[2], and Enhong Chen[3]

[1] USTC-CityU Joint Research Institute
Department of Computer Science, University of Science and Technology of China,
Department of Computer Science, City University of Hong Kong
wweiwei2@cityu.edu.hk
[2] Department of Computer Science, City University of Hong Kong
minmli@cs.cityu.edu.hk
[3] Department of Computer Science, University of Science and Technology of China
cheneh@ustc.edu.cn

Abstract. We study the optimal structure for group broadcast problem where the key tree model is extensively used. The objective is usually to find an optimal key tree to minimize the cost based on certain assumptions. Under the assumption that n members arrive in the initial setup period and only member deletions are allowed after that period, previous works show that when only considering the deletion cost, the optimal tree can be computed in $O(n^2)$ time. In this paper, we first prove a semi-balance property for the optimal tree and use it to improve the running time from $O(n^2)$ to $O(\log \log n)$. Then we study the optimal tree structure when insertion cost is also considered. We show that the optimal tree is such a tree where any internal node has at most degree 7 and children of nodes with degree not equal to 2 or 3 are all leaves. Based on this result we give a dynamic programming algorithm with $O(n^2)$ time to compute the optimal tree.

1 Introduction

Many recent works have researched group broadcast problem due to its cost effectiveness in the applications requiring content security. The applications based on multicast can be divided into two types, one-to-many (e.g., television broadcast, pay per view) and many-to-many (e.g., teleconference, collaborate work, distributed interactive game). All of them require content security which means only authorized users are allowed to access the data broadcasted. Moreover, to

---

[*] This work was supported in part by the National Basic Research Program of China Grant 2007CB807900, 2007CB807901, a grant from the Research Grants Council of the Hong Kong Special Administrative Region, China [Project No. CityU 116907], Program for New Century Excellent Talents in University (No.NCET-05-0549) and National Natural Science Foundation of China (No.60775037).

X. Hu and J. Wang (Eds.): COCOON 2008, LNCS 5092, pp. 521–530, 2008.

deliver messages on insecure channels, we should guarantee confidentiality when users dynamically change in the group with the help of cryptography. Two kinds of confidentiality are usually considered in the literature: future confidentiality (to prevent users deleted from the group from accessing any future keys which will be used to encrypt data) and past confidentiality (to prevent users newly added into the group from accessing any past keys used to encrypt data).

To satisfy these security requirements, the basic strategy is to update the group key whenever a user is deleted from or added into the group. The group controller (GC) will maintain a key structure for the whole group. A recent survey on key management for secure group communications can be found in [2]. However, since encryption is most time consuming, the critical problem is how to decrease the number of encryptions when group members dynamically change. Key tree model proposed by Wong et al. [7] is widely used for key management problem. In this model, it is assumed that a leaf node represents a user and stores his individual key and an internal node stores a key shared by all leaf descendants of that node. The above two assumptions imply that every user possesses all the keys along the path from the leaf node to the root. Whenever a new user is added or deleted, the GC will update the keys along the path in a bottom-up fashion and notify the subset of nodes who share the keys. Wong et al. [7] pointed out that the average measured processing time increases linearly with the logarithm of the group size. Soneyink et al. [5] proved any distribution scheme has a worst-case cost of $\Omega(\log n)$ either for adding or for deleting a user. They also proved that the updating cost of a key tree with n insertions followed by n deletion is $\Theta(n \log n)$. Chen et al. [9] studied the structure of the optimal tree when a deletion sequence is performed. They show that the optimal tree is a tree where any internal node has at most degree 5 and children of nodes with degree not equal to 3 are all leaves. Based on this observation, they designed a dynamic programming algorithm of $O(n^2)$ time to compute an optimal tree. Another scenario where rekeying is done only periodically instead of immediately is studied as batch rekeying strategy in [4]. This strategy is further investigated in [8],[1] and [3] under the assumption that each user has a fixed probability p of being replaced during the batch period.

We further investigate the scenario proposed in [9]. Firstly, we find an important property of the optimal tree when only deletion cost is considered and use it to improve the time for computing the optimal tree from $O(n^2)$ to $O(\log \log n)$. Secondly, we study the optimal tree structure when insertion and deletion cost are simultaneously taken into consideration. Suppose in the initial setup period, the group only accepts membership joins. In the end of this period, a certain key tree is established by multicasting certain encryptions to the users. After that period, the group closes new membership and only accepts membership leaves. Notice that the GC update keys only at the end of the initial setup period and whenever a user leaves afterwards. This is different from the scenario of n insertions followed by n deletions considered in [5] where each insertion triggers an update on the key tree. We show that when considering both key tree establishment cost and deletion cost, the optimal tree is such a tree where any internal node has at most degree 7 and children of nodes with degree not equal to 2 or 3 are all leaves.

The rest of this paper is organized as follows. In Section 2 we review the definition of the key tree model and introduce some related results. In Section 3, we prove an important property of the optimal tree for n deletions and reduce the computation time of the optimal tree from $O(n^2)$ to $O(\log \log n)$. In Section 4 we investigate a more general cost definition where the cost of the key tree establishment is also included. We prove the degree bound for the optimal tree in this scenario and propose an $O(n^2)$ dynamic programming algorithm. Finally, we conclude our work in Section 5. Due to space limit, most of the proofs are omitted in this version.

2 Preliminaries

In this section, we review the key tree model which is referred to in the literature either as key tree [7] or LKH (logical key hierarchy) [6].

In the key tree model, a group controller (GC) maintains the key structure for all the users. A group member holds a key if and only if the key is stored in an ancestor of the member. When a user leaves, GC will update any key that is known by him in a bottom-up fashion, and notify the remaining users who share that key. Take the structure shown in Figure 1 as an example, the user u_1 holds keys (k_1, k_6, k_8), and u_2 holds keys (k_2, k_6, k_8). When u_1 leaves, keys k_6 and k_8 need to be updated. GC will first encrypt the new k_6 with k_2 and multicast the message to the group. Notice that only u_2 can decrypt that message. Then GC encrypts the new k_8 with k_5, k_7 and with the new k_6 separately and multicast them to the group. Notice that all users except u_1 can obtain the new k_8 by decrypting one of those messages. Hence, the GC need 4 encryptions to maintain the key tree structure when u_1 leaves. Based on the updating rule, we introduce the definition of deletion cost.

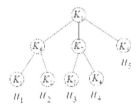

Fig. 1. An example key tree structure for a group with 5 members

Definition 1. *In a key tree T with n leaves, we say a node v has degree d_v if v has d_v children. We denote the set of ancestors of v as $anc(v)$ (not including v itself) and define the ancestor weight of v as $w_v = \sum_{u \in anc(v)} d_u$. The number of leaf descendants of v is denoted as n_v.*

Given a leaf v_i in a key tree T, let $v_i u_1 u_2 \ldots u_k$ be the longest path in T where u_j has only one child for $1 \le j \le k$. We define k as the *exclusive length* of v_i. Notice that when the user v_i is deleted from the group, we need not update any key on the path $v_i u_1 u_2 \ldots u_k$. Hence, we have the following deletion cost (defined in terms of the number of encryptions needed to update the keys after deletions). If not specified otherwise, we abbreviate w_{v_i} and n_{v_i} as w_i and n_i respectively.

Definition 2. *The deletion cost of v_i is $w_i - k - 1$ where k is the exclusive length of v_i, and we denote this deletion cost as c_i.*

Notice that when nodes are deleted, different deletion order may incur different deletion cost. In the tree shown in Figure 1, deletion sequence u_1, u_2, u_3, u_4, u_5 has cost $4 + 2 + 3 + 1 + 0 = 10$, while deletion sequence u_1, u_3, u_2, u_4, u_5 has cost $4 + 4 + 2 + 1 + 0 = 11$.

In our work, we further investigate the scenario where a group only accepts membership joins during the initial setup period. After that period, the only dynamic membership changes are deletions. This requires us to focus on the cost of a sequence instead of a single leaf. We first cite the following definitions from [9].

Definition 3. *In a key tree T with n leaf nodes, we define $\pi = \langle v_1, v_2, ..., v_n \rangle$ as the sequence of all nodes to be deleted in T. Let $\langle c_1, c_2, ..., c_n \rangle$ be the resulting sequence of deletion cost when the deletion sequence was performed. Let $C(T, \pi) = \sum_{i=1}^{n} c_i$ denote the deletion cost of the whole tree T under the deletion sequence π. The worst case deletion cost of the tree T is denoted as $C_{T,deletion} = \max_{\pi} C(T, \pi)$. We define the optimal tree $T_{n,opt}$ as a tree (not necessarily unique) which has the minimum worst case deletion cost over any tree T containing n leaf nodes.*

Definition 4. *Let T be a tree with n leaves. Given a tree T' with r leaves, we call T' a **skeleton** of T if T can be obtained by replacing r leaf nodes $v_1, v_2, \ldots, v_r$ of T' with r trees $T_1, T_2, \ldots, T_r$, where T_i has root v_i for $1 \leq i \leq r$.*

Under this definition, they proved a recursive formula for the worst case deletion cost $C_{T,deletion}$. Let T' be a skeleton of T as defined above. Given a deletion sequence π' for T' as well as a deletion sequence π_i for each T_i $(1 \leq i \leq r)$, they derive a deletion sequence π for T as follows. In the first step, π deletes all leaves in subtree T_i in the order specified by π_i, until there is only 1 leaf left. In the second step, π deletes the sole remaining leaf of each T_i in the order specified by π'. They denote the deletion sequence for T derived this way by $\pi = \langle \pi_1, \ldots, \pi_r, \pi' \rangle$.

Lemma 1. *[9] The sequence $\pi = \langle \pi_1, \pi_2, \ldots, \pi_r, \pi' \rangle$ is a worst-case deletion sequence for T if π_i is a worst-case deletion sequence for T_i and π' is a worst-case deletion sequence for T'. The worst case deletion cost for T is*

$$C_{T,deletion} = C_{T',deletion} + \sum_{i=1}^{r} (C_{T_i,deletion} + (n_i - 1)w_i). \tag{1}$$

In this formula, $C_{T',deletion}$ is the worst-case deletion cost for the skeleton T', and $C_{T_i,deletion}$ is the worst-case deletion cost for the subtree T_i. The values n_i and w_i are the abbreviations of n_{v_i} and w_{v_i} respectively.

3   Semi-balance Property of Key Tree Structure for n Deletions

As [9] shows, to minimize the worst case deletion cost when all n subscribers are deleted from the group one by one, an optimal tree can be found among the trees satisfying the following two conditions: (1) every internal node has degree $d \leqslant 5$ and (2) children of nodes with degree $d \neq 3$ are all leaves. According to this, they gave an algorithm to compute the optimal tree in $O(n^2)$ time. In this section, we prove an important semi-balance property of the optimal tree.

According to the result of [9], if a node v has at least one child being an internal node, it must have degree 3. We further prove the following lemma.

Lemma 2. *There is an optimal tree where the children of any degree 3 node are either all leaves or all internal nodes.*

Proof. Suppose on the contrary in the optimal tree there exists an internal node v with degree 3, which has an internal node child v_1 and a leaf child v_0, then v_1 has at most two leaf descendants. Otherwise, if v_1 has $n_1 \geqslant 3$ leaf descendants, we show in the following that the cost of the tree can be decreased, which contradicts the optimality of the tree. Obviously, v_1 can only have degree $2 \leqslant d_1 \leqslant 5$. Firstly, when $d_1 \geqslant 3$, we can decrease the cost by moving a subtree T_2 (rooted at v_2) of v_1 and combine it with the leaf child (v_0) of v, as shown in Figure 2. Detailed proof is omitted here. On the other hand, when $d_1 = 2$, children of v_1 are all leaves because $d_1 \neq 3$, i.e. v_1 has at most two leaf descendants. Therefore, v has $n_v \leqslant 5$ leaf descendants, and in this case the optimal tree is such a tree where all v's children are leaves.

As a result, if a node has at least one leaf child, its children are all leaves. Furthermore, to make our explanation easier to read, we will give this kind of node a new name in the following.

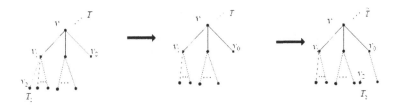

Fig. 2. A degree 3 node with at least one leaf child and one internal node child

Definition 5. *Given a tree T with L levels, we define the pseudo-leaf nodes to be the nodes whose children are all leaves. In addition, we use l_i to denote the level where the pseudo-leaf node u_i is placed and s_i to denote the number of its children.*

Obviously, all pseudo-leaf nodes can only be placed on level $1 \leqslant l_i \leqslant L - 1$ and have $2 \leqslant s_i \leqslant 5$ children.

Lemma 3. *Given a tree T and a pseudo-leaf u_i in T, if we remove one child of u_i, the cost of the resulting tree $\bar{T}$ decreases by $3l_i + s_i - 1$.*

Proof. The lemma can be proved by choosing all the nodes in T except the children of u_i as the skeleton. The details are omitted here.

Lemma 4. *In an optimal tree $T_{n,opt}$, if a pseudo-leaf node u_i satisfies $1 \leqslant l_i \leqslant L - 2$, then we have $s_i = 5$.*

Proof. Suppose $s_i < 5$ and another pseudo-leaf node u_j is on level $l_j = L - 1$. We can get a better tree by moving a child of u_j to be a child of the node u_i when $s_i \leqslant 4$, as shown by Figure 3. The detailed analysis is omitted here.

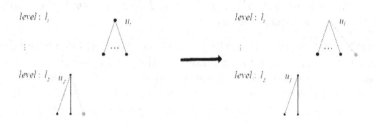

Fig. 3. Transformation of tree T where at least one pseudo-leaf node u_i with $s_i \leqslant 4$ children is on level l_i where $1 \leqslant l_i \leqslant (L - 2)$

Lemma 5. *In an optimal tree $T_{n,opt}$, for any two pseudo-leaf nodes u_i and u_j satisfying $l_j > l_i$, we have $l_j - l_i \leq 1$, which means the pseudo-leaf nodes should only be on level $L - 1$ or $L - 2$.*

Proof. Suppose on the contrary there are two pseudo-leaf nodes u_i and u_j with $l_j - l_i \geqslant 2$ in the optimal tree. According to Lemma 4, we have $s_i = 5$. We can show that the cost decreases at least by 2 after moving a child of u_j to u_i as shown in Figure 4. Details are omitted here.

Lemma 6. *If all pseudo-leaf nodes are on the same level, we have the property that any node u_i, u_j on level $L - 1$ satisfy the inequality $|n_i - n_j| \leqslant 1$ where n_i and n_j are the number of leaf descendants of u_i and u_j respectively.*

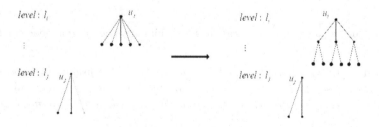

Fig. 4. Transformation of tree T where there are two pseudo-leaf nodes u_i and u_j who was respectively on level l_i and l_j, and $(l_j - l_i) \geqslant 2$

Proof. Because $n_i = s_i$ for all pseudo-leaf nodes, we only need to prove $|s_i - s_j| \leqslant$ 1 for any pseudo-leaf u_i and u_j on the same level. In this case, according to Lemma 5, all of them can only be on level $L - 1$. If $s_i - s_j \geqslant 2$, we can get a better tree by moving a child of u_i to u_j, because the cost decreased is a positive value, which is equal to $s_i - 1 - s_j \geqslant 1$. For the other case when $s_j - s_i \geqslant 2$, we can get a better tree in a similar way. Hence, the lemma is proved.

Lemma 7. *If some of the pseudo-leaf nodes are on different levels, then we have the property that for any node v_i, v_j on the same level $(L - 2$ or $L - 1)$, the inequality $|n_i - n_j| \leqslant 1$ holds.*

Lemma 8. *Given a tree T, for any subtree T_i, T_j whose root v_i, v_j have the same ancestor weight $w_i = w_j$, if we exchange these two subtrees, the cost of the resulting tree does not change.*

Notice that when all pseudo-leaf nodes are on level $L - 1$, the nodes v_i, v_j on level k $(1 \leqslant k \leqslant L - 1)$ have ancestor weight $w_i = w_j = 3k$. Therefore, the total deletion cost will not change when we exchange the subtrees with root nodes on the same level $L - 1$. For the other case when some pseudo-leaf nodes are on level $L - 2$ and others are on level $L - 1$, the nodes v_i, v_j on level $1 \leqslant k \leqslant L - 2$ also have ancestor weight $w_i = w_j = 3k$. According to this observation, for any two nodes v_i, v_j on the upper level, we prove inequality $|n_i - n_j| \leqslant 1$ also holds by exchanging the subtrees.

Lemma 9. *There is an optimal tree $T_{n,opt}$ where for any node v_i, v_j on the same level $1 \leqslant l_i = l_j \leqslant (L - 2)$, the relation $|n_i - n_j| \leqslant 1$ holds.*

Based on these lemmas, we get the following theorem:

Theorem 1. *When the number of leaves $n \geqslant 6$, there is an optimal tree $T_{n,opt}$ where the sizes of its three subtrees differ by at most 1.*

The correctness of the theorem is evident with the support of Lemma 6, 7 and 9. Theorem 1 in fact implies that we can get an optimal tree by distributing the leaves into subtrees in a semi-balanced way until all the pseudo-leaf nodes u_i satisfy $2 \leqslant s_i \leqslant 5$. This can be interpreted as the following optimal tree constructing rule:

(1) When $n \leqslant 5$, optimal tree is a one level tree with all leaves on the same level.
(2) When $n \geqslant 6$, optimal tree is a tree with root degree 3. We distribute n leaves into its three subtrees in a semi-balanced way respectively with $\lceil \frac{n}{3} \rceil, \lceil \frac{n-1}{3} \rceil,$ $\lceil \frac{n-2}{3} \rceil$ leaves.
(3) For each subtree, recursively construct its optimal structure according to (1) and (2).

The rule above implies that the deletion cost can be computed along with the constructing process. In fact, there are $(n \mod 3^k)$ nodes which have $\lceil \frac{n}{3^k} \rceil$ leaf descendants and $(3^k - n \mod 3^k)$ nodes which have $\lceil \frac{n}{3^k} \rceil - 1$ leaf descendants. Notice that when $2 \leqslant \lceil \frac{n}{3^k} \rceil \leqslant 5$ these nodes are pseudo-leaf nodes, while the optimal structure when $\lceil \frac{n}{3^k} \rceil = 6$ is the tree where root degree is 3 and each child has 2 leaves. Hence, we have

Theorem 2. *The worst case deletion cost of the optimal tree $T_{n,opt}$ can be computed in $O(\log \log n)$ time according to the equation below.*

$$T_{n,max} = (n \mod 3^k) \cdot T(\lceil \frac{n}{3^k} \rceil) + (3^k - n \mod 3^k) \cdot T(\lceil \frac{n}{3^k} \rceil - 1) + 3n \cdot k - 3^{k+1} + 3$$

$$where \quad k = \begin{cases} \lceil \log_3 n \rceil - 1 & if \ 2 \leqslant \lfloor \frac{n}{3^{\lceil \log_3 n \rceil - 1}} \rfloor \leqslant 5, \\ \lceil \log_3 n \rceil - 2 & if \ 2 \leqslant \lfloor \frac{n}{3^{\lceil \log_3 n \rceil - 2}} \rfloor \leqslant 5. \end{cases}$$

For basic cases $2 \leqslant n \leqslant 5$, we have $C(2) = 1, C(3) = 3, C(4) = 6, C(5) = 10$. Since it needs $O(\log \log n)$ time to compute the value of $3^{O(\log n)}$, the cost of the optimal tree can be computed in $O(\log \log n)$ time, which is much better than the dynamic programming algorithm with $O(n^2)$ time. Furthermore, to construct the optimal tree, we can distribute the users into subtrees in a semi-balanced way.

4   Optimal Tree Structures for n Insertions Followed by n Deletions

In this section, we investigate a more general setting where the cost of the initial group setup is also considered. In this new setting, the optimal tree we computed in the previous section is probably no longer optimal. We aim to study the optimal tree structure to minimize the cost for the initial setup followed by n deletions.

Lemma 10. *The number of encryptions needed to build the initial tree equals $N - 1$ where N is the number of the nodes in the tree.*

Proof. The tree is built after all the members arrive, we distribute the keys to the users securely in the bottom-up fashion with respect to the tree. Therefore, every key except the key stored in the root will be used once as an encryption key in the whole process, which amounts to $N - 1$ encryptions in total. The lemma is then proved.

To represent the total cost of insertion and deletion in one formula, we modify the cost of *skeleton* a bit as follows.

Definition 6. *Suppose the skeleton T' has t non-root nodes, the worst-case cost for T' is $C_{T',max} = C_{T',deletion} + t$.*

In fact, $C_{T',deletion}$ is the worst-case cost when only deletion is considered, and t is the number of messages encrypted by the keys stored in the t non-root nodes when establishing the initial key tree.

Lemma 11. *If the skeleton T' of T has r leaves, the worst-case cost for T is $C_{T,max} = C_{T',max} + \sum_{i=1}^{r}(C_{T_i,max} + (n_i - 1)w_i)$.*

Proof. By distributing the initial setup cost to every non-root node in the tree, it is easy to see that this lemma is implied by Lemma 1. We omit the detailed proof here.

We then prove the following structural property of the optimal tree which then enables us to find an optimal tree in $O(n^2)$ time. We omit our proof of this property in this version due to space limit.

Lemma 12. *There is an optimal tree $T_{n,opt}$ where every internal node v has degree at most 7 and children of nodes with degree not equal to 2 or 3 are all leaves.*

Based on this lemma, we have the following theorem:

Theorem 3. *Algorithm 1 can compute an optimal tree in $O(n^2)$ time.*

Proof. In Algorithm 1, we use R_i to denote the minimum cost of trees with i leaves and root degree restricted to be 3, D_i to denote the minimum cost of trees with i leaves and root degree restricted to be 2, and C_i to denote the minimum cost of all the trees with i leaves. The analysis is similar with the dynamic programming algorithm in [9] and omitted here.

Algorithm 1. *Sequence_OPT*

---

1. $R_1 = 1$; $R_2 = 3$; $R_3 = 6$; $R_4 = 10$; $R_5 = 15$; $R_6 = 21$; $R_7 = 28$;
2. $D_1 = 1$; $D_2 = 3$; $D_3 = 8$; $D_4 = 13$; $D_5 = 18$; $D_6 = 23$; $D_7 = 29$;
4. $C_1 = 1$; $C_2 = 3$; $C_3 = 6$; $C_4 = 10$; $C_5 = 15$; $C_6 = 21$; $C_7 = 28$;
5.   for $i = 8$ to n
6. $D_i = i^2$; $R_i = i^2$; $C_i = i^2$;
7.   for $k1 = 1$ to $i/2$
8. $k2 = i - k1$;
9.   if $D_i > C_{k1} + C_{k2} + 2 \cdot i - 1$ then
10. $D_i = C_{k1} + C_{k2} + 2 \cdot i - 1$;
11.   if $R_i > C_{k1} + D_{k2} + i + 2 \cdot k1 - 2$ then
12. $R_i = C_{k1} + D_{k2} + i + 2 \cdot k1 - 2$;
13.   end for
14. $C_i = min(D_i, R_i)$;
15. end for

---

5   Conclusion

While many works focus on fixing the cost bound under some multicast protocol, we try to find the optimal structure to minimize the cost. We investigate the scenario where the members all arrive in the initial setup time and then leaves one by one. This can be applied in teleconferencing or applications where the member list can be fixed beforehand. Chen et al. [9] found the optimal tree structure when only deletion cost is considered. We prove an important property of the optimal key tree based on their work. We show that the members can be distributed in a semi-balance way in the optimal tree. Using this property we improve the running time from $O(n^2)$ to $O(\log \log n)$. We then focus on the optimal tree structure when insertion cost for the initial period is simultaneously considered. We obtain a recursive formula and use it to eliminate the impossible degrees in the optimal tree. Based on this observation, we give an algorithm to compute the optimal tree with $O(n^2)$ time.

One possible direction of the future work is to investigate whether there is similar balanced structure for the optimal tree when insertion cost of the initial setup is considered together with the deletion cost.

References

1. Graham, R.L., Li, M., Yao, F.F.: Optimal Tree Structures for Group Key Management with Batch Updates. SIAM Journal on Discrete Mathematics 21(2), 532–547 (2007)
2. Goodrich, M.T., Sun, J.Z., Tamassia, R.: Efficient Tree-Based Revocation in Groups of Low-State Devices. In: Franklin, M. (ed.) CRYPTO 2004. LNCS, vol. 3152, pp. 511–527. Springer, Heidelberg (2004)
3. Li, M., Feng, Z., Graham, R.L., Yao, F.F.: Approximately Optimal Trees for Group Key Management with Batch Updates. In: Proceedings of the Fourth Annual Conference on Theory and Applications of Models of Computation, pp. 284–295 (2007)
4. Li, X.S., Yang, Y.R., Gouda, M.G., Lam, S.S.: Batch Re-keying for Secure Group Communications. In: Proceedings of the Tenth International Conference on World Wide Web, pp. 525–534 (2001)
5. Snoeyink, J., Suri, S., Varghese, G.: A Lower Bound for Multicast Key Distribution. In: Proceedings of the Twentieth Annual IEEE Conference on Computer Communications, pp. 422–431 (2001)
6. Wallner, D., Harder, E., Agee, R.C.: Key Management for Multicast: Issues and Architectures. RFC 2627 (1999)
7. Wong, C.K., Gouda, M.G., Lam, S.S.: Secure Group Communications Using Key Graphs. IEEE/ACM Transactions on Networking (8)(1), 16–30 (2000)
8. Zhu, F., Chan, A., Noubir, G.: Optimal Tree Structure for Key Management of Simultaneous Join/Leave in Secure Multicast. In: Proceedings of Military Communications Conference, pp. 773–778 (2003)
9. Chen, Z.-Z., Feng, Z., Li, M., Yao, F.F.: Optimizing Deletion Cost for Secure Multicast Key Management. Theoretical Computer Science (2008), doi:10.1016/j.tcs.2008.03.016

Throughput Maximization with Traffic Profile in Wireless Mesh Network*

Hejiao Huang and Yun Peng

Harbin Institute of Technology Shenzhen Graduate School, Shenzhen, China
{hjhuang,pengyun}@hitsz.edu.cn

Abstract. Wireless mesh networks (WMNs) are becoming increasingly common for applications, according to its multi-radio and multi-channel advantage over their counterpart, wireless LANs. In WMNs, multiple simultaneous communications over multi-radios using orthogonal channels further improve traffic throughput. Certainly, effective routing and channel assignment are critical, at the same time. Recently, a number of profile-based routing algorithms have emerged, using multi-commodity network flow. Traffic profile used in them records the QoS requirements. But expectant bandwidth requirements may not be assured in wireless condition. Hence, in this paper, we demonstrate a scheme to compute the maximal possible guaranteed bandwidth. Our evaluation demonstrates that our algorithm performs much better than the famous Shortest Path Routing algorithm in routing traffic profile on Grid networks.

1 Introduction

New emerged Wireless Mesh Networks (WMNs) depending on its many excellent performances are being applied as a solution for providing last-mile Internet access for mobile clients located on the edge of the wired network. Many of these networks have been pushed to the market to provide complete commercial solution [11]. A WMN is composed of a network backbone consisting of wireless mesh routers and network clients. As in Ad-Hoc networks, routers of the network backbone can relay messages as an intermediate. But differently mesh routers are rarely mobile that assures infrequent topology changes, limited node failures etc. Also mesh routers have no power constrain. However, the "lazy" backbone does not confine motivation of mobile clients. In WMN, each wireless router may aggregate traffic flows for a large number of mobile clients in its coverage. Since these characters, the traffic load of each mesh router changes infrequently. Under this condition yesterday's traffic may predict today's traffic. Hence traffic profile is introduced as quasi-static information for throughput optimization. Efficacious utilization of it contributes to practical online traffic deliverance [10],[1]. In addition a subset of wireless routers is equipped with a gateway capacity connecting with the wired network. In practice almost all traffic loads come from

---

* This work is financially supported by the National Natural Science Foundation of China with grant No. 10701030 and National High-Tech R&D Program (863 Program) with grant No. 2006AA01Z197.

X. Hu and J. Wang (Eds.): COCOON 2008, LNCS 5092, pp. 531–540, 2008.

the Internet. Thus traffic load is mainly routed between the wireless clients and the gateway nodes. More precisely, traffic load is mainly routed from the gateway nodes to other mesh routers in WMN backbone network. Traffic profile between (source, destination) pairs can be measured by the service provider using a measurement-based mechanism such as policy constrains and service agreements. The main difference between WMN and traditional wireless network is that each router is equipped with multi-radios. Different radios with orthogonal channels allow one router communicates with more than one neighbor at one point of time. In this way traffic throughput may be greatly improved. However, orthogonal channels are limited in number, implying that interference reduce is always a critical issue in WMNs. Links may be assigned channels for communication between neighbors in a static or dynamic fashion. In the first approach, every link between neighbors is combined with a particular channel and this channel does not vary to another over time. However in the second approach a link can be assigned different channels in different time slots. The former is simpler than the latter. Nevertheless, with the growth of network size, any static assignment may likely result in the increase of interference. This phenomenon inevitably degrades the performance of the mesh network. But dynamic channel assignment can alleviate it greatly. In this paper the channel assignment problem is discussed and our work aims at a dynamic channel assignment fashion in WMN. We present a centralized algorithm jointing interference free channel assignment and routing in the next sections. The motivation of our research arises from the problem that given several end-to-end bandwidth demands, there is one design goal to maximize guaranteed bandwidth, while ensuring the demand fairness among (source, destination) pairs. Many previous works regard it as a multi-commodity network flow problem [14],[15]. Flows are allocated to edges and routing is based on this information. However, traffic profile is not utilized efficiently enough. That implies their works only achieve a small fraction of the possible maximal bandwidth. Hence, in this paper, we demonstrate a scheme to compute the maximal guaranteed bandwidth.

The contributions and significance of this paper are as follows.

- A routing and channel assignment algorithm is developed assuring maximal throughput of traffic profile. In other words, guaranteed bandwidth is maximized.
- Routing, channel assignment and scheduling are well integrated while assuring fairness among (source, destination) pairs.
- Resource contention graphs are ingeniously used in our optimal algorithm.
- Our framework is very general and can be extended to multi conditions.

The remainder of this paper is organized as follows. Related works are discussed in Section 2. Problem description and related terminologies are proposed in Section 3. In Section 4 we demonstrate an analytical framework. Multi-commodity flow processing is discussed in Section 5. In Section 6, we propose Interference free channel assignment algorithm. Evaluation of our algorithm is presented in Section 7 and we conclude our work in Section 8.

2   Related Work

In this section we discuss several relative previous works. When traffic condition remains quite stable, traffic profile can improve network performance greatly. Profile-based routing performs much better than load balancing routing and load packing routing [10]. Also, as mentioned before, traffic flows on WMN backbone changes infrequently. Hence, traffic profile contributes to throughput maximization in wireless mesh network in the same way. [13] introduces an iterative channel assignment algorithm aiming at routing a predefined traffic profile. [8] studies topology control algorithms that achieve good network throughput for a given traffic profile. A. Kashyap et al. in [7] study the problem of topology control and multi-path routing for maximizing throughput for a given traffic profile and the single-path routing in [6]. Traditionally, traffic load delivering task is regarded as a multi-commodity network flow problem [14],[15]. Besides this, [16] proposes both an optimal algorithm based on solving a linear programming (LP) and a simple heuristic to route given traffic profile. Distributed channel assignment and routing algorithm is developed in [12]. In [12], each node has a corresponding gateway node and sends all packets for Internet to that node. In addition, traffic costs to reach that gateway vary depending on residual bandwidth to achieve load balancing. In a particular kind of network, Differentiated Services Networks (DiffServ), [9] design an enhanced marking algorithm for assuring TCP and UDP flows rate and fair excess bandwidth distribution. In addition, [5] discusses the traffic profile with infinite amount of flows. [4] studies the problem with dynamic traffic. However, in this paper, we only focus on the finite static traffic profile. Considering channel number limitation, interference-free routing can not be achieved without scheduling. In [2] the authors present a multi-hop packet scheduling mechanism to achieve maximum throughput. Traditionally interference free schedule is transformed into coloring problem (graph theory), which is NP-Complete. In this paper we study it in another point of view.

3   Problem Description and Related Terminologies

In general, a wireless mesh network is modeled as a communication graph $G = (V, E)$, $V \subset \Re \times \Re$ is a set of nodes representing the routers and $E \subset V \times V$ is a set of edges representing the links between routers. $\forall (u, v) \in E$ iff the transmission disk of node u covers node v, vice versa. In addition, we borrow the definition of coverage from [3]. Coverage of edge e, $Cov(e)$ represents the set of nodes that can be affected by edge e. Formally, $Cov(e) = \{w \in V \mid w$ is within $D(u, |uv|)\} \cup \{w \in V \mid w$ is within $D(v, |vu|)\}$, where $|uv|$ is physical length of edge (u, v) and $D(u, |uv|)$ denotes the disk centered at node u with radius $|uv|$. Edge $e_1 = (a, b)$ interferes with edge $e_2 = (u, v)$ iff $u \in Cov(e_1)$ or $v \in Cov(e_2)$, vice versa.

In WMN, multi-orthogonal channels can be used for wireless communication and each node u is equipped with multi-radios. We focus on homogeneous network, which implies that routers are equipped with the same radios. Let R denote the set of radios of a router and C denote the set of orthogonal channels. Without specification, channels mean orthogonal channels for simple. In addition, $\forall e \in E$ has maximal transmission capacity $P(e)$.

As in [15], traffic request is defined as a 4-tuple (i, s_i, d_i, m_i), where i is the request ID, s_i is the ingress router (supply router), d_i is the egress router (demand router), m_i is bandwidth required for the Label Switched Path (LSP). These information (recorded in traffic profile) denotes the flow expected by (ingress, egress) pairs.

The task of *topology control* is to generate a sub-graph (called topology graph) $G' \subset G$ satisfying some constrains and routing is to search for a path on G' for every request recorded in traffic profile. Intuitively, well designed topology graph is a prerequisite of efficient routing. In this paper we focus on routing with splitting flows which assumes that traffic demands are splittable.

Edges must be assigned channels for communication. If we use (e, c) pair to denote that edge e is assigned channel c, *channel assignment* is to compute a subset of $E \times C$, satisfying some constrains. However, due to the limitation of channel number, it is inevitable that two edges interfere with each other but both are assigned the same channel. They can not communicate at the same point of time. Hence, scheduling is needed to allocate them into different time slots. In this way, no interference occurs. Conclusively, if T denotes the set of time slots, interference free channel assignment is to compute a subset of $E \times C \times T$. Tuple (e, c, t) means edge e is assigned channel c in t^{th} time slot. Also for statement simplicity, we assume only one unit of flow is transmitted through an edge during one time slot. The problem studied in this paper is to compute a routing on a given communication graph to maximize guaranteed bandwidth, jointing channel assignment and scheduling. That is to say, traffic requests should be delivered as fast as possible. This problem is described as follows:

Given a graph $G = (V, E)$ and a traffic profile, the Minimum Time Deliverance Problem (MTDP) is to compute a interference free routing, s.t. traffic profile is routed within minimum time.

4   A General Analytical Framework for MTDP

We now present the general analytical framework for modeling MTDP in WMN. It is a 4-steps process. In this paper, MTDP is regarded as a multi-commodity network flow problem, which requires a directed graph. Hence, we need to transfer the undirected communication graph to an auxiliary directed graph without logical lost.

step 1. Given a communication graph $G = (V, E)$, we generate an auxiliary directed graph $G^A = (V, E^A)$. For $\forall (a, b) \in E$, directed edges (a, b) and $(b, a) \in E^A$. This graph is used in ensuing steps. Multi-commodity network flow problem are always solved through linear programming (LP). Through it, commodities are distributed onto edges. In other words, traffic routing is generated. Removing edges having no flows, we get a topology graph.

step 2. Traffic requests are regarded as several kinds of commodities. Applying LP on G^A, we get a topology graph $G' = (V, E')$ and a set F of flows on G'.

Fig. 1. Original Graph

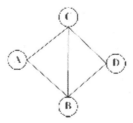

Fig. 2. Interference Graph

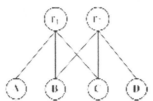

Fig. 3. Resource Contention Graph

As a resulting topology graph, Fig. 1 contains 5 nodes and 4 edges. Capitals on edges denote flows and ellipses reflect interference range. Although the traffic between (ingress, egress) pair is split onto multi-path, packets associated with a network connection still follow a single path to avoid TCP re-ordering.

Through Step2, flows are allocated up. However, they can not be transmitted simultaneously, since interference occurrence. Hence, for maximum throughput we have to design a channel assignment and scheduling scheme. Firstly, we construct an interference graph denoting the interference relationship between flows (Step3). Then from it we construct a resource contention graph, which our channel assignment and scheduling algorithm is based on (Step4).

step 3. From G' and F, we construct an interference graph $G'' = (V'', E'')$. Vertices (called flow vertices) correspond to flows in F. Two vertices are connected if they interfere with each other. Fig. 2 is an interference graph of Fig. 1.

step 4. From the interference graph, we construct a resource contention graph, which is composed of maximum cliques of G' representing distinct contention regions of network. Channel assignment and scheduling are based on this kind of graph. Details of this kind of graph are as follows.

Resource Contention Graph

A resource contention graph, $RCG = (RV, ET, V'')$ is a bipartite graph. RV is a resource vertex set representing contention regions and each one in RV holds multi-resource (channels), V'' is flow vertex set (generated in step 3). A flow vertex is connected with resource vertex r, if it is within contention region r. Size of contention region r, $Size(r)$ is sum of flows within it.

For example, Fig. 2 contains two maximum cliques A, B, C and B, C, D. That is to say the topology graph has two contention regions $r_1 = \{A, B, C\}$ and $r_2 = \{B, C, D\}$. Hence, in Fig. 3 two resource vertices r_1 and r_2 are connected with A, B, C and B, C, D, respectively.

By this way, topology graphs are transferred to RCGs. Problems on topology graphs can be considered through this new kind of graph.

Theorem 1. *Given a resource contention graph $RCG = (RV, ET, V'')$, let T^* denote the time cost for delivering all flows. When $|R| \geq |C|$, we have $T^* \geq \max_{r \in RV} (Size(r_i)/|C|)$, where R is the set of radio of a router and C is available channels set.*

Proof: We select the largest contention region r out. Firstly, we assume $T^* < Size(r)/|C|$. That is to say, flow in contention region r is delivered within $Size(r)/|C|$ time slots. $|R| \geq |C|$ assures that $|C|$ edges simultaneous communication is available in each time slot. Based on our assumption only one unit flow is transferred through an edge in a time slot, $|C|$ edges simultaneous communication can deliver $|C|$ units flow in a time slot at most. That implies $Size(r)$ flows can not be delivered within $Size(r)/|C|$ time slots. It is a contradiction.

Intuitively, if $|R| < |C|$ channel resources may not be fully utilized because of radio limitation. However, when $|R| \geq |C$, it is assured that in each time slot all $|C|$ channels can be slotted onto flows. You may ask if this requirement is available. As we know, IEEE 802.11 can satisfy 3 or 12 channels. Hence, in practice, we can set $|R| \geq |C| = 3$ satisfying IEEE 802.11 requirement.

5   Multi-commodity Flow Processing

Given a communication graph $G = (V, E)$, k traffic requests are regarded as k kinds of commodities. Our goal is to deliver them within minimal time. $\forall u \in V$ is associated a number $b(u)$ indicates its supply or demand depending on whether $b(u) > 0$ or $b(u) < 0$. If node u is a transshipment node, $b(u) = 0$. Let a real valued variable $f_i(e)$ denote the amount of the ith commodity routed through edge e. Then total flow on edge e can not exceed its capacity, $f(e) = \sum f_i(e) \leq P(e)$. Hence, by Theorem 1, we have:

$$\min \max(Size(r_i), r_i \in RV$$

$$\text{subject to} \quad \sum_{(a,b) \in E} \sum_i f_i(a, b) - \sum_{(b,a) \in E} \sum_i f_i(b, a) = b(a) \qquad (1)$$

$$0 \leq \sum_i f_i(a, b) \leq P(a, b), \forall (a, b) \in E. \qquad (2)$$

(1) is a balance constraint. This equation means that flow income (inflow minus outflow) of node a equals flow demand of node a. (2) is load constrain. It assures that total flow of edge e can not exceed capacity of e.

Through this LP, topology graph G' and flow set F is founded. They are used as input of our channel assignment algorithm, which is proposed in next section.

6   Interference Free Channel Assignment

In this section, we demonstrate our Interference Free Channel Assignment (IFCA) algorithm. In this algorithm, $RL(r)$ denotes a descending list of flow vertices within contention region r. Elements are sorted by flow.

IFCA algorithm
Input: A list L recording all contention regions

1. Time=0
2. While Size$(r) \neq 0$ for $\forall r \in L$
3.   For $m = 1$ to $|C|$
4.   DistriChanl(m)
5.   End For
6.   Time++
7.   Update L
8. End While

DistriChanl(c)

9 variant $e_1 = $ an edge of L[1] having the largest flow
10 add e_1 into a temp set
11 For i from 2 to length(L)
12     $j = 1$
13   variant $e_i = $ the jth element in $RL(L[i])$
14   While e_i interfere with any one within the temp set
15   If e_i is not the last element in $RL(L[i])$
16   j++
17             $e_i = $ the jth element in $RL(L[i])$
18   End If
19   Else
20   Go to 16
21   End Else
22   End while
23   add e_i into the temp set
24 End for
25 output (e, c, Time), for $\forall e \in $ the temp set
26 $f(e) = f(e) - 1$, for $\forall e \in $ the temp set
27 clear elements within the temp set

Update L

28 For $i = 1$ to length(L)

29   If Size($L[i]$)=0

30   ReplaceL$[i][i + 1]byL[i + 1][i + 2]\cdots$

31   length(L)=length(L) - 1

32   End If

33 End For

Theorem 2. *IFCA algorithm is correct and output Time*=$\max_{r_i \in RV}$ *($Size(r_i)/|C|$).*

Proof: If each contention region reduces $|C|$ units of flow in one While iteration 2-8, above equation is satisfied. Further more; it is assured if each contention region reduces one units of flow in one For iteration 3-5. Hence, let's consider the sub-process DistriChanl(c) in For iteration. In lines 11-24, from the first contention region to the last one, an edge is selected if it does not interfere with previous "brothers". However, if all edges within a contention region interfere with previous brothers, one previous brother must be within this contention region. That assures one unit flow reduction.

7   Evaluation

In this section, we evaluate the performance of our framework proposed in section IV and compare it with the well-known routing algorithm Shortest-Path (SP). In this simulation, we consider a 4 5 grid topology with 20 nodes. Each node has at most 4 neighbors in the grid. We vary the number of channels and traffic load respectively to study their impacts on the network performance. The capacity of links is fixed as 60 units. There are 5 (ingress, egress) pairs in this network. The ingress and egress nodes are both chosen randomly.

The multi-commodity network flow problem is solved by Matlab 7.0 generating a flow set and a topology graph whose contention regions are minimized. They are input into our IFCA algorithm proposed in section V. Performance comparison with Shortest Path Routing algorithm is presented as follows.

7.1   The Performance Impact of Traffic Load

We assume there are 3 channels for use and traffic load vary from 10 to 60 units. In this experiment, time cost for traffic transmission of our scheme (SP) is 5.0000 (16.1523), 9.8333 (28.9520), 15.0000 (42.0132), 19.6670 (53.2332), 25.0000 (65.9981), and 29.3333 (78.2133), respectively. Correspondingly, the throughput comparison is show in Fig. 4. We can see that IFCA algorithm performs much better than SP and performance is very stable while traffic load growing.

7.2   The Performance Impact of Channels

In this experiment, we assume traffic load is 60. We vary the number of channel from 1 to 6. Time cost for traffic transmission of our scheme (SP) is 88.1024 (232.5461), 44.2013 (116.2745), 30.0106 (77.9680), 23.2500 (59.8762), 19.4000 (48.0116) and 16.8333 (41.2540), respectively. Correspondingly, the throughput comparison is show in Fig. 5. Throughput of IFCA is higher than SP. Further more, it rises faster with channels number growing.

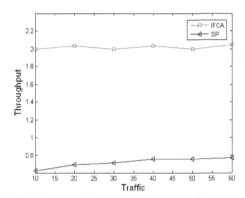

Fig. 4. Throughput Comparison with Traffic Growing

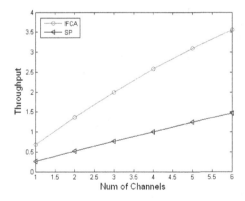

Fig. 5. Throughput Comparison with Channels Num Growing

8   Conclusion

In this paper, we study the problem of integrating routing and channel assignment aiming to maximize guaranteed bandwidth facing given QoS requirements. It is regarded as a multi-commodity network flow problem. Through the solving LP step, flows are distributed into several contention regions. Further more, based on the resource contention graph, IFCA algorithm assigns channels and schedules them such that traffic requests are routed within minimum time. In addition, our evaluation demonstrates that IFCA algorithm performs much better than the famous Shortest Path routing algorithm.

References

1. Alicherry, M., Bhatia, R., Li, L.: Joint Channel Assignment and Routing for Throughput Optimization in Multi-radio Wireless Mesh Networks. In: Proc. of the 11th Annual International Conference on Mobile Computing and Networking, pp. 58–72 (2005)
2. Bahl, P., Chandra, R., Dunagan, J.: SSCH: Slotted Seeded Channel Hopping For Capacity Improvement in IEEE 802.11 Ad-Hoc Wireless Networks. In: ACM MobiCom 2004, pp. 216–230 (2004)
3. Burkhart, M., von Rickenbach, P., Wattenhofer, R., Zollinger, A.: Does Topology Control Reduce Interference? In: MobiCom 2004, pp. 24–26 (2004)
4. Crichigno, J., Wu, M., Shu, W.: Protocols and Architectures for Channel Assignment in Wireless Mesh Networks. Ad Hoc Net (2007)
5. Kalantari, M., Kashyap, A., Lee, K., Shayman, M.: Network Topology Control and Routing under Interface Constraints by Link Evaluation. In: Conference on Information Sciences and Systems (2005)
6. Kashyap, A., Kalantari, M., Lee, K., Shayman, M.: Rollout Algorithms for Topology Control and Routing of Unsplittable Flows in Wireless Optical Backbone Networks. In: Conference on Information Sciences and Systems (2005)
7. Kashyap, A., Khuller, S., Shayman, M.: Topology Control and Routing over Wireless Optical Backbone Networks. In: Conference on Information Sciences and Systems (2004)
8. Kashyap, A., Lee, K., Kalantari, M., Khuller, S., Shayman, M.: Integrated Topology Control and Routing in Wireless Optical Mesh Networks. Computer Networks 51(15), 4237–4251 (2007)
9. Lee, S., Goh, S., Jang, M.: A Traffic Conditioning Algorithm for Enhancing the Fairness Between TCP and UDP Flows in DiffServ. In: Gavrilova, M.L., Gervasi, O., Kumar, V., Tan, C.J.K., Taniar, D., Laganá, A., Mun, Y., Choo, H. (eds.) ICCSA 2006. LNCS, vol. 3981, pp. 204–213. Springer, Heidelberg (2006)
10. Matta, I., Bestavros, A.: A Load Profiling Approach to Routing Guaranteed Bandwidth Flows. In: INFOCOM 1998, pp. 1014–1021 (1998)
11. Mesh Dynamics Inc., http://www.meshdynamics.com
12. Raniwala, A., Chiueh, T.: Architecture and Algorithms for an IEEE 802.11-based Multi-channel Wireless Mesh Network. In: Proc. of IEEE INFOCOM 2005, vol. 3, pp. 2223–2234 (2005)
13. Raniwala, A., Gopalan, K., Chiueh, T.: Centralized Channel Assignment and Routing Algorithms for Multi-Channel Wireless Mesh Networks. ACM Mobile Computing and Communications Review 8(2), 50–65 (2004)
14. Suri, S., Waldvogel, M., Bauer, D., Warkhede, P.R.: Profile-Based Routing and Traffic Engineering. Computer Communications 26, 351–365 (2003)
15. Tabatabaee, V., Kashyap, A., Bhattacharjee, B., La, R.J., Shayman, M.A.: Robust Routing with Unknown Traffic Matrices. In: Proc. of INFOCOM 2007, pp. 2436–2440 (2007)
16. Tang, J., Xue, G., Zhang, W.: Interference-Aware Topology Control and QoS Routing in Multi-Channel Wireless Mesh Networks. In: Proc. of ACM MobiHoc 2005, pp. 68–77 (2005)

Joint Topology Control and Power Conservation for Wireless Sensor Networks Using Transmit Power Adjustment*

Deying Li[1,2], Hongwei Du[3], Lin Liu[2], and Scott C.-H. Huang[3]

[1] Key Laboratory of Data Engineering and Knowledge Engineering
Renmin University of China, MOE, Beijing, China
[2] School of Information, Renmin University of China, Beijing, China
[3] Department of Computer Science, City University of Hong Kong, Hong Kong

Abstract. Topology control with individual node transmission power adjustment in wireless sensor networks has shown to be effective in increasing network capacity and prolonging network lifetime. In this paper, we focus on the strong minimum power restricted topology control problem, which to adjust the limited transmission power for each sensor node to find a power assignment to reserve strong connectivity and achieve minimum energy cost in the wireless sensor network. We proposed three heuristics based on the problem. Simulation results have demonstrated the efficiency of three heuristics.

Keywords: Strong Connectivity, Topology Control, Power Threshold, Wireless Sensor Networks.

1 Introduction

Recent development of the wireless sensor technology has made sensor device (called sensor nodes) widely applied to applications such as health and environmental monitoring, battle-field surveillance and so on. These autonomous devices composed a wireless sensor network (WSN) are inherently resource constrained since they have limited processing speed, storage capacity, communication bandwidth and typically are powered by batteries. In many applications, sensor nodes are deployed in a remote or hazardous area, in this case servicing a node is unpredictable, and recharging a large number of sensor batteries would also be expensive and time consuming, therefore energy constraint is a critical issue in the wireless sensor networks.

A WSN forms a dynamical topology and typically functions as sending data packets back to the data collectors. Basically, topology control is to sustain this data propagation topology and enhance the network connectivity to provide reliable quality of service or prolong the network lifetime. Maintaining the topology control attracts a lot of attentions in recent years such as [6],[3],[1],[2].

* This work was supported in part by the National Natural Science Foundation of China under Grant No. 10671208.

X. Hu and J. Wang (Eds.): COCOON 2008, LNCS 5092, pp. 541–550, 2008.

In a sensor network or any other all-wireless network, all wireless communications among these devices are based on radio propagation, and controlled by their transmit powers. For each order pair (u, v) of sensor nodes, there is a transmission power threshold, denoted by $p(u, v)$. A radio signal transmitted by the sensor u can be received by v only when transmission power of u is at least $p(u, v)$. In practice, the transmission power threshold for a pair of sensors depends on a number of impact factors including the Euclidean distance between the sensors, interference, etc. Note that for any pair $u, v, p(u, v)$ may be not equal to $p(v, u)$.

Given the transmission powers of the sensors, sensor network can be represented by a directed graph. The main goal of topology control is to assign transmission power to each sensor so that the resulting directed graph satisfies some specified properties. Since the battery power of each sensor is an expensive resource, it is important to achieve the goal while minimizing the total power.

Typically, the energy of sensor nodes is not non-limit due to the capacity of the integrated batteries. Thus, it is vital to address topology control problem with limit power levels to adjust for individual sensor nodes. In this paper, we consider topology control problem with a limited transmission power for each sensor.

Definition 1 (Strong Minimum Power Restricted Topology Control Problem). Given n sensors in the Euclidean plane where each sensor has a finite maximum power level and each directed node pair has a power threshold, find a power assignment for each sensor to reach a strongly connected directed graph with minimum total power.

Note that our problem is different from those in [3],[8],[4],[1], which focus on symmetric weighted diagraph G. [3] addresses strong connectivity with bidirectional links. Although [8],[4],[1] address strong connectivity with unidirectional links, but our problem addresses more general weight directed graph G, which G may be a asymmetric weighted directed graph. And the problem which we consider a limited power is also different from [6].

This paper is organized as follows: In Section 2, we briefly summarize some related work. In Section 3, we formulate the network communication model and problem formulation. In Section 4, we present three heuristics according to the problem. Finally, we simulate on proposed heuristics in Section 5 and conclude the paper in Section 6.

2   Related Work

The problem of topology control by adjusting transmission power has been extensively studied in MANETs. One of the classical works in [10] investigates the optimal transmission power of nodes in multi-hop networks to maximize the expected progress of packets in the direction towards the receivers. They assume that the transmission power of each node is on the same range. Li et al propose a QoS topology control strategies for homogeneous and non-homogeneous ad hoc networks in [7],[5], which construct a network topology that meet end users' QoS requirements and achieve the minimal total transmission power. As we are dealing with heuristics of strong minimum power restricted topology control problem in this paper, the closely related works in literature are in [6],[3],[8], [4],[1]. Clementi et al [4] and Chen et al [1] prove the NP-completeness of the broadcast strong connectivity augmentation problem, and [1] reveal that the algorithm based on the minimum spanning tree (MST) has a performance ratio 2 for broadcast strong connectivity augmentation problem for"directed graph model" [8]. [3] prove the NP-completeness of the strong minimum energy topology problem (SMET) and proposes two heuristics: power assignment based on MST and incremental power. And MST has a performance ratio 2. However, Cheng et al [3] used the "undirected graph model" [8]. Although we use the "directed graph model" same as [4],[1], but they look upon symmetric weight directed graph, in contrary, we address the general model-asymmetric weighted directed graph including symmetric weight directed graph as special case.

Another way to improve the performance of topology control of wireless sensor networks is to utilize the connected dominate set (CDS). Ma et al [9] construct network with small number of coordinators while still maintain the topology connectivity. They assume that all wireless links in a sensor network are bidirectional since 802.15.4 MAC has an ACK mechanism for every frame. They consider their problem as minimal connected CDS problem. Three topology algorithms have been proposed with different time complexity and power saving. Cheng et al [2] construct a virtual backbone by computing a CDS in unit-disk graphs, and propose two distributed algorithms to approximate a minimum CDS.

All above works conduct topology control issues or mechanisms in the latest years. But none of them consider utilizing the asymmetric weighted communication model and adjusting transmit powers to minimize the total power consumption for achieving the strong connectivity and enhance the network capacity.

3   Network Communication Model and Problem Formulation

We assume all sensors are randomly deployed in the Euclidean plane. And the transmission power of each sensor is adjustable. For each order pair (u, v) of sensor nodes, there is a transmission power threshold, denoted by $p(u, v)$.

Suppose there are n sensors in plane. $V = \{v_1, v_2, \ldots, v_n\}$ is defined as a set of sensors, Suppose each sensor v has a finite maximum power levels $p_{max}(v)$. Let p be a transmission power assignment on V, i.e. for each node v, $p_{max}(v) \geq p(v)$. A signal transmitted by a sensor u can be received by sensor v only when the transmission power of u is at least $p(u, v)$. Given the transmission powers of the sensors, a sensor networks can be represented by a directed graph $G = (V, E)$. A directed edge (u, v) is in this directed graph if and only if $p_{max}(u) \geq p(u) \geq p(u, v)$. A directed graph is strongly connected if and only if for each directed pair (u, v) of V there is a directed path from u to v. We suppose that each node use its maximum power level, the resulted graph must be strongly connected graph.

Definition 2. Let $G = (V, E)$ be a directed graph resulted by a transmission power assignment p on V. The total power of directed graph G is defined as:

$$TP(G) = \sum_{i=1}^{n} p(v_i)$$

Strong Minimum Power Restricted Topology Control Problem (SM-PRTCP) is formulated as following:

Given n sensors in the Euclidean plane, and each sensor v has its maximum power level $p_{max}(v)$, and each directed pair (u, v) has a power threshold $p(u, v)$, find a power assignment $p(u)$ for each sensor u so that the resulted directed graph $G = (V, E)$ by p is a strongly connected directed graph and total power $TP(G)$ is minimized.

We denote $G_{max} = (V, E_{max})$ as resulted directed graph by maximum power level, i.e. each sensor use its maximum power. V is set of sensors. $u, v \in V$, $(u, v) \in E_{max}$ if and only if $p_{max}(u) \geq p(u, v)$. We assign a weight $p(u, v)$ for (u, v) in E_{max}. Suppose $G = (V, E)$ is a subgraph of $G_{max} = (V, E_{max})$, then the total power of G is

$$TP(G) = \sum_{u \in V} p(u)$$

Where $p(u) = max\{p(u, v) | (u, v) \in E\}$.

The SMPRTCP will be transformed to following problem:

Given a weighted directed graph $G_{max} = (V, E_{max})$ which is strongly connected, to find a spanning subgraph $G = (V, E)$ of G_{max} such that G is a strongly connected and $p(G)$ is minimized.

4   Heuristics

The strong minimum power topology problem is proven to be NP-hard [1]. Therefore, we essay to derive its approximation algorithm. We design three heuristics. Main idea of the first two algorithms is to delete a maximum power such that it still remains strongly connected.

We first give some notations before proposing the algorithm.

Given $G_{max} = (V, E_{max})$ which is resulted directed graph by each sensor node using its maximum power level. For each sensor v, we queue power of v as $w_1^v, w_2^v, ..., w_{k(v)}^v$ from min to max, i.e. We queue all out-directed edges of v along $p(v, u)$ as follows: $p(v, u_1) \leq p(v, u_2) \leq ... \leq p(v, u_{k(v)})$, the different $p(v, u_j)$ is $w_1^v, w_2^v, ..., w_{k(v)}^v$.

We denote that

$$E_{w(j)}^v = \{(v, u) | \text{if } w_j^v \geq p(v, u) > w_{j-1}^v\}$$

Main idea of the algorithm 1 is that we first start G_{max}, then for each time, finding a node with a maximum power, if the resulted graph still keep strong connected after deleting all out-edges under this maximum power, then delete this maximum power of the node, otherwise delete all power levels of the node.

The algorithm 1 is more formally presented as follows.

Algorithm 1. for computing strong connected subgraph

Input: n sensors with their maximum power and threshold powers of node pairs

Output: a strongly connected graph with minimized total power.

Initially, $G \leftarrow G_{max}$, $W \leftarrow \cup_{v \in V} \{w_1^v, w_2^v, ..., w_{k(v)}^v\}$

While $(W \neq \emptyset)$ do

    Choose a node $v \in V$ with maximum power w_j^v in W

      If $G - E_{w(j)}^v$ is a strongly connected, then

        $G \leftarrow G - E_{w(j)}^v$

        $W \leftarrow W - \{w_j^v\}$

      Otherwise $W \leftarrow W - \{w_j^v, w_{j-1}^v, ...w_1^v\}$

The algorithm 2 is similar with algorithm 1, but the rule of selecting a node is different from algorithm 1. The algorithm 2 is more formally presented as follows.

Algorithm 2. for computing strong connected subgraph

Input: n sensors with their maximum power and threshold powers of node pairs

Output: a strongly connected graph with minimized total power.

Initially, $G \leftarrow G_{max}$, $W \leftarrow \cup_{v \in V} \{w_1^v, w_2^v, ..., w_{k(v)}^v\}$

While $(W \neq \emptyset)$ do

    Choose a node $v \in V$ with maximum different $w_j^v - w_{j-1}^v$ in W, where w_j^v is a maximum power of v in W

      If $G - E_{w(j)}^v$ is a strongly connected, then

        $G \leftarrow G - E_{w(j)}^v$

        $W \leftarrow W - \{w_j^v\}$

      Otherwise $W \leftarrow W - \{w_j^v, w_{j-1}^v, ...w_1^v\}$

In the following figures are examples for algorithm 1(A1). We randomly select 8 nodes in a 80×80 area and randomly generate transmission power threshold for each order pair. Then we get a graph showed in Fig.1(a).

Fig.1 illustrates some steps for A1. After some procedure steps, we result in Fig.1(b). In the next step, we pick node 3 which has maximum power, after deleting $(3, 4)$, it is still strong connected and resulted graph is showed in Fig.1(c). The final resulted topology is Fig.1(d).

The main idea of third algorithm is based on minimum Spanning tree. We defined as B-MST. For $G_{max} = (V, E_{max})$, we get a weight undirected graph $G = (V, E)$. $(u, v) \in E$ if and only if $(u, v) \in E_{max}$ and $(v, u) \in E_{max}$, $w(u, v) = \max\{p(u, v), p(v, u)\}$. We find a minimum spanning tree T on graph G, then change each edge in T to two directed edges (u, v) and (v, u), to get a strong connected subgraph of G_{max}.

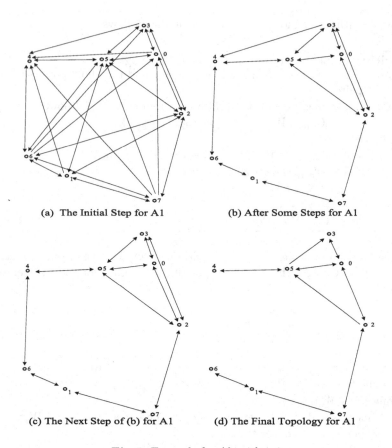

(a) The Initial Step for A1

(b) After Some Steps for A1

(c) The Next Step of (b) for A1

(d) The Final Topology for A1

Fig. 1. Example for Algorithm 1

Algorithm B-MST

Step1: Construct an undirected graph G from G_{max}

Step2: Find a minimum spanning tree T of G, then get a strong connected subgraph of G_{max}.

Note that the B-MST algorithm can be accomplished only when G is a connected graph. Specially, when $p(u,v) = p(v,u) = d(u,v)^{\alpha}$, the algorithm B-MST is similar with MST algorithm in [3].

5   Simulations

In this section, we verify the performance of proposed algorithms via simulation. We simulate a stationary network with sensor nodes randomly located in a 200×200 2-D free-space. We assume the transmission power threshold $p(u,v)$ is equal to $d(u,v)^2 + \rho$ for each directed pair (u,v) of sensor nodes, where ρ is a random value. $p_{max(u)}$ is selected randomly from $[d^2_{min}(u), d^2_{max}(u)]$, where $d_{min}(u)$ is distance from u to the nearest node and $d_{max}(u)$ is the distance from u to the farthest node. In this simulation, we consider the following tunable parameters:

- N, the number of sensor nodes which varies between 10 and 100.
- ρ, the random value which varies between 0 and 200.

We simulate algorithm 1(A1), algorithm 2(A2), and B-MST algorithm, compare their performance, and compute the total power and the average power by varying the number of nodes (N), random value (ρ), then compare our algorithm with MST ([3]) when $\rho = 0$.

The simulation is conducted in a 200×200 2-D free-space by independently allocating N. For any order pair of nodes, the transmission power threshold $p(u, v)$ is equal to $d^2(u, v) + \rho$, ρ is a random value varying from $[0, 20]$, or $[20, 100]$, or $[100, 200]$. We present average of 100 separate runs for each result shown in figures. In each run of the simulations, for given N, we randomly deploy N nodes in the square.

Firstly, base on the three algorithms describe in section 4, we evaluate the algorithms. The simulation results are showed in Fig.2-4. We make the following observation:

(1) The curves in the figures indicate that the performances of three greedy algorithms are close.
(2) The total power increases with the number of sensor nodes. When the number of nodes N enlarges, there are more nodes need to be connected, accordingly, the total power increases.
(3) The average power decreases as the number of sensor nodes increases. This is because when the number of nodes increases, there are more nodes join to relay such that the average power decreases.

In the second experiment, we compare our algorithms (when $\rho = 0$) with the existed algorithm MST [3]. When $\rho = 0$, our model will become the model same as [3]. And our algorithm B-MST is same as MST in this case. The curves in the

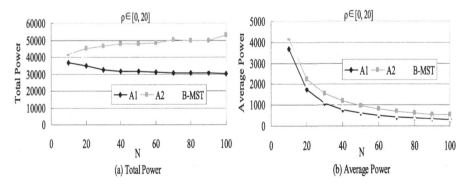

Fig. 2. Total power and average power versus number of nodes ($\rho \in [0, 20]$)

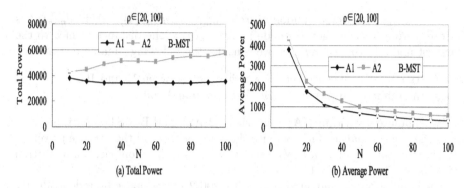

Fig. 3. Total power and average power versus number of nodes ($\rho \in [20, 100]$)

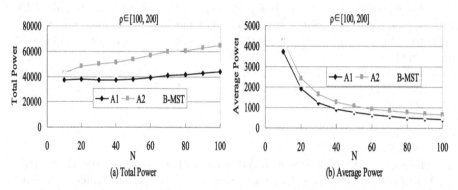

Fig. 4. Total power and average power versus number of nodes ($\rho \in [100, 200]$)

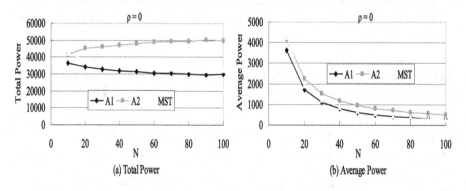

Fig. 5. Comparing A1 and A2 with MST

Fig. 5 indicate that performance of proposed algorithm A1 is better than MST algorithm which is 2-approximation.

Thirdly, we compare our algorithms to optimal solution on smaller number of nodes. We randomly select N nodes from a 30×30 area and result in a topology

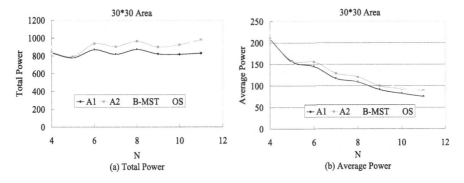

Fig. 6. Comparing our algorithms with optimal solution

of N nodes. We run our algorithms on this topology. We enumerate all possible to get an optimal solution. The simulation result is showed in Fig. 6. From the curves, we get that our algorithms are very close the optimal solution.

6   Conclusion

In this paper, we investigate the strong minimum power restricted topology control problem and present three heuristics. We consider more general case on both bidirectional and unidirectional wireless communication links in WSNs. While we focus on the reality of wireless world, the next step of our research concern will be implementing these algorithms into the real sensor devices to consider the performances of reality.

References

1. Chen, W.T., Huang, N.F.: The Strongly Connecting Problem on Multihop Packet Radio Networks. IEEE Trans. Comm. 3(37), 293–295 (1989)
2. Cheng, X., Ding, M., Du, H., Jia, X.: Virtual Backbone Construction in Multihop Ad Hoc Wireless Networks. Wireless Communications and Mobile Computing 2(6), 183–190 (2006)
3. Cheng, X., Narahari, B., Simha, R., Cheng, X.: Strong Minimum Energy Topology in Wireless Sensor Networks: NP-Completeness and Heuristic. IEEE Transactions on Mobile Computing 3(2), 248–256 (2003)
4. Clementi, A.E.F., Penna, P., Silvestri, R.: Hardness Results for the Power Range Assignment Problem in Packet Radio Networks. In: Hochbaum, D.S., Jansen, K., Rolim, J.D.P., Sinclair, A. (eds.) RANDOM 1999 and APPROX 1999. LNCS, vol. 1671, pp. 197–208. Springer, Heidelberg (1999)
5. Jia, X., Li, D., Du, D.: QoS Topology Control in Ad Hoc Wireless Networks. In: IEEE INFOCOM 2004, vol. 2, pp. 1264–1272 (2004)
6. Li, D., Du, H., Chen, W.: Power Conservation for Strongly Connected Topology Control in Wireless Sensor Networks. In: ACM FOWANC (2008)

7. Li, D., Jia, X., Du, H.: QoS Topology Control for Non-Homogenous Ad Hoc Wireless Networks. In: EURASIP Journal on Wireless Communications and Networking 2006, pp. 1–10 (2006)
8. Lloyd, E.L., Liu, R., Marathe, M.V.: Algorithmic Aspects of Topology Control Problems for Ad Hoc Networks. In: ACM MobiHoc, pp. 123–134 (2002)
9. Ma, J., Gao, M., Zhang, Q., Ni, L.M.: Energy-Efficient Localized Topology Control Algorithms in IEEE 802.15.4-Based Sensor Networks. IEEE Transactions on Parallel and Distributed Systems 5(18), 711–720 (2007)
10. Takagi, H., Kleinrock, L.: Optimal Transmission Ranges for Randomly Distributed Packet Radio Terminals. IEEE Transactions on Communications 32(3), 246–257 (1984)

$(6 + \varepsilon)$-Approximation for Minimum Weight Dominating Set in Unit Disk Graphs*

Xiaofeng Gao[1], Yaochun Huang[1], Zhao Zhang[2,**], and Weili Wu[1]

[1] Department of Computer Science, University of Texas at Dallas, USA
{xxg052000,yxh038100,weiliwu}@utdallas.edu
[2] College of Mathematics and System Sciences, Xingjiang University, China
zhzhao@xju.edu.cn

Abstract. It was a long-standing open problem whether the minimum weight dominating set in unit disk graphs has a polynomial-time constant-approximation. In 2006, Ambühl et al solved this problem by presenting a 72-approximation for the minimum weight dominating set and also a 89-approximation for the minimum weight connected dominating set in unit disk graphs. In this paper, we improve their results by giving a $(6 + \varepsilon)$-approximation for the minimum weight dominating set and a $(10 + \varepsilon)$-approximation for the minimum weight connected dominating set in unit disk graphs where ε is any small positive number.

Keywords: Unit Disk Graph, Approximation algorithm, Dominating Set.

1 Introduction

Consider a graph $G = (V, E)$. A subset A of V is called a *dominating set* if every node in $V - A$ is adjacent to a node in A, and furthermore, A is called a *connected dominating set* if the subgraph $G[A]$ induced by A is connected. A graph is called a *unit disk graph* if every node is associated with unit disk (a disk of diameter one) in the Euclidean plane and there is an edge between two nodes if and only if two corresponding disks have nonempty intersection. Therefore, when we place each node at the center of its associated disk, an edge (u, v) exists if and only if $d(u, v) \leq 1$. The unit disk graph is a mathematical model for wireless sensor networks when all sensors have the same communication range. Both the dominating set and the connected dominating set have important applications in the study of wireless sensor networks [3].

Given a unit-disk graph $G = (V, E)$ with node weight $c : V \to R^+$, find a dominating set with minimum total weight. This is an NP-hard problem [4] and it was open for a long time whether there exists a polynomial-time constant-approximation for this problem. In 2006, Ambühl et al [1] solved this problem by presenting a 72-approximation for the minimum weight dominating set and also a

---

* Support in part by National Science Foundation of USA under grants CCF-9208913 and CCF-0728851; and in part by NSFC (60603003) and XJEDU.
** This work was done when this author visited at University of Texas at Dallas.

X. Hu and J. Wang (Eds.): COCOON 2008, LNCS 5092, pp. 551–557, 2008.
© Springer-Verlag Berlin Heidelberg 2008

89-approximation for the minimum weight connected dominating set in unit disk graphs. In this paper, we improve their results by giving a $(6+\varepsilon)$-approximation for the minimum weight dominating set and a $(10+\varepsilon)$-approximation for the minimum weight connected dominating set in unit disk graphs where ε is arbitrarily small positive number.

2   Preliminaries

Firstly, we set a fixed constant $0 < \mu < \sqrt{2}/2$. Suppose all the nodes of the given unit disk graph are contained in the interior area of a square with edge length $m\mu$. Then, we divide this square into a $m \times m$ grid such that each cell is a square with edge length μ. We may assume that no node lies on any cut-line of the grid since, if such a case occurs, we can always make a little move of the grid to have cut-lines away from nodes. For a cell e and a node subset D, let $D^+(e)$ denote the subset of all nodes each of which is able to dominate (i.e., adjacent to) a node in e and $D(e) = e \cap D$. Ambühl et al [1] studied the following subproblem:

SUBPROBLEM-ON-CELL: Find a minimum weight subset of $V^+(e)$ to dominate $V(e)$.

They found that this subproblem has a polynomial-time 2-approximation, which results in a 72-approximation for minimum weight dominating set in whole unit disk graph. A key lemma in establishing the 2-approximation is as follows.

Lemma 1. *Consider a set P of points lying inside a horizontal strip and a set $\mathcal{D}$ of disks with radius one and with centers either above or below the strip (Fig. 1). Give each disk with a nonnegative weight. Suppose the union of all disks covers P. Then the minimum weight subset of disks covering P can be computed in time $O(m^4 n)$ where $n = |P|$ and $m = |\mathcal{D}|$.*

This lemma is very useful in the whole paper.

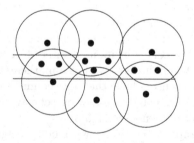

Fig. 1. A covering problem about a strip

3   Main Result

Our new results are based on several new arguments and new ideas.

Firstly, we set $\mu = \sqrt{2}/2$ and establish some new results about a cell e.

Let A, B, C, D be four vertices of e and divide outside of e into eight areas as shown in Fig. 2. For any node $p \in V(e)$, let $\angle p$ be a right angle at p such that two edges intersecting horizontal line AB each at an angle of $\pi/4$. Let $\Delta_{low}(p)$ denote the part of e lying inside of $\angle p$. Similarly, we can define $\Delta_{up}(p)$, $\Delta_{left}(p)$ and $\Delta_{right}(p)$ as shown in Fig. 3.

Lemma 2. *If p is dominated by a node u in area LM then every point in $\Delta_{low}(p)$ can be dominated by u. The similar statement holds for CL and $\Delta_{left}(p)$, CR and $\Delta_{right}(p)$, and UM and $\Delta_{up}(p)$.*

Proof. Since $\Delta_{low}(p)$ is a cover polygon, it sufficient to show that the distance from u to every vertex of $\Delta_{low}(p)$ is at most one.

Suppose v is a vertex of $\Delta_{low}(p)$ on BC (Fig. 4). Draw a line L' perpendicular to pv and equally divide pv. If u is below L', then we have $d(u,v) \leq d(u,p) \leq 1$. If u is above L', then $d(u,v) \geq d(u,p)$, then $\angle uvp < \pi/2$ and hence $\angle uvC < 3\pi/4$. It follows that $d(u,v) < \mu/\cos \pi/4 = 1$.

A similar argument can be applied in the case that the vertex v of $\Delta_{low}(p)$ is on DA or on AB.   □

Consider two nodes $p, p' \in V(e)$. Suppose p is on the left of p'. Extend the left edge of $\angle p$ and the right edge of $\angle p'$ to intersect at point p''. Define $\Delta_{low}(p,p')$ to be the part of e lying inside of $\angle p''$ (Fig. 5). Similarly, we can define $\Delta_{up}(p.p')$.

Lemma 3. *Let K be a subset of $V^+(e) - V(e)$, which dominates $V(e)$. Suppose $p, p' \in V(e)$ are dominated by some nodes in $K \cap LM$ (or $K \cap UM$), but neither p nor p' is dominated by any node in $K \cap (CL \cup CR)$. Then every node in*

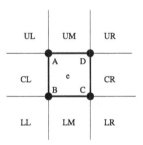

Fig. 2. Outside of e is divided into eight areas

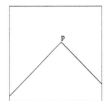 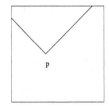 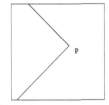 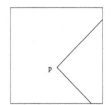

Fig. 3. $\Delta_{low}(p)$, $\Delta_{up}(p)$, $\Delta_{left}(p)$ and $\Delta_{right}(p)$

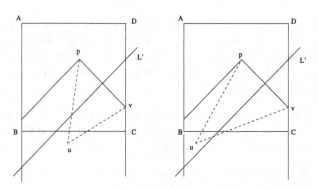

Fig. 4. The proof of Lemma 2

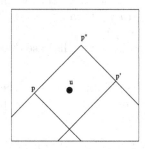

Fig. 5. $\Delta_{low}(p, p')$

$\Delta_{low}(p, p')$ *can be dominated by nodes in* $K \cap (U \cup L)$ *where* $U = ULU \cup UM \cup UR$ *and* $L = LL \cup LM \cup LR$.

Proof. By Lemma 2, it suffices to consider a node u lying in $\Delta_{low}(p, p') \backslash (\Delta_{low}(p) \cup \Delta_{low}(p'))$. For contradiction, suppose u is dominated by a node v in $K \cap (CL \cup CR)$. If $v \in CL$, then $\Delta_{left}(v)$ contains p and by Lemma 2, p is dominated by v, a contradiction. A similar contradiction can result from $v \in CR$. $\square$

Define $\Delta_{low}(\emptyset) = \emptyset$, and $\Delta_{low}(\{p\}) = \Delta_{low}(p)$. Also, define $\Delta_{low}(\{p, p'\}) = \Delta_{low}(p, p')$ if p is on the left of p and $\Delta_{low}(\{p, p'\}) = \Delta_{low}(p)$ if p and p' on a vertical line and p is higher than p'. Similarly, define $\Delta_{up}(W)$ for a subset W of at most two vertices in $V(e)$.

Now, consider an optimal solution Opt for the minimum weight dominating set and its total weight is denoted by opt. Let C be the set of all cells in the grid and $C' = \{e \in C \mid e \cap Opt \neq \emptyset\}$.

For $e \in C - C'$, choose p (p') to be the leftmost (rightmost) node in $V(e)$, which is dominated by a node in $Opt^+(e) \cap LM$, but not by any node in $Opt^+(e) \cap (CL \cup CR)$; if there are more than one such p (p'), then choose the highest one. Choose q (q') to be the leftmost (rightmost) node in $V(e)$, which is dominated by a node in $Opt^+(e) \cap UM$, but not by any node in $Opt^+(e) \cap (CL \cup CR)$; if there are more than one such q (q'), then choose the lowest one.

Define

$$W = \begin{cases} \{p, p'\} & \text{if } p \text{ and } p' \text{ exist} \\ \emptyset & \text{otherwise,} \end{cases}$$

and

$$W' = \begin{cases} \{q, q'\} & \text{if } q \text{ and } q' \text{ exist} \\ \emptyset & \text{otherwise,} \end{cases}$$

By Lemma 3, every node in $V_1(e) = \Delta_{low}(W) \cup \Delta_{up}(W)$ can be dominated by $Opt^+(e) \cup (U \cup L)$ and every node in $V_2(e) = V(e) \setminus V_1(e)$ can be dominated by $Opt^+(e) \cap (L^* \cup R)$ where $L^* = UL \cup CL \cup LL$ and $R = UR \cup CR \cup LR$.

Let $H_1, ..., H_m$ be m horizontal strips and $Y_1, ..., Y_m$ m vertical strips. For every $e \in C'$, choose a node $v_e \in e \cap Opt$. Let $U = \{v_e \mid e \in C'\}$ and Z_U be the set of nodes dominated by U. For each cell e, denote $V_1^+(e) = V^+(e) \cap (U \cup L^*)$ and $V_2^+(e) = V^+(e) \cap (L \cup R)$ and each strip H_i (Y_i), denote $Opt^+(H_i) = (\cup_{e \in (C - C') \cap H_i} Opt_1^+(e)) \setminus U$ $(Opt^+(Y_i) = (\cup_{e \in (C - C') \cap Y_i} Opt_2^+(e)) \setminus U)$.

For each strip H_i, we compute a minimum weight subset $OPT(H_i)$ from set $(\cup_{e \in (C - C') \cap H_i} V_1^+(e)) \setminus U$ to dominate $(\cup_{e \in (C - C') \cap H_i} V_1(e)) \setminus Z_U$ and for each strip Y_i, compute a minimum weight subset of $(\cup_{e \in (C - C') \cap Y_i} V_2^+(e)) \setminus U$ to dominate $(\cup_{e \in (C - C') \cap Y_i} V_2(e)) \setminus Z_U$. Putting $U, OPT(H_i)$ and $OPT(Y_i)$ together, we would obtain a dominating set with total weight at most

$$c(U) + \sum_{i=1}^{m} c(Opt_1^+(H_i)) + \sum_{i=1}^{m} c(Opt_2^+(Y_i)).$$

Note that a node can be in $Opt_1^+(H_i)$ or $Opt_2^+(Y_i)$ for at most six (horizontal or vertical) strips. Therefore,

$$c(U) + \sum_{i=1}^{m} c(Opt_1^+(H_i)) + \sum_{i=1}^{m} c(Opt_2^+(Y_i)) \le 6opt.$$

This means that we obtain a dominating set with total weight at most $6opt$.

However, computing this dominating set has a trouble because C' and U are defined by Opt and for each cell, p, p', q, q' are also determined by Opt. Don't worry! This trouble can be removed by consider all possible C', all possible U and all possible p, p', q, q'. This idea gives the following approximation algorithms.

6-Approximation: Put input unit-disk graph $G = (V, E)$ in the interior of a $m\mu \times m\mu$ square S. Divide the square S into an $m \times m$ grid such that each cell is a $\mu \times \mu$ square. Let C be the set of m^2 cells. Let $C' \subseteq C$. For each cell $e \in C'$, choose a node $v_e \in V(e)$ and let $U = \{v_e \mid e \in C'\}$. For every subset C' and every U, compute a node subset $A(C', U)$ in the following way:

Step 1.   For every cell $e \in C - C'$ and for every $W \subseteq V_{low}(e)$ with $|W| \le 2$ and every $W' \subseteq V_{up}(e)$ with $|W'| \le 2$, let $V_1(e) = \Delta_{low}(W) \cup \Delta_{low}(W')$ and $V_2(e) = V(e) - V_1(e)$.

Step 2.1. For each horizontal strip H, compute a minimum weight subset $OPT(H)$ of $(\cup_{e \in H \setminus C'} V_1^+(e)) \setminus U$ to dominate $(\cup_{e \in H \setminus C'} V_1(e)) \setminus Z_U$.

Step 2.2. For each vertical strip Y, compute a minimum weight subset $OPT(Y)$ of $\cup_{e \in Y \setminus C'} V_2^+(e)$ to dominate $(\cup_{e \in H \setminus C'} V_2(e)) \setminus Z_U$.

Step 2.3. Compute $O = (\cup_H OPT(H)) \cup (\cup_Y OPT(Y))$ to minimize the total weight $c(O)$ over all possible combinations of W, W' for all $e \in C - C'$.

Step 3.   Set $A(C', U) = O(C') \cup U$.

Step 4.   Finally, compute a $A = A(C', U)$ to minimize the total weight $c(A(C', U))$ for C' over all subsets of C and U over all choices.

We now estimate the time for computing A. There are $O(2^{m^2})$ possible subsets of C, $n^{O(m^2)}$ possible choices of U and $O(n^{4m^2})$ possible combinations of W and W' for all cells in $C - C'$. For each combination, computing all $OPT(H)$ and all $OPT(Y)$ needs time $O(n^5)$. Therefore, total computation time is $n^{O(m^2)}$. A good news is that when m is a constant, this is a polynomial-time. A bad news is that in general, m is not a constant. However, we can use partition again to make a constant m!

Theorem 1. *For any $\varepsilon > 0$, there exists $(6+\varepsilon)$-approximation with computation time $n^{O(1/\varepsilon^2)}$ for the minimum weight dominating set in unit disk graph.*

Proof. Choose $m = 12 \max(1, \lceil 1/\varepsilon \rceil)$. Put input unit-disk graph G into a grid with each cell being an $m\mu \times m\mu$ square. For each cell e, solve the problem for subset of nodes within distance one to cell e to dominate the nodes in cell e with the 6-approximation algorithm. Union them together and denote this union by $A(P)$ for the partition P induced by this grid. Shift this grid in diagonal direction with distance one in each time. This results in m partitions $P_1, ..., P_m$. Choose $A = A(P_i)$ to be the one with the minimum weight among $A(P_1), ..., A(P_m)$. We claim $c(A(P_i)) \leq (6 + \varepsilon)opt$.

In fact, each disk with radius one can cross cutlines of at most four partition. When a disk with radius one and center at a vertex u crosses a cutline of a partition P_j, u may involve more than one, but at most four $m\mu \times m\mu$ cells' subproblems. It follows that

$$c(A) \leq (1 + 12/m)opt \leq (1 + \varepsilon)opt.$$

Now, the total computation time is $m \cdot n^{O(m^2)} = n^{O(1/\varepsilon^2)}$.   □

Next, we study the minimum weight connected dominating set problem in unit disk graphs: Given a unit-disk graph $G = (V, E)$ with weight $c : V \to R^+$, find a connected dominating set with minimum total weight.

Theorem 2. *For any $\varepsilon > 0$, there exists a $(10 + \varepsilon)$-approximation with computation time $n^{O(1/\varepsilon^2)}$ for the minimum weight connected dominating set in unit disk graphs.*

Proof. We first compute a $(7 + \varepsilon)$-approximation D for the minimum weight dominating set and then connect D with nodes of total weight at most $4opt$

where *opt* is the minimum weight of connected dominating set. This can be done due to the following:

(1) Let OPT be the minimum weight connected dominating set for G. Then, we can find a minimum-length spanning tree T for $D \cup OPT$ such that every node in $OPT - D$ has degree five by the method in the proof of Lemma 1 in [2].

(2) Using method in [5], we can construct from T a spanning tree T' for D such that each edge (u, v) of T' is a path between u and v in T and each node in $OPT \setminus D$ appears in at most four edges of T'. If we assign the weight of edge (u, v) of T' equal to the total weight of nodes on the path between u and v. The total edge-weight of T' is at most $4opt$.

(3) We can compute a tree T^* with weight as small as that of T' in the following:

Step 1. Construct a complete graph H on D. For each edge (u, v), assign cost $w(u, v)$ with the minimum total weight of internal nodes on a path between u and v in graph G.

Step 2. Compute a minimum spanning tree T'' of H and map T'' back to G in order to obtain a tree T^* for connecting all connected component of D.

Then we can finished the whole proof.   □

4 Conclusion

In this paper we gave an improved constant-factor approximation algorithm to compute minimum weight connected dominating set in unit disk graphs. We divided our algorithm into two parts. The first part selected a minimum weight dominating set for a given unit disk graph, and then the first part connects this dominating set by inserting several disks into the dominating set. The first part is a $(6+\varepsilon)$-approximation, while the whole algorithm is a $(10+\varepsilon)$-approximation.

References

1. Ambühl, C., Erlebach, T., Mihalák, M., Nunkesser, M.: Constant-Approximation for Minimum-Weight (Connected) Dominating Sets in Unit Disk Graphs. In: Díaz, J., Jansen, K., Rolim, J.D.P., Zwick, U. (eds.) APPROX 2006 and RANDOM 2006. LNCS, vol. 4110, pp. 3–14. Springer, Heidelberg (2006)
2. Chen, D., Du, D.-Z., Hu, X.-D., Lin, G.-H., Wang, L., Xue, G.: Approximations for Steiner Trees with Minimum Number of Steiner Points. Theoretical Computer Science 262, 83–99 (2001)
3. Cheng, X., Huang, X., Li, D., Wu, W., Du, D.-Z.: A Polynomial-Time Approximation Scheme for the Minimum-Connected Dominating Set in Ad Hoc Wireless Networks. Networks 42, 202–208 (2003)
4. Clark, B.N., Colbourn, C.J., Johnson, D.S.: Unit Disk Graphs. Discrete Mathematics 86, 165–177 (1990)
5. Mandoiu, I., Zelikovsky, A.: A Note on the MST Heuristic for Bounded Edge-Length Steiner Trees with Minimum Number of Steiner Points. Information Processing Letters 75(4), 165–167 (2000)

Spectrum Bidding in Wireless Networks and Related

Xiang-Yang Li[1,*], Ping Xu[1], ShaoJie Tang[1], and XiaoWen Chu[2]

[1] Illinois Institute of Technology, Chicago, IL, USA
xli@cs.iit.edu, pxu3@iit.edu, stang7@iit.edu
[2] Hong Kong Baptist University, Hong Kong, China
chxw@comp.hkbu.edu.hk

Abstract. In this paper, we study the spectrum assignment problem for wireless access networks. Opportunistic spectrum usage is a promising technology. However, it could suffer from the selfish behavior of secondary users. In order to improve opportunistic spectrum usage, we propose to combine the game theory with wireless modeling. Several versions of problems are formalized under different assumptions. We design PTAS or efficient approximation algorithms for each of these problems such that overall social benefit is maximized. Finally, we show how to design a truthful mechanism based on all these algorithms.

1 Introduction

Wireless technology is expected to play a bigger and more fundamental role in the new Internet than it has today. The radio frequency spectrum has been chronically regulated with static spectrum allocation policies since the early $20th$ century. With the recent fast growing spectrum-based services and devices, remaining spectrum available for future wireless services is being exhausted, known as the *spectrum scarcity problem*. Current fixed spectrum allocation scheme leads to significant spectrum *white spaces* where many allocated spectrum blocks are used only in certain geographical areas and/or in brief periods of time. A huge amount of precious spectrum (below 5GHz), perfect for wireless communications, sit there silently. Recognizing that the traditional spectrum management process can stifle innovation, and it is difficult to provide a certain quality of service (QoS) for systems operated in unlicensed spectrum, the FCC has proposed several new spectrum management models [15].

One promising technology is the opportunistic spectrum usage, secondary users observe channel availability dynamically and explore it opportunistically. While opportunistic spectrum has several advantages, it suffers from selfish behavior of secondary users. Thus, we propose to combine the game theory [13] with wireless

---

* The research of the author was partially supported by National Basic Research Program of China (973 Program) under grant No. 2006CB30300, the National High Technology Research and Development Program of China (863 Program) under grant No. 2007AA01Z180, the RGC under Grant HKBU 2104/06E and CERG under Grant PolyU-5232/07E. Part of the work was done when the author visited MSRA.

X. Hu and J. Wang (Eds.): COCOON 2008, LNCS 5092, pp. 558–567, 2008.
© Springer-Verlag Berlin Heidelberg 2008

communication modeling. More specifically, we study how to share the spectrum and how to charge secondary users such that the *overall social benefit* is maximized even in the presence of selfish behavior, while each secondary user specifies channel, space and time constraints. Note the correlation of time and space constraints introduces high complexity compared with traditional auctions [12],[16],[7],[6],[14],[2].

The main contributions of this paper are as follows. First we design efficient algorithms to allocate channels such that the social efficiency are approximately maximized. Based on these approximation algorithms, we then design strategyproof mechanisms to charge the secondary users. We essentially show that our approximation algorithms satisfy a monotone property.

The rest of the paper is organized as follows. In Section 2, we define the problems to be studied in this paper. From Section 3 to Section 6, we discuss algorithms for several versions of problems described in Section 2. Then we review related results on those spectrum assignment problems in Section 8 and conclude the paper in Section 9 with discussion of some possible future works.

2 Preliminaries

2.1 Network Model

Consider a wireless network system formed by some primary users who hold the right of some spectrum channels, secondary users $\mathcal{V} = \{v_1, v_2, \cdots, v_n\}$ who want to lease the right to use some channels in some region for some time period. For simplicity, we treat all primary users as one unified central authority. In certain applications, each secondary user v_i may provide service to some clients within a geometry region. Let $\mathcal{F} = \{\mathbf{f}_1, \mathbf{f}_2, \cdots, \mathbf{f}_m\}$ be the set of m frequencies that can be used by some secondary users for a given time interval $[0, T]$. For some wireless network systems, it is possible that the primary users will only lease a spectrum frequency for a certain time interval in a certain geographical region. If this is the case, we assume that for each $\mathbf{f}_i \in \mathcal{F}$, we associate it with a region Ω_i and a set of time intervals $\mathcal{T}_i$ that it is available. In this paper, most of our results assume that every channel will be available everywhere and everytime. Our results can easily deal with a general Ω_i and $\mathcal{T}_i$.

We assume that a secondary user v_i may wish to lease a set of channels $\mathcal{F}_i \in \mathcal{F}$. For a bidding, the secondary user will also specify two additional constraints: space condition and time condition. Each user v_i will specify a 2D region exclusively which is typically a disk $D(v_i, r_i)$ centered at node v_i with a radius r_i. User v_i also specifies a time interval $[s_i, e_i]$ or a time duration d_i exclusively. Here it is assumed that $0 \le s_i < e_i \le T$ and $0 < d_i \le T$. Generally, we use T_i to denote the time constraint of user v_i, where T_i is either $[s_i, e_i]$ or a scalar $d_i > 0$.

Two different models of secondary users will be studied in this paper. The first model assumes that every secondary user is *single-minded*: when user v_i bids for $\mathcal{F}_i$, the valuation of v_i over an assignment is 0 if not *all* frequencies in $\mathcal{F}_i$ are allocated. The secondary user v_i will be called *flexible* if it will pay the allocated frequencies separately. For a flexible user v_i, we assume that for each channel $\mathbf{f}_j \in \mathcal{F}_i$, user v_i will bid $b_{i,j}$ for the usage of channel $\mathbf{f}_j$ for a certain time and within certain region.

In this case, we use $b_i = \{b_{i,1}, b_{i,2}, \cdots, b_{i,m}\}$ to denote the bid vector of user i, where $b_{i,j} = 0$ if v_i did not bid for $\mathbf{f}_j$. Thus, a bidding by a user v_i will be written as follows $B_i = [b_i, \mathcal{F}_i, D(v_i, r_i), T_i]$. Upon receiving the bids from secondary users, the central authority decides an allocation method $X = \{x_1, x_2, \cdots, x_n\}$ where $x_i \in \{0, 1\}$ denotes whether user v_i's bid will be satisfied, and also a time-interval $[\mathbf{s}_i, \mathbf{e}_i]$ with $\mathbf{e}_i - \mathbf{s}_i = d_i$ when user v_i required a time-duration d_i in the bid B_i. The allocation must be conflict free among satisfied bids. Here two bids B_i and B_j conflict if $\mathcal{F}_i \cap \mathcal{F}_j \neq \emptyset$, $D(v_i, r_i) \cap D(v_j, r_j) \neq \emptyset$, and $[\mathbf{s}_i, \mathbf{e}_i] \cap [\mathbf{s}_j, \mathbf{e}_j] \neq \emptyset$. The objective of an allocation is to maximize $\sum_{i=1}^{m} x_i b_i$. For simplicity, given a set of bids Y, we use $w(Y)$ to denote the total weight of bids in Y, i.e., $\sum_{B_i \in Y} b_i$.

2.2   Problems Formulation

In this paper, we study several versions of spectrum assignment problems by separately assuming channel, region and time requirements. For notational convenience, we use CRT to denote a problem, where

- C denotes channel requirements. Here C will be either S (denoting that secondary users are single-minded), or F (denoting that secondary users are flexible), or Y (denoting that there is only one channel available).
- R denotes region requirement. Here R will be either O (denoting that required regions overlap) or U (denoting that required regions are unit disks), or G (denoting that required regions are disks with arbitrary radii).
- T denotes time requirement. Here T will be either I (denoting that each required time is an interval) or D (denoting that each required time is a duration) or M (denoting that each required time is an interval or a duration).

For example, problem SUI represents the case that each user v_i will bid for a subset of channels $\mathcal{F}_i$ and is single-minded, will require a unit disk region $D(v_i, 1)$, and a fixed time-interval $[s_i, e_i]$. For each problem where each secondary user bids separately for each channel (C = F), we have a corresponding C = Y problem when considering each user requires k channels as k separate users each requires a different channel. Therefore, we don't discuss these C = F problems as they are special cases of C = Y problems.

Some versions of the problems turn out to be some well-studied problems in the literature and some well-studied problems turns out to be a special case of the above problems. Problem YOD is essentially a knapsack problem, which has a well-known FPTAS [5]. Maximum weighted independent set of a disk graph is a special case of problem YGI with $e_i - s_i \geq T/2$ for each secondary user i. The multi-knapsack problem is also a special case of problem YUD. Due to wireless network applications, we will mainly focus on the problems YOM, YUI, YUD, YUM, SUI.

3   Algorithm for Problem YOM

In this section, we design an approximation algorithm for problem YOM, where there is only one channel available, required regions overlap, and the required time is an interval or a duration. By using FPTAS of knapsack problem and dynamic programming, we can get a simple $(1-\epsilon)/2$ approximation algorithm as follows. We partition the users into two groups: one group of users who required some fixed time intervals and the other group of users who required some time intervals. We solve the assignment problem for each group and take the better assignment as the final solution.

4   PTAS for Problem YUI

In this section, we present a PTAS for the problem YUI, where there is only a single channel available, the required region is a unit disk, and the required time is an interval. The PTAS runs in $O(n^{\frac{1}{\epsilon^2}})$ time and provides approximation factor of $(1 - \epsilon)$ where n is the number of secondary users.

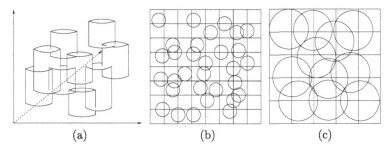

Fig. 1. (a) An illustration of cylinder graph; (b) the network at the end of ith iteration; (c)An illustration of space using hyperplanes. This is a view from the bottom (Z-axis).

Notice that finding the set of bids with the maximum value is equivalent to solve the maximum weighted independent set in the following intersection graph of cylinders. Each bid $B_i = [b_i, \mathcal{F}_i, D(v_i, r_i), T_i]$ defines a cylinder $\mathbf{B}_i = (D(v_i, r_i) \times [s_i, e_i])$ with weight b_i. See Figure 1(a) for illustration. For simplicity, we assume that the three axes are X, Y and Z and the axis Z denotes the time dimension. The disk $D(v_i, r_i)$ is called the base of the cylinder $\mathbf{B}_i$.

Our PTAS is based on the shifting strategy developed in [9]. We partition the space using hyperplanes perpendicular to X-axis and hyperplanes perpendicular to Y-axis. See Figure 1(b) for illustration. In each partition, we throw away some cylinders that intersect some special hyperplanes and then solve the sub-instances of cylinders contained in each cell individually. Let I be the set of all n cylinders. For any give integer $k > 1$, we derive k^2 polynomially solvable sub-instances from the given instance I in polynomial time. The best value of those sub-instance solutions is at least $(1 - \frac{2}{k})w(OPT(I))$, where $OPT(I)$ is the optimal solution on I and $w(OPT(I))$ is the value of the solution. Then, solve these sub-instances using a dynamic programming procedure and return the solution with best value.

4.1 Deriving Sub-instances

For simplicity, we assume that diameter of each disk is 1 and each disk is open disk. Draw a grid consisting of hyperplanes $x = i$ (for $i \in Z$) perpendicular to X-axis and hyperplanes $Y = j$ (for $j \in Z$) perpendicular to Y-axis. The distance between every two parallel neighbor hyperplanes is the diameter of unit disk. So each cylinder will be hit by at most one hyperplane perpendicular to X-axis and at most one hyperplane line perpendicular to Y-axis.

For each i, j belong to $\{0, 1, \cdots, k-1\}$, we compose a sub-instance $I_{i,j}$ containing all cylinders except those being hit by a hyperplane from $\{x = p \mid p \mod k = i\}$ or hit by a hyperplane from $\{y = p \mid p \mod k = j\}$. There are k^2 different sub-instances. For each sub-instance, we calculate its optimal solutions using the dynamic programming in every of the $k \times k$ grids, which will be described in detail in the next subsection. We first establish a technical lemma for the performance guarantee (See [3] for proofs).

Lemma 1. *For at least one sub-instance $I_{i,j}$, $0 \leq i, j < k$, the value of the solution $w(OPT(I_{i,j})) \geq (1 - \frac{2}{k})w(OPT(I))$*

For each sub-instance, each disk is in a $k \times k$ grid. Disks in different grid won't intersect each other. So the union of solutions of all $k \times k$ grids is independent. We show that the problem in a $k \times k$ grid is polynomially solvable as follows.

4.2 Dynamic Programming

We describe a dynamic programming approach to find a maximum weighted independent set for an instance $I_{i,j}$, which only considers cylinders contained in one $k \times k$ cell of $I_{i,j}$. For these cylinders, we sort them in non-decreasing order of their ending time e_i. For simplicity, let $\mathbf{B}_1$, $\mathbf{B}_2$ $\cdots$, $\mathbf{B}_n$ be the n cylinders contained in one cell in the sorted order. In the rest of subsection, we use i to denote the cylinder $\mathbf{B}_i$.

Definition 1. *Pile: A pile is an ordered collection $\langle j_1, j_2, \cdots, j_q \rangle$ of pairwise non-intersecting cylinders that intersect a common hyperplane $z = b$ (for some value b) perpendicular to Z-axis. Here j_i is the ending time of a cylinder $\mathbf{B}_{j_i}$ and $j_t < j_{t+1}$.*

Given a hyperplane $z = b$ for some fixed value b, in a pile that was hit by $z = b$, there are at most $2k^2$ cylinders in the pile. Notice that all cylinders in this pile intersect the hyperplane $z = b$ and are disjoint from each other. Then an area argument implies that the number of cylinders in a pile is at most $k^2 / \frac{\pi}{4} < 2k^2$.

Lemma 2. *The total number of all possible piles in each $k \times k$ grid is polynomial of n.*

Proof. Please see [3].

Our dynamic programming approach will first sort the piles based on an order defined below; then find the optimum solution of all cylinders that are ordered in front of a pile.

Definition 2. *Define an order on the set of piles as following:* $\langle j_1, j_2, \cdots, j_l \rangle \prec \langle i_1, j_2, \cdots, i_m \rangle$ *if (1)* $j_l < i_m$, *or (2)* $j_l = i_m$ *and* $\langle j_1, j_2, \cdots, j_{l-1} \rangle \prec \langle i_1, j_2, \cdots, i_{m-1} \rangle$, *or (3)* $m = 0$.

Definition 3. *We define* $OPT(X \mid j_1, j_2, \cdots, j_l)$ *to be the maximum total weight of pairwise non-overlapping cylinders from set* X *given that cylinders* $j_1, j_2, \cdots, j_l$ *already occupy their places. For a set of non-overlapping cylinders* $\{j_1, j_2, \cdots, j_l\}$ *and a cylinder* t *with* $t < j_1$, *we define the marginal contribution, denoted as* $R_{j_1, j_2, \cdots, j_l}(t)$, *of cylinder* t *to* $OPT(i : i \leq t \mid j_1, j_2, \cdots, j_l)$ *as* $(OPT(i \leq t | j_1, j_2, \cdots, j_l) - OPT(i < t | j_1, j_2, \cdots, j_l))^+$. *Here* $(x)^+ = \max\{0, x\}$.

Based on the above definition, a cylinder t has positive marginal contribution $R_{j_1, j_2, \cdots, j_l}(t)$ will clearly be used in an optimum solution $OPT(i \leq t | j_1, j_2, \cdots, j_l)$. If its marginal contribution is 0, then it means that there is one optimum solution $OPT(i \leq t | j_1, j_2, \cdots, j_l)$ that will not use the cylinder t.

As proved in [11], it is easy to show that

$$OPT(i \leq t | j_1, j_2, \cdots, j_l) = \sum_{i=1}^{t} R_{j_1, j_2, \cdots, j_l}(i)$$

and $R_{j_1, j_2, \cdots, j_l}(t) = b_t + OPT(i < t | j_1, \cdots, j_l, t) - OPT(i < t | j_1, \cdots, j_l)$.

We then present our dynamic programming to find an optimal solution in each $k \times k$ cell as following Algorithm 1.

Algorithm 1. Find Maximum Weighted Independent Set of Cylinders in Each Cell

---

Input: A set S of weighted cylinders contained in a cell.
Output: An optimum maximum weighted independent set in S.

1: Find all the piles $\mathcal{P}$ of cylinders from S and sort them in the order of $\prec$;
2: Take the piles one by one in the order starting from the least one. For each pile $\langle j_1, j_2, \cdots, j_l \rangle$ calculate and store in the memory the value $R_{j_2, \cdots, j_l}(j_1)$ according to the formula: $R_{j_2, \cdots, j_l}(j_1) = \left(b_{j_1} + \sum_{i:i<j_1} R_{j_1, j_2, \cdots, j_l}(i) - \sum_{i:i<j_1} R_{j_2, \cdots, j_l}(i) \right)^+$
3: After all the piles corresponding values of $R_{j_2, j_3, \cdots, j_l}(j_1)$ is calculated for every possible pile $p = \langle j_1, j_2, \cdots, j_l \rangle \in \mathcal{P}$, we schedule the cylinders in the following way. Consider cylinders from S in the order of decreasing number. Take a cylinder j next in that order. Suppose that cylinders $\{j_1, j_2, \cdots, j_l\}$ are already scheduled, then schedule j iff $R_{j_1, j_2, \cdots, j_l}(j)$ is positive.

Theorem 1. *The running time of our dynamic programming is at most* $O(n^{2k^2+1})$.

Proof. Please see [3].

Then by setting $k = 2/\epsilon$, our method implies a PTAS in time $n^{\frac{1}{\epsilon^2}}$.

5   Algorithm for Problem YUD and YUM

In this section, we design approximation algorithms with a constant approximation ratio for problems YUD and YUM respectively.

For YUD, we assume there is only one channel available, the required region is a unit disk and required time is a duration. We can use the same graph of cylinders mentioned above for illustration. The main idea here is to partition cylinders into g groups. Solve the maximum weighted independent set in each group and then take the group with the best solution. By pigeonhole principle, it will give us $1/g$ approximation. We mainly will focus on designing a partition with minimum constant value g, please see [3] for details.

For YUM, where there is only one channel available, the required region is a unit disk and required time is a duration or an interval. The main idea is to partition users into 2 groups. One group G_d includes all users asking for a time duration and the other group G_i includes all users asking for a time-interval. We solve group G_d by the algorithm for YUD, solve group G_i by the PTAS for YUI, and then take the group with the best solution. It is a $1/10$ approximation algorithm for problem YUM, please see [3] for details.

6   Algorithm for Problem SUI

In this section, we design a $\Theta(\sqrt{m})$-approximation algorithm for problem SUI. Notice that the set packing problem is a special case of the problem SUI due to the observation in [3]. Recall that set packing problem is not approximable within $m^{1/2-\varepsilon}$ for any $\varepsilon > 0$, unless NP=ZPP. Thus, we have

Theorem 2. *Problem* SUI *is not approximable within* $m^{1/2-\varepsilon}$ *for any* $\varepsilon > 0$, *unless NP=ZPP.*

Assume that each bidder bids at most k frequencies and single time interval.

When $k = 1$, obviously the optimum solution is the union of the optimum solutions OPT_i, where OPT_i is the optimum solution for the set of users who do bid for frequency $\mathbf{f}_i$. We also use OPT to denote the global optimal solution hereafter. Notice that OPT_i can be solved by dynamic programming.

When $k > 1$, for each frequency $\mathbf{f}_i$, let $OPT|_i$ be the set of users in OPT that bid for frequency $\mathbf{f}_i$. Obviously, we have $\sum_{i=1}^{m} OPT|_i \geq k \times OPT$, since each user in OPT will appear in at least k different $OPT|_i$. Thus, $\max\{OPT|_1, OPT|_2, \cdots, OPT|_m\} \geq \frac{k}{m} \times OPT$.

We partition the bidders into two groups:

1. Z_1 contains all the bidders that bid at least $\sqrt{m}$ frequencies;
2. Z_2 contains all the other bidders, *i.e.*, bid less than $\sqrt{m}$ frequencies.

We approximate optimal solution for each group and return the larger one. We will show $\Theta(\sqrt{m})$-approximate algorithms for both groups respectively. Then, the maximum of these two solution must give us a $\Theta(\sqrt{m})$ approximation.

First for group Z_1, we just use $\max\{OPT_1, OPT_2, \cdots, OPT_m\}$ as our solution. Since each bidders bids at least $k \geq \sqrt{m}$ frequencies, $\max_{i=1}^{m} OPT(Z_1)|_i \geq \frac{\sqrt{m}}{m} OPT(Z_1) = \frac{1}{\sqrt{m}} OPT(Z_1)$. Here $OPT(Z_j)$ is the optimal solution for users in group Z_j for $j = 1, 2$. For bidders in Z_1, we use I_i to denote the set of bidders in Z_1 that bid for frequency $\mathbf{f}_i$. Then we find the maximum weighted independent set OPT_i of I_i using standard dynamic programming. Obviously, $OPT_i \geq OPT(Z_1)|_i$. Thus, $\max_{i=1}^m OPT_i \geq \max_{i=1}^m OPT(Z_1)|_i \geq \frac{1}{\sqrt{m}} OPT(Z_1)$.

For group Z_1, we convert this problem into Scheduling Split Intervals Problem (SSIP) [1]. The ordinary SSIP considers scheduling jobs that are given as groups of non-intersecting segments on the real line. Each job J_i is associated with an interval, I_j, which consists of up to t segments and a positive weight, w_j. Two jobs are in conflict if any of their segments intersect. The objective is to schedule a subset of non-conflicting jobs with maximum total weight.

A $2t$-approximation algorithm for problem SSIP is given in [1]. Here we create a special instance for SSIP problem as follows. Let $[0, T]$ be the original time period where bidders can place their time-interval. We then create a bigger time period $[0, m \cdot T]$ where m is the total number of frequencies. Then a user i will be associated with following $t_i = |\mathcal{F}_i| < \sqrt{m}$ segments in $[0, m \cdot T]$: $\{[(j - 1) \cdot T + s_i, (j - 1) \cdot T + e_i] \mid \mathbf{f}_j \in \mathcal{F}_i, 1 \leq j \leq m\}$ In other words, we duplicate the period $[0, T]$ m times for each of the frequencies in $\mathcal{F}$, and a user i will have a segment in the jth duplication if it bids for frequency $\mathbf{f}_j$. Then there are at most $\sqrt{m}$ segments for every bidder. Then based on algorithms in [1], we get $2\sqrt{m}$-approximation solution, *i.e.*, find a solution with value at least $\frac{OPT(Z_2)}{2\sqrt{m}}$.

Then the maximum of the above two solutions is at least

$$\max \left(\frac{OPT(Z_1)}{\sqrt{m}}, \frac{OPT(Z_2)}{2\sqrt{m}} \right) \geq \frac{1}{3\sqrt{m}} OPT,$$

since either $OPT(Z_1) \geq \frac{1}{3} OPT$, or $OPT(Z_2) \geq \frac{2}{3} OPT$ from $OPT(Z_1) + OPT(Z_2) \geq OPT$.

Notice that by using a different group partition, where group Z_1 contains the bidders that bid for at least $\sqrt{\frac{m}{2}}$ frequencies, we get an algorithm with approximation ration $\frac{\sqrt{2}}{4\sqrt{m}}$.

As a byproduct of our above results, we show that for problem SSIP, there is no approximation algorithm with ratio $O(t^{1-\epsilon})$ for any $\epsilon > 0$ unless $NP = ZPP$.

Theorem 3. *For problem SSIP, there is no polynomial-time approximation algorithm with ratio $O(t^{1-\epsilon})$ for any $\epsilon > 0$ unless $NP = ZPP$.*

Proof. Please see [3].

We can extend the above case. such that each bidder requires at most t time intervals. Obviously, $t \leq \lfloor |\mathcal{F}|/2 \rfloor$, every pair of users have a common requested frequency. Thus, for this case, computing the approximation solution for each frequency is exactly the traditional SSIP. We can get $2t$-approximation solution for this special case. Otherwise, we have a $\Theta(t\sqrt{m})$-approximation algorithm by using similar method above, please see [3] for details.

7   Strategyproof Mechanism Design

In this section, we show how to design a straightforward mechanism, *i.e.*, for each secondary user, reporting its valuation truthfully maximizes its profit, based on the algorithms discussed in previous sections. A strategyproof mechanism can designed by using monotone output algorithm and critical value payment scheme. An output algorithm is monotone if an agent will still participate in the output when it increase its bid. We define a critical value θ_i, *i.e.*, the minimum valuation v_i that makes agent i participate in the output. A critical value payment scheme P^O for an algorithm O such that $p_i = \theta_i$ if agent i is in the output, otherwise, $p_i = 0$. If O is a monotone algorithm and P^O is a critical value payment scheme for O, $M = (O, P^O)$ is a strategyproof mechanism.

In the algorithms mentioned above, we used technology of dynamic programming, grouping and the classic FPTAS for knapsack problem. Dynamic programming and grouping do not affect the monotone property. However, the classic FPTAS for knapsack problem is not monotone. A counterexample is given in [3]. In [4], Briest proposed an alternative rounding scheme that transform a pseudopolynomial algorithm into a monotone FPTAS for knapsack problem. Using this FPTAS for knapsack problem, all algorithms in this paper are monotone. Therefore we can design strategyproof mechanisms $M = (O, P^O)$ for all problems discussed in previous sections.

8   Literature Reviews

The problems we discussed above are at the intersection of a lot of famous problems. Here we review results for some of these problems.

Knapsack problem, which is same as simple problem YOD, has a classic FPTAS by the means of rounding. However, we cannot design a strategyproof mechanism using this FPTAS since it is not monotone. An alternative rounding scheme was proposed by Briest in [4], which gave a new rounding scheme leading to a monotone FPTAS for knapsack problem.

In [10], Jansen and Zhang presented a $(2 + \epsilon)$-approximation algorithm for rectangle packing problem which is similar with our problem YUI. If we don't consider intersections in space, it is a unit-height rectangle packing problem, which is a special case of rectangle packing. Kovaleva described a PTAS for unit-height rectangle packing problem in [11]. We extend this PTAS to a PTAS for problem YUI as described above.

Problem SUI is a special case of set packing problem. In [8], Hastad proved that set packing problem cannot be approximable within $m^{\frac{1}{2}-\epsilon}$, unless $NP = ZPP$. Another special case of set packing problem is Scheduling Split Intervals Problem(SSIP). Bar-Yehuda *et al.* [1] gave a $2t$-approximation algorithm. We show how to convert the problem SUI to SSIP in previous section.

9   Conclusions

In this paper, we combine the game theory with communication modeling to solve the channel assignment problem. We study how to assign the spectrum and charge the secondary users such that the *overall social benefits* is maximized.

More specifically, we formalize several versions of spectrum assignment problems by separately assuming channel, region and time requirements. We also show how to design strategyproof mechanism based those algorithms. We leave it as a future work whether there are efficient approximation algorithms for problem SUD, where single-minded secondary users request unit disk region and a time duration, whether there are PTASs for problems YOM, YUD and YUM.

References

1. Bar-Yehuda, R., Halldórsson, M.M., Naor, J.S., Shachnai, H., Shapira, I.: Scheduling Split Intervals. In: ACM SODA 2002, pp. 732–741 (2002)
2. Bartal, Y., Gonen, R., Nisan, N.: Incentive Compatible Multi Unit Combinatorial Auctions. In: TARK 2003, pp. 72–87 (2003)
3. Li, X.Y., Xu, P., Tang, S.J., Chu, X.W.: Spectrum Bidding in Wireless Networks and Related. Technical Report, IIT (2008)
4. Briest, P., Krysta, P., Vöcking, B.: Approximation Techniques for Utilitarian Mechanism Design. In: ACM STOC 2005, pp. 39–48 (2005)
5. Chekuri, C., Khanna, S.: A PTAS for the Multiple Knapsack Problem. In: ACM SODA 2000, pp. 213–222 (2002)
6. Clarke, E.H.: Multipart Pricing of Public Goods. Public Choice, 17–33 (1971)
7. Groves, T.: Incentives in Teams. Econometrica, 617–631 (1973)
8. Hastad, J.: Clique is Hard to Approximate within $n^{1-\varepsilon}$. Acta Mathematica, 105–142 (1999)
9. Hochbaum, D.S., Maass, W.: Approximation Schemes for Covering and Packing Problems in Image Processing and VLS. Journal of ACM 32, 130–136 (1985)
10. Jansen, K., Zhang, G.: On Rectangle Packing: Maximizing Benefits. In: ACM SODA 2004, pp. 204–213 (2004)
11. Kovaleva, S.: Improved Dynamic Programming Subroutine in the PTAS for the unit-Height Rectangle Packing Problem. In: APPOL II Workshop (2002)
12. Lehmann, D.J., O'Callaghan, L.I., Shoham, Y.: Truth Revelation in Approximately Efficient Combinatorial Auctions. In: ACM Conf. on Electronic Comm, pp. 96–102 (1999)
13. Osborne, M.J., Rubinstein, A.: A Course in Game Theory. The MIT Press, Cambridge (2002)
14. Papadimitriou, A.A.C., Talwar, K., Tardos, E.: An Approximate Truthful Mechanism for Combinatorial Auctions with Single Parameter Agents. In: ACM SODA 2003, pp. 205–214 (2003)
15. Stine, J.A.: Spectrum Management: The Killer Application of ad hoc and Mesh Networking. In: IEEE Symp. on New Frontiers in Dynamic Spectrum Access Net (2005)
16. Vickrey, W.: Counterspeculation, Auctions and Competitive Sealed Tenders. Journal of Finance, 8–37 (1961)

$(1 + \rho)$-Approximation for Selected-Internal Steiner Minimum Tree*

Xianyue Li[1], Yaochun Huang[2], Feng Zou[2], Donghyun Kim[2], and Weili Wu[2]

[1] School of Mathematics and Statistics, Lanzhou University,
Lanzhou, Gansu, 730000, P.R. China
lixianyue@lzu.eud.cn
[2] Department of Computer Science, University of Texas at Dallas,
Richardson, TX 75080, USA
{yaochun.huang,phenix.zou,donghyunkim}@student.utdallas.edu,
weiliwu@utdallas.edu

Abstract. Selected-internal Steiner minimum tree problem is a generalization of original Steiner minimum tree problem. Given a weighted complete graph $G = (V, E)$ with weight function c, and two subsets $R' \subsetneq R \subseteq V$ with $|R - R'| \geq 2$, selected-internal Steiner minimum tree problem is to find a Steiner minimum tree T of G spanning R such that any leaf of T does not belong to R'. In this paper, suppose c is metric, we obtain a $(1+\rho)$-approximation algorithm for this problem, where ρ is the best-known approximation ratio for the Steiner minimum tree problem.

1 Introduction

Given a weighted complete graph $G = (V, E)$ with weight function c, and a subset R, Steiner minimum tree problem is to find a minimum subtree of G spanning R. Steiner tree can be applied in many fields such as VLSI routing [11], network routing [12], phylogeny [5],[10], et.al [2],[4],[8]. In the past years, some generalizations of Steiner minimum tree problem are arisen, such as Steiner minimum tree problem on some special metric space [6],[7], the full Steiner tree problem [3],[13] and the k-size Steiner tree problem [1].

In this paper, we study selected-internal Steiner minimum tree problem, also a generalization of Steiner minimum tree problem. Given a weighted complete graph $G = (V, E)$ with weight function c, and two subsets $R' \subsetneq R \subseteq V$ with $|R - R'| \geq 2$, selected-internal Steiner minimum tree problem is to find a Steiner minimum tree T of G spanning R such that any leaf of T does not belong to R'. Since Steiner minimum tree problem is a special case ($R' = \emptyset$) of this problem, the NP-completeness and MAX SNP-hardness of this problem can immediately follow from the hardness results of Steiner minimum tree problem [9]. Hsieh and Yang [9] gave a 2ρ-approximation algorithm for this problem when c is metric,

---

* Support in part by National Science Foundation under grant CCF-9208013 and CCF-0728851.

X. Hu and J. Wang (Eds.): COCOON 2008, LNCS 5092, pp. 568–576, 2008.

where $\rho = 1 + \frac{\ln 3}{2} \approx 1.55$ is the best-known approximation ratio for Steiner minimum tree problem [14].

In this paper, we present a $(1 + \rho)$-approximation algorithm for this problem when c is metric. Given a complete graph $G = (V, E)$ with weight function c on its edges and a set $R \subseteq V$, let T be a Steiner tree of G spanning R. We call an edge as an *in-edge* if both its endpoints belong to R, otherwise call it an *out-edge*. Denote E_{in}^T and E_{out}^T as the sets of the in-edges and out-edges of T, respectively. For two subsets V_1 and V_2 of $V(G)$, denote $dist(V_1, V_2) = \min_{v_1 \in V_1, v_2 \in V_2} c(v_1 v_2)$ as the distance between V_1 and V_2.

The rest of this paper is organized as follows. In section 2, we present a $(1+\rho)$-approximation algorithm for selected-internal Steiner minimum tree problem by 3 subsections. In subsection 2.1, we firstly give Algorithm 1 to divide R to a pairwise disjoint tree sequence $\mathcal{T} = \{T_1, \ldots T_m\}$. Then, we contract every T_i to v_{T_i} to obtain a new graph G_1 and set $R_1 = \{v_{T_i}\}_{i=1}^m$. Based on Algorithm 1, we present Algorithm 2 to construct a Steiner tree T' of G_1 spanning R_1 with approximation ratio ρ and a Steiner tree T of G spanning R such that there is a Steiner minimum tree T_{opt} of G spanning R, satisfying

$$E(T) = E(T') \bigcup E(\mathcal{T}), \ E(\mathcal{T}) \subseteq E_{in}^T \text{ and } c(E(T) \setminus E(\mathcal{T})) \leq \rho \, c(E(T_{opt}) \setminus E(\mathcal{T})). \quad (1)$$

In subsection 2.2, we modify T' to a new Steiner Tree T'' of G_1 spanning R_1 with approximation ratio $(1 + \rho)$. In T'', every v_{T_i} has degree no less than 2 unless T_i has vertex belonging to $R - R'$. In final subsection, we "blossom" very v_{T_i} of T'' to T_i and modify it to a selected-internal steiner tree T''' of G spanning R with approximation ratio $(1 + \rho)$. Finally, we conclude our results and discuss a special case when $R = V$, which is the selected-internal minimum spanning tree problem.

2 $(1 + \rho)$-Approximation Algorithm

In this section, we will give a $(1+\rho)$-approximation algorithm for selected-internal Steiner minimum tree problem in the following three steps.

2.1   Construction of T' and T

In this subsection, we first give Algorithm 1 to divide R to a disjoint tree sequence $\mathcal{T} = \{T_1, \ldots T_m\}$. Then, we contract every T_i to v_{T_i} to obtain a new graph G_1 and set $R_1 = \{v_{T_i}\}_{i=1}^m$. Based on Algorithm 1, we present Algorithm 2 to construct a Steiner tree T' of G_1 spanning R_1 with approximation ratio ρ and a Steiner tree T of G spanning R satisfying (1).

Algorithm 1
Input: A weighted complete graph $G = (V, E)$ with weight function c and a set $R \subseteq V$.
Output: A sequence of pairwise disjoint subtrees $\mathcal{T} = \{T_1, T_2, \ldots\}$ such that

$\bigcup V(T_i) = R$.

1. Set $S = R$, $i = 1$ and $T = \emptyset$. /*S records the vertices in R not covered by T */

2. **If** $S = \emptyset$, output T, stop;

   Else Choosing a vertex v of S and creating a new tree T_i, set $V(T_i) := \{v\}$, $E(T_i) := \emptyset$ and $S := S - \{v\}$.

3. Search $V - V(T_i)$ with the vertex ordering $S, R - S - V(T_i), V - R$, finding the first vertex u such that $dist(u, T_i) = \min\limits_{w \in V(G - T_i)} dist(w, T_i)$. Let the corresponding edge be ut.

4. **If** $u \in S$, set $V(T_i) := V(T_i) \bigcup \{u\}$, $E(T_i) := E(T_i) \bigcup \{ut\}$ and $S := S - \{u\}$. Goto 3.

   If $u \in R - S - V(T_i)$, then u is contained in some T_j with $1 \leq j < i$. Set $T_j := T_j \bigcup T_i + ut$ and delete T_i. Goto 2.

   If $u \in V - R$ or $u =$NULL, set $T := T \bigcup T_i$ and $i := i + 1$. Goto 2.

At the step 4, if $u \in V - R$, then a new tree T_i will be included in T and holds the following property: there is a vertex $w \in V - R$ such that

$$dist(w, T_i) < dist(v, T_i) \text{ for any } v \in R - V(T_i). \qquad (2)$$

We claim that when the second case of step 4 ($u \in R - S - V(T_i)$) is true, if the old tree T_j follows property (2), the modified tree $T'_j := T_j \bigcup T_i \bigcup \{e'\}$ also follows, where e' is a shortest edge between $V(T_i)$ and $V - V(T_i)$ and one of its endpoint is in T_j. For this purpose, since T_j has property (2), there is $w \in V - R$ such that $dist(w, T_j) < dist(v, T_j)$ for any other $v \in R - T_j$. Denote the corresponding edge as e. It implies that $c(e) < c(e')$. For any edge e'' between $V(T_i)$ and $R - V(T'_j)$, $c(e'') \geq c(e') > c(e)$. Hence, the modified tree T'_j also has the property (2).

By recurrence, we obtain the following result.

Lemma 1. *Let $T_1, T_2, \ldots, T_m$ be the output tree sequence of Algorithm 1. If $R \neq V$, for any T_i, there is a vertex $w \in V - R$ such that $dist(w, T_i) < dist(v, T_i)$ for any $v \in R - V(T_i)$.*

Lemma 2. *For a weighted complete graph $G = (V, E)$ with weight function c and a set $R \subseteq V$, let $T_1, T_2, \ldots, T_m$ be the output of Algorithm 1. Then there exists a Steiner minimum tree T of G spanning R such that every T_i is a subtree of T.*

Proof. Firstly, order edges of $E' = \bigcup\limits_{i=1}^{m} E(T_i)$ as $\{e_1, e_2, \ldots\}$, according to their order of appearance in Algorithm 1. Among all Steiner minimum trees of G spanning R, choose T satisfying the following conditions: (1) $|E(T) \bigcap E'|$ as large as possible; (2) under the condition (1), choosing T such that the index of the first edge in $E' \setminus E(T)$ as large as possible. We shall show that T contains $T_1, \ldots, T_m$ as subtrees.

Suppose to the contrary that the first edge in $E' \setminus E(T)$ is $e_j = uv$. By Algorithm 1, there exists some stage at which there is a subtree T' (which is a subtree of some T_i) such that e_j is a shortest edge between $V(T')$ and $V - V(T')$. Since T is a tree, there is a unique path P between u and v on T. Let e be the unique edge on P between $V(T')$ and $V - V(T')$. Then $c(e_j) \leq c(e)$ and $\widetilde{T} = T + e_j - e$ is also a Steiner tree. Since T is a Steiner minimum tree, we have that $c(e_j) = c(e)$ and $c(T'') = c(T)$. By the choice of T, e must be in E', otherwise $|E(\widetilde{T}) \cap E'| > |E(T) \cap E'|$. Let $e = e_t$. By the structure of T', we have $t > j$. But then, the index of the first edge in $E' \setminus E(\widetilde{T})$ is larger than that of $E' \setminus E(T)$, contradicting to condition (2). So, T contains all edges of $T_1, \ldots, T_m$. $\qquad\square$

Now, we construct a new graph G_1 by contracting every T_i to a new vertex, denoted it by v_{T_i}. Let $R_1 = \{v_{T_1}, \ldots, v_{T_m}\}$. By Lemma 2, we obtain the following corollary.

Corollary 3. *For any Steiner minimum tree T_{R_1} of G_1 spanning R_1, if we "blossom" every v_{T_i} to T_i, then the new tree T_R is a Steiner minimum tree of G spanning R.*

Proof. Suppose to the contrary that the new tree T_R is not a Steiner minimum tree of G spanning R. By Lemma 2, there is a Steiner minimum tree T'_R spanning R such that every T_i is a subtree of T'_R and $c(T'_R) < c(T_R)$. Contract every T_i to a new vertex in T'_R. The resulting tree is a Steiner tree of G_1 spanning R_1 which has less weight than T_{R_1}, a contradiction. $\qquad\square$

Based on Algorithm 1, we present the following algorithm which constructs two Steiner trees T' and T. By Corollary 3, there exists a Steiner minimum tree T_{opt} such that T satisfying (1).

Algorithm 2
Input: A weighted complete graph G with weight function c on $E(G)$ and a set $R \subseteq V$.
Output: Two Steiner trees T' and T.
1. Use Algorithm 1 to obtain $T_1, \ldots, T_m$.
2. Contract T_i to v_{T_i} $(i = 1, \ldots, m)$ to construct G_1. Set $R_1 = \{v_{T_i}\}_{i=1}^m$.
3. Use a ρ-approximation algorithm to obtain a Steiner tree T' of G_1 spanning R_1.
4. "Blossom" v_{T_i} to T_i $(i = 1, \ldots, m)$ to form a Steiner tree T of G spanning R.

2.2 Modification of T'

In subsection 2.1, we obtain a Steiner Tree T' of G_1 spanning R_1 with approximation ratio ρ to the Steiner minimum Tree of G_1 spanning R_1. In this subsection we modify T' to a new Steiner tree T'' with approximation ratio $(1 + \rho)$. In T'', every v_{T_i} has degree no less than 2 unless T_i has vertex belonging to $R - R'$. In the following study, we assume that the weight function c is *metric*.

If $m = 1$, obviously v_{T_1} is the Steiner minimum tree T' of G_1 spanning R_1. Set $T'' = T'$. If $m = 2$, let e be the edge connecting v_{T_1} and v_{T_2}. Then $T' = (\{v_{T_1}, v_{T_2}\}, \{e\})$ is a Steiner minimum tree. Since $|R - R'| \geq 2$, one of T_1 and T_2, say T_1 should contain some vertex of $R - R'$. By Lemma 1, there is a vertex $w \in V - R$ such that $c(wv_{T_2}) = dist(w, T_2) < dist(T_1, T_2) = c(e)$. Set $T'' = T' + \{wv_{T_2}\}$. Then

$$c(T'') = c(T') + c(w, v_{T_2}) < 2c(e) = 2c(T').$$

In the following, suppose $m \geq 3$. Firstly, we give some useful results.

Lemma 4. *Let G be a weighted complete graph on at least two vertices with metric weight function c and T a minimum spanning tree of G. Then, for any two vertices u and v, G has a hamilton cycle C such that edge $uv \in E(C)$ and $c(C) \leq 2c(T)$.*

Proof. We use induction on $|G|$. If $|G| = 2$, the digon through u and v is as desired. Suppose $|G| = k \geq 3$ and the assertion is true for all graphs with less than k vertices. If there is a leaf w different from u and v, let $e = ws$ be the unique edge incident with w and $\widetilde{T} = T - w$. Then $\widetilde{T}$ is a minimum spanning tree of $G[V(\widetilde{T})]$ and $\{u, v\} \subseteq V(\widetilde{T})$. By inductive hypothesis, there is a hamilton cycle C' of $G[V(\widetilde{T})]$ with $c(C') \leq 2c(\widetilde{T})$ and $uv \in E(C')$. Let t be a neighbor of s on C' with $st \neq uv$. Set $C = C' - st + sw + wt$. Then

$$\begin{aligned} c(C) &= c(C') - c(st) + c(sw) + c(wt) \\ &\leq 2c(\widetilde{T}) + 2c(ws) \\ &= 2c(T). \end{aligned}$$

If such a vertex w does not exists, then T is a hamiltion path from u and v, and $C = T + uv$ is as desired.   □

As a consequence, we have

Corollary 5. *Let G be a weighted complete graph with metric weight function c and T a minimum spanning tree of G. For any two vertices u and v, there exists a hamilton path P from u to v such that $c(P) \leq 2c(T)$.*

In the following, we modify T' for the case $m \geq 3$. Let Q be the set of all leaves of T' in R_1. Then $Q \subseteq R_1$. If $|Q|$ is odd, we modify Q as follows. If $Q = R_1$, then there exists a vertex v_{T_i} such that T_i contains some vertex of $R - R'$. In this case, set $Q := Q - v_{T_i}$. If $Q \neq R_1$, then there exists a vertex $v \in R_1 - Q$ and set $Q := Q \bigcup \{v\}$. Now, Q is a subset of R_1 and $|Q|$ is even. Find a minimum perfect matching M of Q in polynomial time. We claim that $c(M) \leq c(T_{opt})$, where T_{opt} is a Steiner minimum tree of G_1 spanning R_1. For this purpose, let C be a hamilton cycle as in Lemma 4. "Short cut" C to a cycle C' containing only vertices of Q. Since c is a metric weight function, we have $c(C') \leq c(C) \leq 2c(T_{opt})$. Then, one of the two perfect matchings of C', say M_1, satisfies $c(M_1) \leq c(T_{opt})$. The claim follows from $c(M) \leq c(M_1)$.

Now, add the edges of M one by one to modify $T^{'}$ such that every v_{T_i}, for which T_i contains only vertices in $R^{'}$, has degree at least 2.

Let e ba an edge in m. Since $T^{'}$ is a tree, $T^{'} + e$ has the unique cycle C. If C contains all vertices of $T^{'}$, then there exists a vertex $v \in R_1 \subseteq V(C)$ such that T_v contains some vertex of $R - R^{'}$. If some neighbor u of v on C with $u \in V - R$, let $T^{'} := T^{'} + e - uv$. Else, the both neighbors of v on C belong to R_1. Let u be one of them. By Lemma 1, there exists a vertex $w \in V(G_1) - R_1$ such that $c(wu) < c(uv)$. If w is not on C, let $T^{'} := T^{'} + e + uw - uv$. If $w \in V(C)$, let s and t be the neighbors of w on C and $T^{'} := T^{'} + e + uw - uv + st - sw - wt$. Then the new tree $T^{'}$ is a Steiner tree of G_1 on R_1 such that the degree of any vertex of R_1 is not less than 2 except v (but it does not matter since T_v contains some vertices in $R - R^{'}$), and the weight of new tree is less than the weight of the old tree plus $c(e)$ (by $c(uw) < c(uv)$ and the assumption that c is metric).

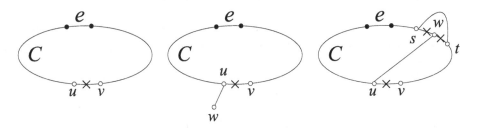

Fig. 1. The three types of new tree

If C does not contain all vertices of $T^{'}$, there exists a vertex v of $V(C)$ such that $d_{T^{'}}(v) \geq 3$. If some neighbor u of v on C has degree more than 2 or $u \in V(G_1) - R_1$, let $T^{'} := T^{'} + e - uv$. Otherwise, let u be a neighbor of v on C. Then $u \in R_1$ with degree 2 in $T^{'}$. By Lemma 1, there exists a vertex $w \in V(G_1) - R_1$ such that $c(wu) < c(uv)$. If $w \notin V(T^{'})$ or $w \in C$, then $T^{'}$ can be modified similarly to the above case. Hence, suppose $w \in V(T^{'} - C)$. In this case, the graph $H_1 = T^{'} + e + uw - uv$ has the unique cycle C_1. Clearly, $c(H_1) \leq c(T^{'}) + c(e)$. If $d_{T^{'}}(w) \geq 2$, let $s \neq u$ be the other neighbor of w on C_1, and t be another neighbor of w on $T^{'}$. Set $T^{'} := H_1 - sw - wt + st$. In all above case, the new tree $T^{'}$ is a Steiner tree satisfying the following conditions:

(1). all vertices of $V(C) \cap R_1$ have degree at least 2,
(2). the degree of any other vertex of R_1 does not alter, and
(3). the weight of the new tree is less than the weight of the old tree plus $c(e)$.

Next, we consider the final case $w \in V(T^{'} - C)$ with $d_{T^{'}}(w) = 1$. Let s be the other neighbor of w on H_1. Set $H_2 := H_1 + us - w$. Then H_1 contains a Steiner tree of G_1 spanning R_1, and $c(H_2) \leq c(H_1) \leq c(T^{'}) + c(e)$. Furthermore, H_2 has unique cycle C_2 and $|V(H_2) - R_1| < |V(H_1) - R_1|$. Repeat the above procedure. Since $|V(G_1) - R_1|$ is finite, there is a stage where we come across a case different

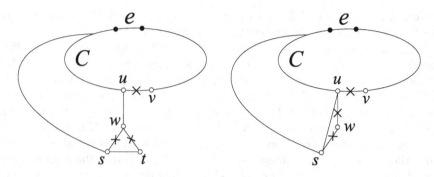

Fig. 2. The two cases of $w \in V(T' - C)$

from the final one, at which a modified tree T' satisfying conditions (1)-(3) is obtained.

After all edges of M having been added and all corresponding modifications having been made, let the final tree be T''. Then T'' is a Steiner tree of G_1 spanning R_1 such that v_{T_i} has degree 1 only when T_i contains vertex in $R - R'$. Furthermore,

$$c(T'') \le c(T') + c(M) \le (1 + \rho)c(T_{opt}).$$

Lemma 6. *We can obtain a Steiner tree T'' of G_1 spanning R_1 with approximation ratio $(1 + \rho)$ in polynomial time. In T'', every v_{T_i} has degree no less than 2 unless T_i contains vertex in $R - R'$.*

2.3 "Blossom" Every v_{T_i} of T'' and Modification

In this subsection, we "blossom" every v_{T_i} to T_i and modify it to a new steiner tree T''' of G spanning R such that every vertex of R' is not a leaf.

If $d_{T''}(v_{T_i}) \ge 2$ and there are at least two vertices u and v of T_i connected to vertices outside of T_i. By Corollary 5, there is a path P from u and v containing all vertices of T_i with $c(P) \le 2c(T_i)$. Replace T_i by P. If $d_{T''}(v_{T_i}) \ge 2$ and there is only one vertex u of T_i connected to vertices outside of T_i. Choose two neighbors of u in $V - V(T_i)$, saying v and w. By Lemma 4, there is a cycle C containing all vertices of T_i with $c(C) \le 2c(T_i)$. Let t be a neighbor of u on C. Replace T_i by $C - ut$ and uw by wt. If $d_{T''}(v_{T_i}) = 1$, then T_i has a vertex $u \in R - R'$. Let v be the unique vertex connected to some vertex outside of T_i. If $u \ne v$, replace T_i by P, where P is a path from v to u containing all vertices of T_i with $c(P) \le 2c(T_i)$. If $u = v$, let w be the unique neighbor of u in $V - V(T_i)$ and t another vertex of T_i. By Lemma 4, there is a cycle C containing all vertices of T_i with $c(C) \le 2c(T_i)$ and $ut \in C$. Replace T_i by $C - vt$ and uw by wt.

After modifying all T_i as above, the resulting graph T''' is a Steiner tree of G spanning R such that every vertex of R' is not a leaf. By Corollary 3, let T_{R_1} be a Steiner minimum tree of G_1 spanning R_1 and T_R be a Steiner minimum tree of G spanning R with $\bigcup_{i=1}^{m} E(T_i) \subseteq E(T_R)$. Then

$$
\begin{aligned}
c(T''') &\leq (1 + \rho)c(T_{R_1}) + 2c(T_1) + \ldots + 2c(T_m) \\
&\leq (1 + \rho)(c(T_{R_1}) + c(T_1) + \ldots + c(T_m)) \\
&\leq (1 + \rho)c(T_R).
\end{aligned}
$$

Hence, we obtain the following theorem.

Theorem 7. *Given a weighted complete graph $G = (V, E)$ with metric weight function c on its edges. Let $R' \subset R \subseteq V$ with $|R - R'| \geq 2$. We can obtain a selected-internal Steiner tree with approximation ratio $(1 + \rho)$ in polynomial time.*

3   Conclusion and Discussion

In this paper, we present a $(1+\rho)$-approximation algorithm for selected-internal Steiner minimum tree problem. It would be interesting and challenging to design a better approximation algorithm or design a approximation algorithm for general wight function c.

Then, we discuss a special case when $R = V$, i.e., the selected-internal minimum spanning tree problem. This problem is NP-complete since when $|R - R'| = 2$, this problem is equivalent to the minimum Hamilton path problem. Similarly, by Corollary 5, we can obtain a 2-approximation algorithm for selected-internal minimum spanning tree problem.

References

1. Borchers, A., Du, D.Z.: The k-Steiner Ratio in Graphs. SIAM Journal on Computing 26(3), 857–869 (1997)
2. Cheng, X., Du, D.Z.: Steiner Trees in Industry. Kluwer Academic Publishers, Dordrecht, Netherlands (2001)
3. Du, D.Z.: On Component-Size Bounded Steiner Trees. Discrete Applied Mathematics 60, 131–140 (1995)
4. Du, D.Z., Smith, J.M., Rubinstein, J.H.: Advance in Steriner Tree. Kluwer Academic Publishers, Dordrecht, Netherlands (2000)
5. Foulds, L., Graham, R.: The Steiner Problem in Phylogeny is NP-complete. Advances in Applied Mathematics 3, 43–49 (1982)
6. Garey, M., Graham, R., Johnson, D.: The Complexity of Computing Steiner Minimal Tree. SIAM Journal on Applied Mathematics 32, 835–859 (1997)
7. Garey, M., Johnson, D.: The Rectilinear Steiner Problem is NP-complete. SIAM Journal on Applied Mathematics 32, 826–834 (1977)
8. Graur, D., Li, W.H.: Fundamentals of Molecular Evolution, 2nd edn. Sinauer Publishers, Sunderland, Massachusetts (2000)
9. Hsieh, S.Y., Yang, S.C.: Approximating the Selected-Internal Steriner Tree. Theoretical Computer Science 381, 288–291 (2007)

10. Hwang, F.K., Richards, D.S., Winter, P.: The Steiner Tree Problem. North-Holland, Amsterdam (1992)
11. Kahng, A.B., Robins, G.: On Optimal Interconnections for VLSI. Kluwer Publishers, Dordrecht (1995)
12. Korte, B., Prömel, H.J., Steger, A.: Steiner Trees in VLSI-Layouts. In: Korte, et al. (eds.) Paths, Flows and VLSI-Layout. Springer, Heidelberg (1990)
13. Lu, C.L., Tang, C.Y., Lee, R.C.T.: The Full Steiner Tree Problem. Theoretical Computer Science 306, 55–67 (2003)
14. Robins, G., Zelikovsky, A.: Improved Steiner Tree Approximation in Graphs. In: Proceedings of the 11th Annual ACM-SIAM Symposium on Discrete Algorithms (SODA), pp. 770–779 (2000)

Computing Maximum Flows in Undirected Planar Networks with Both Edge and Vertex Capacities

Xianchao Zhang[1], Weifa Liang[2], and Guoliang Chen[3]

[1] School of Software, Dalian University of Technology
Dalian, China, 116620
[2] Department of Computer Science, Australian National University
Canberra, ACT 0200, Australia
[3] Department of Computer, University of Science and Technology of China
Hefei, China, 230027
xczhang@dlut.edu.cn, wliang@cs.anu.edu.au, glchen@ustc.edu.cn

Abstract. We study the maximum flow problem in an undirected planar network with both edge and vertex capacities (EVC-network). A previous study reduces the minimum cut problem in an undirected planar EVC-network to the minimum edge-cut problem in another planar network with edge capacity only (EC-network), thus the minimum-cut or the maximum flow value can be computed in $O(n \log n)$ time. Based on this reduction, in this paper we devise an $O(n \log n)$ time algorithm for computing the maximum flow in an undirected general planar EVC-network and an $O(n)$ time algorithm for computing the maximum flow in an undirected (s, t)-planar EVC-network. As a result, the maximum flow problem in undirected planar EVC-networks is as easy as the problem in undirected planar EC-networks in terms of computational complexity.

1 Introduction

The maximum flow problem in a flow network with both edge and vertex capacities (EVC-network) is to find a flow between a pair of vertices such that the value of the flow is maximized. This is a classical combinatorial optimization problem with a wide variety of applications [1]. A special case of the problem is that only the edges in the network have capacities (EC-network), for which extensive studies have been conducted in the past half centuries, and the best algorithms are Goldberg and Tarjan's $O(nm \log(n^2/m))$ time algorithm for real capacity [4] and Goldberg and Rao's $O(\min(n^{2/3}, m^{1/2})m \log(n^2/m) \log U_e)$ time algorithm for integral capacity [5], where n is the number of vertices, m is the number of edges, and U_e is the maximum integral capacity among the edge capacities in the network.

The problem in planar EC-networks has been addressed, and efficient algorithms have been devised by exploiting the network planarity. In particular, Hu [9] transformed the minimum (edge-)cut problem in an (s, t)-planar EC-network into the shortest path problem in the dual network of the network,

X. Hu and J. Wang (Eds.): COCOON 2008, LNCS 5092, pp. 577–586, 2008.

where an (s, t)-planar network is such a planar network that both the source s and the sink t are on the same face. Hassin [7] observed that Hu's algorithm actually computes the maximum flow. Klein *et al* [11] presented a linear algorithm for the single shortest path problem in planar networks. This leads to an $O(n)$ time algorithm for the problem in (s, t)-planar EC-networks [11]. It is not difficult to see that there is an $O(n \log n)$ algorithm for the problem in a general undirected planar EC-network by incorporating the results due to Hassin and Johnson [8] and Klein *et al* [11]. Borradaile and Klein [3] provided an $O(n \log n)$ time algorithm for the problem in a general directed planar EC-network.

It is well known that the maximum flow problem in a general EVC-network can be easily reduced to the maximum flow problem in another EC-network [1]. However, this reduction does not maintain the network planarity if it is applied to a planar network [2]. As a result, the network's planarity can not be exploited if the traditional reduction is applied, and the maximum flow problem in a planar EVC-network takes $O(n^2 \log n)$ time or $O(n^{3/2} \log n \log(\max\{U_e, U_v\})$ time for real or integral capacity respectively, where U_v is the maximum integral capacity among the vertices in the network. Khuler and Naor [10] addressed the planarity-destruction problem of the traditional reduction. By adding some edges and vertices to the dual network of a planar EVC-network, they transformed the minimum cut problem in the primal network into the problem of finding the cut-cycle of the shortest length in the extended dual network, which can be further transformed into the shortest path problem. Their algorithm, incorporating with Klein et al 's [11] shortest path algorithm, can find the minimum cut (or the value of the maximum flow) in $O(n)$ time in an (s, t)-planar EVC-network or in $O(n \log n)$ time in a general planar EVC-network. For an (s, t)-planar EVC-network, they also proposed an algorithm for computing the maximum flow, which can be implemented in $O(n \log \log n)$ time by making use of Han's sorting algorithm [6]. Inspired by Khuler and Naor's transformation, Zhang et al [14] proposed a maintaining-planarity reduction that reduces the minimum cut problem in an undirected planar EVC-network to the minimum edge-cut problem in another planar EC-network. However, finding an algorithm for computing the maximum flow in a general planar EVC-network that takes the advantage of the network's planarity was open until a solution in this paper is proposed.

In this paper we show that the maximum flow in the primal planar EVC-network can be computed by finding a maximum flow in the auxiliary planar EC-network introduced in [14], mapping the flow to a pseudo-flow in the primal EVC-network, and then canceling cycle-flows in the pseudo-flow. Thus, the maximum flow in a general undirected planar EVC-network can be found in $O(n \log n)$ time. We then provide an $O(n)$ time algorithm for canceling cycle-flows in (s, t)-planar networks by showing that the maximum flow in an undirected (s, t)-planar EVC-network can be found in $O(n)$ time. Consequently, the maximum flow problems in undirected planar EVC-networks are as easy as the problems in undirected planar EC-networks in terms of computational complexity.

The rest of the paper is organized as follows. In Section 2 we introduce necessary notations and notions. In Section 3 we show how to compute the maximum flow in an undirected planar EVC-network. In Section 4 we propose an $O(n)$ time algorithm for the problem in an undirected (s,t)-planar EVC-network. In Section 5 we conclude our discussions.

2   Preliminaries

In this section we introduce some notations and concepts which are necessary for the rest of discussions.

Definition 1. *An undirected graph $G = (V, E)$ consists of a set V of vertices and a set of edges E whose elements are pairs of distinct vertices. Denote n and m the number of vertices and edges in N. A path in a graph G is a subgraph of G consisting of a sequence of vertices $v_1 - v_2 - \cdots - v_r$ and the edges between each consecutive pair of vertices in the sequence. A simple path is a path without any repetition of vertices. A cycle is a simple path $v_1 - v_2 - \cdots - v_r$ together with edge (v_1, v_r).*

Definition 2. *An EVC-network $N = (G, c, u, s, t)$ is an undirected graph $G = (V, E)$ with two specified vertices, the source vertex s and the sink vertex t, respectively, an edge capacity function $c : E \mapsto R^+ \cup \{0\}$, an vertex capacity function $u : V \mapsto R^+$ such that $u(s) = u(t) = \infty$. An AC-network with edge capacity only is a special one in which the capacities of all vertices are ∞.*

Definition 3. *A pseudo-flow f in a network N is a function $f : V \times V \to R$ such that:*

$$f(i,j) = f(j,i) = 0, (i,j) \notin E \tag{1}$$

$$0 \le f(i,j) + f(j,i) \le c(i,j) \quad \forall (i,j) \in E \tag{2}$$

$$\sum_{k:(k,i)\in E} f(k,i) = \sum_{j:(i,j)\in E} f(i,j) \quad \forall i \in V \setminus \{s,t\} \tag{3}$$

Formulas (1) and (2) are referred to as the edge-capacity constraint, while Formula (3) is referred to as the flow-conservation constraint. The value of a pseudo-flow f is defined as the net flow f into t, i.e., $val(f) = \sum_{i:(i,t)\in E} f(i,t) - \sum_{j:(t,j)\in E} f(t,j)$. A path from s to t is referred to as an augmenting path if the pseudo-flow in each edge of this path is non-zero.

Definition 4. *A pseudo-flow f in a flow network N is a flow if it meets:*

$$\sum_{j:(i,j)\in E} f(i,j) \le u(i) \quad \forall i \in V - \{s,t\} \tag{4}$$

Formula (4) is the vertex-capacity constraint of the flow. A flow f is the maximum flow if its value $val(f)$ is maximized.

Definition 5. *A path flow f^p in a flow network N on the set $\{P\}$ of all directed paths from s to t is defined as $f^p : \{P\} \rightarrow R^+ \cup \{0\}$. Specifically, a flow on a cycle of a non-simple path is called a cycle-flow. A directed path P from s to t such that $f^p(P) \geq 0$ is called an augmenting path.*

Theorem 1 ([1]). *Flow decomposition theorem: Given a flow network N, every path flow function has a unique representation as a pseudo-flow (defined on the edge set). Conversely, every pseudo-flow function in a network can be represented by at most $n + m$ path-flows, and among them, there are at most m cycle-flows.*

Definition 6. *Canceling cycle-flows within a pseudo-flow f in a network N is to decrease the values of f on edges along some cycles such that there is no cycle-flows in its path-flow representation. If the path flow representation of a pseudo-flow f contains no cycle-flows, f is called an acyclic pseudo-flow.*

Lemma 1 ([1]). *A pseudo-flow is an acyclic pseudo-flow if all its augmenting paths are simple.*

Definition 7. *Given a flow network N, a cut C in N is a minimal collection of vertices and edges whose deletion separates s from t in the resulting network. A cut consisting of only edges is an edge-cut. The sum of the capacities of the terms in a cut C is called the capacity of the cut. A minimum cut is a cut of the minimum capacity among all cuts. A minimum edge-cut is an edge-cut of the minimum capacity among all the edge-cuts.*

Definition 8. *A network is said to be planar if it can be embedded on a plane such that no two edges cross with each other. A planar network partitions the plane into a number of connected regions, and each of this regions is referred to as a face. We say the border $B(F)$ of a face F the set of edges that separate F from other parts of the plane. Two faces are neighbors if their borders share some common edges.*

3   Computing the Maximum Flow in an Undirected Planar EVC-Network

In this section we propose an algorithm for computing the maximum flow in an undirected planar EVC-network N with the aid of an auxiliary planar EC-network. The auxiliary planar EC-network, denoted as N_e, is constructed as follows. (1) Given a vertex $v \in V - \{s, t\}$ of degree d in N, replace v with a cycle consisting of d vertices $v_1, v_2, \cdots, v_d$ and d edges $(v_i, v_{i+1} \mod d), 1 \leq i \leq d$. The edge capacity of each of these edges is $u(v)/2$. (2) The edges in N incident to v are now linked to the vertices of the cycle $v_1, v_2, \cdots, v_d$ one by one in the same clockwise order as they link to v (Fig. 1). Note that if $d = 2$, the "cycle" formed by $v_1 - v_2 - v_1$ degenerates to an edge and its capacity is $u(v)$.

Definition 9. *N_e is called the extended network of N, and the cycle in N_e formed by $v_1, v_2 \cdots, v_d$ is called the corresponding chain-cycle of $v \in N$.*

It is trivial to verify that N_e can be constructed from N in $O(n)$ time.

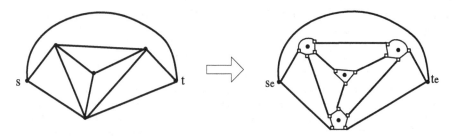

Fig. 1. The construction of an extended network

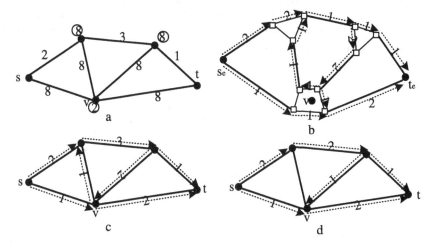

Fig. 2. (a) The original network N, (b) the extended network N_e with a flow, (c) a pseudo-flow in N is obtained from the flow in N_e, (d) a flow is derived by canceling the cycle-flow

Theorem 2. *[14] The capacity of the minimum cut N is equal to the capacity of the minimum edge-cut problem in N_e, i.e., the value of the maximum flow in N is equal to the value of the maximum flow in N_e.*

Our algorithm is based on the above theorem. Let us consider a flow in N_e. If there is a flow f_e in N_e, there is a corresponding function $f \in N, \forall e \in N, f(e) = f_e(e_e)$, where e_e is the corresponding edge of e in N_e. It is not difficult to verify that f follows the edge-capacity and flow-conversation constraints in N, thus, it is a pseudo-flow. However, f may violate the vertex-capacity constraint in N. Fig. 2 illustrates such a violation. Fig. 2(a) is a primal network N, where the value on each edge is its capacity, and the cycled value at each vertex is the capacity of the vertex. Fig. 2(b) is the extended network N_e of N and there is a maximum flow on it. Consider vertex v in N or its corresponding chain-cycle in N_e. The amount of incoming flow to v is 3, exceeding its capacity of 2. Thus, a pseudo-flow f directly obtained from a flow f_e in N_e is not a flow in N. In what follows we consider how to derive a flow from f.

Theorem 3. *If f is an acyclic pseudo-flow, then f follows vertex capacity constraint in N and is a flow.*

Proof. Without loss of generality, consider a chain-cycle in N_e, which corresponds a vertex v in N. Since s and t do not fall inside a chain-cycle, we can draw a line that cross it and separates s from t in the plane. When (a part of) flow f_e goes into the chain-cycle and then out from it, two cases may arise.

Case 1. If the flow f_e goes through the chain-cycle from one side of the plane to the other side (Fig. 3), the amount of the flow through the chain-cycle cannot be greater than the sum of the capacities of two bottleneck edges in the chain-cycle (two thick edges in Fig. 3) due to the edge-capacity constraint. Since the capacity of each edge in the chain-cycle is a half of the capacity of the corresponding vertex v in N. If the chain-cycle is treated as v, the corresponding pseudo-flow f satisfies the vertex-capacity constraint at v.

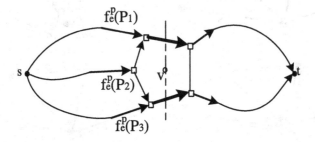

Fig. 3. Case 1. Flow runs through a chain-cycle from one side of the plane to the other side

Case 2. Opposite to case 1, without loss of generality, assume that on the left half plane, there are two path-flows $f_e^p(P_1)$ and $f_e^p(P_2)$ entering the chain-cycle and one path-flow $f_e^p(P_3)$ outgoing from the chain-cycle between the two entering path-flows(Fig. 4). The path-flows $f_e^p(P_1)$ and $f_e^p(P_2)$ and the border of the chain-cycle form a closed region. The path-flow $f_e^p(P_3)$ must go from inside the region to outside of the region because it eventually will reach the sink. Thus, if the chain-cycle is treated as the corresponding vertex v in N, a cycle-flow must be formed in the corresponding pseudo-flow f(Fig. 4).

The above observations indicate that, if f is an acyclic pseudo-flow, case 2 would not arise, and the vertex capacity constraint will not be violated.

If f is not an acyclic pseudo-flow, we can cancel its cycle-flows to make it acyclic. Thus a flow in N can be obtained as follows. First, compute a pseudo-flow f from a flow f_e in N_e, followed by cancelling the cycle-flows of f to get an acyclic pseudo-flow f_a. It is easy to verify that f_a is a flow in N. Fig. 2 depicts this procedure. In Fig. 2(c), a pseudo-flow f in N is obtained from the flow in N_e. Case 2 arises at vertex v and there is a cycle-flow. In Fig. 2(d), the cycle-flow is canceled and a flow f_a in N is obtained.

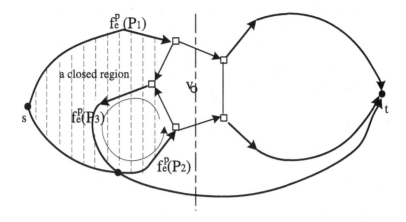

Fig. 4. Case 2. Outgoing path-flows from the chain-cycle exist between path-flows entering into the chain-cycle

Note that canceling cycle-flows does not change the value of a pseudo-flow. So if f_e is a maximum flow in N_e, then f_a is a maximum flow in N.

The algorithm is described as follows.

Algorithm Edge_Vertex_Capacitied_Max_Flow(N)
Begin
1. Construct the extended network N_e;
2. Compute a maximum flow f_e in N_e;
3. Map f_e to the edges in N to get a pseudo-flow f;
4. Cancel cycle flows in f to get a maximum flow f_a in N;
End.

Lemma 2. *[13] A pseudo-flow can be converted into an acyclic pseudo-flow of the same value in $O(m \log n)$ time.*

Theorem 4. *The maximum flow in an undirected planar EVC-network can be computed in $O(n \log n)$ time.*

4   An $O(n)$ Algorithm for Undirected (s,t)-Planar EVC-Networks

In this section, we aim to devise an efficient algorithm by taking advantage of the special properties of (s,t)-planar networks. For an (s,t)-planar network, we assume that s and t are laid in the outer face, a new edge (s,t) is introduced with $c(s,t) = 0$, the new finite face is denoted by s', and the outer face is denoted by t'.

Definition 10. *The dual network $N' = (G'(V', E'), l)$ of a (s,t)-planar network N is a planar network defined as follows. The dual vertex set V' is the set of*

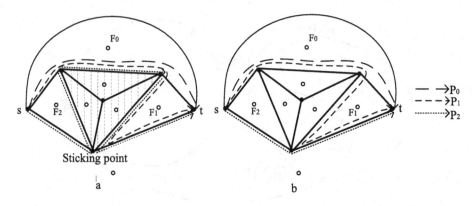

Fig. 5. Check non-simple augmenting path by visiting faces one by one: (a) A non-simple augmenting path P_2 is found, (b) P_2 is made simple by removing its traps

faces in N. For each edge $(i,j) \in E$, let i' be the face to the left of (i,j) when walking from i to j and j' be the face to the right, (i',j') is the dual edge of (i,j) in N', and its length is $l(i',j') = c(i,j)$. For each dual vertex i' in N', the length of the shortest path from s' to i' is called distance of i', denoted as $d(i')$.

Given an undirected (s,t)-planar network N and its dual network N', a label function in N' $h : V' \mapsto R$ with arbitrary value on each vertex has the following property.

Lemma 3. *[7] If two functions $\lambda : V \times V \to R$ and $f : V \times V \to R$ are defined as: $\lambda(i,j) = h(j') - h(i')$, $\lambda(t,s) = h(t') - h(s')$; $f(i,j) = \max\{\lambda(i,j), 0\}$, $f(t,s) = \max\{\lambda(t,s), 0\}$, $f(s,t) = 0$, then f follows the flow-conservation constraint on each vertex, and the value of f is $h(t') - h(s')$. If function $h(\cdot)$ is the distance function $d(\cdot)$, and N is an EC-network, then f is a maximum flow.*

To compute the maximum flow in an undirected planar EVC-network N, we first construct the extended N_e and run a shortest path algorithm in the dual network N_e' of N_e to compute the maximum flow f_e^{max}. Meanwhile, for each dual vertex v_e' that is not a face closed by a chain-cycle in N_e, we assign it an index using the order that it is added to the shortest path tree, and give it a label with its distance in N_e', i.e., $h(v_e') = d(v_e')$. We finally map the index and label of each such vertex in N_e' to its corresponding face in N. A flow f in N can be computed by using the label function $h(\cdot)$ in the way as shown in Lemma 1. It is trivial to verify that f is a pseudo-flow. In what follows we show how to find a maximum flow in N.

The proposed algorithm tranverses the faces in N in increasing order of the indices of the faces, and constructs augmenting paths using the borders of these faces. Non-simple augmenting paths are then made simple by adjusting face labels. Accordingly, the corresponding flow f is acyclic and the maximum flow is found. We refer to this algorithm as **Make_Augmenting_Path_Simple**, which works as follows. During the process, a face is marked as *visited* if it has been visited or its label is updated. A visited face means that it will not be visited again in the future. An edge or vertex is marked as *colored* if it is in the augmenting path, which will be used to update augmenting paths and to check the non-simplicity of augmenting paths.

The algorithm visits the faces in N starting from F_0, which corresponds to the dual vertex s', a simple augmenting path $P_0 = B(F_0) - (s, t)$ is found, F_0 is marked as *visited*, all the edges and vertices in P_0 are marked as *colored*. The algorithm then visits the other faces one by one in increasing order of their indices.

Suppose that F_k has been visited with $k \geq 0$, a simple augmenting path P_j is found, and the edges and vertices in P_j are marked as *colored*, now it visits the next face F_{k+i} which is an unvisited face with the least index. Note that the neighbors of F_{k+i} with smaller indices have already been visited, so some segments of $B(F_{k+i})$ are on P_k and have been marked colored. A new augmenting path P_{j+1} is obtained by replacing the colored segments with those uncolored segments of $B(F_{k+i})$. P_{j+1} may be non-simple, which can be verified by traversing $B(F_{k+i})$ as follows. If two consecutive edges (x, y) and (y, z) on $B(F_{k+i})$ are not colored while y is colored, then P_{j+1} is not simple. Each such vertex y is marked as a *sticking point*. Each non-simple sub-path of P_{j+1} is called a *trap*, and can be easily found by a traversal along P_k that starts from each sticking point, traversing the uncolored edge incident to it first, and ends at it. After this is done, the uncolored edges and vertices in $B(F_{k+i})$ are marked *colored*. Fig. 5(a) gives an illustration of the process, in which when face F_2 is visited, a trap is found.

To make augmenting path P_{j+1} simple, all the labels of faces enclosed by the trap are set to be $h(F_{k+i})$, and they are then marked as *visited* so that they will be not visited again. This label modification results in that the shared edges between $B(F_{k+i})$ and the trap have zero-flows. In the end, the traps in P_{k+1} are removed from it and a simple augmenting path P_{k+1}^* is obtained (Fig. 5(b)).

Lemma 4. *Algorithm Make_Augmenting_Path_Simple runs in $O(n)$ time.*

The proof of Lemma 4 is omitted due to the space limitation. In summary, we have the following theorem.

Theorem 5. *Given an undirected (s, t)-planar ENC-network N, there is an optimal algorithm for computing the maximum flow, which takes $O(n)$ time.*

5   Concluding Remarks

In this paper we considered the general version of the maximum flow problem in an undirected planar EVC-network, and proposed an $O(n \log n)$ time algorithm for the problem. It is still open to whether there is any efficient algorithm for the problems in directed planar networks running in the same amount of time.

References

1. Ahujia, R.K., Magnanti, T.L., Orlin, J.B.: Network Flows: Theory, Algorithms and Applications. Prentice-Hall, New Jersey (1993)
2. Bondy, J.A., Murty, U.S.R.: Graph Theory with Applications. Elsevier, North-Holland (1976)
3. Borradaile, G., Klein, P.: An $O(n \log n)$ Algorithm for Maximum s-t Flow in a Directed Planar Graph. In: Proceedings of the 17th Annual ACM-SIAM Symopsimum on Discrete Algorithms (SODA 2006), pp. 524–533 (2006)
4. Goldberg, A.V., Tarjan, R.E.: A New Approach to the Maximum Flow Problem. Journal of the ACM 35, 921–940 (1988)
5. Goldberg, A.V., Rao, S.: Beyond the Flow Decomposition Barrier. Journal of the ACM 45, 783–797 (1998)
6. Han, Y.: Deterministic Sorting in $O(n \log \log n)$ Time and Linear Space. Journal of Algorithms 50, 96–105 (2004)
7. Hassin, R.: Maximum Flows in (s, t) Planar Networks. Information Processing Letters 13, 107 (1981)
8. Hassin, R., Johnson, D.S.: An $O(n \log^2 n)$ Algorithm for Maximum Flow in Undirected Planar Networks. SIAM Journal on Computing 14, 612–624 (1985)
9. Hu, T.C.: Integer Programing and Network Flows. Addison-Wesley, Reading (1969)
10. Khuler, S., Naor, J.: Flow in Planar Graphs with Vertex Capacities. Algoirthmica 11, 200–225 (1994)
11. Klein, P., Rao, S.B., Rauch-Henzinger, M., Subramanian, S.: Faster Shortest-Path Algorithms for Planar Graphs. Journal of Computer and System Science 55, 3–23 (1997)
12. Reif, J.H.: Minimum $s - t$ Cut of a Planar Undirected Network in $O(n \log^2 n)$ Time. SIAM Journal on Computing 12, 71–81 (1983)
13. Sleator, D.D., Tarjan, R.E.: A Data Structure for Dynamic Tree. Journal of Computer and System Science 26, 362–391 (1983)
14. Zhang, X., Liang, W., Jiang, H.: Flow Equivalent Trees in Node-Edge-Capacitied Undirected Planar Graphs. Information Processing Letters 100, 100–115 (2006)

Spreading Messages*

Ching-Lueh Chang[1] and Yuh-Dauh Lyuu[2]

[1] Department of Computer Science and Information Engineering,
National Taiwan University, Taipei, Taiwan
`d95007@csie.ntu.edu.tw`
[2] Department of Computer Science and Information Engineering,
National Taiwan University, Taipei, Taiwan
`lyuu@csie.ntu.edu.tw`

Abstract. We model a network in which messages spread by a simple directed graph $G = (V, E)$ [8] and a function $\alpha : V \to \mathbb{N}$ mapping each $v \in V$ to a positive integer less than or equal to the indegree of v. The graph G represents the individuals in the network and the communication channels between them. An individual $v \in V$ will be convinced of a message when at least $\alpha(v)$ of its in-neighbors are convinced. Suppose we are to convince a message to all individuals by convincing a subset $S \subseteq V$ of individuals at the beginning and then let the message spread. We study minimum-sized sets S needed to convince all individuals at the end. In particular, our results include a lower bound on the size of a minimum S and the NP-completeness of computing a minimum S. Our lower bound utilizes a technique in [9]. Finally, we analyze the special case where each individual is convinced of a message when more than half of its in-neighbors are convinced.

1 Introduction

Each individual believes to some extent what others believe. A communication channel between two individuals, if it exists, is either bidirectional or unidirectional. Suppose there is an important message that we want to convince all individuals in a group. We start by first convincing some individuals, called the seeds, of the message. Then we sit back and let the message spread among the individuals. The spreading eventually stops when no more individuals remain to be convinced. This paper explores the minimum number of seed individuals that must be convinced initially, so that all individuals are convinced at the end.

Closely related to our work are the issues of broadcasting, accumulation and gossiping, which deal with disseminating information in interconnection networks using as few communication steps as possible [4][5]. A fundamental difference between our work and these is that, in our setting, a non-seed individual is convinced of a message only when sufficiently many others reveal their beliefs to it via direct communication channels. The conviction of non-seed individuals is thus similar to the firing of neurons in neural networks with threshold activation

---

* The authors are supported in part by NSC grant 96-2213-E-002-024.

X. Hu and J. Wang (Eds.): COCOON 2008, LNCS 5092, pp. 587–599, 2008.

functions [6]. Indeed, we may think of individuals as neurons. A convinced individual is one that keeps firing and a non-convinced one becomes convinced when sufficiently many neighboring neurons fire at it. However, our goal of spreading messages is to convince all individuals, whereas neural networks do not aim at making all neurons fire.

The paper is organized as follows. Section 2 gives definitions. Section 3 presents a lower bound on the minimum number of seeds needed to convince all individuals at the end. Section 4 discusses the strict-majority scenario, meaning that each individual is convinced of a message when more than half of its in-neighbors are convinced. The discussion in section 4 is separated into three subsections. Section 4.1 analyzes the spreading of a message in a random graph in the Erdös-Rényi model [1]. Section 4.2 analyzes the spreading of a message beginning from a uniformly and randomly chosen set of seeds. Section 4.3 presents an upper bound on the minimum number of seeds needed to convince all individuals at the end.

2   Definitions

Let $G = (V, E)$ be a simple directed graph (simple digraph for short) [8]. For a vertex $v \in V$, $\deg^{in}(v)$ and $\deg^{out}(v)$ are the indegree and outdegree of v in G, respectively. The set of incoming edges to v and outgoing edges from v are $E^{in}(v)$ and $E^{out}(v)$, respectively. We denote by $N^{in}(v) \subseteq V \setminus \{v\}$ (resp., $N^{out}(v) \subseteq V \setminus \{v\}$) the in-neighbors (resp., out-neighbors) of v, i.e., the set of vertices incident on an edge coming into (resp., going from) v. A simple undirected graph is interpreted as a simple digraph by regarding each undirected edge as two edges with opposite directions. For a vertex v of a simple undirected graph, we write $N(v)$ for both $N^{in}(v)$ and $N^{out}(v)$, and $\deg(v)$ for both $\deg^{in}(v)$ and $\deg^{out}(v)$. Hereafter, undirected and directed graphs are both assumed to be simple unless otherwise specified.

There is also the notion of random graphs. For a positive integer n and a real number $p \in [0, 1]$, the random graph $G(n, p)$ in the Erdös-Rényi model is an undirected graph on the vertex set $[n] \stackrel{\text{def}}{=} \{1, \ldots, n\}$ in which each of the $\binom{n}{2}$ possible edges appears independently with probability p [1].

A finite network $\mathcal{N}(G, \alpha)$ of individuals is modeled by a digraph $G = (V, E)$ and a function $\alpha : V \to \mathbb{N}$ satisfying $1 \leq \alpha(v) \leq \deg^{in}(v)$, $v \in V$. The vertices in V represent the individuals in $\mathcal{N}(G, \alpha)$, and each edge $(u, v) \in E$ represents the communication channel from u to v. The function α reflects the extent to which each individual follows what its in-neighbors believe. Specifically, an individual v will be convinced of a message when at least $\alpha(v)$ of its in-neighbors are convinced. A special case is for each individual to believe a message when more than half of its in-neighbors believe it. In this case, we have $\alpha(v) = \lceil (\deg^{in}(v) + 1)/2 \rceil$ for each $v \in V$, and we call it the strict-majority scenario. For convenience, the strict-majority scenario is also denoted α_{maj}.

The spreading of a message in $\mathcal{N}(G, \alpha)$ proceeds asynchronously. Initially, the message convinces a group $S \subseteq V$ of individuals called the seeds. Thereafter, an individual $v \in V$ becomes convinced after at least $\alpha(v)$ of its in-neighbors are convinced. The spreading ends when no additional individuals remain to be convinced of the message.

For any set $U \subseteq V$, we denote by $c(U) \subseteq V$ the set of individuals that are convinced at the end of the spreading given that U is the set of seeds. We define min-seed(G, α) as $\min_{U \subseteq V, c(U)=V} |U|$, or the minimum number of seeds needed to convince all individuals at the end of the spreading. Define the problem SEED to be that of asking whether min-seed$(G, \alpha) \leq s$ on input $s \in \mathbb{N}$, a digraph $G = (V, E)$ and a function $\alpha : V \to \mathbb{N}$ satisfying $1 \leq \alpha(v) \leq \deg^{\mathrm{in}}(v)$, $v \in V$. The UNDIRECTED-SEED problem is SEED restricted to undirected graphs. The following theorem, whose proof is in the appendix, shows the NP-completeness of UNDIRECTED-SEED.

Theorem 1. *UNDIRECTED-SEED is NP-complete.*

3   A Lower Bound on the Number of Seeds

Let $G = (V, E)$ be a digraph and $\alpha : V \to \mathbb{N}$ satisfy $1 \leq \alpha(v) \leq \deg^{\mathrm{in}}(v)$, $v \in V$. We are interested in min-seed(G, α). For example, if $G = (V, E)$ is undirected and $\alpha(v) = 1$ for every $v \in V$, then min-seed(G, α) equals the number of connected components of G. More elaborate lower bounds can be derived from the following theorem, whose proof uses a technique demonstrated in an example in [9] (the example shows that given n^2 squares placed in an n by n array, if one paints $n - 1$ of the squares black and let any square become black when at least two of its four neighboring squares are black, then at least one square will never be black).

Theorem 2. *Let $G = (V, E)$ be a digraph and $\alpha : V \to \mathbb{N}$ be a function satisfying $1 \leq \alpha(v) \leq \deg^{\mathrm{in}}(v)$, $v \in V$. For any set $S \subseteq V$,*

$$\sum_{v \in S} \left[2\alpha(v) - 2|N^{\mathrm{in}}(v) \setminus N^{\mathrm{out}}(v)| \right] \geq \sum_{v \in c(S)} \left[2\alpha(v) - \deg^{\mathrm{out}}(v) - 2|N^{\mathrm{in}}(v) \setminus N^{\mathrm{out}}(v)| \right].$$

Proof. Consider the spreading of a message beginning from an arbitrary set S of seeds. Whenever two individuals u and v have at least $\alpha(u)$ and $\alpha(v)$ convinced in-neighbors, respectively, it does not matter which of u and v is convinced first for the spreading to succeed in convincing all members of $c(S)$ at the end. Therefore, we may assume without loss of generality that no two individuals become convinced of the message at the same instant.

Denote by $C \subseteq E$ the set of edges going from an individual that is convinced of the message to one that is not. The set C changes during the spreading process. Initially, only the seeds are convinced of the message; hence

$$|C| \leq \sum_{v \in S} \deg^{\mathrm{out}}(v). \tag{1}$$

Let $v \in c(S) \setminus S$ be arbitrary. We know that v is eventually convinced. Let W be the members of $N^{\mathrm{in}}(v) \setminus N^{\mathrm{out}}(v)$ that are convinced before v is and let

$X = (N^{in}(v) \setminus N^{out}(v)) \setminus W$. Denote by Y the members of $N^{out}(v) \cap N^{in}(v)$ that are convinced before v is and let $Z = (N^{out}(v) \cap N^{in}(v)) \setminus Y$. It is easy to see that

$$\left| C \cap (E^{out}(v) \cup E^{in}(v)) \right| = |W| + |Y|$$

right before v is convinced because both count the number of v's in-neighbors that are convinced before v is. Furthermore,

$$\left| C \cap (E^{out}(v) \cup E^{in}(v)) \right| \le |Z| + |N^{out}(v) \setminus N^{in}(v)|$$

right after v is convinced because the left-hand side counts the number of v's out-neighbors that are not convinced before v, whereas the right-hand side may overestimate that by including all of v's out-neighbors in $N^{out}(v) \setminus N^{in}(v)$ that are convinced before v. Therefore, convincing v adds at most

$$
\begin{aligned}
&|Z| + |N^{out}(v) \setminus N^{in}(v)| - (|W| + |Y|) \\
&= \left(|Z| + |Y| + |N^{out}(v) \setminus N^{in}(v)| \right) - |W| - 2|Y| \\
&= \left(|N^{out}(v) \cap N^{in}(v)| + |N^{out}(v) \setminus N^{in}(v)| \right) - |W| - 2|Y| \\
&= \deg^{out}(v) - |W| - 2|Y| \\
&\le \deg^{out}(v) - 2|Y| \qquad\qquad\qquad\qquad\qquad\qquad (2)
\end{aligned}
$$

to $\left| C \cap (E^{out}(v) \cup E^{in}(v)) \right|$.

We have $|Y| \ge \alpha(v) - |N^{in}(v) \setminus N^{out}(v)|$ because right before v is convinced, v has at least $\alpha(v)$ convinced in-neighbors and at most $|N^{in}(v) \setminus N^{out}(v)|$ of them belong in $N^{in}(v) \setminus N^{out}(v)$. This and Eq. (2) imply that convincing v adds at most

$$\deg^{out}(v) - 2\alpha(v) + 2|N^{in}(v) \setminus N^{out}(v)|$$

to $\left| C \cap (E^{out}(v) \cup E^{in}(v)) \right|$. The event that v becomes convinced clearly does not change $C \cap \left(E \setminus (E^{out}(v) \cup E^{in}(v)) \right)$. In summary, each time an individual $v \in c(S) \setminus S$ is convinced,

$$|C| = \left| C \cap (E^{out}(v) \cup E^{in}(v)) \right| + \left| C \cap \left(E \setminus (E^{out}(v) \cup E^{in}(v)) \right) \right|$$

increases by at most $\deg^{out}(v) - 2\alpha(v) + 2|N^{in}(v) \setminus N^{out}(v)|$.

Observe that every $v \in c(S) \setminus S$ becomes convinced exactly once during the spreading process and Eq. (1) holds at the beginning. Hence at the end of the spreading,

$$
\begin{aligned}
0 \\
\le &|C| \\
\le &\sum_{v \in S} \deg^{out}(v) + \sum_{v \in c(S) \setminus S} \left[\deg^{out}(v) - 2\alpha(v) + 2|N^{in}(v) \setminus N^{out}(v)| \right] \\
= &\sum_{v \in c(S)} \left[\deg^{out}(v) - 2\alpha(v) + 2|N^{in}(v) \setminus N^{out}(v)| \right] + \sum_{v \in S} \left[2\alpha(v) - 2|N^{in}(v) \setminus N^{out}(v)| \right],
\end{aligned}
$$

completing the proof. $\qquad\qquad\qquad\qquad\qquad\qquad\qquad\qquad\qquad\qquad\qquad\qquad\qquad$ □

The following corollaries are immediate.

Corollary 1. *Let $G = (V, E)$ be an undirected graph and $\alpha : V \to \mathbb{N}$ be a function satisfying $1 \le \alpha(v) \le \deg(v)$, $v \in V$. For each $S \subseteq V$,*

$$\sum_{v \in S} 2\alpha(v) \ge \sum_{v \in c(S)} [2\alpha(v) - \deg(v)].$$

Corollary 2. *For an undirected graph $G = (V, E)$ and a function $\alpha : V \to \mathbb{N}$ satisfying $1 \le \alpha(v) \le \deg(v)$ for each $v \in V$,*

$$\text{min-seed}(G, \alpha) \ge \frac{\sum_{v \in V} [2\alpha(v) - \deg(v)]}{2 \max_{v \in V} \alpha(v)}.$$

An example where the lower bound on min-seed(G, α) in Corollary 2 is tight is when G is an undirected cycle with an even number of vertices and $\alpha(v) = 2$ for each vertex v of G.

4   The Strict-Majority Scenario

In this section, we explore the case where each individual chooses to believe a message when more than half of its in-neighbors are already convinced.

4.1   Random Graphs

Let n be a positive integer and $p \in (0, 1]$. The following theorem states a lower bound on min-seed$(G(n, p), \alpha_{\text{maj}})$, which also serves as a lower bound of min-seed$(G(n, p), \alpha)$ for any $\alpha : [n] \to \mathbb{N}$ satisfying $\alpha(i) \ge \alpha_{\text{maj}}(i)$, $i \in [n]$.

Theorem 3. *Let n be a positive integer and $p \in (0, 1]$. With probability $1 - o(1)$,*

$$\text{min-seed}(G(n, p), \alpha_{\text{maj}}) = \Omega \left(\min \left\{ n, \frac{1}{p} \right\} \right).$$

Proof. Denote by E the set of edges of $G(n, p)$. Set

$$\epsilon = \frac{1}{1000 \cdot \max\{1, pn\} + 2} \tag{3}$$

and fix an arbitrary $S \subseteq [n]$ of size ϵn. Let the random variable $n_S \in \mathbb{N}$ be the number of edges in E incident on at least one vertex in S. We have

$$\sum_{v \in S} \deg(v) \le 2n_S$$

since each edge in E incident on at least one vertex in S contributes 1 or 2 to the left-hand side and 2 to the right-hand side, whereas any other edge in E contributes none to either side. Therefore,

$$\sum_{v \in S} 2\alpha_{\text{maj}}(v) = \sum_{v \in S} 2 \left\lceil \frac{\deg(v) + 1}{2} \right\rceil \le \sum_{v \in S} [\deg(v) + 2] \le 2n_S + 2\epsilon n. \tag{4}$$

We also have

$$\sum_{v\in[n]} [2\alpha_{\mathrm{maj}}(v) - \deg(v)] = \sum_{v\in[n]} \left\{ 2\left\lceil \frac{\deg(v)+1}{2}\right\rceil - \deg(v)\right\} \geq \sum_{v\in[n]} 1 = n.$$

This and Eq. (4) imply that if $2n_S + 2\epsilon n < n$, then

$$\sum_{v\in S} 2\alpha_{\mathrm{maj}}(v) < \sum_{v\in[n]} [2\alpha_{\mathrm{maj}}(v) - \deg(v)],$$

which in turn implies $c(S) \neq [n]$ by Corollary 1. Therefore,

$$\Pr[c(S) = [n]] \leq \Pr[2n_S + 2\epsilon n \geq n] = \Pr\left[n_S \geq \frac{n - 2\epsilon n}{2}\right]. \qquad (5)$$

Since the complete graph on vertex set $[n]$ has $\binom{|S|}{2}$ edges with both endpoints in S and $|S|(n-|S|)$ edges with exactly one endpoint in S,

$$E[n_S] = \left(\binom{|S|}{2} + |S|(n-|S|)\right) p = \left(\binom{\epsilon n}{2} + \epsilon n(n-\epsilon n)\right) p \leq \epsilon p n^2.$$

This implies

$$500 \times \max\left\{\frac{1}{pn}, 1\right\} E[n_S] \leq 500 \times \max\left\{\frac{1}{pn}, 1\right\} \epsilon p n^2 = \frac{n - 2\epsilon n}{2},$$

where the equality, after some laborious manipulations, follows from Eq. (3). Therefore,

$$\Pr\left[n_S \geq \frac{n - 2\epsilon n}{2}\right] \leq \Pr\left[n_S \geq 500 \max\left\{\frac{1}{pn}, 1\right\} E[n_S]\right] \leq 2^{-n/4} \qquad (6)$$

by the Chernoff bound [2].

By Eqs. (5)–(6) and the fact that there are $\binom{n}{\epsilon n}$ subsets of $[n]$ of size ϵn, we see that with probability at most

$$\binom{n}{\epsilon n} 2^{-n/4},$$

there exists a set $S \subseteq [n]$ of size ϵn satisfying $c(S) = [n]$. The Stirling formula shows that the above probability is $o(1)$. Finally, the proof is completed by verifying that $\epsilon n = \Omega(\min\{n, 1/p\})$.   □

Theorem 3 shows a linear (in n) lower bound on min-seed$(G(n, c/n), \alpha_{\mathrm{maj}})$ for any constant $c > 0$. Consequently, in a network modeled by a random sparse graph and the strict-majority scenario, at least a constant fraction of individuals must be seeds in order to convince all individuals at the end.

4.2   Random Seeds

Now consider the placement of seed individuals in a strict-majority network so huge that we cannot afford to gather information about more than a tiny fraction of individuals. Since we have no information regarding most of the individuals, it is plausible to pick the seed individuals uniformly and randomly. Below we give a lower bound of $(1 - \delta)|V|^2/(2|V| + 2|E|)$ on the number of uniformly and randomly picked seeds needed to convince an expected $1 - \delta$ fraction of individuals in a strict-majority network $\mathcal{N}(G, \alpha_{\text{maj}})$, where $G = (V, E)$ and $\delta \in (0, 1)$. In consideration of the hugeness of the network, our lower bound is practical only if it is accurately estimable by querying only a tiny fraction of all individuals. Fortunately, this is achievable because the average degree in G and therefore the number $|E|$ of communication channels are accurately estimable by querying only a few individuals (assuming that the number $|V|$ of individuals is known) [3].

Theorem 4. *Let $s \in \mathbb{N}$, $\delta \in (0, 1)$, $G = (V, E)$ be an undirected graph and $\alpha : V \to \mathbb{N}$ be a function satisfying $\alpha_{\text{maj}}(v) \le \alpha(v) \le \deg(v)$, $v \in V$. Let S be a set of s uniformly random samples from V, with repetitions removed. Consider the spreading of a message in $\mathcal{N}(G, \alpha)$ beginning with the seed set S. We have*

$$E[\,|c(S)|\,] \le s \cdot \frac{2|E| + 2|V|}{|V|}$$

where the expectation is taken over the randomly chosen elements of S. In particular, if $E[\,|c(S)|\,] \ge (1 - \delta)\,|V|$, then we must have

$$s \ge \frac{(1 - \delta)\,|V|^2}{2|V| + 2|E|}.$$

Proof. We only prove the theorem for $\alpha \equiv \alpha_{\text{maj}}$, which clearly implies the general case. Clearly,

$$\deg(v) + 1 \le 2\alpha_{\text{maj}}(v) \le \deg(v) + 2$$

for each $v \in V$. These inequalities and Corollary 1 imply

$$\sum_{v \in S} [\deg(v) + 2] \ge \sum_{v \in S} 2\alpha_{\text{maj}}(v) \ge \sum_{v \in c(S)} [2\alpha_{\text{maj}}(v) - \deg(v)] \ge |c(S)|. \quad (7)$$

Each uniformly random sample from V has an expected degree of

$$\frac{\sum_{v \in V} \deg(v)}{|V|} = \frac{2|E|}{|V|},$$

where the equality holds because each edge is incident on exactly two vertices. Therefore,

$$E\left[\sum_{v \in S} \deg(v)\right] \le \frac{2|E|\,s}{|V|}$$

by the linearity of expectation and the fact that S is a set of s uniformly random samples from V with repetitions removed. This and Eq. (7) imply

$$E[|c(S)|] \leq E\left[\sum_{v \in S} \deg(v)\right] + 2s \leq s \cdot \frac{2|E| + 2|V|}{|V|}. \qquad \square$$

4.3 An Upper Bound on the Number of Seeds

The following theorem gives a polynomial-time algorithm realizing a $23 \cdot |V|/27$ upper bound on the minimum number of seeds needed to convince all individuals in a network $\mathcal{N}(G, \alpha_{\mathrm{maj}})$, where $G = (V, E)$ is any strongly connected digraph with $|V| \geq 2$. It is not hard to generalize the theorem to digraphs whose strongly connected components all contain at least two vertices.

Theorem 5. *There is a polynomial-time algorithm that, on input a strongly connected digraph $G = (V, E)$ with $|V| \geq 2$ and the strict-majority scenario α_{maj}, outputs a set $S \subseteq V$ of seeds satisfying $c(S) = V$ and*

$$|S| \leq \frac{23 \cdot |V|}{27}.$$

Proof. For two arbitrary disjoint sets $U_1, U_2 \subseteq V$, let $S(U_1, U_2) \subseteq V \setminus (U_1 \cup U_2)$ contain each vertex $v \in V \setminus (U_1 \cup U_2)$ independently with probability $2/3$. Define

$$T(U_1, U_2)$$
$$\stackrel{\text{def}}{=} \left\{ v \in V \setminus (S(U_1, U_2) \cup U_1) \,\Big|\, |N^{\mathrm{in}}(v) \cap S(U_1, U_2)| + |N^{\mathrm{in}}(v) \cap U_1| < \left\lceil \frac{\deg^{\mathrm{in}}(v) + 1}{2} \right\rceil \right\}$$

to be the set of vertices outside of $S(U_1, U_2) \cup U_1$ whose in-neighbors in $S(U_1, U_2) \cup U_1$ do not constitute a strict-majority in $N^{\mathrm{in}}(v)$. Note that the random variable $T(U_1, U_2)$ is completely determined by the random variables used to form $S(U_1, U_2)$.

It is clear that $U_1 \cup S(U_1, U_2) \cup T(U_1, U_2)$ is sufficient as a set of seeds to convince all members of V, that is,

$$c\left(U_1 \cup S(U_1, U_2) \cup T(U_1, U_2)\right) = V, \qquad (8)$$

whatever U_1, U_2 and the realization of $S(U_1, U_2)$ are.

By the linearity of expectation,

$$|S(U_1, U_2)| = \frac{2}{3} \times |V \setminus (U_1 \cup U_2)|. \qquad (9)$$

For each $v \in V \setminus U_1$,

$$\Pr[v \in T(U_1, U_2)]$$
$$= \Pr\left[v \notin S(U_1, U_2) \text{ and } |N^{\mathrm{in}}(v) \cap S(U_1, U_2)| + |N^{\mathrm{in}}(v) \cap U_1| < \left\lceil \frac{\deg^{\mathrm{in}}(v) + 1}{2} \right\rceil\right]$$

$$= \Pr[v \notin S(U_1, U_2)] \cdot \Pr\left[|N^{\text{in}}(v) \cap S(U_1, U_2)| + |N^{\text{in}}(v) \cap U_1| < \left\lceil \frac{\deg^{\text{in}}(v) + 1}{2} \right\rceil\right]$$

$$= \Pr[v \notin S(U_1, U_2)] \cdot \sum_{i \in N, i + |N^{\text{in}}(v) \cap U_1| < \lceil \frac{\deg^{\text{in}}(v)+1}{2} \rceil} \Pr\left[|N^{\text{in}}(v) \cap S(U_1, U_2)| = i\right]$$

$$= \Pr[v \notin S(U_1, U_2)]$$

$$\cdot \sum_{i \in N, i + |N^{\text{in}}(v) \cap U_1| < \lceil \frac{\deg^{\text{in}}(v)+1}{2} \rceil} \binom{|N^{\text{in}}(v) \setminus (U_1 \cup U_2)|}{i} (\frac{2}{3})^i (\frac{1}{3})^{|N^{\text{in}}(v) \setminus (U_1 \cup U_2)| - i},$$

$$\tag{10}$$

where the probabilities are taken over $S(U_1, U_2)$ and $T(U_1, U_2)$. The first equality above follows by the definition of $T(U_1, U_2)$. The second and the fourth equalities hold because $v \notin N^{\text{in}}(v)$ and each vertex in $V \setminus (U_1 \cup U_2)$ is picked into $S(U_1, U_2)$ independently with probability 2/3. For the special case of $U_1 = U_2 = \emptyset$, Eq. (10) shows that

$$\Pr[v \in T(\emptyset, \emptyset)] = \frac{1}{3} \times \sum_{i=0}^{\lceil \frac{\deg^{\text{in}}(v)-1}{2} \rceil} \binom{\deg^{\text{in}}(v)}{i} (2/3)^i (1/3)^{\deg^{\text{in}}(v)-i},$$

$v \in V$. The right-hand side of the above equation can be calculated exactly for $\deg^{\text{in}}(v) \in \{1, \ldots, 5\}$ and bounded from above using the Chernoff bound [2] for $\deg^{\text{in}}(v) > 5$. This yields $\Pr[v \in T(\emptyset, \emptyset)] \leq \frac{5}{27}$ for $v \in V$ (no matter the value of $\deg^{\text{in}}(v)$), which together with Eqs. (9) shows

$$E\left[|S(\emptyset, \emptyset) \cup T(\emptyset, \emptyset)|\right]$$
$$= E[|S(\emptyset, \emptyset)|] + E[|T(\emptyset, \emptyset)|]$$
$$= \frac{2 \cdot |V|}{3} + \sum_{v \in V} \Pr[v \in T(\emptyset, \emptyset)]$$
$$\leq \frac{23 \cdot |V|}{27}. \tag{11}$$

This and Eq. (8) (with $U_1 = U_2 = \emptyset$) guarantee the existence of a $(23 \cdot |V|/27)$-sized set of seeds sufficient to convince all members of V.

To show a deterministic polynomial-time algorithm for computing a set $S \subseteq V$ of seeds satisfying $c(S) = V$ and $|S| \leq 23 \cdot |V|/27$, we use the method of conditional expectation [7]. Let $U_{1,1} = U_{2,1} = \emptyset$. We have

$$E\left[|U_{1,1} \cup S(U_{1,1}, U_{2,1}) \cup T(U_{1,1}, U_{2,1})|\right] \leq \frac{23 \cdot |V|}{27}$$

by Eq. (11). Inductively, suppose we have obtained disjoint sets $U_{1,i}, U_{2,i} \subseteq V$ satisfying

$$E\left[|U_{1,i} \cup S(U_{1,i}, U_{2,i}) \cup T(U_{1,i}, U_{2,i})|\right] \leq \frac{23 \cdot |V|}{27} \tag{12}$$

and $U_{1,i} \cup U_{2,i} \neq V$, $i \geq 1$, we will show how to obtain disjoint sets $U_{1,i+1}, U_{2,i+1} \subseteq V$ with

$$E\left[|U_{1,i+1} \cup S(U_{1,i+1}, U_{2,i+1}) \cup T(U_{1,i+1}, U_{2,i+1})|\right] \leq \frac{23 \cdot |V|}{27} \tag{13}$$

and

$$|U_{1,i+1} \cup U_{2,i+1}| > |U_{1,i} \cup U_{2,i}| \tag{14}$$

in time polynomial in $|V|$. For this purpose, we pick arbitrary a vertex $v \in V \setminus (U_{1,i} \cup U_{2,i})$. We observe that if we take $U_{1,i+1} = U_{1,i} \cup \{v\}$ and $U_{2,i+1} = U_{2,i}$,

$$E\left[|U_{1,i} \cup S(U_{1,i}, U_{2,i}) \cup T(U_{1,i}, U_{2,i})| \big| v \text{ is picked into } S(U_{1,i}, U_{2,i})\right]$$
$$= E\left[|U_{1,i+1} \cup S(U_{1,i+1}, U_{2,i+1}) \cup T(U_{1,i+1}, U_{2,i+1})|\right] \tag{15}$$

because the distribution of $U_{1,i} \cup S(U_{1,i}, U_{2,i})$ conditioned on $v \in S(U_1, U_2)$ is the same as that of $U_{1,i+1} \cup S(U_{1,i+1}, U_{2,i+1})$, and whenever $U_{1,i} \cup S(U_{1,i}, U_{2,i})$ equals $U_{1,i+1} \cup S(U_{1,i+1}, U_{2,i+1})$, it holds that

$$T(U_{1,i}, U_{2,i}) = T(U_{1,i+1}, U_{2,i+1}).$$

Similarly, if we take $U_{1,i+1} = U_{1,i}$ and $U_{2,i+1} = U_{2,i} \cup \{v\}$,

$$E\left[|U_{1,i} \cup S(U_{1,i}, U_{2,i}) \cup T(U_{1,i}, U_{2,i})| \big| v \text{ is not picked into } S(U_{1,i}, U_{2,i})\right]$$
$$= E\left[|U_{1,i+1} \cup S(U_{1,i+1}, U_{2,i+1}) \cup T(U_{1,i+1}, U_{2,i+1})|\right]. \tag{16}$$

Since $v \in V \setminus (U_{1,i} \cup U_{2,i})$ is picked into $S(U_{1,i}, U_{2,i})$ with probability $2/3$,

$$E\left[|U_{1,i} \cup S(U_{1,i}, U_{2,i}) \cup T(U_{1,i}, U_{2,i})|\right]$$
$$= \frac{2}{3} \times E\left[|U_{1,i} \cup S(U_{1,i}, U_{2,i}) \cup T(U_{1,i}, U_{2,i})| \big| v \text{ is picked into } S(U_{1,i}, U_{2,i})\right]$$
$$+ \frac{1}{3} \times E\left[|U_{1,i} \cup S(U_{1,i}, U_{2,i}) \cup T(U_{1,i}, U_{2,i})| \big| v \text{ is not picked into } S(U_{1,i}, U_{2,i})\right].$$

This and Eq. (12) show that at least one of Eqs. (15) and (16) evaluates to at most $23 \cdot |V|/27$. Hence, $(U_{1,i+1}, U_{2,i+1})$ can be selected (to either $(U_{1,i} \cup \{v\}, U_{2,i})$ or $(U_{1,i}, U_{2,i} \cup \{v\})$) in polynomial (in V) time to satisfy Eqs. (13)–(14) if we are able to compute in polynomial (in V) time which of the right-hand sides of Eqs. (15)–(16) evaluates to at most $23 \cdot |V|/27$. The inductive step is therefore complete if $E\left[|U_1' \cup S(U_1', U_2') \cup T(U_1', U_2')|\right]$ is computable in polynomial (in V) time given two arbitrary disjoint sets $U_1', U_2' \subseteq V$. The latter follows by writing

$$E\left[|U_1' \cup S(U_1', U_2') \cup T(U_1', U_2')|\right]$$
$$= |U_1'| + E\left[|S(U_1', U_2')|\right] + E\left[|T(U_1', U_2')|\right]$$
$$= |U_1| + E\left[|S(U_1', U_2')|\right] + \sum_{v \in V \setminus U_1'} \Pr[v \in T(U_1', U_2')]$$

and computing the second and third terms in the last equality using Eqs. (9)–(10).

By inductively computing disjoint sets $U_{1,i}, U_{2,i} \subseteq V$ for $i = 1, 2, \ldots$, we finally obtain disjoints sets $U_{1,k}, U_{2,k} \subseteq V$ satisfying $U_{1,k} \cup U_{2,k} = V$ and

$$E\left[\,|U_{1,k} \cup S(U_{1,k}, U_{2,k}) \cup T(U_{1,k}, U_{2,k})|\,\right] \leq \frac{23 \cdot |V|}{27},$$

implying

$$|U_{1,k} \cup S(U_{1,k}, U_{2,k}) \cup T(U_{1,k}, U_{2,k})| \leq \frac{23 \cdot |V|}{27}$$

as no randomness is involved when $U_{1,k} \cup U_{2,k} = V$. This and Eq. (8) complete the proof. □

References

1. Bollobás, B.: Random Graphs, 2nd edn. Cambridge University Press, Cambridge (2001)
2. Chernoff, H.: A Measure of the Asymptotic Efficiency of Tests of a Hypothesis Based on the Sum of Observations. Annals of Mathematical Statistics 23, 493–507 (1952)
3. Goldreich, O., Ron, D.: On Estimating the Average Degree of a Graph. Technical Report TR04-013. Electronic Colloquium on Computational Complexity (2004)
4. Hedetniemi, S.T., Hedetniemi, S.M., Liestman, A.L.: A Survey of Broadcasting and Gossiping in Communication Networks. Networks 18, 319–349 (1988)
5. Hromkovic, J., Klasing, R., Monien, B., Peine, R.: Dissemination of Information in Interconnection Networks (Broadcasting and Gossiping). In: Du, D.-Z., Hsu, D.F. (eds.) Combinatorial Network Theory, pp. 125–212. Kluwer Academic Publishers, Dordrecht (1996)
6. McCulloch, W.S., Pitts, W.: A Logical Calculus of the Ideas Immanent in Nervous Activity. Neurocomputing: Foundations of Research, 15–27 (1988)
7. Motwani, R., Raghaven, P.: Randomized Algorithms. Cambridge University Press, Cambridge (1995)
8. West, D.B.: Introduction to Graph Theory, 2nd edn. Prentice-Hall, Englewood Cliffs (2001)
9. Yan, Z.-J. (ed.): A Course on High School Mathematics Competition (in Chinese), ch.14, pp. 112–113 (1993), Item 118, http://www.chiuchang.com.tw/catalog/ ISBN: 9576030323

Appendix: Proof of Theorem 1

Proof (Proof of Theorem 1). We reduce 3SAT to SEED. The input to the reduction is a 3-CNF formula F with variables $x_1, \ldots, x_n$. We assume without loss of generality that F contains the clause $(x_i \vee x_i \vee \bar{x}_i)$ for each $1 \leq i \leq n$. The output of the reduction includes a positive integer $s \in \mathbb{N}$, an undirected graph

$G = (V, E)$ and a function α. These three parts of the output are defined as follows. The vertex set of G is

$$V = \{x_i \mid 1 \leq i \leq n\}$$
$$\cup \{\bar{x}_i \mid 1 \leq i \leq n\}$$
$$\cup \{w_{C,j} \mid C \text{ is a clause of } F, 1 \leq j \leq 9n^4\}$$
$$\cup \{u_{C,k} \mid C \text{ is a clause of } F, 1 \leq k \leq 3n^2\}.$$

Here we abuse notation by letting x_i and $\bar{x}_i$ denote literals of F as well as vertices of G, depending on the context. The edges E of G include, for each $1 \leq j \leq 9n^4$, $1 \leq k \leq 3n^2$, $1 \leq h \leq n$ and each clause $C = (\ell_p \vee \ell_q \vee \ell_r)$ of F, the set

$$\{(w_{C,j}, \ell_p)\} \cup \{(w_{C,j}, \ell_q)\} \cup \{(w_{C,j}, \ell_r)\}$$
$$\cup \bigcup_{t=3(k-1)n^2+1}^{3kn^2} \{(u_{C,k}, w_{C,t})\}$$
$$\cup \{(u_{C,k}, x_h)\} \cup \{(u_{C,k}, \bar{x}_h)\}.$$

The function α is given by

$$\alpha(x_i) = \deg(x_i),$$
$$\alpha(\bar{x}_i) = \deg(\bar{x}_i),$$
$$\alpha(w_{C,j}) = 1,$$
$$\alpha(u_{C,k}) = 3n^2,$$

for $1 \leq i \leq n$, $1 \leq j \leq 9n^4$, $1 \leq k \leq 3n^2$ and each clause C of F. Finally we set $s = n$.

We now argue for the correctness of the reduction. Assume that an assignment $\boldsymbol{a}$ satisfies F. For $1 \leq i \leq n$, we choose as a seed the vertex in $\{x_i, \bar{x}_i\}$ which, when seen as a literal, is satisfied by $\boldsymbol{a}$. Under this selection of seeds, it is easy to verify that exactly n seeds are selected and all individuals are eventually convinced.

Now let S be a set of at most n seeds satisfying $c(S) = V$. We need to show that F is satisfiable to complete the proof. Let $C = (\ell_p \vee \ell_q \vee \ell_r)$ be an arbitrary clause of F, where ℓ_p, ℓ_q and ℓ_r are literals and are not necessarily distinct. We first show that

$$\{\ell_p, \ell_q, \ell_r\} \cap S \neq \emptyset. \tag{17}$$

Suppose for contradiction that $\ell_p \notin S, \ell_q \notin S$ and $\ell_r \notin S$. Since $|S| \leq n < 3n^2$, we have $u_{C,\tilde{k}} \notin S$ for some $1 \leq \tilde{k} \leq 3n^2$. Denote

$$U = \{\ell_p, \ell_q, \ell_r\} \cup \{u_{C,\tilde{k}}\} \cup \bigcup_{t=3(\tilde{k}-1)n^2+1}^{3\tilde{k}n^2} \{w_{C,t}\}.$$

Let $S' = (S \cap U) \cup (V \setminus U)$. Simple calculations show that, when S' is used as a seed set, no vertex outside of S' could ever be convinced. Since it is clear that $S \subseteq S'$, we have $c(S) \subseteq c(S') \subsetneq V$, a contradiction.

For each $1 \leq i \leq n$, since F contains the clause $(x_i \vee x_i \vee \bar{x}_i)$, we must have $\{x_i, \bar{x}_i\} \cap S \neq \emptyset$ by Eq. (17). This and the fact that $|S| \leq n$ imply that S contains exactly one of $\{x_i, \bar{x}_i\}$ for $1 \leq i \leq n$, and no other vertices. Let the assignment a assign true to the unique literal in $\{x_i, \bar{x}_i\}$ which, when seen as a vertex of G, belongs to S, $1 \leq i \leq n$. Now Eq. (17) implies that every clause of F is satisfied by a.

On Some City Guarding Problems

Lichen Bao, Sergey Bereg, Ovidiu Daescu*, Simeon Ntafos, and Junqiang Zhou

Department of Computer Science
Erik Jonsson School of Engineering & Computer Science
The University of Texas at Dallas
Richardson, TX 75080, USA
{lxb042000,besp,daescu,ntafos,jxz043000}@utdallas.edu

Abstract. We consider guarding a city of k vertical buildings, each having a rectangular base, by placing guards only at vertices. The aim is to use the smallest number of guards. The problem is a 2.5D variant of the traditional art gallery problem, and finds applications in urban security.

We give upper and lower bounds on the number of guards needed for a few versions of the problem. Specifically, we prove that $\lfloor \frac{2(k-1)}{3} \rfloor + 1$ guards are always sufficient and sometimes necessary to guard all roofs, and $1 + k + \lfloor \frac{k}{2} \rfloor$ guards are always sufficient to guard the roofs, walls, and the ground, while each roof has at least one guard on it.

1 Introduction

Increased concerns about urban security have brought attention to the problem of assigning guards that can see, or *guard*, the buildings of a city. As the guards are expensive, it is desirable to use the smallest number of guards to achieve this goal.

The *city guarding problem* is a 2.5-dimensional (2.5D) variant of the traditional 2D *art gallery problem* posed by Victor Klee in 1973, that asks to determine the minimum number of guards needed to cover the interior of an n-wall art gallery [16]. A large body of work was published since then on art gallery problems. Chvátal [3] proved that $\lfloor \frac{n}{3} \rfloor$ guards are always sufficient and sometimes necessary to guard a simple polygon with n vertices. Lee, Lin, and Aggarwal proved that finding an optimal solution for a simple polygon without holes is NP-hard [1],[9]. Many NP-hardness results for covering and guarding are also known for polygons with holes [4],[10],[12],[13],[17],[18]. More details on previous work can be found in the book by J. O'Rourke [16], the survey by T. Shermer [18], and the recent book chapter by J. Urrutia [19].

Throughout this paper, for simplicity, we make the following assumptions: (a) the city is within some rectangular region, (b) there are k disjoint "box-shaped" buildings in the city, where each building is an axis-parallel box composed of one roof and four walls, (c) the buildings can have arbitrary positive widths, lengths, and heights (in Section 3, we assume that heights can be sorted such

---

* Daescu's research is supported by NSF grant CCF-0635013.

that no cycle exists for roof visibility. This simplifies the problem), and (d) we only consider "vertex guards", i.e. the guards can be placed only at the corners of the buildings or the four points in the city boundary; such vertex guards are called *candidate guards*. A candidate guard placed at a roof corner sees the roof and two incident walls.

In this paper we consider the following three city guarding problems: (1) *Roof Guarding Problem:* determine the minimum number of guards to guard all the roofs, (2) *Ground and Wall Guarding Problem:* determine the minimum number of guards to cover the ground and all the walls, and (3) *City Guarding Problem:* determine the minimum number of guards to cover the whole city, which means all the roofs and walls of the buildings, and the ground.

It is proved that these problems are NP-hard [2]. The purpose of our study is not to give the best solution for a specific city, but to find the bounds for these problems. Thus, the corresponding algorithms serve for it. For the roof guarding problem, we prove a tight bound of $\lfloor \frac{2(k-1)}{3} \rfloor + 1$, i.e., this number of guards is always sufficient and sometimes necessary to guard all roofs. For the ground and wall guarding problem, we prove that $1 + k + \lfloor \frac{k}{4} \rfloor$ guards are always sufficient. For the city guarding problem, we prove that $1 + k + \lfloor \frac{k}{2} \rfloor$ guards are always sufficient. We also give an $O(k \log k)$ time algorithm for this problem.

2 Related Work

Under our assumptions for city guarding, the vertical projection of a city to the ground level results in a rectangle with rectangular holes. Thus, the problem is closely related to guarding orthogonal polygons with holes.

In 1983, O'Rourke [14] conjectured that $\lfloor \frac{n}{4} \rfloor$ point guards suffice to cover any orthogonal polygon of n vertices, independent of the number of holes (h). In 1984, Aggarwal proved the conjecture for $h = 1$ and $h = 2$ [1]. In 1990, Hoffman proved the full conjecture [5].

For vertex guards, it was conjectured in 1982 that any orthogonal polygon with n vertices and h holes can always be guarded by $\lfloor \frac{n+h}{4} \rfloor$ vertex guards (Shermer's Conjecture). According to Kahn, Klawe, and Kleitman [8], $\lfloor n/4 \rfloor$ vertex guards are sometimes necessary and always sufficient to cover the interior of an orthogonal polygon of n vertices, so this conjecture is true for $h = 0$. In 1984, Aggarwal proved Shermer's conjecture is true for $h = 1, 2$ [1].

The first well known vertex guards upper bound for orthogonal polygons with holes was given by O'Rourke [16], who showed that for any orthogonal polygon with n vertices and h holes, $\lfloor \frac{n+2h}{4} \rfloor$ guards suffice. One important technique he used is the L-shaped partitioning of the scene. Some of our solutions for city guarding largely depend on this algorithm.

In 1996, Hoffmann and Kriegel [6] proved that $\lfloor \frac{n}{3} \rfloor$ vertex guards are sufficient to guard an n-vertex orthogonal polygon with holes. In 2005, Zylinski [20] proved that $\lfloor \frac{n+h}{4} \rfloor$ vertex guards are always sufficient to guard an orthogonal polygon with n vertices and an arbitrary number of holes, provided that there exists a quadrilateralization whose dual graph is a cactus. Most recently, Hoffmann,

Kriegel and Tóth [7] proved that $\lfloor \frac{n+h}{4} \rfloor$ vertex guards can guard any orthogonal polygon with n vertices and h holes, thus settling Shermer's conjecture.

Before presenting solutions, we make the following two observations.

Observation 1. If a guard sees one ground edge of another building then this guard can see the corresponding wall.

Observation 2. Each roof except the highest one can be seen by at least one candidate guard position of other building.

3   Roof Guarding Problem

We present a tight bound for the *Roof Guard Problem*. Throughout this section, we do not consider partial visibility, that is, a roof is visible to a guard if and only if the guard sees the whole roof. Since the tight bound is achieved based on the analysis of complete visibility, which is stronger than partial visibility, partial visibility will not change the bound.

We reduce this geometric problem to a graph. Each roof is represented by a node in the graph. If a candidate guard position on roof a can see roof b, we add a directed edge from roof a to roof b in the graph. Since the buildings have ordered heights, the resulting graph is a directed acyclic graph (DAG).

Lemma 1. *Given a city with k roofs, $\lfloor \frac{2(k-1)}{3} \rfloor + 1$ vertex guards are sometimes necessary to cover all roofs.*

Proof. [1]The example in Fig. 1 requires at least $\lfloor \frac{2(k-1)}{3} \rfloor + 1$ guards. The basic setup of this example is as follows: 1) Each "Master" building has two "Slave" buildings on each side. 2) Each "Master" is taller than its "Slaves". 3) The master buildings are in decreasing order of height from top to bottom. 4) Each slave is higher than any master building below it. 5) Non of the roof vertices of the upper group can guard slave roofs in the lower groups.   □

Before claiming a tight bound we prove the following weak upper bound.

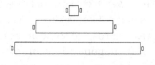

Lemma 2. *Given a city with k roofs, $\lfloor \frac{4(k-1)}{5} \rfloor + 1$ vertex guards are always sufficient to guard all roofs.*

Fig. 1. $\lfloor \frac{2(k-1)}{3} \rfloor + 1$ guards are needed

To obtain a tight bound we need a more detailed analysis. To this end, we classify the edges in the visibility DAG as *blue* and *red*. For an edge of the DAG, if at least two adjacent guard positions on the parent roof can see the lower roof then the edge is a blue edge, else (only one guard position or only two diagonal guard positions can see the lower roof) it is a red edge.

---

[1] Due to the page limitation, we leave the detailed proof in the journal version.

Lemma 3. *Given a parent and its children in the visibility DAG, if all edges from the parent to its children are blue then two diagonal guards are always sufficient to cover the roofs associated with the parent and its children.*

These two guards are marked[2] as *free diagonal guards*, which implies either pair of diagonal guards can be selected to guard the children.

Assuming each node of the graph, with indegree at least one, has at least one blue incoming edge, Lemma 3 and the two-level tree decomposition presented earlier implies at most two diagonal guards are needed for each two-level tree, that is, at most two diagonal guards are needed at a (root) node to guard all its children.

In general, however, it is possible that a node has only red incoming edges. If a node has only red incoming edges, we call the corresponding roof a *trapped roof*. See Fig. 2 for examples. A roof that has a red outgoing edge to a trapped roof is called an *enclosing roof*. We also refer to the group of enclosing roofs and the corresponding trapped roof as a *trap*.

Although traps can assume many geometric structures, and can have recursive structure (a trap can enclose another trap and so on) for our purpose we only need to consider four enclosing roofs for each trapped roof, specifically, the closest one on each side of the trapped roof. Such a *generic trap* has the structure illustrated at the Fig. 2 (a).

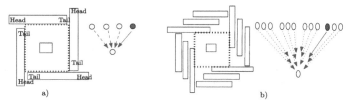

Fig. 2. Trap arrangements (the inner roof is the lowest, and a building's tail blocks the head view of its neighbor)

Enclosing roofs of a generic trap are arranged such that one building's tail blocks the view from the vertices of the next (in circular order) building's head to the trapped roof.

In order to be guarded by other roofs, a trapped roof should have one of its enclosing roofs marked with *fixed diagonal guards*: Unlike for free diagonal guards, after one enclosing roof is marked with fixed diagonal guards, the vertex locations of the two possible guards cannot be changed. We call the enclosing roof marked with fixed diagonal guards an *escaping roof* for the corresponding trapped roof.

Any unmarked roof and any roof marked with free diagonal guards is available for marking with fixed diagonal guards. However, roofs already marked with fixed

---

[2] A pair of marked diagonal guards will be finally assigned with one or two guards only if the corresponding roof is the parent in a resulting two-level tree (discussed in the proof of Theorem 1).

diagonal guards are not available for further marking. Therefore, we need to find an escaping roof for each trapped roof to ensure it can be guarded by another roof.

Lemma 4. *Any roof can be enclosing roof for at most four trapped roofs.*

Proof. Each trapped roof is seen by four vertices from its four closest enclosing roofs respectively. Since each vertex on a closest enclosing roof sees at most one trapped roof, and it has four roof vertices, the enclosing roof can have at most four trapped roofs covered by its four vertices respectively.

Assume there exists an enclosing roof such that three roofs share one of its sides. Then, it follows from the structure of a generic trap that the middle roof is not trapped (see Fig. 3). Therefore, it is enough to show at most two trapped roofs can be on one side.   □

Thus, with respect to its participation in generic traps, a node corresponding to an enclosing roof can have outgoing red edges to at most four other nodes.

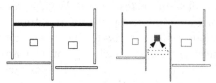

Fig. 3. At most two trapped roofs can share one enclosing roof on one side (a dotted directed edge corresponds to a blue edge in the visibility DAG)

Observe that if two trapped roofs share the same side of an enclosing roof then they conflict with respect to possible positions of fixed diagonal guards.

However, since each trapped roof has (at least) four enclosing roofs, and thus each node with only red incoming edges must have (at least) four incoming edges, we can prove the following lemma.

Lemma 5. *Fixed diagonal guards can always be marked to enclosing roofs such that they guard all trapped roofs.*

Proof (Sketch). Use a vertical scan line and find escaping roofs, then mark fixed diagonal guards in a left-to-right, top-down order (see Fig. 4 for an illustration).   □

Lemma 6. *By finding an escaping roof for each trapped roof, the visibility DAG can be reduced to a tree.*

Proof. We reduce the visibility DAG to a tree as follows. For any node with at least one blue incoming edges, we keep one blue incoming edge and delete all other (blue or red) incoming edges. For any node with only red incoming edges, corresponding to a trapped roof, keep only the incoming edge from its escaping roof and delete all other incoming edges.

After removing those edges every node except the root has exactly one parent. Moreover, the graph connectivity is preserved and thus the resulting acyclic graph is a tree.   □

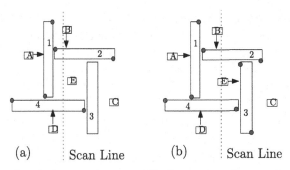

Fig. 4. (a) Scan line before reaching roof E. An arrow points to the marked escape roof, while the dots are the fixed diagonal guards already marked. (b) The scan line reaches E.

Lemma 7. *For any two-level tree in this visibility tree, at most two diagonal guards are required to cover the parent and children.*

Proof. Case 1: If the parent node has all outgoing edges as blue edges, all children and itself will be covered after arbitrarily marking two diagonal guards to the node (by Lemma 3).

Case 2: If the parent node has a red outgoing edge to a child node (there can be only one such edge following our reduction of G to a tree by the scan line method) then it has been selected as an escaping roof for some child node and is marked with two fixed diagonal guards to it.

Notice that the fixed diagonal guards will guard those child nodes that are connected to the current node by a blue edges. Therefore, every parent node needs at most one pair of diagonal guards ("free" or "fixed"), to guard its children and itself (by Lemma 5). □

Theorem 1. *For a city with k buildings, $\lfloor \frac{2(k-1)}{3} \rfloor + 1$ guards are always sufficient and sometimes necessary to guard all roofs.*

Proof. Use the visibility tree from Lemma 6 and the two types of marked diagonal guards from Lemma 3, 5 to assign guards starting from the leaves of the tree, by treating every group of parent and its children as a 2-level subtree. When assigning guards for each parent node we have the following 2 situations: (1) if the node has an out degree $d = 1$, at most d marked diagonal guards are needed, and (2) if the node has an out degree $d \geq 2$, at most 2 marked diagonal guards are needed. The guard assigning process is completed when the whole tree is pruned out. For any parent node in a 2-level subtree, we need $min(d, 2)$ marked diagonal guards, and they totally cover $d + 1$ roofs, corresponding to the parent node and its d children. Since $\frac{min(d,2)}{d+1} \leq \frac{2}{3}$, the ratio of guards/roofs for every 2-level subtree is no more than $\frac{2}{3}$.

Now, let us consider the last step of the partitioning. There can be only two cases: (1) the root is involved in the last 2-level subtree and (2) the root is the only remaining node. Case 1 corresponds to a 2-level subtree, and thus requires at most 2 marked diagonal guards on the parent roof. Since the ratio of guards/roofs for each 2-level subtree is no more than $\frac{2}{3}$, the total number of guards will not exceed $\frac{2k}{3}$. Also due to the fact that the number of guards cannot be a non-integer value, no more than $\lfloor \frac{2k}{3} \rfloor$ guards are needed for this case. For case 2, the remaining root needs one more guard for itself. Therefore, the whole tree needs $\lfloor \frac{2(k-1)}{3} \rfloor + 1$ guards ($\lfloor \frac{2(k-1)}{3} \rfloor$ guards are for the 2-level subtrees and 1 guard is for the root).

Since $\lfloor \frac{2(k-1)}{3} \rfloor + 1 \geq \lfloor \frac{2k}{3} \rfloor$, the upper bound for the roof guarding problem is $\lfloor \frac{2(k-1)}{3} \rfloor + 1$, which matches the lower bound presented earlier. Therefore, $\lfloor \frac{2(k-1)}{3} \rfloor + 1$ is a tight bound for the roof guarding problem. □

4 Ground and Wall Guarding

We show the following result by reduction to guarding orthogonal polygons with holes.

Theorem 2. *The problem of guarding the walls and the ground can be reduced to the art gallery problem for orthogonal polygons with holes in linear time. In particular, the walls and the ground can be guarded by $1 + k + \lfloor \frac{k}{4} \rfloor$ vertex guards on the ground.*

Proof. The reduction step is simply a vertical projection of the city to the ground plane. Each box-shaped building is projected to be a rectangular hole in the ground plane. Walls are reduced to edges of rectangular holes. Because of Observation 1, covering the reduced edges of holes is equivalent to guarding the corresponding walls. All vertices except the four in city boundary are from combining roof corners and corresponding ground corners. The vertex guards for the resulting polygon (a rectangle with rectangular holes) can always be mapped to the ground corners, which are originally candidate vertex guards for the city.

The resulting polygon is a subset of orthogonal polygon with holes. Hoffmann etal [7] proved that an orthogonal polygon with holes can always be guarded by $\lfloor \frac{n+h}{4} \rfloor$ vertex guards. This result maps directly to the ground and wall guarding problem. For our city setup, k is the number of buildings (equivalent to h holes in their bound), and $n = 4 + 4k$. Thus, $1 + k + \lfloor \frac{k}{4} \rfloor$ guards can always guard all walls and the ground. □

5 City Guarding

Given the results from the previous sections, it follows that we can solve the city guarding problem by combining the guards for the roof guarding with those for ground and wall guarding, for a total of at most $\lfloor \frac{2(k-1)}{3} \rfloor + 2 + k + \lfloor \frac{k}{4} \rfloor$ vertex guards. However, we can improve the number of guards to $1 + k + \lfloor \frac{k}{2} \rfloor$ using a strategy based on O'Rourke's theorem [15],[16] for partitioning an orthogonal polygon into L-shaped polygons or *L-polygons*.

L-shaped partition algorithm (Sketch) in [15], page 74 in [16].

a Perform a plane sweep to find all the horizontal cuts[3] of P.
b Partition the polygon at each odd-cut[4]. The resulting pieces are L-polygons
or *histograms*[5].
c Cut each histogram along a vertical line through every other reflex vertex.

Our approach for city guarding is as follows. (i) Project the city to the ground;
use a vertical cut to connect the top right corner of each rectangular hole to the
enclosing rectangle, (ii) find some guards in the plane, and (iii) lift the guards
at vertices of holes to the roofs. This approach works if the polygon with holes
can be guarded by guards selected carefully.

Observation 3. If the projected city can be guarded with t guards such that
at least one guard is assigned to each rectangular hole, then the city can be
guarded by t guards.

Using L-shaped partition algorithm [15],[16], one can hope to achieve $t = 1 + k +$
$\lfloor \frac{k}{2} \rfloor$ guards (one per L-polygon) for city guarding. L-shaped partition algorithm
places the guards at reflex vertices of L-polygons. Note that this will not always
lead to a solution in some cases. Fig. 5 shows an example where the center shaded
roof has no guard assigned. If it is the highest roof, city guarding with this guard
set will fail.

Observation 4. In each L-shaped orthogonal
polygon, a guard can be assigned at one of
the two corners (one reflex vertex, and one
convex vertex as in Fig. 6(a)). We call these
two corners *reflex possible guard position* and
convex possible guard position, respectively.

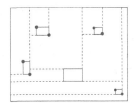

Even though there are two choices for each
L-shaped orthogonal polygon, it is not always

Fig. 5. An counter example

safe to choose the convex one because guard positions on introduced convex
vertices are not legal guard positions at roof corners. Now, the question is: *Can
each rectangular hole always have at least one possible guard position?*

5.1 Partition Properties

To answer the above question, we need to analyze the properties of the partitioning and its outcomes in detail.

---

[3] A horizontal cut is an extension of the horizontal edge incident to a reflex vertex
through the interior of the polygon until it first encounters the boundary of the
polygon. [15],[16].
[4] An odd-cut separates a orthogonal polygon such that at least one resulting part has
odd number of reflex vertices [15],[16].
[5] A horizontal (vertical) histogram is an orthogonal polygon such that there is one
side that can be reached with one vertical (horizontal) line segment from any point
inside the polygon. We have only horizontal histograms here.

Before we classify all the possible partitioning cases for each rectangular hole, we state the following lemma, which is important in cutting histograms.

(a)   (b)

Fig. 6. a. Two possible guard positions. **b.** An example of possible histogram.

Lemma 8. *After cutting all horizontal odd-cuts, there is at most one reflex vertex in the bottom half of each histogram, and can only be at the bottom right.*

Proof. Fig. 6(b) shows a valid histogram after horizontal odd-cutting. The claim is easily verified since every rectangular hole has a vertical cut at its top right corner connecting it to the boundary or another hole.   □

For each **rectangular hole**, we can classify its horizontal odd-cutting into two categories: (1) At most one of the two left vertices of a rectangular hole has one horizontal odd-cut incident to it, and (2) Both left vertices of a rectangular hole have horizontal odd-cuts incident to them. For each category, we further divide it into different classes.

Lemma 9. *After partitioning, at least one vertex of each rectangular hole is a possible guard position, and the corresponding L-polygon is to its left.*

Lemma 10. *By always assigning the guard for each L-shaped orthogonal polygon to its right possible guard position, every rectangular hole has at least one vertex with a guard on it.*

Proof. In all cases, the mentioned possible guarding vertices are all to the right of the L-polygons, which means that they are all right possible guard positions. Therefore, each rectangular hole will get at least one guard using this assigning method.   □

Then, we obtain the **City Guarding Algorithm**.

I Pre-processing
    a Project the city to the ground plane.
    b Using a vertical cut to connect the top right vertex of each hole to the enclosing rectangle, to obtain an orthogonal polygon P without holes.
II Partition using the modified L-shaped partition algorithm in [15],[16].
    a-b The first two steps are same as the original algorithm.
    c For each histogram, index all reflex vertices from **right to left** with consecutive integers starting from 1. Partition each histogram by making a vertical cut through every even-number indexed reflex vertex.
III Assigning Guards
    a The products from the previous steps are all L-shaped orthogonal polygons. Assign guards to right possible guard positions.

Theorem 3. $1 + k + \lfloor \frac{k}{2} \rfloor$ *guards are always sufficient to guard the roofs, walls and the ground of a city with k buildings, and each roof has at least one guard on it.*

Corollary 1. *The city guarding algorithm takes $O(k \log k)$ time.*

6   Conclusion

In this paper we have proposed a new 2.5D guarding problem, and have presented detailed results for each subproblem: *Roof Guarding, Ground and Wall Guarding* and *City Guarding*. We not only have achieved the tight bound for the first problem and upper bounds for the other two, but also have presented a polynomial-time algorithm for *City Guarding*.

Acknowledgments. The authors would like to thank Yuanyi Zhang for kind discussions and advices.

References

1. Aggarwal, A.: The Art Gallery Theorem: Its Variations, Applications, and Algorithmic Aspects. Ph.D. thesis, Johns Hopkins University, Baltimore (1984)
2. Bao, L., Bereg, S., Daescu, O., Ntafos, S., Zhou, J.: The Complexity Study of City Guard Problem. The University of Texas at Dallas (manuscript, 2007)
3. Chvátal, V.: A Combinatorial Theorem in Plan Geometry. J. Combin. Theory Ser. B 18, 39–41 (1975)
4. Garey, R., Johnson, D.: Computers and Intractability. W.H. Freeman, New York (1979)
5. Hoffmann, F.: On the Rectilinear Art Gallery Problem. In: Paterson, M. (ed.) ICALP 1990. LNCS, vol. 443, pp. 717–728. Springer, Heidelberg (1990)
6. Hoffmann, F., Kriegel, K.: A Graph Coloring Result and Its Consequences for Polygon Guarding Problems. SIAM J. Discrete Mathematics 9(2), 210–224 (1996)
7. Hoffmann, F., Kriegel, K., Tóth, C.D.: Vertex Guards in Reclinear Polygons with Holes. In: Hoffmann, F., Kriegel, K., Tóth, C.D. (eds.) Kyoto International Conference on Computational Geometry and Graph Theory (2007)
8. Kahn, J., Klawe, M., Kleitman, D.: Traditional Galleries Require Fewer Watchmen. SIAM J. Algebraic and Discrete Methods 4, 194–206 (1983)
9. Lee, D., Lin, A.: Computational Complexity of Art Gallery Problems. IEEE Trans. Inform. Theory 32, 276–282 (1986)
10. Lenhart, W., Jennings, G.: An Art Gallery Theorem for Line Segments in the Plane, Williams College (manuscript, 1990)
11. Ntafos, S.: On Gallery Watchman in Grid. Info. Proc. Let. 23, 99–102 (1986)
12. Lozano-Pérez, T., Wesley, M.: An Algorithm for Planning Collision-Free Paths among Polyhedral Obstacles. Commun. Ass. Comput. Mach. 22, 560–570 (1979)
13. Lubiw, A.: Orderings and Some Combinatorial Optimization Problems with Geometric Applications. Ph.D. thesis, University of Toronto (1985)
14. O'Rourke, J.: Galleries need Fewer Mobile Guards: A Variation on Chvatal's Theorem. Geometriae Dedicata 14, 273–283 (1983)

15. O'Rourke, J.: An Alternate Proof of the Rectitlinear Art Gallery Theorem. J. of Geometry 21, 118–130 (1983)
16. O'Rourke, J.: Art Gallery Theorems and Algorithms. Oxford University Press, Oxford (1987)
17. O'Rourke, J.: Recovery of Convexity from Visibility Graphs. Tech. Rep. 90.4.6, Dept. Computer Science, Smith College (1990)
18. Shermer, T.: Recent Results in Art Galleries. Tech Report CMPT TR 90-10, Simon Fraser University, Computing Science (1990)
19. Urrutia, J.: Art Gallery and Illumination Problems. In: Sack, J.R., Urrutia, J. (eds.) Handbook on Computational Geometry. Elsevier, Amsterdam (2000)
20. Zylinski, P.: Orthogonal Art Galleries with Holes: a Coloring Proof of Aggarwal's Theorem. The electronic Journal of Combinatorics 13 (2006)

Optimal Insertion of a Segment Highway in a City Metric

Matias Korman and Takeshi Tokuyama

Graduate School of Information Sciences
Tohoku University, Sendai, Japan
{mati,tokuyama}@dais.is.tohoku.ac.jp

Abstract. Given two sets of points in the plane, we are interested in locating a highway h such that an objective function on the *city distance* between points of the two sets is minimized (where the city distance is measured with speed $v > 1$ on a highway and 1 in the underlying metric elsewhere). Extending the results of Ahn et al. ([7]), we consider the option that there are already some built highways. We give a unified approach to this problem to design polynomial-time algorithms for several combinations of objective functions and types of the inserted highway (*turnpike* or *freeway*).

1 Introduction

Geometric optimization related to urban transportation systems is an important topic in computational geometry. For example, when we travel in Tokyo city, we walk to a nearby station, take a train to another station, take a bus, and then walk from a bus stop to the final destination. The nearby station is not necessarily the closest station, since we have Tokyo metro, JR (Japan railway), and several private railways.

There are two major issues on such an urban transportation system. One is the efficient use of a given transportation system, and the other is the design and modification of a given transportation system to increase its usability.

Although the metric given by a real urban transportation system is often quite complicated, simplified mathematical models have been widely studied in order to investigate basic geometric properties of urban transportation systems. A popular formulation is to represent an urban transportation system as a graph. The shortest (i.e. minimum travel time) path query by using such a graph model is well studied. Nowadays there are many commercial systems that give not only the shortest path but also a set of alternative paths that may be useful for the user.

We may also consider the problem as a purely geometrical one: represent highways as polygonal chains consisting of line segments in the plane, giving each line segment an associated speed. Then, the travel time between two points s and t gives a metric. It is common to classify highways into two kinds: A highway that one can only enter/exit at the endpoints of the segment, and a highway that one

X. Hu and J. Wang (Eds.): COCOON 2008, LNCS 5092, pp. 611–620, 2008.

Fig. 1. Sources and their paths to their closest sinks in the city distance

Fig. 2. Updated shortest paths once h is inserted

can enter/exit at any point. In this paper, we call the former a *turnpike*, and the latter a *freeway*. We assume that we can cross a turnpike at any point (imagine an overhead road). We note that turnpike is equivalent to the concept of *walkway* defined by Cardinal et al. [8]. The reader may complain that we cannot enter/exit a real-world freeway in such a flexible way, but this is just a mathematical abstraction. The problem is also important in the design of a working facility of autonomous robots: Robots can move faster on a predetermined route (e.g., a route with guide tape that can be easily recognized by robots) corresponding to a freeway, and also can ride a conveyer corresponding to a turnpike.

The problem of computing the Voronoi diagram and shortest paths under the metric induced by a given transportation network was proposed in [5]. Their algorithm has an $O(nm^2 + m^3 + nm \log n)$ time complexity to compute the Voronoi diagram (where n is the number of sites and m the number of highways), allowing each highway to have different speed. A more simplified transportation model (called the *city distance*) was introduced by Aichholzer et al.([3]). Under the *city distance*, each highway is given as an axis-parallel line segment on which a traveler can move with a speed $v > 1$, and the metric is induced by the L_1 metric. In Figure 1 we can see a sample problem with one freeway (thick segment), one walkway (thin segment), three sources (filled points), each one connected to its closest sink (hollow points). In Figure 2, a new turnpike h has been located and optimal paths have been updated accordingly.

Efficient algorithms for calculating shortest paths and Voronoi diagrams with respect to the city distance $d_{\mathcal{H}}$ were given in [3]. Bae et al.([4]) further improved the time complexities by giving $O((n + m) \log(n + m))$ optimal algorithms for the Voronoi diagram construction under the same distance.

Let us consider the design of an optimal urban system in the city distance model, in which we want to build a highway in order to maximize an objective function representing the user's profit. Cardinal and Langerman ([7]) considered the following problem: given a list of pairs (s_i, t_i) $(i = 1, 2, \dots n)$ of points in the plane, finding the location of a line freeway h (i.e. an infinite length freeway with no endpoint) to minimize $\max_i d_{\{h\}}(s_i, t_i)$. This is a kind of facility location problem in which the user may chose whether or not to use the new facility (i.e.: highway). They gave a linear algorithm (in the number of pairs) to find such a location considering that we can move on h in zero time (i.e., $v = \infty$). Ahn et al.[2] considered the problem of locating a line freeway to reduce the maximum distance among *all* possible pairs of distances of a list of given points S, obtaining a linear time algorithm for locating such highway. They also gave an $O(|S| \log |S|)$ time algorithm for locating an arbitrarily oriented line freeway. Cardinal et al. [8] gave an $O(n \log n)$ algorithm for locating an axis aligned turnpike (and an $\Omega(n \log n)$ lower bound if the maximum diameter is also to be computed). However, in these works, no previously located highways are considered in the environment.

Thus, by applying the above mentioned algorithms we can obtain a transportation system with only one highway. Ahn et al.[2] considered the problem of locating two line freeways simultaneously. However, it has been shown ([11]) that optimally locating k arbitrarily oriented infinite speed line highways is an inapproximable problem by any factor in polynomial time unless $P = NP$. Moreover, in [9] it is conjectured that inapproximability still holds even if we restrict the problem to the case in which highways are axis aligned. Taking this into account, it is natural to consider a method to increment the transportation system by inserting highways one by one.

In this paper, we investigate the computational complexity of the following problem: Given two sets of points S and T called *sources* and *sinks* respectively, and a set of isothetic (i.e.: horizontal or vertical) highways $\mathcal{H}$, we locate another highway h of fixed length that minimizes an objective function related to the *city distance* between sources and sinks. In a real problem in urban planning, source and sink points correspond to houses and working centers, respectively. Naturally, the problem becomes much more involved than the particular case in which $\mathcal{H} = \emptyset$, and to the authors' knowledge, this paper gives the first nontrivial results on this problem (except the authors' preliminary report [10]).

2   Notations and Results

Given sets of sources S, sinks T, and highways $\mathcal{H}$, we call the triplet $(S, T, \mathcal{H})$ a *transportation configuration*. Other than a configuration, an underlying metric d and a constant *speed* $v > 1$ are needed to define the problem. A *usable interval* of a highway is an interval of a highway between an entry point and an exit point; therefore, it is the full highway for a turnpike, and any interval of a freeway.

The travel time of a path $\pi = (p_0, \ldots, p_k)$ is defined by $|\pi| = \sum_{0 \le i < k} \frac{1}{v_i} d(p_i, p_{i+1})$, where $v_i = v$ if the segment $p_i p_{i+1}$ is a usable interval of a highway, and $v_i = 1$ otherwise. The transportation distance between s and t is defined by $d_{\mathcal{H}}(s, t) = \inf_{\pi \in \mathcal{P}(s,t)} |\pi|$, where $\mathcal{P}(s, t)$ is the set of all s-t-paths. Since we consider the L_1 distance as d, we can observe that any path between s and t is a sequence of isothetic segments $p_i p_{i+1}$.

Fixed the location of the new highway h, for each source s in S, we consider the value $cost(s) = \odot_{t \in T} d_{\mathcal{H} \cup \{h\}}(s, t)$, where $\odot$ stands for an operation in $\{\sum, \min, \max\}$, and $d_{\mathcal{H} \cup \{h\}}(s, t)$ is the city distance with the set of highways $\mathcal{H} \cup \{h\}$. Each value $cost(s)$ indicates a statistical measure of how far the sinks are from source s. Our objective function is $\Phi = \odot'_{s \in S} cost(s)$, which shows an aggregated measure of $cost(s)$ for $s \in S$ using an operation $\odot' \in \{\sum, \min, \max\}$.

Without loss of generality we consider that h is a vertical highway of unit length. The insertion of such highway h is parameterized by the location $\beta = (\xi, \eta)$

of its bottom endpoint. The set $\mathcal{H}$ can contain both freeways and turnpikes, but the type of the highway to insert is specified. We may denote by h_β for the inserted highway and also denote $\mathcal{H}_\beta$ for the updated set of highways (i.e.: $\mathcal{H}_\beta = \mathcal{H}_{\xi,\eta} = \mathcal{H} \bigcup \{h_\beta\}$). Regarding as a function of parameters $x = \xi$ and $y = \eta$, we define $F_{s,t}(x,y) = d_{\mathcal{H}_{x,y}}(s,t)$ (abusing notation, we also consider $cost(s)$ and Φ as functions of x and y).

Our basic algorithm design relies on the fact that Φ is described by using the set $\{F_{s,t}(x,y)|s \in \mathcal{S}, t \in \mathcal{T}\}$. The rest of the paper gives the explicit analysis of how to efficiently compute $F_{s,t}(x,y)$ for each pair $s \in \mathcal{S}$, $t \in \mathcal{T}$, and combine them in order to obtain the location β minimizes Φ.

We mainly discuss the case in which $\odot' = \sum$ and $\odot = min$ (i.e.: $\Phi = \sum_{s \in \mathcal{S}} min_{t \in \mathcal{T}} d_{\mathcal{H} \bigcup \{h\}}(s,t)$), since it is probably the most interesting objective function as a facility location problem. The time complexities are $O(S^3H^2 + ST^2)$ and $O((T+H)(S+T+H)^2(SH+T)^2)$ for inserting a turnpike and a freeway, respectively (where the cardinalities of $\mathcal{S}$, $\mathcal{T}$, and $\mathcal{H}$ are denoted by S, T and H, respectively).

The time complexities for optimizing other objective functions are given in Table 1, where the complexities are simplified under the assumption that $S \geq T \geq H$. The shaded entries are results announced in our preliminary report [10] (without formal proceedings), where polynomial factors of the inverse Ackermann function are ignored. Non-shaded entries are further improved complexites updated in the table.

3   Locating a Turnpike

In this section, we are given a transportation configuration $(\mathcal{S}, \mathcal{T}, \mathcal{H})$ and consider the location of a vertical turnpike segment h of unit length (recall that we allow both turnpikes and freeways in $\mathcal{H}$). We use the *shortest path map* [12] that is defined as follows in our context:

Given a point p and $\mathcal{H}$, we define $SPM(p, \mathcal{H}, \delta) = \{q \in \mathbb{R}^2 | d_\mathcal{H}(p,q) = \delta\}$, which is a polygon. Let $V(p, \mathcal{H}, \delta)$ be the set of vertices of the polygon $SPM(p, \mathcal{H}, \delta)$. The straight skeleton with respect to the shortest paths from p is defined as $SK(p, \mathcal{H}) = \{q \in \mathbb{R}^2 | \exists \delta \geq 0, q \in V(p, \mathcal{H}, \delta)\}$. Let $SPM(p, \mathcal{H})$ be the planar subdivision whose cells (i.e., two dimensional faces) are connected components of $\mathbb{R}^2 \setminus SK(p, \mathcal{H})$. Each pair of points in a cell have topologically identical shortest paths from p.

Table 1. Time complexities for the different objective functions

$\odot'$ ＼ $\odot$	freeway			turnpike		
	min	max	$\sum$	min	max	$\sum$
min	$O(STH^3)$	$O^*(ST^2H^6)$	$O(ST^2H^6)$	$O(ST)$	$O^*(ST^2H^4)$	$O(ST^2H^4)$
max	$O(S^4TH)$	$O^*(S^2T^2H^6)$	$O^*(S^2T^4H^{12})$	$O(S^4H^2)$	$O^*(S^2T^2H^4)$	$O^*(S^2T^4H^8)$
$\sum$	$O(S^4TH^2)$	$O^*(S^2T^4H^{12})$	$O(S^3T^2H^2 \log T)$	$O(S^3H^2)$	$O^*(S^2T^4H^8)$	$O(S^3TH^2)$

Fixed a point p, we define $f_p(x, y) = d_{\mathcal{H}}((x, y), p)$ representing the distance between a fixed point p and $q = (x, y)$. Let $\Gamma(x, y) = \min_{t \in \mathcal{T}} \{ f_t(x, y) \}$ be the distance from $q = (x, y)$ to its closest sink. Given a bivariate continuous piecewise linear function $f(x, y)$, its trajectory is a polygonal terrain. The orthogonal projection of the polygonal terrain to the (x, y) plane gives a planar subdivision of the domain of f, which is called the *support complex* of f. By definition, $\Gamma(x, y)$ is a piecewise linear function dependent on $\mathcal{T}$ and $\mathcal{H}$. Let $\mathcal{A}(\mathcal{T}, \mathcal{H})$ be its support complex. The *combinatorial complexity* of a planar subdivision is the total number of cells, edges, and vertices of the subdivision. The following theorems are given by Bae *et al.*([4]):

Theorem 1. $SPM(p, \mathcal{H})$ has $O(H)$ combinatorial complexity, and can be constructed in $O(H \log H)$ time. The support complex $\mathcal{A}(\mathcal{T}, \mathcal{H})$ of $\Gamma(x, y)$ has an $O(T + H)$ combinatorial complexity, and can be computed in $O((T + H) \log (T + H))$ time. Moreover, in each cell c of $SPM(p, \mathcal{H})$, the function f_p is linear (analogously, $\Gamma(x, y)$ is linear in each cell of $\mathcal{A}(\mathcal{T}, \mathcal{H})$).

Note that our environment is slightly more general than the one considered in [4] (we consider two types of highways). Therefore, the modifications needed to adapt the proof can be found in the full version of this paper. Given a region R, we consider the translation of R vertically by -1. Let $e = (0, -1)$, we define $R' = R + e = \{ (x, y) \in \mathbb{R}^2 | (x, y+1) \in R \}$. Accordingly, we define $\mathcal{R}' = \{ R'_1, \ldots R'_k \}$ for a plane subdivision $\mathcal{R} = \{ R_1, \ldots R_k \}$. Thus, we have $\mathcal{A}'(\mathcal{T}, \mathcal{H})$ and $SPM'(s, \mathcal{H})$ by shifting $\mathcal{A}(\mathcal{T}, \mathcal{H})$ and $SPM(s, \mathcal{H})$, respectively.

3.1   Computing Φ from $cost(s)$ Functions

We proceed to compute a subdivision of the plane such that the $cost(s)$ functions can be expressed as the minimum of three linear functions in each cell. Let $\mathcal{A}_{turn}$ be the planar subdivision obtained as the overlay of subdivisions $\mathcal{A}(\mathcal{T}, \mathcal{H})$, $\mathcal{A}'(\mathcal{T}, \mathcal{H})$, and subdivisions $SPM(s, \mathcal{H})$ and $SPM'(s, \mathcal{H})$ for every source $s \in \mathcal{S}$. That is, two points p and q are in the same cell of $\mathcal{A}_{turn}$ if and only if they are in the same cell of each of these planar subdivisions.

Lemma 1. $\mathcal{A}_{turn}$ is a planar subdivision of combinatorial complexity $O(S^2 H^2 + T^2)$ that can be computed in $O(S^2 H^2 + T^2)$ time.

Proof. Simply compute $\mathcal{A}_{turn}$ through the overlay of the regions that form it: two subdivisions of size $O(T + H)$, and S different subdivisions of size $O(H)$ are involved, therefore time and size bounds follow.

Note that, by definition of a turnpike, for any fixed source s and sink t, we have $F_{s,t}(x, y) = \min \{ d_{\mathcal{H}}(s, t), f_s(x, y+1) + 1/v + f_t(x, y), f_s(x, y) + 1/v + f_t(x, y+1) \}$. Therefore, $cost(s)$ can be written as

$$cost(s) = \min \{ \min_{t \in \mathcal{T}} \{ d_{\mathcal{H}}(s, t) \}, f_s(x, y+1) + 1/v + \Gamma(x, y), f_s(x, y) + 1/v + \Gamma(x, y+1) \}$$

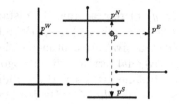

Fig. 3. Neighbouring freeways of a point p

Fig. 4. Subdivision of a shortest path into primitive paths

Which leads to the following theorem:

Theorem 2. *The optimal location of a new isothetic turnpike can be computed in $O(S^3 H^2 + ST^2)$ time.*

Proof. In each cell $c \in A_{turn}$, we have $cost(s) = \min\{g_1^s, g_2^s, g_3^s\}$, where g_2^s and g_3^s are linear functions, and g_1^s is a constant. Therefore, in a fixed cell c, the function $cost(s)$ is concave. By definition, our objective function Φ is the sum of $O(S)$ different concave functions, therefore concave as well within c. Thus, the minimum of Φ in c is located at one of the vertices of c. We will check the value of Φ at all the vertices of A_{turn} to obtain its global minimum. The value of Φ in a single node can be obtained in $O(S)$ time, which iterated among the $O(S^2 H^2 + T^2)$ vertices in A_{turn} proves the theorem.

4   Locating a Freeway

4.1   Characterization of Shortest Paths

In order to adapt the previous method for the freeway case, we first need to characterize the topology of shortest paths under a transportation configuration $(S, T, \mathcal{H}_\beta)$. We define the north neighboring freeway of a point p in $\mathcal{H}_\beta$ as the first freeway that we find along the vertical ray $\{p + (0, t)|t > 0\}$ if it exists. The point found at the intersection is called its North neighbor point p^N (analogously we define p^S, p^E and p^W). Let $N(p, \mathcal{H}_\beta)$ the set of neighboring points of p, V the set of endpoints of highways in $\mathcal{H}_\beta$, and I the set of intersection points of freeways. For any point p, we define $M_p(\mathcal{H}_\beta) = V \bigcup I \bigcup N(p, \mathcal{H}_\beta) \bigcup_{q \in V \cup T} N(q, \mathcal{H}_\beta)$.

A path is called *primitive* if it has at most 1 bend, passes through at most one highway and does not contain any point of $M_s(\mathcal{H}_\beta)$ in its interior. A primitive path with shortest length is called a *shortest* primitive path. The following lemma is given in [4], and implies that even if we allow hopping on and off of a freeway h at any point, there is a finite amount of such points in a shortest path (see figure 4):

Lemma 2. *For any $s \in S$ and $t \in T$ in a transportation configuration, there exists a shortest path π from s to t such that π is a sequence of primitive paths whose endpoints (except s and t) are in $M_s(\mathcal{H}_\beta)$.*

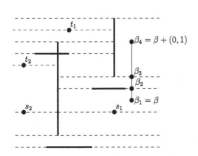

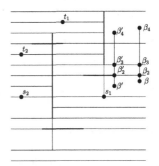

Fig. 5. $\mathcal{H}$ (solid), $T(\mathcal{S},\mathcal{T},\mathcal{H})$ (dashed), and exit points $E_t(\beta)$

Fig. 6. The y coordinates of exit points do not depend on the location of β in a cell of $\mathcal{E}(\mathcal{S},\mathcal{T},\mathcal{H})$ if $1 < i < k(s,\beta)$

4.2 Admissible $\mathcal{S}$-$\mathcal{T}$-Subdivision

For a fixed source s and a point β, we define $E_s(\beta)$ as the set of points in $M_s(\mathcal{H}_\beta)$ that lie in h_β (i.e.: $E_s(\beta) = M_s(\mathcal{H}_\beta) \bigcap h_\beta$). Lemma 2 implies that $E_s(\beta)$ is the set of candidates of the entrance or exit point from h_β in the shortest path from a fixed source s to any sink $t \in \mathcal{T}$. Intuitively, set $E_s(\beta)$ is the set of the endpoints of h_β and points on h_β that are horizontally visible (i.e., there is no freeway blocking the horizontal viewing ray) from $\{s\} \cup V \cup \mathcal{T}$ (see figure 5).

Let $k(s,\beta) = |E_s(\beta)|$. For any $i \in [1, k(s,\beta)]$, let $\beta_i(s)$ be the point in $E_s(\beta)$ whose y coordinate is the $i-th$ lowest one. Note that $\beta_1(s) = \beta$, $\beta_{k(s,\beta)}(s) = \beta + (0,1)$ and $k(s,\beta) \leq T + 2H + 3$.

For $i,j \in [1, k(s,\beta)]$, the length of the shortest s-t path entering h_β at point $\beta_i(s)$ and exiting at point $\beta_j(s)$ equals $d_\mathcal{H}(s, \beta_i(s)) + d(\beta_i(s), \beta_j(s))/v + d_\mathcal{H}(\beta_j(s), t)$. We are interested in the path that starts at s, enters h_β at point $\beta_i(s)$, exits at point $\beta_j(s)$, and from there goes to the nearest sink. The length of such path is: $g_{i,j}^s(\beta) = d_\mathcal{H}(s, \beta_i(s)) + d(\beta_i(s), \beta_j(s))/v + \Gamma(\beta_j(s))$. By definition, for any $s \in \mathcal{S}$, we have $cost(s) = \min\{\min_{t \in \mathcal{T}}\{d_\mathcal{H}(s,t)\}, \min_{i,j \in [1,k(s,\beta)]}\{g_{i,j}^s(\beta)\}\}$.

We can observe that for most of locations for β, the set $E_s(\beta)$ is stable against a small movement of β (except for its top and bottom points). We will thus find a subdivision $\mathcal{E}(\mathcal{S},\mathcal{T},\mathcal{H})$ of the plane such that the y-value of $\beta_i(s)$ (for $1 < i < k(s,\beta)$) is independent of the choice of the point $\beta = (\xi, \eta)$ for each cell c of $\mathcal{E}(\mathcal{S},\mathcal{T},\mathcal{H})$ and every source $s \in \mathcal{S}$.

The subdivision $\mathcal{E}(\mathcal{S},\mathcal{T},\mathcal{H})$ is defined as follows: we draw horizontal rays to both sides from endpoints of each highway and from each $p \in \mathcal{S} \bigcup \mathcal{T}$ until they hit a freeway as in figure 5. Thus we obtain a horizontal trapezoidal map $T(\mathcal{S},\mathcal{T},\mathcal{H})$ that decomposes the plane into rectangles. Let $\mathcal{E}(\mathcal{S},\mathcal{T},\mathcal{H})$ be the subdivision of the plane induced by the arrangement of the set of segments $T(\mathcal{S},\mathcal{T},\mathcal{H}) \bigcup T'(\mathcal{S},\mathcal{T},\mathcal{H})$, illustrated in Figure 6. Recall that $T'(\mathcal{S},\mathcal{T},\mathcal{H})$ is the subdivision obtained by shifting vertically down $T(\mathcal{S},\mathcal{T},\mathcal{H})$ by 1 unit.

We can observe the following:

Lemma 3. *For any cell $c \in \mathcal{E}(\mathcal{S}, \mathcal{T}, \mathcal{H})$ and source $s \in \mathcal{S}$, the value $k(s, \beta)$ is independent of choice of $\beta \in c$, thus, we write $k_c(s)$ for $k(s, \beta)$. For any two points p, q in the same cell c, their corresponding highways h_p and h_q intersect exactly the same horizontal edges in $T(\mathcal{S}, \mathcal{T}, \mathcal{H})$. Thus, the y coordinate of $\beta_i(s)$ is independent of choice of $\beta \in c$ for $i = 2, 3, \ldots, k_c(s) - 1$.*

We observe that, if $i, j \leq k_c(s)$, function $g_{i,j}^s(\beta) = g_{i,j}^s(\xi, \eta)$ is piecewise linear within each cell c. We will define a finer subdivision in which every $g_{i,j}^s$ function is linear in each cell. We say that a subdivision $\mathcal{R}$ is an admissible $\mathcal{S}$-$\mathcal{T}$-subdivision if it satisfies the following:

1. $\mathcal{R}$ is a refinement of subdivision $\mathcal{E}(\mathcal{S}, \mathcal{T}, \mathcal{H})$.
2. Fixed any cell $C \in \mathcal{R}$ (such that $C \subseteq c$ for a cell $c \in \mathcal{E}(\mathcal{S}, \mathcal{T}, \mathcal{H})$), for every $s \in \mathcal{S}$, $\beta, \beta' \in C$, and $i \in [1, k_c(s)]$, points $\beta_i(s)$ and $\beta_i'(s)$ are in the same cells of $SPM(s, \mathcal{H})$ and $\mathcal{A}(\mathcal{T}, \mathcal{H})$.

By definition of admissible $\mathcal{S}$-$\mathcal{T}$-subdivision, we have the following:

Lemma 4. *Function $g_{i,j}^s(\xi, \eta)$ is linear within each cell C of an admissible $\mathcal{S}$-$\mathcal{T}$-subdivision $\mathcal{R}$.*

Once such an admissible subdivision is computed, the minimum of Φ can be found analogously to the turnpike case (as shown later). Thus, our task is to compute an admissible $\mathcal{S}$-$\mathcal{T}$-subdivision of low complexity.

Let $\mathcal{A}_0$ be the arrangement induced by the overlay of regions $\mathcal{A}(\mathcal{T}, \mathcal{H})$, $\mathcal{A}'(\mathcal{T}, \mathcal{H})$, $T(\mathcal{S}, \mathcal{T}, \mathcal{H})$, $T'(\mathcal{S}, \mathcal{T}, \mathcal{H})$ and $SPM(s, \mathcal{H})$ and $SPM'(s, \mathcal{H})$ for every source $s \in \mathcal{S}$. Note that $\mathcal{A}_0$ satisfies property one and, for $i = 1$ and $i = k_c(s)$, property two is also satisfied. We will further refine $\mathcal{A}_0$ to certify the second property for $1 < i < k_c(s)$. From lemma 3, for each $s \in \mathcal{S}$, point $\beta_i(s)$ is on a horizontal segment at a fixed height $\nu_{c,i}(s)$ if $\beta \in c$ and $1 < i < k_c$. We trace the changes of $SPM(s, \mathcal{H})$ and Γ at each height $\nu_{c,i}(s)$: consider the one dimensional functions $f_s(x, \nu_{c,i}(s))$ (for a fixed $s \in \mathcal{S}$) and $\Gamma(x, \nu_{c,i}(s))$. Both functions are one dimensional piecewise linear if $(x, \nu_{c,i}(s)) \in c$. Let $\mathcal{B}_s(c, i)$ be the set of breakpoints (i.e.: non differenciable points) of $f_s(x, \nu_{c,i}(s))$ for a fixed $s \in \mathcal{S}$, c and $1 < i < k_c(s)$ (analogously, let $\mathcal{C}_s(c, i)$ the set of breakpoints of $\Gamma(x, \nu_{c,i}(s))$).

We cut each cell $c \in I$ into subcells by all vertical lines that pass through points in $\bigcup_{1 < i < k_c(s)} \mathcal{B}_t(c, i)(s)$ and $\bigcup_{1 < i < k_c(s)} \mathcal{C}_t(c, i)(s)$ for every source $s \in \mathcal{S}$. Let $\mathcal{A}_{free}$ be the resulting subdivision. We can prove the following lemma:

Theorem 3. *$\mathcal{A}_{free}$ is an admissible $\mathcal{S}$-$\mathcal{T}$-subdivision of combinatorial complexity $O((S + T + H)(SH + T)^2)$ that can be computed in $O((S + T + H)(SH + T)^2)$ time.*

Proof. Such subdivision will be computed through a modified version of the plane sweep algorithm ([6]) of subdivisions $\mathcal{A}(\mathcal{T}, \mathcal{H})$, $\mathcal{A}'(\mathcal{T}, \mathcal{H})$, $T(\mathcal{S}, \mathcal{T}, \mathcal{H})$, $T'(\mathcal{S}, \mathcal{T}, \mathcal{H})$ and $SPM(s, \mathcal{H})$ and $SPM'(s, \mathcal{H})$ for every source $s \in \mathcal{S}$. At each intersection found while computing the overlay, other than the usual operations of a plane sweep, we check whether or not the intersection is caused by a horizontal edge from subdivision $T(\mathcal{S}, \mathcal{T}, \mathcal{H})$ and an edge of either subdivision $SPM(s, \mathcal{H})$ (for any $s \in \mathcal{S}$) or $\mathcal{A}(\mathcal{T}, \mathcal{H})$. Note that each such intersection corresponds to a point in $\mathcal{B}_t(c, i)(s)$ or $\mathcal{C}_t(c, i)(s)$, respectively. If such an intersection is found, we will add a vertical segment of unit length whose upper endpoint is the intersection just found. By construction, we obtain $\mathcal{A}_{free}$. A standard sweep algorithm needs $O(n \log n + i \log n)$ computational time (where n is the size of the input and i is the number of intersections). In our case, we have $n = SH + (T + H) + (S + T + H) \in O(SH + T)$ and $i = O((SH + T)^2)$, giving a cost of $O((SH + T)^2 \log (SH + T))$ to compute $\mathcal{A}_0$.

In order to compute $\mathcal{A}_{free}$, we need to include vertical segments to refine $\mathcal{A}_0$. Those segments are caused by intersections subdivision $T(\mathcal{S}, \mathcal{T}, \mathcal{H})$ (a subdivision of size $O(S + T + H)$) with either a subdivision of size $O(T + H)$ or any of the S different subdivisions of size $O(H)$. That is: there can be $O(S + T + H)(T + H + SH) \in O((S + T + H)(SH + T))$ total intersections. For each such intersection we add a new segment to the subdivision. Being vertical, such segment can only intersect with lines at the current sweep line, therefore $O(SH + T)$ time is needed to treat each new added edge. Thus, $O((S + T + H)(SH + T)^2)$ time extra is needed to process the added segments, therefore proving the theorem.

4.3   Computing Φ for the Freeway Case

In each cell c of $\mathcal{A}_{free}$, we have: $cost(s) = \min\{\min_{t \in \mathcal{T}}\{d_{\mathcal{H}}(s, t)\}, \min_{i,j \in [1, k_c(s)]}\{g_{i,j}^s\}\}$, where each $g_{i,j}^s$ function is linear in each cell c. That is, we again have that Φ is concave in each cell $c \in \mathcal{A}_{free}$. Therefore, evaluating Φ at the vertices of $\mathcal{A}_{free}$ is enough to find its minimum. We only need an efficient method to evaluate Φ at each vertex:

Lemma 5. *For each given β, $\Phi(\beta)$ can be computed in $O((S + T + H)(T + H))$ time.*

Proof. We will see that, fixed β and after a preprocessing of $O(T^2 + H^2)$ time, the value of any $cost(s)$ function can be computed in $O(T + H)$ time as follows:

Fixed $s \in \mathcal{S}$, for every $i \in k(s, \beta)$ we compute the index that minimizes the expression $\min_{j \in k(s, \beta)}\{d(\beta_i(s), \beta_j(s))/v + \Gamma(\beta_j(s))\}$. Let $j_s^*(i)$ be such index and $c_s(i)$ the value (i.e.: $c_s(i) = d(\beta_i(s), \beta_{j^*(i)}(s))/v + \Gamma(\beta_{j^*(i)}(s))$. Recall that $k(s, \beta) \in O(T + H)$, therefore $c_s(i)$ and $j_s^*(i)$ can be computed for every $i \in k(s, \beta)$ in $O(T^2 + H^2)$ time.

Once $c_s(i)$ and $j_s^*(i)$ are known for a particular $s \in \mathcal{S}$, the values of $c_{s'}(i)$ and $j_{s'}^*(i)$ can be computed in $O(1)$ time for any $s' \in \mathcal{S}$. Using $c_s(i)$ we can compute any $cost(s)$ in $O(T + H)$ time as: $cost(s) = \min\{\min_{t \in \mathcal{T}}\{d(s, t)\}, \min_{i \in k(s, \beta)}\{f_s(\beta_i(s)) + c_s(i)\}\}$.

Since it suffices to compute $\Phi(\beta)$ at all $O((S + T + H)(SH + T)^2)$ vertices of $\mathcal{A}_{free}$, we have the following.

Theorem 4. *The optimal location of a new isothetic freeway can be computed in $O((T + H)(S + T + H)^2(SH + T)^2)$ computational time.*

5 Concluding Remarks

We give a list of open problems: an obvious problem is to improve upper bounds of Table 1. Another interesting problem is to modify the algorithm to allow efficient dynamic insertions (and possibly deletions) of highways. Third, in urban transportation design, we need to consider highways with arbitrary directions and different speeds; moreover, there may be obstacles and hazards. Finally, it is an interesting problem is to minimize the length of the matching between S and T, that is, we would like to find the location of a new highway to minimize the cost of the minimum-distance matching of size T (assuming $T \leq S$). This corresponds to the minimization of the total service time of tasks at T by robots (or ambulances) located at S.

References

1. Abellanas, M., Hurtado, F., Icking, C., Klein, R., Langetepe, E., Ma, L., del Río, B.P., Sacristan, V.: Proximity Problems for Time Metrics Induced by the L_1 Metric and Isothetic Networks. In: Proc. Encuentros en Geometría Computacional (2001)
2. Ahn, H.-K., Alt, H., Asano, T., Bae, S.W., Brass, P., Cheong, O., Knauer, C., Na, H.-S., Shin, C.-S., Wolff, A.: Constructing Optimal Highways. In: Computing: The Australian Theory Symposium, vol. 65 (2007)
3. Aicholzer, O., Aurenhammer, F., Palop, B.: Quickest Path, Straight Skeleton and the City Voronoi Diagram. In: ACM Sympos. Comput. Geom., pp. 151–159 (2002)
4. Bae, S.W., Kim, J.-H., Chwa, K.-Y.: Optimal Construction of the City Voronoi Diagram. In: Proc. 17th Annu. Internat. Sympos. Algorithms Comput., pp. 183–192 (2006)
5. Bae, S.W., Chua, K.-Y.: Shortest Paths and Voronoi Diagrams with Transportation Networks under General Distances. In: Proc 16th Annu. Internat. Sympos. Algoritms Comput., pp. 1007–1018 (2005)
6. de Berg, M., van Kreveld, M., Overmars, M., Schwarzkopf, O.: Computational Geometry. Algorithms and Applications. Springer, Heidelberg (1997)
7. Cardinal, J., Langerman, S.: Min-max-min Geometric Facility Location Problems. In: Proc. of the European Workshop on Computational Geometry (2006)
8. Cardinal, J., Collette, S., Hurtado, F., Langerman, S., Palop, B.: Moving Walkways, Escalators, and Elevators. Eprint arXiv:0705.0635 (2007)
9. Hassin, R., Megiddo, N.: Approximation Algorithms for Hitting Objects with Straight Lines. Discrete Applied Mathematics 30(1), 29–42 (1991)
10. Korman, M., Tokuyama, T.: Optimal Insertion of a Segment Highway in a City Metric. In: Proc. of the European Workshop on Computational Geometry, pp. 189–192 (2008)
11. Kranakis, E., Krizanc, D., Meertens, L.: Link Length of Rectilinear Hamiltonian Tours in Grids. Ars Combinatoria 38, 177–192 (1994)
12. Mitchell, J.S.B.: L_1 Shortest Paths among Obstacles in the Plane. Internat. J. Comput. Geom. Appl. 6(3), 309–331 (1996)

Approximating the Generalized Capacitated Tree-Routing Problem

Ehab Morsy and Hiroshi Nagamochi

Department of Applied Mathematics and Physics
Graduate School of Informatics,
Kyoto University
Yoshida Honmachi, Sakyo, Kyoto 606-8501, Japan
{ehab,nag}@amp.i.kyoto-u.ac.jp

Abstract. In this paper, we introduce the *generalized capacitated tree-routing problem* (GCTR), which is described as follows. Given a connected graph $G = (V, E)$ with a sink $s \in V$ and a set $M \subseteq V - \{s\}$ of terminals with a nonnegative demand $q(v)$, $v \in M$, we wish to find a collection $\mathcal{T} = \{T_1, T_2, \ldots, T_\ell\}$ of trees rooted at s to send all the demands to s, where the total demand collected by each tree T_i is bounded from above by a demand capacity $\kappa > 0$. Let $\lambda > 0$ denote a bulk capacity of an edge, and each edge $e \in E$ has an installation cost $w(e) \geq 0$ per bulk capacity; each edge e is allowed to have capacity $k\lambda$ for any integer k, which installation incurs cost $kw(e)$. To establish a tree routing T_i, each edge e contained in T_i requires $\alpha + \beta q'$ amount of capacity for the total demand q' that passes through edge e along T_i and prescribed constants $\alpha, \beta \geq 0$, where α means a fixed amount used to separate the inside of the routing T_i from the outside while term $\beta q'$ means the net capacity proportional to q'. The objective of GCTR is to find a collection $\mathcal{T}$ of trees that minimizes the total installation cost of edges. Then GCTR is a new generalization of the several known multicast problems in networks with edge/demand capacities. In this paper, we prove that GCTR is $(2[\lambda/(\alpha+\beta\kappa)]/\lfloor\lambda/(\alpha+\beta\kappa)\rfloor + \rho_{\mathrm{ST}})$-approximable if $\lambda \geq \alpha+\beta\kappa$ holds, where ρ_{ST} is any approximation ratio achievable for the Steiner tree problem.

1 Introduction

In this paper, we introduce *generalized capacitated tree-routing problem* (GCTR), which is described as follows. Given a connected graph $G = (V, E)$ with a demand capacity $\kappa > 0$, a bulk edge capacity $\lambda > 0$, a sink $s \in V$, and a set $M \subseteq V - \{s\}$ of terminals with a nonnegative demand $q(v)$, $v \in M$, we wish to find a collection $\mathcal{T} = \{T_1, T_2, \ldots, T_\ell\}$ of trees rooted at s to send all the demands to s, where the total demand in the set Z_i of terminals assigned to tree T_i does not exceed the demand capacity κ. Each edge $e \in E$ has an installation cost $w(e) \geq 0$ per bulk capacity; each edge e is allowed to have capacity $k\lambda$ for any integer k, which

X. Hu and J. Wang (Eds.): COCOON 2008, LNCS 5092, pp. 621–630, 2008.

requires installation cost $kw(e)$. To establish a tree routing T_i through an edge e, we assume that e needs to have capacity at least

$$\alpha + \beta \sum_{v \in Z_i \cap D_{T_i}(v_i^e)} q(v)$$

for prescribed coefficients $\alpha, \beta \geq 0$, where v_i^e is the tail of e in T_i and $D_{T_i}(v_i^e)$ denotes the set of descendants of v_i^e in T_i including v_i^e; α means a fixed amount used to separate the inside and outside of the routing T_i while term $\beta \sum_{v \in Z_i \cap D_{T_i}(v_i^e)} q(v)$ means the net capacity proportional to the amount $\sum_{v \in Z_i \cap D_{T_i}(v_i^e)} q(v)$ of demands that passes through edge e along T_i. Hence, given a set $T = \{T_1, T_2, \ldots, T_\ell\}$ of trees, each edge e needs to have capacity $k_e \lambda$ for the least integer k_e such that

$$\sum_{T_i \in T : T_i \text{ contains } e} [\alpha + \beta \sum_{v \in Z_i \cap D_{T_i}(v_i^e)} q(v)] \leq k_e \lambda,$$

and the total installation cost of edges incurred by T is given as $\sum_{e \in E} k_e w(e)$, where $k_e = 0$ if no $T_i \in T$ contains e. The objective of GCTR is to find a set T of trees that minimizes the total installation cost of edges. We formally state GCTR as follows, where we denote the vertex set and edge set of a graph G by $V(G)$ and $E(G)$, respectively, and R^+ denotes the set of nonnegative reals.

Generalized Capacitated Tree-Routing Problem (GCTR)

Input: A graph G, an edge weight function $w : E(G) \to R^+$, a sink $s \in V(G)$, a set $M \subseteq V(G) - \{s\}$ of terminals, a demand function $q : M \to R^+$, a demand capacity $\kappa > 0$, an edge capacity $\lambda > 0$, and prescribed constants $\alpha, \beta \geq 0$.
Feasible solution: A partition $\mathcal{M} = \{Z_1, Z_2, \ldots, Z_\ell\}$ of M and a set $T = \{T_1, T_2, \ldots, T_\ell\}$ of trees of G such that $Z_i \cup \{s\} \subseteq V(T_i)$ and $\sum_{v \in Z_i} q(v) \leq \kappa$ hold for each i.
Goal: Minimize the total installation cost of T, that is,

$$\sum_{e \in E(G)} \lceil \sum_{T_i : e \in E(T_i)} (\alpha + \beta \sum_{v \in Z_i \cap D_{T_i}(v_i^e)} q(v))/\lambda \rceil w(e),$$

where v_i^e is the tail of e in T_i and $D_{T_i}(v_i^e)$ denotes the set of descendants of v_i^e in T_i including v_i^e.

We have a variant of GCTR if it is allowed to purchase edge capacity in any required quantity. In this model, for each edge e of the underlying network, we assign capacity of $\lambda_e = \alpha |T'| + \beta \sum_{T_i \in T'} \sum_{v \in Z_i \cap D_{T_i}(v_i^e)} q(v)$ on e, where T' is the set of trees containing e. That is, the total cost of the constructed trees equals $\sum_{e \in E} \lambda_e w(e)$. We call this variant of GCTR, *the fractional generalized capacitated tree-routing problem* (FGCTR).

We easily see that GCTR and FGCTR contain two classical NP-hard problems, the *Steiner tree problem* and the *bin packing problem* [6]. We see that GCTR with an edge weighted graph G, $\alpha = \lambda = 1$, and $\beta = 0$ is equivalent to the Steiner tree problem in G when $\kappa \geq \sum_{v \in M} q(v)$, and is equivalent to the bin packing problem with bin size κ when G is a complete graph, $w(e) = 1$ for all edges e incident to s and $w(e) = 0$ otherwise. We see that FGCTR also has a similar relationship with the Steiner tree problem and the bin packing problem. The Steiner tree problem is known to be NP-hard even if it has Euclidean or rectilinear costs [5]. A series of approximation algorithms for the Steiner tree problem have been developed over the last two decades [3],[8],[10],[14],[15],[16],[18]. The best known approximation factor for the Steiner tree problem is 1.55 [15].

The characteristic of GCTR and FGCTR is their routing capacity which is a linear combination of the number of trees and the total amount of demands that pass through an edge. Such a general form of capacity constraint can be found in some applications.

Suppose that we wish to find a minimum number of trucks to carry given n items $v_1, v_2, \ldots, v_n$, where each item v_i has size $q(v_i)$ and weight $\beta q(v_i)$, where β is a specific gravity. We also have bins; the weight of a bin is α and the capacity of a bin is κ. Items are first put into several bins, and then the bins are assigned to trucks under capacity constraints. That is, we can put items in a bin B so that the total size $\sum_{v_i \in B} q(v_i)$ of the items does not exceed the bin capacity κ, where the weight of the bin B is given by a linear combination $\alpha + \beta \sum_{v_i \in B} q(v_i)$. We can load packed bins into a truck as long as the total weight of these packed bins does not exceed the truck capacity λ. The objective is to find assignments of items to bins and packed bins to trucks such that the number of required trucks is minimized. This problem can be described as GCTR.

For more possible applications of the problem, refer to the full version of the paper [13].

Similar routing problems in which the objective function is a linear combination of two or more optimization requirements have been studied before [1],[2],[17]. For example, given a *lattice graph* with an edge capacity and a vertex cost function, *the global routing problem in VLSI design* asks to construct a set of trees that spans a given set of *nets* (subsets of the vertex set) under an edge capacity constraint. Terlaky et al. [17] have studied a problem of minimizing an objective function which is defined as a linear combination of the total edge cost and the total number of bends of all trees, where a bend at a vertex corresponds a via in VLSI design, which leads to extra cost in manufacturing.

We here observe that our new problem formulation, GCTR, includes several important routing problems as its special cases such as the *capacitated network design problem* (CND), the *capacitated multicast tree routing problem* (CMTR), and the *capacitated tree-routing problem* (CTR). We refer the readers to [13] for definitions and recent results of these problems.

Table 1 shows a summary of the recent approximation algorithms for CND, CMTR, CTR, and GCTR.

In this paper, we prove that GCTR admits a $(2\lceil \lambda/(\alpha + \beta\kappa) \rceil / \lfloor \lambda/(\alpha + \beta\kappa) \rfloor + \rho_{\mathrm{ST}})$-approximation algorithm if $\lambda \geq \alpha + \beta\kappa$ holds, where ρ_{ST} is any approximation ratio achievable for the Steiner tree problem.. The high-level description of the proposed algorithm resembles our algorithm for CTR problem [12], but we need to derive a new lower bound to the problem. Namely, given an instance $I = (G, w, s, M, q, \alpha, \beta, \kappa, \lambda)$ of GCTR, the main idea of our algorithm is to compute an integer capacity λ' depending on λ, κ, α, and β and then find a feasible tree-routings solution to the instance $I' = (G, w, s, M, q, \kappa, \lambda')$ of CTR. Here such capacity λ' is chosen so that this set of tree-routings is a feasible solution to the original GCTR instance I.

We can show that, with a slight modification, the approximation algorithm proposed for GCTR delivers a $(\alpha + \beta\kappa)(2 + \rho_{ST})$-approximate solution to FGCTR (the details is omitted due to space limitation).

Table 1. Approximation algorithms for CND, CMTR, CTR, and GCTR problems, where $\theta = [\lambda/(\alpha + \beta\kappa)]/\lfloor\lambda/(\alpha + \beta\kappa)\rfloor$

	Problem	unit demands $q \equiv 1$	general demands $q \geq 0$
CND	$\alpha = 0, \beta = 1,$ $\kappa = \lambda \in R^+$	$1 + \rho_{ST}$ [7]	$2 + \rho_{ST}$ [7]
CMTR	$\alpha = 1, \beta = 0,$ $\lambda = 1, \kappa \in R^+$	$2 + \rho_{ST}$ [4], $3/2 + (4/3)\rho_{ST}$ [11]	$2 + \rho_{ST}$ [9]
CTR	$\alpha = 1, \beta = 0$ $\lambda, \kappa \in R^+$	$2 + \rho_{ST}$ [12]	$2 + \rho_{ST}$ [12]
GCTR	$\alpha, \beta, \kappa, \lambda \in R^+$ with $\lambda \geq \alpha + \beta\kappa$	$2\theta + \rho_{ST}$ [this paper]	$2\theta + \rho_{ST}$ [this paper]

The rest of this paper is organized as follows. Section 2 introduces some notations and two lower bounds on the optimal value of GCTR. Section 3 describes some results on tree covers. Section 4 presents our approximation algorithm for GCTR and analyzes its approximation factor. Section 5 makes concluding remarks.

2   Preliminaries

This section introduces some notations and definitions. Let G be a simple undirected graph. We denote by $V(G)$ and $E(G)$ the sets of vertices and edges in G, respectively. For two subgraphs G_1 and G_2 of a graph G, let $G_1 + G_2$ denote the subgraph induced from G by $E(G_1) \cup E(G_2)$. An edge-weighted graph is a pair (G, w) of a graph G and a nonnegative weight function $w : E(G) \to R^+$. The length of a shortest path between two vertices u and v in (G, w) is denoted by $d_{(G,w)}(u, v)$. Given a vertex weight function $q : V(G) \to R^+$ in G, we denote by $q(Z)$ the sum $\sum_{v \in Z} q(v)$ of weights of all vertices in a subset $Z \subseteq V(G)$.

Let T be a tree. A *subtree* of T is a connected subgraph of T. A set of subtrees in T is called a *tree cover* of T if each vertex in T is contained in at least one of the subtrees. For a subset $X \subseteq V(T)$ of vertices, let $T\langle X \rangle$ denote the minimal subtree of T that contains X (note that $T\langle X \rangle$ is uniquely determined).

Now let T be a rooted tree. We denote by $L(T)$ the set of leaves in T. For a vertex v in T, let $Ch(v)$ and $D(v)$ denote the sets of children and descendants of v, respectively, where $D(v)$ includes v. A *subtree* T_v *rooted* at a vertex v is the subtree induced by $D(v)$, i.e., $T_v = T\langle D(v)\rangle$. For an edge $e = (u, v)$ in a rooted tree T, where $u \in Ch(v)$, the subtree induced by $\{v\} \cup D(u)$ is denoted by T_e, and is called a *branch* of T_v. For a rooted tree T_v, the *depth* of a vertex u in T_v is the length (the number of edges) of the path from v to u.

The rest of this section introduces two lower bounds on the optimal value to GCTR. The first lower bound is based on the Steiner tree problem.

Lemma 1. *Given a GCTR instance $I = (G, w, s, M, q, \alpha, \beta, \kappa, \lambda)$, the minimum cost of a Steiner tree to $(G, w, M \cup \{s\})$ is a lower bound on the optimal value to GCTR instance I.*

Proof. Consider an optimal solution $(\mathcal{M}^*, \mathcal{T}^*)$ to GCTR instance I. Let $E^* = \cup_{T' \in \mathcal{T}^*} E(T')$ $(\subseteq E(G))$, i.e., the set of all edges used in the optimal solution. Then the edge set E^* contains a tree T that spans $M \cup \{s\}$ in G. We see that the cost $w(T)$ of T in G is at most that of GCTR solution. Hence the minimum cost of a Steiner tree to $(G, w, M \cup \{s\})$ is no more than the optimal value to GCTR instance I.

The second lower bound is derived from an observation on the distance from vertices to sink s.

Lemma 2. *Let $I = (G, w, s, M, q, \alpha, \beta, \kappa, \lambda)$ be an instance of GCTR. Then,*

$$(\alpha + \beta\kappa)/(\kappa\lambda) \sum_{v \in M} q(v) d_{(G,w)}(s, v)$$

is a lower bound on the optimal value to GCTR instance I.

Proof. Consider an optimal solution $(\mathcal{M}^* = \{Z_1, \ldots, Z_p\}, \mathcal{T}^* = \{T_1, \ldots, T_p\})$ to GCTR instance I. For each edge $e \in E(T_i)$, $i = 1, 2, \ldots, p$, we assume that $e = (u_i^e, v_i^e)$, where $v_i^e \in Ch_{T_i}(u_i^e)$. Let $opt(I)$ denote the optimal value of GCTR instance I. Then we have

$$opt(I) = \sum_{e \in E(G)} \left[[\alpha|\{T_i \mid e \in E(T_i)\}| + \beta \sum_{T_i : e \in E(T_i)} q(Z_i \cap D_{T_i}(v_i^e))]/\lambda \right] w(e)$$

$$\geq \sum_{e \in E(G)} w(e)[\alpha|\{T_i \mid e \in E(T_i)\}| + \beta \sum_{T_i : e \in E(T_i)} q(Z_i \cap D_{T_i}(v_i^e))]/\lambda$$

$$= (\alpha/\lambda) \sum_{e \in E(G)} |\{T_i \mid e \in E(T_i)\}| w(e)$$

$$+ (\beta/\lambda) \sum_{e \in E(G)} w(e) \sum_{T_i : e \in E(T_i)} q(Z_i \cap D_{T_i}(v_i^e))$$

$$= (\alpha/\lambda) \sum_{T_i \in \mathcal{T}^*} w(T_i) + (\beta/\lambda) \sum_{T_i \in \mathcal{T}^*} \sum_{e \in E(T_i)} q(Z_i \cap D_{T_i}(v_i^e)) w(e). \tag{1}$$

Note that, for each tree $T_i \in \mathcal{T}^*$, we have

$$\kappa w(T_i) \geq w(T_i) \sum_{v \in Z_i} q(v) \geq \sum_{v \in Z_i} q(v) d_{(G,w)}(s, v), \tag{2}$$

since $w(T_i) \geq d_{(G,w)}(s,v)$ for all $v \in V(T_i)$. On the other hand, for each tree $T_i \in \mathcal{T}^*$, we have

$$\sum_{e \in E(T_i)} q(Z_i \cap D_{T_i}(v_i^e))w(e) = \sum_{v \in Z_i} q(v)d_{(T_i,w)}(s,v) \geq \sum_{v \in Z_i} q(v)d_{(G,w)}(s,v). \quad (3)$$

Hence by summing (2) and (3) overall trees in $\mathcal{T}^*$ and substituting in (1), we conclude that

$$(\alpha + \beta\kappa)/(\lambda\kappa) \sum_{v \in M} q(v)d_{(G,w)}(s,v) \leq opt(I),$$

which completes the proof.

3   Tree Cover

This section is devoted to present some results on the existence of tree covers, based on which we design our approximation algorithm to GCTR in the next section.

We first review a basic result on tree covers.

Lemma 3. [9] *Given a tree T rooted at r, an edge weight function $w : E(T) \to R^+$, a terminal set $M \subseteq V(T)$, a demand function $q : M \to R^+$, and a demand capacity κ with $\kappa \geq max\{q(v) \mid v \in M\}$, there is a partition $\mathcal{Z} = \mathcal{Z}_1 \cup \mathcal{Z}_2$ of M such that:*

(i) *For each $Z \in \mathcal{Z}$, there is a child $u \in Ch(r)$ such that $Z \subseteq V(T_u)$. Moreover, $|\{Z \in \mathcal{Z}_1 \mid Z \subseteq V(T_u)\}| \leq 1$ for all $u \in Ch(r)$;*
(ii) *$q(Z) < \kappa/2$ for all $Z \in \mathcal{Z}_1$;*
(iii) *$\kappa/2 \leq q(Z) \leq \kappa$ for all $Z \in \mathcal{Z}_2$; and*
(iv) *Let $\mathcal{T} = \{T\langle Z \cup \{r\}\rangle \mid Z \in \mathcal{Z}_1\} \cup \{T\langle Z\rangle \mid Z \in \mathcal{Z}_2\}$. Then $E(T_1) \cap E(T_2) = \emptyset$ for all distinct trees $T_1, T_2 \in \mathcal{T}$.*

Furthermore, such a partition $\mathcal{Z}$ can be obtained in polynomial time.   □

The following corollary is an immediate consequence of the particular construction of a partition $\mathcal{Z}$ in Lemma 3.

Corollary 1. [12] *Let $\mathcal{Z} = \mathcal{Z}_1 \cup \mathcal{Z}_2$ be defined as in Lemma 3 to (T, r, w, M, q, κ). Then:*

(i) *$E(T\langle Z\rangle) \cap E(T\langle \cup_{Z \in \mathcal{Z}_1} Z\rangle) = \emptyset$ for all $Z \in \mathcal{Z}_2$.*
(ii) *Let $Z_0 \in \mathcal{Z}_1$ be a subset such that $Z_0 \subseteq V(T_u)$ for some $u \in Ch(r)$. If $\mathcal{Z}' = \{Z \in \mathcal{Z}_2 \mid Z \subseteq V(T_u)\} \neq \emptyset$, then $\mathcal{Z}'$ contains a subset Z' such that $E(T\langle Z_0 \cup Z'\rangle) \cap E(T\langle Z\rangle) = \emptyset$ for all $Z \in \mathcal{Z} - \{Z_0, Z'\}$.*   □

We now describe a new result on tree covers. For an edge weighted tree T rooted at s, a set $M \subseteq V(T)$ of terminals, and a vertex weight function $d : M \to R^+$, we wish to find a partition $\mathcal{M}$ of M and to construct a set of induced trees $T\langle Z \cup \{t_Z\}\rangle$, $Z \in \mathcal{M}$ by choosing a vertex $t_Z \in V(T)$ for each subset $Z \in \mathcal{M}$, where we call such a vertex t_Z the *hub vertex* of Z. To find a "good" hub vertex t_Z for each $Z \in \mathcal{M}$, we classify a partition $\mathcal{M}$ of M into disjoint collections $\mathcal{C}_1, \mathcal{C}_2, \ldots, \mathcal{C}_f$ and then choose hub vertices t_Z, $Z \in \mathcal{M}$, such that $t_Z = t_j \in \{\mathrm{argmin}_{t \in Z \in \mathcal{C}_j} d(t)\}$ for each $Z \in \mathcal{C}_j$, $j \leq f - 1$, and $t_Z = s$ for each $Z \in \mathcal{C}_f$, as shown in the next lemma.

Lemma 4. *Given a tree T rooted at s, an edge weight function $w : E(T) \to R^+$, a terminal set $M \subseteq V(T)$, a demand function $q : M \to R^+$, a vertex weight function $d : M \to R^+$, a demand capacity κ with $\kappa \geq max\{q(v) \mid v \in M\}$, an edge capacity $\lambda > 0$, and prescribed constants $\alpha, \beta \geq 0$ with $\lambda \geq \alpha + \beta\kappa$, there exist a partition $\mathcal{M} = \cup_{1 \leq j \leq f}\mathcal{C}_j$ of M, and a set $\mathcal{B} = \{t_j \in \{\mathrm{argmin}_{t \in Z \in \mathcal{C}_j} d(t)\} \mid j \leq f - 1\} \cup \{t_f = s\}$ of hub vertices such that:*

  (i) $|\mathcal{C}_j| \leq \lfloor\lambda/(\alpha + \beta\kappa)\rfloor$ *for all* $j = 1, 2, \ldots, f$;
  (ii) $q(Z) \leq \kappa$ *for all* $Z \in \mathcal{M}$;
  (iii) $\sum_{Z \in \mathcal{C}_j} q(Z) \geq \lfloor\lambda/(\alpha + \beta\kappa)\rfloor(\kappa/2)$ *for all* $j = 1, 2, \ldots, f - 1$;
  (iv) $E(T\langle Z\rangle) \cap E(T\langle Z'\rangle) = \emptyset$ *for all distinct* $Z, Z' \in \mathcal{M}$; *and*
  (v) *Let* $T' = \{T\langle Z \cup \{t_j\}\rangle \mid Z \in \mathcal{C}_j, 1 \leq j \leq f\}$, *and let all edges of each* $T\langle Z \cup \{t_j\}\rangle \in T'$, $Z \in \mathcal{C}_j$, $1 \leq j \leq f$ *be directed toward* t_j. *Then for each edge* $e \in E(T)$, *the number of trees in* T' *passing through* e *in each direction is at most* $\lfloor\lambda/(\alpha + \beta\kappa)\rfloor$.

Furthermore, a tuple $(\mathcal{M}, \mathcal{B}, T')$ *can be computed in polynomial time.* □

Before describing the algorithm, we discuss the following lemma.

Lemma 5. *Let* $(\mathcal{M}, \mathcal{B}, T')$ *be a tuple obtained by applying Lemma 4 to* $(T, w, s, M, q, d, \alpha, \beta, \kappa, \lambda)$. *Then we can find a new partition* $\mathcal{C}'_1, \mathcal{C}'_2, \ldots, \mathcal{C}'_f$ *of* M *by swapping subsets between* $\mathcal{C}_1, \mathcal{C}_2, \ldots, \mathcal{C}_f$, *so that each collection* $\mathcal{C}'_j$ *contains at most* $\lfloor\lambda/(\alpha + \beta\kappa)\rfloor$ *subsets from* $\mathcal{M}$, *all of which are assigned to the hub vertex* t_j, $j = 1, 2, \ldots, f$, *and for* $T'' = \{T\langle Z \cup \{t_Z\}\rangle \mid Z \in \mathcal{M}\}$, *it holds* $|\{T' \in T'' \mid e \in E(T')\}| \leq \lfloor\lambda/(\alpha + \beta\kappa)\rfloor$ *for any edge* $e \in E(T)$. □

Proofs of Lemmas 4 and 5 can be found in the full version of the paper [13].

4   Approximation Algorithm to GCTR

This section presents an approximation algorithm for an instance $I = (G, w, s, M, q, \alpha, \beta, \kappa, \lambda)$ of GCTR problem based on results on tree covers in the previous section. Our algorithm begins by computing an approximate Steiner tree T in $(G, w, M \cup \{s\})$. We then find a tree cover T'' of the tree T such that, for each $e \in E(T)$, $|\{T' \in T'' \mid e \in E(T')\}| \leq \lfloor\lambda/(\alpha + \beta\kappa)\rfloor$ and hence $\sum_{T' \in T'':e \in E(T')}(\alpha + \beta q(D_{T'}(v^e) \cap M)) \leq (\alpha + \beta\kappa)|\{T' \in T'' \mid e \in E(T')\}| \leq (\alpha + \beta\kappa)\lfloor\lambda/(\alpha + \beta\kappa)\rfloor \leq \lambda$, where $e = (u^e, v^e) \in E(T')$ with $v^e \in Ch_{T'}(u^e)$. Finally, we connect each tree in T'' to s in order to get a tree-routings $\mathcal{T}$ in the instance I.

Algorithm APPROXGCTR
Input: An instance $I = (G, w, s, M, q, \alpha, \beta, \kappa, \lambda)$ of GCTR.
Output: A solution $(\mathcal{M}, \mathcal{T})$ to I.

Step 1. Compute a ρ_{ST}-approximate solution T to the Steiner tree problem in G that spans $M \cup \{s\}$ and then regard T as a tree rooted at s.
Define a function $d : M \to R^+$ by setting

$$d(t) := d_{(G,w)}(s, t), \quad t \in M.$$

Step 2. Apply Lemma 4 to $(T, w, s, M, q, d, \alpha, \beta, \kappa, \lambda)$ to get a partition $\mathcal{M} = \cup_{1 \leq j \leq f} \mathcal{C}_j$ of M, a set $\mathcal{B} = \{t_1, t_2, \ldots, t_f\}$ of hub vertices, where $t_Z = t_j$ for each $Z \in \mathcal{C}_j$, $j = 1, 2, \ldots, f$, and a set $T' = \{T\langle Z \cup \{t_Z\}\rangle \mid Z \in \mathcal{M}\}$ of subtrees of T that satisfy Conditions (i)-(v) of the lemma.
Step 3. Apply Lemma 5 to the tuple $(\mathcal{M}, \mathcal{B}, T')$ output from Step 2 to get a new partition $\mathcal{C}'_1, \mathcal{C}'_2, \ldots, \mathcal{C}'_f$ of $\mathcal{M}$ and a set $T'' = \{T\langle Z \cup \{t_Z\}\rangle \mid Z \in \mathcal{M}\}$ of subtrees of T that satisfy the conditions of the lemma.
Step 4. For each $j = 1, 2, \ldots, f - 1$, choose a shortest path $SP(s, t_j)$ between s and t_j in (G, w) and join t_j to s by installing a copy of each edge in $SP(s, t_j)$. Let $\mathcal{T} := \{T_Z = T\langle Z \cup \{t_Z\}\rangle + SP(s, t_Z) \mid Z \in \mathcal{M}\}$ and output $(\mathcal{M}, \mathcal{T})$.

Now we show the feasibility and analyze the approximation factor of the approximate solution $(\mathcal{M}, \mathcal{T})$ output by algorithm APPROXGCTR.

Theorem 1. *For an instance $I = (G, w, s, M, q, \alpha, \beta, \kappa, \lambda)$ of GCTR, algorithm* APPROXGCTR *delivers a $(2\lceil \lambda/(\alpha + \beta\kappa)\rceil / \lfloor \lambda/(\alpha + \beta\kappa)\rfloor + \rho_{\mathrm{ST}})$-approximate solution $(\mathcal{M}, \mathcal{T})$, where ρ_{ST} is the approximation ratio of solution T to the Steiner tree problem.*

Proof. Lemma 4(ii) implies that $(\mathcal{M}, \mathcal{T})$ satisfies the demand capacity constraint on each tree.

Now we show that $\mathcal{T}$ satisfies the edge capacity constraint. Let $\mathcal{M} = \cup_{1 \leq j \leq f} \mathcal{C}'_j$ and T'' be output from Step 3 of algorithm APPROXGCTR. Note that each tree $T_Z \in \mathcal{T}$ is a tree $T\langle Z \cup \{t_Z\}\rangle \in T''$ plus the shortest path $SP(s, t_Z)$ between s and t_Z in (G, w). By Lemma 5, $|\{T' \in T'' \mid e \in E(T')\}| \leq \lfloor \lambda/(\alpha + \beta\kappa)\rfloor$ for any $e \in E(T)$. On the other hand, each collection $\mathcal{C}'_j$, $j \leq f$, contains at most $\lfloor \lambda/(\alpha + \beta\kappa)\rfloor$ subsets from $\mathcal{M}$, all of which are assigned to a common hub vertex t_j. Thus, by installing one copy of each edge of the Steiner tree T and each edge in a shortest path $SP(s, t_j)$ between s and t_j in (G, w), $j \leq f - 1$ ($t_f = s$), we get a set $\mathcal{T}$ of tree-routings such that $|\{T_Z \in \mathcal{T} \mid e \in E(T_Z)\}| \leq k_e \lfloor \lambda/(\alpha + \beta\kappa)\rfloor$ for any $e \in E(G)$, where k_e is the number of installed copies of e. Consequently, for any $e \in E(G)$, we observe that

$$\sum_{T_Z \in \mathcal{T} : e \in E(T_Z)} (\alpha + \beta q(D_{T_Z}(v^e) \cap Z)) \leq \sum_{T_Z \in \mathcal{T} : e \in E(T_Z)} (\alpha + \beta q(Z))$$

$$\leq (\alpha + \beta\kappa) k_e \lfloor \lambda/(\alpha + \beta\kappa)\rfloor \leq k_e \lambda,$$

where $e = (u^e, v^e) \in E(T_Z)$ with $v^e \in Ch_{T_Z}(u^e)$. Thereby $(\mathcal{M}, \mathcal{T})$ is feasible to I and the total weight of the installed edges on the network is bounded by

$$w(T) + \sum_{1 \leq j \leq f-1} d(t_j).$$

For a minimum Steiner tree T^* that spans $M \cup \{s\}$, we have $w(T) \leq \rho_{\text{ST}} \cdot w(T^*)$ and $w(T^*) \leq opt(I)$ by Lemma 1, where $opt(I)$ denotes the weight of an optimal solution to GCTR. Hence $w(T) \leq \rho_{\text{ST}} \cdot opt(I)$ holds. To prove the theorem, it suffices to show that

$$\sum_{1 \leq j \leq f-1} d(t_j) \leq 2[\lambda/(\alpha + \beta\kappa)]/\lfloor \lambda/(\alpha + \beta\kappa) \rfloor opt(I). \qquad (4)$$

Consider a collection $\mathcal{C}_j$, $j \leq f - 1$ obtained by applying Lemma 4 to $(T, w, s, M, q, d, \alpha, \beta, \kappa, \lambda)$ in Step 2. Note that even if some subsets of $\mathcal{C}_j$ are applied by swapping in Step 3, the hub vertex of the new collection $\mathcal{C}'_j$ remains unchanged. That is, the set $\mathcal{B}$ of hub vertices computed in Step 2 remains unchanged throughout the algorithm. The choice of t_j and Lemma 4(iii) imply that

$$\sum_{t \in Z \in \mathcal{C}_j} q(t)d(t) \geq d(t_j) \sum_{t \in Z \in \mathcal{C}_j} q(t) \geq \lfloor \lambda/(\alpha + \beta\kappa) \rfloor (\kappa/2)d(t_j). \qquad (5)$$

By summing inequality (5) overall $\mathcal{C}_j$'s, $j \leq f - 1$, we have

$$(\alpha + \beta\kappa)\lfloor \lambda/(\alpha + \beta\kappa) \rfloor/(2\lambda) \sum_{1 \leq j \leq f-1} d(t_j) \leq (\alpha + \beta\kappa)/(\kappa\lambda) \sum_{1 \leq j \leq f-1} \sum_{t \in Z \in \mathcal{C}_j} q(t)d(t)$$

$$\leq (\alpha + \beta\kappa)/(\kappa\lambda) \sum_{t \in M} q(t)d(t).$$

Hence Lemma 2 completes the proof of (4).

5   Conclusion

In this paper, we have studied the generalized capacitated tree-routing problem (GCTR), a new routing problem formulation under a multi-tree model with a general routing capacity, which unifies several important routing problems such as the capacitated network design problem (CND), the capacitated multicast tree routing problem (CMTR), and the capacitated tree-routing problem (CTR). We have proved that GCTR with $\lambda \geq \alpha + \beta\kappa$ is $(2[\lambda/(\alpha + \beta\kappa)]/\lfloor \lambda/(\alpha + \beta\kappa) \rfloor + \rho_{\text{ST}})$-approximable based on a new lower bound to the problem and some new results on tree covers, where ρ_{ST} is any approximation factor achievable for the Steiner tree problem. Future work may include design of approximation algorithms for GCTR in the case of $\lambda < \alpha + \beta\kappa$. Also, it will be interested to obtain a better approximation algorithm for the fractional generalized capacitated tree-routing problem (FGCTR). We remark that GCTR with a very small λ compared with $\alpha + \beta\kappa$ is closely related with FGCTR.

References

1. Behjat, L.: New Modeling and Optimization Techniques for the Global Routing Problem. Ph.D. Thesis, University of Waterloo (2002)
2. Behjat, L., Vannelli, A., Rosehart, W.: Integer Linear Programming Models for Global Routing. Informs Journal on Computing 18(2), 137–150 (2002)
3. Berman, P., Ramaiyer, V.: Improved Approximations for the Steiner Tree Problem. J. Algorithms 17, 381–408 (1994)
4. Cai, Z., Lin, G.-H., Xue, G.: Improved Approximation Algorithms for the Capacitated Multicast Routing Problem. In: Wang, L. (ed.) COCOON 2005. LNCS, vol. 3595, pp. 136–145. Springer, Heidelberg (2005)
5. Garey, M.R., Johnson, D.S.: The Rectilinear Steiner Tree Problem is NP-Complete. SIAM J. Appl. Math. 32, 826–843 (1977)
6. Garey, M.R., Johnson, D.S.: Computers and Intractability, a Guide to the Theory of NP-completeness. Freeman, San Francisco (1978)
7. Hassin, R., Ravi, R., Salman, F.S.: Approximation Algorithms for a Capacitated Network Design Problem. Algorithmica 38, 417–431 (2004)
8. Hougardy, S., Prömmel, H.J.: A 1.598-Approximation Algorithm for the Steiner Problem in Graphs. In: Proceedings of the 9th Annual ACM-SIAM Symposium on Discrete Algorithms (SODA 1999), pp. 448–453 (1999)
9. Jothi, R., Raghavachari, B.: Approximation Algorithms for the Capacitated Minimum Spanning Tree Problem and its Variants in Network Design. In: Díaz, J., Karhumäki, J., Lepistö, A., Sannella, D. (eds.) ICALP 2004. LNCS, vol. 3142, pp. 805–818. Springer, Heidelberg (2004)
10. Karpinsky, M., Zelikovsky, A.: New Approximation Algorithms for the Steiner Tree Problem. J. Combin. Optim. 1, 47–65 (1997)
11. Morsy, E., Nagamochi, H.: An Improved Approximation Algorithm for Capacitated Multicast Routings in Networks. Theoritical Computer Sceince 390, 81–91 (2008)
12. Morsy, E., Nagamochi, H.: Approximating Capacitated Tree-Routings in Networks. In: Cai, J.-Y., Cooper, S.B., Zhu, H. (eds.) TAMC 2007. LNCS, vol. 4484, pp. 342–353. Springer, Heidelberg (2007)
13. Morsy, E., Nagamochi, H.: Approximating the Generalized Capacitated Tree-Routing Problem. Technical Report, 2008-001, Discrete Mathematics Lab., Graduate School of Informatics, Kyoto University, http://www.amp.i.kyoto-u.ac.jp/tecrep/TR2008.html
14. Prömmel, H.J., Steger, A.: RNC-Approximation Algorithms for the Steiner Problem. In: Proceedings of the 14th Annual Symposium on Theoritical Aspects of Computer Science, pp. 559–570 (1997)
15. Robins, G., Zelikovsky, A.Z.: Improved Steiner Tree Approximation in Graphs. In: Proceedings of the 11th Annual ACM-SIAM Symposium on Discrete Algorithms (SODA 2000), pp. 770–779 (2000)
16. Takahashi, H., Matsuyama, A.: An Approximate Solution for the Steiner Problem in Graphs. Math. Japon. 24, 573–577 (1980)
17. Terlaky, T., Vannelli, A., Zhang, H.: On Routing in VLSI Design and Communication Networks. In: Deng, X., Du, D.-Z. (eds.) ISAAC 2005. LNCS, vol. 3827, pp. 1051–1060. Springer, Heidelberg (2005)
18. Zelikovsky, A.: An 11/6-approximation Algorithm for the Network Steiner Problem. Algorithmica 9, 463–470 (1993)

Column Generation Algorithms for the Capacitated m-Ring-Star Problem

Edna A. Hoshino[1,*] and Cid C. de Souza[2,**]

[1] University of Mato Grosso do Sul, Department of Computing and Statistic,
Campo Grande MS, Brazil
eah@dct.ufms.br
[2] University of Campinas, Institute of Computing, Campinas SP, Brazil
cid@ic.unicamp.br

Abstract. In this paper we propose an integer programming formulation for the capacitated m-ring-star problem (CmRSP) based on a *set covering* model and develop an exact *branch-and-price* (BP) algorithm to solve it exactly. The CmRSP is a variant of the classical one-depot capacitated vehicle routing problem in which a customer is either on a route or is connected to another customer or to some *connection point* present in a route. The set of potential connection points and the number m of vehicles are given *a priori*. Routing and connection costs are also known and the goal is to minimize the sum of routing and connection costs. To our knowledge, the only exact approach for the CmRSP is a *branch-and-cut* (BC) proposed in [2]. Extensive experimentation reported here shows that our BP algorithm is competitive with the BC algorithm. This performance was achieved after a profound investigation of the alternatives for column generation relaxation and a careful implementation of the pricing algorithm.

1 Introduction

In the capacitated m-ring-star problem (CmRSP) a set of customers has to be visited by m vehicles initially located at a central depot. Each vehicle performs a route or *ring* that starts and ends at the depot and is characterized by an ordered set of customers and *connection points*. These connection points are selected among a set of predefined sites called the *Steiner* points. Besides, there is also a *star* associated to each vehicle. The *star* of vehicle t is a set of pairs of the form (u, v) where u is a customer and v is a customer or *Steiner* point belonging to the *ring* of t. In the latter situation, we say that u is *connected* to v. Customers in the *ring-star*(i.e., *ring* or *star*) of t are said to be covered by t and their quantity is limited by the capacity of the vehicle which is assumed to be the same for the entire fleet. Now, a solution for the CmRSP can be viewed as

---

* Supported by Capes/PICDT scholarship.
** Supported by grants 301732/2007-8, 478470/2006-1, 472504/2007-0 and 473726/2007-6 from *Conselho Nacional de Desenvolvimento Científico e Tecnológico* and grant 03/09925-5 from *Fundação de Amparo à Pesquisa do Estado de São Paulo*.

X. Hu and J. Wang (Eds.): COCOON 2008, LNCS 5092, pp. 631–641, 2008.
© Springer-Verlag Berlin Heidelberg 2008

a set of *m ring-stars* covering all customers. Routing costs incur for every pair of consecutive sites in a *ring*, while connection costs incur for every connection defined by a *star*. The cost of a solution is then given by the sum of all routing costs plus all the connection costs induced by its *m ring-stars*. The C*m*RSP asks for a solution with minimum cost and can be easily shown to be $\mathcal{NP}$-hard since it generalizes the *Traveling Salesman Problem*.

The C*m*RSP was introduced by Baldacci et al. [2] who describe an application in the design of a large fiber optics networks. The authors proposed a *branch-and-cut* (BC) algorithm for the problem and reported experiments where moderated-size instances were solved in reasonable time. To the best of our knowledge, this is the only exact algorithm available for the C*m*RSP. On the heuristic side, Mauttone et al. [14] proposed an algorithm combining GRASP and Tabu Search which obtained good solutions for the same instances tested in [2]. The literature also exhibits results for the *Ring-Star Problem* (RSP) and its variations [11]. The RSP can be viewed as restricted case of the C*m*RSP where a single uncapacitated vehicle is available. On the other hand, the *Vehicle Routing-Allocation Problem* (VRAP) presented by Beasley and Nascimento in [3] is a generalization of the C*m*RSP where customers must remain unattended, though this situation is penalized in the objective function. Special attention has been paid to instances of the VRAP where only one vehicle is at hand, the so-called *Single Vehicle Routing-Allocation Problem*.

Motivation and Our Contribution. The C*m*RSP can be interpreted as a generalization of the classical Capacitated Vehicle Problem (CVRP). In particular, the C*m*RSP is suited to cope with CVRP applications where customers can be served indirectly by displacing themselves to a site covered by one of the *m* routes. The best results reported in the literature concerning the exact solution of the CVRP were obtained by a robust *branch-and-cut-price* (BCP) algorithm proposed in [7]. BCP algorithms embed cutting planes and column generation in a standard branch-and-bound procedure for solving Integer Programming (IP) problems. One of the key ingredients of the BCP algorithm described in [7] refers to the relaxation to the pricing (column generation) problem. Encouraged by the success of this approach for the CVRP we decided to investigate the adequacy of column generation with relaxed pricing to the C*m*RSP.

This paper proposes an IP *set covering* model for the C*m*RSP based on column generation and describes ways to relax the pricing problem. The approach is validated by computational experiments who showed that the *branch-and-price* algorithm we implemented is at least as good as the best existing exact algorithms for the problem.

Organization of the Text. In the next section we present some basic notation and an IP *set covering* formulation for C*m*RSP. Section 3 describes the relaxations for the pricing problem we have considered. Details of the implementation issues as well the presentation and analysis of the and computational results are the subject of Section 4. Finally, Section 5 is devoted to our conclusions and to the future directions we envision for this work.

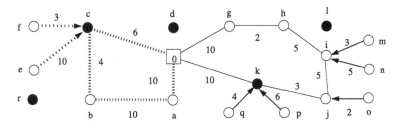

Fig. 1. Two *ring-stars* $(\{(0, g), (g, h), (h, i), (i, j), (j, k), (k, 0)\}, \{mi, ni, oj, pk, qk\})$ and $(\{(0, a), (a, b), (b, c), (c, 0)\}, \{(ec, fc)\})$

2   Notation, Definitions and Problem Formulation

A mixed graph is one that admits both undirected and directed arcs. Let $G = (V, E \cup A)$ be a mixed graph where E and A are the sets of the undirected and directed arcs, respectively. Given $R \subseteq E \cup A$, $V[R]$ represents the subset of the vertices which are extremities of arcs in R. The set of the incident arcs in a vertex i is denoted by $\delta(i)$. A *path* P is a sequence of vertices $(v_0, v_1, \ldots, v_p)$ of G such as each (v_{i-1}, v_i), $1 \leq i \leq k$, is an arc of G. The *length* of P is the number of arcs in P. A *cycle* is a path $(v_0, v_1, \ldots, v_p)$ of length $p > 1$ having $v_0 = v_p$. When the length of a cycle C is less or equal to k, C is called a k-*cycle*. A path P is k-*cycle free* if P does not contain a k-cycle as a subpath. Paths and cycles having no vertex repetitions are said to be *elementary*.

Let us define the $CmRSP$ according to the graph terminology. To this end, we consider a mixed graph $G = (\{0\} \cup U \cup W, E \cup A)$, where 0 denotes the central depot, U is the set of the customers and W is the set of the *Steiner points*. Assume that the set of undirected arcs is complete and that the set of directed arcs represents the possible connections for the vertices in U. Denote by C_i the set of the nodes to which customer i can be connected. In other words, we have that $A = \{ij : i \in U, j \in C_i\}$. Each undirected arc e is associated with a non-negative routing cost c_e and, for the purpose of this paper, assume that these costs satisfy the triangle inequalities. Besides, assume that each directed arc ij is associated with a non-negative connection cost w_{ij}. As an additional notation, given $S \subseteq A$, we denote by $C(S) = \{i \in U : ij \in S\}$ and by $C'(S) = \{j \in U \cup W : ij \in S\}$. We also define a Q-*ring-star* as a pair of arc subsets (R, S) where: $R \subseteq E$ and $S \subseteq A$; R defines an elementary cycle containing the central depot; $C'(S) \subseteq V[R]$; $V[R] \cap C(S) = \emptyset$ and $|V[R] \cap U| + |C(S)| \leq Q$. The cost of a *ring-star* $p = (R, S)$ is $\sigma(p) = \sum_{e \in R} c_e + \sum_{ij \in S} w_{ij}$. Figure 1 shows two 10-*ring-stars* with costs 43 and 55.

Let Q and m be non-negative integer values. With G defined as above, the $CmRSP$ can be formulated as the problem of finding m Q-*ring-stars* in G, say $p_1 = (R_1, S_1)$, ..., $p_m = (R_m, S_m)$, satisfying: (i) $V[R_i \cup S_i] \cap V[R_j \cup S_j] = \emptyset$, $\forall i \neq j \in [1 \ldots m]$; (ii) $U \subseteq \bigcup_{i=1}^m V[R_i \cup S_i]$; and (iii) $\sum_{i=1}^m \sigma(p_i)$ is minimized. Since all costs are non-negative and the routing costs define a metric, it can be shown that a *ring-star* (R, S) of any optimum solution for $CmRSP$ satisfies the following properties: (i) R is a minimum cost Hamiltonian circuit in $G[R]$; (ii) a Steiner node w belongs to the $V[R]$ if, and only if exists uw in S, for some $u \in U$ and (iii) an arc (i, j) belongs to S if and only if $j = \arg\min_{k \in V[R]} w_{ik}$.

We now turn our attention to the mathematical formulation of the problem. First, let P be the set of all Q-*ring-stars* of G. Given a *ring-star* $p = (R, S)$ in P, let $r^p \in \mathbb{Z}_{+}^{|U|+|W|+|E|}$ and $s^p \in \mathbb{Z}_{+}^{|U|+|A|}$ be, respectively, the characteristic vector of the *ring* R and of the *star* S. That is, for all $i \in U \cup W$, $r_i^p = 1$ if and only if $i \in V[R]$ and, for all $e \in E$, r_e^p denotes the number of times edge e occurs in R. On the other hand, for all $i \in U$, let $s_i^p = 1$ if and only if $i \in C(S)$ and $s_{ij}^p = 1$ if and only if $ij \in S$. Thus, a *set covering* model for CmRSP is:

$$(F) \quad \min \sum_{p \in P} c_p \lambda_p$$

$$\text{subject to} \quad \sum_{p \in P} \lambda_p = m \tag{1}$$

$$\sum_{p \in P} (r_i^p + s_i^p)\lambda_p \geq 1, \ \forall i \in U \tag{2}$$

$$\sum_{p \in P} \left(\sum_{e \in \delta(i)} r_e^p \right)\lambda_p \leq 2, \ \forall i \in V \setminus \{0\} \tag{3}$$

$$\sum_{p \in P} r_e^p \lambda_p \leq u_e, \ \forall e \in E \tag{4}$$

$$\sum_{p \in P} s_{ij}^p \lambda_p \leq 1, \ \forall ij \in A \tag{5}$$

$$\lambda_p \in \{0, 1\}, \forall p \in P \tag{6}$$

where $c_p = \sigma(p)$ and $u_e = 1$, if $e \in E \setminus \delta(0)$ or $u_e = 2$, if $e \in \delta(0)$. Constraint (1) fixes the number of *ring-stars* to be selected from P. The covering constraints in (2) force each customer to be in a *ring* or connected to one of them. Constraints (3) forbids a customer from being simultaneously in two or more *rings*. Constraints (4) bound the use of an arc to one *ring* while constraints (5) to the same for *stars*. Finally, constraints (6) restrict the λ variables to assume 0–1 values depending on whether or not the associated *ring-star* is in the solution. The number of variables in this model grows exponentially with the size of G. Column generation is a handy way of tackling this problem. The technique is described in many textbooks on IP (cf., [16]) and is not discussed here.

The Pricing Problem. Let π, μ, ν,β and α be the dual variables related to constraints (1), (2),(3), (4) and (5), respectively. Given a solution of the linear relaxation of (F), let $\tilde{c}_e = c_e - \beta_e - \nu_i - \nu_j$, $\tilde{w}_{ij} = w_{ij} - \mu_i - \alpha_{ij}$ and $\tilde{p}_i = -\mu_i$. Then, the reduced cost of a *ring-star* p is $\bar{c}_p = \sum_{e \in E} \tilde{c}_e r_e^p + \sum_{ij \in A} \tilde{w}_{ij} s_{ij}^p + \sum_{i \in U} \tilde{p}_i r_i^p + \pi$. The pricing problem requires the computation of $\min_{p \in P} \bar{c}_p$.

Through polynomial reductions involving a variant of Traveling Salesman Problem (TSP) with Profits [6], called Profitable Tour Problem (PTSP) [5], one can easily show that the pricing problem defined above is $\mathcal{NP}$-hard.

3   Relaxed Pricing Problem

Since the pricing problem is $\mathcal{NP}$-hard, a relaxed problem is considered. This approach is similar to that using Q-routes to relax the pricing problem for the CVRP [7]. A Q-route is a closed walk containing the vertex 0 and such as the sum of the demands of costumers in the walk is limited by the value Q. Thus, a Q-route is a relaxation to the route that admits repetition of vertices. In a similar fashion, let us define a *relaxed Q-ring-star* as an ordinary Q-ring-star except that vertices are allowed to be repeated in the *ring* and/or in the *star*. In the context of the pricing problem, we named by *weight* the cost of a relaxed *ring-star* (R, S). Thus, the weight of (R, S) is given by $\tilde{\sigma}_p = \sum_{e \in E} \tilde{c}_e x_e + \sum_{ij \in A} \tilde{w}_{ij} z_{ij} + \sum_{i \in V'} \tilde{p}_i y_i$, where x_e, z_{ij} and y_i are the number of times e, ij and i occurs in R, S, and in $V[R]$, respectively. The *relaxed pricing problem* consists in finding a minimum weight relaxed Q-ring-star. Notice that the formulation (F) is also valid for relaxed pricing problem.

Now, consider a *ring-star* (R, S) having $q \leq Q$ customers and where one of the arcs incident to the depot is $(j, 0)$. If we remove the arc $(j, 0)$ from (R, S) we get a (q, j)-*walk-star*, denoted by $p(q, j)$. Given j and q, let $F(j, q)$ be the minimum weight among all (q, j)-*walk-stars*. Then, the weight of an optimum *relaxed ring-star* for the pricing problem is $\min_{j \in V', q \in [1..Q]} F(j, q) + \tilde{c}_{j0}$.

The construction of $p(j, q)$, for given j and q, can be done iteratively by three operations: adding a connection arc $kj \in A$, adding a routing arc $(i, j) \in E$, and adding both routing and connection arcs. The latter operation can only be done if vertex j is a *Steiner* point whilst the second one requires j to be a costumer. Once we have introduced these operations, it is not difficult to see that the values in the cells of the bidimensional matrix F can be calculated by applying the recursive formulas below. These computations may take place after the cells in the first row and column of F have been initialized by setting $F(j, q)$ to *zero*, if $j = q = 0$, and to *infinity*, otherwise. Now, recalling that $V = \{0\} \cup U \cup W$, the computation of each cell of F spends $O(|V|)$ time. Therefore, as F is a $|V| \times Q$ matrix, the relaxed pricing problem can be solved in $O(|V|^2 Q)$.

$$F(j, q) = \min \begin{cases} \min\limits_{k \in U : j \in C_k} F(j, q-1) + \tilde{w}_{kj}, \\ \begin{cases} \min\limits_{i \in V, i \neq j} F(i, q-1) + \tilde{c}_{ij} + \tilde{p}_j & \text{, if } j \in U \\ \min\limits_{i \in V, i \neq j, k \in U : j \in C_k} F(i, q-1) + \tilde{w}_{kj} + \tilde{c}_{ij} & \text{, if } j \in W. \end{cases} \end{cases}$$

The dual bound obtained by relaxing *ring-stars* can be improved by avoid the occurrence of k-*cycles* inside the *rings*. The prohibition of k-*cycles* in paths used in the relaxation of discrete optimization problems is not a novelty. Examples where this idea was applied can be found in [8] for the TSP and in [10] for the *Resource Constrained Shortest Path Problem*. The elimination of k-*cycles* in q-routes was used to solve the *Vehicle Routing Problem* in [4]. The success of this idea relies on the fact that, for small values of k, the elimination of k-cycles can be done without changing the complexity of the dynamic programming algorithm that computes the paths or q-routes. Below, we extend this idea to prohibit cycles and a more complex structure to appear on *relaxed ring-stars*.

k-Cycle Elimination. In this presentation we adopt the same notation as in [10]. Irnich and Villeneuve used the *label setting algorithm* (LSA) [1] to solve the *Non-elementary Shortest Path Problem with Resource Constraints and k-cycle Elimination* (SPPRC-k-CYC) [9,10]. We restrict our discussion to the case with only one resource and comment on how it can be adapted to solve the CmRSP. Given some limited resource, assume that there is a resource consumption r_{ij} associated to each arc ij in the graph. A path is *resource feasible* if the sum of resource consumptions of its arcs, denoted by $res(P)$, is less than a certain fixed limit. The goal in the SPPRC-k-CYC is to find a minimum cost path that is k-cycle free and resource feasible. The chief fact about using the LSA to solve SPPRC-k-CYC is that the resource consumption function must be monotonically increasing in the path length. In the context of CMRSP, a resource represents the capacity Q and each arc on the *ring* consumes one resource unit, except those incident to a *Steiner* vertex, and each arc ij in the *star* increases by one the consumption related to the arc incident in j on the *ring*.

Let S to be the set of all k-cycle free paths and $\mathcal{F}(s,w)$ the set of resource feasible paths starting in s and ending in vertex w. Given paths $P = (v_0, \ldots, v_p)$ and $Q = (w_0, \ldots, w_q)$, we say P extends Q if the path $(v_0, \ldots, v_p = w_0, \ldots, w_q)$, denoted by (P, Q), belongs to S. The set of paths that extends a path P will be denoted by $\mathcal{E}(P)$. Path P dominates path Q ($P \prec_{\text{dom}} Q$) if $res(P) \leq res(Q)$ and the cost of P is less or equal to the cost of Q. Irnich and Villeneuvue [10] proved that a path $P \in S \cap \mathcal{F}(s,w)$ is *useful* and can not be discarded during the execution of LSA if $\mathcal{E}(P) \not\subseteq \bigcup_{Q \in S \cap \mathcal{F}(s,w): Q \prec_{\text{dom}} P} \mathcal{E}(Q)$. They proposed the so-called *Intersection Algorithm* (IA) which computes the complement of $\mathcal{E}(P)$ to identify useful paths. It is based on the definition of a *set form*, a set whose elements are paths with a finite set of positions fixed *a priori*. A *set form* is represented by a string s of elements in $V \cup \{\cdot\}$, where $s_i \neq \cdot$ means that the i-th vertex of the path is necessarily s_i, otherwise, it can be any vertex. A *self-hole set* of a path P, denoted by $H(P)$, is a set of *set forms*, representing all the paths that can not extend P to produces a path in S. For example, if S is the set of 3-cycle free paths, the self-hole set of the path $P = (a, b, c, d)$ is $H(P) = \{(b..), (c..), (.c.), (d..), (.d.), (..d)\}$. Now, we can redefine a *useful path* in terms of *self-hole set*: P is a *useful path* if $\bigcap_{Q \in S \cap \mathcal{F}(s,w): Q \prec_{\text{dom}} P} H(Q) \subseteq H(P)$. The IA consists of a sequence of intersection and inclusion operations applied to self-hole sets. Further details of the algorithm can be found in [9],[10].

In [10], the authors also proved an upper bound on the number of useful paths and conjectured that this number is $k!$. They have proved this conjecture for $k = 2$ and 3 using a digraph, called the *intersection digraph*. Consider useful paths $P_1, \ldots, P_t$ where $P_1 \prec_{\text{dom}} \ldots \prec_{\text{dom}} P_t$ and $I_1 \supset \ldots \supset I_t$, where $I_i = \cap_{l=1}^i H(P_l)$. In the intersection digraph, each vertex represents one intersection of self-hole sets, I_i, and an arc (I_i, I_j) exists if $I_i \cap H(P) = I_j$ and $I_i \neq I_j$, for some useful path P. Moreover, the digraph has a unique source vertex which represents the set form set $I_0 = \{(\cdot)\}$ and a single sink vertex representing the self-hole set $\emptyset$.

Thus, a dipath in this digraph represents a set of useful paths in the original graph and the length of the longest dipath in the intersection digraph is equal to the total number of useful paths.

k-Stream Elimination. By prohibiting k-cycles we avoid some vertex repetitions in the *ring*, but not in the *star*. To get rid of some repetitions in the *ring-star* (R, S) we can do as follows. First we put all vertices in $V[R] \cup C[S]$ in a string in some predefined order. We name such string a *stream* and, as if the string was representing a path, we forbid the occurrence of k-cycles in the stream. As this structure is not actually a k-cycle we call it a k-stream. We represent the stream related to a *ring-star* (R, S) in the same order in which the vertices in $V[R]$ occur in R. Moreover, the vertices $i \in U$ such as $ij \in S$ are inserted inside the stream immediately before the position of j. For example, the streams related to the *ring-stars* in Figure 1 are `OabefcO` and `OghmniojpqkO`. Since the three operations that extend a *ring-star* are conform to the definition of stream, the same rules to prohibit k-cycle are valid to prohibit k-stream.

4   Computational Experiments

This section describes the computational experiments we carried out with the the BP algorithm we developed for the CmRSP. Before we report and analyze our results, we give below some basic implementation details.

Initial basis. A crucial step in column generation procedures is how one obtain an initial basis for the first linear relaxation of the IP model. Here we introduced artificial variables with huge costs, so that they will be eliminated later, and whose columns form an identity matrix. To accelerate convergence, we also included a set of columns corresponding to *ring-stars* that are part of a feasible solution generated by a naïve heuristic inspired in [14].

Branching rule. Following the idea discussed in [13] and later used in [2], we branch on valid constraints of the form: $\sum_{e \in \delta(S)} \sum_{p \in P} r_e^p \lambda_p \geq 2 \left\lceil \frac{|\{i \in U : C_i \subseteq S\}|}{Q} \right\rceil$, $S \subseteq V \setminus \{0\}, S \neq \emptyset$. Roughly speaking, the LHS of this inequality computes the number of arcs in the cutset of S that are in some *ring*. Clearly, this quantity must be even. Given a fractional solution of the current linear relaxation, let $f(S)$ denote the difference between the LHS and the RHS of the inequality above. The branching is done for a set S satisfying $0 < f(S) < 2$. Thus, the following cuts are used to branch: $f(S) = 0$ and $f(S) \geq 2$.

Node selection. The classical best-bound strategy is used to select the next node to be explored during the enumeration procedure.

Bounding. No heuristics were developed to compute feasible solutions during the enumeration. Therefore, primal bounds must correspond to IP solutions. On the other hand, dual bounds are directly obtained from the linear relaxations or from Lasdon's formula (cf., [12]).

Pricing. After solving the relaxed pricing problem, the following columns are added to the current linear program: (i) all columns with negative reduced cost associated to non relaxed *ring-stars* and (ii) for each $j \in V \setminus \{0\}$, the column related to the minimum weight relaxed (q, j)-*walk-star*, if the corresponding reduced cost is negative. In our experiments we used two ways to identify the useful paths in LSA. The first one applies the IA, discussed in Section 3. The other one uses a deterministic finite automaton (DFA), whose diagram is the intersection digraph defined in Section 3. The DFA processes a string of paths, each one of them representing a k-cycle free (q, j)-*walk-star*. The paths are organized in the string in non decreasing order of cost and are treated as symbols of the DFA's alphabet. Two situations characterize a non-useful path: either the current state is final, i.e, is the sink vertex of the digraph, or no transition is defined for this path in the current state.

Computational Results. We have conducted empirical analysis using the same instance classes reported in [2], but adapted so as to satisfy the triangle inequalities. In particular, we used the TSPLIB [15] instances named ei151.tsp, ei176.tsp and ei1101.tsp. Moreover, we generated instances called ei140.tsp and ei164.tsp consisting of the first 40 and 64 vertices in ei151.tsp and ei1101.tsp instances, respectively. For a given number of vertices, there are 9 different instances. All our coding were done in C language using gcc 4.1.2 compiler. We implemented a *branch-and-cut* algorithm, herein after denoted by BC, following the ideas presented in [2]. To this end, we use the *branch-and-cut* framework provided by the libraries of XPRESS-MP [17] version 17.01.01. In the *branch-and-price* algorithm the XPRESS solver was used solely to compute linear relaxations. The experiments were ran on a Pentium IV 3.4 GHz and 1Gb of RAM. In the table that follow, running times are reported in seconds. We implemented three algorithms to solve the pricing problem: BPr, BPkc and BPks. The first one solves the original relaxed pricing problem discussed in Section 3, with vertex repetitions allowed. The second one prohibits k-cycles in the *rings*, while the last one forbids k-streams in *ring-stars*. The version of BPks that identifies useful paths via a DFA is denoted by BPksA. Table 1 summarizes the comparative performance of these implementations of BP and of BC. The running time of each execution of any code was limited to 1800 seconds. Codes are compared pairwisely. The group of columns under the header *code A × code B* contain the following data. Column GAP refers to the average percentage reduction in the duality gap obtained by code B with respect do code A taken over all the instances not solved by both codes. This gap was calculated relative to the best primal solution found by all the implemented codes, including the naïve heuristic used to populate the initial linear relaxation. Now, considering only the instances whose optimum was proved by both codes, column TIME shows the average speed-up rate defined as the running time of code A divided by that of code B. Finally, for the group of instances associated to the table row, the contents of column OPT are of the form x/y where y ($y - x$) is the number of instances solved to optimality by code B (code A). With these definitions, it is clear that the value in some cells may be undefined, which is denoted by a "*".

First we concentrate on the three different versions of the BP algorithm: BPr, BP3c and BP3s. Inspecting columns OPT from the 2 first groups in Table 1,

Table 1. Comparison between BPr, BP3c, BP3s, BP3sA and BC codes

	BPr × BP3c			BP3c × BP3s			BP3s × BP3sA			BC × BP3sA		
instance	GAP	TIME	OPT	GAP	TIME	OPT	GAP	TIME	OPT	GAP	TIME	OPT
eil40.tsp	1.0	2.8	2/8	*	1.6	1/9	*	3.3	0/9	*	0.3	0/9
eil51.tsp	1.0	1.4	0/2	0.4	19.9	4/6	0.2	3.5	0/6	-0.6	0.7	0/6
eil64.tsp	1.4	*	0/0	1.3	*	4/4	0.1	2.3	0/4	0.1	1.2	0/4
eil76.tsp	0.6	*	1/1	2.5	48.7	0/1	0.2	2.4	0/1	11.8	*	1/1

one can see that the more complex the structures forbidden in the relaxation, the larger is the number of instances that are solved to optimality. The gain is even more evident when we start eliminating 3-streams rather than 3-cycles. If we turn our attention to columns TIME, we see that more stringent relaxations lead to higher speedups in running times. This can be explained by the fact that the dual bounds from the linear relaxations become tighter, which permits the early pruning of the enumeration tree. This is confirmed by the reductions in duality gaps expressed in columns GAP, which are small but always occur. From these results, we can conclude that BP3s dominates the two other codes analyzed so far. Therefore, it is important to have an efficient implementation of this relaxation. The use of a DFA is a step towards this direction. The gain on running time achieved by using a DFA can be appreciated by checking column TIME in the group associated to codes BP3s and BP3sA in Table 1. Further evidences of the effectiveness of BP3sA can be seen in Figure 2. In this figure we compare the number of nodes explored during the enumeration (in (a)) and the time spent by the procedure in solving the relaxed pricing problem (in (b)) for all the 36 instances in our dataset. The fact that BP3sA solves the pricing problem faster than BP3s allows it to explore larger portions of the solution space.

Once we have identified BP3sA as our best BP implementation, we compare this code with the BC code. Here, the picture is less clear. The last group of columns in Table 1 shows that, on average, BC is faster than BP3sA. As for the duality gaps, it seems that BC performs slightly better for small instances while BP3sA becomes more attractive for bigger ones. In fact, looking at column GAP,

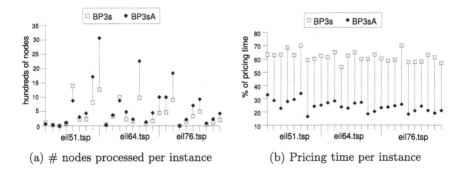

(a) # nodes processed per instance   (b) Pricing time per instance

Fig. 2. Comparison between BP3s and BP3sA codes

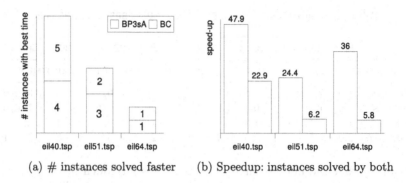

(a) # instances solved faster   (b) Speedup: instances solved by both

Fig. 3. Comparison between BP3sA and BC

we see that BP3sA was the only code that could solve one of the larger instances to optimality. A closer inspection of the results reveals that these two codes behave quite differently. First, as it is often the case when a discrete optimization problem admits a set covering formulation, the dual bound provided by the latter is better than that obtained with a "natural" model for the problem. This was precisely what happened in our tests where we have observed that the average improvement in the lower bound generated by BP3sA with respect to that of BC was 3.6%, 3,8%, 5.0% and 6.6% for the instance classes eil40, eil51, eil64 and eil76, respectively. In all but one instance in class eil40 and one in class eil51 the lower bound of BP3sA was strictly better than that of BC. The numbers in the last column OPT of Table 1 show that, except for class eil76, both algorithms solved the same amount of instances. However, the instances in each class that were actually solved differ. In Figure 3(a) we can see how many instances were solved by both codes in each class and in how many of them each code was the fastest one. On the other hand, in Figure 3(b), we display the speedup obtained by each code in the total time spent in processing those instances where they ran faster. From these charts and the previous discussion, it is clear that none of the codes dominates the other. There are instances which are far better suited to be tackled by BP3sA and also others for which the opposite is true.

5   Conclusion and Further Works

In this paper we proposed a *set covering* IP model for C*m*RSP and *branch-and-price* algorithms to solve this model. The key point in developing such algorithms is how we solve the pricing problem, which is $\mathcal{NP}$-hard. In order to get a faster code, we relaxed the pricing problem in different ways and, through experimentation, we investigate which of these variants of the algorithm leads to better performance. We end up with a *branch-and-price* code for the C*m*RSP, called BP3sA, which we proved to be competitive with a *branch-and-cut* algorithm proposed earlier. This was achieved through a careful implementation of the algorithm that solves the relaxed pricing problem by means of a deterministic finite automaton. We also noticed that the BP3sA and BC do not dominate each other and some instances are better suited for one or the other algorithm. This suggests that the next step in this research should be the development of a *branch-and-cut-and-price* algorithm combining the two techniques, a direction we start pursuing now.

References

1. Ahuja, R., Magnanti, T., Orlin, J.: Network Flows: Theory, Algorithms, and Applications. Prentice-Hall, Englewood Cliffs (1993)
2. Baldacci, R., Dell'Amico, M., Salazar, J.: The Capacitated m-Ring Star Problem. Operations Research 55, 1147–1162 (2007)
3. Beasley, J., Nascimento, E.: The Vehicle Routing-Allocation Problem: A Unifying Framework. Trabajos de OPerativa 4, 65–86 (1996)
4. Christofides, N., Mingozzi, A., Toth, P.: Exact Algorithms for the Vehicle Routing Problem, Based on Spanning Tree and Shortest Path Relaxations. Mathematical Programming 20, 255–282 (1981)
5. Dell'Amico, M., Maffioli, F., Varbrand, P.: On Prize-Collecting Tours and the Asymmetric Travelling Salesman Problem. International Transactions in Operational Research 2(3), 297–308 (1995)
6. Feillet, D., Dejax, P., Gendreau, M.: Traveling Salesman Problems with Profits. Transportation Science 39(2), 188–205 (2005)
7. Fukasawa, R., Longo, H., Lysgaard, J., de Aragão, M.P., Reis, M., Uchoa, E., Werneck, R.: Robust Branch-and-Cut-and-Price for the Capacitated Vehicle Routing problem. Mathematical Programming 106(3), 491–511 (2006)
8. Houck, D., Picard, J., Queyranne, M., Vemuganti, R.: The Travelling Salesman Problem as a Constrained Shortest Path Problem: Theory and Computational Experience. Opsearch 17, 93–109 (1980)
9. Irnich, S., Desaulniers, G.: Shortest Path Problems with Resource Constraints. In: Column Generation, pp. 33–65. Springer, Heidelberg (2005)
10. Irnich, S., Villeneuve, D.: The Dhortest Path Problem with Tesource Vonstraints and k-Cycle Elimination for k ≥ 3. INFORMS J. on Computing 18(3), 391–406 (2006)
11. Labbé, M., Laporte, G., Martín, I.R., González, J.S.: The Ring-Star Problem: Polyhedral Analysis and Exact Algorithm. Networks 43, 117–189 (2004)
12. Lasdon, L.: Optimization Theory for Large Systems. Macmillan, Basingstoke (1970)
13. Lysgaard, J., Letchford, A., Eglese, R.: A New Branch-and-Cut Algorithm for the Capacitated Vehicle Routing problem. Mathematical Programming 100(2), 423–445 (2004)
14. Mauttone, A., Nesmachnow, S., Olivera, A., Robledo, F.: A Hybrid Metaheuristic Algorithm to Solve the Capacitated m-Ring Star Problem. In: International Network Optimization Conference (2007)
15. TSPLIB,
   http://www.iwr.uni-heidelberg.de/groups/comopt/software/ TSPLIB95/tsp/
16. Wolsey, L.A.: Integer Programming. Wiley-Interscience, Chichester (1998)
17. Xpress-Optimizer. Dash Optimization (2007)

Two-Agent Scheduling with Linear Deteriorating Jobs on a Single Machine*

Peng Liu and Lixin Tang

The Logistics Institute, Northeastern University, Shenyang, China
liup7802@163.com,
qhjytlx@mail.neu.edu.cn

Abstract. This paper considers the two-agent scheduling problems with linear deteriorating jobs to be processed on a single machine. By a deteriorating job we mean that the processing time of the job is a function of its starting time. Two agents compete for the usage of a common single machine and each agent has his own criterion to optimize. There are four objective functions: makespan, maximum lateness, maximum cost, and total completion time. Some basic properties of two different scheduling problems to minimize the objective function for one agent with a constraint on the other agent's objective function are proved. Based on these properties, the optimal algorithms with polynomial time are presented for two different scheduling problems, respectively.

1 Introduction

In the classical scheduling problems it is assumed that the job processing times are constant and independent of the starting time, but that is not proper for all actual situations. Job processing times are not constant because jobs deteriorate as they wait to be processed. The phenomenon of the processing time of job depends on its starting time is called deterioration of processing time, and this problem is known as the deteriorating job scheduling problem. However, researches on the deteriorating job scheduling problem assume that all jobs to be processed on a common machine are belonged to a single agent. In real production settings there are several agents, each with a set of jobs. The agents have to schedule their jobs on a common processing resource, and each agent wishes to minimize an objective function that depends on the completion times of his own set of jobs. In this paper, we study the two-agent scheduling problems with linear deteriorating jobs to be processed on a single machine. To the best of our knowledge, this type of problem has not been discussed in the literature.

Scheduling problems with deterioration and multi-agent have been studied separately in the most of previous literatures. In this paper we combine scheduling, deterioration and multi-agent decisions. We consider not only the scheduling

---

* This work is supported by NSFC (Grant No. 70425003 and Grant No. 60674084), National 863 High-Tech Research and Development Program of China through approved No.2006AA04Z174 .

X. Hu and J. Wang (Eds.): COCOON 2008, LNCS 5092, pp. 642–650, 2008.

with deteriorating jobs to be processed on a single machine, but also the scheduling with two agents to compete for the usage of the common single machine. We will present a brief review on the related scheduling problems as follows.

The work of this paper is related to machine scheduling problem with deteriorating jobs. The deteriorating job scheduling problem is first introduced independently by Gupta and Gupta [10] and Browne and Yechiali [5]. An extensive survey of different models and problems concerning jobs with starting time-dependent processing times can be found in Alidaee and Womer [3] and Cheng et al. [6]. Mosheiov [11] considers the single machine scheduling problems of simple linear deterioration where jobs have a fixed job-dependent growth rate but no basic processing time. He show that most commonly used performance, such as the makesapn, the total flow time, the total weighted completion time, the total lateness, the maximum lateness, the maximum tardiness, and the number of tardy jobs, remain polynomially solvable. Ng et al. [12] consider three scheduling problems with different linearly decreasing function to minimize the total completion time on a single machine. Two of the problems are solved optimally and a pseudo-polynomial time algorithm is constructed to solve the third problem using dynamic programming. Wu and Lee [13] study the problem that scheduling linear deteriorating jobs to minimize makespan on a single machine with an availability constraint. They show that the linear deteriorating model can be solved using the 0-1 integer programming technique. The literatures mentioned above cover the scheduling with deteriorating jobs, however, they do not consider the scheduling with multiple agents.

Another line of research related our problem focuses on multi-agent scheduling problems. Cheng et al. [7] study the multi-agent scheduling problem with minimizing the total weighted number of tardy jobs on a single machine. They show that the problem is strongly NP-complete in general, and present a fully polynomial-time approximation scheme for the case of the fixed number of agents. Cheng et al. [8] consider the multi-agent scheduling problem on a single machine, where the agent's objective functions are of the max-form. The problem of the feasibility model can be solved in polynomial time. They show that the minimality model is strongly NP-hard and identify some polynomial time solvable cases. Agnetic et al. [1] consider the two-agent scheduling problem on a single machine and the shop environment. They study constrained optimization problem and Pareto-optimization problem. They analyze the complexity of these problems and propose solution algorithms. In this paper one of the problems of Agnetic et al. [1] is extended to incorporate the situation where deteriorating jobs are allowed at the single machine. Agnetic et al. [2] investigate the complexity of some scheduling problems in which several agents have to negotiate the usage of a common processing resource. They find a schedule attaining minimum quality requirements specified by each single agent. Baker and Smith [4] consider a multiple-criterion scheduling problem involving a single machine being utilized by two or more customers. They provide some dominance properties and demonstrate where the problem becomes computationally difficult. Yuan et al. [14] show that two dynamic programming recursions in Baker and Smith

[4] are incorrect and give a polynomial-time algorithm for the same problems. Different from the above scheduling models, we are in addition to taking into account scheduling with deteriorating jobs to be processed on a single machine.

The remainder of this paper is organized as follows. In Section 2, we describe the proposed problems. In Section 3, we develop polynomial time algorithms for two different scheduling problems. Section 4 gives some concluding remarks.

2 Problem Description

We now describe our problems formally. There are two families of independent and non-preemptive jobs $J^A = \{J_1^A, J_2^A, ..., J_{n_A}^A\}$ and $J^B = \{J_1^B, J_2^B, ..., J_{n_B}^B\}$ to be processed on a common single machine. The jobs in J^A and J^B are called A-agent's jobs and B-agent's jobs, respectively. The processing time of the job depends on its starting time. The processing time p_h^A of job J_h^A is a linear function of its starting time, i.e., $p_h^A = \alpha_h^A t, h = 1, 2, ..., n_A$, where $\alpha_h^A > 0$ is the deterioration rate and $t \geq t_0$ is the starting time. All jobs are available for processing at time $t_0 > 0$. The assumption $t_0 > 0$ is made here to avoid the trivial case of $t_0 = 0$ (when $t_0 = 0$, the completion time of each job will be 0). Similarly, the processing time p_k^B of job J_k^B is $\alpha_k^B t$, i.e., $p_k^B = \alpha_k^B t, k = 1, 2, ..., n_B$. Let $X \in \{A, B\}$ be given and let S indicate a feasible schedule of the $n = n_A + n_B$ jobs. The completion time of job J_j^X in S is denoted as $C_j^X(S)$. The job J_j^X has a positive due date d_j^X, and the lateness of J_j^X is $L_j^X(S) = C_j^X(S) - d_j^X$. The makespan of agent X is $C_{\max}^X = \max_{j=1,2,...,n_X} \{C_j^X(S)\}$, and the maximum lateness of agent X is $L_{\max}^X = \max_{j=1,2,...,n_X} \{L_j^X(S)\}$. For each job J_j^X, let $f_j^X(\cdot)$ be a non-decreasing function of the completion time of job J_j^X. The objective function of agent X takes one of the following four forms: makespan $C_{\max}^X(S)$, maximum lateness $L_{\max}^X(S)$, maximum cost $f_{\max}^X(S) = \max_{j=1,2,...,n_X} \{f_j^X(C_j^X(S))\}$, and the total completion time $\sum C_j^X(S)$. Note that $C_{\max}^X$ and $L_{\max}^X$ are the special case of $f_{\max}^X$.

The goal is to find a schedule such that the objective function value of agent A is minimized, subject to the requirement that the objective function value of agent B is bounded by an upper bound U. An instance of the scheduling problem may not have feasible solutions (e.g., if U is too small). If there is at least one feasible solution, we say that the instance is feasible. By using the three-field notation $\Psi_1|\Psi_2|\Psi_3$ of Graham et al. [9], we denote two considered scheduling problems as $1|p_h^A = \alpha_h^A t, p_k^B = \alpha_k^B t|L_{\max}^A : C_{\max}^B \leq U$ and $1|p_h^A = \alpha_h^A t, p_k^B = \alpha_k^B t|\sum C_h^A : f_{\max}^B \leq U$, respectively.

3 Main Results

In subsection 3.1 and 3.2, we consider problem $1|p_h^A = \alpha_h^A t, p_k^B = \alpha_k^B t|L_{\max}^A : C_{\max}^B \leq U$ and $1|p_h^A = \alpha_h^A t, p_k^B = \alpha_k^B t|\sum C_h^A : f_{\max}^B \leq U$, respectively.

3.1 Problem $1|p_h^A = \alpha_h^A t, p_k^B = \alpha_k^B t|L_{\max}^A : C_{\max}^B \leq U$

In this subsection, we show that the problem $1|p_h^A = \alpha_h^A t, p_k^B = \alpha_k^B t|L_{\max}^A : C_{\max}^B \leq U$ is polynomially solvable. Before proposing the algorithm, we first present some properties of the problem.

Lemma 1. *In the scheduling problem that we consider, there exists an optimal schedule without idle time between jobs on the machine.*

Proof. Since the objective function $f_{\max}$ and $\sum C_j$ are non-decreasing function of the completion time C_j, they are the regular measure of performance. If there exists idle time, the subsequent jobs can be moved earlier without increasing the objective value. □

Lemma 2. *For the problem $1|p_h^A = \alpha_h^A t, p_k^B = \alpha_k^B t|C_{\max}$, the makespan is equal to $t_0 \prod_{h=1}^{n_A} (1+\alpha_h^A) \prod_{k=1}^{n_B} (1+\alpha_k^B)$.*

Proof. For job set $J^A = \{J_1^A, J_2^A, ..., J_{n_A}^A\}$ and $J^B = \{J_1^B, J_2^B, ..., J_{n_B}^B\}$, we re-index the $n = n_A + n_B$ jobs such that job set $J = J^A \cup J^B = \{J_1, J_2, ..., J_{n_A}, J_{n_A+1}, J_{n_A+2}, ..., J_{n_A+n_B}\}$ with deterioration rate α_j, where $\alpha_j = \alpha_j^A$ for $j = 1, 2, ..., n_A$ and $\alpha_j = \alpha_{j-n_A}^B$ for $j = n_A + 1, j = n_A + 2, ..., j = n_A + n_B$. We consider the schedule S_0 that bases on the re-indexed permutation $(1, 2, ..., n_A, n_A + 1, ..., n)$ and starts at t_0. From Lemma 1 and Mosheiov [11], we know that the makespan is sequence independent for any schedule of job set J and $C_n = t_0 \prod_{j=1}^{n} (1+\alpha_j)$. So we have $C_{\max} = t_0 \prod_{j=1}^{n} (1+\alpha_j)$ for job set J. Therefore, substituting α_j by α_h^A or α_k^B, the makespan is obtained: $C_{\max} = t_0 \prod_{j=1}^{n_A} (1+\alpha_j) \prod_{j=n_A+1}^{n_A+n_B} (1+\alpha_j) = t_0 \prod_{j=1}^{n_A} (1 + \alpha_j^A) \prod_{j=n_A+1}^{n_A+n_B} (1 + \alpha_{j-n_A}^B) = t_0 \prod_{h=1}^{n_A} (1 + \alpha_h^A) \prod_{k=1}^{n_B} (1 + \alpha_k^B)$. This completes the proof. □

Lemma 3. *For the problem $1|p_h^A = \alpha_h^A t, p_k^B = \alpha_k^B t|L_{\max}$, there exists an optimal schedule such that all A-agent's jobs are processed in the non-decreasing order of d_h^A.*

Proof. For job set $J^A = \{J_1^A, J_2^A, ..., J_{n_A}^A\}$ and $J^B = \{J_1^B, J_2^B, ..., J_{n_B}^B\}$, we re-index the $n = n_A + n_B$ jobs such that job set $J = J^A \cup J^B = \{J_1, J_2, ..., J_{n_A}, J_{n_A+1}, J_{n_A+2}, ..., J_{n_A+n_B}\}$ with deterioration rate α_j and due date d_j, where $\alpha_j = \alpha_j^A$ and $d_j = d_j^A$ for $j = 1, 2, ..., n_A$, and $\alpha_j = \alpha_{j-n_A}^B$ and $d_j = d_{j-n_A}^B$ for $j = n_A + 1, j = n_A + 2, ..., j = n_A + n_B$. A pairwise job interchange argument is used in the following proof. Consider a schedule S' for job set J, suppose which is not in the non-decreasing order of d_j, is optimal. In this schedule there must be at least two adjacent jobs J_i and J_j, J_i followed by J_j, such that J_i starts at time T and $d_i > d_j$. Schedule S^* is obtained from schedule S' by interchanging jobs J_i and J_j. Hence, under S', we have $L_i(S') = T + \alpha_i T - d_i$

and $L_j(S') = T + \alpha_i T + \alpha_j(T + \alpha_i T) - d_j$, whereas under S^* it is $L_j(S^*) = T + \alpha_j T - d_j$ and $L_i(S^*) = T + \alpha_j T + \alpha_i(T + \alpha_j T) - d_i$. Since $d_i > d_j$, then $\max\{L_i(S^*), L_j(S^*)\} < \max\{L_i(S'), L_j(S')\}$. Hence, interchanging the position of jobs J_i and J_j cannot increase the value of $L_{\max}$. A finite number of such changes transform S' into the non-decreasing order of d_j, showing that the non-decreasing order of d_j is optimal. Therefore, all A-agent's jobs are also processed in the non-decreasing order of d_h^A in an optimal schedule, even though they may not all appear consecutively in the optimal schedule.   □

Lemma 4. *For the problem* $1|p_h^A = \alpha_h^A t, p_k^B = \alpha_k^B t|C_{\max}$, *there exists an optimal schedule such that all B-agent's jobs are consecutively processed.*

Proof. By Lemma 2, the makespan is sequence independent. Use a pairwise job interchange argument, we can consecutively process B-agent's jobs and consolidate all B-agent's jobs into a block.   □

Now we consider a feasible instance of $1|p_h^A = \alpha_h^A t, p_k^B = \alpha_k^B t|L_{\max}^A : C_{\max}^B \leq U$. From Lemmas 3 and 4, we know that the A-agent's jobs can be sequenced according to the non-decreasing order of d_h^A and the B-agent's jobs can be consolidated into a block. Without loss of generality, we denote the A-agent's jobs as $\{J_1^A, J_2^A, ..., J_{n_A}^A\}$ according to the non-decreasing order of d_h^A and denote all B-agent's jobs as a dummy job B1. Let $u = t_0 \prod_{h=1}^{n_A}(1 + \alpha_h^A) \prod_{k=1}^{n_B}(1 + \alpha_k^B)$. Define schedule $S = \{J_1^A, J_2^A, ..., J_{n_A-1}^A, J_{n_A}^A, B1\}$ by placing the A-agent's jobs at the beginning of the schedule, followed by the dummy job B1. If there exists a B-agent's job $J_{k_0}^B$ such that $C_{k_0}^B = u \leq U$, then schedule S is an optimal schedule. Otherwise, we arrange job $J_{n_A}^A$ in the last position. Consequently, we obtain schedule $S_1 = \{J_1^A, J_2^A, ..., J_{n_A-1}^A, B1, J_{n_A}^A\}$ and $u_1 = t_0 \prod_{h=1}^{n_A-1}(1+\alpha_h^A) \prod_{k=1}^{n_B}(1+\alpha_k^B)$. If there exists a B-agent's job $J_{k_1}^B$ such that $C_{k_1}^B = u_1 \leq U$, then S_1 is an optimal schedule. Otherwise, we arrange job $J_{n_A-1}^A$ in the second last position, so we obtain schedule $S_2 = \{J_1^A, J_2^A, ..., J_{n_A-2}^A, B1, J_{n_A-1}^A, J_{n_A}^A\}$ and $u_2 = t_0 \prod_{h=1}^{n_A-2}(1 + \alpha_h^A) \prod_{k=1}^{n_B}(1 + \alpha_k^B)$. By the similar argument, if there exists a B-agent's job $J_{k_i}^B$ such that $C_{k_i}^B = u_i = t_0 \prod_{h=1}^{n_A-i}(1 + \alpha_h^A) \prod_{k=1}^{n_B}(1 + \alpha_k^B) \leq U$, then $S_i = \{J_1^A, J_2^A, ..., J_{n_A-i}^A, B1, J_{n_A-i+1}^A, ..., J_{n_A}^A\}$ is an optimal schedule. Therefore, the problem $1|p_h^A = \alpha_h^A t, p_k^B = \alpha_k^B t|L_{\max}^A : C_{\max}^B \leq U$ is solved by finding the best location of the dummy job B1 among the A-agent's jobs sequenced by the non-decreasing order of d_h^A. The computational complexity is $O(n_A \log n_A)$.

3.2   Problem $1|p_h^A = \alpha_h^A t, p_k^B = \alpha_k^B t|\sum C_h^A : f_{\max}^B \leq U$

In this subsection, we derive an polynomial time algorithm to solve the problem $1|p_h^A = \alpha_h^A t, p_k^B = \alpha_k^B t|\sum C_h^A : f_{\max}^B \leq U$. It is easy to see that Lemma 1 and

Lemma 2 still hold for the problem $1|p_h^A = \alpha_h^A t, p_k^B = \alpha_k^B t| \sum C_h^A : f_{\max}^B \leq U$. Now we give Lemma 5 and Lemma 6.

Lemma 5. *For the problem* $1|p_h^A = \alpha_h^A t, p_k^B = \alpha_k^B t| \sum C_h^A : f_{\max}^B \leq U$, *let*

$$u = t_0 \prod_{h=1}^{n_A} (1+\alpha_h^A) \prod_{k=1}^{n_B} (1+\alpha_k^B).$$ *If for all B-agent's jobs* J_k^B *such that* $f_k^B(u) > U$, *then for any optimal schedule the A-agent's job with the largest deterioration rate is scheduled in the last position.*

Proof. Use a pairwise job interchange argument. Assume there is an optimal schedule S', such that the A-agent's job with the largest deterioration rate, say $J_{h_0}^A$, is not scheduled in the last position. Since the last job in schedule S' is an A-agent's job, we may let it be J_l^A such that $\alpha_l^A < \alpha_{h_0}^A$. By interchanging J_l^A and $J_{h_0}^A$, and leaving the other jobs unchanged, we obtain a new schedule S^* from S'. Since there is no change for any job preceding $J_{h_0}^A$ in schedule S', we may assume that job $J_{h_0}^A$ in S' starts its processing at time T. Therefore, we have $C_{h_0}^A(S') = T + \alpha_{h_0}^A T$ and $C_l^A(S^*) = T + \alpha_l^A T$. It is easy to see that $C_l^A(S^*) < C_{h_0}^A(S')$, so the starting time of each job between J_l^A and $J_{h_0}^A$ in S^* is earlier than the starting time of the corresponding job between $J_{h_0}^A$ and J_l^A in S'. Hence, $C_i^X(S^*) < C_i^X(S')$ for any job J_i^X between $J_{h_0}^A$ and J_l^A. The completion time of the last job $J_{h_0}^A$ in schedule S^* is equal to the completion time of the last job J_l^A in schedule S', i.e., $C_{h_0}^A(S^*) = C_l^A(S') = u = t_0 \prod_{h=1}^{n_A} (1+\alpha_h^A) \prod_{k=1}^{n_B} (1+\alpha_k^B)$. Since $f_k^B(\cdot)$ is a non-decreasing function of the completion time of job J_k^B, $f_{\max}^B(S^*) \leq f_{\max}^B(S')$. And we have $\sum C_h^A(S^*) < \sum C_h^A(S')$. It shows that the new schedule S^* is better than the optimal schedule S', which contradicts our assumption. This completes the proof. $\square$

Lemma 6. *For the problem* $1|p_h^A = \alpha_h^A t, p_k^B = \alpha_k^B t| \sum C_h^A : f_{\max}^B \leq U$, *let* $u = t_0 \prod_{h=1}^{n_A} (1+\alpha_h^A) \prod_{k=1}^{n_B} (1+\alpha_k^B)$. *If there is a B-agent's job* $J_{k_0}^B$ *such that* $f_{k_0}^B(u) \leq U$, *then there exists an optimal schedule such that* $J_{k_0}^B$ *is scheduled in the last position and there exists no optimal schedule such that an A-agent's job is scheduled in the last position.*

Proof. Assume there is an optimal schedule S', such that $J_{k_0}^B$ is not scheduled in the last position. Schedule S^* is obtained from S' by moving $J_{k_0}^B$ to the last position. There is no change for any job preceding $J_{k_0}^B$ in schedule S'. Since job $J_{k_0}^B$ in schedule S' is moved to the last position, the starting time of each job following $J_{k_0}^B$ in S' is larger than the starting time of the corresponding job in the schedule S^*. As a consequence, we have $C_i^X(S^*) \leq C_i^X(S')$ for any job J_i^X other than $J_{k_0}^B$. The completion time of the last job in schedule S' is equal to the completion time of the last job $J_{k_0}^B$ in S^*. Consequently, $\sum C_h^A(S^*) \leq \sum C_h^A(S')$.

The completion time of job $J_{k_0}^B$ in schedule S^* is $u = t_0 \prod_{h=1}^{n_A} (1+\alpha_h^A) \prod_{k=1}^{n_B} (1+\alpha_k^B)$

and $f_{k_0}^B(u) \leq U$. The schedule S^* is still feasible and optimal because $f_k^B(\cdot)$ is a non-decreasing function of the completion time of job J_k^B. If an A-agent's job is in the last position in schedule S', then $\sum C_h^A(S^*) < \sum C_h^A(S')$. This contradicts the optimality of schedule S'. The proof is complete. $\qquad\square$

Using Lemmas 1, 2, 5 and 6, an algorithm to determine an optimal schedule of the problem $1|p_h^A = \alpha_h^A t, p_k^B = \alpha_k^B t| \sum C_h^A : f_{\max}^B \leq U$ is developed as follows.

Let E denote the set of jobs to be scheduled, and let F denote the set of jobs already scheduled. $C_{\max}(E)$ is the makespan of jobs in E.

Algorithm 1

Step 1: Let $E = J^A \cup J^B = \{J_1^A, J_2^A, ..., J_{n_A}^A, J_1^B, J_2^B, ..., J_{n_B}^B\}$, $F = \phi$.

Step 2: Calculate $u = C_{\max}(E)$ according to Lemma 2. If there exists an unscheduled B-agent's job J_k^B such that $f_k^B(u) \leq U$, then arrange job J_k^B in the last position among unscheduled jobs and let $J_{j^*} = J_k^B$. Otherwise, let J_h^A be the unscheduled A-agent's job with the largest deterioration rate, arrange job J_h^A in the last position among unscheduled jobs and let $J_{j^*} = J_h^A$. If all A-agent's jobs have been scheduled and no B-agent's job can be scheduled in the last position, the instance is not feasible.

Step 3: Set $E = E - \{J_{j^*}\}$, $F = F + \{J_{j^*}\}$. If $E = \phi$, stop; otherwise, go to Step 2.

Theorem 1. *Algorithm 1 generates an optimal schedule for the problem $1|p_h^A = \alpha_h^A t, p_k^B = \alpha_k^B t| \sum C_h^A : f_{\max}^B \leq U$ in $O(n_A \log n_A + n_B \log n_B)$ time.*

Proof. The proof of optimality follows from Lemmas 1, 2, 5 and 6. We now turn to time complexity. The time to sequence the jobs in job set J^A according to the non-decreasing order of deterioration rate α_h^A requires $O(n_A \log n_A)$. For each B-agent's job J_k^B, we can define a deadline D_k^B such that $f_k^B(C_k^B) \leq U$ for $C_k^B \leq D_k^B$ and $f_k^B(C_k^B) > U$ for $C_k^B > D_k^B$. So the time to sequence the jobs in job set J^B according to the non-decreasing order of deadline D_k^B also requires $O(n_B \log n_B)$. At each step there is only an A-agent's job from job set J^A or a B-agent's job from job set J^B to be considered. Therefore, the overall time complexity of Algorithm 1 is bounded by $O(n_A \log n_A + n_B \log n_B)$. $\qquad\square$

Furthermore, to illustrate the application of Algorithm 1, we consider the following numerical example.

Example 1. We assume that $t_0 = 1, U = 150, 75, 10$, and $f_k^B(C_k^B) = C_k^B$. Job set $J^A = \{J_1^A, J_2^A, J_3^A\}$: $n_A = 3, \alpha_1^A = 1, \alpha_2^A = 3, \alpha_3^A = 0.2$; and job set $J^B = \{J_1^B, J_2^B, J_3^B\}$: $n_B = 3, \alpha_1^B = 0.5, \alpha_2^B = 4, \alpha_3^B = 1$.

Case 1: $U = 150$. First, let $E = J^A \cup J^B = \{J_1^A, J_2^A, J_3^A, J_1^B, J_2^B, J_3^B\}$, $F = \phi$. Second, calculate $u = C_{\max}(E) = t_0 \prod_{h=1}^{n_A} (1 + \alpha_h^A) \prod_{k=1}^{n_B} (1 + \alpha_k^B) = 144$ according to Lemma 2. Since $f_3^B(u) = 144 < U$, arrange J_3^B in last position, and go to Step 3. By Step 3 we have $E = \{J_1^A, J_2^A, J_3^A, J_1^B, J_2^B\}$ and $F = \{J_3^B\}$, calculate $u = C_{\max}(E) = 72, f_2^B(u) = 72 < U$, arrange J_2^B in last position, and go to

Step 3. By Step 3 we have $E = \{J_1^A, J_2^A, J_3^A, J_1^B\}$ and $F = \{J_3^B, J_2^B\}$. Similarly, calculate $u = C_{\max}(E) = 14.4$, $f_1^B(u) = 14.4 < U$, arrange J_1^B in last position, and go to Step 3.

By Step 3 we have $E = \{J_1^A, J_2^A, J_3^A,\}$ and $F = \{J_3^B, J_2^B J_1^B\}$. According to Step 2, calculate $u = C_{\max}(E) = 9.6$. Since there is no unscheduled B-agent's job, arrange the largest deterioration rate job J_2^A in the last position, and go to Step 3. By Step 3 we have $E = \{J_1^A, J_3^A\}$ and $F = \{J_3^B, J_2^B, J_1^B, J_2^A\}$, calculate $u = C_{\max}(E) = 2.4$. Since there is no unscheduled B-agent's job, arrange the largest deterioration rate job J_1^A in the last position, and go to Step 3. By Step 3 we have $E = \{J_3^A\}$ and $F = \{J_3^B, J_2^B, J_1^B, J_2^A, J_1^A\}$. Similarly, calculate $u = C_{\max}(E) = 1.2$. Since there is no unscheduled B-agent's job, arrange the largest deterioration rate job J_3^A in the last position, and go to Step 3. By Step 3 we have $E = \phi$ and $F = \{J_3^B, J_2^B, J_1^B, J_2^A, J_1^A, J_3^A\}$.

So we obtain an optimal schedule $J_3^A \rightarrow J_1^A \rightarrow J_2^A \rightarrow J_1^B \rightarrow J_2^B \rightarrow J_3^B$, and $\sum C_h^A = 1.2 + 2.4 + 9.6 = 13.2$.

Case 2: $U = 75$. First, let $E = J^A \cup J^B = \{J_1^A, J_2^A, J_3^A, J_1^B, J_2^B, J_3^B\}$, $F = \phi$. Second, calculate $u = C_{\max}(E) = 144$. Since for all B-agent's jobs $f_k^B(u) = 144 > U$, arrange the largest deterioration rate job J_2^A in the last position, and go to Step 3. By Step 3 we have $E = \{J_1^A, J_3^A, J_1^B, J_2^B, J_3^B\}$ and $F = \{J_2^A\}$. According to Step 2, calculate $u = C_{\max}(E) = 36$, $f_3^B(u) = 36 < U$, arrange job J_3^B in the last position, and go to Step 3. By Step 3 we have $E = \{J_1^A, J_3^A, J_1^B, J_2^B\}$ and $F = \{J_2^A, J_3^B\}$. The remainder of solution process is similar to Case 1, we finally have $E = \phi$ and $F = \{J_2^A, J_3^B, J_2^B, J_1^B, J_1^A, J_3^A\}$.

So we obtain an optimal schedule $J_3^A \rightarrow J_1^A \rightarrow J_1^B \rightarrow J_2^B \rightarrow J_3^B \rightarrow J_2^A$, and $\sum C_h^A = 1.2 + 2.4 + 144 = 147.6$.

Case 3: $U = 10$. First, let $E = J^A \cup J^B = \{J_1^A, J_2^A, J_3^A, J_1^B, J_2^B, J_3^B\}$, $F = \phi$. Second, calculate $u = C_{\max}(E) = 144$. Since for all B-agent's jobs $f_k^B(u) = 144 > U$, arrange the largest deterioration rate job J_2^A in the last position, and go to Step 3. By Step 3 we have $E = \{J_1^A, J_3^A, J_1^B, J_2^B, J_3^B\}$ and $F = \{J_2^A\}$, calculate $u = C_{\max}(E) = 36$. Since for all unscheduled B-agent's jobs $f_k^B(u) = 36 > U$, arrange the largest deterioration rate job J_1^A in the last position, and go to Step 3. By Step 3 we have $E = \{J_3^A, J_1^B, J_2^B, J_3^B\}$ and $F = \{J_2^A, J_1^A\}$. Similarly, calculate $u = C_{\max}(E) = 18$. Since for all unscheduled B-agent's jobs $f_k^B(u) = 18 > U$, arrange the largest deterioration rate job J_3^A in the last position, and go to Step 3. By Step 3 we have $E = \{J_1^B, J_2^B, J_3^B\}$ and $F = \{J_2^A, J_1^A, J_3^A\}$, calculate $u = C_{\max}(E) = 15$, for all unscheduled B-agent's jobs $f_k^B(u) = 15 > U$. Therefore, we know that all A-agent's jobs have been scheduled and no B-agent's job can be scheduled in the last position, the instance is not feasible (U is too small).

4   Conclusions

In this paper, we combine two important issues in scheduling that recently have received increasing attention from researchers: deteriorating jobs and multiple agents. We study the problem that deteriorating jobs belonged to two agents are scheduled on a single machine. The goal is to find a schedule such that the objective function value of one agent is minimized, subject to the requirement that the other agent's objective is bounded by an upper bound U. We show that two scheduling problems $1|p_h^A = \alpha_h^A t, p_k^B = \alpha_k^B t|L_{\max}^A : C_{\max}^B \leq U$ and $1|p_h^A = \alpha_h^A t, p_k^B = \alpha_k^B t|\sum C_h^A : f_{\max}^B \leq U$ are polynomially solvable.

In the future research, other scheduling objectives such as number of late jobs, the total weighted completion time and tardiness, can be tested. Extending this problem to other time-dependent deteriorating model is also an interesting issue.

References

1. Agnetis, A., Mirchandani, P.B., Pacciarelli, D., Pacifici, A.: Scheduling Problems with Two Competing Agents. Operations Research 52, 229–242 (2004)
2. Agnetis, A., Pacciarelli, D., Pacifici, A.: Multi-agent Single Machine Scheduling. Annals of Operations Research 150, 3–15 (2007)
3. Alidaee, B., Womer, N.K.: Scheduling with Time Dependent Processing Times: Review and Extensions. Journal of the Operational Research Society 50, 711–720 (1999)
4. Baker, K.R., Smith, J.C.: A multiple-criterion Model for Machine Scheduling. Journal of Scheduling 6, 7–16 (2003)
5. Browne, S., Yechiali, U.: Scheduling Deteriorating Jobs on a Single Processor. Operations Research 38, 495–498 (1990)
6. Cheng, T.C.E., Ding, Q., Lin, B.M.T.: A Concise Survey of Scheduling with Time-Dependent Processing Times. European Journal of Operational Research 152, 1–13 (2004)
7. Cheng, T.C.E., Ng, C.T., Yuan, J.J.: Multi-agent Scheduling on a Single Machine to Minimize Total Weighted Number of Tardy Jobs. Theoretical Computer Science 362, 273–281 (2006)
8. Cheng, T.C.E., Ng, C.T., Yuan, J.J.: Multi-agent Scheduling on a Single Machine with Max-Form Criteria. European Journal of Operational Research 188, 603–609 (2008)
9. Graham, R.L., Lawler, E.L., Lenstra, J.K., Rinnooy Kan, A.H.G.: Optimization and Approximation in Deterministic Sequencing and Scheduling Theory: a Survey. Annals of Discrete Mathematics 5, 287–326 (1979)
10. Gupta, J.N.D., Gupta, S.K.: Single Facility Scheduling with Nonlinear Processing Times. Computers and Industrial Engineering 14, 387–393 (1988)
11. Mosheiov, G.: Scheduling Jobs under Simple Linear Deterioration. Computers and Operations Research 21, 653–659 (1994)
12. Ng, C.T., Cheng, T.C.E., Bachman, A., Janiak, A.: Three Scheduling Problems with Deteriorating Jobs to Minimize the Total Completion Time. Information Processing Letters 81, 327–333 (2002)
13. Wu, C.C., Lee, W.C.: Scheduling Linear Deteriorating Jobs to Minimize Makespan with an Availability Constraint on a Single Machien. Information Processing Letters 87, 89–93 (2003)
14. Yuan, J.J., Shang, W.P., Feng, Q.: A Note on the Scheduling with Two Families of Jobs. Journal of Scheduling 8, 537–542 (2005)

A Two-Stage Flexible Flowshop Problem with Deterioration

Hua Gong and Lixin Tang*

The Logistics Institute, Northeastern University, Shenyang, China
gonghua1018@sina.com,
lixintang@mail.neu.edu.cn

Abstract. We study a scheduling problem under considering deterioration on a two-stage flexible flowshop of particular structure (parallel machines in the first stage and a single batching machine in the second stage). The deterioration of a job means that its processing time on the batching machine is dependent on its waiting time, i.e., the time between the completion of the job in the first stage and the start of the job in the second stage. The objective is to minimize the makespan plus the total penalty cost of batching-machine utilization ratio. First, we prove the problem is strongly NP-hard. An efficient heuristic algorithm for the general problem is constructed and its worst-case bound is analyzed. Computational experiments show that the heuristic algorithm performs well on randomly generated problem instances.

1 Introduction

Job scheduling with waiting time constraint in a flexible flowshop environment is an important research topic. The motivation of this paper comes from the steel and iron industry, and in particular, steel ingot process. Teeming and soaking are two key operations in steel ingot process, which correspond to two stages of a flexible flowshop. A soaking pit is expensive that can simultaneously heat several ingots in a batch. If the waiting time of an ingot for the soaking pit is more than limited waiting time, then the soaking time of the ingot increases. In order to improve the utilization of the soaking pit and its benefit-to-cost ratio, it has to be scheduled carefully. This is the motivation for the considered problem.

In this paper, we address a scheduling problem that combines waiting-time dependent deterioration and a two-stage flexible flowshop with parallel machines in the first stage and a single batching machine in the second stage. The deterioration feature means that the processing time of each job on the batching machine is dependent on its waiting time between two stages. In the second stage, each job has a basic constant processing time if its waiting time is no more than limited waiting time, it is called hot job; otherwise it has an extended

---

* This work is partly supported by NSFC (Grant No. 70425003 and Grant No. 60674084), National 863 High-Tech Research and Development Program of China through approved No.2006AA04Z174 .

X. Hu and J. Wang (Eds.): COCOON 2008, LNCS 5092, pp. 651–660, 2008.

processing time and is called deteriorating job. The batching machine can process a number of jobs at a time. The processing time of one batch is equal to the largest processing time among all the jobs in the batch. The objective is to sequence and partition the jobs to minimize the sum of makespan and total penalty cost of batching-machine utilization ratio.

Flexible flowshop scheduling problems have been studied by a number of researchers. Riane et al. [6] study a three-stage hybrid flowshop problem with one machine in first and third stages and two dedicated machines in stage two. Soewandi and Elmaghraby [7] consider a three-stage flexible flowshop problem with identical machines in each stage and propose shop-partitioning approach and auxiliary process-based approach to develop heuristic procedures. Ahmadi et al. [1] study a class of scheduling problems defined in a two- or three-machine flowshop with at least one batching machine incorporated. Sung and Kim [8] consider a scheduling problem for a two-machine flowshop where a discrete processing machine is followed by a batching machine and a finite number of jobs arrives dynamically at the first machine.

Deterioration effect has also been observed in many production situations. Surveys on scheduling deteriorating jobs are given by Cheng et al. [2]. Leung et al. [4] consider the scheduling problem on identical parallel machines, where the jobs are processed in batches and the processing time of each job is a step function of its waiting time. This paper studies parallel machines rather than a batching machine.

However, to the best of our knowledge, no work has been done on models with both a two-stage flexible flowshop (identical parallel machines and single batching machine) and waiting-time-dependent deterioration. We feel that this is a probable scenario in real-life scheduling environments, and hence deserves some attention. When we consider batch together a set of jobs for processing in order to improve utilization ratio of the batching machine, it is likely to incur a long makespan. On the other hand, if we create too many batches, then we will end up with large penalty cost of machine utilization ratio. So batching is important. Hence, we have to find a feasible schedule that achieves some balance between makespan and total penalty cost of batching-machine utilization ratio.

This paper is organized as follows. In the next section we introduce the notation to be used. In section 3, we show the strong NP-hardness result. In section 4, we develop a heuristic algorithm for the general problem and analyze its worst-case performance. Computational results to demonstrate the effectiveness of the heuristic algorithm are reported in section 5. Finally, some conclusive remarks are made in last section.

2   Notation and Formulation

As mentioned above, we consider a two-stage flowshop schedule with parallel machines in the first stage and a single batching machine in the second stage, and the processing time of each job is dependent on its waiting time between two stages. The flowing assumptions are made for the flexible flowshop schedule.

(1) Every job has to be processed in the same technological order: first, the first stage and the second stage.

(2) At stage 1, a machine can process only one job at a time.

(3) At stage 2, the batching machine can process a batch of jobs simultaneously, and the processing time of a batch is equal to the longest processing time of the jobs assigned to it. Once the processing of a batch is initiated, it cannot be interrupted, nor can other jobs be introduced into the batch.

(4) All jobs are available simultaneously at time zero.

(5) Jobs are not preemptive.

(6) Each job completed at the first stage moves to the batching machine immediately. Transfer time between two stages is negligible.

(7) The waiting time of a job is the time that elapses between the completion of the job at stage 1 and the start of processing at stage 2.

To describe the problem, we introduce the following notation:

n:number of jobs;

m: number of machines in the first stage, it is fixed and not part of problem input;

$J=\{J_1, J_2, \ldots, J_n\}$: job set to be processed;

B: capacity of the batching machine;

p_{j1}: processing time of job J_j in the first stage;

w_j: waiting time of job J_j between two stages;

W: limited waiting time;

p_{j2}: processing time of job J_j in the second stage, $p_{j2} = \begin{cases} p_h, & w_j \leq W, \\ p_c, & w_j > W. \end{cases}$

b: number of batches to be processed in the second stage;

B_l: batch l;

b_l: number of jobs in batch B_l;

C_{ji}: completion time of job J_j at stage i , $i=1,2$;

$C_{\max}$: completion time of all jobs in the flowshop, i.e., makespan;

$\alpha(b)$: penalty cost of machine utilization ratio about the batching machine,

$$\alpha(b) = \beta \cdot \sum_{l=1}^{b} (1 - b_l/B);$$

Objective function: $F=C_{\max} + \alpha(b)$. Adopting the three-field notation introduced by Pinedo [5], we denote our problem as $Pm \rightarrow B|w_j|C_{\max} + \alpha(b)$.

3   Analysis of NP-Hardness

In this section, we consider the complexity issues of the problem. First, the following lemmas establish several properties for an optimal schedule for the problem. We show that the problem in general is strongly NP-hard by a reduction from 3-PARTITION, which is known to be NP-hard in the strong sense (see Garey and Johnson [3]).

Lemma 1. *There exists an optimal schedule for $Pm \rightarrow B|w_j|C_{\max} + \alpha(b)$ that satisfies the flowing properties at some time t:*

(1) If there are some jobs whose waiting time is no more than limited waiting time, then at stage 2 their schedule precedes the schedule of deteriorating jobs.

(2) The sequence of deteriorating jobs is immaterial at stage 2.

(3) The jobs whose waiting time is no more than limited waiting time are sequenced in non-decreasing order of their completion times at stage 1.

Proof: (1) The processing time of a deteriorating job has not change as its waiting time increases. On the other hand, for a hot job, the increase of the waiting time may result in deterioration as the job is not processed in time. Hence, we can perform simple job interchange without increasing the objective value.

(2) The property is proved by simple job interchange arguments.

(3) We know that the job sequence of non-decreasing arrival time order on the batching machine is same as the sequence of non-decreasing completion time order in the first stage. Let C_{j1} denote the completion time of job j in the first stage, for $j=1,\ldots,n$. Consider a schedule S^*, suppose it is not sequenced in non-decreasing completion time order to the hot jobs, then we have a job i is processed in batch B_{l+1} and another job j is processed in batch B_l such that $C_{i1} < C_{j1}$ and job i and job j are hot jobs at time t. Based on Lemma 1, we assume that the remaining jobs of batch B_l are the jobs whose waiting times are not more than the limited waiting time. Without loss of generality, suppose that the starting time of batch B_l on the batching machine is $S_{B_l} = t$. Then the completion time of batch B_l is $C_{B_l} = t + p_h$. This implies that the earliest possible starting time of batch B_{l+1} on the batching machine is $t+p_h$. We assume that $S_{B_{l+1}} =t+p_h$. It is easy to see that $w_i = t + p_h - C_{i1}$ and $w_j = t - C_{j1}$.

Consider a schedule S' that is created by interchanging jobs i and j, and forming $i \in B_l$ and $j \in B_{l+1}$. All other jobs are identical at the same time in S' as in S^*. We have the waiting time of job j in S' is $w'_j = t + p_h - C_{j1} < w_i$. If $w_i \leq W$ in S^*, then the objective value associated with S' is equal to the objective value associated with S^*. If $w_i > W$, then the completion time of batch B_{l+1} is $C_{B_{l+1}} = t + p_h + p_c$ in S^*. We prove the lemma by considering the two cases that arise through the possible relations of w'_j and W. Case 1: $w'_j = t + p_h - C_{j1} \leq W$. When all other jobs of the batch are hot jobs, then $C'_{B_{l+1}} = t+2p_h < C_{B_{l+1}}$. The makespan may be decreased in S'. The completion times of all jobs processed in S' are not affected by this interchange when batch contains at least a cold job. Case 2: $w'_j = t + p_h - C_{j1} > W$. In this case, the makespan in S' is equal to the makespan in S^* due to the deterioration of job j.

Hence, interchanging the jobs i and j will not increase the objective value. So we have shown that there exists an optimal schedule in which the hot jobs are sequenced in non-decreasing order of the arrival times on the batching machine at some time t.   □

Next we consider the complexity issue of the problem. First, it is interesting to note that, when the second stage is negligible, $Pm \rightarrow B|w_j|C_{\max} + \alpha(b)$ simply reduces to the classical parallel machine scheduling problem $Pm \parallel C_{\max}$. Since $Pm \parallel C_{\max}$ has been shown to be NP-complete in the ordinary sense when m is fixed [8]. We have $Pm \rightarrow B|w_j|C_{\max} + \alpha(b)$ is at least NP-hard. The following theorem states the computational complexity of the problem.

Theorem 1. *Problem $Pm \to B|w_j|C_{\max} + \alpha(b)$ is NP-hard in the strong sense even if* m=2.

Proof. To prove the theorem, we reduce the 3-Partition problem to decision version of problem $Pm \to B|w_j|C_{\max} + \alpha(b)$.

3-Partition problem: Given a set of $3h$ items $H = \{1, 2, \ldots, 3h\}$ and an integer a such that each item $j \in H$ has an integer size a_j satisfying $a/4 < a_j < a/2$ and $\sum\limits_{j=1}^{3h} a_j = ha$. Do there exist h disjoint subsets $H_1, H_2, \ldots, H_h$ of H such that each subset contains exactly three items and its total size $\sum\limits_{a_j \in H_i} a_j = a$ for $i = 1, 2, \ldots, h$?

Given an instance of 3-Partition problem, we construct the following instance of the proposed problem. There are $n = 6h$ jobs split into two groups: the X-jobs (partition jobs) denoted by X_j, $j = 1, 2, \ldots, 3h$, the Y-jobs (auxiliary jobs) denoted by Y_j, $j = 1, 2, \ldots, 3h$. The capacity of the batching machine is denoted by $B=6$. Their processing times are given by the formulas:

$$p_{X_j,1} = a_j \;, \; p_{Y_j,1} = a \;, p_{X_j,2} = p_{Y_j,2} = \begin{cases} 2a, & w_j \le a, \\ 2ha, & w_j > a. \end{cases}, j = 1, 2, \ldots, 3h.$$

The penalty cost is denoted by $\beta = 2a(h + 1)$, and threshold value is $Z = 2a(h + 1)$.

Note that, in above instance, it is easily seen that the smallest total processing time on each machine in the first stage is $2ha$ and the smallest processing time of each job on the batching machine in the second stage is $2a$. This implies that $2(h+1)a$ is a lower bound of the optimal objective value. Furthermore, there is no idle time on any one of machines in two stages. We show that for the constructed instance of our problem, a schedule S^* with $C_{\max} + \beta \sum\limits_{l=1}^{b}(1 - b_l/B) \le 2(h+1)a$ exists if and only if 3-Partition problem has a solution. This will then imply that the proposed scheduling problem is strongly NP-hard.

$\to$ Given a solution to the 3-Partition problem instance, $H_1, H_2, \ldots, H_h$, we construct a schedule for our problem as shown.

$$machine1 : Y_1, H_1(X), Y_2, H_2(X), \ldots, Y_h, H_h(X)$$

$$machine2 : Y_{h+1}, Y_{2h+1}, Y_{h+2}, Y_{2h+2}, \ldots, Y_{2h}, Y_{3h}.$$

Machine 1 of the first stage processes jobs Y_j in time-slots $[2(j-1)a, (2j-1)a]$ and jobs $H_j(X)$ in time-slots $[(2j-1)a, 2ja], j = 1, \ldots, h$. Machine 2 of the first stage processes in series jobs $Y_j, j = h+1, \ldots, 3h$. The batching machine processes batch $B_l = \{H_l(X), Y_l, Y_l + h, Y_l + 2h\}$ in time-slots $[2la, 2(l+1)a], l = 1, \ldots, h$. It is easy to see that the above schedule is feasible and the objective value is

$$C_{\max} + \beta \sum\limits_{l=1}^{h}(1 - b_l/B) = 2(h+1)a.$$

$\leftarrow$ Now suppose that there exists an optimal schedule S^* for the proposed problem whose objective value is no more than $2(h+1)a$. For schedule S^*, we know that the waiting time of any job is not exceed the limited waiting time a. Otherwise, the total completion time on the batching machine is greater than $2(h+1)a$, which is a contradiction. This means that if an auxiliary job and a partition job are in the same batch and assigned to the same machine in the first stage, then the auxiliary job precedes the partition job. Furthermore, the number of batches on the batching machine must be equal to h. First, the number of batches is at least h due to the machine capacity. Second, assume there are $h+1$ batches in schedule S^*. We obtain:

$$C_{\max} + \beta \sum_{l=1}^{h+1}(1 - b_l/B) > 2ha + 2(h+1)a.$$

Thus the jobs are grouped into exactly h batches such that each batch contains exactly six jobs.

With the above argument, we have that schedule S^* must contain exactly h batches, say $B_1, B_2, ..., B_h$, such that there are three partition jobs and three auxiliary jobs in each batch. Assume batch B_l is the first batch containing four or more auxiliary jobs. Then the waiting time of jobs of batch B_l is more than a, a contradiction. Besides, we conclude that each batch has the following schedule in the first stage: one auxiliary job and three partition jobs are processed on one machine, the auxiliary job precedes three partition jobs; two auxiliary jobs are processed on another machine. Now we prove that $\sum_{X_j \in B_l} p_{X_j,1} = a$, $l = 1, ..., h$.

If $\sum_{X_j \in B_l} p_{X_j,1} < a$, then $\sum_{X_j \in B_k} p_{X_j,1} > a$ for some $k \neq l$. This implies that the waiting time of the auxiliary job in batch B_k is greater than the limited waiting time. From the earlier discussion, we know this will happen. Thus $\sum_{X_j \in B_l} p_{X_j,1} = a$

must be true. This completes the proof.   □

Theorem 1 indicates that the existence of a polynomial time algorithm to solve our scheduling problem is unlikely. Therefore, developing fast heuristic algorithm for yielding near-optimal solutions is of great interest.

4   Heuristics and Performance Analysis

In this section, we give a heuristic algorithm for problem $Pm \rightarrow B|w_j|C_{\max}+\alpha(b)$. The heuristic algorithm is based on the following idea: schedule all jobs in longest processing time order at stage 1 and then a dynamic programming algorithm at stage 2.

Algorithm H

Step 1. Arrange the jobs in non-increasing order of their processing times of the first stage, and index the jobs in such a way that $p_{11} \geq p_{21} \geq ... \geq p_{n1}$. Then we get a list S_1.

Step 2. At stage 1, use the first available machine rule for S_1, where the first unscheduled job is scheduled to the earliest available machine of identical parallel machines until all the jobs are scheduled. For job j , computer the completion time C_{j1} at stage 1, $j = 1, ..., n$. Re-index the jobs in non-decreasing order of C_{j1}, then we get a list S_2.

Step 3. Let $b = \lceil n/B \rceil$. For S_2, form b batches to be processed on the batching machine at stage 2 and computer $\alpha(b) = \beta(\lceil n/B \rceil - n/B)$.

Step 4. Define $f(j, l)$ as the minimum makespan of a partial schedule containing the first j jobs $\{1, .., j\}$, where the last batch B_l contains jobs $i + 1, ..., j$ for $l = 1, ..., b$. Let $f(0, 0) = 0$.

Recurrence relations:

$$f(j, l) = \min_{\substack{0 < j - i \leq B \\ \lceil i/B \rceil \leq l - 1 \leq i \leq n}} \begin{cases} \max\{f(i, l - 1), C_{j1}\} + p_h, & w_{i+1} \leq W, \\ \max\{f(i, l - 1), C_{j1}\} + p_c, & w_{i+1} > W. \end{cases}$$

where $w_{i+1} = \max\{f(i, l - 1), C_{j1}\} - C_{i+1,1}$.

Step 5. Computer $C_{\max}^H = f(n, b)$ and $F^H = C_{\max}^H + \alpha(b)$.

Let $F^* = C_{\max}^* + \alpha(b^*)$ denote the objective value of the optimal schedule. Based on step 1, we assume that $P_1 = P_{n1} + P_{n-1,1} + ... + P_{n-\lceil n/m \rceil + 1,1}$. The following lemma provides a lower bound on $C_{\max}^*$.

Lemma 2. *The lower bound is derived as*

$$LB = \max \left\{ \max \left\{ \max_j \{p_{j1}\}, \frac{1}{m} \sum_{j=1}^{n} p_{j1}, P_1 \right\} + p_h, \min_j \{p_{j1}\} + \lceil n/B \rceil p_h \right\}.$$

Proof. Because the time required to complete all jobs at stage 1 is at least $\max \left\{ \max_j \{p_{j1}\}, \frac{1}{m} \sum_{j=1}^{n} p_{j1}, P_1 \right\}$, the first lower bound can be derived as the sum of the smallest completion time and the processing time of one batch. At stage 2, when the waiting time constraint is ignored, the processing time of each job on the batching machine is equal to p_h. So the second lower bound is derived as $\min_j \{p_{j1}\} + \lceil n/B \rceil p_h$. $\square$

Theorem 2. *The worst-case performance ratio of Algorithm H is less than or equal to $4/3 - 1/3m + p_c/p_h$.*

Proof. Ignoring the second stage, the flexible flowshop scheduling problem is reduced to problem $Pm \parallel C_{\max}$. Let $C_{\max}^0$ denote the minimum makespan for the problem with ignoring the second stage and $C_{\max}^*(P)$ denote the makespan of optimal schedule for problem $Pm \parallel C_{\max}$. We have

$$C_{\max}^0 / C_{\max}^*(P) \leq 4/3 - 1/3m.$$

For any schedule generated by Algorithm H, we modify this schedule by keeping the batching machine idle until all the jobs finish processing on parallel machines at stage 1. We consider a situation where each batch has at least one deteriorating job in any job schedule generated by Algorithm H. That is, the processing time of each batch is p_c. It is easy to see that the makespan of this schedule is equal to $C_{max}^0 + \lceil n/B \rceil p_c$. By Lemma 2, we have

$$\frac{C_{max}^H}{C_{max}^*} \le \frac{C_{max}^0 + \lceil n/B \rceil p_c}{C_{max}^*} \le \frac{C_{max}^0}{C_{max}^*(P)} + \frac{\lceil n/B \rceil p_c}{\lceil n/B \rceil p_h} \le \frac{4}{3} - \frac{1}{3m} + \frac{p_c}{p_h}.$$

In this heuristics, the number of batches is $b = \lceil n/B \rceil$, then the total penalty cost satisfies that $\alpha(b) = \beta(\lceil n/B \rceil - n/B) \le \alpha(b^*)$. We can obtain that

$$\frac{F^H}{F^*} \le \frac{C_{max}^H + \alpha(b)}{C_{max}^* + \alpha(b)} \le \frac{4}{3} - \frac{1}{3m} + \frac{p_c}{p_h},$$

where the last inequality comes from

$$\frac{x+z}{y+z} \le \frac{x}{y},$$

for any $z \ge 0$ and $0 < y \le x$. □

The worst case analysis we presented, provides an upper bound on the performance ratio of the heuristic algorithm. However, from an actual point of view it may be too pessimistic, since the worst case may be rarely occur.

5   Computational Experiments

To test the performance of our proposed heuristics, a set of computational experiments is conducted. The heuristic scheduling algorithm is coded in Visual C++ language and implemented on the computer with 512MB RAM and 256KB L2 cache. Test problems are randomly generated as follows:

(1) Number of jobs $n \in \{20, 50, 100, 200\}$, number of machines in the first stage $m \in \{2, 4, 6\}$, and capacity of the batching machine $B \in \{2, 4, 6, 10\}$.

(2) The processing times of jobs and the limited waiting time are generated from the following discrete uniform distributions:

$$p_{j1} \sim U[1, 10], \ p_{j2} \sim U[1, 20], \ W \sim U[20, 30] \text{ and } \beta = 10.$$

For each combination of number of jobs, number of machines in the first stage and capacity of batching machine, we randomly generate 10 problem instances. Total 480 instances are generated with 48 combinations. We report both average and maximum relative gaps (denoted by $avg(r)$ and $\max(r)$, respectively) on test problems, which are measured over the derived lower bound of the objective value, are used for the performance test of Algorithm H. The relative gap on a test problem is defined as $r = 100 * (F^H - F^{LB})/F^{LB}$, where $F^H = C_{max}^H + \alpha(b)$ and $F^{LB} = LB + \alpha(b)$.

Based on the results in Table 1, the average value of r among all test instances is 3.66, although the maximum value is 16.78. We observe that the effectiveness of proposed heuristics increases as n increases. The heuristic solutions seem

Table 1. Computational results

n	B	$m{=}2$ avg(r)	$m{=}2$ max(r)	$m{=}4$ avg(r)	$m{=}4$ max(r)	$m{=}6$ avg(r)	$m{=}6$ max(r)
20	2	5.05	6.42	4.67	6.40	3.78	5.78
	4	3.14	7.89	8.92	10.22	8.10	10.38
	6	1.19	4.20	6.11	16.78	8.63	11.84
	10	1.35	6.02	1.17	4.80	4.88	12.08
50	2	2.29	2.77	2.16	2.79	2.10	2.74
	4	3.33	15.03	4.48	5.84	4.31	5.12
	6	2.49	6.60	9.48	15.00	6.01	8.13
	10	0.33	1.49	3.55	7.90	3.46	6.97
100	2	1.24	1.68	1.26	1.72	1.03	1.43
	4	1.58	5.88	2.40	3.28	2.14	2.56
	6	2.57	4.99	3.49	5.56	3.19	4.88
	10	0.38	1.39	0.46	1.38	0.74	1.91
200	2	0.64	0.87	0.72	0.88	0.64	0.88
	4	6.70	9.04	1.26	1.56	1.32	1.71
	6	1.97	7.80	2.63	5.77	1.69	2.57
	10	0.26	1.04	0.89	1.43	0.73	1.15

to be greatly dependent upon the relation between m and B. In addition, for $B{=}2$, we observe that the maximum relative gap is close to the average relative gap. These results demonstrate that the heuristics is capable of generating near-optimal solutions within a reasonable amount of CPU time.

6   Conclusions

In this paper, we study a two-stage flexible flowshop scheduling problem with deterioration in which there are identical parallel machines in the first stage and a single batching machine in the second stage, and the processing time of a job on the batching machine is dependent on its waiting time between two stages. We show that the problem is strongly NP-hard. We develop an efficient heuristics along with a worst-case performance ratio for the general problem. The computational experiments show that the proposed heuristic algorithm is effective in practice.

A number of important issues remain open for future investigation. One extension is to analyze the problems include multiple batching machines and other objective functions. Another interesting research direction is to develop problem with the linear deteriorating processing times.

References

1. Ahmadi, J.H., Ahmadi, R.H., Dasu, S., Tang, C.S.: Batching and Scheduling Jobs on Batch and Discrete Processors. Operations Research 39, 750–763 (1992)
2. Cheng, T.C.E., Ding, Q., Lin, B.M.T.: A Concise Survey of Scheduling with Time-Dependent Processing Times. European Journal of Operational Research 152, 1–13 (2004)

3. Garey, M.R., Johnson, D.S.: Computers and Intractability: A Guide to the Theory of NP-Completeness. W. H. Freeman and Company, New York (1979)
4. Leung, T.T., Ng, C.T., Cheng, T.C.E.: Minimizing Sum of Completion Times for Batch Scheduling of Jobs with Deteriorating Processing Times. European Journal of Operational Research 187, 1090–1099 (2008)
5. Pinedo, M.: Scheduling: Theory, Algorithm and Systems. Prentice-Hall, Englewood Cliffs (1995)
6. Riane, F., Artiba, A., Elmaghraby, S.E.: A Hybrid Three-stage Flowshop Problem: Efficient Heuristics to Minimize Makespan. European Journal of Operational Research 109, 321–329 (1998)
7. Soewandi, H., Elmaghraby, S.E.: Sequencing Three-stage Flexible Flowshops with Identical Machines to Minimize Makespan. IIE Transactions 33, 985–993 (2001)
8. Sung, C.S., Kim, Y.H.: Minimizing Makespan in a Two-machine Flowshop with Dynamic Arrivals Allowed. Computers Operations Research 29, 275–294 (2002)

A Lower Bound for the On-Line Preemptive Machine Scheduling with ℓ_p Norm

Tianping Shuai[1],[*] and Donglei Du[2],[**]

[1] School of Science, Beijing University of Posts and Telecommunications, China
shuaitp@gmail.com
[2] Faculty of Business Administration, University of New Brunswick, Canada
ddu@unb.ca

Abstract. We consider the on-line version of the preemptive scheduling problem that minimizes the machine completion time vector in the ℓ_p norm (a direct extension of the ℓ_∞ norm: the makespan) on m parallel identical machines. We present a lower bound on the competitive ratio of any randomized on-line algorithm with respect to the general ℓ_p norm. This lower bound amounts to calculating a (non-convex) mathematical program and generalizes the existing result on makespan. While similar technique has been utilized to provide the best possible lower bound for makespan, the proposed lower bound failed to achieve the best possible lower bound for general ℓ_p norm (though very close), and hence revealing intricate and essential difference between the general ℓ_p norm and the makespan.

1 Introduction

We consider the following scheduling problem. We are given m machines and a sequence of jobs. In the variation with preemption any job may be divided into several pieces that may be processed on several machines; in addition, the time slots assigned to different pieces must be disjoint. The objective is to find a schedule which minimizes the ℓ_p norm of the machines' completion times, that is, $\sqrt[p]{L_1^p + L_2^p + \cdots + L_m^p}$, where $L_i, i = 1, 2, ..., m$ are the completion times of the last job on M_i. We denote this problem as $Pm|pmtn, online - list|\ell_p$. The off-line version of problem, where the full information (number of jobs and sizes of jobs) on all the jobs is known in advance, can be denoted as $Pm|pmtn|\ell_p$ using the three-field notation in [14].

The makespan, the most studied objective function in scheduling, is just a special ℓ_p norm—that is—the ℓ_∞. Scheduling in the *general* ℓ_p norm has also been widely studied in the literature [1],[2],[3],[4],[5],[6],[7],[9],[12],[13],[15],[16], [18], and the ℓ_p norm of the completion time vector is one of the basic and fundamental objectives investigated in scheduling theory. While the makespan

---

[*] This work was done while the author was visiting the Faculty of Business Administration, University of New Brunswick. Supported in part by NSFC grant 10726058.
[**] Supported in part by NSERC grant 283103 and URF, FDF at UNB.

X. Hu and J. Wang (Eds.): COCOON 2008, LNCS 5092, pp. 661–669, 2008.

only characterizes the latest completion time among all machines, the general ℓ_p norm is more appropriate when we are interested in the average behavior of the machine completion times rather than the worst-case scenario. A particular application of ℓ_p norm scheduling in disk storage allocation problem is illustrated in [7].

The main focus of this work is on the lower bound analysis of the competitive ratio of the on-line scheduling problem introduced in the beginning. The quality of an on-line algorithm is measured by its *competitive ratio* r_{ON}, which is defined to be the superimum of ratio C_{ON}/C_{OFF} over all problem instances, where C_{ON} and C_{OFF} denote respectively the ℓ_p norm of machine completion time vector of the on-line schedule constructed by ON and that of the corresponding (off-line) optimal schedule.

Related work. Scheduling in the ℓ_p norm has been investigated from both on-line (and semi-on-line) [2],[7],[9],[16],[15],[18] and off-line [1],[3],[4],[5],[6],[12], [13] points of view. But we only review the results related to the on-line case as it is the main focus of this work. Readers who are interested in the off-line case are referred to the references listed here and further pointers therein.

Directly related to the current work is [9], in which a best possible on-line algorithm for $P2|pmtn, online - list|\ell_p$ is developed. Moreover, several semi-online models of $P2||\ell_p$, and $Q2||\ell_p$ are studied (both preemptive and non-preemptive versions) in [7],[15],[18].

Our results. The main contribution of this work is to provide a lower bound for the competitive ratio of any randomized on-line algorithm for the problem $Pm|pmtn, online - list|\ell_p$, generalizing existing result for makepsan [8]. We adopt a similar lower bounding technique developed in [11],[10] for the uniform machines scheduling $Qm|pmtn, online - list|C_{\max}$. The proposed lower bound amounts to calculating a (non-convex) mathematical program. While the technique of [11],[10] has been utilized to provide the best possible lower bound for makespan, our lower bound derived from the similar technique (with more involved technical development) fails to achieve the best possible lower bound for general ℓ_p norm as shown in this paper (though very close), and hence revealing intricate and essential difference between the general ℓ_p norm and the makepsan. This leaves the obvious open question on finding the best possible lower bound for the ℓ_p norm problem by either extending the existing lower bounding techniques or developing more powerful new methods.

The rest of the paper is organized as follows. After some preliminary results in Section 2, we present our lower bound in Section 3. Finally, a solution for $m = 2$ is illustrated in Section 4.

2   Preliminaries and Notations

We will need the following closed-form formula of the off-line optimal objective value for the problem $Pm|pmtn|\ell_p$. This formula is explicitly implied by a general result in [12]. For the special case of the identical parallel machines, a much simplified proof is possible, which is offered here for completeness.

Lemma 1. [12] *For any given instance of the problem $Pm|pmtn|\ell_p$ with m machines and n jobs $\mathcal{J} = \{p_n, \ldots, p_1\}$ such that $p_n \leq \ldots \leq p_1$, the optimal objective value $\mathrm{OPT}(m, \mathcal{J})$ satisfies:*

$$\mathrm{OPT}(m, J) = \left(\max\left\{ p_1, \frac{\sum_{i=1}^n p_i}{m} \right\}^p + \max\left\{ \min\left\{ \frac{\sum_{i=1}^n p_i}{m}, \frac{\sum_{i=2}^n p_i}{m-1} \right\}, p_2 \right\}^p \right.$$

$$+ \ldots + \max\left\{ \min\left\{ \frac{\sum_{i=1}^n p_i}{m}, \frac{\sum_{i=2}^n p_i}{m-1}, \ldots, \frac{\sum_{i=m-1}^n p_i}{2} \right\}, p_{m-1} \right\}^p$$

$$\left. + \min\left\{ \frac{\sum_{i=1}^n p_i}{m}, \frac{\sum_{i=2}^n p_i}{m-1}, \ldots, \frac{\sum_{i=m-1}^n p_i}{2}, \sum_m^n p_i \right\}^p \right)^{1/p} \quad (1)$$

Proof. The proof is by induction on m, the number of machines. The formula is obviously true for $m = 1$. For $m = 2$, if $p_1 \leq \sum_{i=1}^n p_i/2$, then the optimal solution given by McNaughton's Warp-around algorithm [17] for makespan is also an optimal solution in the ℓ_p norm due to the convexity of ℓ_p norm. For this case, all the machine loads are equal to $\sum_{i=1}^n p_i/2$, implying that the objective value is $\sqrt[p]{2}(\sum_{i=1}^n p_i/2)$ and

$$\mathrm{OPT}(2, J) = \sqrt[p]{\left(\max\left\{ \frac{1}{2}\sum_{i=1}^n p_i, p_1 \right\} \right)^p + \left(\min\left\{ \frac{1}{2}\sum_{i=1}^n p_i, \sum_{i=2}^n p_i \right\} \right)^p}$$

$$= \sqrt[p]{2}\left(\frac{1}{2}\sum_{i=1}^n p_i \right).$$

If $p_1 > \sum_{i=1}^n p_i/2$, then in any optimal solution, p_1 must be assigned entirely to one machine and all remaining jobs are assigned to another machine, implying that

$$\sqrt[p]{p_1^p + \left(\sum_{i=2}^n p_i \right)^p} = \sqrt[p]{\left(\max\left\{ p_1, \frac{1}{2}\sum_{i=1}^n p_i \right\} \right)^p + \left(\min\left\{ \frac{1}{2}\sum_{i=1}^n p_i, \sum_{i=2}^n p_i \right\} \right)^p}.$$

Next suppose that the formula were true for $m - 1$ machines, we shall show that it is also true for m machines.

(i) If $p_1 \leq \sum_{i=1}^n p_i/m$, then the optimal solution given by McNaughton for makespan is also the optimal solution in the ℓ_p norm by the convexity of the objective function. For this case, all the machine loads are equal to $\sum_{i=1}^n p_i/m$, implying that the objective value is $\sqrt[p]{m}(\sum_{i=1}^n p_i/m)$. Next, consider each term in the formula (1). It is easy to prove that $\sum_{i=1}^n p_i/m \geq p_i, i = 1, 2, \ldots, n$, and $\sum_{i=1}^n p_i/m \leq \sum_{i=k}^n p_i/(m - k + 1), k = 1, 2, \ldots, m - 1$ since $p_1 \leq \sum_{i=1}^n p_i/m$ and $p_1 \geq p_2 \geq \cdots \geq p_n$. So each term in (1) is equal to $(\sum_{i=1}^n p_i/m)^p$, implying that (1) is just equal to $\sqrt[p]{m}(\sum_{i=1}^n p_i/m)$ and we are done.

(ii) If $p_1 > \sum_{i=1}^n p_i/m$, then in any optimal solution, p_1 must be assigned entirely on one machine, and $\text{OPT}^p(J) = p_1^p + \text{OPT}^p(m-1, J')$, where $J' = J \setminus \{p_1\}$ and $\text{OPT}(m-1, J')$ is an optimal value for $m-1$ machines. The rest of argument is similar to (i). Consider each term in the formula (1). The first term is p_1^p, and deleting $\sum_{i=1}^n p_i/m$ from each term results in the sum of the last $m-1$ terms being equal to $\text{OPT}^p(m-1, J')$ due to $\sum_{i=1}^n p_i/m \geq \sum_{i=k}^n p_i/(m-k+1), k = 1, 2, ..., m-1$. We are done.

3   Lower Bound

Consider the problem $Pm|pmtn, online - list|\ell_p$. Let $J = (p_n, \ldots, p_1)$ be a sequence of jobs arriving in that order. Denote $\text{OPT}(J)$ as the optimal (off-line) objective value for input J. For a given randomized algorithm A, let $A(J)$ be the objective value of the schedule generated on input J by A. Algorithm A is σ-competitive if, $E[A(J)] \leq \sigma \text{OPT}(J)$ for any sequence J, where $E[A(J)]$ denotes the expected objective value of the schedule generated by A. The best possible competitive ratio among all randomized algorithm will be denoted as R for the problem $Pm|pmtn, online - list|\ell_p$.

For any $i = 1, \ldots, n$, we denote $J_i = (p_n, \ldots, p_i)$. Let $\text{OPT}(J_i)$ be the optimal objective value of the instance of the problem $Pm|pmtn, online - list|\ell_p$ with job sequence J_i. Let $L = \sum_{i=1}^n p_i$.

Theorem 1. *For any job sequence $J_n = \{p_n, \ldots, p_1\}$ such that $p_{m-1} \leq \ldots \leq p_1$, the best possible competitive ratio R among all randomized algorithm must satisfy*

$$\sqrt[p]{L^p + (L - p_1)^p + (L - p_1 - p_2)^p + \cdots + \left(L - \sum_{i=1}^{m-1} p_i\right)^p} \leq R \sum_{i=1}^m \text{OPT}(J_i) \quad (2)$$

Proof. Fix a sequence of random bits used by random algorithm A. Since we are considering the best possible randomized algorithm, we can assume that there is no unenforced idle time before the last job in the schedule generated by A. Suppose that L_k $(k = 1, \ldots, m)$ is the load on machine k after all jobs are scheduled by algorithm A, where $L_k \leq L_{k+1}$. Let T_i $(i = 1, \ldots, m)$ be the resultant objective value by algorithm A on job sequence J_i.

First we claim that, for any i $(i = 1, \ldots, m)$,

$$\sqrt[p]{\sum_{j=1}^{m-i+1} L_j^p + (L_{m-i+2} - p_{i-1})^p + (L_{m-i+3} - p_{i-2})^p + \cdots + (L_m - p_1)^p} \leq T_i. \quad (3)$$

Since the algorithm is on-line, the schedule for J_i is obtained from the schedule for J_n by removing the last $i-1$ jobs. After all jobs are assigned, the machines loads are $L_1 \leq L_2 \leq \cdots \leq L_m$. The last $m-1$ jobs being nondecreasing and the objective function being convex imply that the objective value produced by the algorithm is at least $L_1^p + L_2^p + \cdots + L_{m-i+1}^p + (L_{m-i+2} - p_{i-1})^p + (L_{m-i+3} - p_{i-2})^p + \cdots + (L_m - p_1)^p$, and hence (3) holds.

By (3), we have

$$\sum_{i=1}^{m} \sqrt[p]{\sum_{j=1}^{m-i+1} L_j^p + (L_{m-i+2} - p_{i-1})^p + (L_{m-i+3} - p_{i-2})^p + \cdots + (L_m - p_1)^p} \leq \sum_{i=1}^{m} T_i.$$

Using Minkovski's inequality, we have

$$\sum_{i=1}^{m} \sqrt[p]{\sum_{j=1}^{m-i+1} L_j^p + (L_{m-i+2} - p_{i-1})^p + (L_{m-i+3} - p_{i-2})^p + \cdots + (L_m - p_1)^p}$$
$$\geq \sqrt[p]{L^p + (L - p_1)^p + (L - p_1 - p_2)^p + \cdots + (L - \sum_{i=1}^{m-1} p_i)^p}.$$

Algorithm A being R-competitive implies that $E[A(J)] \leq R \cdot \text{OPT}(J)$ and the theorem follows.

Theorem 1 implies that a lower bound on R can be obtained by formulating a mathematic program, which is explained below. Note that

$$R \geq \frac{\sqrt[p]{L^p + (L - p_1)^p + (L - p_1 - p_2)^p + \cdots + (L - \sum_{i=1}^{m-1} p_i)^p}}{\sum_{i=1}^{m} \text{OPT}(J_i)}$$

We can normalize the jobs such that $\sum_{i=1}^{m} \text{OPT}(J_i) = 1$, and therefore

$$R^p \geq L^p + (L - p_1)^p + (L - p_1 - p_2)^p + \cdots + \left(L - \sum_{i=1}^{m-1} p_i\right)^p$$

Now it is easy to given a mathematical program to compute the above lower bound. The program has variables $q_1, q_2, ..., q_m, O_1, O_2, ..., O_m$. Variable q_1 corresponds to the sum of all processing times in J_m, variables $q_2, ..., q_m$ to the processing times of the last $m-1$ jobs, and variables O_i correspond to $\text{OPT}(J_{m-i+1})$.

Definition 1. *Let r be the value of the objective function of the optimal solution of the following program:*

$$\max \quad r^p = q_1^p + (q_1 + q_2)^p + \cdots + (q_1 + q_2 + \cdots + q_m)^p$$

$$\frac{q_1^p}{m^{p-1}} \leq O_1^p$$

$$\max\left\{\frac{q_1 + q_2}{m}, q_2\right\}^p + \min\left\{\frac{q_1 + q_2}{m}, \frac{q_1}{m-1}\right\}^p (m-1) \leq O_2^p$$

$$\max\left\{\frac{q_1 + q_2 + q_3}{m}, q_3\right\}^p + \max\left\{\min\left\{\frac{q_1 + q_2 + q_3}{m}, \frac{q_1 + q_2}{m-1}\right\}, q_2\right\}^p$$
$$+ \min\left\{\frac{q_1 + q_2 + q_3}{m}, \frac{q_1 + q_2}{m-1}, \frac{q_1}{m-2}\right\}^p (m-2) \leq O_3^p$$

$$\vdots$$

$$\max\left\{\frac{q_1 + \cdots + q_m}{m}, q_m\right\}^p$$

$$+\max\left\{\min\left\{\frac{q_1 + \cdots + q_m}{m}, \frac{q_1 + \cdots + q_{m-1}}{m-1}\right\}, q_{m-1}\right\}^p + \cdots$$

$$+\max\left\{\min\left\{\frac{q_1 + \cdots + q_m}{m}, \frac{q_1 + \cdots + q_{m-1}}{m-1}, \cdots, \frac{q_1 + q_2}{2}\right\}, q_2\right\}^p$$

$$+\min\left\{\frac{q_1 + \cdots + q_m}{m}, \frac{q_1 + \cdots + q_{m-1}}{m-1}, \cdots, \frac{q_1 + q_2}{2}, q_1\right\}^p \le O_m^p$$

$$O_1 + O_2 + \cdots + O_m = 1$$
$$O_i \ge 0, i = 1, 2, \cdots, m$$
$$q_i \le q_{i+1}, i = 2, 3, \cdots, m$$
$$q_i \ge 0, i = 1, 2, \cdots, m$$

The above program has a feasible solution with the only non-zero variable $O_m = 1$. It is also easy to see that the objective function is bounded. The constraints imply that $(q_1 + q_2 + \cdots + q_i)^p \le m^{p-1}O_i^p$ for each i and $q_1^p + (q_1 + q_2)^p + \cdots + (q_1 + q_2 + \cdots + q_m)^p \le m^{p-1}\sum_{i=1}^m O_i^p \le m^{p-1}(O_1 + O_2 + \cdots + O_m)^p = m^{p-1}$. Thus the value r is well defined.

Theorem 2. *Any randomized on-line algorithm for m identical machines has competitive ratio at least r.*

Proof. There exist optimal solutions $q_1^*, q_2^*, ..., q_m^*, O_1^*, O_2^*, ..., O_m^*$. We create instance I as follows: The first m jobs have processing times $p_1 = p_2 = \cdots = p_m = q_1^*/m$. The remaining $m-1$ jobs have processing times $p_{m+1} = q_2^*, p_{m+2} = q_3^*, \cdots, p_{2m-1} = q_m^*$. By lemma 1,

$$\text{OPT}(I_{m-i+1}) = \max\left\{\frac{p_1 + \cdots + p_{m+i}}{m}, p_{m+i}\right\}^p$$

$$+ \max\left\{\min\left\{\frac{p_1 + \cdots + p_{m+i}}{m}, \frac{p_1 + \cdots + p_{m+i-1}}{m-1}\right\}, p_{m+i-1}\right\}^p$$

$$+ \cdots$$

$$+ \max\left\{\min\left\{\frac{p_1 + \cdots + p_{m+i}}{m}, \cdots, \frac{p_1 + \cdots + p_{m+1}}{m-i+1}\right\}, p_{m+1}\right\}^p$$

$$+ \min\left\{\frac{p_1 + \cdots + p_{m+i}}{m}, \cdots, \frac{p_1 + \cdots + p_m}{m-i}\right\}^p (m-i)$$

and the constraints of the program implies that $O_i^* \ge \text{OPT}(I_{m-i+1})$. Hence, by Theorem 1, we have

$$\frac{\sqrt[p]{q_1^{*p} + q_2^{*p} + \cdots + q_m^{*p}}}{\sum_{i=1}^m C_{\text{OPT}}(I_{m-i+1})} \ge \frac{\sqrt[p]{q_1^{*p} + q_2^{*p} + \cdots + q_m^{*p}}}{\sum_{i=1}^m O_i^*} = r$$

4 Solutions for $m = 2$

The mathematical program in Definition 1 is non-convex for general m. Below we shall show that, when $m = 2$, we can decompose the original program into two convex sub-programs, and hence the best bound is obtained by $r = \max\{r_1, r_2\}$:

$$\max \ r_1^p = q_1^p + (q_1 + q_2)^p$$
$$q_1^p \le 2^{p-1}O_1^p$$
$$(q_1 + q_2)^p \le 2^{p-1}O_2^p$$
$$O_1 + O_2 = 1$$
$$q_1 \ge q_2$$
$$O_i \ge 0, \quad i = 1, 2.$$
$$q_i \ge 0, \quad i = 1, 2.$$

$$\max \ r_2^p = q_1^p + (q_1 + q_2)^p$$
$$q_1^p \le 2^{p-1}O_1^p$$
$$q_1^p + q_2^p \le O_2^p$$
$$O_1 + O_2 = 1$$
$$q_1 \le q_2$$
$$O_i \ge 0, \quad i = 1, 2.$$
$$q_i \ge 0, \quad i = 1, 2.$$

Note that for each of the above two programs, the first two inequalities must be *tight* (satisfied as equality) in any optimal solutions, otherwise we can increase the objective value by either increasing q_2 when the second inequality is non-tight, or decreasing O_1, increasing q_2 and O_2 when the first inequality is non-tight. So, the above programs can reformulate as follows.

$$\max \ r_1^p = q_1^p + (q_1 + q_2)^p$$
$$q_1^p = 2^{p-1}O_1^p$$
$$(q_1 + q_2)^p = 2^{p-1}O_2^p$$
$$O_1 + O_2 = 1$$
$$q_1 \ge q_2$$
$$O_i \ge 0, \quad i = 1, 2.$$
$$q_i \ge 0, \quad i = 1, 2.$$

$$\max \ r_2^p = q_1^p + (q_1 + q_2)^p$$
$$q_1^p = 2^{p-1}O_1^p$$
$$q_1^p + q_2^p = O_2^p$$
$$O_1 + O_2 = 1$$
$$q_1 \le q_2$$
$$O_i \ge 0, \quad i = 1, 2.$$
$$q_i \ge 0, \quad i = 1, 2.$$

For the first program, it is easy to see that the optimal solution is $q_1 = 1, q_2 = 0, O_1 = O_2 = 1/2$ and the objective value is 1. For the second program, letting $t = O_1$, leads to a two-stage problem

$$\max_{t \in [0,1]} f(t), \tag{4}$$

where $f(t)$ is the objective function of the following parameterized problem:

$$f(t) = \max \ q_1^p + (q_1 + q_2)^p$$
$$q_1^p = 2^{p-1}t^p$$
$$q_1^p + q_2^p = (1 - t)^p$$
$$q_1 \le q_2$$
$$q_i \ge 0, \ i = 1, 2.$$

Case 1: $q_1 = q_2$. By direct computation, $t = 1/3$ and $f(t) = (2^p + 1)2^{p-1}/3^p$.

Case 2: $q_1 < q_2$. Then $q_1 = \sqrt[p]{2^{p-1}t}, q_2 = \sqrt[p]{(1-t)^p - 2^{p-1}t^p}$. So

$$f(t) = 2^{p-1}t^p + (2^{p-1}t^p + \sqrt[p]{(1-t)^p - 2^{p-1}t^p})^p.$$

Consider the equation $f'(t) = 0$; that is,

$$(2t)^{p-1}\left((1-t)^p - 2^{p-1}t^p\right)\left(1 + \left(2^{p-1}t^p + \sqrt[p]{(1-t)^p - 2^{p-1}t^p}\right)^{p-1}\right) =$$

$$\left(2^{p-1}t^p + \sqrt[p]{(1-t)^p - 2^{p-1}t^p}\right)^{p-1}\left((1-t)^{p-1} - (2t)^{p-1}\right)\sqrt[p]{(1-t)^p - 2^{p-1}t^p}. \quad (5)$$

The solution of the desired problem (4) is obtained by first calculating the roots of $f'(t) = 0$ and choose the best one.

As an example, when $p = 2$, the equation (5) is

$$2(3t - 1) + \frac{2\sqrt{2}(1 - 3t - 2t^2)}{\sqrt{1 - 2t - t^2}} = 0, \quad (6)$$

which can be simplified as follows:

$$17t^4 + 36t^3 - 10t^2 - 4t + 1 = 0,$$

resulting in four roots: $t_1 = -0.3413, t_2 = 0.2535, t_3 = 0.2927, t_4 = -2.3226$. Among which, only t_1 and t_2 are the solution to the equation (6), and t_2 is the maximizer of the problem (4).

For $p = 2, 3, 4, 5$, we obtain the following results. Compared with optimal competitive ratio from [9], we found that this bound is very close to the optimal bound and approaches the best bound when $p \to \infty$.

Table 1. Results on lower bounds and optimal bounds

	$p = 2$	$p = 3$	$p = 4$	$p = 5$
This paper	1.07483	1.1267	1.16358	1.19055
Optimal [9]	1.0758	1.1272	1.1638	1.19066

From the above analysis, we know that the bound given by the above program is not tight in general, and hence revealing intricate and essential difference between the general ℓ_p norm and the makepsan, and moreover, leaving the obvious open question on finding the best possible lower bound for the ℓ_p norm problem by either extending the existing lower bounding techniques or developing more powerful new methods.

References

1. Alon, N., Azar, Y., Woeginger, G.J., Yadid, T.: Approximation Schemes for Scheduling. In: SODA 1997, pp. 493–500 (1997)
2. Avidor, A., Azar, Y., Sgall, J.: Ancient and New Algorithms for Load Balancing in the ℓ_p Norm. Algorithmica 29, 422–441 (2001)

3. Azar, Y., Epstein, A.: Convex programming for Scheduling Unrelated Parallel Machines. In: STOC 2005, pp. 331–337 (2005)
4. Azar, Y., Epstein, A., Epstein, L.: Load Balancing of Temporary Tasks in the ℓ_p Norm. Theoretical Computer Science 361(2-3), 314–328 (2006)
5. Azar, Y., Epstein, L., Richter, Y., Woeginger, G.J.: All-Norm Approximation Algorithms. In: Penttonen, M., Schmidt, E.M. (eds.) SWAT 2002. LNCS, vol. 2368, pp. 288–297. Springer, Heidelberg (2002)
6. Azar, Y., Taub, S.: All-Norm Approximation for Scheduling on Identical Machines. In: Hagerup, T., Katajainen, J. (eds.) SWAT 2004. LNCS, vol. 3111, pp. 298–310. Springer, Heidelberg (2004)
7. Chandra, A.K., Wong, C.K.: Worst-Case Analysis of a Placement Algorithm Related to Storage Allocation. SIAM Journal on Computing 1, 249–263 (1975)
8. Chen, B., van Vliet, A., Woeginger, G.J.: An Optimal Algorithm for Preemptive On-Line Scheduling. Operations Research Letters 18(3), 127–131 (1995)
9. Du, D.-L., Jiang, X., Zhang, G.: Optimal Preemptive Online Scheduling to Minimize ℓ_p Norm on Two Processors. Journal of Manufacturing and Management Optimization 1(3), 345–351 (2005)
10. Ebenlendr, T., Jawor, W., Sgall, J.: Preemptive Online Scheduling: Optimal Algorithms for All Speeds. In: Azar, Y., Erlebach, T. (eds.) ESA 2006. LNCS, vol. 4168, pp. 327–339. Springer, Heidelberg (2006)
11. Epstein, L., Sgall, J.: A Lower Bound for On-Line Scheduling on Uniformly Related Machines. Operations Research Letters 26(1), 17–22 (2000)
12. Epstein, L., Tassa, T.: Optimal Preemptive Scheduling for General Target Functions. Journal of Computer and System Sciences 72(1), 132–162 (2006)
13. Kumar, V.S.A., Marathe, M.V., Parthasarathy, S., Srinivasan, A.: Approximation Algorithms for Scheduling on Multiple Machines. In: FOCS 2005, pp. 254–263 (2005)
14. Lawler, E.L., Lenstra, J.K., Rinnooy Kan, A.H.G., Shmoys, D.B.: Sequencing and Scheduling: Algorithms and Complexity. In: Graves, S.C., Rinnooy Kan, A.H.G., Zipkin, P.H. (eds.) Logistics of Production and Inventory, pp. 445–522. North-Holland, Amsterdam (1993)
15. Lin, L.: Semi-Online Scheduling Algorithm under the ℓ_p Norm on Two Identical Machines. Journal of Zhejiang University (Science Edition) 34(2), 148–151 (2007) (in Chinese)
16. Lin, L., Tan, Z.Y., He, Y.: Deterministic and Randomized Scheduling Problems under the ℓ_p Norm on Two Identical Machines. Journal of Zhejiang University Science 6(1), 20–26 (2005)
17. McNaughton, R.: Scheduling with Deadlines and Loss Functions. Management Science 6(1), 1–12 (1959)
18. Tan, Z., He, Y., Epstein, L.: Optimal On-line Algorithms for the Uniform Machine Scheduling Problem with Ordinal Data. Information and Computation 196(1), 57–70 (2005)

The Coordination of Two Parallel Machines Scheduling and Batch Deliveries

Hua Gong[1,2] and Lixin Tang[1,*]

[1] The Logistics Institute, Northeastern University, Shenyang, China
lixintang@mail.neu.edu.cn
[2] College of Science, Shenyang Ligong University, Shenyang, China
gonghua1018@sina.com

Abstract. In this paper, we consider a two-identical-parallel-machine scheduling problem with batch delivery. Although NP-hardness of this problem has been already proved by Hall and Potts, but they did not give the pseudo-polynomial-time algorithm for the problem. We prove the problem is NP-hard in the ordinary sense by constructing a pseudo-polynomial-time algorithm for the problem. We also give a polynomial-time algorithm to solve the case when the job assignment to parallel machines is given.

1 Introduction

In this paper, we investigate a coordinated scheduling problem with two identical parallel machines and batch delivery, which is first proposed by Hall and Potts [4], and can be described as follows. We are given n jobs $\{J_1, J_2, ..., J_n\}$ which must be first non-preemptively processed on two identical parallel machines ($M1$ and $M2$) and then delivered to the customer in batches. Job J_j needs a processing time of p_j on the machine where we let $P = \sum_{j=1}^{n} p_j$. All jobs delivered together in one shipment are defined as a delivery batch. Let D denote the nonintegative cost of delivering a batch and y denote the number of deliveries in the schedule. Define T as the minimum time between any two consecutive deliveries, where we assume that $T \leq P$. The value T of is expressed as the time amount that it takes to load a transporter if there is a fleet of transporters, or as the transportation time to and from the customers plus loading time if there is only one single transporter. We also define c_j as the completion time of job J_j on the parallel machines, and C_j as the time at which job J_j is delivered to its customer. The goal is to find a schedule for processing jobs on parallel machines and delivering finished jobs to the customer such that the sum of total delivery cost and job completion times is minimized. Adopting the notation introduced by Hall and Potts [4], we denote the problem as $P2|T| \sum C_j + Dy$.

---

* This work is partly supported by NSFC (Grant No. 70425003 and Grant No. 60674084), National 863 High-Tech Research and Development Program of China through approved No.2006AA04Z174.

X. Hu and J. Wang (Eds.): COCOON 2008, LNCS 5092, pp. 670–677, 2008.

As we know, the coordination between production and delivery in most manufacturing system has become more practical and become one of the most important topics. Lee and Chen [7] study two-type transportation problems in a single machine, a set of parallel machines or a series of flowshop machines. They consider constraints on both transportation capacity and transportation times, where they do not consider the transportation cost. Chang and Lee [1] have extended one of Lee and Chen's problems to the situation in which every job occupies a different amount of storage space in the vehicle, in which they minimize the makespan. For the problems in which jobs are processed on a single machine and on two parallel machines, they have presented polynomial time algorithms with a worst-case ratio of 5/3 and 2, respectively. For single machine of Chang and Lee's problems, He et al. [5] and Zhong et al. [9] give further improved algorithms for two problems of Chang and Lee, respectively.

We briefly dicuss some works related to integrated scheduling and transportation cost decisions. Wang and Cheng [8] consider a parallel machine scheduling problem with batch delivery. The batch delivery date is equal to the completion time of the last job in a batch. The objective is to minimize the sum of the total flow time and delivery cost. Hall et al. [2] study an environment in which delivery dates are fixed in advance of detailed schedule planning. Hall and Potts [3] analyze the complexity of some single machine scheduling problems with deliveries in supply chain, but without a transporter availability constraint. Hall and Potts [4] also consider single machine or identical parallel machine scheduling problems where deliveries are made in batches with each batch delivered to the customer in a single shipment. They prove that problem $P2|T|\sum C_j + Dy$ is NP-hard by a reduction from Partition problem. The problem remains open as to pseudo- polynomial time solvability.

In this paper, we give a pseudo-polynomial time algorithm to solve problem $P2|T|\sum C_j + Dy$ by dynamic programming and prove the problem is ordinarily NP-hard. We also give a polynomial-time algorithm to solve the special case when the job assignment to two parallel machines is given.

This paper is organized as follows. In section 2, we give some properties and present a pseudo-polynomial-time algorithm for problem $P2|T|\sum C_j + Dy$. In section 3, we show that the problem with given job assignment is solvable in polynomial time. Finally, we present our conclusions in the last section. .

2   Problem $P2|T|\sum C_j + Dy$

In this section, we consider the solvability of problem $P2|T|\sum C_j + Dy$. We first give several properties for an optimal schedule for the problem and find all possible departure times of each batch delivery by Procedure Partition in Kovalyov and Kubiak [6]. Then by dint of the possible departure times, we present a dynamic programming algorithm to solve the problem.

Lemma 1. *(Hall and Potts [4]) For problem $P2|T|\sum C_j + Dy$, there exists an optimal schedule without idle time between jobs on any machine.*

Lemma 2. *(Hall and Potts [4]) For problem $P2|T| \sum C_j + Dy$, there exists an optimal schedule in which all deliveries are made either at job completion times or immediately the transporter becomes available.*

Lemma 3. *For problem $P2|T| \sum C_j + Dy$, there exists an optimal schedule in which: (1) all jobs assigned to the same machine are scheduled in the shortest processing time (SPT) order. (2) the lth delivery batch contains all jobs which finish processing in the time interval $(s_{l-1}, s_l]$ where s_l is the departure time of the lth delivery batch.*

Proof. (1) It can be easily proved by a pairwise job interchange argument.

(2) Index all jobs in the non-decreasing order of their completion times on parallel machines, i.e., $c_1 \leq c_2 \leq ... \leq c_n$. Denote the lth delivery batch by B_l. Without loss of generality, assume that $J_k \in B_l$ and $J_{k+1} \in B_{l+1}$ such that $c_k \leq c_{k+1} \leq s_l$. By definition, we know that $s_{l+1} \geq s_l + T$. Assign J_{k+1} to batch delivery B_l and the remaining jobs are not changed. It is obvious that the objective value decreases at least T. Hence, all jobs which finish processing in the time interval $(s_{l-1}, s_l]$ are assigned to the same batch delivery. □

Based on Lemma 2, when a transporter is available, it will either transport a batch of jobs immediately or wait until the completion time of a job before transporting the batch that contains that job. Hence, the departure time of each delivery needs to be decided. Before we start investigating the algorithm of problem $P2|T| \sum C_j + Dy$, we first study all possible departure times by the procedure Partition.

Let us index all the jobs such that $p_1 \leq p_2 \leq ... \leq p_n$. We introduce 0-1 variable x_j, $j =1,2,...,n$, where $x_j = 1$ if job J_j is scheduled to complete on $M1$ and $x_j = 0$ if job J_j is scheduled to complete on $M2$. Let X be the set of all 0-1 vectors $x = (x_1, x_2, .., x_n)$. Let $G_j^i(x)$ be the total processing time on Mi when the jobs $\{J_1, ..., J_j\}$ have been processed, $i=1,2$. We define the following initial and recursive functions on X:

$$G_0^i(x) = 0, i = 1, 2;$$

$$G_j^1(x) = G_{j-1}^1(x) + x_j p_j;$$

$$G_j^2(x) = G_{j-1}^2(x) + (1 - x_j)p_j.$$

Thus, due to the definition of $G_j^1(x)$ and $G_j^2(x)$, we conclude that $G_j^1(x)$ and $G_j^2(x)$ are not only possible completion times but also possible departure times.

We introduce the Procedure Partition (A, f, δ) presented in Kovalyov and Kubiak [6], where $A \subseteq X$, f is a nonnegative integer function on X, and δ is a number. This procedure partitions A into the disjoint subsets $A_1^f, A_2^f, ..., A_{r_f}^f$, such that $|f(x) - f(x')| \leq \delta \min\{f(x), f(x')\}$ for any x, x' from the same subset A_l^f, $l = 1, 2, ..., r_f$. The following present gives details of Partition (A, f, δ).

Procedure Partition (A, f, δ)

Arrange the vectors $x \in A$ in the order $x^{(1)}, x^{(2)}, ..., x^{(|A|)}$, where $0 \leq f(x^{(1)}) \leq f(x^{(2)}) \leq ... \leq f(x^{(|A|)})$. Assign the vectors $x^{(1)}, x^{(2)}, ..., x^{(i_1)}$ to set A_1^f until i_1 is found such that $f(x^{(i_1)}) \leq (1 + \delta)f(x^{(1)})$ and $f(x^{(i_1+1)}) > (1 + \delta)f(x^{(1)})$. If such i_1 does not exist, then take $A_1^f = A$ and stop.

Assign $x^{(i_1+1)}, x^{(i_1+2)}, ..., x^{(i_2)}$ to set A_2^f until i_2 is found such that $f(x^{(i_2)}) \leq (1 + \delta)f(x^{(i_1+1)})$ and $f(x^{(i_2+1)}) > (1 + \delta)f(x^{(i_1+1)})$. If such i_2 does not exist, then take $A_2^f = A - A_1^f$ and stop.

Continue the above process until $x^{(|A|)}$ is included in $A_{r_f}^f$ for some r_f.

Procedure Partition requires $O(|A| \log |A|)$ operations to arrange the vectors of A in nondecreasing order of $f(x)$ and $O(|A|)$ operations to provide a partition. Kovalyov and Kubiak [6] give the main property in the following.

Lemma 4. *(Kovalyov and Kubiak [6]).* $r_f \leq \log f(x^{|A|})/\delta + 2$ *for* $0 < \delta \leq 1$ *and* $1 \leq f(x^{|A|})$.

Now we present the following algorithm to give birth to all possible departure times.

Algorithm A

Step 1. (Initialization) Number the jobs such that $p_1 \leq p_2 \leq ... \leq p_n$. Set $X_0 = \{(0, 0, ..., 0)\}$, $j = 1$ and $G_0^i(x) = 0$, for $i = 1, 2$.

Step 2. (Generation of $X_1, X_2, ..., X_n$) From the set X_{j-1}, generate the set X_j' by adding 0 and 1 in position j of each vector from X_{j-1}, i.e., $X_j' = X_{j-1} \cup \{x + (0, ..., 0, x_j = 1, 0, ..., 0)|x \in X_{j-1}\}$. Calculate the following for any $x \in X_j'$:

$$G_j^1(x) = G_{j-1}^1(x) + x_j p_j;$$

$$G_j^2(x) = G_{j-1}^2(x) + (1 - x_j)p_j.$$

If $j = n$, then the set $X_n = X_n'$ and go to step 3.

If $j < n$, then set $\delta = 1/(2(n + 1))$, and perform the following computations.

Call Partition (X_j', G_j^i, δ) $(i=1,2)$ to partition set X_j' into disjoint subsets $X_1^{G^i}, X_2^{G^i}, ..., X_{r_j}^{G^i}$. Divide set X_j' into disjoint subsets $X_{ab} = X_a^{G^1} \cap X_b^{G^2}$, $a = 1, 2, ..., r_{f_1}$, $b = 1, 2, ..., r_{f_2}$, for each nonempty subset X_{ab}, choose a vector $x^{(ab)}$ such that $G_j^1(x^{(ab)}) = \min\{G_j^1(x)|x \in X_{ab}\}$.

Set $X_j := \{x^{(ab)}|a = 1, 2, ..., r_{f_1}, b = 1, 2, ..., r_{f_2}$ and $X_a^{G^1} \cap X_b^{G^2} \neq \varPhi\}$ and $j := j + 1$, go to step 2.

Step 3. (Solution) Find $G_j^1(x)$ and $G_j^2(x)$, for $j = 1, 2, ..., n$.

Lemma 5. *Algorithm A finds all the possible departure times* $G_j^1(x)$ *and* $G_j^2(x)$ *in time* $O(n^3(\log P)^3)$, *for* $j = 1, 2, ..., n$.

Proof. Note that the most time-consuming operation is iteration j in step 2, i.e., a call of Procedure Partition, which requires $O(|X_j'| \log |X_j'|)$ time to complete. By Algorithm A, we have $|X_{j+1}'| \leq 2|X_j'| \leq 2r_{f_1}r_{f_2}$. By Lemma 4, we have

$r_{f_1} \leq 2(n+1)\log P + 2$, and the same for r_{f_2}. Thus $|X'_j| = O(n^2(\log P)^2)$, and the time complexity of Algorithm A is $O(n^3(\log P)^3)$.   □

In the following, we propose a dynamic programming algorithm to solve problem $P2|T|\sum C_j + Dy$. In summary, all the possible departure times of each delivery can be $G_j^1(x)$, $G_j^2(x)$, $G_j^1(x)+T$, $G_j^1(x)+2T$, ..., $G_j^1(x)+qT$, $G_j^2(x)+T$, $G_j^2(x)+2T$, ..., $G_j^2(x)+qT$, for some $q \leq n-j$ and $j = 1,2,...,n$. Thus, there are at most n^4 candidate departure time points. For ease of presentation, we assume $G(G \leq n^4)$ candidate departure time points that are indexed as 1, 2,..., G such that earlier time points have smaller indices. Let $s_l(k)$ denote the actual time of batch B_l corresponding to departure time point k, for $k = 1,2,...,G$. Also, let $s_0(0) = 0$. Since T is denoted as a constraint on the minimum time interval between consecutive deliveries, we have $s_l(k) \geq s_{l-1}(h) + T$.

Algorithm DP

   Step 1. Perform Algorithm A and find all possible departure times.

   Step 2. Define $f(j, t_1, t_2, s_l(k))$ as the minimum objective value if we have scheduled jobs J_1 up to J_j such that the total processing time of jobs assigned to machine u is t_u, $u = 1, 2$, and the departure time is $s_l(k)$ for batch B_l, $k = 1, 2, ..., G$.

   Initial conditions: $f(0, 0, 0, s_0(0)) = 0$.

   Recursive equations:

$$f(j, t_1, t_2, s_l(k)) = \min_{1 \leq l \leq j} \begin{cases} f(j-1, t_1-p_j, t_2, s_l(k)) + s_l(k) + T, \\ f(j-1, t_1, t_2-p_j, s_l(k)) + s_l(k) + T, \\ f(j-1, t_1-p_j, t_2, s_{l-1}(h)) + s_l(k) + T + D, \\ f(j-1, t_1, t_2-p_j, s_{l-1}(h)) + s_l(k) + T + D, \end{cases}$$

where $s_l(k) \geq s_{l-1}(h) + T$ and $s_l(k) \geq \max\{t_1, t_2\}$.

   Step 3. Optimal solution:

$$\min\{f(n, t_1, t_2, s_l(k)) | \texttt{all possible states}(n, t_1, t_2, s_l(k))\}.$$

Theorem 1. *Algorithm DP finds an optimal schedule for problem* $P2|T|\sum C_j + Dy$ *in* $O(n^6 P^2 + n^3(logP)^3)$.

Proof. As discussed above, the optimality of the algorithm can be proved. By definition, $j, l \leq n$. t_1 and t_2 are independent and $t_1, t_2 \leq P$. For $s_l(k)$, there are at most n^4 candidate departure time points. Therefore, based on Lemma 4, the overall time complexity of Algorithm DP is $O(n^6 P^2 + n^3(logP)^3)$.   □

The existence of such a pseudo-polynomial algorithm means that the problem $P2|T|\sum C_j + Dy$ is NP-hard in the ordinary sense.

Theorem 2. *Problem* $P2|T|\sum C_j + Dy$ *is NP-hard in the ordinary sense.*

3   A Special Case

In this section, we assume that the job assignment to parallel machines is given, i.e, it is not part of the problem decision. This special case characterizes the actual situation where each machine is devoted to a special group of jobs. It is evident that the special case reduces to sequence jobs on each machine and partition delivery batches. We show that the problem with given job assignment is polynomially solvable by a dynamic programming algorithm. It is obvious that the special case reduces to an optimal batching problem.

By Lemma 1 and 3 (1), schedule all jobs in the SPT order on each machine without idle time. Then the completion time of job J_j on parallel machines is determined. Let the jobs be reindexed such that $c_1 \leq c_2 \leq ... \leq c_n$. By Lemma 2, the departure time of each delivery will be either the completion time of some job or the time back to the machines. In the first case, the departure time is c_j. In the second case, the departure time is traced back and expressed as $c_j + qT$ for some $q \leq n - j$ and some job J_j. As discussed above, the possible departure times can be $c_j, c_j + T, ..., c_j + qT$ for some $q \leq n - j$ and $j = 1, 2, ..., n$. Thus, there are at most n^2 candidate departure time points. Define s_l as the departure time of batch B_l that satisfy $s_l \geq \max_{j \in B_l}\{c_j\}$. Note that s_l depends on the previous delivery and satisfies $s_l \geq s_{l-1} + T$.

Define $f(j, k, s_l)$ as the minimum objective value of a partial schedule containing the first k jobs $J_1, ..., J_k$, provided that the current last batch B_l contains $J_{j+1}, ..., J_k$ and is delivered at time s_l. We present the dynamic programming algorithm as follows.

Algorithm G
Schedule the jobs on each machine in the SPT order, and then reindex all the jobs in accordance with the job completion time on the machines such that $c_1 \leq c_2 \leq ... \leq c_n$.
    Initial conditions: $f(0, 0, 0) = 0$.
    Recursive relations:

$$f(j, k, s_l) = \min_{0 \leq i < j, 1 \leq l \leq j} \{f(i, j, s_{l-1}) + (k - j)(s_l + T) + D\}$$

where $s_l \geq c_k$ and $s_l \geq s_{l-1} + T$.
    Optimal solution:

$$\min_{0 \leq j < n, 1 \leq l \leq n} \{f(j, n, s_l)\}$$

Theorem 3. *Algorithm G finds an optimal schedule for the problem with given job assignment in $O(n^4)$.*

Proof. The proof of optimality follows from Lemmas 1, 2 and 3. By definition, $j, l \leq n$. The departure time s_l requires at most $O(n^2)$ time. Therefore, the overall time complexity of Algorithm G is $O(n^4)$. □

We now demonstrate the above solution method with a numerical example.

Example. Consider the instance with $\{J_1, J_2, J_3, J_4\}$, $p_1 = 1, p_2 = 2, p_3 = p_4 = 4, T = 2$ and $D = 5$. Assume that J_1 and J_3 are assigned to $M1$, J_2 and J_4 are assigned to $M2$. From above method, all jobs are sequenced in SPT order on each machine. We can obtain completion times of four jobs: $c_1 = 1, c_2 = 2, c_3 = 5, c_4 = 6$.

We use the above dynamic programming algorithm to solve the instance, we have the following results:

$f(0, 1, s_1) = f(0, 0, 0) + c_1 + T + D = 8,$

$f(0, 2, s_1) = f(0, 0, 0) + 2(c_2 + T) + D = 13,$

$f(0, 3, s_1) = f(0, 0, 0) + 3(c_3 + T) + D = 26,$

$f(0, 4, s_1) = f(0, 0, 0) + 4(c_4 + T) + D = 37.$

$f(1, 2, s_2) = f(0, 1, s_1) + \max\{(s_1 + T), c_2\} + T + D = 18,$

$f(1, 3, s_2) = f(0, 1, s_1) + 2(\max\{(s_1 + T), c_3\} + T) + D = 27,$

$f(1, 4, s_2) = f(0, 1, s_1) + 3(\max\{(s_1 + T), c_4\} + T) + D = 37,$

$f(2, 3, s_2) = f(0, 2, s_1) + \max\{(s_1 + T), c_3\} + T + D = 25,$

$f(2, 4, s_2) = f(0, 2, s_1) + 2(\max\{(s_1 + T), c_4\} + T) + D = 34,$

$f(3, 4, s_2) = f(0, 3, s_1) + \max\{(s_1 + T), c_4\} + T + D = 40,$

$f(2, 3, s_3) = f(1, 2, s_2) + \max\{(s_2 + T), c_3\} + T + D = 30,$

$f(2, 4, s_3) = f(1, 2, s_2) + 2(\max\{(s_2 + T), c_4\} + T) + D = 39,$

$f(3, 4, s_3) = \min \begin{cases} f(1, 3, s_2) + \max\{(s_2 + T), c_4\} + T + D = 41, \\ f(2, 3, s_2) + \max\{(s_2 + T), c_4\} + T + D = 39. \end{cases}$

$f(3, 4, s_4) = f(2, 3, s_3) + \max\{(s_3 + T), c_4\} + T + D = 44.$

Thus

$$\min_{0 \leq j < 4, \ 1 \leq l \leq 4} \{f(j, 4, s_l)\} = f(2, 4, s_2) = 34.$$

Hence, we can obtain an optimal schedule $\{\{J_1, J_2\}, \{J_3, J_4\}\}$ with two delivery batches.

4   Concluding Remarks

In this paper, we study a parallel machine scheduling problem with batch delivery. Hall and Potts have shown the problem to minimize the sum of total completion time and delivery cost is NP-hard. We provide a pseudo-polynomial-time to solve the problem and prove the problem is ordinarily NP-hard. When the job assignment on parallel machines is given, we give a polynomial-time algorithm to solve the special case. We only solve the problem in theory. Hence, a future interesting issue is to develop effective heuristic to solve the problem.

References

1. Chang, Y.-C., Lee, C.-Y.: Machine Scheduling with Job Delivery Coordination. European Journal of Operational Research 158, 470–487 (2004)
2. Hall, N.G., Lesaoana, M.A., Potts, C.N.: Scheduling with Fixed Delivery Dates. Operations Research 49, 854–865 (2001)
3. Hall, N.G., Potts, C.N.: Supply Chain Scheduling: Batching and Delivery. Operations Research 51, 566–584 (2003)
4. Hall, N.G., Potts, C.N.: The Coordination of Scheduling and Batch Deliveries. Annals of Operations Research 135, 41–64 (2005)

5. He, Y., Zhong, W., Gu, H.: Improved Algorithms for Two Single Machine Scheduling Problems. Theoretical Computer Science 363, 257–265 (2006)
6. Kovalyov, M.Y., Kubiak, W.A.: A Fully Polynomial Approximation Scheme for Minimizing Makespan of Deteriorating Jobs. Journal of Heuristics 3, 287–297 (1998)
7. Lee, C.-Y., Chen, Z.-L.: Machine Scheduling with Transportation Considerations. Journal of Scheduling 4, 3–24 (2001)
8. Wang, G.Q., Cheng, T.C.E.: Parallel Machine Scheduling with Batch Delivery Costs. International Journal of Production Economics 68, 177–183 (2000)
9. Zhong, W., Dsa, G., Tan, Z.: On the Machine Scheduling Problem with Job Delivery Coordination. European Journal of Operational Research 182, 1057–1072 (2007)

Author Index

Lecture Notes in Computer Science

Sublibrary 1: Theoretical Computer Science and General Issues

For information about Vols. 1–4739
please contact your bookseller or Springer